NEW STATES OF MATTER IN HADRONIC INTERACTIONS

Other Related Titles from AIP Conference Proceedings

619 Hadron Spectroscopy: Ninth International Conference on Hadron Spectroscopy; HADRON2001
Edited by Dmitry Amelin and Alexander M. Zaitsev, June 2002, 0-7354-0067-9

602 QCD@Work: International Workshop on Quantum Chromodynamics: Theory and Experiment
Edited by Pietro Colangelo and Giuseppe Nardulli, December 2001, 0-7354-0046-6

594 Hadrons and Nuclei: First International Symposium
Edited by Il-Tong Cheon, Taekeun Choi, Seung-Woo Hong, and Su Houng Lee, November 2001, 0-7354-0037-7

583 Advanced Computing and Analysis Techniques in Physics Research: VII International Workshop; ACAT 2000
Edited by Pushpalatha C. Bhat and Matthias Kasemann, August 2001, 0-7354-0023-7

549 Intersections of Particle and Nuclear Physics: 7th Conference, CIPANP2000
Edited by Zohreh Parsa and William J. Marciano, December 2000, 1-56396-978-5

482 RHIC Physics and Beyond: Kay Kay Gee Day
Edited by Berndt Müller and Robert Pisarski, July 1999, 1-56396-878-9

To learn more about these titles, or the AIP Conference Proceedings Series, please visit the webpage **http://proceedings.aip.org**

NEW STATES OF MATTER IN HADRONIC INTERACTIONS

Pan-American Advanced Study Institute

São Paulo, Brazil 7–18 January 2002

SPONSORING ORGANIZATIONS
U.S. Department of Energy
U.S. National Science Foundation
CLAF (Latin American Physics Center)
CNPq (Conselho Nacional de
  Desenvolvimento Científico e
  Tecnológico)
FAPESP (Fundação de Amparo à
  Pesquisa do Estado de São Paulo)
FAPERJ (Fundação Carlos Chagas
  Filho de Amparo à Pesquisa do
  Estado do Rio de Janeiro)

EDITORS
H.-Thomas Elze
Erasmo Ferreira
Takeshi Kodama
UFRJ, Rio de Janeiro, Brazil
Jean Letessier
LPTHE, University of Paris 7, France
Johann Rafelski
Robert L. Thews
*University of Arizona
Tuscon, Arizona*

Melville, New York, 2002
AIP CONFERENCE PROCEEDINGS ■ VOLUME 631

Editors:

H.-Thomas Elze
Erasmo Ferreira
Takeshi Kodama

Instituto de Física
Universidade Federal
  do Rio de Janeiro
Caixa Postal 68.528
21945-970 Rio de Janeiro
BRAZIL

E-mail: thomas@if.ufrj.br
         erasmo@if.ufrj.br
         tkodama@if.ufrj.br

Johann Rafelski
Robert L. Thews

Department of Physics
University of Arizona
1118 East 4th Street
Tuscon, AZ 85721
USA

E-mail: rafelski@physics.arizona.edu
        thews@physics.arizona.edu

Jean Letessier
LPTHE
Université Paris 7 - Denis Diderot
2, place Jussieu
75251 Paris Cedex 05
FRANCE

E-mail: jletes@lpthe.jussieu.fr

Cover: Particle production observed in STAR (top right) and PHENIX (bottom left) detectors at the Brookhaven National Laboratory (BNL) Relativistic Heavy Ion Collider (RHIC).

Thousands of newly made charged particles are produced and are observed in these detectors when two beams of gold ions collide head on at maximum energy of RHIC. Particle tracks emanate outwards from the center, where beams traveling out of picture plane collided. The gentle bend in the tracks is due to a magnetic field which helps to identify individual particles.

STAR image courtesy of Brookhaven National Laboratory and the STAR experiment. The PHENIX image courtesy of the PHENIX experiment and Jeff Mitchell.

Authorization to photocopy items for internal or personal use, beyond the free copying permitted under the 1978 U.S. Copyright Law (see statement below), is granted by the American Institute of Physics for users registered with the Copyright Clearance Center (CCC) Transactional Reporting Service, provided that the base fee of $19.00 per copy is paid directly to CCC, 222 Rosewood Drive, Danvers, MA 01923. For those organizations that have been granted a photocopy license by CCC, a separate system of payment has been arranged. The fee code for users of the Transactional Reporting Service is: 0-7354-0086-5/02/$19.00.

© 2002 American Institute of Physics

Individual readers of this volume and nonprofit libraries, acting for them, are permitted to make fair use of the material in it, such as copying an article for use in teaching or research. Permission is granted to quote from this volume in scientific work with the customary acknowledgment of the source. To reprint a figure, table, or other excerpt requires the consent of one of the original authors and notification to AIP. Republication or systematic or multiple reproduction of any material in this volume is permitted only under license from AIP. Address inquiries to Office of Rights and Permissions, Suite 1NO1, 2 Huntington Quadrangle, Melville, N.Y. 11747-4502; phone: 516-576-2268; fax: 516-576-2450; e-mail: rights@aip.org.

L.C. Catalog Card No. 2002111919
ISBN 0-7354-0086-5
ISSN 0094-243X
Printed in the United States of America

Contents

Preface ix
Photos xiii

STRONG INTERACTIONS

Introduction to Relativistic Gases 3
    T. Kodama
High-Energy Nuclear Interactions and QCD: An Introduction 27
    D. E. Kharzeev and J. Raufeisen
A QCD Primer 70
    G. Altarelli
Lattice QCD at Finite Temperature 112
    F. Karsch
Quark-Gluon Plasma and Strangeness 142
    J. Rafelski and J. Letessier
QCD Vacuum and Confinement 168
    A. Di Giacomo
Introduction to Effective Field Theories in QCD 191
    U. van Kolck, L. J. Abu-Raddad, and D. M. Cardamone
The Nucleon and the Nuclear Force: Effective Theory and Path-Integral Methods 221
    L. J. Abu-Raddad
Relativistic Quantum Transport Theory 229
    H.-T. Elze

EXPERIMENTS

RHIC Physics Overview 255
    J. W. Harris, C. Beckmann, J. Gans, and K. H. Gulbrandsen
Introduction to Ultrarelativistic Heavy-Ion Physics at the CERN-SPS 294
    F. Antinori, A. Billmeier, and J. Zaranek
PHENIX for Beginners 325
    W. A. Zajc (for the PHENIX Collaboration) and P. Fachini
Survey of Recent Results from the PHOBOS Experiment at RHIC 349
    C. Roland (for the PHOBOS Collaboration)
$K^*(892)^0$ Production in Relativistic Heavy-Ion Collisions 358
    P. Fachini (for the STAR Collaboration)
Azimuthal Anisotropy of K_S^0 and Λ Production at Mid-rapidity from Au+Au Collisions at $\sqrt{S_{NN}}$=130 GeV 366
    P. Sorensen (for the STAR Collaboration)
Multi-strange Baryon Production at Mid-rapidity in Au+Au Collisions at $\sqrt{S_{NN}}$=130 GeV 371
    J. Castillo (for the STAR Collaboration)

Particle Tracking. 377
    K. Šafařík, I. Kraus, J. Newby, and P. Sorensen

RELATIVISTIC HEAVY-ION COLLISIONS

Physics of Event Generators 397
    K. Werner
Baryon Production in Proton-Proton Collisions. 422
    F. M. Liu and K. Werner
Hydrodynamics for Modeling Ultrarelativistic Heavy-Ion Reactions 428
    L. P. Csernai
Nonequilibrium Hadrochemistry in QGP Hadronization 460
    J. Rafelski and J. Letessier
Nonlinear Behavior of Quarkonium Formation and
Deconfinement Signals 490
    R. L. Thews
Nuclear Effects in the Drell-Yan Process. 525
    J. Raufeisen
Strange Hadron Resonances: Freeze-Out Probes in
Heavy-Ion Collisions. 533
    C. Markert, G. Torrieri, and J. Rafelski
Critical Review of Quark Gluon Plasma Signatures. 553
    H. Stöcker, J. Berger, U. Eichmann, S. Salur, S. Scherer, and D. Zschiesche
Hot Hadronic Matter with the NJL Model. 580
    J. Aichelin, G. Torrieri, and F. Gastineau
Purpose and Physics of Event-by-Event Analysis 599
    T. A. Trainor

LATIN-AMERICAN RESEARCH

QCD/QGP in Latin America 2002 615
    E. Ferreira
Λ^0 Polarization and QGP Production in High-Energy
Heavy-Ion Collisions. 620
    A. Ayala, E. Cuautle, G. Herrera, and L. M. Montaño
Soft and Hard QCD in Charmonium Production 631
    C. Brenner Mariotto, M. B. Gay Ducati, and G. Ingelman
Unitarity Corrections and Structure Functions 637
    M. B. Gay Ducati and M. V. T. Machado
Particle Emission in Hydrodynamics: A Problem Not Yet Solved 646
    F. Grassi
The F_2^D Logarithmic Slope and the Saturation Phenomena 655
    M. V. T. Machado
Small Quark Stars in the Chromodielectric Model. 658
    M. Malheiro, E. O. Azevedo, L. G. Nuss, M. Fiolhais, and A. Taurines

Strangelets and Their Possible Astrophysical Origin 666
    L. Masperi

Is Nuclear Multifragmentation Related with the Liquid-Gas Phase Transition? 672
    D. P. Menezes and C. Providência

QCD Sum Rules Applications to the J/ψ Dissociation 676
    M. Nielsen, F. S. Navarra, R. S. Marques de Carvalho, G. Krein, and
    M. E. Bracco

Hydrodynamical Evolution from Event-by-Event Fluctuating Initial Conditions 686
    T. Osada, C. E. Aguiar, Y. Hama, and T. Kodama

Finite Temperature and Density in Nonlocal Chiral Quark Models 695
    N. N. Scoccola and D. Gomez Dumm

Front Form Approach to $q\tilde{Q}$ Mesons with Harmonic Confinement 704
    C. A. Z. Vasconcellos, M. Dillig, E. F. Lütz, F. G. Pilotto, and
    G. F. Marranghello

Crossing Symmetry Violation of Inverse Amplitude Method 713
    I. P. Cavalcante and J. Sá Borges

Bosons Production $W^\pm$ via e^-p-Collisions at CERN LEP/LHC Energies with a $W^\pm$ Anomalous Magnetic Moment 715
    A. Gutiérrez-Rodríguez

Nucleon-Nucleon Potential: Relativistic and Heavy Baryon Formalism 717
    R. Higa

Models of Black Hole Production at LHC 719
    R. Klippert and C. J. Solano Salinas

Proton-Proton Total Cross-Section Scenarios at Cosmic-Ray Energies 721
    E. G. S. Luna and M. J. Menon

Charmonium-Pion Cross Section from QCD Sum Rules 723
    F. S. Navarra, M. Nielsen, R. S. Marques de Carvalho, and G. Krein

Coulomb Gauge Quark Model, Vacuum Condensates, and High-Density Quark Matter 725
    P. K. Panda and G. Krein

Author Index 727

Preface

For more than two decades nuclear and particle physicists have been searching for experimental evidence of a phase transition between hadronic matter and a state of deconfined color charge, generically known as the quark-gluon plasma. This behavior is predicted by numerical solutions of QCD at finite temperature, and as such would be a dramatic manifestation of the theory which governs strong interactions. To create the required conditions for this phase transition to occur in the laboratory, several experimental facilities have been developed to collide heavy ions at high energy. Such collisions have the potential to provide, albeit for a very short time duration, the energy density, temperature and extended volume which would be required for the phase transition to occur. The CERN SPS program of heavy ion beams on fixed targets has been in operation for over a decade, and has produced a variety of very interesting results spanning a wide range of experimental measurables. The RHIC heavy ion collider at Brookhaven National Laboratory initiated operations in the latter half of 2000, and the first two physics runs have already provided exciting new results in this high energy regime. The next generation experiments will occur at CERN, using the heavy ion colliding beam capability of the Large Hadron Collider. Present plans anticipate the start of experimental operations in 2007. To understand these data and extract physical insights is not a trivial task, requiring highly sophisticated methods and concepts, both experimental and theoretical.

In order to ensure maximal utilization of these facilities, the high energy and nuclear physics communities must continue to recruit new and active participants. It is essential to provide opportunities to obtain the basic knowledge and to learn the advanced techniques which are required for the study of this fascinating and complex area of physics. Many summer schools and workshops have already been organized, especially in the US and Europe, to fulfill these needs. It is of fundamental importance to offer young students chances to participate in advanced schools with a scientific program of high standards, where they will also absorb the spirit of international collaboration and competition. However, little has been done to extend these efforts to involve physicists from many of the Latin American countries.

The idea of organizing the present PASI (Pan American Advanced Study Institute) originated early in the year 2000, when some of us realized that such a school could benefit the community in a unique way. One could build upon the existence of several emerging research groups in Latin American countries which already collaborate with BNL and CERN. Since the experiments are concentrated in few big international laboratories, new collaborations and research links are vital to bring in new participants from widely-scattered and distant locations in Latin America. Existing research collaborations between several of us at Arizona and UFRJ led to the formation of a Board of

Directors, which was soon expanded into an Organizing Committee with both international and local components. A series of preliminary meetings led to the determination of various parameters of the meetings. It was decided that the core subject of this PASI would be dedicated to the study of the quark-gluon plasma. A grant application was submitted to the U.S. NSF for support of a meeting in Brazil with this focus, and funds were approved in late 2000. In addition, we agreed that it would be important to offer additional program elements covering closely-related subfields of physics. We decided to devote about 1/4 of the final program to these topics. These extended activities were to be funded primarily with resources made available by Brazilian funding agencies.

This Institute was planned as an advanced school, neither a conference nor a workshop. The academic objective of this Institute is primarily pedagogical, but of comparable importance is the development of new personal relationships among the participants. Thus we made every effort to ensure that all Institute participants were present at the meeting for the full period. The selection of the meeting site was done with this in mind.

The final selection of the location in Campos do Jordao (Sao Paulo State, Brazil), a resort town near Sao Paulo, was influenced by easy access from Rio de Janeiro and Sao Paulo, where most international participants arrived. Choosing the Hotel Leao da Montanha (The Mountain Lion) was both a wise and fortunate choice which contributed immensely to the success and atmosphere of the meeting. The owner Arie Yaari has been always present and willing to assist in solving problems of participants. His staff was dedicated and understanding of the needs of our meeting. The meals were superb in both quantity and quality, consistently from the beginning to the end of the meeting.

The weather was quite variable, with some periods which were quite miserable indeed, with most of Brazil suffering from a prolonged and unseasonal rainy weather. However, this turned out to be very conducive to full attendance at all lectures, barring most participants from adventurous hiking trips and excursions into the most beautiful mountain area with the extensions of the coastal tropical rain-forest. Ample time during meals and diverse social activities in the evenings encouraged informal discussions concerning possible future reinforcements of international collaboration. Many U.S. and Latin American students and postdocs developed contacts which will allow future collaborations and exchanges to be facilitated.

Among 128 participants there were: 20 US/European and 28 Latin American physicists; 5 US/European and 16 Latin American postdocs; 23 US/European and 36 Latin American students. The enthusiasm shown by the junior participants was stimulating for the lecturers and very promising for the future of the field. For the production of lecture notes for these proceedings, we paired groups of students with each lecturer to provide note-taking and other assistance. In several of these cases, this lead to a listing of students as co-authors on the written versions.

The following presentations have been divided into 4 categories for the convenience of the reader. In addition to these presentations there were several very introductory tutorials and discussions. Both oral and poster (2 pages) contributions by students are included in this volume.

We are pleased to recognize the contributions of the International Advisory Committee, which is composed of many eminent physicists active in this field. They did a wonderful job providing recommendations and advice concerning the choice of primary lecturers and specific topics. Many of them were also active participants at the PASI itself. The success of this meeting is due in no small part to their efforts.

The primary lecturers did a superb job. Many of them had to make serious sacrifices in time and resources to be away from their home institutions. They delivered hot-off-the-press scientific material, with careful and kind pedagogical competence, and in elegant style. The motivation they transferred to the students, from whom they received clear and immediate response, will remain as a legacy of their efforts. Daily discussion sessions have allowed for an unusually intense interaction between the principal lecturers and all participants of the school.

The primary founding financial support for the meeting was provided by the U.S. Department of Energy and U.S. National Science Foundation through the Americas Program. Brazilian and CLAF (Latin American Physics Center) funds made up about 36 percent of the total operating budget. Many Brazilian agencies and institutions (CNPq, Universidade Federal do Rio de Janeiro, Universidade de Sao Paulo, FAPESP, FAPERJ) are gratefully acknowledged for their most generous and ample support, which has been delivered in due time, which has helped to make this unique PASI school become a reality. Several participants were supported by an individual travel grant provided to them by the Heavy-Ion Laboratory, GSI-Darmstadt.

We express our sincere hopes that this successful collaborative effort in science and funding will lead to additional jointly-funded activities in the future between the countries of the Americas. Rapidly-expanding fields of study such as ours will reap the benefits of such activities for years to come.

H.-T. Elze, E. Ferreira, T. Kodama, J. Letessier, J. Rafelski and R. L. Thews

New States of Matter in Hadronic Interactions, Pan-American Advanced Study Institute, Campos do Jordão, January 7–18, 2002

STRONG INTERACTIONS

Introduction to Relativistic Gases

Takeshi Kodama

Instituto de Física - Universidade Federal do Rio de Janeiro

INTRODUCTION

Some of the observed quantities in relativistic heavy-ion collisions, such as ratios of produced particles, transverse spectra, etc. [1], are often well described in terms of simple thermal models. However, this fact does not necessarily mean that the whole system is in thermal equilibrium, and the interpretation of these analyses should be done with care. In this lecture, I would like to introduce some fundamental concepts of statistical physics to derive thermodynamical properties of a relativistic gas, with the aim of improving the understanding of the forthcoming lectures on advanced models discussed in this school. Several expressions for thermodynamical quantities frequently used in thermal models are derived, and it is shown how these quantities can be calculated in practice. Therefore, this lecture note is a kind of short digest of more complete books on statistical physics [2, 3].

GRAND CANONICAL ENSEMBLE

First, consider a small portion of hadronic matter (or quark-gluon plasma) formed in a relativistic heavy-ion collision. When the time scale of space-time evolution of the system is relatively slow compared to the microscopic time scale (such as the mean collision time), a lot of different physical configurations appear within a significant global time scale of the system. Thus, physical properties of this segment are basically determined by the statistical average over all the *microstates* which appear within the macroscopic time-scale.

Let us suppose that a microstate of our system is specified by the occupation numbers

$$\alpha \equiv \{n_1, n_2, \ldots, n_i, \ldots\ldots\}, \tag{1}$$

where n_i is the number of particles which are in the $i-th$ "single particle state". For an ideal gas in a box of volume V, we can take plane wave states, and the index i can be identified as the wavenumber vector,

$$i \to \vec{k}$$

Note that neither α nor i are simple numbers, but (infinite) sets of numbers.

For one microstate α, we can write the energy E_α and the total number of particles N_α as

$$E_\alpha = \sum_i n_i \varepsilon_i, \qquad (2)$$

$$N_\alpha = \sum_i n_i, \qquad (3)$$

where the sum extends over all single-particle states of the system.

The macroscopic state of the system, which is determined by a small number of parameters, such as temperature, pressure, total energy, etc., usually does not specify the microstate α. Microscopically, a lot of interactions among particles will take place within a macroscopic time-lapse τ_h. Almost an infinite set of microstates appear and disappear in this time scale. This (infinite) set of microstates relevant for one macroscopic state is called *ensemble*. The basic question in equilibrium statistical mechanics is to determine the probability of a given microstate α to appear in the ensemble.

Let p_α be the probability corresponding to a microstate α. To determine p_α, we argue as follows. First, let us build the ensemble by preparing $\mathcal{M}$ systems of the same gas, all of them in the same *macroscopic* state. Here, we assume $\mathcal{M} \gg 1$ (in fact, we take $\mathcal{M} \to \infty$). Each system should be in some *microstate* α. Let $\mathcal{M}_\alpha$ be the total number of systems which are in the microstate α (also $\mathcal{M}_\alpha \gg 1$). Obviously,

$$\mathcal{M} = \sum_\alpha \mathcal{M}_\alpha, \qquad (4)$$

and the probability p_α is then given by

$$p_\alpha = \frac{\mathcal{M}_\alpha}{\mathcal{M}}. \qquad (5)$$

We have

$$\sum_\alpha p_\alpha = 1. \qquad (6)$$

Depending the way such an ensemble of systems is prepared, they are not always identical. We can construct an ensemble, for example, in which every system is exactly in the same microstate $\alpha = \alpha_0$. In this case, every $\mathcal{M}_\alpha$ is equal to zero except $\mathcal{M}_{\alpha_0} = \mathcal{M}$. Another extreme is to prepare the ensemble in such a way that every microstate occurs exactly the same number of times. In this case, we have the same number $\mathcal{M}_\alpha$ for all α. Thus, the distribution of $\mathcal{M}_\alpha$ depends on how the system is prepared. We may specify the state of the ensemble by this distribution of $\mathcal{M}_\alpha$,

$$\left\{ \mathcal{M}_{\alpha_1}, \mathcal{M}_{\alpha_2}, \cdots, \mathcal{M}_{\alpha_i}, \cdots \right\},$$

where the ensemble contains $\mathcal{M}_{\alpha_1}$ systems in the microstate α_1, $\mathcal{M}_{\alpha_2}$ systems in the microstate α_2, etc.

Now, let us introduce a basic hypothesis. We suppose that in the equilibrium state of the system, every microstate has the same a-priori probability. That is, every microstate can appear in the same way as others, just as the numbers of an unbiased die. So, we are considering a kind of die that has an infinite number of surfaces, each surface corresponding to a microstate α. As we know very well, even if the probability is equal for any number of a die, when we throw 2 dice, then the probabilities for the sum of numbers of the 2 dice are not uniform. The probability of having a total number 7 is larger than the probability of having 2. This is because with two numbers from 1 to 6 there are more ways to compose 7 than 2. Certain configurations can occur more frequently than others just because there are more ways to realize such configurations. In our case, the number of different ways to have a specific partition $\{\mathcal{M}_{\alpha_1}, \mathcal{M}_{\alpha_2}, \cdots\}$ of an integer $\mathcal{M}$ is given by

$$W = \frac{\mathcal{M}!}{\mathcal{M}_{\alpha_1}! \mathcal{M}_{\alpha_2}! \cdots \mathcal{M}_{\alpha_i}! \cdots}. \tag{7}$$

We then expect that the equilibrium should correspond to the configuration of the ensemble for which W is maximum. This state of the ensemble is more probable than others.

- In equilibrium, the configuration $\{\mathcal{M}_{\alpha_1}, \mathcal{M}_{\alpha_2}, \cdots, \mathcal{M}_{\alpha_i}, \cdots\}$ is obtained by maximizing W.

It is easy to see that if we don't have any more physical constraints, the maximum of W is given simply by

$$\mathcal{M}_{\alpha_1} = \mathcal{M}_{\alpha_2} = \cdots = \mathcal{M}_{\alpha_i} = \cdots,$$

that is, every microstate appears with the same probability. However, if we are considering a system where the average value of the total energy and the number of particles are fixed, for example, then we have to maximize W (or equivalently $\ln W$) taking into account two constraints. We have

$$\delta \ln W = 0, \tag{8}$$

together with

$$\delta \langle E \rangle = 0, \tag{9}$$
$$\delta \langle N \rangle = 0, \tag{10}$$

where the average total energy is

$$\langle E \rangle = \frac{1}{\mathcal{M}} \sum_\alpha E_\alpha \mathcal{M}_\alpha, \tag{11}$$

and the average total number of particles is

$$\langle N \rangle = \frac{1}{\mathcal{M}} \sum_\alpha N_\alpha \mathcal{M}_\alpha. \tag{12}$$

Furthermore, we have to fix the total number of systems in the ensemble,

$$\mathcal{M} = \sum_\alpha \mathcal{M}_\alpha. \tag{13}$$

Using the Lagrangian multiplier method, the configuration $\{\mathcal{M}_{\alpha_1}, \mathcal{M}_{\alpha_2}, \cdots\}$ is determined by the condition

$$\delta\left(\frac{1}{\mathcal{M}} \ln W\right) - \beta\, \delta\langle E\rangle + \lambda\, \delta\langle N\rangle = 0, \tag{14}$$

or equivalently

$$\delta\left[\ln W - \beta \sum_\alpha E_\alpha \mathcal{M}_\alpha + \lambda \sum_\alpha N_\alpha \mathcal{M}_\alpha - \gamma \mathcal{M}\right] = 0, \tag{15}$$

where β, λ and γ are Lagrange multipliers, and the variation should be taken with respect to the change of the configuration $\{\mathcal{M}_{\alpha_1}, \mathcal{M}_{\alpha_2}, \cdots\}$, that is, with respect to the numbers of the systems in each microstate α in the ensemble.

Now, for $\mathcal{M}_\alpha \gg 1$, we can make an approximation,

$$\begin{aligned}\ln W &= \ln \mathcal{M}! - \sum_\alpha \ln \mathcal{M}_\alpha! \\ &\simeq C - \sum_\alpha \mathcal{M}_\alpha (\ln \mathcal{M}_\alpha - 1),\end{aligned} \tag{16}$$

where $C = \mathcal{M}(\ln \mathcal{M} - 1)$ is a constant and we have used the Stirling formula, $\ln N! \simeq N(\ln N - 1)$. We get

$$\sum_\alpha \delta \mathcal{M}_\alpha [\ln \mathcal{M}_\alpha + \beta E_\alpha - \lambda N_\alpha + \gamma] = 0 \tag{17}$$

for all $\delta \mathcal{M}_\alpha$ so that

$$\ln \mathcal{M}_\alpha + \beta E_\alpha - \lambda N_\alpha + \gamma = 0. \tag{18}$$

Equivalently, we have

$$\mathcal{M}_\alpha = \mathcal{M}_0 e^{-\beta E_\alpha + \lambda N_\alpha}, \tag{19}$$

where

$$\mathcal{M}_0 = e^{-\gamma}. \tag{20}$$

Since

$$\sum_\alpha \mathcal{M}_\alpha = \mathcal{M}, \tag{21}$$

we have

$$\mathcal{M}_0 \sum_\alpha e^{-\beta E_\alpha + \lambda N_\alpha} = \mathcal{M}. \tag{22}$$

We define the *partition function* by

$$Z(V,\beta,\mu) = \sum_\alpha e^{-\beta E_\alpha + \lambda N_\alpha}. \tag{23}$$

Here, the quantity V is the volume of the system. The dependence on V appears, because the number of single particle-states depends on the volume of the system. In the thermodynamical limit, $V \to \infty$, this number is proportional to V (see the later discussion).

We can determine β and λ from the conditions

$$\frac{\mathscr{M}_0}{\mathscr{M}} \sum_\alpha N_\alpha e^{-\beta E_\alpha + \lambda N_\alpha} = \langle N \rangle, \tag{24}$$

$$\frac{\mathscr{M}_0}{\mathscr{M}} \sum_\alpha E_\alpha e^{-\beta E_\alpha + \lambda N_\alpha} = \langle E \rangle. \tag{25}$$

In this way, in equilibrium, the probability of finding a given microstate α in the ensemble is given by

$$p_\alpha = \frac{\mathscr{M}_\alpha}{\mathscr{M}} = \frac{1}{Z} e^{-\beta(E_\alpha - \mu N_\alpha)}, \tag{26}$$

where

$$\mu = \frac{\lambda}{\beta}. \tag{27}$$

Remember that α is the index for distinguishing a microstate as introduced in Eq.(1), and not a single number. In the equations above, the sum over α is in fact a sum over all single particle configurations,

$$\{n_1, n_2, \cdots, n_i, \cdots\}.$$

Consequently, we have

$$\sum_\alpha = \sum_{n_1} \sum_{n_2} \sum_{n_3} \cdots \sum_{n_i} \cdots. \tag{28}$$

Equation (14) can be rewritten as

$$\delta\langle E \rangle = \frac{1}{\beta} \delta\left(\frac{1}{\mathscr{M}} \ln W\right) + \mu \, \delta\langle N \rangle. \tag{29}$$

The above relation should hold for any change $\{\mathscr{M}_\alpha\} \to \{\mathscr{M}'_\alpha\}$, fixing the parameters β and μ. Therefore, this equation should be compared with the well-known thermodynamical law

$$\delta\langle E \rangle = T \, \delta\langle S \rangle + \mu \, \delta\langle N \rangle, \tag{30}$$

for fixed volume ($\delta V = 0$). We thus identify the temperature T, the chemical potential μ, and the entropy S, respectively. Thus,

$$kT = \frac{1}{\beta}, \tag{31}$$

$$S = k \left(\frac{1}{\mathscr{M}} \ln W\right), \tag{32}$$

where k is the Boltzmann constant (here, it is introduced to adjust the dimension of T). If we use the relation

$$p_\alpha = \frac{\mathscr{M}_\alpha}{\mathscr{M}}, \tag{33}$$

we have

$$S \simeq k\frac{1}{\mathscr{M}}\left[\mathscr{M}(\ln\mathscr{M} - 1) - \sum_\alpha p_\alpha \mathscr{M}[\ln(p_\alpha \mathscr{M}) - 1]\right]$$
$$= -k\sum_\alpha p_\alpha \ln p_\alpha + \text{Const}. \tag{34}$$

We may take the constant above as zero, so that if the system is in a particular microstate $\alpha = \alpha_0$, that is,

$$p_\alpha = \begin{cases} 1, & \alpha = \alpha_0 \\ 0, & \text{otherwise} \end{cases}, \tag{35}$$

we have zero entropy.

We can reformulate the variational principle in terms of the probability distribution p_α. Eq.(29) reads

$$\delta \sum_\alpha \{p_\alpha E_\alpha + kT p_\alpha \ln p_\alpha - \mu N_\alpha p_\alpha + \gamma p_\alpha\} = 0, \tag{36}$$

for arbitrary variation δp_α. The last term in this equation is added to incorporate the restriction

$$\sum_\alpha p_\alpha = 1. \tag{37}$$

Eq.(36) is read as

$$\delta\langle E\rangle - T\delta\langle S\rangle - \mu\delta\langle N\rangle + \gamma\delta\langle 1\rangle = 0, \tag{38}$$

where

$$\langle E\rangle = \sum_\alpha p_\alpha E_\alpha,$$

etc., and the last term

$$\delta\langle 1\rangle \equiv \delta\left(\sum_\alpha p_\alpha\right)$$

is the change of the overall normalization of the total probability. We may interpret Eq.(38) as minimizing the free energy,

$$F = E - TS - \mu N,$$

under the fixed temperature and chemical potential and the normalization. Of course, the original form of Eq.(38) is

$$\delta\langle S\rangle - \beta[\delta\langle E\rangle + \mu\delta\langle N\rangle - \gamma\delta\langle 1\rangle] = 0, \tag{39}$$

that is, to maximize the entropy fixing the total average energy and total average number of particles, together with the constraint of conserving the total probability.

Expressing the average values in terms of p_α, we obtain from Eq.(38)

$$\sum_\alpha \delta p_\alpha \{E_\alpha - \mu N_\alpha + \gamma + kT(\ln p_\alpha + 1)\} = 0, \qquad (40)$$

and consequently,

$$p_\alpha = \frac{1}{Z} e^{-(E_\alpha - \mu N_\alpha)/kT}, \qquad (41)$$

where

$$\frac{1}{Z} \equiv e^{\gamma/kT + 1} \qquad (42)$$

is related to the normalization of the probability. Using Eq.(37), we obtain

$$Z = Z(V, T, \mu) = \sum_\alpha e^{-(E_\alpha - \mu N_\alpha)/kT}. \qquad (43)$$

This function is known as the partition function for the grand canonical ensemble.

Once we know the partition function Z in terms of μ and T, we can calculate all the thermodynamical quantities. We have

$$\langle E \rangle = -\left.\frac{\partial \ln Z}{\partial \beta}\right|_{\mu\beta}, \qquad (44)$$

$$\langle N \rangle = \frac{1}{\beta} \left.\frac{\partial \ln Z}{\partial \mu}\right|_\beta, \qquad (45)$$

$$\langle S \rangle = -k \langle \ln p_\alpha \rangle \qquad (46)$$

$$= \frac{1}{T} \sum_\alpha (E_\alpha - \mu N_\alpha) p_\alpha + k \ln Z \qquad (47)$$

$$= \frac{1}{T} (\langle E \rangle - \mu \langle N \rangle) + k \ln Z. \qquad (48)$$

Note that the partial derivative in Eq.(44) should be performed by fixing the quantity $\lambda = \mu\beta$. When the system is large enough and the general extensive thermodynamical relation

$$\langle S \rangle = \frac{1}{T} \langle E \rangle - \frac{\mu}{T} \langle N \rangle + \frac{1}{T} PV, \qquad (49)$$

is valid, then we identify

$$\frac{1}{\beta} \ln Z = PV, \qquad (50)$$

which is in fact the thermodynamical potential for a grand canonical ensemble.

Ideal Fermi Gas

Let us apply the above results to an ideal gas of Fermi-Dirac particles. For a Fermi gas, due to the Pauli exclusion principle, the occupation number of particles n_i for each state i is limited to be 0 or 1. Thus, the partition function becomes

$$
\begin{aligned}
Z(V,\beta,\mu) &= \sum_\alpha e^{-\beta(E_\alpha - \mu N_\alpha)} \\
&= \sum_{n_1}\sum_{n_2}\sum_{n_3}\cdots\sum_{n_i}\cdots \exp\left\{-\beta \sum_i n_i(\varepsilon_i - \mu)\right\} \\
&= \prod_i \sum_{n_i=0,1} e^{-\beta n_i(\varepsilon_i - \mu)} \\
&= \prod_i \left\{1 + e^{-\beta(\varepsilon_i - \mu)}\right\} \\
&= \exp \sum_i \ln\left[1 + e^{-\beta(\varepsilon_i - \mu)}\right].
\end{aligned}
\quad (51)
$$

Reminding that the single particle state i for an ideal gas is taken as a plane wave state of momentum $\vec{k}$, we may replace the sum over states i by an integral in $\vec{k}$,

$$\sum_i \to \frac{gV}{(2\pi\hbar)^3} \int d^3k,$$

where g is the statistical factor of the particle[1]. For simplicity, from now on, we switch to the system of units where $\hbar = c = 1$. For a spin 1/2 particle, this factor is 2. We get

$$\ln Z(V,T,\mu) = \frac{gV}{(2\pi)^3} \int d^3k \ln\left[1 + e^{-\beta(\varepsilon_k - \mu)}\right], \quad (52)$$

where ε_k is the energy of the state with momentum $\vec{k}$.

The total energy of the system is

$$
\begin{aligned}
\langle E \rangle &= -\left.\frac{\partial}{\partial \beta} \ln Z(V,T,\mu)\right|_{\beta\mu} \\
&= \frac{gV}{(2\pi)^3} \int d^3k \frac{\varepsilon_k e^{-\beta(\varepsilon_k - \mu)}}{1 + e^{-\beta(\varepsilon_k - \mu)}} \\
&= \frac{gV}{(2\pi)^3} \int d^3k \frac{\varepsilon_k}{e^{\beta(\varepsilon_k - \mu)} + 1}
\end{aligned}
\quad (53)
$$

and the total number of particles of the system becomes

$$
\begin{aligned}
\langle N \rangle &= \left.\frac{1}{\beta}\frac{\partial}{\partial \mu} \ln Z(V,T,\mu)\right|_\beta \\
&= \frac{gV}{(2\pi)^3} \int d^3k \frac{1}{e^{\beta(\varepsilon_k - \mu)} + 1}.
\end{aligned}
\quad (54)
$$

---

[1] Note that this is true in the thermodynamical limit, $V \to \infty$. See Ref[5].

The above Eqs.(53) and (54) show that the average of occupation number for the energy level ε_k in the Fermi gas is given by

$$f(\varepsilon_k) = \frac{1}{e^{\beta(\varepsilon_k-\mu)}+1}, \qquad (55)$$

which is known as the Fermi distribution. The pressure is given by

$$P = \frac{g}{(2\pi)^3}\frac{1}{\beta}\int d^3k \ln\left[1+e^{-\beta(\varepsilon_k-\mu)}\right]. \qquad (56)$$

Finally, the entropy is calculated from

$$T\langle S\rangle = \langle E\rangle - \mu\langle N\rangle + PV.$$

Ideal Bose-Einstein Gas

For bosons, the sum over states differs from that of a Fermi gas. There is no restriction for the occupation numbers n_i, so we have to sum over all non-negative integers:

$$\begin{aligned}
Z(V,\beta,\mu) &= \sum_\alpha e^{-\beta(E_\alpha-\mu N_\alpha)} \\
&= \sum_{n_1}\sum_{n_2}\sum_{n_3}\cdots\sum_{n_i}\cdots \exp\left\{-\beta\sum_i n_i(\varepsilon_i-\mu)\right\} \\
&= \prod_i \sum_{n_i=0}^{\infty} e^{-\beta n_i(\varepsilon_i-\mu)} \\
&= \prod_i \frac{1}{1-e^{-\beta(\varepsilon_i-\mu)}} \\
&= \exp\left\{-\sum_i \ln\left[1-e^{-\beta(\varepsilon_i-\mu)}\right]\right\},
\end{aligned} \qquad (57)$$

where we have assumed

$$\varepsilon_i - \mu > 0 \qquad (58)$$

to assure the convergence. Introducing again the integral over plane-wave states, we get

$$\ln Z(V,T,\mu) = -\frac{gV}{(2\pi)^3}\int d^3k \ln\left[1-e^{-\beta(\varepsilon_k-\mu)}\right]. \qquad (59)$$

In an analogous way as in the case of the Fermi gas, the expressions for the energy and the particle number of a boson gas are found to be

$$\langle E\rangle = \frac{gV}{(2\pi)^3}\int d^3k \frac{\varepsilon_k}{e^{\beta(\varepsilon_k-\mu)}-1}, \qquad (60)$$

$$\langle N\rangle = \frac{gV}{(2\pi)^3}\int d^3k \frac{1}{e^{\beta(\varepsilon_k-\mu)}-1}. \qquad (61)$$

The pressure is given by

$$P = -\frac{g}{(2\pi)^3}\frac{1}{\beta}\int d^3k \ln\left[1 - e^{-\beta(\varepsilon_k - \mu)}\right], \tag{62}$$

and again

$$T\langle S\rangle = \langle E\rangle - \mu\langle N\rangle + PV. \tag{63}$$

The probability of occupation of the energy level ε_k for a boson gas is

$$f(\varepsilon_k) = \frac{1}{e^{\beta(\varepsilon_k - \mu)} - 1}. \tag{64}$$

Note that we should have $\varepsilon_k > \mu$ for all k, so that

$$m > \mu \tag{65}$$

where m is the mass of the boson. In the limit $\mu \to m$, $\langle E\rangle/V$ and $\langle N\rangle/V$ diverge, and the behavior of the equation of state changes qualitatively. This is known as Bose-Einstein condensate (see the later discussion).

RELATIVISTIC IDEAL GASES

In the final stage of relativistic heavy ion collisions, a lot of hadrons are produced. Let us assume that such a state can approximately be described as an ideal gas of hadrons. Such an approximation will be valid if the thermal energy is sufficiently large compared to the interaction energies between hadrons. This means that for low mass particles, we have to treat their kinematics relativistically. We have to evaluate the integral in Eq.(59) with[2]

$$\varepsilon_k = \sqrt{k^2 + m^2}, \tag{66}$$

where m is the mass of the particle. Expressions for number density n, energy density ε and pressure P are given by

$$n = \frac{g}{2\pi^2}\int_0^\infty dk\, k^2 \frac{1}{e^{\beta(\sqrt{k^2+m^2}-\mu)} \pm 1}, \tag{67}$$

$$\varepsilon = \frac{g}{2\pi^2}\int_0^\infty dk\, k^2 \frac{\sqrt{k^2+m^2}}{e^{\beta(\sqrt{k^2+m^2}-\mu)} \pm 1}, \tag{68}$$

$$P = \pm\frac{g}{2\pi^2}\frac{1}{\beta}\int_0^\infty dk\, k^2 \ln\left[1 \pm e^{-\beta(\sqrt{k^2+m^2}-\mu)}\right], \tag{69}$$

where the double sign $\pm$ correspond to the case of Fermions and Bosons, respectively.

---

[2] We use the natural unit, $\hbar = c = 1$.

Non-degenerate Case

First, let us evaluate the pressure. When

$$e^{-\beta(m-\mu)} < 1, \tag{70}$$

we can expand the integrand as

$$\ln\left[1 \pm e^{-\beta(\varepsilon_k - \mu)}\right] = \pm \sum_{n=1}^{\infty} \frac{(\mp 1)^{n-1}}{n} e^{-\beta n(\varepsilon_k - \mu)}, \tag{71}$$

so that

$$P = \frac{g}{2\pi^2} \frac{1}{\beta} \sum_{n=1}^{\infty} \frac{(\mp 1)^{n-1}}{n} e^{\beta n \mu} \Phi(\beta n, m), \tag{72}$$

where

$$\begin{aligned}
\Phi(\beta n, m) &\equiv \int_0^\infty dk\, k^2\, e^{-\beta n \varepsilon_k} \\
&= \int_0^\infty dk\, k^2\, e^{-\beta n \sqrt{k^2 + m^2}} \\
&= m^3 \int_0^\infty dx\, x^2\, e^{-z\sqrt{x^2 + 1}}
\end{aligned} \tag{73}$$

with $z = \beta n m$. Using the integral representation of modified Bessel functions,

$$K_\nu(z) = \frac{\sqrt{\pi}}{\Gamma(\nu + \frac{1}{2})} \left(\frac{z}{2}\right)^\nu \int_0^\infty e^{-z\sqrt{x^2+1}} \frac{x^{2\nu}}{\sqrt{x^2+1}} dx,$$

which holds for $z > 0$, Re $\nu > 1/2$, we identify

$$\begin{aligned}
\Phi(\beta n, m) &= -m^3 \frac{2\Gamma(\frac{3}{2})}{\sqrt{\pi}} \left[\frac{d}{dz}\left(\frac{1}{z} K_1(z)\right)\right]_{z=\beta n m} \\
&= m^3 \left[\frac{1}{z} K_2(z)\right]_{z=\beta n m}.
\end{aligned} \tag{74}$$

Finally, we get

$$P = \frac{g}{2\pi^2} \frac{m^3}{\beta} \sum_{n=1}^{\infty} \frac{(\mp 1)^{n-1}}{n} e^{n\beta\mu} \left[\frac{1}{z} K_2(z)\right]_{z=\beta n m}. \tag{75}$$

Once P is expressed as a function of β and μ, we obtain the number density

$$\begin{aligned}
n &= \left.\frac{\partial P}{\partial \mu}\right|_\beta \\
&= \frac{g m^3}{2\pi^2} \sum_{n=1}^{\infty} (\mp 1)^{n-1} e^{n\beta\mu} \left[\frac{1}{z} K_2(z)\right]_{z=\beta n m}
\end{aligned} \tag{76}$$

and the energy density

$$\begin{aligned}
\varepsilon &= -\left.\frac{\partial(\beta P)}{\partial \beta}\right|_{\beta\mu} \\
&= -\frac{gm^4}{2\pi^2}\sum_{n=1}^{\infty}(\mp 1)^{n-1}e^{n\beta\mu}\left[\frac{d}{dz}\left(\frac{1}{z}K_2(z)\right)\right]_{z=\beta nm} \\
&= \frac{gm^4}{2\pi^2}\sum_{n=1}^{\infty}(\mp 1)^{n-1}e^{n\beta\mu}\left[\frac{1}{z}\left(\frac{3}{z}K_2(z)+K_1(z)\right)\right]_{z=\beta nm},
\end{aligned} \qquad (77)$$

while the entropy density can be calculated as

$$Ts = \varepsilon + P - \mu n. \qquad (78)$$

The above expressions are only valid for

$$e^{-\beta(m-\mu)} < 1. \qquad (79)$$

The series converges very slowly for $e^{-\beta(m-\mu)} \to 1$, and for $e^{-\beta(m-\mu)} \geq 1$ the sum does not converge. For bosons, this last situation does not happen, but for fermions it can occur for rather high density and low temperature.

For practical applications, it is important to know the limit of validity of the series expansion. In the figure 1 below, we show the number density of a typical baryon ($m = m_n = 938$ MeV) as a function of temperature T, for $\mu = m$. The series representation of integrals shown above are valid only for the domain below this curve.

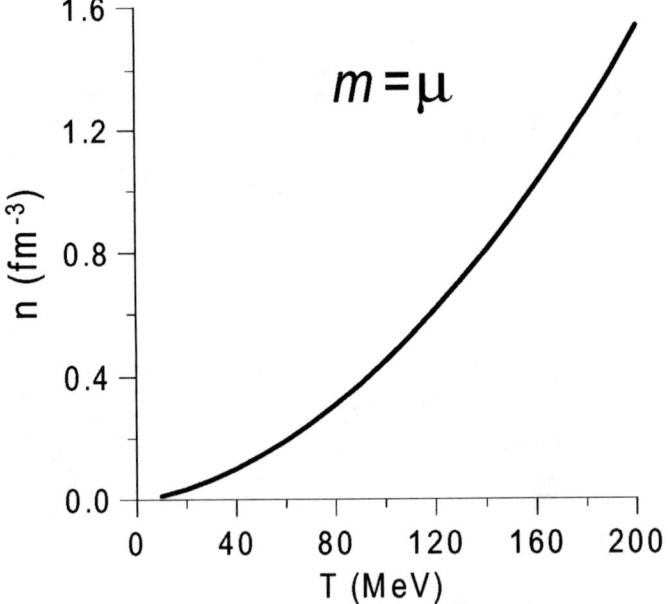

Figure 1: Nuclear density as function of temperature T for $m = \mu$.

Boltzmann Limit

When
$$e^{-\beta(m-\mu)} \ll 1,$$
or equivalently $m - \mu$ is sufficiently larger than T, the above series expansion converges very rapidly and in practice, only the first term gives a good approximation. In this limit, there is no difference between bosons and fermions. Explicitly, we have

$$P \to P_{Boltz} = \frac{g}{2\pi^2} m^2 T^2 e^{\frac{\mu}{T}} K_2\left(\frac{m}{T}\right), \tag{80}$$

$$n \to n_{Boltz} = \frac{g}{2\pi^2} m^2 T e^{\frac{\mu}{T}} K_2\left(\frac{m}{T}\right), \tag{81}$$

$$\varepsilon \to \varepsilon_{Boltz} = \frac{g}{2\pi^2} m^3 T e^{\frac{\mu}{T}} \left[K_1\left(\frac{m}{T}\right) + \frac{3T}{m} K_2\left(\frac{m}{T}\right)\right]. \tag{82}$$

We can immediately see
$$P_{Boltz} = n_{Boltz} T,$$
which is the well-known classical ideal gas equation of state. Furthermore, for $m \gg T$, we can express the mean energy per particle

$$\left(\frac{\varepsilon}{n}\right)_{Boltz} = m \left[\frac{K_1(z)}{K_2(z)} + \frac{3}{z}\right]_{z=\frac{m}{T}}$$

$$= m \left[1 + \frac{3}{2z} + \frac{15}{8z^2} + \cdots\right]_{z=\frac{m}{T}}$$

$$= m + \frac{3}{2} T + \frac{15}{8m} T^2 + \cdots$$

where we have used the asymptotic expansion of Bessel functions. The first term is the rest mass and the second term is the classical formula for the mean kinetic energy of ideal gas. Higher terms are relativistic corrections.

In Figs. 2a, b, and c, we show how the series expansions for the energy density, pressure and entropy density converge to the exact values when the number of terms N is increased. Here, we take for a boson gas with mass = 150 MeV, corresponding to typically π-mesons. The temperature is taken to be 200 MeV. In these figures, the curves indicated as $N = 1$ (Boltzmann) correspond to the Boltzmann approximation. For boson gas, at $\mu \to m$, the energy density and particle density diverge, corresponding to the Bose-Einstein condensate. However, as a function of the number density, the pressure and entropy density tend to constant, and the energy per particle decreases, tending to the rest mass energy. This is because, the increase of particle number of the system after certain amount is just consumed up to fill lowest energy states and does not contribute to the total energy and entropy. For more details of Bose-Einstein condensation, see the standard text books [2].

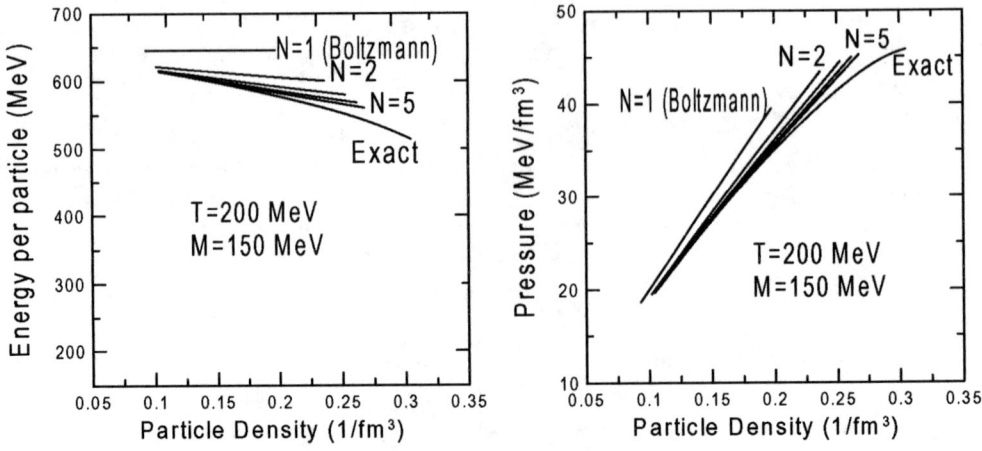

Figure 2a: Energy per particle of a boson gas.   Figure 2b: Pressure of a boson gas

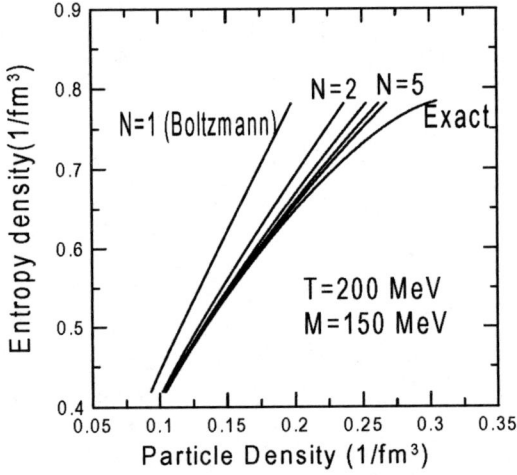

Figure 2c: Entropy density of a boson gas

For boson gas, at $\mu \to m$, the energy density and particle density diverge, corresponding to the Bose-Einstein condensate. However, as a function of the number density, the pressure and entropy density tend to constant, and the energy per particle decreases, tending to the rest mass energy. This is because, the increase of particle number of the system after certain amount is just consumed up to fill lowest energy states and does not contribute to the total energy and entropy. For more details of Bose-Einstein condensation, see the standard text books[2].

In Figs. 3a, b and c, we plotted the behaviors of the energy per particle, pressure and entropy density of a fermion gas as functions of particle density. Here, we show the example of a typical baryon gas, with mass $m = 900$ MeV at the temperature 200 MeV.

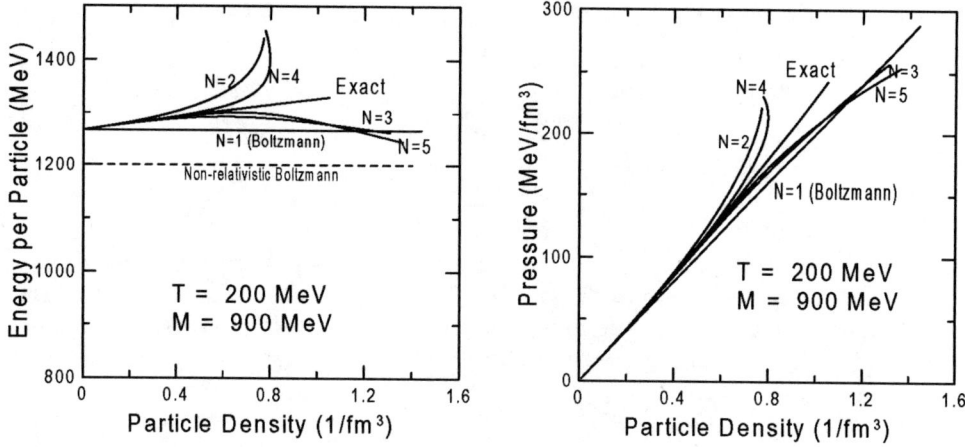

Figure 3a: Energy per particle of a fermion gas   Figure 3b: Pressure of a fermion gas.

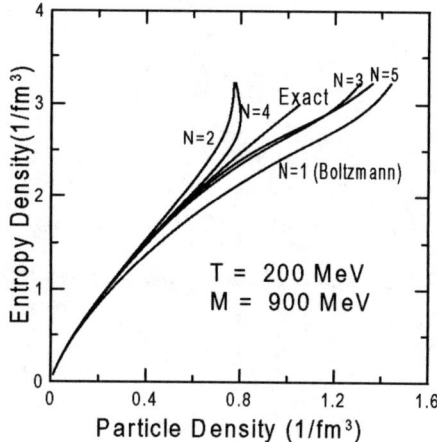

Figure 3c: Entropy of a fermion gas.

As we see from these figures, the Boltzmann approximation is not so bad at these temperature and density values but the convergence of the series expansion becomes catastrophic for particle density greater than $0.8 fm^{-3}$. For very high density, see the section

Mixture of Particles and Chemical Equilibrium

It is easy to extend the formulation of the previous section to a system which contains more than one kind of particles. Let us denote baryon number, strangeness, and electric charge of the type t particle as $b_t, s_t,$ and e_t, respectively. The total baryon number Q_b, total strangeness Q_s and total electric charge Q_e of the system are

$$Q_b = \sum_t b_t N_t, \tag{83}$$

$$Q_s = \sum_t s_t N_t, \tag{84}$$

$$Q_e = \sum_t e_t N_t, \tag{85}$$

where N_t is the total number of particles of the type t. Suppose that these are the only conserved quantum numbers of the system. Then we may ask "what is the probability distribution of microstates at equilibrium, given the numbers of conserved quantum numbers?" To find the answer, we extend Eq.(39) as

$$\delta\langle S\rangle - \beta\left[\delta\langle E\rangle + \mu_b\,\delta\langle Q_b\rangle + \mu_s\delta\langle Q_s\rangle + \mu_e\delta\langle Q_e\rangle - \sum_t \gamma_t\delta\langle 1\rangle_t\right] = 0, \tag{86}$$

where

$$\langle S\rangle = \sum_t \langle S_t\rangle, \tag{87}$$

$$\langle E\rangle = \sum_t \langle E_t\rangle, \tag{88}$$

are the entropy and total energy of the system. Here, $\langle S_t\rangle$ and $\langle E_t\rangle$ represent the entropy and energy of the particle t. Variation should be taken with respect to the probability distribution $\left\{p_\alpha^{(t)}\right\}$ for each particle of type t, so that the last term in Eq.(86) represents the constraints of the normalization of each $\left\{p_\alpha^{(t)}\right\}$. Substituting Eqs.(83,84,85,87,88) into Eq.(86), we have

$$\sum_t \delta\left\{\langle S_t\rangle - \beta\left[\delta\langle E_t\rangle + \mu_t\,\delta\langle N_t\rangle - \gamma_t\delta\langle 1\rangle_t\right]\right\} = 0,$$

where

$$\mu_t \equiv b_t\mu_b + s_t\mu_s + e_t\mu_e.$$

Since $\left\{p_\alpha^{(t)}\right\}$ are independent for each particle type t, we get

$$\delta\left\{\langle S_t\rangle - \beta\left[\delta\langle E_t\rangle + \mu_t\,\delta\langle N_t\rangle - \gamma_t\delta\langle 1\rangle_t\right]\right\} = 0, \tag{89}$$

for each t. This last equation shows that all the results for the unique particle-type case can readily be generalized to a mixture of many different kinds of particles, just substituting the chemical potential of the type t particle by

$$\mu \to \mu_t = b_t\mu_b + s_t\mu_s + e_t\mu_e,$$

thus introducing chemical potentials for each conserved quantity. The resulting formulas describe the chemical and thermal equilibrium among particles.

SOME USEFUL APPROXIMATIONS

As we see from the above figure, the domain of applicability of the series expansion for Fermi integrals is rather small. Particularly for light mass fermions, the situation becomes worse. For example, if we extend our ideal gas description to quarks, then the series expansion does not apply for most regions of interest. In this section, we will see some useful analytical approximations of the above Fermi integrals[4].

When a fermion gas is in thermal equilibrium, its antiparticle also appears due to pair production or other possible reaction channels. The chemical potential of the antiparticle is just the opposite of the chemical potential of the particle. Since all thermodynamical quantities in the Grand Canonical Ensemble are deduced from the thermodynamical potential, let us consider here only the pressure. The pressure of a system of particles and antiparticles in equilibrium is then

$$P(\mu,T) = \frac{gT}{(2\pi)^3}\int d^3k \left[\ln\left(e^{-(E-\mu)/T}+1\right) + \ln\left(e^{-(E+\mu)/T}+1\right)\right]$$

$$= \frac{g}{6\pi^2}\int_m^\infty dE\, (E^2-m^2)^{3/2} \left[\frac{1}{e^{(E-\mu)/T}+1} + \frac{1}{e^{(E+\mu)/T}+1}\right]. \qquad (90)$$

Let us consider the evaluation of the integral

$$F(a,b) = \int_a^\infty dx\, (x^2-a^2)^{3/2} \left[\frac{1}{e^{(x-b)}+1} + \frac{1}{e^{(x+b)}+1}\right]. \qquad (91)$$

The pressure P is proportional to a function of this form,

$$P = \frac{gT^4}{6\pi^2} F(a,b), \qquad (92)$$

where a and b are related to the mass and the chemical potential by

$$a = \frac{m}{T}, \qquad (93)$$

$$b = \frac{\mu}{T}, \qquad (94)$$

with T measured in units of energy ($k=1$).

1) Degenerate case

At extremely high densities, the pressure of a Fermi gas is determined essentially by the density and less dependent on the temperature. Such a situation occurs often in astrophysical processes, for example, in the core of advanced stage of heavy stars, white dwarfs, and neutron stars. Several peculiar processes like supernova explosion are intimately associated to the degeneracy of Fermi gas. This is due to the change of behavior of the pressure with respect to the temperature [6]. In our expression the degeneracy corresponds a large chemical potential compared to the temperature

$$b \gg 1, \qquad (95)$$

and for this case, we can safely approximate

$$\left[\frac{1}{e^{(x-b)}+1} + \frac{1}{e^{(x+b)}+1}\right] \simeq \frac{1}{e^{(x-b)}+1}, \qquad (96)$$

for all x between zero and infinity. Then, writing

$$x = bu, \qquad (97)$$

we get

$$F(a,b) \simeq b^4 \int_{a/b}^{\infty} du \left(u^2 - (a/b)^2\right)^{3/2} \frac{1}{e^{b(u-1)}+1}. \qquad (98)$$

In general, for any function $f(u)$, we can write

$$\int_{a/b}^{\infty} du\, f(u) \frac{1}{e^{b(u-1)}+1}$$
$$= \int_{a/b}^{\infty} du\, f(u) \left(\theta(1-u) + \frac{1}{e^{b(u-1)}+1} - \theta(1-u)\right)$$
$$= \int_{a/b}^{1} du\, f(u) + \int_{a/b}^{\infty} du\, f(u) \left(\frac{1}{e^{b(u-1)}+1} - \theta(1-u)\right)$$
$$\equiv I_1 + I_2. \qquad (99)$$

Here the second term is

$$I_2 \equiv \int_{a/b}^{\infty} du\, f(u) \left(\frac{1}{e^{b(u-1)}+1} - \theta(1-u)\right)$$
$$= \int_{a/b}^{1} du\, f(u) \left(\frac{1}{e^{b(u-1)}+1} - 1\right) + \int_{1}^{\infty} du\, f(u) \frac{1}{e^{b(u-1)}+1}$$
$$= -\int_{a/b}^{1} du\, f(u) \frac{e^{b(u-1)}}{e^{b(u-1)}+1} + \int_{0}^{\infty} du\, f(u+1) \frac{1}{e^{bu}+1}$$
$$= -\int_{a/b}^{1} du\, f(u) \frac{1}{1+e^{-b(u-1)}} + \int_{0}^{\infty} du\, f(u+1) \frac{1}{e^{bu}+1}$$
$$= -\int_{0}^{1-a/b} du\, f(1-u) \frac{1}{1+e^{bu}} + \int_{0}^{\infty} du\, f(u+1) \frac{1}{e^{bu}+1} \qquad (100)$$

For $a \ll b$, we may safely approximate (with an error the order of $\simeq e^{-(b-a)}$)

$$I_2 \simeq -\int_{0}^{\infty} du\, f(1-u) \frac{1}{1+e^{bu}} + \int_{0}^{\infty} du\, f(u+1) \frac{1}{e^{bu}+1}$$
$$= \int_{0}^{\infty} du \frac{1}{1+e^{bu}} \left[f(u+1) - f(1-u)\right] \qquad (101)$$

Expanding the function f in a power series of u, we get

$$f(u+1) = f(1) + \frac{1}{1!}f^{(1)}(1)u + \frac{1}{2!}f^{(2)}(1)u^2 + \cdots, \quad (102)$$

$$f(1-u) = f(1) - \frac{1}{1!}f^{(1)}(1)u + \frac{1}{2!}f^{(2)}(1)u^2 - \cdots, \quad (103)$$

and

$$I_2 \simeq \sum_{k=0}^{\infty} \frac{2}{(2k+1)!} f^{(2k+1)}(1) \int_0^{\infty} du \frac{u^{2k+1}}{1+e^{bu}} \quad (104)$$

In addition, we have

$$\int_0^{\infty} du \frac{u^{2k+1}}{1+e^{bu}} = \int_0^{\infty} du\, u^{2k+1} \sum_{n=1}^{\infty} (-1)^{n-1} e^{-bnu}$$

$$= \frac{1}{b^{2k+2}} \sum_{n=1}^{\infty} \frac{(-1)^{n-1}}{n^{2k+2}} \int_0^{\infty} dx\, x^{2k+1} e^{-x}$$

$$= \frac{(2k+1)!}{b^{2k+2}} \sum_{n=1}^{\infty} \frac{(-1)^{n-1}}{n^{2k+2}}$$

$$= \frac{(2k+1)!}{b^{2k+2}} \left(1 - \frac{1}{2^{2k+1}}\right) \zeta(2k+2), \quad (105)$$

so that

$$I_2 \simeq \sum_{k=0}^{\infty} \frac{2}{b^{2k+2}} \left(1 - \frac{1}{2^{2k+1}}\right) \zeta(2k+2) f^{(2k+1)}(1)$$

$$= \frac{1}{b^2} \zeta(2) f^{(1)}(1) + \frac{7}{4b^4} \zeta(4) f^{(3)}(1) + \cdots \quad (106)$$

Finally, for $b \gg 1$, a,

$$\int_{a/b}^{\infty} du\, f(u) \frac{1}{e^{b(u-1)}+1}$$

$$\simeq \int_{a/b}^{1} du\, f(u) + \frac{\pi^2}{6b^2} f^{(1)}(1) + \frac{7\pi^4}{360 b^4} f^{(3)}(1) + \cdots \quad (107)$$

This is known as Sommerfeld expansion. For

$$f = (u^2 - z^2)^{3/2}, \quad (108)$$

where $z \equiv a/b$, we have

$$f^{(1)}(1) = 3\sqrt{1-z^2} \quad (109)$$

and

$$f^{(3)}(1) = 3\frac{2 - 3z^2}{(1-z^2)^{3/2}}, \quad (110)$$

so that

$$F(a,b) \simeq \frac{1}{8}\left[b\left(2b^2 - 5a^2\right)\left(b^2 - a^2\right)^{1/2} + 3a^4 \ln\frac{b+\sqrt{b^2-a^2}}{a}\right]$$
$$+ \frac{\pi^2 b\sqrt{b^2-a^2}}{2} + \frac{7\pi^4}{120}\frac{b(2b^2 - 3a^2)}{(b^2-a^2)^{3/2}} + \cdots \qquad (111)$$

This is valid for $b \gg a, 1$. The approximation contains errors of the order of

$$\sim e^{-(b-a)}.$$

2) Ultrarelativistic Limit

For $a \to 0$, we can evaluate the integral as shown below. We have

$$F(a,b) = \int_0^\infty dx \left(x^3 - \frac{3}{2}a^2 x\right)\left[\frac{1}{e^{(x-b)}+1} + \frac{1}{e^{(x+b)}+1}\right] + O(a^3) \qquad (112)$$

In general,

$$\int_0^\infty dx\, f(x)\left[\frac{1}{e^{(x-b)}+1} + \frac{1}{e^{(x+b)}+1}\right]$$
$$= \int_{-b}^\infty dx\, f(x+b)\frac{1}{e^x+1} + \int_b^\infty dx\, f(x-b)\frac{1}{e^x+1}$$
$$= \int_0^\infty dx\,[f(x+b) + f(x-b)]\frac{1}{e^x+1}$$
$$+ \int_{-b}^0 dx\, f(x+b)\frac{1}{e^x+1} - \int_0^b dx\, f(x-b)\frac{1}{e^x+1} \qquad (113)$$

However,

$$\int_{-b}^0 dx\, f(x+b)\frac{1}{e^x+1} = \int_0^b dx\, f(-x+b)\frac{1}{e^{-x}+1}$$
$$= \int_0^b dx\, f(-x+b)\frac{e^x}{e^x+1}$$
$$= \int_0^b dx\, f(-x+b)\left[1 - \frac{1}{e^x+1}\right], \qquad (114)$$

so that

$$\int_0^\infty dx\, f(x)\left[\frac{1}{e^{(x-b)}+1} + \frac{1}{e^{(x+b)}+1}\right]$$
$$= \int_0^\infty dx\,[f(x+b)+f(x-b)]\frac{1}{e^x+1} + \int_0^b dx\, f(-x+b)$$
$$- \int_0^b dx\,[f(x-b) + f(-x+b)]\frac{1}{e^x+1}. \qquad (115)$$

In our case, $f(x) = x^3 - (3a^2/2)x$ is an odd function of x, so that the last term just vanishes. Following same steps as above,

$$\int_0^\infty dx \frac{x^k}{e^x+1} = k!\left(1 - \frac{1}{2^k}\right)\zeta(k+1),$$

and

$$\begin{aligned}
F(a,b) &\simeq \frac{1}{4}b^4 - \frac{3}{4}a^2b^2 + 2\int_0^\infty dx\left[x^3 + 3\left(b^2 - \frac{1}{2}a^2\right)x\right]\frac{1}{e^x+1} \\
&= \frac{1}{4}b^4 - \frac{3}{4}a^2b^2 + 2\cdot 3\left(b^2 - \frac{1}{2}a^2\right)\frac{1}{2}\zeta(2) + 2\cdot 3!\cdot\frac{7}{8}\zeta(4) \\
&= \frac{1}{4}b^4 - \frac{3}{4}a^2b^2 + \frac{\pi^2}{2}\left(b^2 - \frac{1}{2}a^2\right) + \frac{7\pi^4}{60}.
\end{aligned} \quad (116)$$

Note that the first terms in the above expression coincide with the Taylor expansion in a of the degenerate limit, Eq.(111). Therefore, we can express both cases, the ultra-relativistic and extreme degeneracy limits as

$$F(a,b) \simeq b^4 F_0(z) + \frac{\pi^2}{2}b^2\sqrt{1-z^2} + \frac{7\pi^4}{120}\frac{(2-3z^2)}{(1-z^2)^{3/2}}, \quad (117)$$

where $z = a/b = m/\mu$ and

$$F_0(z) = \frac{1}{8}\left[(2-5z^2)(1-z^2)^{1/2} + 3z^4 \ln\left(\frac{1+\sqrt{1-z^2}}{z}\right)\right]. \quad (118)$$

This approximation is valid in the whole domain ($0 < b < \infty$) as far as z is sufficiently smaller than unity. Finally, we get the expression for the pressure of an ultra-relativistic fermion gas including the antiparticles as

$$\begin{aligned}
P(\mu,T) &\simeq \frac{g}{6\pi^2(\hbar c)^3}\left[\frac{\mu^4}{8}\left((2-5z^2)(1-z^2)^{1/2} + 3z^4 \ln\left(\frac{1+\sqrt{1-z^2}}{z}\right)\right)\right. \\
&\left. + \frac{\pi^2}{2}\mu^2 T^2 \sqrt{1-z^2} + \frac{7\pi^4 T^4}{120}\frac{(2-3z^2)}{(1-z^2)^{3/2}}\right],
\end{aligned} \quad (119)$$

and the corresponding number density is

$$n \simeq \frac{g}{6\pi^2(\hbar c)^3}\left[\mu^3(1-z^2)^{3/2} + \frac{\pi^2}{2}\mu T^2 \frac{2-z^2}{\sqrt{1-z^2}} + \frac{7\pi^4 T^4}{40\mu}\frac{z^4}{(1-z^2)^{5/2}}\right].$$

The approximation is valid for

$$z = \frac{m}{\mu} \ll 1, \quad (120)$$

and they are exact for both of $T = 0$ and $m = 0$ cases. In Fig. 4, we show the equation of state $p = p(n, T)$ for a Fermi gas with $m = 900$ MeV, for two different temperatures. These curves are obtained by Eqs.(119,120), eliminating the chemical potential μ. For comparison, we also show the corresponding curves using the exact integrals. For higher densities, the pressure tends to independent of the temperature T and in this region, the approximation becomes asymptotically exact.

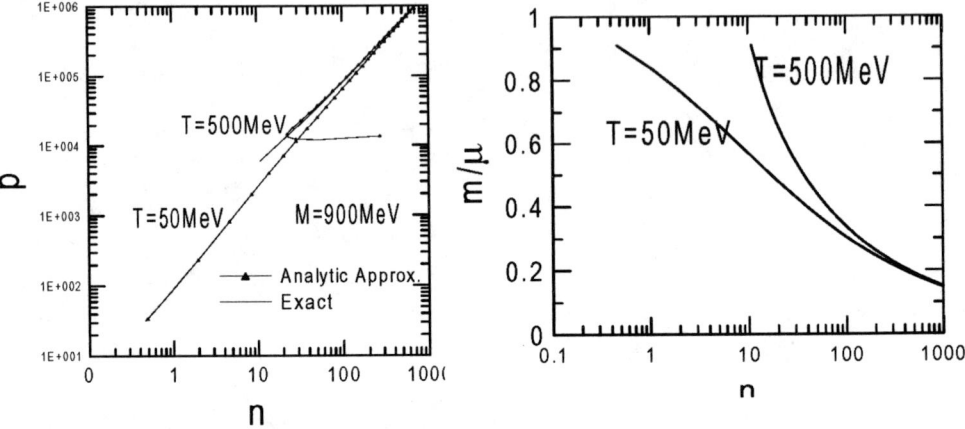

Figure 4: Comparison of the analytic approximation with the exact values for the degenerate Fermi gas.

Figure 5: Behavior of chemical potential as a function of particle density.

As we see, for the lower temperature, the approximation extends much more to the domain of low particle densities compared to high temperature case. This is because, when T becomes smaller, the value of $z = m/\mu$ stay smaller than unity for lower densities as we can see from Fig. 5.

- Exercise: Derive the expression for the pressure in ultra-relativistic regime of a boson gas.

Numerical Method

Although the analytical expressions above are useful to discuss the general properties of a relativistic gas, none of them can cover the whole region of parameters μ and T. For practical calculations, it is desirable to possess a simple and efficient method to obtain precise values of thermodynamical quantities. In this sense, it is more effective to evaluate directly the integrals Eqs.(67,68,69) using Gauss' quadrature method. To do this, we can rewrite the integrals as

$$n = \frac{gT^3}{2\pi^2 (\hbar c)^3} \int_0^\infty dx \frac{x^{1/2} (x+a)(x+2a)^{1/2}}{e^{a-b} \pm e^{-x}} e^{-x}, \qquad (121)$$

$$\varepsilon = \frac{gT^4}{2\pi^2(\hbar c)^3} \int_0^\infty dx \frac{x^{1/2}(x+a)^2(x+2a)^{1/2}}{e^{a-b} \pm e^{-x}} e^{-x}, \qquad (122)$$

$$P = \frac{gT^4}{6\pi^2(\hbar c)^3} \int_0^\infty dx \frac{x^{3/2}(x+2a)^{3/2}}{e^{a-b} \pm e^{-x}} e^{-x}, \qquad (123)$$

where as before, $a = m/T$ and $b = \mu/T$. We note that all these integrals have the form

$$\int_0^\infty dx\, x^\alpha e^{-x} f(x),$$

which can well be evaluated by the Gauss-Laguerre quadrature methods as

$$\int_0^\infty dx\, x^\alpha e^{-x} f(x) \simeq \sum_{i=1}^N \omega_i f(x_i),$$

where x_i and ω_i are calculated in terms of orthogonal polynomials associated to confluent hypergeometric functions. A subtle detail here is that, for the n-integration Eq.(121), for example, we can take,

$$\alpha \to \frac{1}{2},$$

$$f(x) \to \frac{(x+a)(x+2a)^{1/2}}{e^{a-b} \pm e^{-x}},$$

but if a is much smaller than the first x value of Gauss-Laguerre quadrature, that is,

$$a \ll x_1,$$

then the approximation becomes better choosing

$$\alpha \to 2,$$

$$f(x) \to \frac{(1+a/x)(1+2a/x)^{1/2}}{e^{a-b} \pm e^{-x}}.$$

Similarly, for the ε-integral,

$$\alpha \to \frac{1}{2},$$

$$f(x) \to \frac{(x+a)^2(x+2a)^{1/2}}{e^{a-b} \pm e^{-x}}$$

but if

$$a \ll x_1,$$

then

$$\alpha \to 3,$$

$$f(x) \to \frac{(1+a/x)^2(1+2a/x)^{1/2}}{e^{a-b} \pm e^{-x}}.$$

and similarly for the pressure. The Gauss-Laguerre quadrature abscissa $\{x_i\}$ and weights $\{\omega_i\}$ can readily be calculated by, for example, the program, "gaulag" in the Numerical Recipes [7]. For practical uses, the number of abscissa points N can be taken to the order of 30 to 40, within relative errors of the order of 10^{-6} for the most of necessary a and b values, as far as $b \ll x_N$. For extremely degenerate region of a Fermi gas, a larger value of N is required.

The author acknowledges Drs. D. Gomez Dumm and A. Mihara for careful revision of the text and useful suggestions. He also thanks J.Rafelski and H-Th. Elze for critical reading of the manuscript and suggestions. This work is supported by PRONEX #41.96.0886.00, CNPq and FAPERJ.

REFERENCES

1. For general reference on Relativistic Heavy Ion collisions, see for example: C. Y. Wong, *Introduction to High Energy Heavy-Ion Collisions* (World Scientific Publishing, 1994), and in particular, a detailed description of hydrodynamic models, see: L. P. Csernai, *Introduction to Relativistic Heavy-Ion Collisions* (John Wiley & Sons, 1994). See also: J. Letessier and J. Rafelski, *Hadrons and Quark Gluon Plasma* (Cambridge U. Press, 2002) for the most recent developments.
2. L.D. Landau and E.M. Lifshitz, *Statistical Physics* (Pergamon Press, 1959).
3. W. Greiner, L. Neise and H. Stöcker, *Thermodynamics and Statistical Mechanics* (Springer-Verlag, 1995).
4. H-Th. Elze, W. Greiner and J. Rafelski, J. Phys. G6(1980) L149.
5. See R.K.Pathria, Statistical Mechanics (Butterworth-Heinemann, 1996). For the application to QGP, H-Th. Elze and W. Greiner, Phys. Lett. **B179** (1986) 385.
6. N.K.Glendenning, Compact Stars, Springer-Verlag, 2000.
7. W.H. Press, B. P. Flannery, S.A. Teukolsky and W.T. Vetterling, *Numerical Recipes* (Cambridge University Press,1986).

High Energy Nuclear Interactions and QCD: an Introduction

D. E. Kharzeev* and J. Raufeisen†

*Physics Department, Brookhaven National Laboratory, Upton, New York 11973, USA
†Los Alamos National Laboratory, MS H846, Los Alamos, NM 87545, USA

Abstract. The goal of these lectures, oriented towards the students just entering the field, is to provide an elementary introduction to QCD and the physics of nuclear interactions at high energies. We first introduce the general structure of QCD and discuss its main properties. Then we proceed to Glauber multiple scattering theory which lays the foundation for the theoretical treatment of nuclear interactions at high energies. We introduce the concept of Gribov's inelastic shadowing, crucial for the understanding of quantum formation effects. We outline the problems facing Glauber approach at high energies, and discuss how asymptotic freedom of QCD helps to resolve them, introducing the concepts of parton saturation and color glass condensate.

1. QUANTUM CHROMO-DYNAMICS – THE THEORY OF STRONG INTERACTIONS

1.1. What is QCD?

Strong interaction is, indeed the strongest force of nature. It is responsible for over 80% of the baryon masses, and thus for most of the mass of everything on Earth. Strong interactions bind nucleons in nuclei which, being then dressed with electrons and bound into molecules by the much weaker electro-magnetic force, give rise to the variety of the physical world.

Quantum Chromodynamics (QCD) is *the* theory of strong interactions. The fundamental degrees of freedom of QCD, quarks and gluons, are already well established even though they cannot be observed as free particles, but only in color neutral bound states (confinement). Today, QCD has firmly occupied its place as part of the Standard Model. However, understanding the physical world does not only mean understanding its fundamental constituents; it means mostly understanding how these constituents interact and bring into existence the entire variety of physical objects composing the universe. In these lectures, we try to explain why high energy nuclear physics offers us unique tools to study QCD.

1.1.1. The QCD Lagrangian

So what is QCD? QCD emerges when the naïve quark model is combined with local SU(3) gauge invariance. Quark model classifies the large number of hadrons in terms of a few, more fundamental constituents. Baryons consist of three quarks, while mesons are made of a quark and an antiquark. For example, the proton is made of two up-quarks and one down quark, $|p\rangle = |uud\rangle$, and the π^+-meson contains one up and one anti-down quark, $|\pi^+\rangle = |u\bar{d}\rangle$. However, the quark model in this naïve form is not complete, because the Pauli exclusion principle would not allow for a particle like the Δ isobar $|\Delta^{++}\rangle = |uuu\rangle$ with spin 3/2. The only way to construct a completely antisymmetric wavefunction for the Δ^{++} is to postulate an additional quantum number, which may be called "color". Quarks can then exist in three different color states; one may choose calling them red, green and blue. Correspondingly, we can define a quark-state "vector" with three components,

$$q(x) = \begin{pmatrix} q^{\text{red}}(x) \\ q^{\text{green}}(x) \\ q^{\text{blue}}(x) \end{pmatrix}. \quad (1)$$

The transition from quark model to QCD is made when one decides to treat color similarly to the electric charge in electrodynamics. As is well known, the entire structure of electrodynamics emerges from the requirement of local gauge invariance, i.e. invariance with respect to the phase rotation of electron field, $exp(i\alpha(x))$, where the phase α depends on the space–time coordinate. One can demand similar invariance for the quark fields, keeping in mind that while there is only one electric charge in QED, there are three color charges in QCD.

To implement this program, let us require the free quark Lagrangian,

$$\mathscr{L}_{\text{free}} = \sum_{q=u,d,s...} \sum_{\text{colors}} \bar{q}(x) \left(i\gamma_\mu \frac{\partial}{\partial x_\mu} - m_q \right) q(x), \quad (2)$$

to be invariant under rotations of the quark fields in color space,

$$U: \quad q^j(x) \rightarrow U_{jk}(x) q^k(x), \quad (3)$$

with $j, k \in \{1...3\}$ (we always sum over repeated indices). Since the theory we build in this way is invariant with respect to these "gauge" transformations, all physically meaningful quantities must be gauge invariant.

In electrodynamics, there is only one electric charge, and gauge transformation involves a single phase factor, $U = exp(i\alpha(x))$. In QCD, we have three different colors, and U becomes a (complex valued) unitary 3×3 matrix, i.e. $U^\dagger U = UU^\dagger = 1$, with determinant Det $U = 1$. These matrices form the fundamental representation of the group $SU(3)$ where 3 is the number of colors, $N_c = 3$. The matrix U has $N_c^2 - 1 = 8$ independent elements and can therefore be parameterized in terms of the 8 generators T^a_{kj}, $a \in \{1...8\}$ of the fundamental representation of $SU(3)$,

$$U(x) = \exp\left(-i\phi_a(x) T^a\right) \quad (4)$$

By considering a transformation U that is infinitesimally close to the **1** element of the group, it is easy to see that the matrices T^a must be Hermitian ($T^a = T^{a\dagger}$) and traceless (tr $T^a = 0$). The T^a's do not commute; instead one defines the $SU(3)$ structure constants

f_{abc} by the commutator

$$[T^a, T^b] = \mathrm{i} f_{abc} T^c. \tag{5}$$

These commutator terms have no analog in QED which is based on the abelian gauge group $U(1)$. QCD is based on a non-abelian gauge group $SU(3)$ and is thus called a non-abelian gauge theory.

The generators T^a are normalized to

$$\mathrm{tr}\, T^a T^b = \frac{1}{2} \delta_{ab}, \tag{6}$$

where δ_{ab} is the Kronecker symbol. Useful information about the algebra of color matrices, and their explicit representations, can be found in many textbooks (see, e.g., [2]).

Since U is x-dependent, the free quark Lagrangian (2) is not invariant under the transformation (3). In order to preserve gauge invariance, one has to introduce, following the familiar case of electrodynamics, the gauge (or "gluon") field $A^\mu_{kj}(x)$ and replace the derivative in (2) with the so-called *covariant derivative*,

$$\partial^\mu q^j(x) \rightarrow D^\mu_{kj} q^j(x) \equiv \left\{ \delta_{kj} \partial^\mu - \mathrm{i} A^\mu_{kj}(x) \right\} q^j(x). \tag{7}$$

Note that the gauge field $A^\mu_{kj}(x) = A^\mu_a T^a_{kj}(x)$ as well as the covariant derivative are 3×3 matrices in color space. Note also that Eq. (7) differs from the definition often given in textbooks, because we have absorbed the strong coupling constant in the field A^μ. With the replacement given by Eq. (7), all changes to the Lagrangian under gauge transformations cancel, provided A^μ transforms as

$$U: \quad A^\mu(x) \rightarrow U(x) A^\mu(x) U^\dagger(x) + \mathrm{i} U(x) \partial^\mu U^\dagger(x). \tag{8}$$

(From now on, we will often not write the color indices explicitly.)

The QCD Lagrangian then reads

$$\mathscr{L}_{\mathrm{QCD}} = \sum_q \bar{q}(x) \left(\mathrm{i} \gamma_\mu D^\mu - m_q \right) q(x) - \frac{1}{4g^2} \mathrm{tr}\, G^{\mu\nu}(x) G_{\mu\nu}(x), \tag{9}$$

where the first term describes the dynamics of quarks and their couplings to gluons, while the second term describes the dynamics of the gluon field. The strong coupling constant g is the QCD analog of the elementary electric charge e in QED. The gluon field strength tensor is given by

$$G^{\mu\nu}(x) \equiv \mathrm{i} [D^\mu, D^\nu] = \partial^\mu A^\nu(x) - \partial^\nu A^\mu(x) - \mathrm{i} [A^\mu(x), A^\nu(x)]. \tag{10}$$

This can also be written in terms of the color components A^μ_a of the gauge field,

$$G^{\mu\nu}_a(x) = \partial^\mu A^\nu_a(x) - \partial^\nu A^\mu_a(x) + f_{abc} A^\mu_b(x) A^\nu_c(x). \tag{11}$$

For a more complete presentation, see [1] and modern textbooks like [2, 3, 4].

FIGURE 1. Due to the non-abelian nature of QCD, gluons carry color charge and can therefore interact with each others via these vertices.

The crucial, as will become clear soon, difference between electrodynamics and QCD is the presence of the commutator on the *r.h.s.* of Eq. (10). This commutator gives rise to the gluon-gluon interactions shown in Fig. 1 that make the QCD field equations non-linear: the color fields do not simply add like in electrodynamics. These non-linearities give rise to rich and non-trivial dynamics of strong interactions.

1.1.2. Asymptotic Freedom

Let us now turn to the discussion of the dynamical properties of QCD. To understand the dynamics of a field theory, one necessarily has to understand how the coupling constant behaves as a function of distance. This behavior, in turn, is determined by the response of the vacuum to the presence of external charge. The vacuum is the ground state of the theory; however, quantum mechanics tells us that the "vacuum" is far from being empty – the uncertainty principle allows particle-antiparticle pairs to be present in the vacuum for a period time inversely proportional to their energy. In QED, the electron-positron pairs have the effect of screening the electric charge, see Fig. 2. Thus, the electromagnetic coupling constant increases toward shorter distances. The dependence of the charge on distance is given by

$$e^2(r) = \frac{e^2(r_0)}{1 + \frac{2e^2(r_0)}{3\pi} \ln \frac{r}{r_0}}, \qquad (12)$$

which can be obtained by resumming (logarithmically divergent, and regularized at the distance r_0) electron–positron loops dressing the virtual photon propagator.

The formula (12) has two surprising properties: first, at large distances r away from the charge which is localized at r_0, $r \gg r_0$, where one can neglect unity in the denominator, the "dressed" charge $e(r)$ becomes independent of the value of the "bare" charge $e(r_0)$ – it does not matter what the value of the charge at short distances is. Second, in the local limit $r_0 \to 0$, if we require the bare charge $e(r_0)$ be finite, the effective charge vanishes at any finite distance away from the bare charge! This is the celebrated Landau's zero charge problem [5]: the screening of the charge in QED does not allow to reconcile the presence of interactions with the local limit of the theory. This is a fundamental problem of QED, which shows that i) either it is not a truly fundamental theory, or ii)

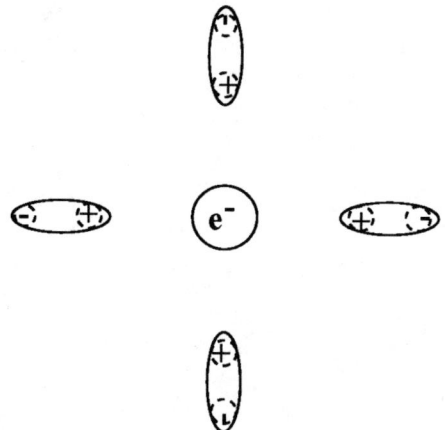

FIGURE 2. In QED, virtual electron-positron pairs from the vacuum screen the bare charge of the electron. The larger the distance, the more pairs are present to screen the bare charge and the electromagnetic coupling decreases. Conversely, the coupling is larger when probed at short distances.

Eq. (12), based on perturbation theory, in the strong coupling regime gets replaced by some other expression with a more acceptable behavior. The latter possibility is quite likely since at short distances the electric charge becomes very large and its interactions with electron–positron vacuum cannot be treated perturbatively. A solution of the zero charge problem, based on considering the rearrangement of the vacuum in the presence of "super–critical", at short distances, charge was suggested by Gribov [6].

Fortunately, because of the smallness of the physical coupling $\alpha_{em}(r) = e^2(r)/(4\pi) = 1/137$, this fundamental problem of the theory manifests itself only at very short distances $\sim \exp(-3/[8\alpha_{em}])$. Such short distances will probably always remain beyond the reach of experiment, and one can safely apply QED as a truly effective theory.

In QCD, as we are now going to discuss, the situation is qualitatively different, and corresponds to *anti*-screening – the charge is small at short distances and grows at larger distances. This property of the theory, discovered by Gross, Wilczek, and Politzer [7], is called asymptotic freedom.

While the derivation of the running coupling is conventionally performed by using field theoretical perturbation theory, it is instructive to see how these results can be illustrated by using the methods of condensed matter physics. Indeed, let us consider the vacuum as a continuous medium with a dielectric constant ε. The dielectric constant is linked to the magnetic permeability μ and the speed of light c by the relation

$$\varepsilon \mu = \frac{1}{c^2} = 1. \tag{13}$$

Thus, a screening medium ($\varepsilon > 1$) will be diamagnetic ($\mu < 1$), and conversely a paramagnetic medium ($\mu > 1$) will exhibit antiscreening which leads to asymptotic freedom. In order to calculate the running coupling constant, one has to calculate the magnetic permeability of the vacuum. We follow [8] in our discussion, where this has been done in a framework very similar to Landau's theory of the diamagnetic properties of a free electron gas. In QED one has

$$\varepsilon_{QED} = 1 + \frac{2e^2(r_0)}{3\pi} \ln \frac{r}{r_0} > 1 \qquad (14)$$

So why is the QCD vacuum paramagnetic while the QED vacuum is diamagnetic? The energy density of a medium in the presence of an external magnetic field $\vec{B}$ is given by

$$u = -\frac{1}{2} 4\pi \chi \vec{B}^2 \qquad (15)$$

where the magnetic susceptibility χ is defined by the relation

$$\mu = 1 + 4\pi\chi. \qquad (16)$$

When electrons move in an external magnetic field, two competing effects determine the sign of magnetic susceptibility:

- The electrons in magnetic field move along quantized orbits, so-called Landau levels. The current originating from this movement produces a magnetic field with opposite direction to the external field. This is the diamagnetic response, $\chi < 0$.
- The electron spins align along the direction of the external $\vec{B}$-field, leading to a paramagnetic response ($\chi > 0$).

In QED, the diamagnetic effect is stronger, so the vacuum is screening the bare charges. In QCD, however, gluons carry color charge. Since they have a larger spin (spin 1) than quarks (or electrons), the paramagnetic effect dominates and the vacuum is anti-screening.

Let us explain this in more detail. Basing on the considertaions given above, the energy density of the QCD vacuum in the presence of an external color-magnetic field can be calculated by using the standard formulas of quantum mechanics, see *e.g.* [9], by summing over Landau levels and taking account of the fact that gluons and quarks give contributions of different sign. Note that a summation over all Landau levels would lead to an infinite result for the energy density. In order to avoid this divergence, one has to introduce a cutoff Λ with dimension of mass. Only field modes with wavelength $\lambda \gtrsim 1/\Lambda$ are taken into account. The upper limit for λ is given by the radius of the largest Landau orbit, $r_0 \sim 1/\sqrt{gB}$, which is the only dimensionful scale in the problem; the summation thus is made over the wave lengths satisfying

$$\frac{1}{\sqrt{|gB|}} \gtrsim \lambda \gtrsim \frac{1}{\Lambda}, \qquad (17)$$

The result is [8]

$$u_{vac}^{QCD} = -\frac{1}{2} B^2 \frac{11N_c - 2N_f}{48\pi^2} g^2 \ln \frac{\Lambda^2}{|gB|}, \qquad (18)$$

where N_f is the number of quark flavors, and $N_c = 3$ is the number of flavors. Comparing this with Eqs. (15) and (16), one can read off the magnetic permeability of the QCD vacuum,

$$\mu_{vac}^{QCD}(B) = 1 + \frac{11N_c - 2N_f}{48\pi^2} g^2 \ln \frac{\Lambda^2}{|gB|} > 1. \qquad (19)$$

The first term in the denominator ($11N_c$) is the gluon contribution to the magnetic permeability. This term dominates over the quark contribution ($2N_f$) as long as the number of flavors N_f is less than 17 and is responsible for asymptotic freedom.

The dielectric constant as a function of distance r is then given by

$$\varepsilon_{vac}^{QCD}(r) = \frac{1}{\mu_{vac}^{QCD}(B)} \bigg|_{\sqrt{|gB|} \to 1/r}. \qquad (20)$$

The replacement $\sqrt{|gB|} \to 1/r$ follows from the fact that ε and μ in Eq. (20) should be calculated from the same field modes: the dielectric constant $\varepsilon(r)$ could be calculated by computing the vacuum energy in the presence of two static colored test particles located at a distance r from each other. In this case, the maximum wavelength of field modes that can contribute is of order r so that

$$r \gtrsim \lambda \gtrsim \frac{1}{\Lambda}. \qquad (21)$$

Combining Eqs. (17) and (21), we identify $r = 1/\sqrt{|gB|}$ and find

$$\varepsilon_{vac}^{QCD}(r) = \frac{1}{1 + \frac{11N_c - 2N_f}{24\pi^2} g^2 \ln(r\Lambda)} < 1. \qquad (22)$$

With $\alpha_s(r_1)/\alpha_s(r_2) = \varepsilon_{vac}^{QCD}(r_2)/\varepsilon_{vac}^{QCD}(r_1)$ one finds to lowest order in α_s

$$\alpha_s(r_1) = \frac{\alpha_s(r_2)}{1 + \frac{11N_c - 2N_f}{6\pi} \alpha_s(r_2) \ln\left(\frac{r_2}{r_1}\right)}. \qquad (23)$$

Apparently, if $r_1 < r_2$ then $\alpha_s(r_1) < \alpha_s(r_2)$. The running of the coupling constant is shown in Fig. 3, $Q \sim 1/r$. The intuitive derivation given above illustrates the original field–theoretical result of [7].

At high momentum transfer, corresponding to short distances, the coupling constant thus becomes small and one can apply perturbation theory, see Fig. 3. There is a variety of processes that involve high momentum scales, e.g. deep inelastic scattering, Drell-Yan dilepton production, e^+e^--annihilation into hadrons, production of heavy quarks/quarkonia, high p_T hadron production .... QCD correctly predicts the Q^2 dependence of these, so-called "hard" processes, which is a great success of the theory.

1.2. Challenges in QCD

1.2.1. Confinement

While asymptotic freedom implies that the theory becomes simple and treatable at short distances, it also tells us that at large distances the coupling becomes very strong.

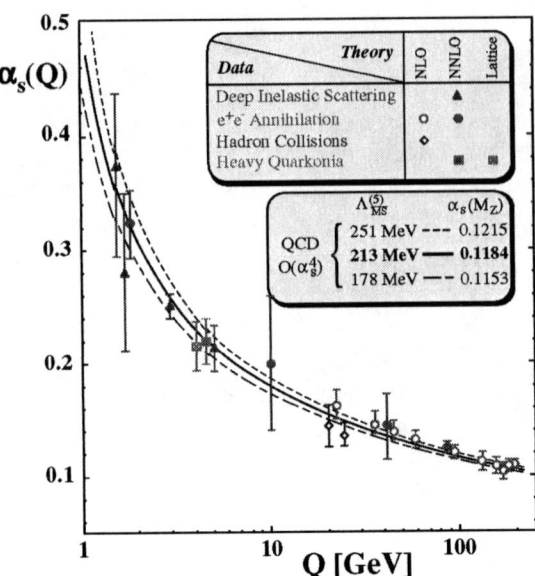

FIGURE 3. The running coupling constant $\alpha_s(Q^2)$ as a function of momentum transfer Q^2 determined from a variety of processes. (Reprinted with permission from [10], Copyright 2000, Institute of Physics (UK), IOP Publishing Limited, and courtesy of S. Bethke).

In this regime we have no reason to believe in perturbation theory. In QED, as we have discussed above, the strong coupling regime starts at extremely short distances beyond the reach of current experiments – and this makes the "zero–charge" problem somewhat academic. In QCD, the entire physical World around us is defined by the properties of the theory in the strong coupling regime – and we have to construct accelerators to study it in the much more simple, "QED–like", weak coupling limit.

We do not have to look far to find the striking differences between the properties of QCD at short and large distances: the elementary building blocks of QCD – the "fundamental" fields appearing in the Lagrangean (9), quarks and gluons, do not exist in the physical spectrum as asymptotic states. For some, still unknown to us, reason, all physical states with finite energy appear to be color–singlet combinations of quarks and gluons, which are thus always "confined" at rather short distances on the order of 1 fm. This prevents us, at least in principle, from using well–developed formal S-matrix approaches based on analyticity and unitarity to describe quark and gluon interactions.

The property of confinement can be explored by looking at the propagation of heavy quark–antiquark pair at a distance R propagating in time a distance T. An object which describes the behavior of this system is the Wilson loop [11]

$$W(R,T) = Tr\left[P\, exp\left[i\int_C A^a_\mu T^a dx^\mu\right]\right], \qquad (24)$$

where A^a_μ is the gluon field, T^a is the generator of $SU(3)$, and the contour C is chosen as a rectangle with side R in one of the space dimensions and T in the time direction. It can be shown that at large T the asymptotics of the Wilson loop is

$$\lim_{T \to \infty} W(R,T) = exp[-TV(R)], \qquad (25)$$

where $V(R)$ is the static potential acting between the heavy quarks. At large distances, this potential grows as

$$V(R) = \sigma R, \qquad (26)$$

where $\sigma \sim 1$ GeV/fm is the string tension. We thus conclude that at large T and R the Wilson loop should behave as

$$W(R,T) \simeq exp[-\sigma TR], \qquad (27)$$

The formula (27) is the celebrated "area law", which signals confinement.

It should be noted, however, that the introduction of dynamical quarks leads to the string break–up at large distances, and the potential $V(R)$ saturates at a constant. The presence of light dynamical quarks is most important in Gribov's confinement scenario [6], in which the color charges at large distances behave similarly to the "supercritical" charge in electrodynamics, polarizing the vacuum and producing copious quark–antiquark pairs which screen them. In this scenario, in the physical world with light quarks there is never a confining force acting on color charges at large distances, just quark–antiquark pair production ("soft confinement"). This may explain why the spectra of jets, for example, computed in perturbative QCD, appear to be consistent with experiment; this fact would be difficult to reconcile with the existence of strong confining forces. There exists a special situation, however, when the law (27) should be appropriate even in the presence of light quarks – the heavy quarkonium. The sizes of heavy quarkonia are quite small, and their masses are below the threshold to produce a pair of heavy mesons. This is why heavy quarkonia are especially useful probes of confinement.

At high temperatures, the long–range interactions responsible for confinement become screened away – instead of the growing potential (26), we expect

$$V(R) \sim -\frac{g^2(T)}{R} exp(-m_D R), \qquad (28)$$

where $m_D \sim gT$ is the Debye mass. Mathematically, this transition to the deconfined phase can again be studied by looking at the properties of the Wilson loop. At finite temperature, the theory is defined on a cylinder: Euclidean time τ varies within $0 \le \tau \le \beta = 1/T$, and the gluon fields satisfy the periodic boundary conditions:

$$A_\mu^a(\vec{x}, \beta) = A_\mu^a(\vec{x}, \beta). \qquad (29)$$

Let us now consider the Wilson loop wrapped around this cylinder (the Polyakov loop), and choose a gauge where A_0^a is time–independent:

$$P(\vec{x}) = Tr \, exp[ig\beta A_0^a(\vec{x}) t^a]; \qquad (30)$$

the correlation function of these objects can be defined as

$$C_T(\vec{x}) = <P(\vec{x}) P^*(\vec{x})>_T. \qquad (31)$$

Again, it can be shown that this correlation function is related to the free energy, and thus static potential $V(R)$, of the heavy quark–antiquark pair. Assuming, as before, that the heavy quarks are separated by the spatial distance $R = |\vec{x}|$, one finds

$$C_T(R) \sim exp[-\beta V(R)]. \tag{32}$$

Again, if we define the limit value $L(T)$ of the correlation function,

$$\lim_{R \to \infty} C_T(R) \equiv L(T) \tag{33}$$

it would have to vanish in the confined phase in the absence of dynamical quarks, since $V(R)$ tends to infinity in this case: $L(T) = 0$. In the deconfined phase, on the other hand, because of the screening $V(R)$ should tend to a constant, and this implies a finite value $L(T) \neq 0$. The correlation function of Polyakov loops therefore can be used as an order parameter of the deconfinement. The behavior of $L(T)$ as a function of temperature has been measured on the lattice; one indeed observes a transition from the confined phase with $L(T) = 0$ to the deconfined phase with $L(T) \neq 0$ at some critical temperature T_c. In the presence of light quarks, as we have already discussed above, the potential would tend to a constant even in the confined phase, and $L(T)$ ceases to be a rigorous order parameter.

1.2.2. Chiral symmetry breaking

The decades of experience with "soft pion" techniques and current algebra convinced physicists that the properties of the world with massless pions are quite close to the properties of our physical World. The existence of massless particles is always a manifestation of a symmetry of the theory – photons, for example, appear as a consequence of local gauge invariance of the electrodynamics. However, unlike photons, pions have zero spin and cannot be gauge bosons of any symmetry. The other possibility is provided by the Goldstone theorem, which states that the appearance of massless modes in the spectrum can also reflect a spontaneously broken symmetry, i.e. the symmetry of the theory which is broken in the ground state. Because of the great importance of this theorem, let us briefly sketch its proof.

Suppose that the Hamiltonian H of the theory is invariant under some symmetry generated by operators Q_i, so that

$$[H, Q_i] = 0. \tag{34}$$

Spontaneous symmetry breaking in the ground state of theory implies that for some of the generators Q_i

$$Q_i|0> \neq 0. \tag{35}$$

Since Q_i commute with the Hamiltonian, this means that this new state $Q_i|0>$ has the same energy as the ground state. The vacuum is therefore degenerate, and in a relativistically invariant theory this implies the existence of massless particles – Goldstone bosons. A useful example of that is provided by the phonons in a crystal, where the continuous translational symmetry of the QED Lagrangean is spontaneously broken by the existence of the fixed period of the crystal lattice.

Even though all six quark flavors enter the Lagrangean, it is intuitively clear that at small scales $Q << M_c, M_b, M_t$, heavy quarks should not have any influence on the dynamics. In a rigorous way this statement is formulated in terms of decoupling theorems, which we will discuss in detail later. At the moment let us just assume that we are interested in the low–energy behavior, and that only light quarks are relevant for that purpose. Then it makes sense to consider the approximate symmetry, which becomes exact when the quarks are massless. In fact, in this limit, the Lagrangean does not contain any terms which connect the right– and left–handed components of the quark fields:

$$q_R = \frac{1}{2}(1+\gamma_5)q; \qquad q_L = \frac{1}{2}(1-\gamma_5)q. \qquad (36)$$

The Lagrangean of QCD (9) is therefore invariant under the independent transformations of right– and left–handed fields ("chiral rotations"). In the limit of massless quarks, QCD thus possesses an additional symmetry $U_L(N_f) \times U_R(N_f)$ with respect to the independent transformation of left– and right–handed quark fields $q_{L,R} = \frac{1}{2}(1 \pm \gamma_5)q$:

$$q_L \to V_L q_L; \quad q_R \to V_R q_R; \quad V_L, V_R \in U(N_f); \qquad (37)$$

this means that left– and right–handed quarks are not correlated.

Even a brief look into the Particle Data tables, or simply in the mirror, can convince anyone that there is no symmetry between left and right in the physical World. One thus has to assume that the symmetry (37) is spontaneously broken in the vacuum.

The presence of the "quark condensate" $<\bar{q}q>$ in QCD vacuum signals spontaneous breakdown of this symmetry, since

$$<\bar{q}q> = <\bar{q}_L q_R> + <\bar{q}_R q_L>, \qquad (38)$$

which means that left– and right–handed quarks and antiquarks can transform into each other. Quark condensate therefore can be used as an order parameter of chiral symmetry. Lattice calculations show that around the deconfinement phase transition, quark condensate dramatically decreases, signalling the onset of the chiral symmetry restoration.

This spontaneous breaking of $U_L(3) \times U_R(3)$ chiral symmetry, by virtue of the Goldstone theorem presented above, should give rise to $3^2 = 9$ Goldstone particles. The flavor composition of the existing eight candidates for this role (3 pions, 4 kaons, and the η) suggests that the $U_A(1)$ part of $U_L(3) \times U_R(3) = SU_L(3) \times SU_R(3) \times U_V(1) \times U_A(1)$ does not exist. This constitutes the famous "$U_A(1)$ problem".

1.2.3. The origin of mass

There is yet another problem with the chiral limit in QCD. Indeed, as the quark masses are put to zero, the Lagrangian (9) does not contain a single dimensionful scale – the only parameters are pure numbers N_c and N_f. The theory is thus apparently invariant with respect to scale transformations, and the corresponding scale current is conserved: $\partial_\mu s_\mu = 0$. However, the absence of a mass scale would imply that all physical states in the theory should be massless!

1.2.4. Quantum anomalies

Both apparent problems – the missing $U_A(1)$ symmetry and the origin of hadron masses – are related to quantum anomalies. A symmetry of a classical theory can be broken when that theory is quantized, due to the requirements of regularization and renormalization. This is called anomalous symmetry breaking. Regularization of the theory on the quantum level brings in a dimensionful parameter – remember the cutoff Λ of Eq. (17) we had to impose on the wavelength of quarks and gluons.

Once the theory is quantized, we already know that the coupling constant is scale dependent and therefore scale invariance is broken (note that the four-divergence of the scale current in field theory is equal to the trace of the energy momentum tensor Θ^μ_μ). One finds

$$\partial^\mu s_\mu = \Theta^\mu_\mu = \sum_q m_q \bar{q} q + \frac{\beta(g)}{2g^3} \mathrm{tr} G^{\mu\nu} G_{\mu\nu}, \tag{39}$$

where $\beta(g)$ is the QCD β-function, which governs the behavior of the running coupling:

$$\mu \frac{dg(\mu)}{d\mu} = \beta(g); \tag{40}$$

note that as discussed in Section *1.1.1* we include coupling g in the definition of the gluon fields. As we already discussed, at small coupling g, the β function is negative, which means that the theory is asymptotically free. The leading term in the perturbative expansion is (compare with Eq. (23))

$$\beta(g) = -b \frac{g^3}{(4\pi)^2} + O(g^5), \qquad b = 11 N_c - 2 N_f, \tag{41}$$

where N_c and N_f are the numbers of colors and flavors, respectively.

Hadron masses are related to the forward matrix element of trace of the QCD energy-momentum tensor, $2 m_h^2 = \langle h | \Theta^\mu_\mu | h \rangle$. Apparently, light hadron masses must receive dominant contributions from the G^2-term in Eq. (39). Note also that the flavor sum in Eq. (39) includes heavy flavors, too. This would lead to the unphysical picture that e.g. the proton mass is dominated by heavy quark masses. However, the heavy flavor contribution to the sum (39) is exactly canceled by a corresponding heavy flavor contribution to the β-function.

Similar thing happens with the axial current, $j^5_\mu = \bar{q} \gamma_\mu \gamma^5 q$, genrated by the $U_A(1)$ group. The corresponding axial charge is not conserved because of the contribution of

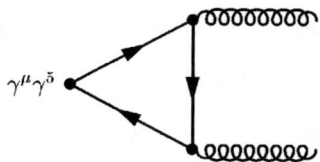

FIGURE 4. The triangle graph that leads to the $U_A(1)$-anomaly. The corresponding graph with the two gluons interchanged in the final state is not shown.

the triangle graph in Fig. 4, and the four-divergence of the axial current is given by [12]

$$\partial^\mu j_\mu^5 = \sum_q 2im_q \bar{q}\gamma^5 q + \frac{N_f}{8\pi^2} \text{tr}\, G^{\mu\nu}\widetilde{G}_{\mu\nu}. \tag{42}$$

Since the gluonic part on the *rhs* of this equation is a surface term (a full divergence), there would be no physical effect, if the QCD vacuum were "empty".

1.2.5. Classical solutions

However, it appears that due to non–trivial topology of the $SU(3)$ gauge group, QCD equations of motion allow classical solutions even in the absence of external color source, i.e. in the vacuum. The well–known example of a classical solution is the *instanton*, corresponding to the mapping of a three–dimensional sphere S^3 onto the $SU(2)$ subgroup of color $SU(3)$ (for reviews, see [13, 14]. As a result, the ground state of classical Chromodynamics is not unique. There is an enumerable infinite number of gauge field configurations with different topologies (corresponding to different winding number in the $S^3 \to SU(2)$ mapping, and the ground state looks like a periodic potential, see Fig. 5.

In a quantum theory, however, the system will not stay in one of the minima, like the classical system would. Instead, there will be tunneling processes between different minima. These tunneling processes, in Minkowski space, correspond to instantons. Since tunneling, in general, lowers the ground state energy of the system, one expects the QCD vacuum to have a complicated structure.

Instantons, through the anomaly relation (42), lead to the explicit violation of the $U_A(1)$ symmetry and thus solve the mystery of the missing ninth Goldstone boson - the η'. Physically, axial symmetry $U_A(1)$ is broken because the tunneling processes between topologically different vacua are accompanied by the change in quark helicity – even in the vacuum, left quarks periodically turn into right and *vice versa*.

1.2.6. Strong CP problem

The vacuum structure shown in Fig. 5 immediately leads to a puzzle known as the *strong CP problem:* When one calculates the expectation value of an observable in the

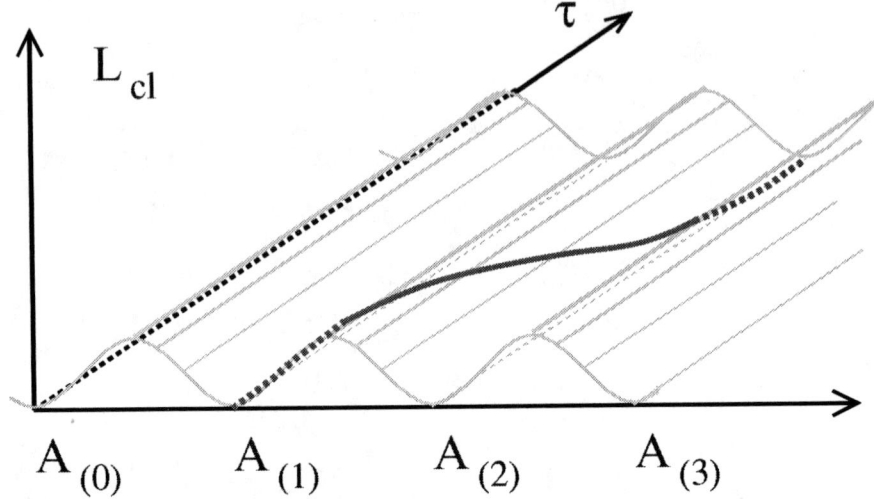

FIGURE 5. Topological structure of QCD vacuum. The minima correspond to classical ground states with topologically different gauge field configurations $A_{(n)}$. Also shown is an instanton trajectory interpolating between the classical vacua $A_{(1)}$ and $A_{(2)}$. The third axis shows the Euclidean time τ. From [14]; courtesy of H. Forkel.

vacuum, one has to average over all topological sectors of the vacuum. This is equivalent to adding an additional term to the QCD-Lagrangian,

$$\mathscr{L}_{\mathrm{QCD}} \to \mathscr{L}_{\mathrm{QCD}} - \frac{\theta}{16\pi^2} \, \mathrm{tr} \, G^{\mu\nu} \widetilde{G}_{\mu\nu}, \qquad (43)$$

where $\widetilde{G}_{\mu\nu} = \varepsilon_{\mu\nu\kappa\lambda} G^{\kappa\lambda}/2$ is the dual field strength tensor and $\theta \in [0, 2\pi]$ is a parameter of the theory which has to be determined from experiment. Since the θ-term in Eq. (43) is CP violating, a non-zero value of θ would have immediate phenomenological consequences, *e.g.* an electric dipole moment of the neutron. However, precision measurements of this dipole moment constrain θ to $\theta < 10^{-9}$. The fact that θ is so unnaturally small constitutes the strong CP problem. The most likely solution to this problem [15] implies the existence of a light pseudoscalar meson, the *axion*. However, despite many efforts, axions remain unobserved in experiment.

1.2.7. Phase structure

As was repeatedly stated above, the most important problem facing us in the study of all aspects of QCD is understanding the structure of the vacuum, which, in a manner of saying, does not at all behave as an empty space, but as a physical entity with a complicated structure. As such, the vacuum can be excited, altered and modified in physical processes [16].

Collisions of heavy ions are the best way to create high energy density in a "macroscopic" (on the scale of a single hadron) volume. It thus could be possible to create and to study a new state of matter, the *Quark-Gluon Plasma*(QGP), in which quarks and gluons are no longer confined in hadrons, but can propagate freely. The search for QGP is one of the main motivations for the heavy ion research.

Lattice calculations predict that QCD at high temperatures undergoes phase transitions in which confinement property is lost and chiral symmetry is restored. The critical temperature for the chiral phase transition is similar (or maybe even equal) to the critical temperature for deconfinement.

Heavy ion collisions at RHIC may also give us the possibility to study the θ angle dependence of the QCD phase diagram. In a heavy ion collision, bubbles containing a metastable vacuum with $\theta \neq 0$ may be produced, and reveal themselves through their unusual decay pattern [17].

2. NUCLEAR INTERACTIONS AT HIGH ENERGIES

2.1. Glauber-Gribov Theory

It is intuitively clear that heavy ion collisions are governed by multiple scattering effects. As a short introduction to the basics of multiple scattering theory, we introduce here the eikonal approximation to high energy scattering processes and the Glauber multiple scattering theory [18]. We also discuss Gribov's inelastic corrections [19] to Glauber's theory.

2.1.1. The Eikonal Approximation

The eikonal approximation is the classical approximation to the angular momentum l. In partial wave expansion, *i.e.* in an expansion in angular momentum eigenstates, the scattering amplitude $f(s,t)$ reads [9]

$$f(s,t) = \frac{1}{2\mathrm{i}p} \sum_l (2l+1) \left[e^{2\mathrm{i}\delta_l} - 1 \right] P_l(\cos\theta), \tag{44}$$

where s and t are the usual Mandelstam variables (center-of-mass energy squared and invariant momentum transfer, respectively), p is the momentum of the projectile and P_l are the Legendre functions, which depend on the cosine of the scattering angle θ. All information about the interaction is contained in the scattering phases δ_l.

High energy scattering is of course a process that is far from being spherically symmetric. Therefore, very large values of l will dominate the sum Eq. (44) and we can treat the angular momentum classically. Since the angular momentum is given by pb, one replaces the variable l by the impact parameter b,

$$pb = l + \frac{1}{2}. \tag{45}$$

Note that b is now a continuous variable, so angular momentum is no longer quantized.

At large l and for small scattering angles θ, the Legendre functions can be expressed to good approximation as

$$P_l(\cos\theta) \approx \int_0^{2\pi} \frac{d\phi}{2\pi} e^{i(2l+1)\sin(\theta/2)\cos(\phi)} = \int_0^{2\pi} \frac{d\phi}{2\pi} e^{i\vec{q}\cdot\vec{b}}, \tag{46}$$

where $\vec{q} = \vec{p} - \vec{p}\,'$ is the momentum transfer in the scattering process ($t \approx -|\vec{q}|$) and $|\vec{p}| = |\vec{p}\,'|$ for elastic scattering. At high energy, $\vec{q}$ lies in the impact parameter plane. We have used the relation

$$(2l+1)\sin(\theta/2)\cos(\phi) = 2p\sin(\theta/2)\frac{l+1/2}{p}\cos(\phi) = \vec{q}\cdot\vec{b} \tag{47}$$

to obtain the second equality in Eq. (46).

Thus, the scattering amplitude in eikonal approximation reads

$$f(s,t) = \frac{ip}{2\pi}\int d^2b\, e^{i\vec{q}\cdot\vec{b}}\left[1 - e^{i\chi(s,\vec{b})}\right], \tag{48}$$

where the phase shift of the projectile is related to the scattering phase δ_l by

$$\chi(s,\vec{b}) \equiv 2\delta(s,b). \tag{49}$$

In the case of scattering off a potential $V(\vec{r})$, this phase shift is simply given by

$$\chi(\vec{b}) = -\frac{1}{v}\int_{-\infty}^{\infty} V(\vec{r})dz, \tag{50}$$

where v is the velocity of the projectile. The scattering amplitude then reads

$$f(s,t) = \frac{ip}{2\pi}\int d^2b\, e^{i\vec{q}\cdot\vec{b}}\left[1 - \exp\left(-\frac{i}{v}\int V(\vec{r})dz\right)\right]. \tag{51}$$

The total cross section can now be obtained from the forward scattering amplitude via the optical theorem,

$$\sigma_{tot} = \frac{4\pi}{p}\mathrm{Im}\, f(s,t=0) = 2\int d^2b\left(1 - \mathrm{Re}\, e^{i\chi(b)}\right). \tag{52}$$

For completeness, we also give the expressions for the elastic and inelastic cross sections. The elastic cross section is obtained by squaring the elastic scattering amplitude and integrating over the solid angle,

$$\sigma_{el} = \int d\Omega_{p'}\, |f(\theta,\phi)|^2. \tag{53}$$

With the approximation $d\Omega_{p'} \approx d^2p'/p'^2$, which assumes that scattering takes place predominantly in forward direction, one obtains

$$\sigma_{el} = \int d^2b\left|1 - e^{i\chi(b)}\right|^2 \tag{54}$$

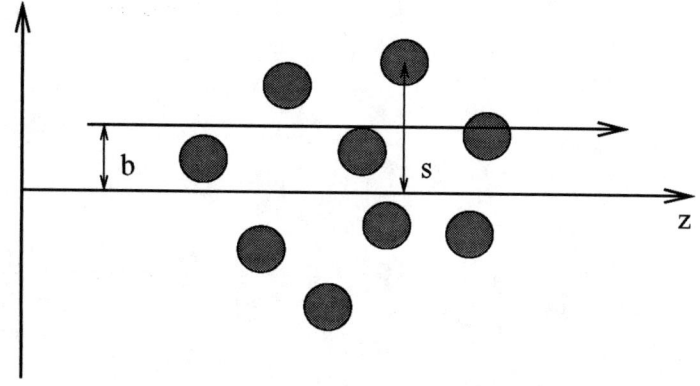

FIGURE 6. Scattering off a composite system. The impact parameter of the projectile is denoted by $\vec{b}$, while the impact parameters of the scattering centers are denoted by $\vec{s}_j$.

Finally, the inelastic cross section is

$$\sigma_{inel} = \sigma_{tot} - \sigma_{el} = \int d^2 b \left(1 - \left|e^{i\chi(b)}\right|^2\right). \tag{55}$$

For potential scattering, the inelastic cross section, of course, vanishes because $\chi(b)$ is real. In general, however, $\chi(b)$ will have an imaginary part.

The expressions Eqs. (52), (54) and (55) could have been obtained directly from the partial wave decomposition of the total, elastic and inelastic cross section, as well. The conditions under which the eikonal approximation is applicable are investigated in detail in [18].

2.1.2. Multiple Scattering Theory

Based on the eikonal approximation, it is quite straightforward to develop a theory for scattering off a composite system. In this section, we explain the basic features of the multiple scattering theory developed by Glauber [18]. A much more detailed presentation of this subject can be found in [18].

Assuming that the scatterings on different nucleons are independent, the phase shifts from each scattering simply add up,

$$\chi(\vec{b}, \vec{r}_1, \vec{r}_2, \ldots \vec{r}_a) = \sum_{j=1}^{A} \chi_j(\vec{b} - \vec{s}_j). \tag{56}$$

Here, $\vec{b}$ is the impact parameter of the projectile and $\vec{s}_j$, $j = 1 \ldots A$ are the impact parameters of the A nucleons in the nucleus, see Fig. 6. The amplitude for scattering off a nuclear target then can be written as

$$F_{fi}^A = \frac{ip}{2\pi} \int d^2b\, e^{i\vec{q}\cdot\vec{b}} \langle f | 1 - \prod_{j=1}^A e^{i\chi_j(\vec{b}-\vec{s}_j)} | i \rangle \tag{57}$$

$$= \frac{ip}{2\pi} \int d^2b\, e^{i\vec{q}\cdot\vec{b}} \langle f | 1 - \prod_{j=1}^A [1 - \gamma_j(\vec{b}-\vec{s}_j)] | i \rangle, \tag{58}$$

where $|f\rangle$ and $|i\rangle$ are the final and initial state of the target, respectively. In the second step, we introduced the profile function $\gamma(\vec{b})$, which is related to the single-scattering amplitude $f(\vec{q})$ by

$$\gamma(\vec{b}) = \frac{1}{2\pi i p} \int d^2q\, e^{-i\vec{q}\cdot\vec{b}} f(\vec{q}). \tag{59}$$

Thus, we have expressed the nuclear scattering amplitude in terms of the amplitude for scattering off a single nucleon.

In the case of a purely imaginary $f(\vec{q})$, $\gamma(\vec{b})$ is the probability of absorption of the projectile by a nucleon and the nuclear scattering amplitude, Eq. 58 has a simple probabilistic interpretation. Namely, $1 - \gamma_j(\vec{b}-\vec{s}_j)$ is the probability of not being absorbed by nucleon number j. Taking the product over all $j \in \{1 \ldots A\}$ yields the probability of not being absorbed by any nucleon in the target. Finally, $1 - \prod_{j=1}^A [1 - \gamma_j(\vec{b}-\vec{s}_j)]$ is the probability that the projectile is absorbed by any of the nucleons.

Also, if one in addition assumes that all nucleons in the target are identical, the nuclear cross section can be expressed in terms of the cross section for scattering on a single nucleon,

$$\sigma_{tot}^A = \frac{4\pi}{p} \text{Im} F_{ii}^A(t=0) \tag{60}$$

$$= 2 \int d^2b \left(1 - \left(1 - \frac{\sigma_{tot}^N T(\vec{b})}{2A} \right)^A \right) \tag{61}$$

$$\approx 2 \int d^2b \left(1 - \exp\left(-\frac{\sigma_{tot}^N T_A(\vec{b})}{2} \right) \right), \tag{62}$$

where the nuclear thickness function $T_A(\vec{b})$ is the integral over the nuclear density,

$$T_A(\vec{b}) = \int_{-\infty}^{\infty} dz\, \rho_A(\vec{b},z). \tag{63}$$

The simple expression, Eq. (61), resums all multiple scattering terms. We stress that the probabilistic interpretation of Eq. (58) as well as Eqs. (61) and (62) only hold for a purely imaginary $f(\vec{q})$.

The meaning of the nuclear scattering amplitude, Eq. (58), is further explained by expanding the probability of particle absorption by any of the nucleons in powers of

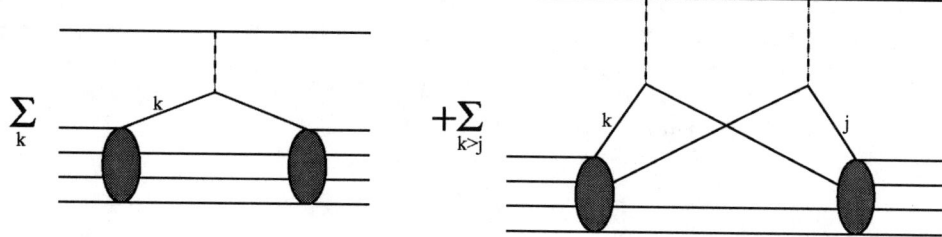

FIGURE 7. Illustration of the single and double scattering terms in Eq. (65). The coherent sum over all graphs leads to interferences that reduce the total cross section.

$\gamma(\vec{b})$,

$$\Gamma(\vec{b},\vec{r}_1,\vec{r}_2,\ldots\vec{r}_a) \equiv 1 - \prod_{j=1}^{A}[1 - \gamma_j(\vec{b}-\vec{s}_j)] \quad (64)$$

$$= \sum_{k=1}^{A}\gamma_k(\vec{b}-\vec{s}_k) - \sum_{k>j}\gamma_k(\vec{b}-\vec{s}_k)\gamma_j(\vec{b}-\vec{s}_j) + \ldots. \quad (65)$$

The first two terms in this expansion are illustrated in Fig. 7. The first term in Eq. (65) is just the sum of single scattering amplitudes. However, different nucleons in the nucleus compete to interact with the projectile. This effect is contained in the second term in Eq. (65), which reduces the cross section. This reduction is an interference effect that appears because the amplitudes for scattering on different nucleons have to be added coherently. This destructive interference can be observed in experiment as shadowing in hadron-nucleus interactions (eclipse effect in deuterium). Note, however, that shadowing is not completely explained by Glauber theory, as will be explained in the following section.

The easiest application of Glauber multiple scattering theory to nuclear systems is the calculation of the inelastic nucleus-nucleus (AB) cross section, which can be written as

$$\sigma_{AB}^{in} = \int d^2 b(1 - P_0(b)). \quad (66)$$

here, $P_0(b)$ is the probability that no interaction takes place,

$$P_0(b) = (1 - \sigma_{NN}^{in} T_{AB}(b))^{AB}, \quad (67)$$

where the nuclear overlap function is given by

$$T_{AB}(\vec{b}) = \int d^2 s T_A(\vec{s}) T_B(\vec{b}-\vec{s}). \quad (68)$$

(Obviously, $1 - P_0(b)$ is then the probability of an inelastic interaction, and the meaning of Eq. (66) becomes very transparent.) As it is common, we have labeled the two nuclei by their atomic mass numbers A and B.

Another application is the calculation of inclusive particle spectra. With the help of crossing symmetry, the cross section for production of a particle of type a in an AB collision, $AB \to aX$, can be calculated from the total cross section of the process $\bar{a}AB \to X$, where $\bar{a}$ is the antiparticle of a. According to so-called AGK cutting rules [21], the nuclear cross section for this process is given by

$$E\frac{d^3\sigma^a_{AB}}{d^3p} = T_{AB}(\vec{b})E\frac{d^3\sigma^a_{NN}}{d^3p}. \tag{69}$$

Integration over impact parameter b yields

$$E\frac{d^3\sigma^a_{AB}}{d^3p} = AB\, E\frac{d^3\sigma^a_{NN}}{d^3p}, \tag{70}$$

and correspondingly the charged particle multiplicity would scale proportional to AB,

$$\frac{dn_{ch}}{d\eta} = AB\, \frac{1}{\sigma^{in}_{AB}}\frac{d\sigma^{ch}_{NN}}{d\eta}, \tag{71}$$

the meaning of which is obvious – if collisions are truly independent, the resulting multiplicity should scale with the number of collisions, AB.

However, the relation (71) appears to be badly violated in experiment. What went wrong? It appears that the disagreement between the result Eq. (71) and experimental data is due to the fact that there are important corrections to the Glauber multiple scattering theory, which we neglected so far. These corrections are known as Gribov's inelastic shadowing [19] and will be the subject of the next section.

2.1.3. Gribov's "Inelastic Shadowing"

The assumed independence of nucleon–nucleon collisions is violated by the diagrams of the type of Fig. 8, where the projectile is excited into a state $|n\rangle$ by the interaction. The diagram in Fig. 8 does not describe independent collisions, and a high energies it will interfere with the double scattering graph in Fig. 7.

The excitation of an inelastic state in the scattering is accompanied by a longitudinal momentum transfer

$$\Delta p_L = \frac{M_f^2 - M_i^2}{2p}, \tag{72}$$

where M_f is the invariant mass of the excited system and M_i is the invariant mass of the projectile in the initial state. The diagram in Fig. 8 is only important if it can make a significant contribution to the forward scattering amplitude F^A_{ii}. This requires that the longitudinal momentum transfer must be so small that the nucleus has a chance to remain intact, i.e.

$$\Delta p_L R_A \lesssim 1, \tag{73}$$

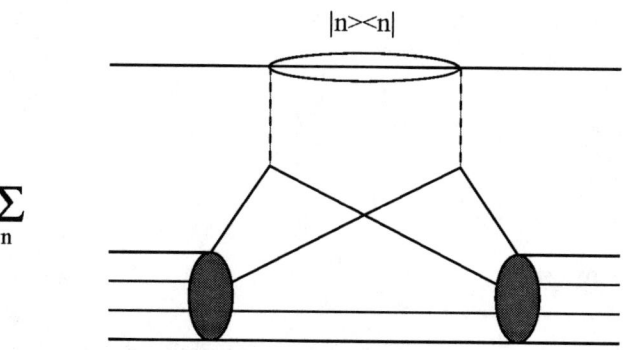

FIGURE 8. If the projectile is a composite particle, it can be excited by the interaction. Therefore, this graph will interfere with the double scattering graph in Fig. 7.

where R_A is the nuclear radius. Apparently, this condition is fulfilled for sufficiently large values of the projectile momentum p in Eq. (72). Thus, as it was first found by Gribov in [19], Glauber theory receives important corrections at high energy.

The condition, Eq. (73), which determines, whether Gribov's inelastic shadowing becomes relevant, leads us to the important quantum mechanical concept of *formation time*, or *formation length*. The formation time is the lifetime of the excitation $|n\rangle$ in Fig. 8 in the target rest frame and the formation length is the longitudinal distance over which the excited state $|n\rangle$ lives. At high energy, of course, both quantities are identical. The formation time/length can be determined in a time-dependent and in a time-independent approach.

In the time dependent formulation, one starts from the energy-time uncertainty relation,

$$\Delta E \Delta t \gtrsim 1. \tag{74}$$

The lifetime of the excitation in rest frame of the projectile is given by

$$\tau_f \approx \frac{1}{M_f - M_i}. \tag{75}$$

In order to obtain the formation time, we have to transform τ_f to the target rest frame, by multiplying τ_f with the relativistic γ-factor,

$$t_f \equiv \Delta t = \gamma \tau_f = \frac{p}{\bar{M}} \tau_f, \tag{76}$$

where

$$\bar{M} = \frac{M_f + M_i}{2}. \tag{77}$$

We finally obtain for the formation time

$$t_f \approx \frac{2p}{M_f^2 - M_i^2}. \tag{78}$$

In the time-independent approach, one starts from the coordinate-momentum uncertainty relation,

$$\Delta p_L \Delta z \gtrsim 1. \qquad (79)$$

The longitudinal momentum transfer was already given in Eq. 72. It is calculated in the following way,

$$\Delta p_L = \sqrt{E^2 - M_i^2} - \sqrt{E^2 - M_f^2} \approx \frac{M_f^2 - M_i^2}{2p}. \qquad (80)$$

According to the uncertainty relation Eq. (79), the excited state lives over the longitudinal extension

$$l_f \equiv \Delta z \approx \frac{1}{\Delta p_L} \approx \frac{2p}{M_f^2 - M_i^2}. \qquad (81)$$

As expected, the formation length is identical to the formation time given in Eq. (78).

We see from Eqs. (78) and (81) that for large initial projectile momentum, the process develops at large longitudinal distances in the target rest frame. At the high center of mass energies of RHIC and LHC, the coherence length will be much larger than the nuclear radius and all scattering processes will be governed by coherence effects (the coherence length becomes as long as several hundreds fm).

2.2. Elementary hadron–hadron scattering at high energies

All of the formalism presented above is completely independent of the underlying interaction. Before concluding this section, we will briefly discuss the main properties of hadron–hadron scattering at high energies. Let us begin by listing some empirical facts about hadronic cross sections:

- Total hadronic cross sections are approximately constant at cm energies of order $\sqrt{s} \sim 20\,\text{GeV}$ and slowly rise, $\sigma_{tot} \sim s^{0.08}$, up to the highest energies accessible in experiment (Tevatron energy, $\sqrt{s} = 1.8\,\text{TeV}$).
- The diffraction cone shrinks as energy increases, indicating that the size of the hadron increases with energy.
- The mean transverse momentum of produced particles is approximately constant or increases only slowly with energy, respectively.

The (approximate) constancy of the total cross section in the framework of QCD implies that high energy hadronic scattering is dominated by two gluon exchange [22], see Fig. 9 (left). The two gluon exchange model also yields a purely imaginary forward scattering amplitude. In order to explain the increase of the total cross section with energy, one has to take the radiation of additional gluons into account, see Fig. 9 (right). The probability of gluon emission is proportional to $\alpha_s y \sim \alpha_s \ln s$, where y is rapidity. Thus, each gluon radiation in Fig. 9 (right) contributes a factor $\ln s$ to the total cross section. Resumming an infinite number of gluon emissions ordered in rapidity, one finds

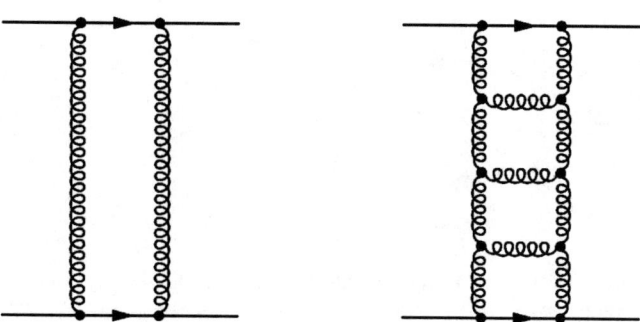

FIGURE 9. Double gluon exchange (left), yields an imaginary scattering amplitude and a constant cross section. The rise of hadronic cross sections and the shrinkage of the diffraction cone at high energy is due to radiation of additional gluons (right).

that the total cross section behaves like

$$\sigma_{tot} \propto \sum_n \frac{(\ln s)^n}{n!} a^n = s^a, \tag{82}$$

where $a \propto \alpha_s$.

This gluon radiation also explains the shrinkage of the diffraction cone. At high energy, the t-differential cross section in hadronic collisions behaves like

$$\frac{d\sigma}{dt} = \left.\frac{d\sigma}{dt}\right|_{t=0} e^{-B(s)|t|}, \tag{83}$$

where $B(s) \propto \ln s$ increases with energy. Such a behavior emerges, if the elastic scattering amplitude in impact parameter space is given by

$$f_{el} \propto \exp\left(\frac{b}{R_h^2(s)}\right), \tag{84}$$

where the effective hadron radius $R_h^2(s)$ increases as a function of energy. Therefore, the shrinkage of the diffraction peak suggests an increase of hadronic sizes with energy. In QCD, this can be understood in the following way: Gluons are radiated off the projectile with different transverse momenta. As rapidity, or energy, increases, these gluons perform a random walk in the impact parameter plane and correspondingly, the transverse size of the gluon cloud surrounding the projectile increases. This can be regarded as a diffusion process in the impact parameter plane, in which rapidity plays the role of time.

The slow increase of the mean transverse momentum with energy is likely to be related to asymptotic freedom. Indeed, at large transverse momentum, $p_\perp \gg \Lambda_{QCD}$, the strong coupling constant becomes small, $\alpha_s(p_\perp) \ll 1$, which suppresses the production of high $p_\perp$ particles.

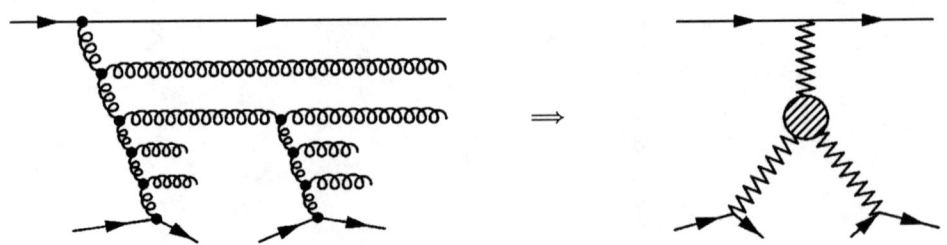

FIGURE 10. At sufficiently high energy, emitted gluons can themselves develop showers. In the squared amplitude (right), the gluons combine to ladders, which are denoted by zigzag lines. Fusion of gluon ladders is the mechanism behind gluon shadowing. A resummation of fan diagrams like in the left figure, corresponds to classical solutions [24] of Reggeon Field Theory [25].

Eventually, the power-like growth of the total hadronic cross section will violate the Froissart-Martin bound [23], which states that as a consequence of unitarity and analyticity, total cross sections cannot rise faster than

$$\sigma_{tot} \lesssim C \ln^2 s, \qquad (85)$$

where C is a constant. At sufficiently high energy, emitted gluons can develop showers themselves, see Fig. 10. Due to this process, the projectile sees a reduced gluon density in the target and the growth of the cross section is slowed down. This effect, in the squared amplitude, realizes a QCD realization of Gribov's inelastic shadowing (see Fig. 10.) Even though such unitarity corrections might already be present in proton-antiproton scattering at Tevatron, they will be much more pronounced in nuclear collisions at RHIC.

As the magnitude of these effects increases with energy and/or atomic number of the colliding nuclei, the classification of diagrams in terms of individual nucleon–nucleon amplitudes (or parton ladders) rapidly starts to lose sense – the non–linear effects become extremely important. The treatment of nuclear interactions in this high–density regime will be considered in the following section.

3. CLASSICAL CHROMODYNAMICS OF RELATIVISTIC HEAVY ION COLLISIONS

3.1. QCD in the classical regime

Most of the applications of QCD so far have been limited to the short distance regime of high momentum transfer, where the theory becomes weakly coupled and can be linearized. While this is the only domain where our theoretical tools based on perturbation theory are adequate, this is also the domain in which the beautiful non–linear structure of QCD does not yet reveal itself fully. On the other hand, as soon as we decrease the momentum transfer in a process, the dynamics rapidly becomes non–linear, but our understanding is hindered by the large coupling. Being perplexed by this problem, one is tempted to dream about an environment in which the coupling is weak,

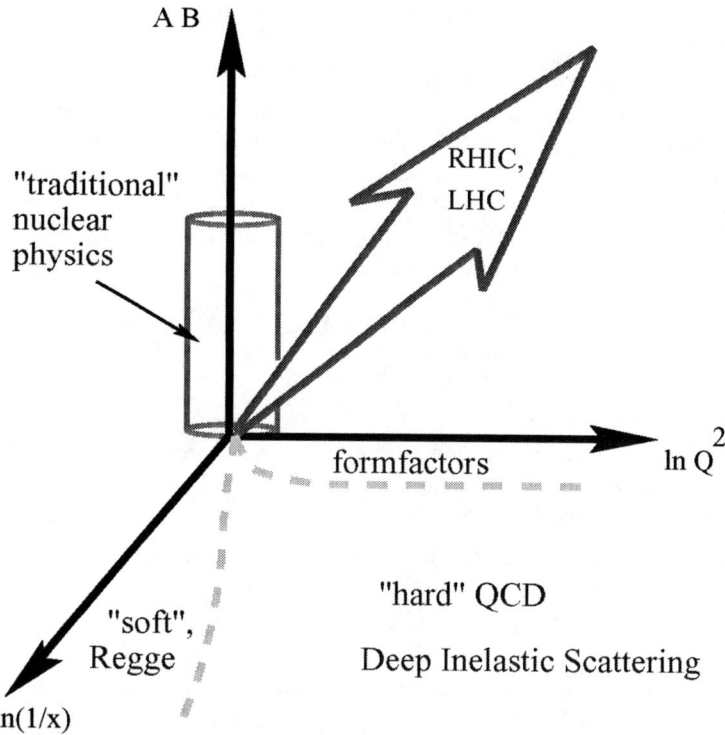

FIGURE 11. The place of relativistic heavy ion physics in the study of QCD; the vertical axis is the product of atomic numbers of projectile and target, and the horizontal axes are the momentum transfer Q^2 and rapidity $y = \ln(1/x)$ (x is the Bjorken scaling variable).

allowing a systematic theoretical treatment, but the fields are strong, revealing the full non–linear nature of QCD. We are going to argue now that this environment can be created on Earth with the help of relativistic heavy ion colliders. Relativistic heavy ion collisions allow to probe QCD in the non–linear regime of high parton density and high color field strength, see Fig. 11.

It has been conjectured long time ago that the dynamics of QCD in the high density domain may become qualitatively different: in parton language, this is best described in terms of *parton saturation* [27, 28, 29], and in the language of color fields – in terms of the *classical* Chromo–Dynamics [30]; see the lectures [31] and [32] and references therein. In this high density regime, the transition amplitudes are dominated not by quantum fluctuations, but by the configurations of classical field containing large, $\sim 1/\alpha_s$, numbers of gluons. One thus uncovers new non–linear features of QCD, which cannot be investigated in the more traditional applications based on the perturbative approach. The classical color fields in the initial nuclei (the "color glass condensate" [31]) can be thought of as either perturbatively generated, or as being a topologically non–trivial superposition of the Weizsäcker-Williams radiation and the quasi–classical vacuum fields [33, 34, 35].

3.1.1. Geometrical arguments

Let us consider an external probe J interacting with the nuclear target of atomic number A. At small values of Bjorken x, by uncertainty principle the interaction develops over large longitudinal distances $z \sim 1/mx$, where m is the nucleon mass. As soon as z becomes larger than the nuclear diameter, the probe cannot distinguish between the nucleons located on the front and back edges of the nucleus, and all partons within the transverse area $\sim 1/Q^2$ determined by the momentum transfer Q participate in the interaction coherently. The density of partons in the transverse plane is given by

$$\rho_A \simeq \frac{xG_A(x,Q^2)}{\pi R_A^2} \sim A^{1/3}, \tag{86}$$

where we have assumed that the nuclear gluon distribution scales with the number of nucleons A. The probe interacts with partons with cross section $\sigma \sim \alpha_s/Q^2$; therefore, depending on the magnitude of momentum transfer Q, atomic number A, and the value of Bjorken x, one may encounter two regimes:

- $\sigma \rho_A \ll 1$ – this is a familiar "dilute" regime of incoherent interactions, which is well described by the methods of perturbative QCD;
- $\sigma \rho_A \gg 1$ – in this regime, we deal with a dense parton system. Not only do the "leading twist" expressions become inadequate, but also the expansion in higher twists, i.e. in multi–parton correlations, breaks down here.

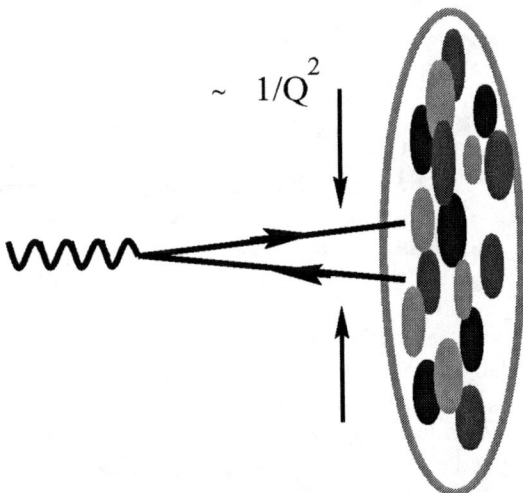

FIGURE 12. Hard probe interacting with the nuclear target resolves the transverse distance $\sim 1/\sqrt{Q}$ (Q^2 is the square of the momentum transfer) and, in the target rest frame, the longitudinal distance $\sim 1/(mx)$ (m is the nucleon mass and x – Bjorken variable).

The border between the two regimes can be found from the condition $\sigma \rho_A \simeq 1$; it determines the critical value of the momentum transfer ("saturation scale"[27]) at which the parton system becomes to look dense to the probe[1]:

$$Q_s^2 \sim \alpha_s \frac{xG_A(x, Q_s^2)}{\pi R_A^2}. \tag{87}$$

In this regime, the number of gluons from (87) is given by

$$xG_A(x, Q_s^2) \sim \frac{\pi}{\alpha_s(Q_s^2)} Q_s^2 R_A^2, \tag{88}$$

where $Q_s^2 R_A^2 \sim A$. One can see that the number of gluons is proportional to the *inverse* of $\alpha_s(Q_s^2)$, and becomes large in the weak coupling regime. In this regime, as we shall now discuss, the dynamics is likely to become essentially classical.

3.1.2. Saturation as the classical limit of QCD

Indeed, the condition (87) can be derived in the following, rather general, way. As a first step, let us note that the dependence of the action corresponding to the Lagrangian (9) on the coupling constant is given by

$$S \sim \int \frac{1}{g^2} G_{\mu\nu}^a G_{\mu\nu}^a \, d^4x. \tag{89}$$

Let us now consider a classical configuration of gluon fields; by definition, $G_{\mu\nu}^a$ in such a configuration does not depend on the coupling, and the action is large, $S \gg \hbar$. The number of quanta in such a configuration is then

$$N_g \sim \frac{S}{\hbar} \sim \frac{1}{\hbar g^2} \rho_4 V_4, \tag{90}$$

where we re-wrote (89) as a product of four–dimensional action density ρ_4 and the four–dimensional volume V_4.

Note that since (90) depends only on the product of the Planck constant $\hbar$ and the coupling g^2, the classical limit $\hbar \to 0$ is indistinguishable from the weak coupling limit $g^2 \to 0$. The weak coupling limit of small $g^2 = 4\pi\alpha_s$ therefore corresponds to the semi-classical regime.

The effects of non–linear interactions among the gluons become important when $\partial_\mu A_\mu \sim A_\mu^2$ (this condition can be made explicitly gauge invariant if we derive it from the expansion of a correlation function of gauge-invariant gluon operators, e.g., G^2). In momentum space, this equality corresponds to

$$Q_s^2 \sim (A_\mu)^2 \sim (G^2)^{1/2} = \sqrt{\rho_4}; \tag{91}$$

---

[1] Note that since $xG_A(x, Q_s^2) \sim A^{1/3}$, which is the length of the target, this expression in the target rest frame can also be understood as describing a broadening of the transverse momentum resulting from the multiple re-scattering of the probe.

Q_s is the typical value of the gluon momentum below which the interactions become essentially non–linear.

Consider now a nucleus A boosted to a high momentum. By uncertainty principle, the gluons with transverse momentum Q_s are extended in the longitudinal and proper time directions by $\sim 1/Q_s$; since the transverse area is πR_A^2, the four–volume is $V_4 \sim \pi R_A^2/Q_s^2$. The resulting four–density from (90) is then

$$\rho_4 \sim \alpha_s \frac{N_g}{V_4} \sim \alpha_s \frac{N_g Q_s^2}{\pi R_A^2} \sim Q_s^4, \tag{92}$$

where at the last stage we have used the non–linearity condition (91), $\rho_4 \sim Q_s^4$. It is easy to see that (92) coincides with the saturation condition (87), since the number of gluons in the infinite momentum frame $N_g \sim xG(x, Q_s^2)$.

In view of the significance of saturation criterion for the rest of the material in these lectures, let us present yet another argument, traditionally followed in the discussion of classical limit in electrodynamics [48]. The energy of the gluon field per unit volume is $\sim \vec{E}^{a2}$. The number of elementary "oscillators of the field", also per unit volume, is $\sim \omega^3$. To get the number of the quanta in the field we have to divide the energy of the field by the product of the number of the oscillators $\sim \omega^3$ and the average energy $\hbar\omega$ of the gluon:

$$N_{\vec{k}} \sim \frac{\vec{E}^{a2}}{\hbar \omega^4}. \tag{93}$$

The classical approximation holds when $N_{\vec{k}} \gg 1$. Since the energy ω of the oscillators is related to the time Δt over which the average energy is computed by $\omega \sim 1/\Delta t$, we get

$$\vec{E}^{a2} \gg \frac{\hbar}{(\Delta t)^4}. \tag{94}$$

Note that the quantum mechanical uncertainty principle for the energy of the field reads

$$\vec{E}^{a2} \, \omega^4 \sim \hbar, \tag{95}$$

so the condition (94) indeed defines the quasi–classical limit.

Since $\vec{E}^{a2}$ is proportional to the action density ρ_4, and the typical time is $\Delta t \sim 1/k_\perp$, using (92) we finally get that the classical description applies when

$$k_\perp^2 < \alpha_s \frac{N_g}{\pi R_A^2} \equiv Q_s^2. \tag{96}$$

3.1.3. The absence of mini–jet correlations

When the occupation numbers of the field become large, the matrix elements of the creation and annihilation operators of the gluon field defined by

$$\hat{A}^\mu = \sum_{\vec{k},\alpha} (\hat{c}_{\vec{k}\alpha} A^\mu_{\vec{k}\alpha} + \hat{c}^\dagger_{\vec{k}\alpha} A^{\mu*}_{\vec{k}\alpha}) \tag{97}$$

become very large,
$$N_{\vec{k}\alpha} = \langle \hat{c}^\dagger_{\vec{k}\alpha} \hat{c}_{\vec{k}\alpha} \rangle \gg 1, \tag{98}$$
so that one can neglect the unity on the r.h.s. of the commutation relation
$$\hat{c}_{\vec{k}\alpha} \hat{c}^\dagger_{\vec{k}\alpha} - \hat{c}^\dagger_{\vec{k}\alpha} \hat{c}_{\vec{k}\alpha} = 1 \tag{99}$$
and treat these operators as classical c-numbers.

This observation, often used in condensed matter physics, especially in the theoretical treatment of superfluidity, has important consequences for gluon production – in particular, it implies that the correlations among the gluons in the saturation region can be neglected:
$$\langle A(k_1)A(k_2)...A(k_n)\rangle \simeq \langle A(k_1)\rangle\langle A(k_2)\rangle...\langle A(k_n)\rangle. \tag{100}$$
Thus, in contrast to the perturbative picture, where the produced mini-jets have strong back-to-back correlations, the gluons resulting from the decay of the classical saturated field are uncorrelated at $k_\perp \lesssim Q_s$.

Note that the amplitude with the factorization property (100) is called point–like. However, the relation (100) cannot be exact if we consider the correlations of final–state hadrons – the gluon mini–jets cannot transform into hadrons independently. These correlations caused by color confinement however affect mainly hadrons with close three–momenta, as opposed to the perturbative correlations among mini–jets with the opposite three–momenta.

It will be interesting to explore the consequences of the factorization property of the classical gluon field (100) for the HBT correlations of final–state hadrons. It is likely that the HBT radii in this case reflect the universal color correlations in the hadronization process.

Another interesting property of classical fields follows from the relation
$$\langle (\hat{c}^\dagger_{\vec{k}\alpha} \hat{c}_{\vec{k}\alpha})^2 \rangle - \langle \hat{c}^\dagger_{\vec{k}\alpha} \hat{c}_{\vec{k}\alpha} \rangle^2 = \langle \hat{c}^\dagger_{\vec{k}\alpha} \hat{c}_{\vec{k}\alpha} \rangle, \tag{101}$$
which determines the fluctuations in the number of produced gluons. We will discuss the implications of Eq. (101) for the multiplicity fluctuations in heavy ion collisions later.

3.2. Classical QCD in action

3.2.1. Centrality dependence of hadron production

In nuclear collisions, the saturation scale becomes a function of centrality; a generic feature of the quasi–classical approach – the proportionality of the number of gluons to the inverse of the coupling constant (90) – thus leads to definite predictions [37] on the centrality dependence of multiplicity.

Let us first present the argument on a qualitative level. At different centralities (determined by the impact parameter of the collision), the average density of partons (in the transverse plane) participating in the collision is very different. This density ρ is proportional to the average length of nuclear material involved in the collision, which in turn approximately scales with the power of the number N_{part} of participating nucleons, $\rho \sim N_{part}^{1/3}$. The density of partons defines the value of the saturation scale, and so we expect

$$Q_s^2 \sim N_{part}^{1/3}. \tag{102}$$

The gluon multiplicity is then, as we discussed above, is

$$\frac{dN_g}{d\eta} \sim \frac{S_A Q_s^2}{\alpha_s(Q_s^2)}, \tag{103}$$

where S_A is the nuclear overlap area, determined by atomic number and the centrality of collision. Since $S_A Q_s^2 \sim N_{part}$ by definitions of the transverse density and area, from (103) we get

$$\frac{dN_g}{d\eta} \sim N_{part} \ln N_{part}, \tag{104}$$

which shows that the gluon multiplicity shows a logarithmic deviation from the scaling in the number of participants.

To quantify the argument, we need to explicitly evaluate the average density of partons at a given centrality. This can be done by using Glauber theory, which allows to evaluate the differential cross section of the nucleus–nucleus interactions. The shape of the multiplicity distribution at a given (pseudo)rapidity η can then be readily obtained by using the formulae introduced in section 2:

$$\frac{d\sigma}{dn} = \int d^2b \, \mathscr{P}(n;b) \, (1 - P_0(b)), \tag{105}$$

where $P_0(b)$ is the probability of no interaction among the nuclei at a given impact parameter b:

$$P_0(b) = (1 - \sigma_{NN} T_{AB}(b))^{AB}; \tag{106}$$

σ_{NN} is the inelastic nucleon–nucleon cross section, and $T_{AB}(b)$ is the nuclear overlap function for the collision of nuclei with atomic numbers A and B; we have used the three–parameter Woods–Saxon nuclear density distributions [39].

The correlation function $\mathscr{P}(n;b)$ is given by

$$\mathscr{P}(n;b) = \frac{1}{\sqrt{2\pi a \bar{n}(b)}} \exp\left(-\frac{(n-\bar{n}(b))^2}{2a\bar{n}(b)}\right), \tag{107}$$

here $\bar{n}(b)$ is the mean multiplicity at a given impact parameter b; the formulae for the number of participants and the number of binary collisions can be found in [38]. The parameter a describes the strength of fluctuations; for the classical gluon field, as follows from (101), $a = 1$. However, the strength of fluctuations can be changed by the subsequent evolution of the system and by hadronization process. Moreover, in a real experiment, the strength of fluctuations strongly depends on the acceptance. In describing the PHOBOS distribution [46], we have found that the value $a = 0.6$ fits the data well.

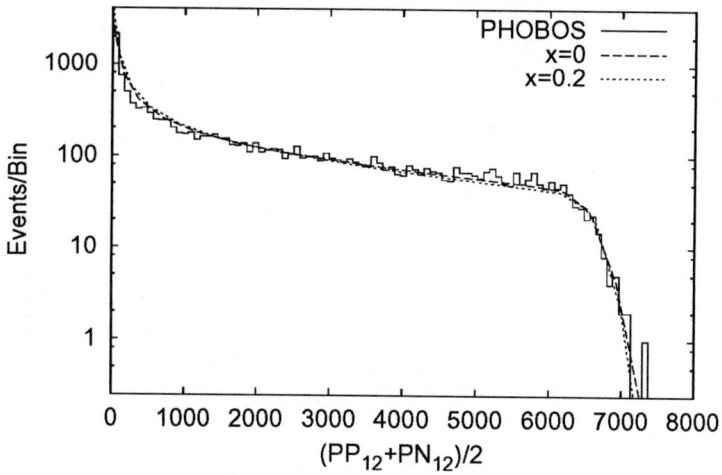

FIGURE 13. Charged multiplicity distribution at $\sqrt{s} = 130$ A GeV; solid line (histogram) – PHOBOS result; dashed line – distribution corresponding to participant scaling ($x = 0$); dotted line – distribution corresponding to the 37% admixture of "hard" component in the multiplicity; see text for details.

In Fig. 13, we compare the resulting distributions for two different assumptions about the scaling of multiplicity with the number of participants to the PHOBOS experimental distribution, measured in the interval $3 < |\eta| < 4.5$. One can see that almost independently of theoretical assumptions about the dynamics of multiparticle production, the data are described quite well. At first this may seem surprising; the reason for this result is that at high energies, heavy nuclei are almost completely "black"; unitarity then implies that the shape of the cross section is determined almost entirely by the nuclear geometry. We can thus use experimental differential cross sections as a reliable handle on centrality. This gives us a possibility to compute the dependence of the saturation scale on centrality of the collision, and thus to predict the centrality dependence of particle multiplicities, shown in Fig. 14. (see [37] for details).

3.2.2. Energy dependence

Let us now turn to the discussion of energy dependence of hadron production. In semi–classical scenario, it is determined by the variation of saturation scale Q_s with Bjorken $x = Q_s/\sqrt{s}$. This variation, in turn, is determined by the $x-$ dependence of the gluon structure function. In the saturation approach, the gluon distribution is related to the saturation scale by Eq.(87). A good description of HERA data is obtained with saturation scale $Q_s^2 = 1 \div 2$ GeV^2 with W - dependence ($W \equiv \sqrt{s}$ is the center-of-mass energy available in the photon–nucleon system) [49]

$$Q_s^2 \propto W^\lambda, \qquad (108)$$

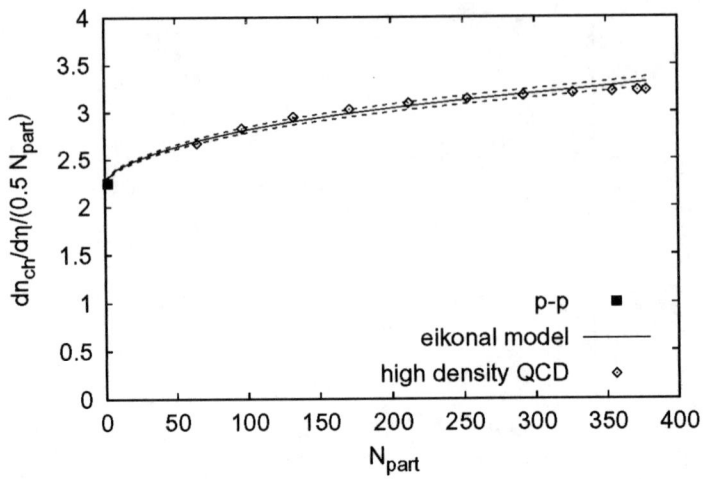

FIGURE 14. Centrality dependence of the charged multiplicity per participant pair near $\eta = 0$ at $\sqrt{s} = 130$ A GeV; the curves represent the prediction based on the conventional eikonal approach, while the diamonds correspond to the high density QCD prediction (see text). The square indicates the pp multiplicity.

where $\lambda \simeq 0.25 \div 0.3$. In spite of significant uncertainties in the determination of the gluon structure functions, perhaps even more important is the observation [49] that the HERA data exhibit scaling when plotted as a function of variable

$$\tau = \frac{Q^2}{Q_0^2}\left(\frac{x}{x_0}\right)^\lambda, \qquad (109)$$

where the value of λ is again within the limits $\lambda \simeq 0.25 \div 0.3$. In high density QCD, this scaling is a consequence of the existence of dimensionful scale [27, 30])

$$Q_s^2(x) = Q_0^2 \, (x_0/x)^\lambda. \qquad (110)$$

Using the value of $Q_s^2 \simeq 2.05$ GeV^2 extracted [37] at $\sqrt{s} = 130$ GeV and $\lambda = 0.25$ [49] used in [40], equation (120) leads to the following approximate formula for the energy dependence of charged multiplicity in central $Au - Au$ collisions:

$$\left\langle \frac{2}{N_{part}} \frac{dN_{ch}}{d\eta} \right\rangle_{\eta<1} \approx 0.87 \left(\frac{\sqrt{s}\,(\text{GeV})}{130}\right)^{0.25} \times$$

$$\times \left[3.93 + 0.25 \ln\left(\frac{\sqrt{s}\,(\text{GeV})}{130}\right)\right]. \qquad (111)$$

At $\sqrt{s} = 130$ GeV, we estimate from Eq.(111) $2/N_{part} \, dN_{ch}/d\eta \,|_{\eta<1} = 3.42 \pm 0.15$, to be compared to the average experimental value of 3.37 ± 0.12 [46, 44, 45, 47]. At $\sqrt{s} = 200$ GeV, one gets 3.91 ± 0.15, to be compared to the PHOBOS value [46] of

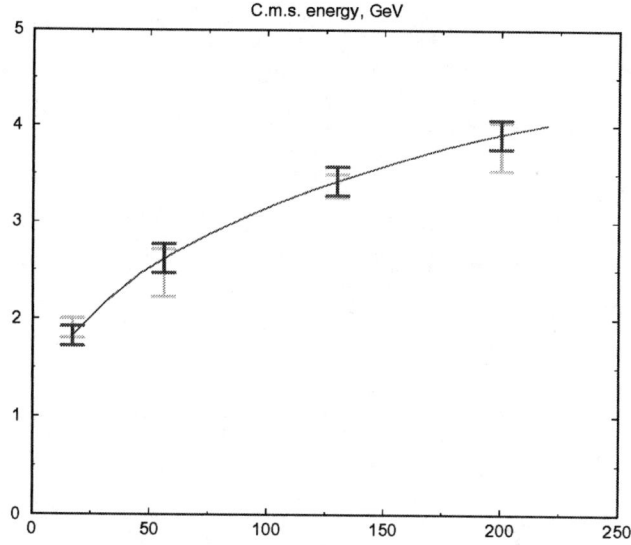

FIGURE 15. Energy dependence of charged multiplicity per participant pair at RHIC energies; solid line is the result (111).

3.78 ± 0.25. Finally, at $\sqrt{s} = 56$ GeV, we find 2.62 ± 0.15, to be compared to [46] 2.47 ± 0.25. It is interesting to note that formula (111), when extrapolated to very high energies, predicts for the LHC energy a value substantially smaller than found in other approaches:

$$\left\langle \frac{2}{N_{part}} \frac{dN_{ch}}{d\eta} \right\rangle_{\eta<1} = 10.8 \pm 0.5; \quad \sqrt{s} = 5500 \text{ GeV}, \tag{112}$$

corresponding only to a factor of 2.8 increase in multiplicity between the RHIC energy of $\sqrt{s} = 200$ GeV and the LHC energy of $\sqrt{s} = 5500$ GeV (numerical calculations show that when normalized to the number of participants, the multiplicity in central $Au - Au$ and $Pb - Pb$ systems is almost identical). The energy dependence of charged hadron multiplicity per participant pair is shown in Fig.15.

One can also try to extract the value of the exponent λ from the energy dependence of hadron multiplicity measured by PHOBOS at $\sqrt{s} = 130$ GeV and at at $\sqrt{s} = 56$ GeV; this procedure yields $\lambda \simeq 0.37$, which is larger than the value inferred from the HERA data (and is very close to the value $\lambda \simeq 0.38$, resulting from the final–state saturation calculations [51]).

3.2.3. Radiating the classical glue

Let us now proceed to the quantitative calculation of the (pseudo-) rapidity and centrality dependences [50]. We need to evaluate the leading tree diagram describing

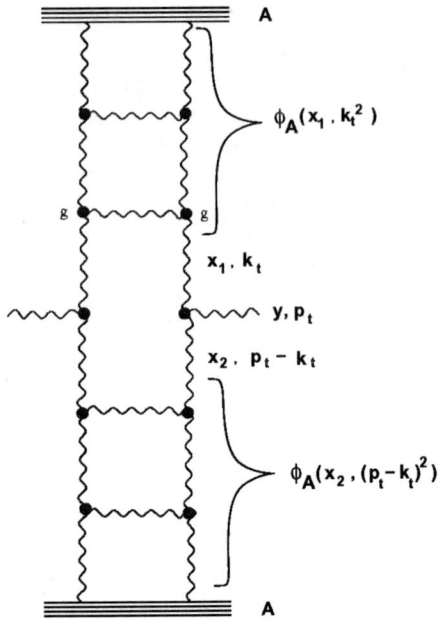

FIGURE 16. The Mueller diagram for the classical gluon radiation.

emission of gluons on the classical level, see[2] Fig. 16.

Let us introduce the unintegrated gluon distribution $\varphi_A(x, k_t^2)$ which describes the probability to find a gluon with a given x and transverse momentum k_t inside the nucleus A. As follows from this definition, the unintegrated distribution is related to the gluon structure function by

$$xG_A(x, p_t^2) = \int^{p_t^2} dk_t^2 \, \varphi_A(x, k_t^2); \tag{113}$$

when $p_t^2 > Q_s^2$, the unintegrated distribution corresponding to the bremsstrahlung radiation spectrum is

$$\varphi_A(x, k_t^2) \sim \frac{\alpha_s}{\pi} \frac{1}{k_t^2}. \tag{114}$$

In the saturation region, the gluon structure function is given by (88); the corresponding unintegrated gluon distribution has only logarithmic dependence on the transverse momentum:

$$\varphi_A(x, k_t^2) \sim \frac{S_A}{\alpha_s}; \quad k_t^2 \leq Q_s^2, \tag{115}$$

where S_A is the nuclear overlap area, determined by the atomic numbers of the colliding nuclei and by centrality of the collision.

---

[2] Note that this "mono–jet" production diagram makes obvious the absence of azimuthal correlations in the saturation regime discussed above, see eq (100).

The differential cross section of gluon production in a AA collision can now be written down as [27, 41]

$$E\frac{d\sigma}{d^3p} = \frac{4\pi N_c}{N_c^2-1} \frac{1}{p_t^2} \int dk_t^2 \; \alpha_s \; \varphi_A(x_1,k_t^2) \; \varphi_A(x_2,(p-k)_t^2), \tag{116}$$

where $x_{1,2} = (p_t/\sqrt{s})\exp(\pm\eta)$, with η the (pseudo)rapidity of the produced gluon; the running coupling α_s has to be evaluated at the scale $Q^2 = max\{k_t^2,(p-k)_t^2\}$. The rapidity density is then evaluated from (116) according to

$$\frac{dN}{dy} = \frac{1}{\sigma_{AA}} \int d^2p_t \left(E\frac{d\sigma}{d^3p}\right), \tag{117}$$

where σ_{AA} is the inelastic cross section of nucleus–nucleus interaction.

Since the rapidity y and Bjorken variable are related by $\ln 1/x = y$, the x– dependence of the gluon structure function translates into the following dependence of the saturation scale Q_s^2 on rapidity:

$$Q_s^2(s;\pm y) = Q_s^2(s;y=0) \; \exp(\pm\lambda y). \tag{118}$$

As it follows from (118), the increase of rapidity at a fixed $W \equiv \sqrt{s}$ moves the wave function of one of the colliding nuclei deeper into the saturation region, while leading to a smaller gluon density in the other, which as a result can be pushed out of the saturation domain. Therefore, depending on the value of rapidity, the integration over the transverse momentum in Eqs. (116),(117) can be split in two regions: i) the region $\Lambda_{QCD} < k_t < Q_{s,min}$ in which the wave functions are both in the saturation domain; and ii) the region $\Lambda \ll Q_{s,min} < k_t < Q_{s,max}$ in which the wave function of one of the nuclei is in the saturation region and the other one is not. Of course, there is also the region of $k_t > Q_{s,max}$, which is governed by the usual perturbative dynamics, but our assumption here is that the rôle of these genuine hard processes in the bulk of gluon production is relatively small; in the saturation scenario, these processes represent quantum fluctuations above the classical background. It is worth commenting that in the conventional mini–jet picture, this classical background is absent, and the multi–particle production is dominated by perturbative processes. This is the main physical difference between the two approaches; for the production of particles with $p_t \gg Q_s$ they lead to identical results.

To perform the calculation according to (117),(116) away from $y=0$ we need also to specify the behavior of the gluon structure function at large Bjorken x (and out of the saturation region). At $x \to 1$, this behavior is governed by the QCD counting rules, $xG(x) \sim (1-x)^4$, so we adopt the following conventional form: $xG(x) \sim x^{-\lambda}(1-x)^4$.

We now have everything at hand to perform the integration over transverse momentum in (117), (116); the result is the following [50]:

$$\frac{dN}{dy} = const \; S_A \; Q_{s,min}^2 \; \ln\left(\frac{Q_{s,min}^2}{\Lambda_{QCD}^2}\right) \times$$

$$\times \left[1+\frac{1}{2}\ln\left(\frac{Q^2_{s,max}}{Q^2_{s,min}}\right)\left(1-\frac{Q_{s,max}}{\sqrt{s}}e^{|y|}\right)^4\right], \tag{119}$$

where the constant is energy–independent, S_A is the nuclear overlap area, $Q^2_s \equiv Q^2_s(s; y = 0)$, and $Q^2_{s,min(max)}$ are defined as the smaller (larger) values of (118); at $y = 0$, $Q^2_{s,min} = Q^2_{s,max} = Q^2_s(s) = Q^2_s(s_0) \times \times (s/s_0)^{\lambda/2}$. The first term in the brackets in (119) originates from the region in which both nuclear wave functions are in the saturation regime; this corresponds to the familiar $\sim (1/\alpha_s) Q^2_s R^2_A$ term in the gluon multiplicity. The second term comes from the region in which only one of the wave functions is in the saturation region. The coefficient $1/2$ in front of the second term in square brackets comes from k_t ordering of gluon momenta in evaluation of the integral of Eq.(116).

The formula (119) has been derived using the form (115) for the unintegrated gluon distributions. We have checked numerically that the use of more sophisticated functional form of φ_A taken from the saturation model of Golec-Biernat and Wüsthoff [49] in Eq.(116) affects the results only at the level of about 3%.

Since $S_A Q^2_s \sim N_{part}$ (recall that $Q^2_s \gg \Lambda^2_{QCD}$ is defined as the density of partons in the transverse plane, which is proportional to the density of participants), we can re–write (119) in the following final form [50]

$$\frac{dN}{dy} = c\, N_{part} \left(\frac{s}{s_0}\right)^{\frac{\lambda}{2}} e^{-\lambda|y|}\left[\ln\left(\frac{Q^2_s}{\Lambda^2_{QCD}}\right) - \lambda|y|\right] \times$$

$$\times \left[1+\lambda|y|\left(1-\frac{Q_s}{\sqrt{s}}e^{(1+\lambda/2)|y|}\right)^4\right], \tag{120}$$

with $Q^2_s(s) = Q^2_s(s_0)(s/s_0)^{\lambda/2}$. This formula is the central result of our paper; it expresses the predictions of high density QCD for the energy, centrality, rapidity, and atomic number dependences of hadron multiplicities in nuclear collisions in terms of a single scaling function. Once the energy–independent constant $c \sim 1$ and $Q^2_s(s_0)$ are determined at some energy s_0, Eq. (120) contains no free parameters. At $y = 0$ the expression (119) coincides exactly with the one derived in [37], and extends it to describe the rapidity and energy dependences.

3.2.4. Converting gluons into hadrons

The distribution (120) refers to the radiated gluons, while what is measured in experiment is, of course, the distribution of final hadrons. We thus have to make an assumption about the transformation of gluons into hadrons. The gluon mini–jets are produced with a certain virtuality, which changes as the system evolves; the distribution in rapidity is thus not preserved. However, in the analysis of jet structure it has been found that the *angle* of the produced gluon is remembered by the resulting hadrons; this property of "local parton–hadron duality" (see [43] and references therein) is natural if one assumes that the hadronization is a soft process which cannot change the direction of the emitted radiation. Instead of the distribution in the angle θ, it is more convenient to use the distribution in pseudo–rapidity $\eta = -\ln\tan(\theta/2)$. Therefore, before we can compare (119) to the data, we have to convert the rapidity distribution (120) into the gluon distribution in pseudo–rapidity. We will then assume that the gluon and hadron distributions are dual to each other in the pseudo–rapidity space.

To take account of the difference between rapidity y and the measured pseudo-rapidity η, we have to multiply (119) by the Jacobian of the $y \leftrightarrow \eta$ transformation; a simple calculation yields

$$h(\eta; p_t; m) = \frac{\cosh \eta}{\sqrt{\frac{m^2+p_t^2}{p_t^2} + \sinh^2 \eta}}, \tag{121}$$

where m is the typical mass of the produced particle, and p_t is its typical transverse momentum. Of course, to plot the distribution (120) as a function of pseudo-rapidity, one also has to express rapidity y in terms of pseudo-rapidity η; this relation is given by

$$y(\eta; p_t; m) = \frac{1}{2} \ln \left[\frac{\sqrt{\frac{m^2+p_t^2}{p_t^2} + \sinh^2 \eta} + \sinh \eta}{\sqrt{\frac{m^2+p_t^2}{p_t^2} + \sinh^2 \eta} - \sinh \eta} \right]; \tag{122}$$

obviously, $h(\eta; p_t; m) = \partial y(\eta; p_t; m)/\partial \eta$.

We now have to make an assumption about the typical invariant mass m of the gluon mini–jet. Let us estimate it by assuming that the slowest hadron in the mini–jet decay is the ρ-resonance, with energy $E_\rho = (m_\rho^2 + p_{\rho,t}^2 + p_{\rho,z}^2)^{1/2}$, where the z axis is pointing along the mini-jet momentum. Let us also denote by x_i the fractions of the gluon energy q_0 carried by other, fast, i particles in the mini-jet decay. Since the sum of transverse (with respect to the mini-jet axis) momenta of mini-jet decay products is equal to zero, the mini-jet invariant mass m is given by

$$m_{jet}^2 \equiv m^2 = \left(\sum_i x_i q_0 + E_\rho\right)^2 - \left(\sum_i x_i q_z + p_{\rho,z}\right)^2 \simeq$$

$$\simeq 2 \sum_i x_i q_z \cdot (m_{\rho,t} - p_{\rho,z}) \equiv 2 Q_s \cdot m_{eff}, \tag{123}$$

where $m_{\rho,t} = (m_\rho^2 + p_{\rho,t}^2)^{1/2}$. In Eq. (123) we used that $\sum_i x_i = 1$ and $q_0 \approx q_z = Q_s$. Taking $p_{\rho,z} \approx p_{\rho,t} \approx 300$ MeV and ρ mass, we obtain $m_{eff} \approx 0.5$ GeV.

We thus use the mass $m^2 \simeq 2 Q_s m_{eff} \simeq Q_s \cdot 1$ GeV in Eqs.(121,122). Since the typical transverse momentum of the produced gluon mini-jet is Q_s, we take $p_t = Q_s$ in (121). The effect of the transformation from rapidity to pseudo–rapidity is the decrease of multiplicity at small η by about 25 – 30%, leading to the appearance of the $\approx 10\%$ dip in the pseudo-rapidity distribution in the vicinity of $\eta = 0$. We have checked that the change in the value of the mini–jet mass by two times affects the Jacobian at central pseudo-rapidity to about $\simeq 10\%$, leading to $\sim 3\%$ effect on the final result.

The results for the $Au - Au$ collisions at $\sqrt{s} = 130$ GeV are presented in Figs 17 and 18. In the calculation, we use the results on the dependence of saturation scale on the

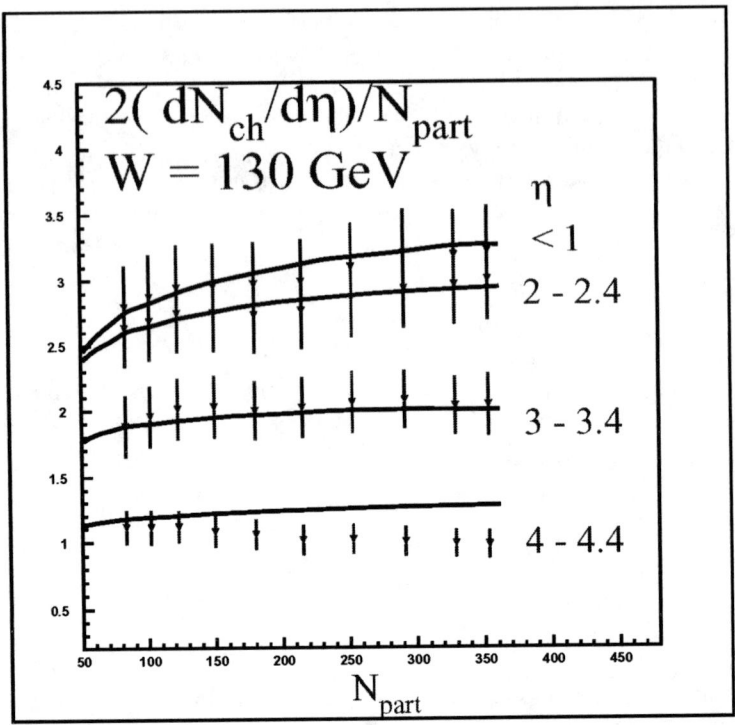

FIGURE 17. Centrality dependence of charged hadron production per participant at different pseudo-rapidity η intervals in $Au-Au$ collisions at $\sqrt{s} = 130$ GeV; the data are from [46] (Reprinted from Ref. [50], Copyright 2001, with permission from Elsevier Science).

mean number of participants at $\sqrt{s} = 130$ GeV from [37], see Table 2 of that paper. The mean number of participants in a given centrality cut is taken from the PHOBOS paper [46]. One can see that both the centrality dependence and the rapidity dependence of the $\sqrt{s} = 130$ GeV PHOBOS data are well reproduced below $\eta \simeq \pm 4$. The rapidity dependence has been evaluated with $\lambda = 0.25$, which is within the range $\lambda = 0.25 \div 0.3$ inferred from the HERA data [49]. The discrepancy above $\eta \simeq \pm 4$ is not surprising since our approach does not properly take into account multi–parton correlations which are important in the fragmentation region.

Our predictions for $Au-Au$ collisions at $\sqrt{s} = 200$ GeV are presented in [50]. The only parameter which governs the energy dependence is the exponent λ, which we assume to be $\lambda \simeq 0.25$ as inferred from the HERA data. The absolute prediction for the multiplicity, as explained above, bears some uncertainty, but there is a definite feature of our scenario which is distinct from other approaches. It is the dependence of multiplicity on centrality, which around $\eta = 0$ is determined solely by the running of the QCD strong coupling [37]. As a result, the centrality dependence at $\sqrt{s} = 200$ GeV is somewhat less steep than at $\sqrt{s} = 130$. While the difference in the shape at these two energies is quite small, in the perturbative mini-jet picture this slope should increase, reflecting the growth

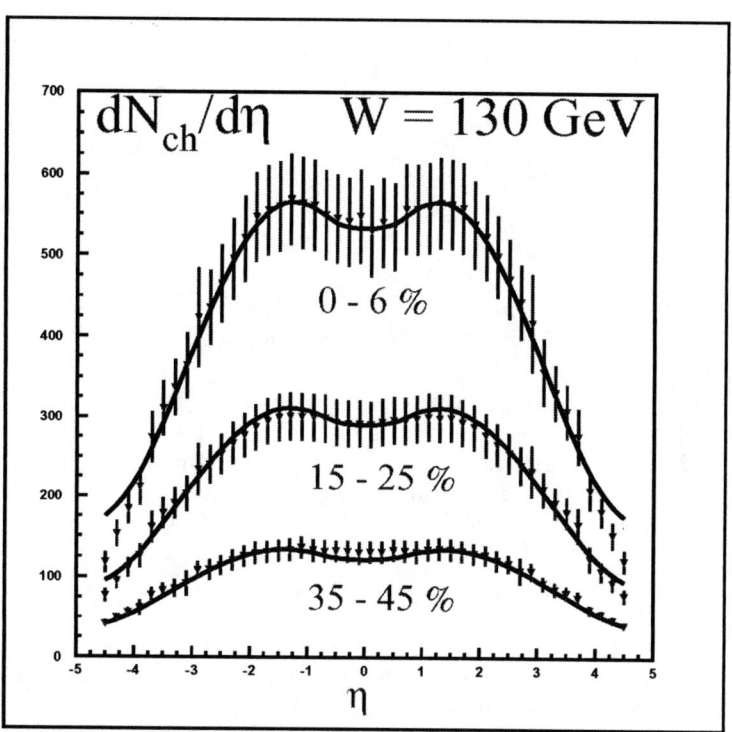

FIGURE 18. Pseudo–rapidity dependence of charged hadron production at different cuts on centrality in $Au-Au$ collisions at $\sqrt{s} = 130$ GeV; the data are from [46] (Reprinted from Ref. [50], Copyright 2001, with permission from Elsevier Science.)

of the mini-jet cross section with energy [42].

3.2.5. Further tests

Checking the predictions of the semi–classical approach for the centrality and pseudo–rapidity dependence at $\sqrt{s} = 200$ GeV is clearly very important. What other tests of this picture can one devise? The main feature of the classical emission is that it is coherent up to the transverse momenta of about $\sqrt{2}\, Q_s$ (about $\simeq 2$ GeV/c for central $Au-Au$ collisions). This means that if we look at the centrality dependence of particle multiplicities above a certain value of the transverse momentum, say, above 1 GeV/c, it should be very similar to the dependence without the transverse momentum cut-off. On the other hand, in the two–component "soft plus hard" model the cut on the transverse momentum would strongly enhance the contribution of hard mini–jet production processes, since soft production mechanisms presumably do not contribute to particle production at high transverse momenta. Of course, at sufficiently large value of the cutoff all of the observed particles will originate from genuine hard processes, and the centrality dependence will become steeper, reflecting the scaling with the number of collisions. It will be very interesting to explore the transition to this hard scattering regime experimentally.

Another test, already discussed above (see eq.(100)) is the study of azimuthal correlations between the produced high p_t particles. In the saturation scenario these correlations should be very small below $p_t \simeq 2$ GeV/c in central collisions. At higher transverse momenta, and/or for more peripheral collisions (where the saturation scale is smaller) these correlations should be much stronger.

3.3. Does the vacuum melt?

The approach described above allows us to estimate the initial energy density of partons achieved at RHIC. Indeed, in this approach the formation time of partons is $\tau_0 \simeq 1/Q_s$, and the transverse momenta of partons are about $k_t \simeq Q_s$. We thus can use the Bjorken formula and the set of parameters deduced above to estimate [37]

$$\varepsilon \simeq \frac{<k_t>}{\tau_0} \frac{d^2N}{d^2bd\eta} \simeq Q_s^2 \frac{d^2N}{d^2bd\eta} \simeq 18 \text{ GeV/fm}^3 \qquad (124)$$

for central $Au - Au$ collisions at $\sqrt{s} = 130$ GeV. This value is well above the energy density needed to induce the QCD phase transition according to the lattice calculations. However, the picture of gluon production considered above seems to imply that the gluons simply flow from the initial state of the incident nuclei to the final state, where they fragment into hadrons, with nothing spectacular happening on the way. In fact, one may even wonder if the presence of these gluons modifies at all the structure of the physical QCD vacuum.

To answer this question theoretically, we have to possess some knowledge about the non–perturbative vacuum properties. While in general the problem of vacuum structure still has not been solved (and this is one of the main reasons for the heavy ion research!), we do know one class of vacuum solutions – the instantons. It is thus interesting to investigate what happens to the QCD vacuum in the presence of strong external classical fields using the example of instantons [35].

The problem of small instantons in a slowly varying background field was first addressed in [52, 53] by introducing the effective instanton Lagrangian $L_{eff}^{I(\bar{I})}(x)$

$$L_{eff}^{I}(x_0) = \int d\rho \, n_0(\rho) \, dR \exp\left(-\frac{2\pi^2}{g^2} \rho^2 \, \overline{\eta}_{a\mu\nu}^{M} \, R^{aa'} \, G_{\mu\nu}^{a'}(x_0)\right) \qquad (125)$$

in which $n_0(\rho)$ is the instanton size distribution function in the vacuum, $\overline{\eta}_{a\mu\nu}^{M}$ is the 't Hooft symbol in Minkowski space, and $R^{aa'}$ is the matrix of rotations in color space, with dR denoting the averaging over the instanton color orientations.

The complete field of a single instanton solution could be reconstructed by perturbatively resumming the powers of the effective instanton lagrangian which corresponds to perturbation theory in powers of the instanton size parameter ρ^2. In our case here the

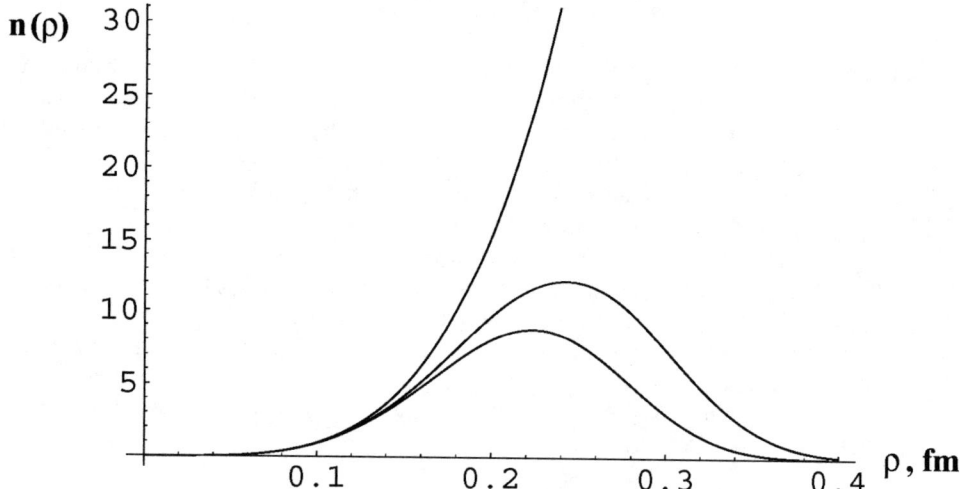

FIGURE 19. Distributions of instanton sizes in vacuum for QCD with three light flavors (upper curve) versus the distribution of instanton sizes in the saturation environment produced by a collision of two identical nuclei for $c = 1$ (middle curve) and $c = 2\ln 2$ (lower curve) with $Q_s^2 = 2\,\text{GeV}^2$. (Reprinted from Ref. [35], Copyright 2002, with permission from Elsevier Science.).

background field arises due to the strong source current J_μ^a. The current can be due to a single nucleus, or resulting from the two colliding nuclei. Perturbative resummation of powers of the source current term translates itself into resummation of the powers of the classical field parameter $\alpha_s^2 A^{1/3}$ [30, 54]. Thus the problem of instantons in the background classical gluon field is described by the effective action in Minkowski space

$$S_{eff} = \int d^4x \left(-\frac{1}{4g^2} G_{\mu\nu}^a(x) G_{\mu\nu}^a(x) + L_{eff}^I(x) + L_{eff}^{\bar{I}}(x) + J_\mu^a A_\mu^a(x) \right). \tag{126}$$

The problem thus is clearly formulated; by using an explicit form for the radiated classical gluon field, it was possible to demonstrate [35] that the distribution of instantons gets modified from the original vacuum one $n_0(\rho)$ to

$$n_{sat}^{AA}(\rho) = n_0(\rho) \exp\left(-\frac{c\rho^4 Q_s^4}{8\alpha_s^2 N_c (Q_s \tau_0)^2} \right), \tag{127}$$

where τ_0 is the proper time. The result Eq. (127) shows that large size instantons are suppressed by the strong classical fields generated in the nuclear collision (see Fig. 19)[3]. The vacuum does melt!

---

[3] Of course, at large proper times $\tau_0 \to \infty$ the vacuum "cools off", and the instanton distribution returns to the vacuum one.

ACKNOWLEDGMENTS

This manuscript is based on lectures given by one of us (D. K.) at the School on "New States of Matter in Hadronic Interactions" of the Pan American Advanced Studies Institute in Campos do Jordão, São Paulo, Brazil, on January 7-18, 2002. We thank the Organizers for the invitation to this stimulating meeting.

D. K. is grateful to his collaborators – Yuri Kovchegov, Eugene Levin, and Marzia Nardi – with whom most of the original results presented here were obtained. The work of D. K. was supported by the U.S. Department of Energy under Contract No. DE-AC02-98CH10886. He also wishes to acknowledge the hospitality of the Kavli Institute for Theoretical Physics in Santa Barbara, where this work was completed.

J. R. wishes to thank the Gesellschaft für Schwerionenforschung (GSI), Darmstadt, Germany for financial support during this summer school. The work of J. R. was supported in part by the U.S. Department of Energy at Los Alamos National Laboratory under Contract No. W-7405-ENG-38.

REFERENCES

1. G. Altarelli, *in this volume*.
2. R. D. Field, *Applications of Perturbative QCD*, Addison-Wesley, New York, USA, 1989.
3. R. K. Ellis, W. J. Stirling, and B. R. Webber, *QCD and Collider Physics*, Cambridge University Press, Cambridge, UK, 1996.
4. T. Muta, *Foundations of Quantum Chromodynamics*, World Scientific, Singapore, Singapore, 1987.
5. L. D. Landau and I. Y. Pomeranchuk, Dokl. Akad. Nauk Ser. Fiz. **102**, 489 (1955); also published in *The Collected Papers of L.D. Landau*. Edited by D. Ter Haar, Pergamon Press, 1965. pp. 654-658.
6. V. N. Gribov, Eur. Phys. J. C **10**, 71 (1999) [arXiv:hep-ph/9807224];
7. D. J. Gross and F. Wilczek, Phys. Rev. Lett. **30**, 1343 (1973); H. D. Politzer, Phys. Rev. Lett. **30**, 1346 (1973).
8. N. K. Nielsen, Am. J. Phys. **49**, 1171 (1981).
9. L. D. Landau and E. M. Lifshits, *Quantum Mechanics: Nonrelativistic Theory*, Landau-Lifshits physics course vol. 3, Pergamon Press, Oxford, UK, 1965.
10. S. Bethke, J. Phys. G **26**, R27 (2000) [arXiv:hep-ex/0004021].
11. K. Wilson, Phys.Rev.**D10**, 2445 (1974).
12. S. L. Adler, Phys. Rev. **177**, 2426 (1969); J. S. Bell and R. Jackiw, Nuovo Cim. A **60**, 47 (1969).
13. T. Schäfer and E. V. Shuryak, Rev. Mod. Phys. **70**, 323 (1998) [arXiv:hep-ph/9610451].
14. H. Forkel, "A primer on instantons in QCD," arXiv:hep-ph/0009136.
15. R. D. Peccei and H. R. Quinn, Phys. Rev. Lett. **38**, 1440 (1977); Phys. Rev. D **16**, 1791 (1977). F. Wilczek, Phys. Rev. Lett. **40**, 279 (1978). S. Weinberg, Phys. Rev. Lett. **40**, 223 (1978).
16. T. D. Lee and G. C. Wick, Phys. Rev. D **9**, 2291 (1974).
17. D. Kharzeev, R. D. Pisarski and M. H. Tytgat, Phys. Rev. Lett. **81**, 512 (1998) [arXiv:hep-ph/9804221]; D. Kharzeev and R. D. Pisarski, Phys. Rev. D **61**, 111901 (2000) [arXiv:hep-ph/9906401].
18. R. J. Glauber "High Energy Collision theory" in *Lectures in Theoretical Physics*, vol. 1, W. E. Brittin and L. G. Duham (eds.) Interscience, New York, 1959.
19. V. N. Gribov, Sov. Phys. JETP **29**, 483 (1969) [Zh. Eksp. Teor. Fiz. **56**, 892 (1969)].
20. V. N. Gribov, arXiv:hep-ph/0006158.
21. V. A. Abramovsky, V. N. Gribov and O. V. Kancheli, Yad. Fiz. **18**, 595 (1973) [Sov. J. Nucl. Phys. **18**, 308 (1973)].
22. F. E. Low, Phys. Rev. D **12**, 163 (1975); S. Nussinov, Phys. Rev. Lett. **34**, 1286 (1975).
23. M. Froissart, Phys. Rev. **123**, 1053 (1961); A. Martin, Phys. Rev. **129**, 1432 (1963).
24. A. Schwimmer, Nucl. Phys. B **94**, 445 (1975).

25. V. N. Gribov, Sov. Phys. JETP **26**, 414 (1968) [Zh. Eksp. Teor. Fiz. **53**, 654 (1968)]. H. D. Abarbanel, J. B. Bronzan, R. L. Sugar and A. R. White, Phys. Rept. **21**, 119 (1975).
26. D. Kharzeev, Nucl. Phys. A **699**, 95 (2002) [arXiv:nucl-th/0107033].
27. Gribov, L.V., Levin, E.M. and Ryskin, M.G., Phys. Rept. 100:1, 1983.
28. Mueller, A.H. and Qiu, J.-W., Nucl. Phys. B268:427, 1986.
29. Blaizot, J.P. and Mueller, A.H., Nucl. Phys. B289:847, 1987.
30. McLerran, L.D. and Venugopalan, R., Phys. Rev. D49:2233, 1994; D49:3352, 1994.
31. Iancu, I., Leonidov, A., and McLerran, L. hep-ph/0202270.
32. Mueller, A.H., hep-ph/0111244.
33. Kharzeev, D., Kovchegov, Yu. and Levin, E., Nucl. Phys. A690:621, 2001;
34. Nowak, M., Shuryak, E., and Zahed, I., Phys.Rev. D64:034008, 2001.
35. Kharzeev, D., Kovchegov, Yu. and Levin, E., Nucl.Phys.A699:745, 2002.
36. Berestetskii, V. B., Lifshitz, E.M. and Pitaevskii, L.P., *"Quantum electrodynamics"*, Oxford New York, Pergamon Press, 1982.
37. Kharzeev, D. and Nardi, Phys. Lett. B507:121, 2001.
38. Kharzeev, D., Lourenço, C., Nardi, M. and Satz, H., Z.Phys. C74:307, 1997.
39. De Jager, C., De Vries, H. and De Vries, C., Atom. Nucl. Data Tabl. 14:479, 1974.
40. Kharzeev, D. and Levin, E., Nucl. Phys. B578:351, 2000.
41. Gyulassy, M. and McLerran, L., Phys. Rev. C56:2219, 1997.
42. Wang, X.N. and Gyulassy, M., Phys. Rev. Lett. 86:3496, 2001.
43. Dokshitzer, Yu.L., hep-ph/9812252.
44. Adcox, K. et al., (The PHENIX Collaboration), Phys. Rev. Lett. 86:3500, 2001; Phys. Rev. Lett. 87:052301, 2001; Milov, A. et al., (The PHENIX Collaboration), nucl-ex/0107006.
45. Adler, C. et al., (The STAR Collaboration), Phys. Rev. Lett. 87:112303, 2001; Phys. Rev. Lett. 87:082301, 2001.
46. Back, B. et al., PHOBOS Coll., Phys. Rev. Lett. 85:3100, 2000; Phys. Rev. Lett. 87:102303, 2001; nucl-ex/0105011; nucl-ex/0108009.
47. Bearden I.G. et al., (The BRAHMS Collaboration), Phys. Rev. Lett. 87:112305, 2001; nucl-ex/0102011; nucl-ex/0108016
48. Berestetskii, V. B., Lifshitz, E.M. and Pitaevskii, L.P., *"Quantum electrodynamics"*, Oxford New York, Pergamon Press, 1982.
49. Golec-Biernat, K. and Wüsthof, M., Phys. Rev. D59:014017, 1999; Phys. Rev. D60:114023, 1999; Stasto, A., Golec-Biernat, K. and Kwiecinski, J., Phys. Rev. Lett. 86:596, 2001.
50. Kharzeev, D. and Levin, E., Phys. Lett. B523:79, 2001.
51. Eskola, K.J., Kajantie, K. and Tuominen, K., Phys. Lett. B497:39, 2001;
    Eskola, K.J., Kajantie, K., Ruuskanen, P.V. and Tuominen, K., Nucl. Phys. B 570:379, 2000.
52. Callan, C.G., Dashen, R. and Gross, D.J., Phys. Rev. D19:1826, 1979.
53. Shifman, M.A., Vainshtein, A.I., Zakharov, V.I., Nucl. Phys. B165:45, 1980.
54. Kovchegov, Yu.V., Phys. Rev. D54:5463, (1996); 55:5445, 1997.

A QCD Primer

G. Altarelli

CERN, Geneva, Switzerland

Abstract.
Introduction
Massless QCD and Scale Invariance
The Renormalisation Group and Asymptotic Freedom
More on the Running Coupling
Application to Hard Processes
  $R_{e^+e^-}$ and Related Processes
  The Final State in e^+e^- Annihilation
  Deep Inelastic Scattering
  Factorisation and the QCD Improved Parton Model
Measurements of α_s
  α_s from e^+e^- Colliders
  α_s from Deep Inelastic Scattering
Conclusion
Appendix: The Formalism of Gauge Theories

INTRODUCTION

These four lectures are devoted to an elementary introduction to Quantum Chromo-Dynamics (QCD), the theory of strong interactions. Four lectures are not much. So after a general introduction I will concentrate on the basic principles and the main applications of perturbative QCD (for reviews of the subject see for example [1], [2], [3]). I will try to put the main emphasis on ideas with only a minimum of technicalities.

At present most of particle physics is well understood in terms of the Standard Model (SM), which is a gauge theory of strong and electroweak interactions based on the group $SU(3) \otimes SU(2) \otimes U(1)_Y$. The $SU(3)$ factor is the colour group of QCD, while $SU(2) \otimes U(1)_Y$ is the Glashow-Weinberg-Salam electroweak symmetry group. The electroweak symmetry is spontaneously broken down to $U(1)_Q$, the phase group of the electric charge Q_e (that is, Q is the charge operator in units of e, the proton charge), different from the $U(1)_Y$ of weak hypercharge: $Q = t_3 + Y/2$, where t_3 is the third component of the weak isospin generator of $SU(2)$. The group $SU(3) \otimes U(1)_Q$ is believed to be an exact gauge symmetry of nature. The corresponding gauge bosons, the eight gluons and the photon are massless. Matter fields include three families of coloured quarks and colourless leptons. Quarks and gluons (q and g) are the fields that have strong interactions described by QCD. The statement that QCD is a gauge theory based on the group $SU(3)$ with colour triplet quark matter fields fixes the QCD lagrangian density to be (a summary of the general formalism of gauge theories is presented in Appendix):

$$\mathcal{L} = -\frac{1}{4}\sum_{A=1}^{8} F^{A\mu\nu} F^A_{\mu\nu} + \sum_{j=1}^{n_f} \bar{q}_j(i\slashed{D} - m_j)q_j \tag{1}$$

Here: q_j are the quark fields (of n_f different flavours) with mass m_j; $\slashed{D} = D_\mu \gamma^\mu$, where γ^μ are the Dirac matrices and D_μ is the covariant derivative:

$$D_\mu = \partial_\mu - ie_s \mathbf{g}_\mu; \tag{2}$$

e_s is the gauge coupling, later we will mostly use, in analogy with QED

$$\alpha_s = \frac{e_s^2}{4\pi}; \tag{3}$$

$\mathbf{g}_\mu = \sum_A t^A g^A_\mu$ where g^A_μ, $A = 1, 8$, are the gluon fields and t^A are the $SU(3)$ group generators in the triplet representation of quarks (i.e. t_A are 3x3 matrices acting on q); the generators obey the commutation relations $[t^A, t^B] = iC_{ABC} t^C$ where C_{ABC} are the complete antisymmetric structure constants of $SU(3)$ (the normalisation of C_{ABC} and of e_s is specified by $Tr[t^A t^B] = \delta^{AB}/2$);

$$F^A_{\mu\nu} = \partial_\mu g^A_\nu - \partial_\nu g^A_\mu - e_s C_{ABC} g^B_\mu g^C_\nu \tag{4}$$

The Feynman rules of QCD are listed in Fig 1. The physical vertices in QCD include the gluon-quark-antiquark vertex, analogous to the QED photon-fermion-antifermion coupling, but also the 3-gluon and 4-gluon vertices, of order e_s and e_s^2 respectively, which have no analogue in an abelian theory like QED. In fact the QED lagrangian density is given by:

$$\mathcal{L} = -\frac{1}{4} F^{\mu\nu} F_{\mu\nu} + \sum_\psi \bar{\psi}(i\slashed{D} - m_\psi)\psi \tag{5}$$

with the covariant derivative given in terms of the photon field A_μ and the charge operator Q by:

$$D_\mu = \partial_\mu - ieA_\mu Q \tag{6}$$

and

$$F_{\mu\nu} = \partial_\mu A_\nu - \partial_\nu A_\mu \tag{7}$$

In QED the photon is coupled to all electrically charged particles but itself is neutral. In QCD the gluons are coloured hence self-coupled. This is reflected in the fact that in QED $F_{\mu\nu}$ is linear in the gauge field, so that the term $F^2_{\mu\nu}$ in the lagrangian is a pure kinetic term, while in QCD $F^A_{\mu\nu}$ is quadratic in the gauge field so that in $F^{A2}_{\mu\nu}$ we find cubic and quartic vertices beyond the kinetic term. Also instructive is to consider the case of scalar QED:

$$\mathcal{L} = -\frac{1}{4} F^{\mu\nu} F_{\mu\nu} + (D_\mu \phi)^\dagger (D^\mu \phi) - m^2(\phi^\dagger \phi) \tag{8}$$

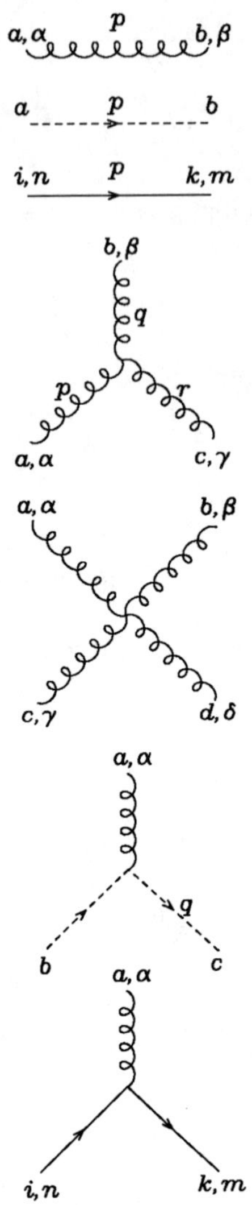

$$a,\alpha \underset{p}{} b,\beta \quad = \quad \delta^{ab}\frac{-ig^{\alpha\beta}}{p^2+i\epsilon} \quad \text{(Feynman gauge)}$$

$$a \underset{p}{\text{-----}} b \quad = \quad \delta^{ab}\frac{i}{p^2+i\epsilon}$$

$$i,n \underset{p}{} k,m \quad = \quad \delta^{ik}\frac{i}{\not{p}-m+i\epsilon}\bigg|_{mn}$$

$$= gf^{abc}\left[g^{\alpha\beta}(p-q)^\gamma + g^{\beta\gamma}(q-r)^\alpha + g^{\gamma\alpha}(r-p)^\beta\right]$$

$$\begin{aligned}&= -ig^2 f^{xac}f^{xbd}\left(g^{\alpha\beta}g^{\gamma\delta}-g^{\alpha\delta}g^{\beta\gamma}\right)\\ & -ig^2 f^{xad}f^{xbc}\left(g^{\alpha\beta}g^{\gamma\delta}-g^{\alpha\gamma}g^{\beta\delta}\right)\\ & -ig^2 f^{xab}f^{xcd}\left(g^{\alpha\gamma}g^{\beta\delta}-g^{\alpha\delta}g^{\beta\gamma}\right)\end{aligned}$$

$$= -gf^{abc}\, q^\alpha$$

$$= ig\,\lambda^a_{ki}\,\gamma^\alpha_{mn}$$

FIGURE 1. Feynman rules for QCD. The solid lines represent the femions, the curly lines the gluons, and the dotted lines represent the ghosts.

For $Q=1$ we have:

$$(D_\mu\phi)^\dagger(D^\mu\phi) = (\partial_\mu\phi)^\dagger(\partial^\mu\phi) + ieA_\mu[(\partial^\mu\phi)^\dagger\phi - \phi^\dagger(\partial^\mu\phi)] + e^2 A_\mu A^\mu \phi^\dagger\phi \quad (9)$$

We see that for a charged boson in QED, given that the kinetic term for bosons is quadratic in the derivative, there is a two-gauge vertex of order e^2. Thus in QCD the

3-gluon vertex is there because the gluon is coloured and the 4-gluon vertex because the gluon is a boson.

The QCD lagrangian in eq.(1) has a simple structure but a very rich dynamical content. It gives rise to a complex spectrum of hadrons, it implies the striking properties of confinement and asymptotic freedom, is endowed with an approximate chiral symmetry which is spontaneously broken, has a highly non trivial topological vacuum structure (instantons, $U(1)_A$ symmetry breaking, strong CP violation (?)...), an intriguing phase transition diagram (colour deconfinement, quark-gluon plasma, chiral symmetry restauration, colour superconductivity, ...).

Confinement is the property that no isolated coloured charge can exist but only colour singlet particles. For example, the potential between a quark and an antiquark has been studied on the lattice and it has a Coulomb part at short distances and a linearly rising term at long distances:

$$V_{q\bar{q}} \approx C_F[\frac{\alpha_s(r)}{r} + ... + \sigma r] \qquad (10)$$

where

$$1_3\, C_F = \sum_A t^A t^A = \frac{N_C^2 - 1}{2N_C}\, 1_3 \qquad (11)$$

with N_C the number of colours ($N_C = 3$ in QCD). The scale dependence of α_s (the distance r is Fourier-conjugated to momentum transfer) will be explained in detail in this course. The linearly rising term makes it energetically impossible to separate a $q - \bar{q}$ pair. If the pair is created at one space-time point, for example in e^+e^- annihilation, and then the quark and the antiquark start moving away from each other in the center of mass frame, it soon becomes energetically favourable to create additional pairs, smoothly distributed in rapidity between the two leading charges, which neutralise colour and allow the final state to be reorganised into two jets of colourless hadrons, that communicate in the central region by a number of "wee" hadrons with small energy. It is just like the familiar example of the broken magnet: if you try to isolate a magnetic pole by stretching a dipole, the magnet breaks down and two new poles appear at the breaking point. Very often in QCD one computes inclusive rates for partons (the fields in the lagrangian, that is, in QCD, quarks and gluons) and takes them as equal to rates for hadrons. Partons and hadrons are considered as two equivalent sets of complete states. This is called "global duality" and it is rather safe in the rare instance of a totally inclusive final state. It is less so for distributions, like distributions in the invariant mass M ("local duality") where it can be reliable only if smeared over a sufficiently wide bin in M.

Confinement is essential to explain why nuclear forces have very short range while massless gluon exchange would be long range. Nucleons are colour singlets and they cannot exchange colour octet gluons but only colourless states. The lightest colour singlet hadronic particles are pions. So the range of nuclear forces is fixed by the pion mass $r \simeq m_\pi^{-1} \simeq 10^{-13}\ cm : V \approx \exp(-m_\pi r)/r$.

Why $SU(N_C = 3)_{colour}$? The selection of $SU(3)$ as colour gauge group is unique in view of a number of constraints. (a) The group must admit complex representations because it must be able to distinguish a quark from an antiquark. In fact there are meson states made up of $q\bar{q}$ but not analogous qq bound states. Among simple groups

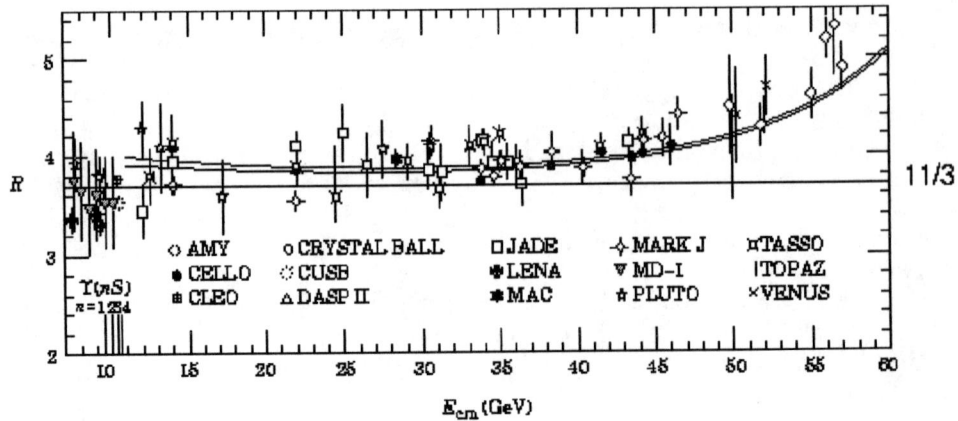

FIGURE 2.

this restricts the choice to $SU(N)$ with $N \geq 3$, $SO(4N+2)$ with $N \geq 2$ (taking into account that $SO(6)$ has the same algebra as $SU(4)$) and $E(6)$. (b) The group must admit a completely antisymmetric colour singlet baryon made up of 3 quarks: qqq. From the study of hadron spectroscopy we know that the low lying baryons, completing an octet and a decuplet of (flavour) $SU(3)$ (the approximate symmetry that rotate the 3 light quarks u, d and s), are made up of three quarks and are colour singlets. The qqq wave function must be completely antisymmetric in colour in order to agree with Fermi statistics. Indeed if we consider, for example, a N^{+++} with spin z-component $+3/2$, this is made up of $(u \Uparrow u \Uparrow u \Uparrow)$ in an s-state. Thus its wave function is totally symmetric in space, spin and flavour so that complete antisymmetry in colour is required by Fermi statistics. In QCD this requirement is very simply satisfied by $\epsilon_{abc} q^a q^b q^c$ where a, b, c are $SU(3)_{colour}$ indices. (c) The choice of $SU(N_C = 3)_{colour}$ is confirmed by many processes that directly measure N_C. Some examples are listed here. The total rate for hadronic production in e^+e^- annihilation is linear in N_C. Precisely if we consider $R = \sigma(e^+e^- \to hadrons)/\sigma_{point}(e^+e^- \to \mu^+\mu^-)$ above $b\bar{b}$ threshold and below m_Z and we neglect small computable radiative corrections (that will be discussed later) we have a sum of individual contributions (proportional to Q^2) from $q\bar{q}$ final states with $q = u, c, d, s, b$:

$$R \approx N_C[2 \cdot \frac{4}{9} + 3 \cdot \frac{1}{9}] \approx N_C \frac{11}{9} \quad (12)$$

The data neatly indicate $N_C = 3$ as seen from Fig. 2 [4]. The slight excess of the data with respect to the value 11/3 is due to the QCD radiative corrections. Similarly we can consider the branching ratio $B(W^- \to e^- \bar{\nu})$, again in Born approximation. The possible fermion-antifermion $(f\bar{f})$ final states are for $f = e^-, \mu^-, \tau^-, d, s$ (there is no $f = b$ because the top quark is too heavy for $b\bar{t}$ to occur). Each channel gives the same contribution, except that for quarks we have N_C colours:

$$B(W^- \to e^- \bar{\nu}) \approx \frac{1}{3 + 2N_C} \quad (13)$$

For $N_C = 3$ we obtain $B = 11\%$ and the experimental number is $B = 10.7\%$. Another analogous example is the branching ratio $B(\tau^- \to e^- \bar{\nu}_e \nu_\tau)$. From the final state channels with $f = e^-$, μ^-, d we find

$$B(\tau^- \to e^- \bar{\nu}_e \nu_\tau) \approx \frac{1}{2 + N_C} \qquad (14)$$

For $N_C = 3$ we obtain $B = 20\%$ and the experimental number is $B = 18\%$ (the less accuracy in this case is explained by the larger radiative and phase-space corrections because the mass of τ^- is much smaller than m_W). An important process that is quadratic in N_C is the rate $\Gamma(\pi^0 \to 2\gamma)$. This rate can be reliably calculated from a solid theorem in field theory which has to do with the chiral anomaly:

$$\Gamma(\pi^0 \to 2\gamma) \approx (\frac{N_C}{3})^2 \frac{\alpha^2 m_{\pi^0}^3}{32\pi^3 f_\pi^2} = (7.73 \pm 0.04)(\frac{N_C}{3})^2 \; eV \qquad (15)$$

where the prediction is obtained for $f_\pi = (130.7 \pm 0.37) \; MeV$. The experimental result is $\Gamma = (7.7 \pm 0.5) \; eV$ in remarkable agreement with $N_C = 3$. There are many more experimental confirmations that $N_C = 3$: for example the rate for Drell-Yan processes (see section 5.4) is inversely proportional to N_C.

How do we get testable predictions from QCD? On the one hand there are non perturbative methods. The most important at present is the technique of lattice simulations: it is based on first principles, it has produced very valuable results on confinement, phase transitions, bound states, hadronic matrix elements and so on, and it is by now an established basic tool. The main limitation is computing power and therefore there is continuous progress and a lot of good perspectives for the future. Another class of approaches is based on effective lagrangians which provide simpler approximations than the full theory, valid in some definite domain of physical conditions. Chiral lagrangians are based on soft pion theorems and are valid for suitable processes at energies below 1 GeV. Heavy quark effective theories are obtained from expanding in inverse powers of the heavy quark mass and are mainly important for the study of b and, to less accuracy, c decays. The approach of QCD sum rules has led to interesting results but appears to offer not much potential for further development. Similarly specific potential models for quarkonium have a limited range of application. On the other hand, the perturbative approach, based on asymptotic freedom, still remains the main quantitative connection to experiment, due to its wide range of applicability to all sorts of "hard" processes. Perturbative QCD will be the main subject in the following. I will discuss its foundations and main applications in the next sections.

MASSLESS QCD AND SCALE INVARIANCE

The QCD lagrangian in eq.(1) only specifies the theory at the classical level. The procedure for quantisation of gauge theories involves a number of complications that arise from the fact that not all degrees of freedom of gauge fields are physical because of the constraints from gauge invariance which can be used to eliminate the dependent variables. This is already true for abelian theories and we are familiar with the QED case. One introduces a gauge fixing term (an additional term in the lagrangian density that acts as a Lagrange multiplier in the action extremisation). One can choose to preserve manifest Lorentz invariance. In this case, one adopts a covariant gauge, like the Lorentz gauge, and in QED one proceeds according to the formalism of Gupta-Bleuler. Or one can give up explicit formal covariance and work in a non covariant gauge, like the Coulomb or the axial gauges, and only quantise the physical degrees of freedom (the transverse components of the photon field). While this is all for an abelian gauge theory, in the non-abelian case some additional complications arise, in particular the necessity to introduce ghosts for the formulation of Feynman rules. There are in general as many ghost fields as gauge bosons and they appear in the form of a transformation Jacobian in the Feynman diagram functional integral. Ghosts only propagate in closed loops and their vertices with gluons can be included as additional terms in the lagrangian density which are fixed once the gauge fixing terms and their infinitesimal gauge transformations are specified. We skip the detailed derivation of the complete Feynman rules in a given gauge as they appear in Fig 1.

Once the Feynman rules are derived we have a formal perturbative expansion but loop diagrams generate infinities. First a regularisation must be introduced, compatible with gauge symmetry and Lorentz invariance. This is possible in QCD. In principle one can introduce a cut-off Λ (with dimensions of energy), for example, a' la Pauli-Villars. But at present the universally adopted regularisation procedure is dimensional regularisation that we will briefly describe later on. After regularisation the next step is renormalisation. In a renormalisable theory (like for all gauge theories in 4 spacetime dimensions and for QCD in particular) the dependence on the cutoff can be completely reabsorbed in a redefinition of particle masses, of gauge coupling(s) and of wave function normalisations. After renormalisation is achieved the perturbative definition of the quantum theory that corresponds to a classical lagrangian like in eq.(1) is completed.

In the QCD Lagrangian of eq.(1) quark masses are the only parameters with physical dimensions (we work in the natural system of units $\hbar = c = 1$). Naively we would expect that massless QCD is scale invariant. This is actually true at the classical level. Scale invariance implies that dimensionless observables should not depend on the absolute scale of energy but only on ratios of energy-dimensional variables. The massless limit should be relevant for the asymptotic large energy limit of processes which are non singular for $m \to 0$.

The naive expectation that massless QCD should be scale invariant is false in the quantum theory. The scale symmetry of the classical theory is unavoidably destroyed by the regularisation and renormalisation procedure which introduce a dimensional parameter in the quantum version of the theory. When a symmetry of the classical theory is necessarily destroyed by quantisation, regularisation and renormalisation one talks of an "anomaly". So, in this sense, scale invariance in massless QCD is anomalous.

While massless QCD is finally not scale invariant, the departures from scaling are asymptotically small, logarithmic and computable. In massive QCD there are additional mass corrections suppressed by powers of m over the energy scale (for non singular processes in the limit $m \to 0$). At the parton level (q and g) we can conceive to apply the asymptotics from massless QCD to processes and observables (we use the word "processes" for both) with the following properties ("hard processes"). (a) All relevant energy variables must be large:

$$E_i = z_i Q, \qquad Q >> m_j; \qquad z_i: \text{scaling variables } o(1) \qquad (16)$$

(b) There should be no infrared singularities (one talks of "infrared safe" processes). (c) The processes concerned must be finite for $m \to 0$ (no mass singularities). To possibly satisfy these criteria processes must be as "inclusive" as possible: one should include all final states with massless gluon emission and add all mass degenerate final states (given that quarks are massless also $q - \bar{q}$ pairs can be massless if "collinear", that is moving together in the same direction at the common speed of light).

Let us discuss more in detail infrared and collinear safety. Consider, for example, a quark virtual line that ends up into a real quark plus a real gluon (Fig. 3).

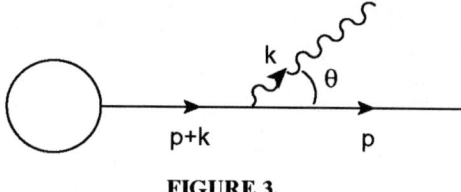

FIGURE 3.

For the propagator we have:

$$\text{propagator} = \frac{1}{(p+k)^2 - m^2} = \frac{1}{2(p \cdot k)} = \frac{1}{2 E_k E_p} \cdot \frac{1}{1 - \beta_p \cos\theta} \qquad (17)$$

Since the gluon is massless, E_k can vanish and this corresponds to an infrared singularity. Remember that we have to take the square of the amplitude and integrate over the final state phase space, or all together, dE_k/E_k. So we get $1/E_k^2$ from the squared amplitude and $d^3k/E_k \sim E_k dE_k$ from the phase space. Also, for $m \to 0$, $\beta_p = \sqrt{1 - m^2/E_p^2} \to 1$ and $(1 - \beta_p \cos\theta)$ vanishes at $\cos\theta = 1$. This leads to a collinear mass singularity.

There are two very important theorems on infrared and mass singularities. The first one is the Bloch-Nordsieck theorem: infrared singularities cancel between real and virtual diagrams (see Fig. 4) when all resolution indistinguishable final states are added up. For example, for each real detector there is a minimum energy of gluon radiation that can be detected. For the cancellation of infrared divergences, one should add all possible gluon emission with a total energy below the detectable minimum. The second one is the Kinoshita-Lee-Nauenberg theorem: mass singularities connected with an external particle of mass m are canceled if all degenerate states (that is with the same mass) are summed up. That is for a final state particle of mass m we should add all final states that in the limit $m \to 0$ have the same mass, also including gluons and massless pairs. If a completely inclusive final state is taken, only the mass singularities from the initial state particles remain (we shall see that they will be absorbed inside the non perturbative parton densities, which are probability densities of finding the given parton in the initial hadron).

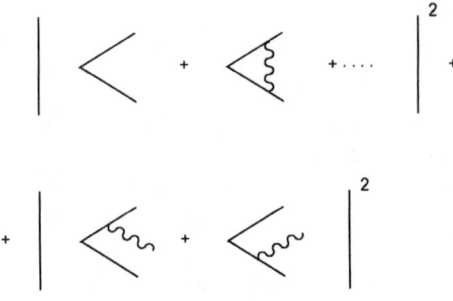

FIGURE 4.

Hard processes to which the massless QCD asymptotics can possibly apply must be infrared and collinear safe, that is must satisfy the requirements from the Bloch-Nordsieck and the Kinoshita-Lee-Nauenberg theorems. We give now some examples of important hard processes. One of the simplest hard processes is the totally inclusive cross section for hadron production in e^+e^- annihilation, Fig. 5, parameterised in terms of the already mentioned dimensionless observable $R = \sigma(e^+e^- \to hadrons)/\sigma_{point}(e^+e^- \to \mu^+\mu^-)$. The pointlike cross section in the denominator is given by $\sigma_{point} = 4\pi\alpha^2/3s$, where $s = Q^2 = 4E^2$ is the squared total center of mass energy and Q is the mass of the exchanged virtual gauge boson. At parton level the final state is $(q\bar{q} + n\, g + n'\, q'\bar{q}')$ and n and n' are limited at each order of perturbation theory. It is assumed that the conversion of partons into hadrons does not affect the rate (it happens with probability 1). We have already mentioned that in order for this to be true within a given accuracy an averaging over a sufficiently large bin of Q must be understood. The binning width is larger in the vicinity of thresholds: for example when one goes across the charm $c\bar{c}$ threshold the physical cross-section shows resonance bumps which are absent in the smooth partonic counterpart which however gives an average of the cross-section.

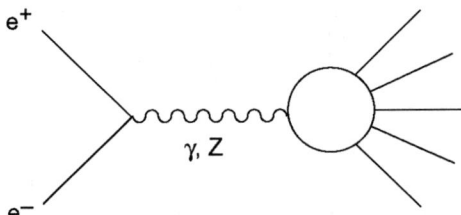

FIGURE 5.

A very important class of hard processes is Deep Inelastic Scattering (DIS)

$$l + N \to l' + X \qquad l = e^\pm, \mu^\pm, \nu, \bar{\nu} \qquad (18)$$

which has played and still plays a very important role for our understanding of QCD and of nucleon structure. For the processes in eq.(18), shown in Fig 6, we have, in the lab

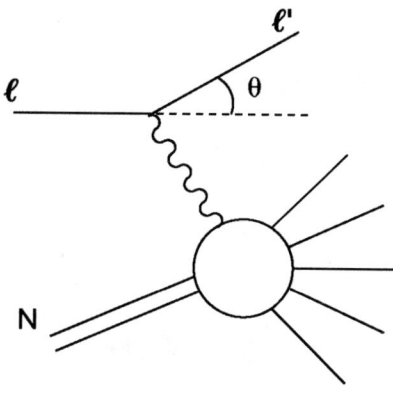

FIGURE 6.

system where the nucleon of mass m is at rest:

$$Q^2 = -q^2 = -(k-k')^2 = 4EE'\sin^2\theta/2; \qquad m\nu = (p.q); \qquad x = \frac{Q^2}{2m\nu} \qquad (19)$$

In this case the virtual momentum q of the gauge boson is spacelike. x is the familiar Bjorken variable.

THE RENORMALISATION GROUP AND ASYMPTOTIC FREEDOM

In this section we aim at providing a reasonably detailed introduction to the renormalisation group formalism and the concept of running coupling which leads to the result that QCD has the property of asymptotic freedom. We start with a summary on how renormalisation works.

In the simplest conceptual situation imagine that we implement regularisation of divergent integrals by introducing a dimensional cut-off Λ that respects gauge and Lorentz invariance. The dependence of renormalised quantities on Λ is eliminated by absorbing it into a redefinition of m (the quark mass: for simplicity we assume a single flavour here), the gauge coupling e (can be e in QED or e_s in QCD) and the wave function renormalisation factors $Z_{q,g}^{1/2}$ for q and g, using suitable renormalisation conditions (that is precise definitions of m, g and Z). For example we can define the renormalised mass m as the position of the pole in the quark propagator and, similarly, the normalisation Z_q as the residue at the pole:

$$\text{Propagator} = \frac{Z_q}{p^2 - m^2} + \text{no}-\text{pole terms} \qquad (20)$$

The renormalised coupling e can be defined in terms of a renormalised 3-point vertex at some specified values of the external momenta. We now become more specific by

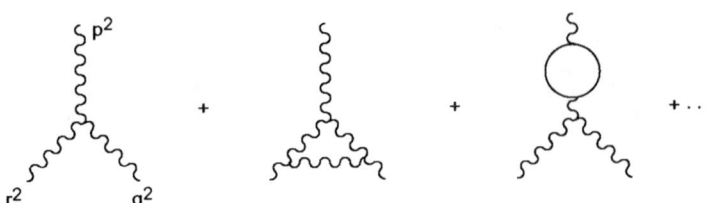

FIGURE 7.

concentrating in the case of massless QCD. If we start from a vanishing mass at the classical (or "bare") level, $m_0 = 0$, the mass is not renormalised because it is protected by a symmetry, chiral symmetry. The conserved currents of chiral symmetry are axial currents: $\bar{q}\gamma_\mu\gamma_5 q$. The divergence of the axial current gives, by using the Dirac equation, $\partial^\mu(\bar{q}\gamma_\mu\gamma_5 q) = 2m\bar{q}\gamma_5 q$. So the axial current and corresponding axial charge are conserved in the massless limit. Since QCD is a vector theory we have not to worry about chiral anomalies in this respect. So one can choose a regularisation that preserves chiral symmetry besides gauge and Lorentz symmetry. Then the renormalised mass remains zero. The renormalised propagator has the form in eq.(20) with $m = 0$.

The renormalised coupling e_s can be defined from the renormalised 3-gluon vertex at a scale $-\mu^2$ (Fig. 7):

$$V_{bare}(p^2, q^2, r^2) = Z V_{ren}(p^2, q^2, r^2), \quad Z = Z_g^{3/2} Z_V, \quad e_s = V_{ren}(-\mu^2, -\mu^2, -\mu^2) \quad (21)$$

Here V_{bare} is what is obtained from computing the Feynman diagrams including, for example, the 1-loop corrections at the lowest non trivial order (V_{bare} is defined so that it coincides with e_{s0} in lowest order). It contains the cut-off Λ but does not know μ. Z is a factor that depends on the cut-off. Because of infrared singularities the defining scale μ cannot vanish. The negative value $-\mu^2 < 0$ is chosen to stay away from physical cuts (a gluon with negative virtual mass cannot decay). Similarly we can define Z_g from the massless gluon propagator at the same scale $-\mu^2$ (the vanishing mass of the gluon is guaranteed by gauge invariance).

After computing all 1-loop diagrams in Fig. 7 we have:

$$\begin{aligned}V_{bare}(p^2, p^2, p^2) &= e_{0s}[1 + c\alpha_{0s}\cdot\log\frac{\Lambda^2}{p^2} + ...] \\ &= [1 + c\alpha_s\cdot\log\frac{\Lambda^2}{-\mu^2} + ...]e_s[1 + c\alpha_s\cdot\log\frac{-\mu^2}{p^2} + ...] = Z V_{ren}\end{aligned} \quad (22)$$

Note the replacement of e_0 with e in the second step. The definition of e_s demands that one precisely specifies what is included in Z. For this, in a given renormalisation scheme, a prescription is fixed to specify the finite terms that go into Z (i.e. the terms of order α_s that accompany $\log\Lambda^2$). Then V_{ren} is specified and the renormalised coupling is defined from it according to eq.(21). For example, in the momentum subtraction scheme we define $V_{ren}(p^2, p^2, p^2) = e_s + V_{bare}(p^2, p^2, p^2) - V_{bare}(-\mu^2, -\mu^2, -\mu^2)$, which is equivalent to say, at 1-loop, that all finite terms that do not vanish at $p^2 = -\mu^2$ are included in Z.

A crucial observation is that V_{bare} depends on Λ but not on μ, which is only introduced when Z, V_{ren} and hence α_s are defined. (From here on, for shorthand, we write α to indicate either the QED coupling or the QCD coupling α_s). More in general for a generic Green function G, we similarly have:

$$G_{bare}(\Lambda^2, \alpha_0, p_i^2) = Z G_{ren}(\mu^2, \alpha, p_i^2) \tag{23}$$

so that we have:

$$\frac{dG_{bare}}{d\log\mu^2} = \frac{d}{d\log\mu^2}[ZG_{ren}] = 0 \tag{24}$$

or

$$Z[\frac{\partial}{\partial \log\mu^2} + \frac{\partial \alpha}{\partial \log\mu^2}\frac{\partial}{\partial \alpha} + \frac{1}{Z}\frac{\partial Z}{\partial \log\mu^2}]G_{ren} = 0 \tag{25}$$

Finally the renormalisation group equation (RGE) can be written as:

$$[\frac{\partial}{\partial \log\mu^2} + \beta(\alpha)\frac{\partial}{\partial \alpha} + \gamma(\alpha)]G_{ren} = 0 \tag{26}$$

where

$$\beta(\alpha) = \frac{\partial \alpha}{\partial \log\mu^2} \tag{27}$$

and

$$\gamma(\alpha) = \frac{\partial \log Z}{\partial \log\mu^2} \tag{28}$$

Note that $\beta(\alpha)$ does not depend on which Green function G we are considering, but it is a property of the theory and the renormalisation scheme adopted, while $\gamma(\alpha)$ also depends on G.

Assume that we want to apply the RGE to some hard process at a large scale Q, related to a Green function G that we can always take as adimensional (by multiplication by a suitable power of Q). Since the interesting dependence on Q will be logarithmic we introduce the variable t as :

$$t = \log \frac{Q^2}{\mu^2} \tag{29}$$

Then we can write $G_{ren} \equiv F(t, \alpha, x_i)$ where x_i are scaling variables (we often omit to write them in the following). In the naive scaling limit F should be independent of t. To find the actual dependence on t, we want to solve the RGE

$$[-\frac{\partial}{\partial t} + \beta(\alpha)\frac{\partial}{\partial \alpha} + \gamma(\alpha)]G_{ren} = 0 \tag{30}$$

with a given boundary condition at $t = 0$ (or $Q^2 = \mu^2$): $F(0, \alpha)$.

We first solve the RGE in the simplest case that $\gamma(\alpha) = 0$. This is not an unphysical case: for example, it applies to $R_{e^+e^-}$ where the vanishing of γ is related to the non renormalisation of the electric charge in QCD (otherwise the proton and the electron

charge would not exactly compensate: this will be better explained later). So we consider the equation:
$$[-\frac{\partial}{\partial t} + \beta(\alpha)\frac{\partial}{\partial \alpha}]G_{ren} = 0 \qquad (31)$$

The solution is simply
$$F(t,\alpha) = F[0,\alpha(t)] \qquad (32)$$
where the "running coupling" $\alpha(t)$ is defined by:
$$t = \int_\alpha^{\alpha(t)} \frac{1}{\beta(\alpha')}d\alpha' \qquad (33)$$

Note that from this definition it follows that $\alpha(0) = \alpha$, so that the boundary condition is also satisfied. To prove that $F[0,\alpha(t)]$ is indeed the solution, we first take derivatives with respect to t and α (the two independent variables) of both sides of eq.(33). By taking d/dt we obtain
$$1 = \frac{1}{\beta(\alpha(t))}\frac{\partial \alpha(t)}{\partial t} \qquad (34)$$

We then take $d/d\alpha$ and obtain
$$0 = -\frac{1}{\beta(\alpha)} + \frac{1}{\beta(\alpha(t))}\frac{\partial \alpha(t)}{\partial \alpha} \qquad (35)$$

These two relations make explicit the dependence of the running coupling on t and α:
$$\frac{\partial \alpha(t)}{\partial t} = \beta(\alpha(t)) \qquad (36)$$
$$\frac{\partial \alpha(t)}{\partial \alpha} = \frac{\beta(\alpha(t))}{\beta(\alpha)}$$

Using these two equations one immediately checks that $F[0,\alpha(t)]$ is indeed the solution.

Similarly, one finds that the solution of the more general equation with $\gamma \neq 0$, eq.(30), is given by:
$$F(t,\alpha) = F[0,\alpha(t)]\exp\int_\alpha^{\alpha(t)} \frac{\gamma(\alpha')}{\beta(\alpha')}d\alpha' \qquad (37)$$

In fact the sum of the two derivatives acting on the factor $F[0,\alpha(t)]$ vanishes and the exponential is by itself a solution of the complete equation. Note that the boundary condition is also satisfied.

The important point is the appearance of the running coupling that determines the asymptotic departures from scaling. The next step is to study the functional form of the running coupling. From eq.(36) we see that the rate of change with t of the running coupling is determined by the β function. In turn $\beta(\alpha)$ is determined by the μ dependence of the renormalised coupling through eq.(27). Clearly there is no dependence on μ of the basic 3-gluon vertex in lowest order (order e). The dependence starts at 1-loop, that is at order e^3 (one extra gluon has to be emitted and reabsorbed). Thus we obtain that in perturbation theory:

$$\frac{\partial e}{\partial \log \mu^2} \propto e^3 \tag{38}$$

Recalling that $\alpha = e^2/4\pi$, we have:

$$\frac{\partial \alpha}{\partial \log \mu^2} \propto 2e \frac{\partial e}{\partial \log \mu^2} \propto e^4 \propto \alpha^2 \tag{39}$$

Thus the behaviour of $\beta(\alpha)$ in perturbation theory is as follows:

$$\beta(\alpha) = \pm b\alpha^2 [1 + b'\alpha + ...] \tag{40}$$

Since the sign of the leading term is crucial in the following discussion, we stipulate that always $b > 0$ and we make the sign explicit in front. By direct calculation at 1-loop one finds:

$$\text{QED}: \quad \beta(\alpha) \sim +b\alpha^2 + ... \qquad b = \sum_i \frac{N_C(Q^2)_i}{3\pi} \tag{41}$$

where $N_C = 3$ for quarks and $N_C = 1$ for leptons and the sum runs over all fermions of charge Qe that are coupled. Also, one finds:

$$\text{QCD}: \quad \beta(\alpha) \sim -b\alpha^2 + ... \qquad b = \frac{11 N_C - 2n_f}{12\pi} \tag{42}$$

where, as usual, n_f is the number of coupled flavours of quarks (we assume here that $n_f \leq 16$ so that $b > 0$ in QCD). If $\alpha(t)$ is small we can compute $\beta(\alpha(t))$ in perturbation theory. The sign in front of b then decides the slope of the coupling: $\alpha(t)$ increases with t (or Q^2) if β is positive at small α (QED), or $\alpha(t)$ decreases with t (or Q^2) if β is negative at small α (QCD). A theory like QCD where the running coupling vanishes asymptotically at large Q^2 is called (ultraviolet) "asymptotically free". An important result that can be proven is that in 4 spacetime dimensions all and only non-abelian gauge theories can be asymptotically free.

Going back to eq.(33) we replace $\beta(\alpha) \sim \pm b\alpha^2$, do the integral and perform a simple algebra. We find

$$\text{QED}: \quad \alpha(t) \sim \frac{\alpha}{1 - b\alpha t} \tag{43}$$

and

$$\text{QCD}: \quad \alpha(t) \sim \frac{\alpha}{1 + b\alpha t} \tag{44}$$

A slightly different form is often used in QCD. Defining $1/\alpha = b \log \mu^2 / \Lambda_{QCD}^2$ we can write:

$$\alpha(t) \sim \frac{1}{\frac{1}{\alpha} + bt} = \frac{1}{b \log \frac{\mu^2}{\Lambda_{QCD}^2} + b \log \frac{Q^2}{\mu^2}} = \frac{1}{b \log \frac{Q^2}{\Lambda_{QCD}^2}} \tag{45}$$

We see that $\alpha(t)$ decreases logarithmically with Q^2 and that one can introduce a dimensional parameter Λ_{QCD} that replaces μ. Often in the following we will simply write Λ

for Λ_{QCD}, assuming that confusion with Λ = ultraviolet cut-off is avoided by the reader. Note that it is clear that Λ depends on the particular definition of α, not only on the defining scale μ but also on the renormalisation scheme (see, for example, the discussion in the next session). Through the parameter b, and in general through the β function, it also depends on the number n_f of coupled flavours. It is very important to note that QED and QCD are theories with "decoupling": up to the scale Q only quarks with masses $m \ll Q$ contribute to the running of α. This is clearly very important, given that all applications of perturbative QCD so far apply to energies below the top quark mass m_t. For the validity of the decoupling theorem it is necessary that the theory where all the heavy particle internal lines are eliminated is still renormalisable and that the coupling constants do not vary with the mass. These requirements are true for the mass of heavy quarks in QED and QCD, but are not true in the electroweak theory where the elimination of the top would violate $SU(2)$ symmetry (because the t and b left quarks are in a doublet) and the quark couplings to the Higgs multiplet (hence to the longitudinal gauge bosons) are proportional to the mass. In conclusion, in QED and QCD, quarks with $m \gg Q$ do not contribute to n_f in the coefficients of the relevant β function. The effects of heavy quarks are power suppressed and can be taken separately into account. For example, in e^+e^- annihilation for $2m_c < Q < 2m_b$ the relevant asymptotics is for $n_f = 4$, while for $2m_b < Q < 2m_t$ $n_f = 5$. Going accross the b threshold the β function coefficients change, so the $\alpha(t)$ slope changes. But $\alpha(t)$ is continuous, so that Λ changes so as to keep constant $\alpha(t)$ at the matching point at $Q \sim o(m_b)$. The effect on Λ is large: approximately $\Lambda_5 \sim 0.65\Lambda_4$.

Note the presence of a pole in eqs.(43),(44) at $\pm b\alpha t = 1$, called the Landau pole, who realised its existence in QED already in the '50's. For $\mu \sim m_e$ (in QED) the pole occurs beyond the Planck mass. In QCD the Landau pole is located for negative t or at $Q < \mu$ in the region of light hadron masses. Clearly the issue of the definition and the behaviour of the physical coupling in the region around the Landau pole is a problem that lies outside the domain of perturbative QCD.

The non leading terms in the asymptotic behaviour of the running coupling can in principle be evaluated going back to eq.(40) and computing b' at 2-loops and so on. But in general the perturbative coefficients of $\beta(\alpha)$ depend on the definition of the renormalised coupling α (the renormalisation scheme), so one wonders whether it is worthwhile to do a complicated calculation to get b' if then it must be repeated for a different definition or scheme. In this respect it is interesting to remark that actually both b and b' are independent of the definition of α, while higher order coefficients do depend on that. Here is the simple proof. Two different perturbative definitions of α are related by $\alpha' \sim \alpha(1 + c_1\alpha + ...)$. Then we have:

$$\begin{aligned}\beta(\alpha') = \frac{d\alpha'}{d\log\mu^2} &= \frac{d\alpha}{d\log\mu^2}(1 + 2c_1\alpha + ...) \\ &= \pm b\alpha^2(1 + b'\alpha + ...)(1 + 2c_1\alpha + ...) \\ &= \pm b\alpha'^2(1 + b'\alpha' + ... \end{aligned} \quad (46)$$

which shows that, up to the first subleading order, $\beta(\alpha')$ has the same form as $\beta(\alpha)$.

In QCD ($N_C = 3$) one has calculated:

$$b' = \frac{153 - 19n_f}{2\pi(33 - 2n_f)} \quad (47)$$

By taking b' into account one can write the expression of the running coupling at next to the leading order (NLO):

$$\alpha(Q^2) = \alpha_{LO}(Q^2)[1 - b'\alpha_{LO}(Q^2) \log\log\frac{Q^2}{\Lambda^2} + ...] \qquad (48)$$

where $\alpha_{LO}^{-1} = b\log Q^2/\Lambda^2$ is the LO result.

Summarizing, we started from massless classical QCD which is scale invariant. But we have seen that the procedure of quantisation, regularisation and renormalisation necessarily breaks scale invariance. In the quantum QCD theory there is a scale of energy, Λ_{QCD}, which from experiment is of the order of a few hundred MeV, its precise value depending on the definition, as we shall see in detail. Adimensional quantities depend on the energy scale through the running coupling which is a logarithmic function of Q^2/Λ^2. In QCD the running coupling decreases logarithmically at large Q^2 (asymptotic freedom), while in QED the coupling has the opposite behaviour.

MORE ON THE RUNNING COUPLING

In the previous section we have introduced the renormalised coupling α in terms of the 3-gluon vertex at $p^2 = -\mu^2$ (momentum subtraction). The Ward identities of QCD then ensure that the coupling defined from other vertices like the $\bar{q}qg$ vertex are renormalised in the same way and the finite radiative corrections are related. But at present the universally adopted definition of α_s is in terms of dimensional regularisation because of computational simplicity which is essential given the great complexity of present day calculations. So we now briefly review the principles of dimensional regularisation and the definition of Minimal Subtraction (MS) and Modified Minimal Subtraction ($\overline{MS}$). The $\overline{MS}$ definition of α_s is the one most commonly adopted in the literature and a value quoted for it is nomally referring to this definition.

Dimensional Regularisation (DR) is a gauge and Lorentz invariant regularisation that consists in formulating the theory in $D < 4$ spacetime dimensions in order to make loop integrals ultraviolet finite. In DR one rewrites the theory in D dimensions (D is integer at the beginning, but then we will see that the expression of diagrams makes sense at all D except for isolated singularities). The metric tensor is extended into a $D \times D$ matrix $g_{\mu\nu} = diag(1, -1, -1, ..., -1)$ and 4-vectors are given by $k^\mu = (k^0, k^1, ..., k^{D-1})$. The Dirac γ^μ are $f(D) \times f(D)$ matrices and it is not important what is the precise form of the function $f(D)$. It is sufficient to extend the usual algebra in a straightforward way like $\gamma^\mu\gamma^\nu\gamma_\mu = -(D-2)\gamma^\nu$ or $Tr(\gamma^\mu\gamma^\nu) = f(D)g_{\mu\nu}$.

The physical dimensions of fields change in D dimensions and, as a consequence, the gauge couplings become dimensional $e_D = \mu^\epsilon e$, where e is adimensional, $D = 4 - 2\epsilon$ and μ is a scale of mass (this is how a scale of mass is introduced in the DR of massless QCD!). The dimension of fields is determined by requiring that the action $S = \int d^D x \mathcal{L}$ is adimensional. By inserting for $\mathcal{L}$ terms like $m\bar{\Psi}\Psi$ or $m^2\phi^\dagger\phi$ or $e\bar{\Psi}\gamma^\mu\Psi A_\mu$ the dimensions of the fields and coupling are determined as: $m, \Psi, \phi, A_\mu, e = 1, (D-1)/2, (D-2)/2, (D-2)/2, (4-D)/2$, respectively. The formal expression of loop integrals can be written for any D. For example:

$$\int \frac{d^D k}{(2\pi)^D} \frac{1}{(k^2-m^2)^2} = \frac{\Gamma(2-D/2)(-m^2)^{D/2-2}}{(4\pi)^{D/2}} \qquad (49)$$

For $D = 4 - 2\epsilon$ one can expand using:

$$\Gamma(\epsilon) = \frac{1}{\epsilon} - \gamma_E + o(\epsilon), \qquad \gamma_E = 0.5772..... \qquad (50)$$

For some Green function G, normalised to 1 in lowest order, (like V/e with V the 3-g vertex function at the symmetric point $p^2 = q^2 = r^2$, considered in the previous section) we typically find at 1-loop:

$$G_{bare} = 1 + \alpha_0 (\frac{-\mu^2}{p^2})^\epsilon [c(\frac{1}{\epsilon} + \log 4\pi - \gamma_E) + d + o(\epsilon)] \qquad (51)$$

In $\bar{MS}$ one rewrites this at 1-loop accuracy (diagram by diagram: this is a virtue of the method):

$$G_{bare} = Z G_{ren}$$
$$Z = 1 + \alpha [c(\frac{1}{\epsilon} + \log 4\pi - \gamma_E)]$$
$$G_{ren} = 1 + \alpha [c \log \frac{-\mu^2}{p^2} + d]$$

In the original MS prescription only $1/\epsilon$ was subtracted (that clearly plays the role of a cutoff) and not also $\log 4\pi$ and γ_E. Later, since these constants always appear from the expansion of Γ functions it was decided to modify MS into $\overline{MS}$. Note that the $\overline{MS}$ definition of α is different than that in the momentum subtraction scheme because the finite terms (those beyond logs) are different. In particular here δG_{ren} does not vanish at $p^2 = -\mu^2$.

The third coefficient of the QCD β function is also known in the $\overline{MS}$ prescription (recall that only the first two coefficients are scheme independent). Translated in numbers, for $n_f = 5$ one obtains [5]:

$$\beta(\alpha) = -0.610\alpha^2 [1 + 1.261... \frac{\alpha}{\pi} + 1.475...(\frac{\alpha}{\pi})^2 + ...] \qquad (52)$$

It is interesting to remark that the expansion coefficients are all of order 1, so that the $\bar{MS}$ expansion looks well behaved.

APPLICATION TO HARD PROCESSES

$R_{e^+e^-}$ and Related Processes

The simplest hard process is $R_{e^+e^-}$ that we have already started to discuss. R is adimensional and in perturbation theory is given by $R = N_C \sum_i Q_i^2 F(t, \alpha_s)$, where

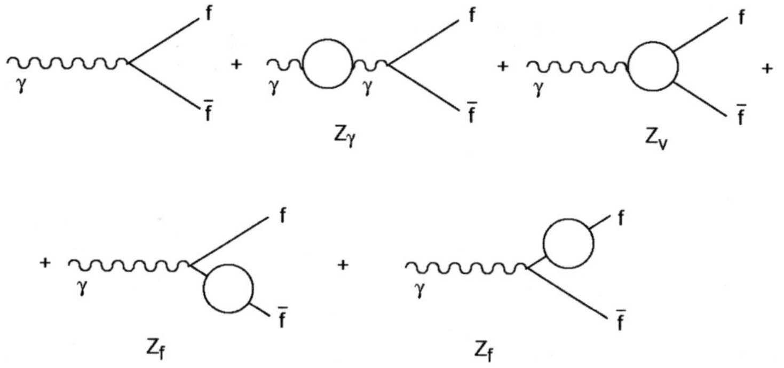

FIGURE 8.

$F = 1 + o(\alpha_S)$. We have already mentioned that for this process the "anomalous dimension" function vanishes: $\gamma(\alpha_s) = 0$ because of electric charge non renormalisation by strong interactions. Let us review how this happens in detail. The diagrams that are relevant for charge renormalisation in QED at 1-loop are shown in Fig. 8. The Ward identity that follows from gauge invariance in QED imposes that the vertex (Z_V) and the self-energy (Z_f) renormalisation factors cancel and the only divergence remains in Z_γ, the vacuum polarization of the photon. So the charge is only renormalised by the photon blob, hence it is universal (the same factor for all fermions, independent of their charge) and is not affected by QCD at 1-loop. It is true that at higher orders the photon vacuum polarization diagram is affected by QCD (for example, at 2-loops we can exchange a gluon between the quarks in the photon loop) but the renormalisation induced by the vacuum polarisation diagram remains independent of the nature of the fermion to which the photon line is attached. The gluon contributions to the vertex (Z_V) and to the self-energy (Z_f) cancel because they have exactly the same structure as in QED, and there is no gluon contribution to the lowest order photon vacuum polarisation blob. So $\gamma(\alpha_s) = 0$.

At 1-loop the diagrams relevant for the computation of R are shown in Fig. 9. There are virtual diagrams and real diagrams with one additional gluon in the final state. Infrared divergences cancel between the interference term of the virtual diagrams and the absolute square of the real diagrams, according to the Bloch-Nordsieck theorem. Similarly there are no mass singularities, in agreement with the Kinoshita-Lee-Nauenberg theorem, because the initial state is purely leptonic and all degenerate states that can appear at the given order are included in the final state. Given that $\gamma(\alpha_s) = 0$ the RGE prediction is simply given, as we have already seen, by $F(t,\alpha_s) = F[0,\alpha_s(t)]$. This means that if we do, for example, a 2-loop calculation, we must obtain a result of the form:

$$F(t,\alpha_s) = 1 + c_1\alpha_s(1 - b\alpha_s t) + c_2\alpha_s^2 + o(\alpha_s^3) \tag{53}$$

In fact we see that this form, taking into account that from eq.(44) we have:

$$\alpha_s(t) \sim \frac{\alpha_s}{1 + b\alpha_s t} \sim \alpha_s(1 - b\alpha_s t + ...) \tag{54}$$

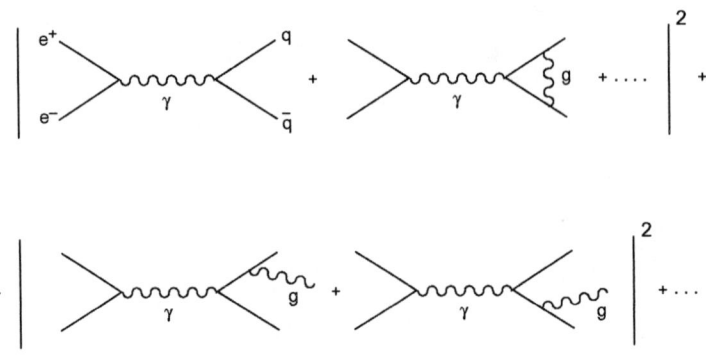

FIGURE 9.

can be rewritten as

$$F(t, \alpha_s) = 1 + c_1 \alpha_s(t) + c_2 \alpha_s^2(t) + o(\alpha_s^3(t)) = F[0, \alpha_s(t)] \tag{55}$$

The content of the RGE prediction is, at this order, that there are no $\alpha_s t$ and $(\alpha_s t)^2$ terms (the leading log sequence must be absent) and the term of order $\alpha_s^2 t$ has the coefficient that allows to reabsorb it in the transformation of α_s into $\alpha_s(t)$.

At present the first 3 coefficients have been computed in the $\bar{MS}$ scheme [6]. Clearly $c_1 = 1/\pi$ does not depend on the definition of α_s but c_2 and c_3 do. The subleading coefficients also depend on the scale choice: if instead of expanding in $\alpha_s(Q)$ we decide to choose $\alpha_s(Q/2)$ the coefficients c_2 and c_3 change. In the $\bar{MS}$ scheme, for γ-exchange and $n_f = 5$, which are good approximations for $2m_b \ll Q \ll m_Z$, one has:

$$F[0, \alpha_s(t)] = 1 + \frac{\alpha_s(t)}{\pi} + 1.409...(\frac{\alpha_s(t)}{\pi})^2 - 12.8...(\frac{\alpha_s(t)}{\pi})^3 + ... \tag{56}$$

Similar perturbative results at 3-loop accuracy also exist for $R_Z = \Gamma(Z \to hadrons)/\Gamma(Z \to leptons)$, $R_\tau = \Gamma(\tau \to \nu_\tau + hadrons)/\Gamma(\tau \to \nu_\tau + leptons)$, etc. We will discuss these results later when we deal with measurements of α_s.

The perturbative expansion in powers of $\alpha_s(t)$ takes into account all contributions that are suppressed by powers of logarithms of the large scale Q^2 ("leading twist" terms). In addition there are corrections suppressed by powers of the large scale Q^2 ("higher twist" terms). The pattern of power corrections is controlled by the light-cone Operator Product Expansion (OPE) which (schematically) leads to:

$$F = \text{pert.} + r_2 \frac{m^2}{Q^2} + r_4 \frac{<0|\text{Tr}[\mathbf{F}_{\mu\nu}\mathbf{F}^{\mu\nu}]|0>}{Q^4} + ... + r_6 \frac{<0|O_6|0>}{Q^6} + ... \tag{57}$$

Here m^2 generically indicates mass corrections, notably from b quarks, for example (t quark mass corrections only arise from loops, vanish in the limit $m_t \to \infty$ and are included in the coefficients as those in eq.(56) and the analogous ones for higher twist terms), $\mathbf{F}_{\mu\nu} = \Sigma_A \mathbf{F}_{\mu\nu}^A \mathbf{t}^A$, O_6 is typically a 4-fermion operator, etc. For each possible gauge invariant operator the corresponding power of Q^2 is fixed by dimensions.

We now consider the light-cone OPE in some more detail. $R_{e^+e^-} \sim \Pi(Q^2)$ where $\Pi(Q^2)$ is the scalar spectral function related to the hadronic contribution to the imaginary part of the photon vacuum polarization $T_{\mu\nu}$:

$$T_{\mu\nu} = (-g_{\mu\nu}Q^2 + q_\mu q_\nu)\Pi(Q^2) = \int \exp iqx <0|J_\mu^\dagger(x)J_\nu(0)|0> dx =$$
$$= \sum_n <0|J_\mu^\dagger(0)|n><n|J_\nu(0)|0> (2\pi)^4 \delta^4(q-p_n)$$

For $Q^2 \to \infty$ the $x^2 \to 0$ region is dominant. To all orders in perturbation theory the OPE can be proven. Schematically, dropping Lorentz indices, for simplicity, near $x^2 \sim 0$ we have:

$$J_\mu^\dagger(x) J_\nu(0) = I(x^2) + E(x^2) \sum_{n=0}^{\infty} c_n(x^2) x^{\mu_1}...x^{\mu_n} \cdot O_{\mu_1...\mu_n}^n(0) +$$
$$+ \text{ less sing. terms}$$

Here $I(x^2)$, $E(x^2)$,..., $c_n(x^2)$ are c-number singular functions, O^n is a string of local operators. $E(x^2)$ is the singularity of free field theory, $I(x^2)$ and $c_n(x^2)$ contain powers of $\log \mu x$ in interaction. Some O^n are already present in free field theory, other ones appear when interactions are switched on. $\Pi(Q^2)$ is related to the Fourier transform. Less singular terms in x^2 lead to power suppressed terms in $1/Q^2$. The perturbative terms come from $I(x^2)$ which is the leading twist term. The logarithmic scaling violations induced by the running coupling are the logs in $I(x^2)$.

The Final State in e^+e^- Annihilation

Experiments on e^+e^- annihilation at high energy provide a remarkable possibility of systematically testing the distinct signatures predicted by QCD for the structure of the final state averaged over a large number of events. Typical of asymptotic freedom is the hierarchy of configurations emerging as a consequence of the smallness of $\alpha_s(Q^2)$. When all corrections of order $\alpha_s(Q^2)$ are neglected one recovers the naive parton model prediction for the final state: almost collinear events with two back-to-back jets with limited transverse momentum and an angular distribution as $(1+\cos^2\theta)$ with respect to the beam axis (typical of spin 1/2 parton quarks: scalar quarks would lead to a $\sin^2\theta$ distribution). At order $\alpha_s(Q^2)$ a tail of events is predicted to appear with large transverse momentum $p_T \sim Q/2$ with respect to the thrust axis (the axis that maximizes the sum of the absolute values of the longitudinal momenta of the final state particles). This small fraction of events with large p_T mostly consists of three-jet events with an almost planar topology. The skeleton of a three-jet event, at leading order in $\alpha_s(Q^2)$, is formed by three hard partons $q\bar{q}g$, the third being a gluon emitted by a quark or antiquark line. The distribution of three-jet events is given by:

$$\frac{1}{\sigma} \frac{d\sigma}{dx_1 dx_2} = \frac{2\alpha_s}{3\pi} \frac{x_1^2 + x_2^2}{(1-x_1)(1-x_2)} \qquad (58)$$

here $x_{1,2}$ refer to energy fractions of massless quarks: $x_i = 2E_i/\sqrt{s}$ with $x_1+x_2+x_3=2$. At order $\alpha_s^2(Q^2)$ a hard perturbative non planar component starts to build up and a small fraction of four-jet events $q\bar{q}gg$ or $q\bar{q}q\bar{q}$ appear, and so on.

For precise testing and for measuring α_s a quantitatively specified definition of jet counting must be introduced which must be infrared safe (i.e. not altered by soft particle emission or collinear splittings of massless particles) in order to be computable at parton level and as much as possible insensitive to the transformation of partons into hadrons. One introduces a resolution parameter y_{cut} and a suitable pair variable, for example:

$$y_{ij} = \frac{min(E_i^2, E_j^2)(1-\cos\theta_{ij})}{s} \qquad (59)$$

The particles i,j belong to different jets for $y_{ij} > y_{cut}$. Clearly the number of jets becomes a function of y_{cut}: there are more jets for smaller y_{cut}. Measurements of $\alpha_s(Q^2)$ have been performed starting from jet multiplicities, the largest error coming from the necessity of correcting for non-perturbative hadronisation effects.

Deep Inelastic Scattering

Deep Inelastic Scattering (DIS) processes have played and still play a very important role for our understanding of QCD and of nucleon structure. This set of processes actually provides us with a rich laboratory for theory and experiment. There are several structure functions that can be studied, $F_i(x, Q^2)$, each a function of two variables. This is true separately for different beams and targets and different polarizations. Depending on the charges of l and l' we can have neutral currents (γ,Z) or charged currents in the l'-l channel (Fig. 6). In the past DIS processes were crucial for establishing quarks and gluons as partons and QCD as the theory of strong interactions. At present DIS is very important for quantitative studies and tests of QCD. The theory of scaling violations for totally inclusive DIS structure functions, based on operator expansions and renormalization group techniques, is crystal clear and the predicted Q^2 dependence can be tested at each value of x. The measurement of quark and gluon densities in the nucleon, as functions of x at some reference value of Q^2, which is an essential starting point for the calculation of all relevant hadronic hard processes, is performed in DIS processes. At the same time one measures $\alpha_s(Q^2)$ and the DIS values can be compared with those obtained from other processes. At all times new theoretical challenges arise from the study of DIS processes. Recent examples are the so-called "spin crisis" in polarized DIS and the behaviour of singlet structure functions at small x as revealed by HERA data. In the following we will review the past successes and the present open problems in the physics of DIS.

The cross-section $\sigma \sim L^{\mu\nu}W_{\mu\nu}$ is given in terms of the product of a leptonic ($L^{\mu\nu}$) and a hadronic ($W_{\mu\nu}$) tensor. While $L^{\mu\nu}$ is simple and easily obtained from the lowest order electroweak (e-w) vertex plus QED radiative corrections, the complicated strong interaction dynamics is contained in $W_{\mu\nu}$. The latter is proportional to the Fourier transform of the forward matrix element between the nucleon target states of the product of two e-w currents:

$$W_{\mu\nu} = \int dx \, \exp iqx \, <p|J_\mu^\dagger(x)J_\nu(0)|p> \tag{60}$$

Structure functions are defined starting from the general form of $W_{\mu\nu}$ given Lorentz invariance and current conservation. For example, for e-w currents between unpolarized nucleons we have:

$$W_{\mu\nu} = (-g_{\mu\nu} + \frac{q_\mu q_\nu}{q^2}) W_1(\nu, Q^2) + (p_\mu - \frac{m\nu}{q^2} q_\mu)(p_\nu - \frac{m\nu}{q^2} q_\nu) \frac{W_2(\nu, Q^2)}{m^2} -$$
$$- \frac{i}{2m^2} \epsilon_{\mu\nu\lambda\rho} p^\lambda q^\rho \, W_3(\nu, Q^2)$$

W_3 is absent for pure vector currents. In the limit $Q^2 >> m^2$, x fixed, the structure functions obey approximate Bjorken scaling which in reality is broken by logarithmic corrections that can be computed in QCD:

$$\begin{aligned} mW_1(\nu, Q^2) &\to F_1(x) \\ \nu W_{2,3}(\nu, Q^2) &\to F_{2,3}(x) \end{aligned} \tag{61}$$

The $\gamma - N$ cross-section is given by ($W_i = W_i(Q^2, \nu)$):

$$\frac{d\sigma^\gamma}{dQ^2 d\nu} = \frac{4\pi\alpha^2 E'}{Q^4 E} \cdot [2\sin^2\theta/2 W_1 + \cos^2\theta/2 W_2] \tag{62}$$

while for the $\nu - N$ or $\bar{\nu} - N$ cross-section one has:

$$\frac{d\sigma^{\nu,\bar{\nu}}}{dQ^2 d\nu} = \frac{G_F^2 E'}{2\pi E} (\frac{m_W^2}{Q^2 + m_W^2})^2 \cdot [2\sin^2\theta/2 W_1 + \cos^2\theta/2 W_2 \pm \frac{E+E'}{m}\sin^2\theta/2 W_3] \tag{63}$$

(W_i for photons, ν and $\bar{\nu}$ are all different, as we shall see in a moment).

In the scaling limit the longitudinal and transverse cross sections are given by:

$$\begin{aligned} \sigma_L &= \frac{1}{s}[\frac{F_2(x)}{2x} - F_1(x)] \\ \sigma_{RH,LH} &\sim \frac{1}{s}[F_1(x) \pm F_3(x)] \\ \sigma_T &= \sigma_{RH} + \sigma_{LH} \end{aligned} \tag{64}$$

where L, RH, LH refer to the helicity 0, 1, -1, respectively, of the exchanged gauge vector boson.

In the '60's the demise of hadrons from the status of fundamental particles to that of bound states of constituent quarks was the breakthrough that made possible the construction of a renormalisable field theory for strong interactions. The presence of an unlimited number of hadrons species, many of them with large spin values, presented an obvious dead-end for a manageable field theory. The evidence for constituent quarks emerged clearly from the systematics of hadron spectroscopy. The complications of the hadron spectrum could be explained in terms of the quantum numbers of spin 1/2, fractionally charged, u, d and s quarks. The notion of colour was introduced to reconcile the observed spectrum with Fermi statistics. But confinement that forbids the observation of free quarks was a clear obstacle towards the acceptance of quarks as real constituents and not just as fictitious entities describing some mathematical pattern (a doubt expressed even by Gell-Mann at the time). The early measurements at SLAC of DIS dissipated all doubts: the observation of Bjorken scaling and the success of the naive (not so much after all) parton model of Feynman imposed quarks as the basic fields for describing the nucleon structure (parton quarks).

In the language of Bjorken and Feynman the virtual γ (or, in general, any gauge boson) sees the quark partons inside the nucleon target as quasi-free, because the (Lorentz dilated) QCD interaction time is much longer than $\tau_\gamma \sim 1/Q$. Since the virtual photon 4-momentum is spacelike, we can go to a Lorentz frame where $E_\gamma = 0$ (Breit frame). In this frame $q = (E_\gamma = 0; 0, 0, Q)$ and the nucleon momentum, neglecting the mass $m \ll Q$, is $p = (Q/2x; 0, 0, -Q/2x)$ (note that this correctly gives $x = Q^2/2(p \cdot q)$). Consider (Fig. 10) the interaction of the photon with a quark carrying a fraction y of the nucleon 4-momentum: $p_q = yp$ (we are neglecting the transverse components of p_q which are of order m). The incoming parton with $p_q = yp$ absorbs the photon and the final parton has 4-momentum p'_q. Since in the Breit frame the photon carries no energy but only a longitudinal momentum Q, the photon can only be absorbed by those partons with $y = x$: then the longitudinal component of $p_q = yp$ is $-yQ/2x = -Q/2$ and can be flipped into $+Q/2$ by the photon. As a result, the photon longitudinal momentum $+Q$ disappears, the parton quark momentum changes of sign from $-Q/2$ into $+Q/2$ and the energy is not changed. So the structure functions are proportional to the density of partons with fraction x of the nucleon momentum, weighted with the squared charge. Also, recall that the helicity of a massless quark is conserved in a vector (or axial vector) interaction. So when the momentum is reversed also the spin must flip. Since the process is collinear there is no orbital contribution and only a photon with helicity ± 1 (transverse photon) can be absorbed. If partons were spin zero only longitudinal photons would instead contribute.

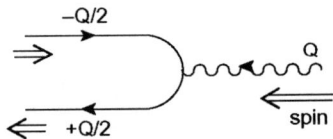

FIGURE 10.

Using these results, which are maintained in QCD at leading order, the quantum numbers of the quarks were confirmed by early experiments. The observation that $R = \sigma_L/\sigma_T \to 0$ implies that the charged partons have spin 1/2. The quark charges were derived from the data on the electron and neutrino structure functions:

$$F_{ep} = 4/9 u(x) + 1/9 d(x) + ...; \qquad F_{en} = 4/9 d(x) + 1/9 u(x) + ...$$
$$F_{\nu p} = F_{\bar\nu n} = 2d(x) + ...; \qquad F_{\nu n} = F_{\bar\nu p} = 2u(x) + ... \qquad (65)$$

where $F \sim 2F_1 \sim F_2/x$ and $u(x)$, $d(x)$ are the parton number densities in the proton (with fraction x of the proton longitudinal momentum), which, in the scaling limit, do not depend on Q^2. The normalisation of the structure functions and the parton densities are such that the charge relations hold:

$$\int_0^1 [u(x) - \bar{u}(x)]dx = 2, \quad \int_0^1 [d(x) - \bar{d}(x)]dx = 1, \quad \int_0^1 [s(x) - \bar{s}(x)]dx = 0 \quad (66)$$

Also it was proven by experiment that at values of Q^2 of a few GeV^2, in the scaling region, about half of the nucleon momentum, given by the momentum sum rule:

$$\int_0^1 [\sum_i (q_i(x) + \bar{q}_i(x)) + g(x)]x dx = 1 \quad (67)$$

is carried by neutral partons (gluons).

In QCD there are calculable log scaling violations induced by $\alpha_s(t)$. The parton rules just introduced can be summarised in the formula:

$$F(x,t) = \int_x^1 dy \frac{q_0(y)}{y} \sigma_{point}(\frac{x}{y}, \alpha_s(t)) + o(\frac{1}{Q^2}) \quad (68)$$

Before QCD corrections $\sigma_{point} = e^2 \delta(x/y - 1)$ and $F = e^2 q_0(x)$ (here we denote by e the charge of the quark in units of the positron charge, i.e. $e = 2/3$ for the u quark). QCD modifies σ_{point} at order α_s via the diagrams of Fig. 11. Note that the integral is from x to 1, because the energy can only be lost by radiation before interacting with the photon (which eventually wants to find a fraction x, as we have explained). From a direct computation of the diagrams one obtains a result of the following form:

$$\sigma_{point}(z, \alpha_s(t)) \simeq e^2[\delta(z-1) + \frac{\alpha_s}{2\pi}(t \cdot P(z) + f(z))] \quad (69)$$

For $y > x$ the correction arises from diagrams with real gluon emission. Only the sum of the two real diagrams in Fig. 11 is gauge invariant, so that the contribution of one given diagram is gauge dependent. There is a special form of axial gauge, called physical gauge, where, among real diagrams, the diagram of Fig. 11(c) gives the whole t-proportional term. It is obviously not essential to go to this gauge, but this diagram has a direct physical interpretation: a quark in the proton has a fraction $y > x$ of the parent 4-momentum; it then radiates a gluon and looses energy down to a fraction x before interacting with the photon. The log arises from the virtual quark propagator, according to the discussion of collinear mass singularities in eq.(17). In fact in the massless limit one has:

$$\text{propagator} = \frac{1}{r^2} = \frac{1}{(k-h)^2} = \frac{-1}{2E_k E_h} \cdot \frac{1}{1-\cos\theta}$$

$$= \frac{-1}{4E_k E_h} \cdot \frac{1}{\sin^2\theta/2} \propto \frac{-1}{p_T^2} \quad (70)$$

where p_T is the transverse momentum of the virtual quark. So the square of the propagator goes like $1/p_T^4$. But there is a p_T^2 factor in the numerator, because in the collinear limit, when $\theta = 0$ and the initial and final quarks and the emitted gluon are all aligned,

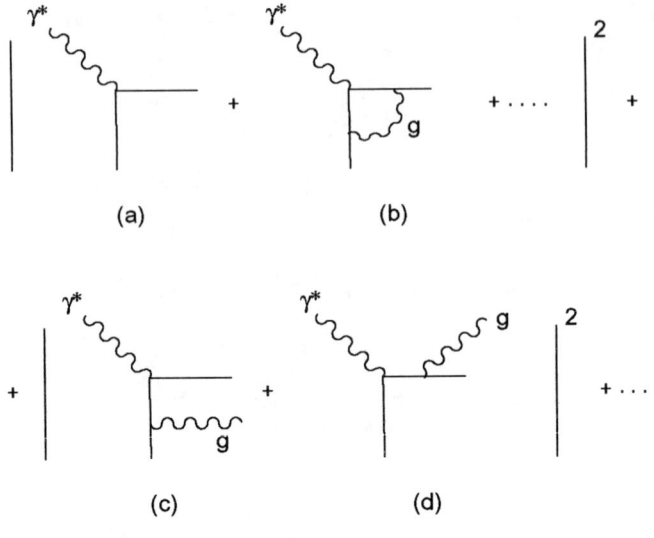

FIGURE 11.

the quark helicity cannot flip (vector interaction) and the real gluon cannot have but ±1 helicity. So the cross-section behaves as:

$$\sigma \sim \int^{Q^2} \frac{1}{p_T^2} dp_T^2 \sim \log Q^2 \qquad (71)$$

Actually the log should be read as $\log Q^2/m^2$ because in the massless limit a genuine mass singularity appears. In fact the mass singularity connected with the initial quark line is not cancelled because we do not have the sum of all degenerate initial states, but only a single quark. But in correspondence to the initial quark we have the (bare) quark density $q_0(y)$ that appear in the convolution integral. This is a non perturbative quantity that is determined by the nucleon wave function. So we can factorize the mass singularity in a redefinition of the quark density: we replace $q_0(y) \to q(y,t) = q_0(y) + \Delta q(y,t)$ with:

$$\Delta q(y,t) = \frac{\alpha_s}{2\pi} t \int_x^1 dy \frac{q_0(y)}{y} \cdot P(\frac{x}{y}) \qquad (72)$$

Here the factor of t is a bit symbolic: it stands for $\log Q^2/km^2$ and what we exactly put below Q^2 depends on the definition of the renormalised quark density, that also determines the exact form of the finite term $f(z)$ in eq.(69).

The effective parton density $q(y,t)$ that we have defined is now scale dependent. In terms of this scale dependent density we have the following relations, where we have also replaced the fixed coupling with the running coupling according to the prescription derived from the RGE:

$$F(x,t) = \int_x^1 dy \frac{q(y,t)}{y} e^2 [\delta(\frac{x}{y} - 1) + \frac{\alpha_s(t)}{2\pi} f(\frac{x}{y}))] = e^2 q(x,t) + o(\alpha_s(t))$$

$$\frac{d}{dt}q(x,t) = \frac{\alpha_s(t)}{2\pi} \int_x^1 dy \frac{q(y,t)}{y} \cdot P(\frac{x}{y}) + o(\alpha_s(t)^2)$$

We see that in lowest order we reproduce the naive parton model formulae for the structure functions in terms of effective parton densities that are scale dependent. The evolution equations for the parton densities are written down in terms of kernels (the "splitting functions") that can be expanded in powers of the running coupling. At leading order, we can interpret the evolution equation by saying that the variation of the quark density at x is given by the convolution of the quark density at y times the probability of emitting a gluon with fraction x/y of the quark momentum.

It is interesting that the integro-differential QCD evolution equation for densities can be transformed into an ordinary differential equation for Mellin moments. The moment f_n of a density $f(x)$ is defined as:

$$f_n = \int_0^1 dx\, x^{n-1} f(x) \tag{73}$$

By taking moments of both sides of the second of eqs.(73) one finds, with a simple interchange of the integration order, the simpler equation for the nth moment:

$$\frac{d}{dt}q_n(t) = \frac{\alpha_s(t)}{2\pi} \cdot P_n \cdot q_n(t) \tag{74}$$

To solve this equation we observe that:

$$\log \frac{q_n(t)}{q_n(0)} = \frac{P_n}{2\pi} \int_0^t \alpha_s(t) dt = \frac{P_n}{2\pi} \int_{\alpha_s}^{\alpha_s(t)} \frac{d\alpha'}{-b\alpha'} \tag{75}$$

where we used eq.(36) to change the integration variable from dt to $d\alpha(t)$ (denoted as $d\alpha'$) and $\beta(\alpha) \simeq -b\alpha^2 + ...$ Finally the solution is:

$$q_n(t) = [\frac{\alpha_s}{\alpha_s(t)}]^{\frac{P_n}{2\pi b}} \cdot q_n(0) \tag{76}$$

The connection of these results with the RGE general formalism occurs via the light cone OPE (recall eq.(60) for $W_{\mu\nu}$ and eq.(58) for the OPE of two currents). In the case of DIS the c-number term $I(x^2)$ does not contribute, because we are interested in the connected part $<p|...|p> - <0|...|0>$. The relevant terms are:

$$J_\mu^\dagger(x) J_\nu(0) = E(x^2) \sum_{n=0}^\infty c_n(x^2) x^{\mu_1}...x^{\mu_n} \cdot O_{\mu_1...\mu_n}^n(0) + \text{less sing. terms} \tag{77}$$

A formally intricate but conceptually simple argument (Ref.[3], page 28) based on the analiticity properties of the forward virtual Compton amplitude shows that the Mellin moments M_n of structure functions are related to the individual terms in the OPE, precisely to the Fourier transform $c_n(Q^2)$ (we will write it as $c_n(t, \alpha)$) of the coefficient $c_n(x^2)$ times a reduced matrix element h_n from the operators O^n: $<p|O_{\mu_1...\mu_n}^n(0)|p> = h_n p_{\mu_1}...p_{\mu_n}$:

$$c_n <p|O^n|p> \to M_n = \int_0^1 dx\, x^{n-1} F(x) \tag{78}$$

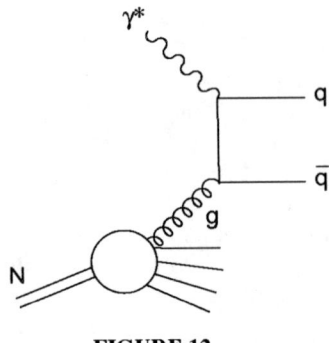

FIGURE 12.

Since the matrix element of the products of currents satisfy the RGE so do the moments M_n. Hence the general form of the Q^2 dependence is given by the RGE solution (see eq.(37)):

$$M_n(t,\alpha) = c_n[0,\alpha(t)] \exp \int_\alpha^{\alpha(t)} \frac{\gamma_n(\alpha')}{\beta(\alpha')} d\alpha' \cdot h_n(\alpha) \tag{79}$$

In lowest order, identifying in the simplest case M_n with q_n, we have:

$$\gamma_n(\alpha) = \frac{P_n}{2\pi}\alpha + ..., \qquad \beta(\alpha) = -b\alpha^2 + ... \tag{80}$$

and

$$q_n(t) = q_n(0) \exp \int_\alpha^{\alpha(t)} \frac{\gamma_n(\alpha')}{\beta(\alpha')} d\alpha' = [\frac{\alpha_s}{\alpha_s(t)}]^{\frac{P_n}{2\pi b}} \cdot q_n(0) \tag{81}$$

which exactly coincides with eq.(76).

Up to this point we have implicitely restricted our attention to non-singlet (under the flavour group) structure functions. The Q^2 evolution equations become non diagonal as soon as we take into account the presence of gluons in the target. In fact the quark which is seen by the photon can be generated by a gluon in the target (Fig. 12).

The quark evolution equation becomes:

$$\frac{d}{dt}q_i(x,t) = \frac{\alpha_s(t)}{2\pi}[q_i \otimes P_{qq}] + \frac{\alpha_s(t)}{2\pi}[g \otimes P_{qg}] \tag{82}$$

where we introduced the shorthand notation:

$$[q \otimes P] = [P \otimes q] = \int_x^1 dy \frac{q(y,t)}{y} \cdot P(\frac{x}{y}) \tag{83}$$

(it is easy to check that the convolution, like an ordinary product, is commutative). At leading order, the interpretation of eq.(82) is simply that the variation of the quark density is due to the convolution of the quark density at a higher energy times the probability of finding a quark in a quark (with the right energy fraction) plus the gluon density at a higher energy times the probability of finding a quark (of the given flavour

i) in a gluon. The evolution equation for the gluon density, needed to close the system, can be obtained by suitably extending the same line of reasoning to a gedanken probe sensitive to colour charges, for example a virtual gluon. The resulting equation is of the form:

$$\frac{d}{dt}g(x,t) = \frac{\alpha_s(t)}{2\pi}[\sum_i (q_i+\bar{q}_i) \otimes P_{gq}] + \frac{\alpha_s(t)}{2\pi}[g \otimes P_{gg}] \quad (84)$$

The explicit form of the splitting functions in lowest order can be explicitly derived from the QCD vertices. They are a property of the theory and do not depend on the particular process the parton density is taking part into. The results are:

$$P_{qq} = \frac{4}{3}[\frac{1+x^2}{(1-x)_+} + \frac{3}{2}\delta(1-x)] + o(\alpha_s)$$

$$P_{gq} = \frac{4}{3}\frac{1+(1-x)^2}{x} + o(\alpha_s)$$

$$P_{qg} = \frac{1}{2}[x^2+(1-x)^2] + o(\alpha_s)$$

$$P_{gg} = 6[\frac{x}{(1-x)_+} + \frac{1-x}{x} + x(1-x)] + \frac{33-2n_f}{6}\delta(1-x) + o(\alpha_s) \quad (85)$$

The "+" distribution is defined as, for a generic non singular weight function $f(x)$:

$$\int_0^1 \frac{f(x)}{(1-x)_+}dx = \int_0^1 \frac{f(x)-f(1)}{1-x}dx \quad (86)$$

The $\delta(1-x)$ terms arise from the virtual corrections to the tree diagram. Their coefficient can be simply obtained by imposing the validity of charge and momentum sum rules. In fact, from the request that the charge sum rules in eq.(66) are not affected by the Q^2 dependence one derives that

$$\int_0^1 P_{qq}(x)dx = 0 \quad (87)$$

which can be used to fix the coefficient of the $\delta(1-x)$ terms of P_{qq}. Similarly, by taking the t-derivative of the momentum sum rule in eq.(67) and imposing its vanishing for generic q_i and g, one obtains:

$$\int_0^1 [P_{qq}(x) + P_{gq}(x)]xdx = 0, \quad \int_0^1 [2n_f P_{qg}(x) + P_{gg}(x)]xdx = 0, \quad (88)$$

At higher orders the evolution equations are easily generalised but the calculation of the splitting functions rapidly becomes very complicated. The splitting functions are completely known at NLO accuracy: $\alpha_s P \sim \alpha_s P_1 + \alpha_s^2 P_2 + ...$. More recently the NNLO results P_3 have been derived in analytic form for the first few moments [7]. The full NNLO calculation is in progress and could be finished soon.

The scaling violations are clearly observed by experiment and their pattern is very well reproduced by QCD fits at NLO. Examples are seen in Figs. 13(a-d) [8]. These fits provide an impressive confirmation of a quantitative QCD prediction, a measurement of

$q_i(x, Q_0^2)$ and $g(x, Q_0^2)$ at some reference value Q_0^2 of Q^2 and a precise measurement of $\alpha_s(m_Z^2)$. At small x and large but fixed Q^2 when terms of order $(\alpha_s \log 1/x)^n$ become of order 1 and cannot be neglected, the validity of the NLO or NNLO approximations should breakdown. However, the small x data collected by HERA can be fitted reasonably even at the smallest measured values of x by the NLO QCD evolution equations, so that there is no neat evidence in the data for departures. But the extracted gluon density (which is dominant at small x) and the fitted value of $\alpha_s(m_Z^2)$ could be biassed by this effect if data at too small values of x are included.

Factorisation and the QCD Improved Parton Model

The parton densities defined and measured in DIS are used to compute hard processes initiated by hadronic collisions via the Factorisation Theorem (FT). Suppose you have a hadronic process of the form $h_1 + h_2 \to X + all$ where X is some triggering particle or pair of particles which specify the large scale Q^2 relevant for the process, in general somewhat but not much smaller than s, the total c.o.m. squared mass. For example, in pp or $p\bar{p}$ collisions, X can be a W or a Z or a virtual photon with large Q^2, or a jet at large transverse momentum p_T, or a pair of heavy quark-antiquark of mass M. By "all" we mean a totally inclusive collection of gluons and light quark pairs. The FT implies that for the cross-section or some other sufficiently inclusive distribution we can write the expression:

$$\sigma(s,\tau) = \sum_{AB} \int dx_1 dx_2 p_{1A}(x_1, Q^2) p_{2B}(x_2, Q^2) \sigma_{AB}(x_1 x_2 s, \tau) \qquad (89)$$

Here $\tau = Q^2/s$ is a scaling variable, p_{iC} are generic parton-C densities inside the hadron h_i, σ_{AB} is the partonic cross-section for parton-A + parton-B$\to X + all'$. This result is based on the fact that the mass singularities that are associated with the initial legs are of universal nature, so that one can reproduce the same modified parton densities, by absorbing these singularities into the bare parton densities, as in deep inelastic scattering. Once the parton densities and α_s are known from other measurements the prediction of the rate for a given hard process is obtained with no free parameters. The NLO calculation of the reduced partonic cross-section is needed in order to correctly specify the scale and in general the definition of the parton densities and of the running coupling in the leading term. The residual scale and scheme dependence is often the most important source of theoretical error. In the following we consider a few examples.

A comparison of data and predictions on the production of jets at large $\sqrt{s}$ and p_T in pp or $p\bar{p}$ collisions is shown in Fig. 14 [4].

This is a particularly significant test because the rates at different c.o.m. energies and, for each energy, at different values of p_T span over many orders of magnitude. Also the corresponding values of $\sqrt{s}$ and p_T are large enough to be well inside the perturbative region. The overall agreement of the data from ISR, UA1,2 and CDF and D0 is spectacular. Only at very large p_T there might be some problem according to CDF and, to a lesser extent, to D0. A harder gluon at large x and inclusion of systematic errors can alleviate or eliminate the problem. This issue will be clarified in the near

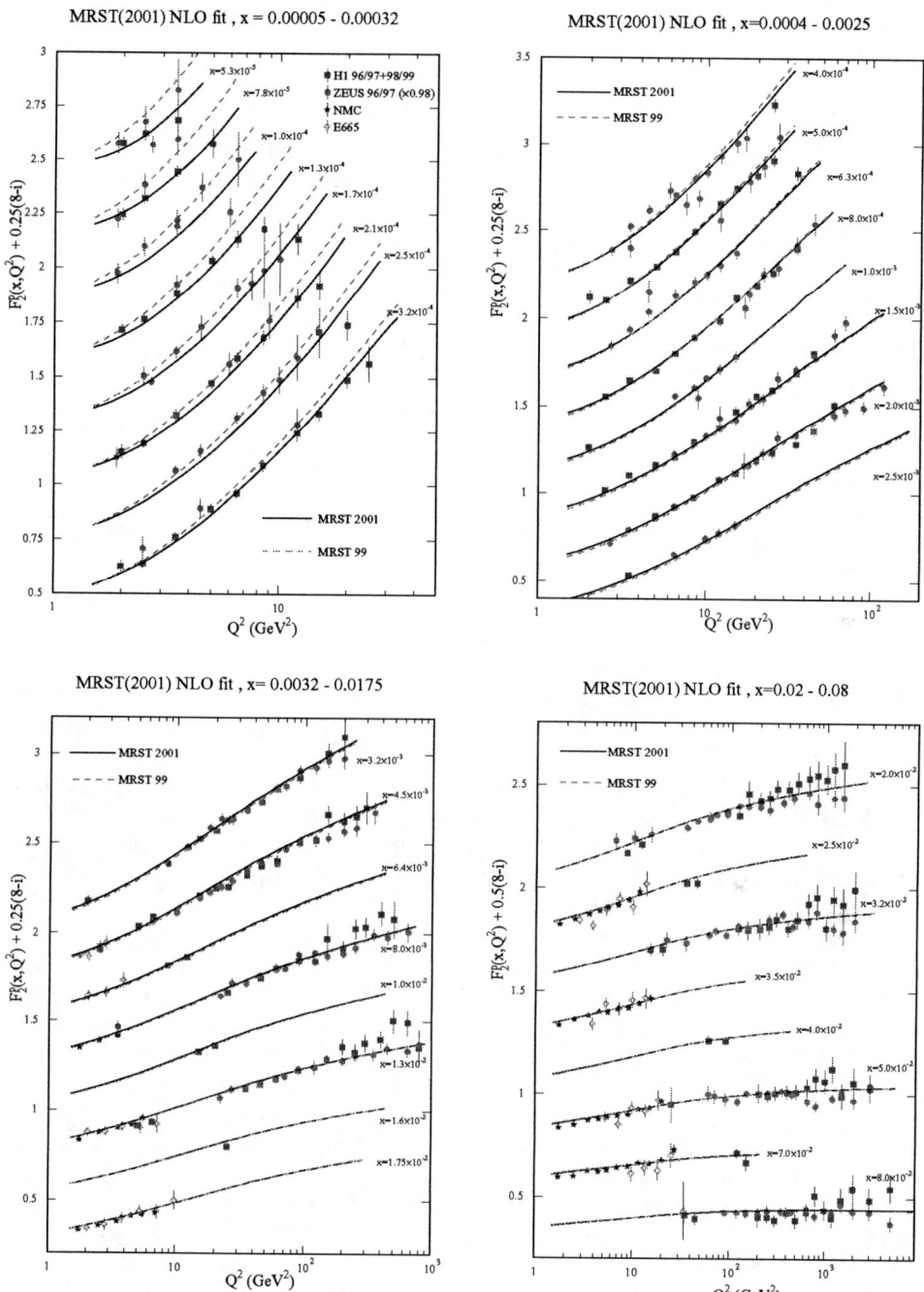

FIGURE 13. A recent NLO fit of scaling violations for different x ranges, as functions of Q^2. Reprinted with permission of Authors and European Journal of Physics C, Springer Verlag Heidelberg, see Ref. [8].

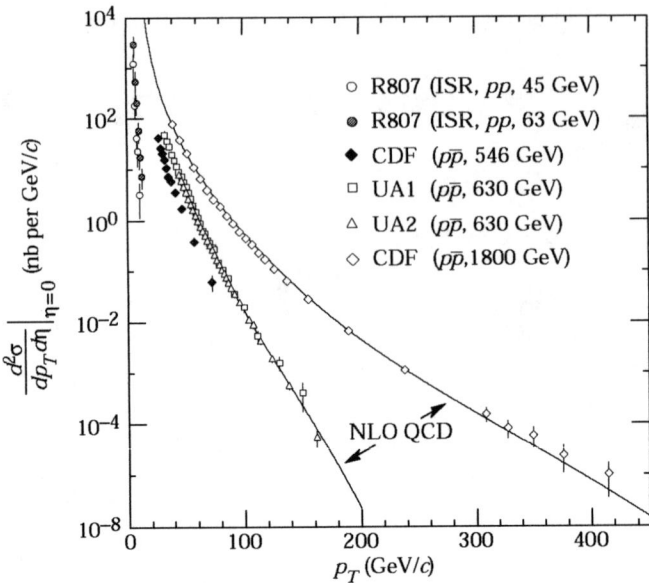

FIGURE 14. Jet production cross-section at pp or $p\bar{p}$ colliders, as function of p_T

future by the RunII results at the Tevatron. Similar results also hold for the production of photons at large p_T. The collider data, shown in Fig. 15 [4], are in fair agreement with the theoretical predictions. A less clear situation is found with fixed target data. Here, first of all, the experimental results show some internal discrepancies. Also, the p_T accessible values being smaller, the theoretical uncertainties are larger. But it is true that the agreement is poor, so that the necessity of an "intrinsic" transverse momentum of partons inside the hadron of over 1 GeV has been claimed, which I do not think one is free to introduce.

For heavy quark production at colliders the agreement is very good for top production at the Tevatron (Fig. 16). To present a rare example of a puzzle, the bottom production at the Tevatron remains problematic. The total rate and the p_T distribution of b quarks observed at CDF is in excess of the prediction, up to the largest measured values of p_T (fig.17). It is true that this is a complicated case, with different scales being present at the same time: $\sqrt{s}$, p_T, m_b. Also some non-perturbative ingredients, like fragmentation functions, are present in the calculation and could be in part reponsible for the effect. But no clear explanation of this phenomenon is available. Probably it occurs through an unfavourable conspiracy of several small effects.

Drell-Yan processes, including lepton pair production via virtual γ, W or Z exchange, offer a good opportunity to test QCD. The process is quadratic in parton densities, and the final state is totally inclusive, while the large scale is specified and measured by the invariant mass squared Q^2 of the lepton pair which itself is not strongly interacting. The QCD improved parton model leads directly to a prediction for the total rate as a function of Q^2. The value of the LO cross-section is inversely proportional to the number of

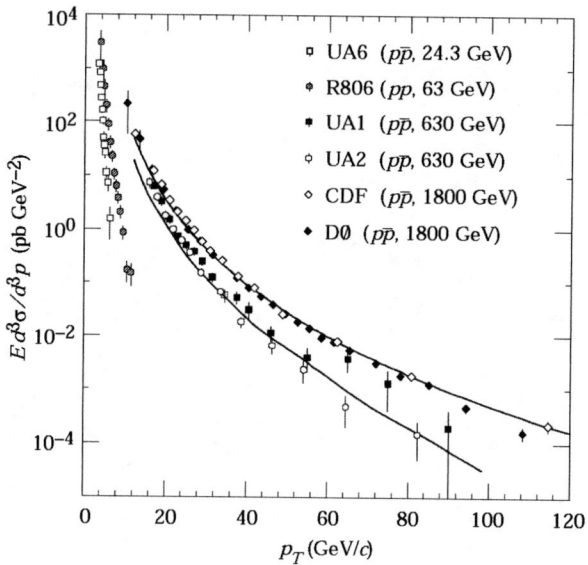

FIGURE 15. Single photon production in $p\bar{p}$ colliders as function of p_T

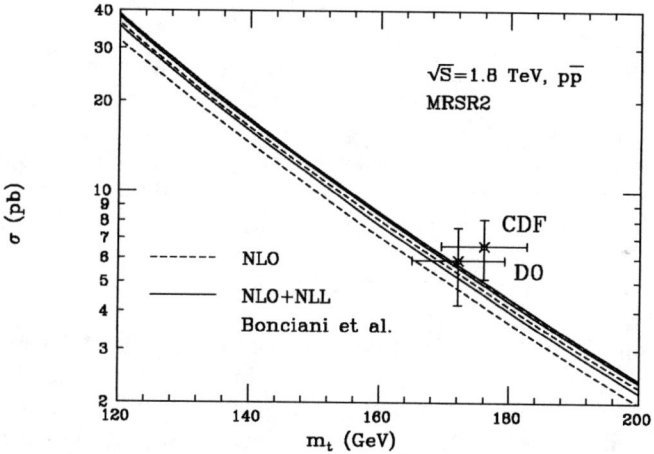

FIGURE 16. The t production cross-section at the Tevatron $p\bar{p}$ collider

colours N_C because a quark of given colour can only annihilate with an antiquark of the same colour to produce a colourless lepton pair. The order $\alpha_s(Q^2)$ corrections to the total rate were computed long ago and found to be particularly large, when the quark densities are defined from the structure function F_2 measured in DIS at $q^2 = -Q^2$. The ratio $\sigma_{corr}/\sigma_{LO}$ of the corrected and the Born cross-sections, was called K-factor (by me), because it is almost a constant in rapidity. More recently also the NLO full

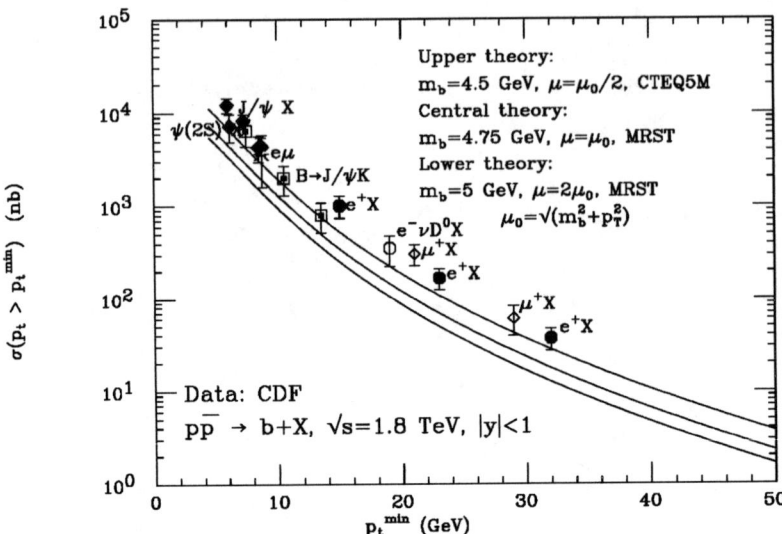

FIGURE 17. The b production p_T distribution at the Tevatron $p\bar{p}$ collider (CDF collaboration). The cross-section for $p_T > p_T^{min}$ is plotted as function of p_T^{min}.

calculation of the K-factor was completed, a very remarkable calculation [9]. The QCD predictions can be best tested for W and Z production at CERN and Tevatron energies. $Q \sim m_{W,Z}$ is large enough to make the prediction reliable (a not too large K-factor) and the ratio $\sqrt{\tau} = Q/\sqrt{s}$ is not too small. Recall that in lowest order $x_1 x_2 s = Q^2$ so that the parton densities are probed at x values around $\sqrt{\tau}$. We have $\sqrt{\tau} = 0.13 - 0.15$ at $\sqrt{s} = 630$ GeV (CERN $Sp\bar{p}S$ Collider) and $\sqrt{\tau} = 0.041 - 0.052$ at the Tevatron. In this respect the prediction is more delicate at the LHC, where $\sqrt{\tau} \sim 5.9 - 6.5 \cdot 10^{-3}$. One comparison of the experimental total rates at the Tevatron with the QCD predictions is shown in Fig. 18, together with the expected rates at the LHC (based on the structure functions obtained in [8]).

The calculation of the W/Z p_T distribution has been a classic problem in QCD. For large p_T, for example $p_T \sim o(m_W)$, the p_T distribution can be reliably computed in perturbation theory, which was done up to NLO in the late '70's and early '80's. A problem arises in the intermediate range $\Lambda_{QCD} << p_T << m_W$, where the bulk of the data is concentrated, because terms of order $\alpha_s(p_T^2) \log m_W^2/p_T^2$ become of order 1 and should included to all orders. At order α_s we have:

$$\frac{1}{\sigma_0}\frac{d\sigma_0}{dp_T^2} = (1+A)\delta(p_T^2) + \frac{B}{p_T^2}\log\frac{m_W^2}{p_T^2}\bigg|_+ + \frac{C}{(p_T^2)_+} + D(p_T^2) \qquad (90)$$

where A, B, C, D are coefficients of order α_s. The "+" distribution is defined in complete analogy with eq.(86):

$$\int_0^{p_{TMAX}^2} g(z)f(z)_+ dz = \int_0^{p_{TMAX}^2} [g(z) - g(0)]f(z)dz \qquad (91)$$

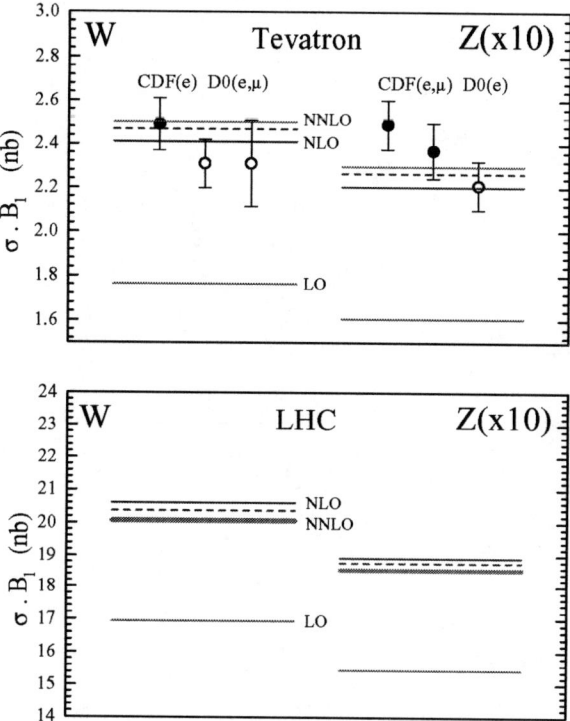

FIGURE 18. Data vs. theory for W and Z production at the Tevatron ($\sqrt{s} = 1.8$ TeV) together with the corresponding predictions for the LHC ($\sqrt{s} = 1.4$ TeV)

The content of this at first sight mysterious definition is that the singular "+" terms do not contribute to the total cross-section. In fact for the cross-section the weight function $g(z) = 1$ and we obtain:

$$\sigma = \sigma_0[(1+A) + \int_0^{p_{TMAX}^2} D(z)dz] \qquad (92)$$

The singular terms, of infrared origin, are present at the non completely inclusive level but disappear in the total cross-section. These singularities are proven to exponentiate and lead to the following expression:

$$\frac{1}{\sigma_0}\frac{d\sigma_0}{dp_T^2} = \int \frac{d^2b}{4\pi} \exp(-ib \cdot p_T)(1+A)\exp S(b) \qquad (93)$$

with:

$$S(b) = \int_0^{p_{TMAX}} \frac{d^2k_T}{2\pi}[\exp ik_T \cdot b - 1][\frac{B}{k_T^2}\log\frac{m_W^2}{k_T^2} + \frac{C}{k_T^2}] \qquad (94)$$

At large p_T the LO perturbative expansion is recovered. At intermediate p_T the infrared p_T singularities are resummed (the Sudakov log terms, which are typical of vector glu-

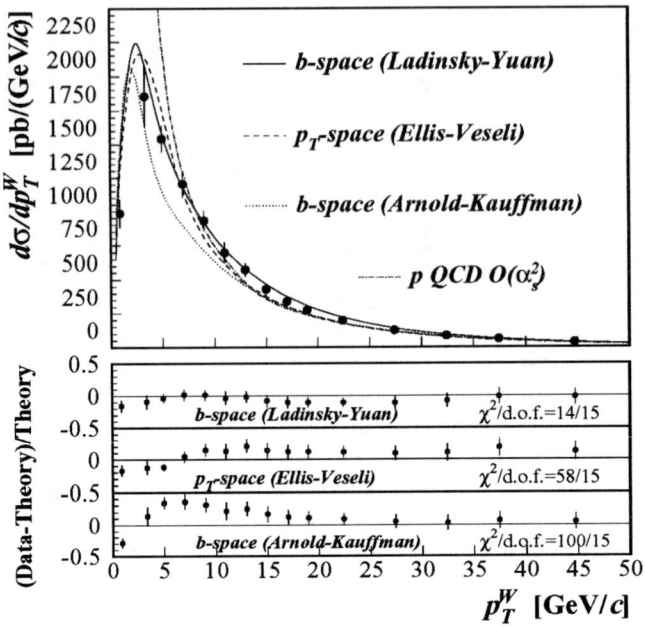

FIGURE 19. QCD predictions for the W p_T distribution compared with recent D0 data at the Tevatron ($\sqrt{s}$ = 1.8 TeV)

ons, are related to the fact that for a charged particle in acceleration it is impossible not to radiate, so that the amplitude for no soft gluon emission is exponentially suppressed). However this formula has problems at small p_T, for example, because of the presence of α_s under the integral for $S(b)$: presumably the relevant scale is of order k_T^2. So it must be completed by some non perturbative ansatz or an extrapolation into the soft region. All the exercise has been extended to NLO accuracy, where one starts from the perturbative expansion at order α_s^2, and generalises the resummation to also include NLO terms of order $\alpha_s(p_T^2)^2 \log m_W^2/p_T^2$. The comparison with the data is very impressive. In Fig. 19 we see the p_T distribution as predicted in QCD (with a number of variants that mainly differ in the approach to the soft region) compared with some recent data from the D0 Collaboration at the Tevatron [10].

MEASUREMENTS OF α_S

The most precise and reliable measurements of $\alpha_s(m_Z^2)$ in the $\bar{M}S$ definition are obtained from e^+e^- colliders (in particular LEP) and from Deep Inelastic Scattering.

α_s from e^+e^- colliders

The main methods at e^+e^- colliders are: a) Inclusive hadronic Z decay, R_l, σ_h, Γ_Z. b) Inclusive hadronic τ decay. c) Event shapes and jet rates.

As we have seen, for a quantity like R_l we can write a general expression of the form:

$$R_l = \frac{Z,\tau \to hadrons}{Z,\tau \to leptons} \sim R^{EW}(1 + \delta_{QCD} + \delta_{NP}) + ... \tag{95}$$

where R^{EW} is the electroweak-corrected Born approximation, δ_{QCD}, δ_{NP} are the perturbative (logarithmic) and non perturbative (power suppressed) QCD corrections. If we consider measurement at the Z, from R_l only, assuming the standard electroweak theory, one finds [11]:

$$\alpha_s(m_Z) = 0.123 \pm 0.004 \tag{96}$$

Better, one can use all info from R_l, $\Gamma_Z = 3\Gamma_l + \Gamma_h$ and $\sigma_h = 12\pi\Gamma_l\Gamma_h/(m_Z^2\Gamma_Z^2)$. From these observables together one obtains:

$$\alpha_s(m_Z) = 0.120 \pm 0.003 \tag{97}$$

By adding all other electroweak precision electroweak tests (in particular m_W) one finds:

$$\alpha_s(m_Z) = 0.118 \pm 0.003 \tag{98}$$

The final error is predominantly theoretical and is dominated by our ignorance on m_H and from higher orders in the QCD expansion.

We now consider the measurement of $\alpha_s(m_Z)$ from τ decay. R_τ has a number of advantages that, at least in part, tend to compensate for the smallness of $m_\tau = 1.777 \ GeV$. First, R_τ is maximally inclusive, more than $R_{e^+e^-}(s)$, because one also integrates over all values of the invariant hadronic squared mass:

$$R_\tau = \frac{1}{\pi}\int_0^{m_\tau^2} \frac{ds}{m_\tau^2}(1 - \frac{s}{m_\tau^2})^2 Im\Pi_\tau(s) \tag{99}$$

Analyticity can be used to transform the integral into one on the circle at $|s| = m_\tau^2$:

$$R_\tau = \frac{1}{2\pi i}\oint_{|s|=m_\tau^2} \frac{ds}{m_\tau^2}(1 - \frac{s}{m_\tau^2})^2 \Pi_\tau(s) \tag{100}$$

Also, the factor $(1 - \frac{s}{m_\tau^2})^2$ is important to kill the sensitivity the region $Re s = m_\tau^2$ where the physical cut and the associated thresholds are located. Still the quoted result (a combination of ALEPH and OPAL analyses) looks a bit too precise:

$$\alpha_s(m_Z) = 0.1181 \pm 0.0007(exp) \pm 0.003(th) \tag{101}$$

This precision is obtained by taking for granted that corrections suppressed by $1/m_\tau^2$ are negligible. This is because, in the massless theory, the light cone expansion is given by:

$$\delta_{NP} = \frac{ZERO}{m_\tau^2} + c_4 \cdot \frac{<O_4>}{m_\tau^4} + c_6 \cdot \frac{<O_6>}{m_\tau^6} + \cdots \tag{102}$$

In fact there are no dim-2 Lorentz and gauge invariant operators. For example, $g_\mu g^\mu$ is not gauge invariant. In the massive theory, the ZERO is replaced by light quark mass-squared m^2. This is still negligible if m is taken as a lagrangian mass of a few MeV. But would not at all be negligible, actually would very much affect the result, if it is taken as a constituent mass of order $m \sim \Lambda_{QCD}$. Most people believe the optimistic version. I am not convinced that the gap is not filled up by ambiguities of $0(\Lambda_{QCD}^2/m_\tau^2)$ from δ_{pert}. In any case, one can discuss the error, but it is true and remarkable, that the central value from τ decay, obtained at very small Q^2, is in perfect agreement with all other precise determinations of α_s at more typical LEP values of Q^2.

α_s from Deep Inelastic Scattering

QCD predicts the Q^2 dependence of $F(x,Q^2)$ at each fixed x, not the x shape. But the Q^2 dependence is related to the x shape by the QCD evolution equations. For each x-bin the data allow to extract the slope of an approximately stright line in $dlogF(x,Q^2)/dlogQ^2$: the log slope. The Q^2 span and the precision of the data are not much sensitive to the curvature, for most x values. A single value of Λ_{QCD} must be fitted to reproduce the collection of the log slopes. For the determination of α_s the scaling violations of non-singlet structure functions would be ideal, because of the minimal impact of the choice of input parton densities. We can write the non-singlet evolution equations in the form:

$$\frac{d}{dt}logF(x,t) = \frac{\alpha_s(t)}{2\pi} \int_x^1 \frac{dy}{y} \frac{F(y,t)}{F(x,t)} P_{qq}(\frac{x}{y},\alpha_s(t)) \tag{103}$$

where P_{qq} is the NLO splitting function. It is clear from this form that, for example, the normalisation of the input density drops away, and the dependence on the input is reduced to a minimum (also there is a single density, while in general there are quark and gluon densities). Unfortunately the data on non-singlet structure functions are not very accurate. If we take the difference of data on protons and neutrons, $F_p - F_n$, experimental errors add up in the difference and finally are large. The $F_{3\nu N}$ data are directly non-singlet but are not very precise. A determination of α_s from the CCFR data on $F_{3\nu N}$ has led to [12]:

$$\alpha_s(m_Z) = 0.118 \pm 0.006 \tag{104}$$

When one measures α_s from scaling violations on F_2 from e or μ beams, the data are abundant, the errors small but there is an increased dependence on input parton densities and especially a strong correlation between the result on α_s and the input on the gluon density. There are two most complete and accurate derivations of α_s from scaling violations in F_2. In the first analysis, by Santiago and Yndurain [13], the data on protons from SLAC, BCDMS, E665 and HERA are used with NLO kernels plus the NNLO first few moments. The analysis is based on an original method that uses projections on a specially selected basis of orthogonal polynomials. The quoted result is given by:

$$\alpha_s(m_Z) = 0.117 \pm 0.003 \tag{105}$$

A different analysis by Alekhin [14] of a similar collection of proton data from SLAC, BCDMS, NMC and HERA with NLO kernels and a more conventional method leads to

$$\alpha_s(m_Z) = 0.116 \pm 0.003 \qquad (106)$$

In both analysis the dominant error is theoretical and could be somewhat larger than quoted.

If we compare these results on α_s from DIS with the findings at the Z, given by eq.(98), we see that the agreement looks perfect. But, in my opinion, the situation of α_s from DIS is not yet completely satisfactory (while it is so for α_s from e^+e^-). The data have shown large fluctuations in the recent past. For example, the original result obtained by combining BCDMS and SLAC data was [15]:

$$\alpha_s(m_Z) = 0.113 \pm 0.005 \qquad (107)$$

This low value appeared to be confirmed by the CCFR result from F_2 and F_3 data combined:

$$\alpha_s(m_Z) = 0.111 \pm 0.005 \qquad (108)$$

But later the same data were corrected by a new energy calibration performed by the collaboration and the result become:

$$\alpha_s(m_Z) = 0.119 \pm 0.005 \qquad (109)$$

So the experimental errors are perhaps larger than quoted and there is a problem for matching the systematics of different experiments. On the theory side the analysis methods are perhaps still not completely optimised.

Summarising: there is very good agreement among many different measurements of α_s (see Fig. 20, [4]). This is a very convincing, quantitative test of QCD. The average value quoted by PDG 2000 is

$$\alpha_s(m_Z) = 0.118 \pm 0.002 \qquad (110)$$

The corresponding value of Λ (for $n_f = 5$) is:

$$\Lambda_5 = 209 \pm 24 \; MeV \qquad (111)$$

Λ is the scale of mass that finally appears in massless QCD. It is the scale where $\alpha_s(\Lambda)$ is of order 1. Hadron masses are determined by Λ. Actually the ρ mass or the nucleon mass receive little contribution from the quark masses (the case of pseudoscalar mesons is special, as they are the pseudo Goldstone bosons of broken chiral invariance). Hadron masses would be almost the same in massless QCD.

CONCLUSION

We have seen that perturbative QCD based on asymptotic freedom offers a rich variety of tests and we have described some examples in detail. QCD tests are not as precise as

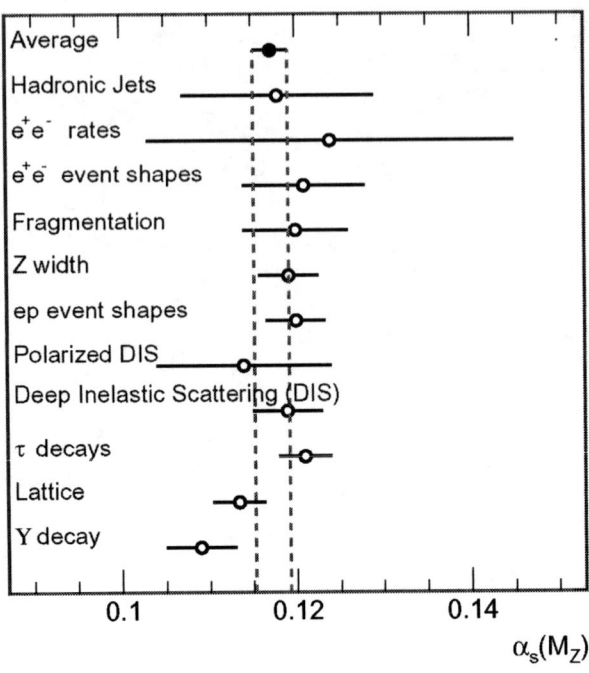

FIGURE 20. Compilation (see [4]) of the existing precise measurements of $\alpha_s(M_2)$

for the electroweak sector. But the number and diversity of such tests has established a very firm experimental foundation for QCD as a theory of strong interactions.

The field is still very much in movement. There are areas of continuous development: I list here some of these domains where work is in progress and which would deserve an expanded discussion.

- Higher order calculations: important examples are the effort to complete the 3-loop analytic determination of the splitting functions, and the calculation of 4-jet distributions at NLO in e^+e^- annihilation.
- Resummations in problems with 2 (or more) different large scales: if y=scale1/scale2 one needs to resum $(\alpha_s \log^a y)^n$, with $a = 1, 2$, at all orders in n. We have seen an example on the W, Z p_T distribution in pp or $p\bar{p}$ collisions for $\Lambda_{QCD} \ll p_T \ll m_W$ where all terms of order $(\alpha_s(p_T^2) \log m_W^2/p_T^2)^n$ must be taken into account. Other examples involve structure functions at small x $((\alpha_s(Q^2) \log 1/x)^n)$ or at $x \sim 1$ $((\alpha_s(Q^2) \log 1/(1-x))^n)$, or thrust distributions near $T \sim 1$ etc.
- Renormalons and power suppressed corrections (for an introduction see, for example, Ref [16]).. The QED and QCD perturbative series, after renormalisation, have all their coefficients finite, but it is well known that the expansion does not converge. Actually the perturbative series is not even Borel summable. After Borel resummation for a given process one is left with a result which is ambiguous by

terms typically down by $-\exp(-n/b\alpha)$, with n an integer and b the first β function coefficient. In QED these corrective terms are extremely small and not very important in practice. On the contrary in QCD $\alpha = \alpha_s(Q^2) \sim 1/(b\log Q^2/\Lambda^2)$ and the ambiguous terms are of order $(1/Q^2)^n$, that is are power suppressed. The problem arises of the precise relation between the ambiguities of the perturbative expansion and the higher twist corrections, which is a very interesting field of research.
- New areas explored by experiment. Particularly interesting physics cases are present in polarized deep inelastic scattering, or in the phenomenology of structure functions at small x as measured at HERA.
- An important domain of activity is the QCD event simulation for the preparation of LHC experiment which poses highly non trivial theoretical and technical problems.

APPENDIX: THE FORMALISM OF GAUGE THEORIES

We summarize here the definition and the structure of a gauge Yang–Mills theory.

Consider a Lagrangian density $\mathcal{L}[\phi, \partial_\mu \phi]$ which is invariant under a D dimensional continuous group of transformations:

$$\phi' = U(\theta^A)\phi \qquad (A = 1, 2, ..., D) . \tag{112}$$

For θ^A infinitesimal, $U(\theta^A) = 1 + ig\sum_A \theta^A T^A$, where T^A are the generators of the group Γ of transformations eq.(112) in the (in general reducible) representation of the fields ϕ. Here we restrict ourselves to the case of internal symmetries, so that T^A are matrices that are independent of the space–time coordinates. The generators T^A are normalized in such a way that for the lowest dimensional non-trivial representation of the group Γ (we use t^A to denote the generators in this particular representation) we have

$$\text{tr}(t^A t^B) = \frac{1}{2}\delta^{AB} . \tag{113}$$

The generators satisfy the commutation relations

$$[T^A, T^B] = iC_{ABC}T^C . \tag{114}$$

In the following, for each quantity V^A we define

$$\mathbf{V} = \sum_A T^A V^A . \tag{115}$$

If we now make the parameters θ^A depend on the space–time coordinates $\theta^A = \theta^A(x_\mu)$, $\mathcal{L}[\phi, \partial_\mu \phi]$ is in general no longer invariant under the gauge transformations $U[\theta^A(x_\mu)]$, because of the derivative terms. Gauge invariance is recovered if the ordinary derivative is replaced by the covariant derivative:

$$D_\mu = \partial_\mu + ig\mathbf{V}_\mu , \tag{116}$$

where V_μ^A are a set of D gauge fields (in one-to-one correspondence with the group generators) with the transformation law

$$\mathbf{V}'_\mu = U\mathbf{V}_\mu U^{-1} - (1/ig)(\partial_\mu U)U^{-1} \, . \tag{117}$$

For constant θ^A, $\mathbf{V}$ reduces to a tensor of the adjoint (or regular) representation of the group:

$$\mathbf{V}'_\mu = U\mathbf{V}_\mu U^{-1} \simeq \mathbf{V}_\mu + ig[\theta, \mathbf{V}_\mu] \, , \tag{118}$$

which implies that

$$V'^C_\mu = V^C_\mu - gC_{ABC}\theta^A V^B_\mu \, , \tag{119}$$

where repeated indices are summed up.

As a consequence of Eqs. (116) and (117), $D_\mu \phi$ has the same transformation properties as ϕ:

$$(D_\mu \phi)' = U(D_\mu \phi) \, . \tag{120}$$

Thus $\mathcal{L}[\phi, D_\mu \phi]$ is indeed invariant under gauge transformations. In order to construct a gauge-invariant kinetic energy term for the gauge fields V^A, we consider

$$[D_\mu, D_\nu]\phi = ig\{\partial_\mu \mathbf{V}_\nu - \partial_\nu \mathbf{V}_\mu + ig[\mathbf{V}_\mu, \mathbf{V}_\nu]\}\phi \equiv ig\mathbf{F}_{\mu\nu}\phi \, , \tag{121}$$

which is equivalent to

$$F^A_{\mu\nu} = \partial_\mu V^A_\nu - \partial_\nu V^A_\mu - gC_{ABC}V^B_\mu V^C_\nu \, . \tag{122}$$

From Eqs. (112), (120) and (121) it follows that the transformation properties of $F^A_{\mu\nu}$ are those of a tensor of the adjoint representation

$$\mathbf{F}'_{\mu\nu} = U\mathbf{F}_{\mu\nu}U^{-1} \, . \tag{123}$$

The complete Yang–Mills Lagrangian, which is invariant under gauge transformations, can be written in the form

$$\mathcal{L}_{\text{YM}} = -\frac{1}{4}\sum_A F^A_{\mu\nu} F^{A\mu\nu} + \mathcal{L}[\phi, D_\mu \phi] \, . \tag{124}$$

For an Abelian theory, as for example QED, the gauge transformation reduces to $U[\theta(x)] = \exp[ieQ\theta(x)]$, where Q is the charge generator. The associated gauge field (the photon), according to Eq. (117), transforms as

$$V'_\mu = V_\mu - \partial_\mu \theta(x) \, . \tag{125}$$

In this case, the $F_{\mu\nu}$ tensor is linear in the gauge field V_μ so that in the absence of matter fields the theory is free. On the other hand, in the non-Abelian case the $F^A_{\mu\nu}$ tensor contains both linear and quadratic terms in V^A_μ, so that the theory is non-trivial even in the absence of matter fields.

REFERENCES

1. G. Altarelli, *Phys. Rep.* **81**,1, (1982).
2. G. Altarelli, *Ann. Rev. Nucl. Part. Sci.* **39**, 357, (1989).
3. G. Altarelli, "The Development of Perturbative QCD", World Scientific, 1994.
4. K. Hagiwara, *et al.*, Particle Data Group, *Phys.Rev.* D **66**, 010001, (2002).
5. O.V. Tarasov, A.A. Vladimirov and A. Yu. Zharkov, *Phys. Lett.* B **93**, 429, (1980).
6. S.G. Gorishny, A.L. Kataev and S.A. Larin, *Phys. Lett.* B **259**,144, (1991); L. R. Surguladze and M. A. Samuel, *Phys. Rev. Lett.* **66**, 560, (1991).
7. for a review, see, for example, S. Moch, J. A. M. Vermaseren and M. Zhou, hep-ph/0108033.
8. A. D. Martin, R. G. Roberts, W. J. Stirling and R. S. Thorne, *Phys. Lett.* B **531**, 216, (2002); *The Europ. Phys. J.* C **23**, 73, (2002).
9. R. Hamberg, W. L. van Neerven and T. Matsuura, *Nucl. Phys.* B **359**, 343, (1991); W. L. van Neerven and E. B. Zijlstra, *Nucl. Phys.* B **382**, 11, (1992).
10. V.M.Abazov, *et al.*, The D0 collaboration, *Phys. Lett.* B **513**, 292 (2001).
11. For up-to-date results see, The LEP Electroweak Group, http://lepewwg.web.cern.ch/LEPEWWG/.
12. A.L. Kataev, G. Parente and A.V. Sidorov, *Nucl. Phys.* B **573**, 405, (2000).
13. J. Santiago and F.J. Yndurain *Nucl. Phys.* B **563**, 45, (1999).
14. S.I. Alekhin, *Phys.Rev.* D **63**, 094022, (2001).
15. M. Virchaux and A. Milsztajn, *Phys. Lett.* B **274**, 221, (1992).
16. G. Altarelli, Proceedings of the E. Majorana Summer School, Erice, 1995, Plenum Press, ed. by A. Zichichi.

Lattice QCD at Finite Temperature

Frithjof Karsch

Fakultät für Physik, Universität Bielefeld, D-33615 Bielefeld, Germany

Abstract. After a brief introduction into basic aspects of the formulation of lattice regularized QCD at finite temperature and density we discuss our current understanding of the QCD phase diagram at finite temperature, present results on the QCD equation of state, the transition temperature and the heavy quark free energy. Furthermore, we discuss in-medium properties of hadrons and present first results from a calculation of thermal dilepton rates resulting from quark anti-quark annihilation in the QCD plasma phase.

INTRODUCTION

Almost immediately after the ground-breaking demonstration that the numerical analysis of lattice regularized quantum field theories [1] can also provide quantitative results on fundamental non-perturbative properties of QCD [2] it has been realized that this approach will also allow to study the QCD phase transition [3, 4] and the equation of state of the quark-gluon plasma [5]. During the last 20 years we have learned a lot from lattice calculations about the phase structure of QCD at finite temperature. However, it is only now that numerical calculations start to reach a level of accuracy that allows us to seriously consider quantitative studies of QCD with a realistic light quark mass spectrum. An important ingredient in the preparation of such calculations has been the development of new regularization schemes in the fermion sector of the QCD Lagrangian, which reduce discretization errors and also improve the flavour symmetry of the lattice actions. The currently performed investigations of QCD thermodynamics provide first results with such improved actions and prepare the ground for calculations with a realistic light quark mass spectrum.

One of the central goals in analyzing the properties of QCD under extreme conditions is to reach a quantitative description of the behaviour of matter at high temperature and density. This does provide important input for a quantitative description of experimental signatures for the occurrence of a phase transition in heavy ion collisions and should also help to better understand the phase transitions that occurred during the early times of the evolution of the universe. Eventually it also may allow to answer the question whether a quark-gluon plasma can exist in the interior of dense neutron stars or did exist in early stages of supernovae explosions. To answer such questions requires a quantitative understanding of the QCD equation of state, the determination of critical parameters such as the critical temperature and the critical energy density and an analysis of possible thermal modifications of basic hadron properties (masses, decay widths).

In the next section we give a short introduction into the lattice formulation of QCD thermodynamics. In Section 3 we discuss basic lattice results on the structure of the QCD phase diagram at finite temperature. Sections 4 and 5 are devoted to a discussion of the quark mass and flavour dependence of the QCD equation of state and the phase transition temperature. In Section 6 we analyze the temperature dependence of the heavy quark free energy and discuss consequences for the formation of heavy quark bound states. Thermal modifications of the light quark spectrum, on the other hand, are discussed in Section 7. Here we also present results from a calculation of dilepton rates in the high temperature plasma phase. Finally we give our conclusions in Section 8.

THE LATTICE FORMULATION OF QCD THERMODYNAMICS

Starting point for the discussion of the equilibrium thermodynamics of QCD is the Euclidean path integral formulation of the partition function. This is given as an integral over the fundamental quark ($\bar{\psi}$, ψ) and gluon (A_v) fields. In addition to its dependence on the bare couplings of QCD, *i.e.* the gauge coupling g^2 and the quark masses m_f for $f = 1,..,n_f$ different quark flavours, the partition function also depends explicitly on the thermodynamic parameters, volume (V) and temperature (T)[1],

$$Z(V,T) = \int \mathcal{D}A_v \mathcal{D}\bar{\psi}\mathcal{D}\psi \, e^{-S_E(V,T)} \quad . \tag{1}$$

Here A_v and $\bar{\psi}$, ψ obey periodic and anti-periodic boundary conditions in Euclidean time, respectively. The Euclidean action $S_E \equiv S_G + S_F$ contains a purely gluonic contribution (S_G) expressed in terms of the field strength tensor, $F_{\mu\nu} = \partial_\mu A_v - \partial_v A_\mu - ig[A_\mu,A_v]$, and a fermionic part ($S_F$), which couples the gauge and fermions field through the standard minimal substitution,

$$S_E(V,T) \equiv S_G(V,T) + S_F(V,T) \tag{2}$$

$$S_G(V,T) = \int_0^{1/T} dx_4 \int_V d^3\mathbf{x} \, \frac{1}{2} \text{Tr} \, F_{\mu\nu}F_{\mu\nu} \tag{3}$$

$$S_F(V,T) = \int_0^{1/T} dx_4 \int_V d^3\mathbf{x} \, \sum_{f=1}^{n_f} \bar{\psi}_f \left(\gamma_\mu[\partial_\mu - igA_\mu] + m_f \right) \psi_f. \tag{4}$$

The path integral appearing in Eq. 1 is regularized by introducing a four dimensional space-time lattice of size $N_\sigma^3 \times N_\tau$ with a lattice spacing a. Volume and temperature are then related to the number of points in space and time directions, respectively,

$$V = (N_\sigma \, a)^3 \quad , \quad T^{-1} = N_\tau \, a \quad . \tag{5}$$

The pure gauge sector: The most important step towards a lattice formulation of QCD that also could be handled in computer simulations was to provide a gauge invari-

---

[1] We will discuss here only the thermodynamics for vanishing baryon number density or equivalently vanishing baryon chemical potential.

ant, discretized version of the field strength tensor in terms of compact *link variables* $U_{x,\mu}$ which are elements of the $SU(3)$ group colour [1]. The link variables $U_{x,\mu}$ are associated with the link between two neighbouring sites of the lattice and describe the parallel transport of the field A_μ from site x to $x + \hat{\mu}a$,

$$U_{x,\mu} = P\exp\left(ig \int_x^{x+\hat{\mu}a} dx^\mu A_\mu(x)\right) \quad , \tag{6}$$

where $\hat{\mu}$ is a unit vector in the μ-direction of the 4-dimensional lattice and P denotes the path ordering. A product of link variables around an elementary plaquette (Wilson loop of length 4) may be used to define an approximation to the continuum gauge action,

$$\begin{aligned}W^{(1,1)}_{n,\mu\nu} = 1 - \frac{1}{3}\mathrm{Re}\,\square_{n,\mu\nu} &\equiv \mathrm{Re}\,\mathrm{Tr}\,U_{n,\mu}U_{n+\hat{\mu},\nu}U^\dagger_{n+\hat{\nu},\mu}U^\dagger_{n,\nu}\\ &= \frac{g^2 a^4}{2}F^\alpha_{\mu\nu}F^\alpha_{\mu\nu} + \mathcal{O}(a^6) \quad .\end{aligned} \tag{7}$$

A discretized version of the Euclidean gauge action, which reproduces the continuum version up to cut-off errors of order a^2, thus is given by the *Wilson action* [1],

$$\beta S_G = \beta \sum_{\substack{n\\ 1\le\mu<\nu\le 4}} W^{(1,1)}_{n,\mu\nu} \implies \int d^4x\,\frac{1}{2}\mathrm{Tr}F_{\mu\nu}F_{\mu\nu} + \mathcal{O}(a^2) \quad , \tag{8}$$

where we have introduced the lattice gauge coupling $\beta = 6/g^2$. This suffices to reproduce the correct continuum physics in the gluonic sector of QCD in the limit $a \to 0$. In fact, in many cases one is interested in physical quantities that are related to long distance properties of lattice correlation functions, e.g. hadron masses or the asymptotic behaviour of the heavy quark potential (string tension). In these cases cut-off effects in typical lattice calculations are generally found to be on the 10% level or less. However, in particular in thermodynamic calculations the cut-off effects can become large and the use of better discretization schemes becomes mandatory.

One may use more involved discretization schemes, which also eliminate the leading $\mathcal{O}(a^2)$ errors. In general this can be achieved by adding in Eq. 8 larger Wilson loops. The simplest extension of the Wilson one-plaquette action thus is to include an additional contribution from a planar six-link Wilson loop,

$$W^{(1,2)}_{n,\mu\nu} = 1 - \frac{1}{6}\mathrm{Re}\left(\square\!\square + \begin{array}{c}\square\\\square\end{array}\right)_{n,\mu\nu} \quad . \tag{9}$$

A suitable combination of these six-link loops and the four-link plaquette term eliminates the leading $\mathcal{O}(a^2)$ corrections and yields a formulation that reproduces the continuum action up to $\mathcal{O}(a^4)$ corrections,

$$\beta S_G = \beta \sum_{\substack{n\\ 0\le\mu<\nu\le 3}} c_{1,1}W^{(1,1)}_{n,\mu\nu} + c_{1,2}W^{(1,2)}_{n,\mu\nu} \quad , \tag{10}$$

with $c_{1,1} = 5/3$ and $c_{1,2} = -1/6$. This choice of coefficients defines the so-called tree-level improved (1×2)-action. It insures that the leading cut-off effects are removed on the classical level, *i.e.* at $\mathcal{O}(g^0)$. The action may be further improved perturbatively by either eliminating the leading lattice cut-off effects also at $\mathcal{O}(g^2)$, *i.e.* $c_{i,j} \Rightarrow c_{i,j}^{(0)} + g^2 c_{i,j}^{(1)}$, or by introducing non-perturbative modifications of the coefficients $c_{i,j}$ (*tadpole improvement*). Another well-studied gluon action with non-perturbative corrections is the RG-improved action introduced by Y. Iwasaki [6]. The *RG-action* also has the structure of Eq. 10 but with coefficients $c_{1,1}^{RG} = 3.648$ and $c_{1,2}^{RG} = -0.662$. This choice of parameters is motivated by an analysis of scaling properties of the pure gauge action at zero temperature and finite values of the cut-off. In the continuum limit it, however, still leads to $\mathcal{O}(a^2)$ discretization errors.

The fermion sector: Although the discretization of the fermion sector of QCD does at first look much simpler, it turns out that additional problems arise. In the lattice formulation it is by no means easy to reproduce the correct number of fermionic degrees of freedom without violating essential symmetries (chiral symmetry) of the continuum action. The naive discretization of the fermionic part of the action, which is obtained by introducing a simple finite difference scheme to discretize the derivative appearing in the fermion Lagrangian, *i.e.* $\partial_\mu \psi_f(x) = (\psi_{n+\hat{\mu}} - \psi_{n-\hat{\mu}})/2a$, does not reproduce the particle content one started with in the continuum. The massless lattice fermion propagator has poles not only at zero momentum but also at all other corners of the Brillouin zone and thus generates 16 rather than a single fermion species in the continuum limit ($a \to 0$). One therefore faces a severe species doubling problem. The way out has been to either introduce a higher derivative term, $a \partial_\mu^2 \psi_f(x)$, which is proportional to the lattice spacing a and thus vanishes in the continuum limit (Wilson fermions [1]), or to distribute the components of the fermion Dirac spinors over several lattice sites (staggered fermions) [7]. While the Wilson formulation removes all the unwanted doublers in the continuum limit by giving them a mass of $\mathcal{O}(a^{-1})$ on the lattice, the staggered fermion formulation aims at a reduction of the number of fermion doublers. It does not eliminate the problem completely. One still obtains four degenerate fermion species in the continuum limit. On the other hand, the staggered fermion formulation has the advantage that it preserves a continuous subgroup of the original global chiral symmetry, while this symmetry is explicitly broken in the Wilson formulation. The former thus has the advantage that it still provides a true order parameter for critical behaviour at finite temperature even for non-zero values of the lattice spacing. In the massless limit the chiral condensate still is an order parameter.

In the following we will present results from calculations based on the Wilson as well as staggered fermion formulations. Progress has also been made in formulating lattice QCD with chiral fermion actions which do avoid the species doubling and at the same time preserve the chiral symmetry of the QCD Lagrangian [8]. At present, however, very little has been done to study QCD thermodynamics on the lattice with these actions [9]. As an example for fermion actions we will discuss in a bit more detail the staggered fermion formulation [7]. This action has been used extensively in studies of the QCD phase transition. It can be written as

$$S_F^{KS} = \sum_{nm} \bar{\chi}_n Q_{nm}^{KS} \chi_m \quad, \tag{11}$$

where the staggered fermion matrix Q^{KS} is given by

$$\begin{aligned}Q_{nm}^{KS}(m_q) &= \frac{1}{2}\sum_{\mu=1}^{3}(-1)^{x_1+\ldots x_\mu}(\delta_{n+\hat{\mu},m}U_{n,\mu} - \delta_{n,m+\hat{\mu}}U_{m,\mu}^\dagger) \\ &+ \frac{1}{2}(\delta_{n+\hat{4},m}U_{n,4} - \delta_{n,m+\hat{4}}U_{m,4}^\dagger) + \delta_{nm}m_q \quad .\end{aligned} \tag{12}$$

As the fermion action is quadratic in the Grassmann valued quark fields $\bar{\chi}$ and χ we can integrate them out in the partition function and finally arrive at a representation of $Z(V,T)$ on a 4-dimensional lattice of size $N_\sigma^3 \times N_\tau$,

$$Z(N_\sigma, N_\tau, \beta, m_q) = \int \prod_{n,\nu} dU_{n,\nu} (\det Q^{KS}(m_q))^{n_f/4} e^{-\beta S_G} \quad . \tag{13}$$

We have made explicit the fact that the staggered fermion action does lead to four degenerate fermion flavours in the continuum limit, *i.e.* taking the continuum limit with the action given in Eqs. 11 and 12 corresponds to $n_f = 4$ in Eq. 13. As the number of fermion species does appear only as an appropriate power of the fermion determinant, which is true also in the continuum limit, one also may choose $n_f \neq 4$ in Eq. 13. This is the approach used to perform simulations with different number of flavours in the staggered fermion formulation.

Within the framework of staggered fermions one again is free to choose other discretization schemes, e.g. higher order difference schemes, which may lead to a smoother approach towards the continuum limit. Unlike in the pure gauge sector it, however, is difficult to systematically eliminate $\mathcal{O}(a^2)$ errors beyond the tree level. A particular form of action, which improves the rotational symmetry of the fermion propagator and reduces cut-off effects in bulk thermodynamic observables, is the *p4-action* [10]. In addition to the standard one-link term that appears in Eq. 12 it also includes a set of bended three-link terms,

$$\begin{aligned}S_F(m_q) &= c_1^F S_{1-link,fat}(\omega) + c_3^F S_{3-link} + m_q \sum_x \bar{\chi}_x^f \chi_x^f \\ &\equiv \sum_x \bar{\chi}_x^f \sum_\mu \eta_\mu(x) \left(\frac{3}{8} \left[\begin{array}{c} y\text{—}x\text{—}y \end{array} + \omega \sum_{\nu \neq \mu} \begin{array}{c} \text{(bended link diagrams)} \end{array} \right] \right. \\ &\left. + \frac{1}{96} \sum_{\nu \neq \mu} \left[\begin{array}{c} \text{(four bended three-link diagrams)} \end{array} \right] \right) \chi_y^f\end{aligned}$$

116

$$+m_q \sum_x \bar{\chi}_x^f \chi_x^f \quad . \tag{14}$$

The action contains an additional free parameter ω which can be optimized to improve the flavour symmetry of the action.

Continuum limit: The lattice discretized QCD actions discussed above reproduce the continuum action in the limit $a \to 0$. It depends on dimensionless couplings (β, m_q) and boundary conditions (V, T) which fix the thermal environment and are defined through Eg. 5. The actual value of the lattice spacing a is controlled through the bare couplings appearing in the action. In order to reproduce continuum physics at constant temperature without errors emerging from systematic cut-off effects, we will have to take the limit $(a \to 0, N_\tau \to \infty)$ with $T = 1/N_\tau a$ fixed. Eventually we thus have to analyze our observables on different size lattices and extrapolate results to $N_\tau \to \infty$ at fixed temperature. Unless we perform calculations at a well defined temperature, e.g. the critical temperature, we will have to determine the temperature scale from an independent (zero-temperature) calculation of an observable for which we know its physical value (in MeV). This requires a calculation at the same value of the cut-off, i.e. for the same values of bare couplings. Of course, we know such a quantity only for the physical case realized in nature, i.e. QCD with two light up and down quark flavours and a heavier strange quark. Nonetheless, we have good reason to believe that certain observables are quite insensitive to changes in the quark masses, e.g. quenched hadron masses[2] ($\tilde{m}_H$) or the string tension ($\tilde{\sigma}$) are believed to be suitable observables to set a physical scale. Even in the limit of infinite quark masses, i.e. in calculations within the pure $SU(3)$ gauge theory (quenched QCD), they differ from experimentally known values only on the 10% level. We thus may use calculations of these quantities to define a temperature scale,

$$T/\sqrt{\sigma} = 1/\sqrt{\tilde{\sigma}} N_\tau \quad \text{or} \quad T/m_H = 1/\tilde{m}_H N_\tau \quad . \tag{15}$$

THE QCD PHASE DIAGRAM AT FINITE TEMPERATURE

At vanishing baryon number density the properties of the QCD phase transition depend on the number of quark flavours and their masses. While it is a detailed quantitative question at which temperature the transition to the high temperature plasma phase occurs, we do expect that qualitative aspects of the transition, e.g. its order and details of the critical behaviour, are controlled by global symmetries of the QCD Lagrangian. Such symmetries only exist in the limits of either infinite or vanishing quark masses. For any non-zero, finite value of quark masses the global symmetries are explicitly broken. In fact, the explicit symmetry breaking induced by the finite quark masses is very much similar to that induced by an external ferromagnetic field in spin models. We thus expect that a continuous phase transition, which may exist in the zero or infinite

---

[2] A physical observable O is calculated on the lattice as dimensionless quantity, which we denote here by $\tilde{O}$. Quite often, however, we will also adopt the customary lattice notation, which explicitly specifies the cut-off dependence in the continuum limit, e.g. $\tilde{m}_H \equiv m_H a$ or $\tilde{\sigma} \equiv \sigma a^2$.

quark mass limit, will turn into a non-singular crossover behaviour for any finite value of the quark mass. First order transitions, on the other hand, may persist for some time before they end in a continuous transition. However, whether in QCD with the physical quark mass spectrum a true phase transition exists or whether in this case the transition is just a (rapid) crossover, again becomes a quantitative question which we have to answer through direct numerical calculations.

Our current understanding of the qualitative aspects of QCD phase diagram is based on universality arguments for the symmetry breaking patterns in the heavy [11] as well as the light quark mass regime [12, 13]. In the limit of infinitely heavy quarks, the pure $SU(3)$ gauge theory, the large distance behaviour of the heavy quark free energy, $F_{\bar{q}q}$, provides a unique characterization of confinement below T_c and deconfinement above T_c. On a lattice of size $N_\sigma^3 \times N_\tau$ the heavy quark free energy[3] can be calculated from the expectation value of the Polyakov loop correlation function

$$\exp\left(-\frac{F_{\bar{q}q}(r,T)}{T}\right) = \langle \mathrm{Tr} L_{\vec{x}} \mathrm{Tr} L_{\vec{y}}^\dagger \rangle \quad , \quad rT = |\vec{x}-\vec{y}|/N_\tau \qquad (16)$$

where the Polyakov loops $L_{\vec{x}}$ and $L_{\vec{y}}^\dagger$ represent static quark and anti-quark sources located at the spatial points $\vec{x}$ and $\vec{y}$, respectively,

$$L_{\vec{x}} = \prod_{x_4=1}^{N_\tau} U_{n,4} \quad , \quad n \equiv (\vec{x}, x_4) \quad . \qquad (17)$$

For large separations ($r \to \infty$) the correlation function approaches $|\langle L \rangle|^2$, where $\langle L \rangle = N_\sigma^{-3} \langle \sum_{\vec{x}} \mathrm{Tr} L_{\vec{x}} \rangle$ denotes the Polyakov loop expectation value, which therefore characterizes the behaviour of the heavy quark free energy at large distances and is an order parameter for deconfinement in the $SU(3)$ gauge theory,

$$\langle L \rangle \begin{cases} = 0 \Leftrightarrow \text{confined phase}, & T < T_c \\ > 0 \Leftrightarrow \text{deconfined phase}, & T > T_c \end{cases} . \qquad (18)$$

The effective theory for the order parameter is a 3-dimensional spin model with global $Z(3)$ symmetry. Universality arguments then suggest that the phase transition is first order in the infinite quark mass limit [11].

In the limit of vanishing quark masses the classical QCD Lagrangian is invariant under chiral symmetry transformations; for n_f massless quark flavours the symmetry is

$$U_A(1) \times SU_L(n_f) \times SU_R(n_f).$$

The $SU_{L/R}(n_f)$ flavour part of this symmetry is spontaneously broken in the vacuum. This gives rise to $(n_f^2 - 1)$ massless Goldstone particles, the pions. The axial $U_A(1)$

---

[3] In the $T \to 0$ limit this is just the heavy quark potential; at non-zero temperature $F_{\bar{q}q}$ does, however, also include a contribution resulting from the overall change of entropy that arises from the presence of external quark and anti-quark sources.

symmetry, however, only is a symmetry of the classical Lagrangian. It is explicitly broken due to quantum corrections in the QCD partition function and the true symmetry therefore only is a discrete $Z(n_f)$ symmetry. The basic observable which reflects the chiral properties of QCD is the chiral condensate,

$$\langle \bar{\chi}\chi \rangle = \frac{1}{N_\sigma^3 N_\tau} \frac{\partial}{\partial m_q} \ln Z \quad . \tag{19}$$

In the limit of vanishing quark masses the chiral condensate stays non zero as long as chiral symmetry is spontaneously broken. The chiral condensate thus is an obvious order parameter in the chiral limit,

$$\langle \bar{\chi}\chi \rangle \begin{cases} > 0 \Leftrightarrow \text{symmetry broken phase,} & T < T_c \\ = 0 \Leftrightarrow \text{symmetric phase,} & T > T_c \end{cases} . \tag{20}$$

For light quarks the global chiral symmetry is expected to control the critical behaviour of the QCD phase transition. In particular, the order of the transition is expected to depend on the number of light or massless flavours. The basic aspects of the n_f-dependence of the phase diagram have been derived by Pisarski and Wilczek [12] from an effective, 3-dimensional Lagrangian for the order parameter[4],

$$\begin{aligned}\mathscr{L}_{eff} = & -\frac{1}{2}\text{Tr}(\partial_\mu \Phi^\dagger \partial^\mu \Phi) - \frac{1}{2}m^2 \text{Tr}(\Phi^\dagger \Phi) + \frac{\pi^2}{3}g_1 \left(\text{Tr}(\Phi^\dagger \Phi)\right)^2 \\ & +\frac{\pi^2}{3}g_2 \text{Tr}\left((\Phi^\dagger \Phi)^2\right) + c\left(\det \Phi + \det \Phi^\dagger\right) \quad , \end{aligned} \tag{21}$$

with $\Phi \equiv (\Phi_{ij})$, $i,j = 1,...,n_f$. $\mathscr{L}_{eff}$ has the same global symmetry as the QCD Lagrangian. A renormalization group analysis of this Lagrangian suggests that the transition is first order for $n_f \geq 3$ and second order for $n_f = 2$. The latter, however, is expected to hold only if the axial $U_A(1)$ symmetry breaking, related to the $\det \Phi$ terms in Eq. 21, does not become too weak at T_c so that the occurrence of a fluctuation induced first order transition would also become possible.

This basic pattern indeed is supported by lattice calculations. So far no indication for a discontinuous transition has been observed for $n_f = 2$. The transition is found to be first order for $n_f \geq 3$. Moreover, the transition temperature is decreasing with increasing n_f and there are indications that chiral symmetry is already restored in the vacuum above a critical number of flavours [15].

The anticipated phase diagram of 3-flavour QCD at vanishing baryon number density is shown in Fig. 1.

An interesting aspect of the phase diagram is the presence of a second order transition line in the light quark mass regime which forms the boundary of the region of first order phase transitions. On this line the transition is controlled by an effective 3-dimensional theory with global $Z(2)$ symmetry [13]. As this is not a symmetry of the

---

[4] It should be noted that this ansatz assumes that chiral symmetry is broken at low temperatures. Instanton model calculations suggest that the vacuum, in fact, is chirally symmetric already for $n_f \geq 5$ [14].

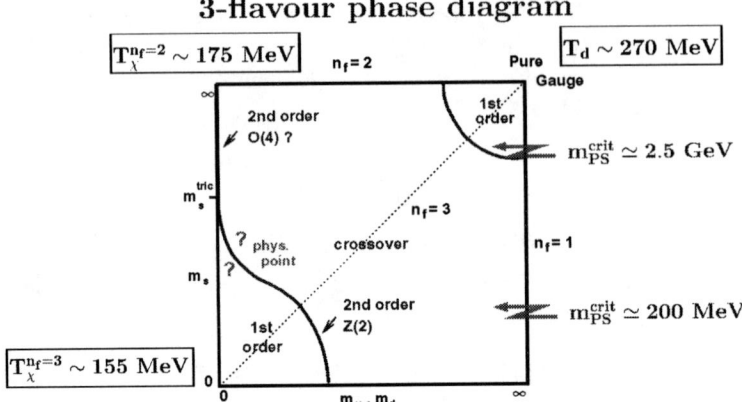

FIGURE 1. The QCD phase diagram of 3-flavour QCD with degenerate (u,d)-quark masses and a strange quark mass m_s.

QCD Lagrangian neither the chiral condensate nor the Polyakov loop will be the order parameter. The critical behaviour (large fluctuations) caused by this transition may, however, equally important for the critical or crossover behaviour of QCD with a realistic quark mass spectrum as the nearby critical point in the chiral limit. It therefore is important to determine in detail the chiral critical line in the QCD phase diagram and the location of the physical point relative to it. A first step along this line has been the verification that the chiral point in three flavour QCD indeed belongs to the universality class of the 3-d Ising model [16]. The determination of the pseudo-scalar meson mass at this critical point indicates that the first order transition regime is rather small and only persists for $m_{PS} \lesssim 200$ MeV.

THE QCD EQUATION OF STATE

The most fundamental quantity in equilibrium thermodynamics is, of course, the partition function itself, or the free energy density,

$$f = -\frac{T}{V} \ln Z(T,V) \; . \tag{22}$$

All basic bulk thermodynamic observables can be derived from the free energy density. In the thermodynamic limit we obtain directly the pressure, $p = -f$ and subsequently also other quantities like the energy (ε) and entropy (s) densities,

$$\frac{\varepsilon - 3p}{T^4} = T \frac{d}{dT}\left(\frac{p}{T^4}\right) \quad , \quad \frac{s}{T^3} = \frac{\varepsilon + p}{T^4} \; . \tag{23}$$

In the limit of infinite temperature asymptotic freedom suggests that these observables approach the ideal gas limit for a gas of free quarks and gluons, $\varepsilon_{SB} = 3p_{SB} = -3f_{SB}$

with $p_{SB}/T^4 = (16 + 10.5n_f)\pi^2/90$. Deviations from this ideal gas value have been studied in high temperature perturbation theory. However, it was well-known that this expansion is no longer calculable perturbatively at $\mathcal{O}(g^6)$ [17]. By now all calculable orders up to $\mathcal{O}(g^5 \ln g)$ have been calculated [18, 19]. Unfortunately it turned out that the information gained from this expansion is rather limited. The expansion has bad convergence properties and suggests that it is of use only at temperatures several orders of magnitude larger than the QCD transition temperature. In analytic approaches one thus has to go beyond perturbation theory which currently is being attempted by either using hard thermal loop resummation techniques [20, 21] or perturbative dimensional reduction combined with numerical simulations of the resulting effective 3-dimensional theory [22].

Lattice calculations allow to calculate the basic thermodynamic quantities non-perturbatively. However, in order to make use of the basic thermodynamic relations, Eqs. 22 and 23, in numerical calculations one has to go through an additional intermediate step. The free energy density itself is not directly accessible in Monte Carlo calculations; e.g. only expectation values can be calculated easily. One thus proceeds by calculating differences of the free energy density at two different temperatures. These are obtained by taking a suitable derivative of $\ln Z$ followed by an integration, e.g.

$$\left. \frac{f}{T^4} \right|_{T_o}^{T} = -\frac{1}{V} \int_{T_o}^{T} dx \frac{\partial x^{-3} \ln Z(x,V)}{\partial x} \quad . \tag{24}$$

This ansatz readily translates to the lattice. Taking derivatives with respect to the gauge coupling, $\beta = 6/g^2$, rather than the temperature as was done in Eq. 24, we obtain expectation values of the Euclidean action which can be integrated again to give the free energy density,

$$\left. \frac{f}{T^4} \right|_{\beta_o}^{\beta} = N_\tau^4 \int_{\beta_o}^{\beta} d\beta' (\langle \tilde{S} \rangle - \langle \tilde{S} \rangle_{T=0}) \quad . \tag{25}$$

Here

$$\langle \tilde{S} \rangle \equiv -\frac{1}{N_\sigma^3 N_\tau} \frac{\partial \ln Z}{\partial \beta} = \langle S_E + \frac{\partial S_E}{\partial \beta} \rangle \quad , \tag{26}$$

is calculated on a lattice of size $N_\sigma^3 \times N_\tau$ and $\langle ... \rangle_{T=0}$ denotes expectation values calculated on zero temperature lattices, which usually are approximated by symmetric lattices with $N_\tau \equiv N_\sigma$. The lower integration limit is chosen at low temperatures so that f/T_o^4 is small and may be ignored[5].

A little bit more involved is the calculation of the energy density as we have to take derivatives with respect to the temperature, $T = 1/N_\tau a$. On lattices with fixed

---

[5] In the gluonic sector the relevant degrees of freedom at low temperature are glueballs. Even the lightest ones calculated on the lattice have large masses, $m_G \simeq 1.5$ GeV. The free energy density thus is exponentially suppressed already close to T_c. In QCD with light quarks the dominant contribution to the free energy density comes from pions. As long as we are dealing with massive quarks also this contribution gets suppressed exponentially. However, in the massless limit clearly some care has to be taken with the normalization of the free energy density.

temporal extent N_τ we rewrite this in terms of a derivative with respect to the lattice spacing a which in turn is controlled through the bare couplings of the QCD Lagrangian, $a \equiv a(\beta, m_q)$. We thus find for the case of degenerate quark flavours of mass m_q

$$\frac{(\varepsilon - 3p)}{T^4} = N_\tau^4 \left[\left(\frac{d\beta(a)}{d\ln a}\right) \left(\langle \tilde{S} \rangle - \langle \tilde{S} \rangle_{T=0}\right) - \left(\frac{dm_q(a)}{d\ln a}\right) \left(\langle \bar{\chi}\chi \rangle - \langle \bar{\chi}\chi \rangle_{T=0}\right) \right] . \quad (27)$$

An evaluation of the energy density e.g. requires the knowledge of the two β-functions appearing in Eq. 27. These may be determined by calculating two physical observables in lattice units for given values of β and m_q; for instance, the string tension, $\tilde{\sigma}$ and a ratio of hadron masses, $m_{PS}/m_V \equiv m_\pi/m_\rho$. These quantities have to be calculated at zero temperature. Using Eq. 15 one then can determine also a temperature scale in physical units.

The numerical calculation of thermodynamic quantities is done on finite lattices with spatial extent N_σ and temporal extent N_τ. In order to perform calculations close to the thermodynamic limit we want to use a large spatial extent of the lattice. In general it has been found that lattices with $N_\sigma \gtrsim 4N_\tau$ provide a good approximation to the infinite volume limit. In addition, we want to get close to the continuum limit in order to eliminate discretization errors and at the same time keep the temperature constant; we thus have to combine the limit of vanishing lattice spacing ($a \to 0$) with the limit of infinite lattice size ($N_\tau \to \infty$). In order to perform this limit in a controlled way we have to analyze in how far lattice calculations of bulk thermodynamic observables are influenced by the introduction of a finite lattice cut-off ($a > 0$). These cut-off effects are largest in the high (infinite) temperature limit which can be analyzed analytically in weak coupling lattice perturbation theory. In Fig. 2 we show the results of such a calculation for lattices with different temporal extent. This clearly shows the importance of improved actions for the reduction of cut-off effects on finite lattices.

Using the tree level improved gauge action in combination with the improved staggered fermion action in numerical simulations at finite temperature it is possible to perform calculations with small systematic cut-off errors already on lattices with small temporal extent, e.g. $N_\tau = 4$ or 6. In actual calculations performed with various actions in the pure gauge sector one finds that for temperatures $T \lesssim 5T_c$ the cut-off dependence of thermodynamic observables shows the pattern predicted by the weak coupling perturbative calculation. At these temperatures the absolute magnitude of the cut-off effects, however, is smaller by about a factor of two. Although such a detailed systematic study of the cut-off dependence does not exist for QCD with light quarks it is reasonable to expect that at least at high temperature the situation will be similar to that in the pure gauge sector.

Thermodynamics of the SU(3) gauge theory

Before entering a discussion of bulk thermodynamics in two and three flavour QCD it is worthwhile to discuss some results on the equation of state in the heavy quark mass

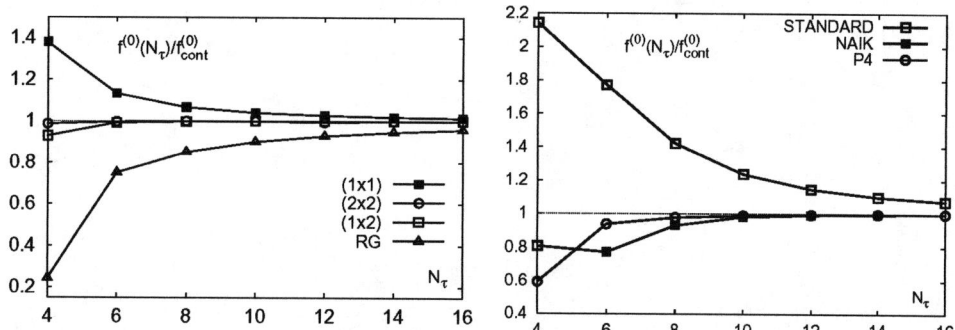

FIGURE 2. Cut-off dependence of the ideal gas pressure or free energy density ($p \equiv -f$) for the $SU(3)$ gauge theory (left) and several staggered fermion actions (right). Cut-off effects for the Wilson fermion action are compatible with those of the standard staggered fermion action defined by Eqs.11 and 12. The p4-action is defined by Eq. 14 and the Naik action has a similar structure with the bended 3-link terms replaced by straight 3-link terms ([23]).

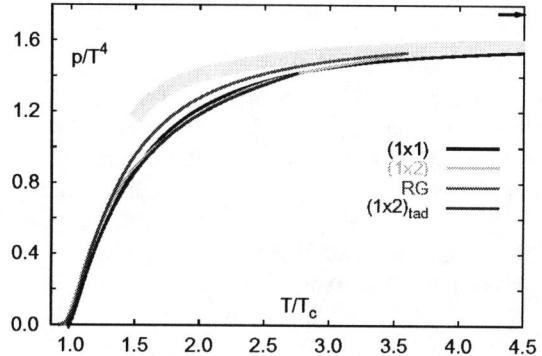

FIGURE 3. Pressure of the SU(3) gauge theory calculated on lattices with different temporal extent and extrapolated to the continuum limit. Shown are results from calculations with the standard Wilson (1×1)-action [24] and several improved actions [26, 27]. The broad band shows the approximately self-consistent HTL calculation of [28].

limit of QCD – the SU(3) gauge theory. In this case the temperature dependence of the pressure and energy density has been studied in great detail, calculations with the standard action [24] and various improved actions [25, 26, 27] have been performed, the cut-off dependence has explicitly been analyzed through calculations on lattices with varying temporal extent N_τ and results have been extrapolated to the continuum limit. In Fig. 3 we show some results for the pressure obtained from such detailed analyzes with different actions [24, 26, 27]. This figure shows the basic features of the temperature dependence of bulk thermodynamic quantities in QCD, which also carry over to the case of QCD with light quarks. The pressure stays small for almost all temperatures below T_c; this is expected, as the only degrees of freedom in the low temperature phase

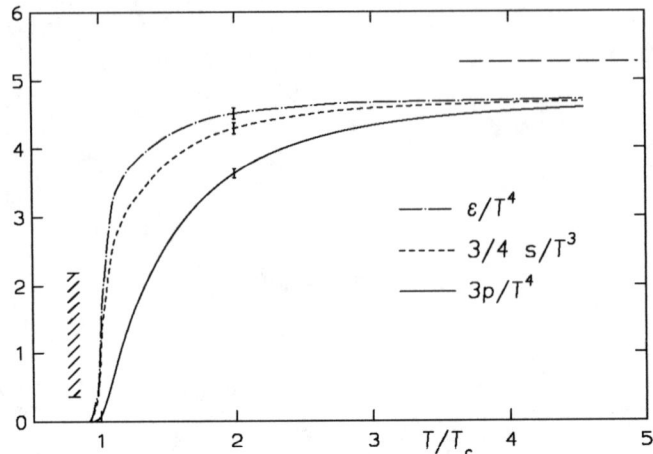

FIGURE 4. Energy density, entropy density and pressure of the SU(3) gauge theory calculated on lattices with different temporal extent and extrapolated to the continuum limit. The dashed band indicates the size of the latent heat gap in energy and entropy density

are glueballs which are rather heavy and thus lead to an exponential suppression of pressure and energy density at low temperature. Above T_c the pressure rises rapidly and reaches about 70% of the asymptotic ideal gas value at $T = 2\, T_c$. For even larger temperatures the approach to this limiting value proceeds rather slowly. In fact, even at $T \simeq 4\, T_c$ deviations from the ideal gas value are larger than 10%. This is still too much to be described in terms of weakly interacting massless gluons as it is done in ordinary high temperature perturbation theory [18, 19]. Even at these high temperatures non-perturbative effects have to be taken into account which may be described in terms of interactions among quasi-particles. The broad band in Fig. 3 shows the result of a self-consistent HTL resummation [28], which leads to good agreement with the lattice calculations for $T \gtrsim 3T_c$. Other approaches [22, 29] reach a similarly good agreement in the high temperature regime.

Compared to the pressure the energy density rises much more rapidly in the vicinity of T_c. In fact, as the transition is first order in the $SU(3)$ gauge theory the energy density is discontinuous at T_c with a latent heat of about $1.5T_c^4$ [30]. In Fig. 4 we show results for the energy density, entropy density and the pressure obtained from calculations with the Wilson action which have been extrapolated to the continuum limit [24]. The delayed rise of the pressure compared to that of the energy density has consequences for the velocity of sound in the QCD plasma. In the vicinity of T_c it is substantially smaller than in the high temperature ideal gas limit.

Flavour dependence of the QCD equation of state

The calculation of the QCD pressure based on Eq. 25 is independent of the quark mass. The analysis of the pressure in QCD with light quarks thus proceeds along the

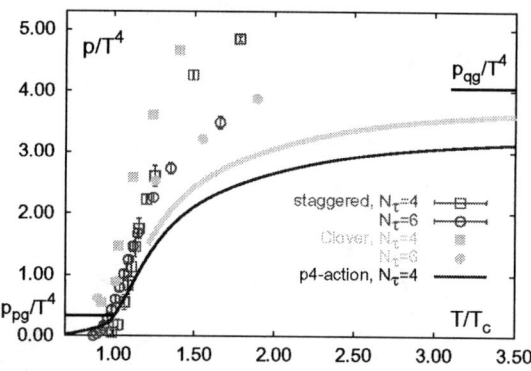

FIGURE 5. The pressure in two flavour QCD calculated with unimproved gauge and staggered fermion actions (open symbols) [32], RG-improved gauge and clover improved Wilson action (full symbols) [33] and the p4-action (improved gauge and improved staggered fermions (full line) [31]. The broad line estimates the results in the continuum limit as described in the text. The horizontal lines to the left and right show the Stefan-Boltzmann values for an ideal pion gas (pg) and a free quark-gluon gas (qg), respectively.

same line as in the pure gauge sector. Unlike in the pure gauge case it, however, will be difficult to perform calculations on lattices with large temporal extent. In fact, at present all calculations of the equation of state are restricted to lattices with $N_\tau = 4$ and 6 [31, 32, 33]. The use of an improved fermion action thus seems to be even more important in this case. Of course, an additional problem arises from insufficient chiral properties of staggered and Wilson fermion actions. This will mainly be of importance in the low temperature phase and in the vicinity of the transition temperature. The continuum extrapolation thus will be more involved in the case of QCD with light quarks than in the pure gauge theory and we will have to perform calculations closer to the continuum limit. Nonetheless, in particular for small number of flavours, we may expect that the flavour symmetry breaking only has a small effect on the overall magnitude of bulk thermodynamic observables. After all, for $n_f = 2$, the pressure of an ideal massless pion gas contributes less than 10% of that of an ideal quark-gluon gas in the high temperature limit. For our discussion of bulk thermodynamic observables the main source for lattice artifacts thus still seems to arise from the short distance cut-off effects, which we have to control. Additional confidence in the numerical results can be gained by comparing simulations performed with different fermion actions.

The importance of an improved lattice action, which leads to small cut-off errors at least in the high temperature ideal gas limit is apparent from Fig. 5, where we compare the results of a calculation of the pressure in 2-flavour QCD performed with unimproved gauge and staggered fermion actions [32], the RG-improved gauge action combined with improved Wilson fermions (clover action) [33] and an improved staggered fermion action (p4-action) [31]. At temperatures above $T \simeq 2 T_c$ these actions qualitatively reproduce the cut-off effects calculated analytically in the infinite temperature limit. In particular, it is evident that also the clover improved Wilson action leads to an

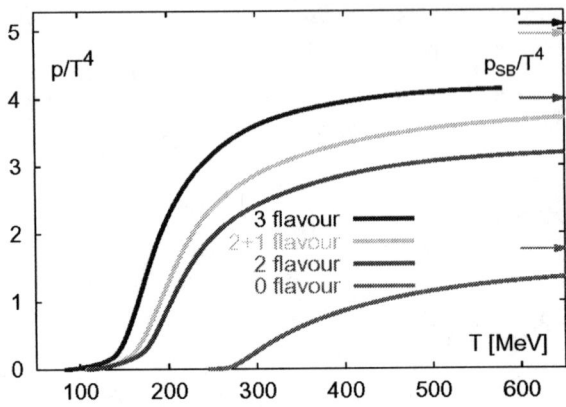

FIGURE 6. The pressure in QCD with different number of degrees of freedom as a function of temperature. The curve labeled (2+1)-flavour corresponds to a calculation with two light and a four times heavier strange quark mass [31].

overshooting of the continuum ideal gas limit. This is expected as the clover term in the Wilson action does eliminate $\mathcal{O}(ag^2)$ cut-off effects but does not improve the high temperature ideal gas limit, which is $\mathcal{O}(g^0)$. The clover improved Wilson action thus leads to the same large $\mathcal{O}(a^2)$ cut-off effects as the unimproved Wilson action. The influence of cut-off effects in bulk thermodynamic observables thus is similar in calculations with light quarks and in the $SU(3)$ gauge theory. This observation may also help to estimate the cut-off effects still present in current calculations with light quarks. In particular, we know from the analysis performed in the pure gauge sector that in the interesting temperature regime of a few times T_c the cut-off dependence seems to be about a factor two smaller than calculated analytically in the infinite temperature limit; we may expect that this carries over to the case of QCD with light quarks. This is the basis for the estimated continuum extrapolation of the $n_f = 2$ results shown as a broad line in Fig. 5.

In Fig. 6 we show results for the pressure obtained in calculations with different number of flavours. This figure clearly shows that the transition region shifts to smaller temperatures as the number of degrees of freedom is increased. Such a conclusion, of course, requires the determination of a temperature scale that is common to all *QCD-like* theories which have a particle content different from that realized in nature. We have determined this temperature scale by assuming that the string tension is flavour and quark mass independent. This assumption is supported by the observation that already in the heavy quark mass limit the string tension calculated in units of quenched hadron masses, e.g. $m_\rho/\sqrt{\sigma} = 1.81$ (4) [34], is in good agreement with values required in QCD phenomenology, $\sqrt{\sigma} \simeq 425$ MeV.

At high temperature the magnitude of p/T^4 clearly reflects the change in the number of light degrees of freedom present in the ideal gas limit. When we rescale the pressure by the corresponding ideal gas values it becomes, however, apparent that the overall

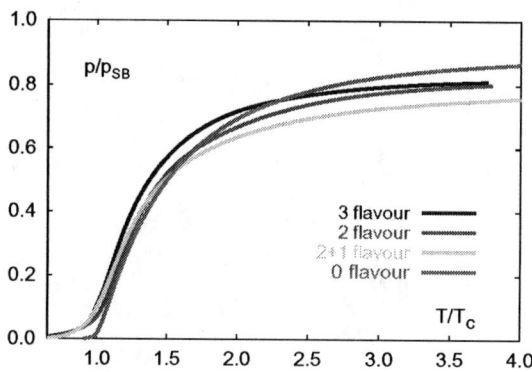

FIGURE 7. The pressure in units of the ideal gas pressure for the $SU(3)$ gauge theory and QCD with various number of flavours. The latter calculations have been performed on lattices with temporal extent $N_\tau = 4$ using the p4-action. Results are not yet extrapolated to the continuum limit.

pattern of the temperature dependence of p/T^4 is quite similar in all cases. This is shown in Fig. 7. In particular, when one takes into account that a proper continuum extrapolation in QCD with light quarks is still missing this agreement achieved with improved staggered fermions is quite remarkable.

We also note that the pressure at low temperature is enhanced in QCD with light quarks compared to the pure gauge case. This is an indication for the contribution of hadronic states, which are significantly lighter than the heavy glueballs of the $SU(3)$ gauge theory. Their contribution is even more clearly visible in the behaviour of the energy density. In Fig. 8 we show the energy density of two and three flavour QCD obtained with improved staggered[6] fermions. These calculations yield as an estimate for the energy density at T_c,

$$\varepsilon_c \simeq (6 \pm 2) T_c^4 \quad . \tag{28}$$

Similar results for the critical energy density can be deduced from calculations with Wilson [33] fermions and unimproved staggered fermions[32]. The estimate for ε_c/T_c^4 is an order of magnitude larger than the critical value on the hadronic side of the transition in the pure gauge theory (see Fig. 4). It is, however, interesting to note that this difference gets to a large extent compensated by the shift in T_c to smaller values. When we convert the result for ε_c in units of [MeV/fm^3] the transitions in the infinite quark mass limit and in QCD with light quarks seem to take place at compatible values of the energy density, $\varepsilon_c \simeq (0.3 - 1.3) \text{GeV/fm}^3$. At present the largest uncertainty on this number arises from uncertainties on the value of T_c (see next section). However, also the magnitude of ε_c/T_c^4 still has to be determined more accurately. Here two competing effects will be relevant. On the one hand we expect ε_c/T_c^4 to increase with decreasing quark masses, *i.e.* closer to the chiral limit. On the other hand, it is likely that finite volume effects are similar to

---

[6] The figure for staggered fermions is based on data from Ref. [31]. Here a contribution to ε/T^4 which is proportional to the bare quark mass and vanishes in the chiral limit is not taken into account.

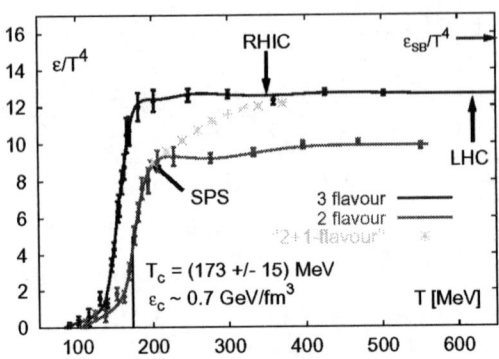

FIGURE 8. The energy density in QCD for two and three flavour QCD. The calculations have been performed on a lattice with temporal extent $N_\tau = 4$ and a pseudo-scalar to vector meson mass ratio of $m_{PS}/m_V = 0.7$. The crosses give a guess for the temperature dependence of QCD with a realistic strange mass value (see text). Also indicated are energy densities expected to be reached at the SPS at CERN, the RHIC at Brookhaven as well as the future LHC at CERN.

those in the pure gauge sector, which suggests that ε_c/T_c^4 will still decrease closer to the thermodynamic limit, *i.e.* for $N_\sigma \to \infty$.

In the case of 2-flavour QCD simulations have been performed also with improved Wilson fermions [33]. Here the quark mass dependence of bulk thermodynamic observables has been analyzed in a wide range of masses, which is controlled by the ratio of the pseudo-scalar to vector meson mass m_{PS}/m_V. The results show no significant quark mass dependence up to $m_{PS}/m_V \simeq 0.9$, which corresponds to pseudo-scalar meson masses of about 1.5 GeV. As this mass is somewhat larger than the Φ-meson mass the corresponding quark mass is compatible with that of the strange quark. The approximate quark mass independence of the equation of state observed in the high temperature phase thus is consistent with our expectation that quark mass effects should become significant only when the quark masses get larger than the temperature. This observation led to the estimate for the temperature dependence of the energy density in QCD with realistic (u, d, s)-quark masses which is indicated with crosses in Fig. 8. While the strange quarks will not contribute to the thermodynamics close to T_c they rapidly will contribute like an additional massless degree of freedom above T_c. The pressure and energy density will then be close to that of massless 3-flavour QCD.

THE CRITICAL TEMPERATURE OF THE QCD TRANSITION

As discussed in Section 3 the transition to the high temperature phase is continuous and non-singular for a large range of quark masses. Nonetheless, for all quark masses the transition proceeds rather rapidly in a small temperature interval. A definite transition point thus can be identified, for instance through the location of maxima of the susceptibilities of the Polyakov loop or the chiral condensate,

$$\chi_L = N_\sigma^3 \left(\langle L^2 \rangle - \langle L \rangle^2 \right) \quad , \quad \chi_m = \frac{\partial}{\partial m_q} \langle \bar{\psi}\psi \rangle \quad . \tag{29}$$

For a given value of the quark mass this defines a pseudo-critical coupling, $\beta_{pc}(m_q)$, on a lattice with temporal extent N_τ. An additional calculation of an experimentally or phenomenologically known observable at zero temperature, e.g. a hadron mass or the string tension, is still needed to determine the transition temperature from Eq. 15. In the pure gauge theory the transition temperature, again has been analyzed in great detail and the influence of cut-off effects has been examined through calculations on different size lattices and with different actions. From this one finds for the critical temperature of the first order phase transition,

$$\underline{\text{SU(3) gauge theory}}: \quad T_c/\sqrt{\sigma} = 0.637 \pm 0.005$$
$$T_c = (271 \pm 2) \text{ MeV} \tag{30}$$

Already the early calculations for the transition temperature in QCD with dynamical quark degrees of freedom [35, 36] indicated that the inclusion of light quarks leads to a significant decrease of the transition temperature. However, these early calculations, which have been performed with standard Wilson [35] and staggered [36] fermion actions, also led to significant discrepancies in the results for T_c as well as the order of the transition. These differences strongly diminished in the newer calculations which are based on improved Wilson fermions (clover action) [36, 37, 38], domain wall fermions as well as improved staggered fermions (p4-action) [39]. A compilation of these newer results is shown in Fig. 9 for various values of the quark masses. In order to compare calculations performed with different actions the results are presented in terms of a *physical observable*, the meson mass ratio m_{PS}/m_V. In Fig. 9a we show T_c/m_V obtained for 2-flavour QCD while Fig. 9b gives a comparison of results obtained with improved staggered fermions [39] for 2 and 3-flavour QCD. Also shown there is a result for the case of (2+1)-flavour QCD, *i.e.* for two light and one heavier quark flavour degree of freedom. Unfortunately the quark masses in this latter case are still too large to be compared directly with the situation realized in nature. We note however, that the results obtained so far suggest that the transition temperature in (2+1)-flavour QCD is close to that of 2-flavour QCD. The 3-flavour theory, on the other hand, leads to consistently smaller values of the critical temperature, $T_c(n_f = 2) - T_c(n_f = 3) \simeq 20$ MeV. Extrapolations of the transition temperatures to the chiral limit gave

$$\underline{2-\text{flavour QCD}}: \quad T_c = \begin{cases} (171 \pm 4) \text{ MeV}, & \text{clover-improved Wilson fermions [37]} \\ (173 \pm 8) \text{ MeV}, & \text{improved staggered fermions [39]} \end{cases}$$

$$\underline{3-\text{flavour QCD}}: \quad T_c = (154 \pm 8) \text{ MeV}, \quad \text{improved staggered fermions[38]}$$

Here m_ρ has been used to set the scale for T_c. Although the agreement between results obtained with Wilson and staggered fermions is striking, one should bear in mind that

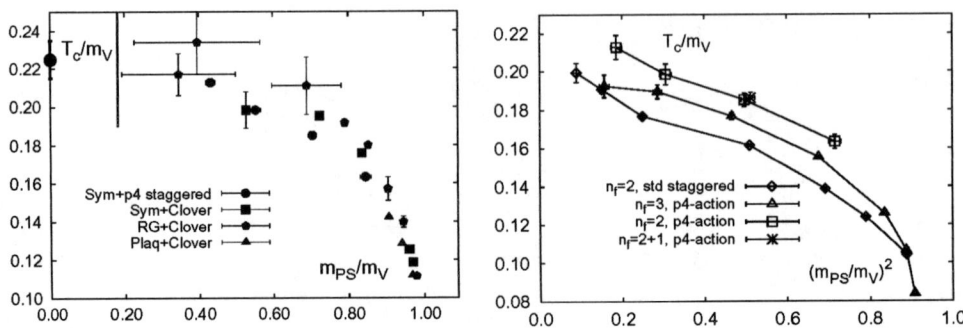

FIGURE 9. Transition temperatures in units of m_V. The left figure shows a collection of results obtained for 2-flavour QCD with various fermion actions (for references see [40]) while in the right figure we compare results obtained in 2 and 3-flavour QCD with the p4-action. All results are from simulations on lattices with temporal extent $N_\tau = 4$. The large dot drawn for $m_{PS}/m_V = 0$ indicates the result of chiral extrapolations based on calculations with improved Wilson [37] as well as improved staggered [39] fermions. The vertical line in the upper figure shows the location of the physical limit, $m_{PS} \equiv m_\pi = 140$ MeV.

all these results have been obtained on lattice with temporal extent $N_\tau = 4$, i.e. at rather large lattice spacing, $a \simeq 0.3$ fm. Moreover, there are uncertainties involved in the ansatz used to extrapolate to the chiral limit. We thus estimate that the systematic error on the value of T_c/m_ρ still is of similar magnitude as the purely statistical error quoted above.

We note from Fig. 9 that T_c/m_V drops with increasing ratio m_{PS}/m_V, i.e. with increasing quark mass. This may not be too surprising as m_V, of course, does not take on the physical ρ-meson mass value as long as m_{PS}/m_V did not reach its physical value (vertical line in Fig. 9a). In fact, m_V will increase with increasing quark mass. The ratio T_c/m_V thus will approach zero for $m_{PS}/m_V = 1$ as T_c will stay finite and take on the value calculated in the pure $SU(3)$ gauge theory. Fig. 9 thus does not yet allow to quantify how T_c depends on the quark mass. A simple percolation picture for the QCD transition would suggest that $T_c(m_q)$ or better $T_c(m_{PS})$ will increase with increasing m_q; with increasing m_q also the hadron masses increase and it becomes more difficult to excite the low lying hadronic states. It thus becomes more difficult to create a sufficiently high particle/energy density in the hadronic phase that can trigger a phase (percolation) transition. Such a picture also follows from chiral model calculations [41, 42].

As argued previously we should express T_c in units of an observable, which itself is not dependent on m_q; the string tension (or also a quenched hadron mass) seems to be suitable for this purpose. In fact, this is what tacitly has been assumed when one converts the critical temperature of the SU(3) gauge theory, $T_c/\sqrt{\sigma} \simeq 0.63$, into physical units as it has been done in Eq. 30.

To quantify the quark mass dependence of the transition temperature one may express T_c in units of $\sqrt{\sigma}$. This ratio is shown in Fig. 10 as a function of $m_{PS}/\sqrt{\sigma}$. As can be seen the transition temperature starts deviating from the quenched values for $m_{PS} \lesssim (6-7)\sqrt{\sigma} \simeq 2.5$ GeV. We also note that the dependence of T_c on $m_{PS}/\sqrt{\sigma}$ is almost linear in the entire mass interval. Such a behaviour might, in fact, be expected for light quarks in

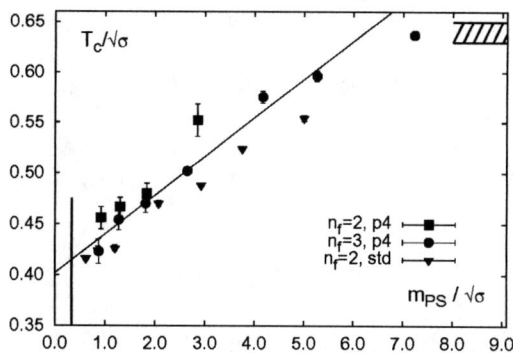

FIGURE 10. The transition temperature in 2 (filled squares) and 3 (circles) flavour QCD versus $m_{PS}/\sqrt{\sigma}$ using an improved staggered fermion action (p4-action). Also shown are results for 2-flavour QCD obtained with the standard staggered fermion action (open squares). The dashed band indicates the uncertainty on $T_c/\sqrt{\sigma}$ in the quenched limit. The straight line is the fit given in Eq. 32.

the vicinity of a 2nd order chiral transition where the dependence of the pseudo-critical temperature on the mass of the Goldstone-particle follows from the scaling relation

$$T_c(m_\pi) - T_c(0) \sim m_\pi^{2/\beta\delta} \ . \tag{31}$$

For 2-flavour QCD the critical indices β and δ are expected to belong to the universality class of 3-d, $O(4)$ symmetric spin models and one thus indeed would expect to find $2/\beta\delta \simeq 1.1$. However, this clearly cannot be the origin for the quasi linear behaviour which is observed for rather large hadron masses and seems to be independent of n_f. Moreover, unlike in chiral models [41, 42] the dependence of T_c on m_{PS} turns out to be rather weak. The line shown in Fig. 10 is a fit to the 3-flavour data, which gave

$$\left(\frac{T_c}{\sqrt{\sigma}}\right)_{m_{PS}/\sqrt{\sigma}} = \left(\frac{T_c}{\sqrt{\sigma}}\right)_0 + 0.04(1)\left(\frac{m_{PS}}{\sqrt{\sigma}}\right) \ . \tag{32}$$

It seems that the transition temperature does not react strongly to changes of the lightest hadron masses. This favours the interpretation that the contributions of heavy resonance masses are equally important for the occurrence of the transition. In fact, this also can explain why in the heavy quark mass limit the transition still sets in at quite low temperatures even when all hadron masses, including the pseudo-scalars, attain masses of the order of 1 GeV or more. Such an interpretation also is consistent with the weak quark mass dependence of the critical energy density which we found from the analysis of the QCD equation of state in the previous section.

For the quark masses currently used in lattice calculations a resonance gas model combined with a percolation criterion thus provides an appropriate description of the thermodynamics close to T_c. It remains to be seen whether the role of the light meson sector becomes more dominant when we get closer to the chiral limit.

THE HEAVY QUARK FREE ENERGY

The heavy quark free energy defined in Eq. 16 plays a central role in our understanding of the QCD phase transition and the thermal properties of the high temperature plasma phase. As discussed in Section 2 the correlation function for static quark anti-quark sources defines the excess free energy due the presence of these sources in a thermal medium [3],

$$\frac{F_{\bar{q}q}(r,T)}{T} = -\ln\left(\langle \text{Tr}L_{\vec{x}}\text{Tr}L_{\vec{y}}^{\dagger}\rangle\right) + c(T) \quad , \quad rT = |\vec{x}-\vec{y}|/N_\tau \quad , \tag{33}$$

with a so far unspecified normalization constant $c(T)$ related to the self-energy of the quark anti-quark sources. The heavy quark free energy probes the confining properties of the thermal medium. From the large distance behaviour of $F_{\bar{q}q}$ one can extract a screening mass. Its T-dependence at has extensively been used in the analysis of heavy quark bound state formation in the plasma phase and the discussion of J/ψ-suppression [43]. Equally important for this discussion, however, is to understand the short distance properties of $F_{\bar{q}q}$ which is an statistical average of contributions arising from $(\bar{q}q)$-pairs in singlet and octet states (colour averaged free energy) while the potential relevant for heavy quark bound state formation refers to the singlet potential only. For this reason it is important to understand in detail the structure of $F_{\bar{q}q}$ and firmly establish its connection to the heavy quark potential.

The long distance behaviour of $F_{\bar{q}q}$ is given by $-\ln|\langle L\rangle|^2$ (see Eqs. 17 and 18). In the pure SU(3) gauge theory this provides an order parameter for the deconfinement phase transition as indicated in Eq. 18. In the presence of dynamical quarks, *i.e.* for QCD with arbitrary but finite quark masses, the heavy quark free energy stays finite at all distances and all temperatures,

$$F_\infty(T) = \lim_{r\to\infty} F_{\bar{q}q}(r,T) \quad . \tag{34}$$

We would like to interpret F_∞ as the change in free energy due to the presence of two well separated quarks, each of which is screened by a cloud of quarks and gluons. In order to do so we have to clarify the normalization condition for $F_{\bar{q}q}(r,T)$ which unambiguously fixes the T-dependent constant in Eq. 33. This can be achieved through an analysis of the short distance behaviour of $F_{\bar{q}q}(r,T)$. At short distances the quark anti-quark pair interacts through the exchange of gluons, which may be calculated perturbatively. More importantly we expect that this interaction is essentially T-independent for separations r which are smaller than the average separation between partons in the thermal medium. We thus would like to fix the normalization constant $c(T)$ through a matching of the short distance behaviour of the heavy quark free energy with that of the zero temperature heavy quark potential.

We may split $F_{\bar{q}q}$ in contributions arising from quark anti-quark pairs in singlet (F_1) and octet (F_8) states [3],

$$e^{-F_{\bar{q}q}(r,T)/T} = \frac{1}{9}e^{-F_1(r,T)/T} + \frac{8}{9}e^{-F_8(r,T)/T} \quad . \tag{35}$$

In zero temperature perturbation theory one finds that the singlet free energy is attractive whereas the octet free energy is repulsive. To leading order (1-gluon exchange) they are

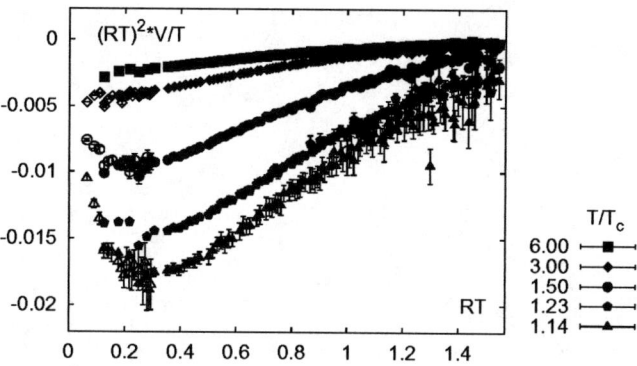

FIGURE 11. The heavy quark free energy at various temperatures in the deconfined phase of the $SU(3)$ gauge theory. The asymptotic value at infinite quark anti-quark separation has been subtracted. Calculations have been performed on lattices of size $32^3 \times 8$ (filled symbols) and $32^3 \times 16$ (open symbols).

given by

$$F_1(r,T) = -\frac{g^2}{3\pi}\frac{1}{r} \quad , \quad F_8(r,T) = \frac{g^2}{24\pi}\frac{1}{r} \quad . \quad (36)$$

The relative strength of F_1 and F_8 is such that the 1-gluon exchange contribution cancels in the colour averaged heavy quark free energy $F_{\bar{q}q}(r,T)$. The leading contribution then arises from the exchange of two gluons. This leads to the well known perturbative result $F_{\bar{q}q}(r,T)/T \sim (F_1(r,T)/T)^2 \sim 1/(rT)^2$, which can also directly be obtained from a weak coupling expansion of the Polyakov loop correlation function. From Eq. 35 it is, however, apparent, that the cancellation of singlet and octet contributions only occurs at large distances and/or temperatures ($rT \gg 1$) where it is justified to expand the exponentials appearing in Eq. 35. At short distances the repulsive octet term is exponentially suppressed and the contribution from the attractive singlet channel will dominate the heavy quark free energy,

$$\begin{aligned}\frac{F_{\bar{q}q}(r,T)}{T} &= \frac{F_1(r,T)}{T} - \ln 9 \\ &= -\frac{g^2(T)}{3\pi}\frac{1}{rT} - \ln 9 \quad \text{for} \quad rT \ll 1 \quad . \end{aligned} \quad (37)$$

We thus expect that at short distances the colour averaged free energy coincides with the colour singlet free energy and shows a Coulomb-like $1/r$ singularity. Such a behaviour indeed has been found in recent numerical calculations which aimed at an analysis of the short distance structure of $F_{\bar{q}q}$. Some results of this analysis which has been performed in the $SU(3)$ gauge theory [44] are shown in Fig. 11. In order to eliminate the anticipated power-like behaviour at large distances we show $(rT)^2 V(r,T)/T$, where $V(r,T) \equiv F_{\bar{q}q}(r,T) - F_\infty(T)$. The rise at short distances, which is particularly pronounced at temperatures close to T_c, indicates that $F_{\bar{q}q}(r,T)$ drops less rapid than $1/(rT)^2$; in fact, it is consistent with a Coulomb-like behaviour. Furthermore, one can establish that the

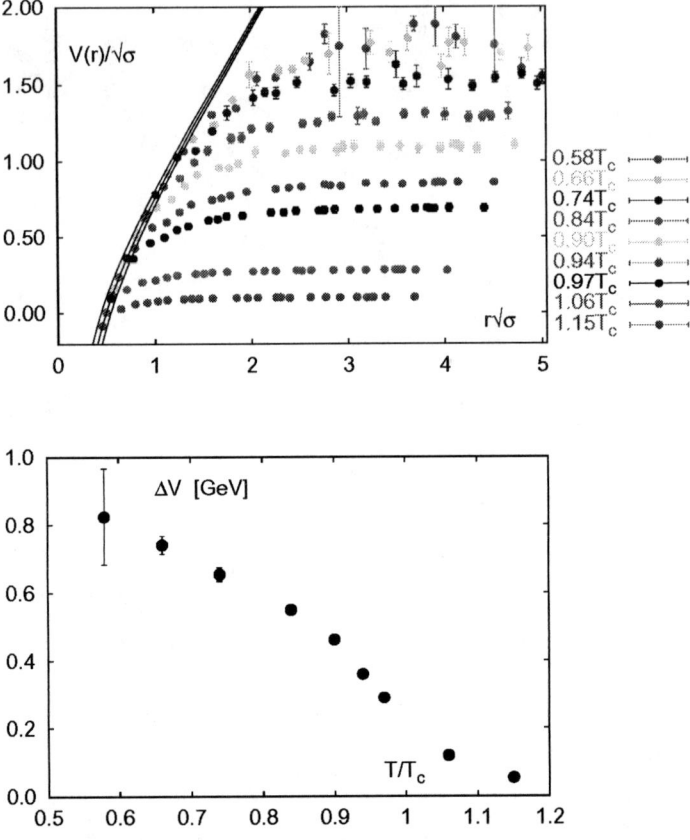

FIGURE 12. The heavy quark free energy at various temperatures in the high temperature phase of three flavour QCD (upper figure). The free energy has been normalized at the shortest distance available ($r = 1/T$) to the zero temperature Cornell potential, $V(r)/\sqrt{\sigma} = -\alpha/r\sqrt{\sigma} + r\sqrt{\sigma}$ with $\alpha = 0.25 \pm 0.05$ (solid band). The lower figure shows the temperature dependence of the change in free energy when separating the quark anti-quark pair from a distance $\bar{r} = 0.5/\sqrt{\sigma}$ to infinity.

singlet and colour averaged free energies differ by the statistical factor $T \ln 9$ at short distances [44].

At large distances the heavy quark free energy is exponentially screened. As can be seen the change from the Coulomb-like behaviour at short distances to the exponential screening at large distances is well localized. For $T_c \lesssim T \lesssim 2T_c$ it occurs already for $rT \simeq 0.2$ or $r \simeq 0.15\,(T_c/T)$ fm and shifts slightly to smaller rT with increasing temperature. A consequence of this efficient screening at short distances is that heavy quark bound states get destroyed close to T_c in the plasma phase ([43, 45, 46]).

As we have established that the short distance part of the colour averaged free energy is dominated by the singlet contribution we may proceed and try to normalize the finite-T heavy quark free energy by matching it to the zero temperature heavy quark potential.

This is shown in Fig. 12 for QCD with two degenerate, light quark flavours [39]. We notice that the asymptotic, large distance value of the colour averaged heavy quark free energy rapidly decreases in the vicinity of T_c. In the lower part of Fig. 12 we show the change in free energy needed to separate a quark anti-quark pair from a distance typical for the J/ψ radius, i.e. $\bar{r} = 0.5/\sqrt{\sigma} \simeq 0.2$ fm, to infinity. At T_c the change in free energy is only about 200 MeV, which is of the same magnitude as typical parton momenta in the thermal medium. It thus is likely that a thermal break-up of $\bar{c}c$-bound states is possible at this temperature and provides the explanation for the experimentally observed suppression of J/ψ bound states [47].

THERMAL MODIFICATIONS OF HADRON PROPERTIES

Spectral functions: A central issue in the discussion of the QCD phase transition and possible signatures for its occurrence in heavy ion collisions is to understand the thermal modifications of hadron properties. In the previous section we have discussed modifications of the heavy quark free energy which indicate drastic changes of the heavy quark potential in the QCD plasma phase. As a consequence heavy quark bound states cannot form above certain critical temperatures. Similarly it is expected that the QCD plasma cannot support the formation of light quark bound states. In the pseudo-scalar sector the disappearance of the light pions clearly is related to the vanishing of the chiral condensate at T_c. For $T > T_c$ the pions are no longer (nearly massless) Goldstone bosons. In the plasma phase one thus may expect to find only massive quasi-particle excitations in the pseudo-scalar quantum number channel. However, also below T_c it is expected that the gradual disappearance of the spontaneous breaking of chiral flavour symmetry as well as the weakening of the explicit breaking of the axial $U_A(1)$ symmetry will lead to thermal modifications of hadron properties, e.g. their masses and widths. While the breaking of the $SU_L(n_f) \times SU_R(n_f)$ flavour symmetry leads, for instance, to the splitting of scalar (f_0) and pseudo-scalar (pion) particle masses, the $U_A(1)$ symmetry breaking is visible in the splitting of the pion and the scalar a_0 meson.

The effect of the breaking/restoration of chiral symmetries does not only lead to a temperature dependence of the splitting of physical particle states. The symmetry transformations can also be performed directly on the hadronic currents[7], $H(\tau,\vec{x}) = \bar{q}(\tau,\vec{x})\Gamma q(\tau,\vec{x})$, and the hadron correlation functions constructed from them,

$$G_H(\tau,\vec{x}) = \langle H(\tau,\vec{x}) H^\dagger(0) \rangle \quad , \tag{38}$$

or equivalently the correlation functions at fixed momentum,

$$G_H(\tau,\vec{p}) = \int d^3x \, \exp(i\,\vec{p}\,\vec{x})\, G_H(\tau,\vec{x}) \quad . \tag{39}$$

One thus may analyze directly the temperature dependence of hadronic correlation functions. Such a comparison is given in Fig. 13 for scalar (a_0) and pseudo-scalar

---

[7] Here we denote with Γ a combination of Dirac and flavour matrices appropriate for the given quantum number channel H.

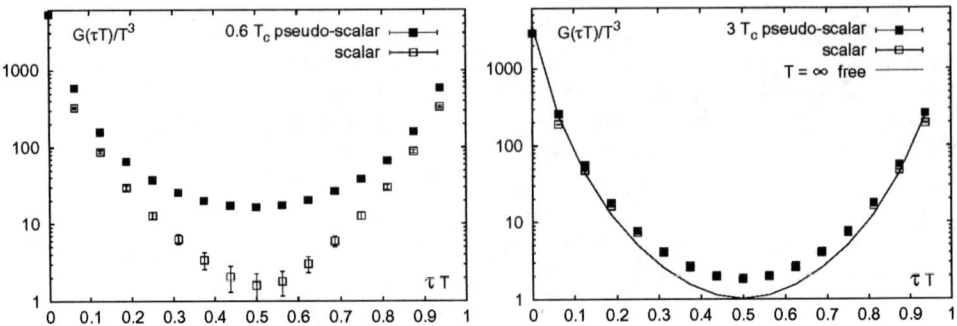

FIGURE 13. Zero momentum projected temporal correlation functions, $G_H(\tau T) \equiv G_H(\tau T, \vec{p}=0)$, for scalar ($a_0$) and pseudo-scalar mesons at $T = 0.6T_c$ and $3T_c$. The correlation functions have been calculated in the quenched approximation of QCD using clover improved Wilson fermions on $32^3 \times 16$ and $64^3 \times 16$ lattices, respectively.

correlation functions calculated in the pure $SU(3)$ gauge theory (quenched QCD) at a temperature below ($T = 0.6T_c$) and above ($T = 3T_c$) the deconfinement phase transition. The effective restoration of the chiral $U_A(1)$ symmetry is clearly visible in this figure; scalar and pseudo-scalar correlators coincide above T_c within statistical accuracy.

Similar results hold for vector and pseudo-vector correlation functions. The crucial question, of course, is to what extent the obvious temperature dependence of hadronic correlation functions also results in modifications of hadron masses. The answer to this can be obtained through a study of the spectral representation of the Euclidean time correlation functions,

$$G_H(\tau, \vec{p}) = \int_0^\infty d\omega\, \sigma_H(\omega, \vec{p}) \frac{\cosh(\omega(\tau - 1/2T))}{\sinh(\omega/2T)} \quad . \tag{40}$$

Here we have introduced the spectral function σ_H. On the other hand $G_H(\tau, \vec{p})$ can be represented as the Fourier transform of the momentum space correlation function,

$$G_H(\tau, \vec{p}) = T \sum_n e^{-i\omega_n \tau} \tilde{G}_H(\omega_n, \vec{p}) \quad , \tag{41}$$

with Matsubara frequencies $\omega_n \equiv 2\pi n T$. From Eqs. 40 and 41 one finds

$$\tilde{G}_H(\omega_n, \vec{p}) = \int_{-\infty}^\infty d\omega \frac{\sigma_H(\omega, \vec{p}, T)}{\omega - i\omega_n} \quad , \tag{42}$$

which shows that the spectral function appearing in the Euclidean correlation function indeed coincides with the spectral function introduced in Minkowski space through the retarded correlation function [48], $\sigma_H(\omega, \vec{p}, T) = \pi^{-1} \text{Im}\, \tilde{G}_H^R(\omega, \vec{p})$, with $\tilde{G}_H^R(\omega, \vec{p}) = \tilde{G}_H(-i\omega, \vec{p})$.

In the infinite temperature limit the normalized scalar and pseudo-scalar correlation functions shown in Fig. 13 approach the correlation function for a freely propagating

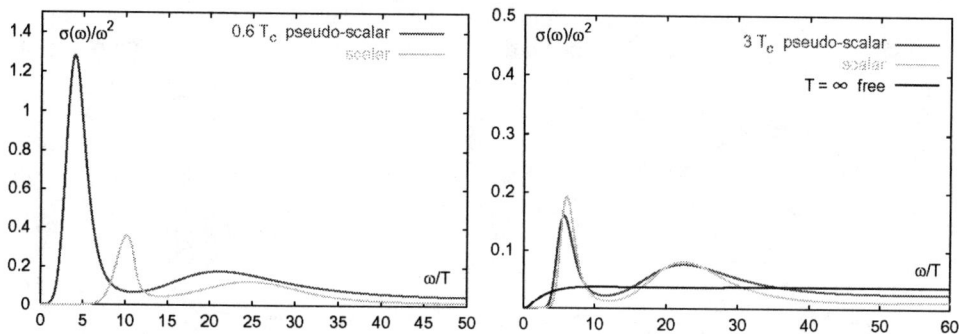

FIGURE 14. Scalar and pseudo-scalar spectral functions at $T = 0.6T_c$ and $3T_c$ reconstructed from the correlation functions shown in Fig. 13.

quark anti-quark pair. Introducing dimensionless variables, $\tilde{\omega} = \omega/T$ and $\tilde{\tau} = \tau T$ the free correlator is given by

$$\frac{G_H^{free}(\tau,\vec{p})}{T^3} \equiv \lim_{T \to \infty} \frac{G_H(\tau,\vec{p})}{T^3}$$

$$= \int_0^\infty d\tilde{\omega}\, \sigma_H^{free}(\tilde{\omega},\vec{p}) \frac{\cosh(\tilde{\omega}(\tilde{\tau}-1/2))}{\sinh(\tilde{\omega}/2)} \quad , \quad (43)$$

where the free spectral function can be constructed from the quark spectral functions projected to the pair-momentum $\vec{p}$ [49, 50]. It takes on a simple form for vanishing quark masses and $\vec{p} = 0$; $\sigma_H^{free}(\tilde{\omega},\vec{0}) = 0.375 a_H \pi^{-2} \tilde{\omega}^2 \tanh(\tilde{\omega}/4)$ with $a_H = 1$ or 2 in the pseudo-scalar and vector channels, respectively. In the free (infinite temperature) limit, the spectral function only receives contributions from the two particle continuum. In general, however, σ_H will also receive contributions from T-dependent bound states, which show up as δ-function singularities or resonance peaks in addition to a T-dependent continuum which, moreover, may only contribute beyond a certain T-dependent threshold.

In order to control these different T-dependent features of the spectral function and, in particular, in order to separate modifications of the pole and resonance structure from thermal modifications of the continuum, it will not be sufficient to analyze only the correlation functions. A direct determination of σ_H, on the other hand, requires the inversion of the integral equation, Eq. 40. This can, in general, quite efficiently be achieved with statistical techniques like the maximum entropy method (MEM) [51, 52]. An additional problem arises in lattice calculations where the correlation functions can only be determined at a discrete set of lattice grid points and, moreover, temperature is controlled through the finite lattice extent in temporal direction. At finite temperature the temporal correlation function are thus only determined for Euclidean times $\tau T = k/N_\tau$, $k = 0, 1, ..., N_\tau - 1$. Nonetheless, recent attempts [53, 54] to reconstruct hadronic spectral functions using MEM have shown that this becomes possible already with a moderate number of grid points in the temporal direction of the lattice. The reconstruction of the spectral functions corresponding to the correlators shown in Fig. 13 is shown in Fig. 14. This clearly reflects the drastic, qualitative change which occurs in the pseudo-scalar channel when one crosses T_c. The large peak which corresponds to the Goldstone below T_c pole[8] drastically reduces in size above T_c. In the high temperature phase the spectral functions reconstructed from the scalar and pseudo-scalar correlation functions coincide. This, of course, simply reflects the almost perfect agreement of the correlation functions themselves (Fig. 13). The peak, still visible in the spectral functions shifts with increasing temperature. This suggests that the pion pole indeed disappears above T_c.

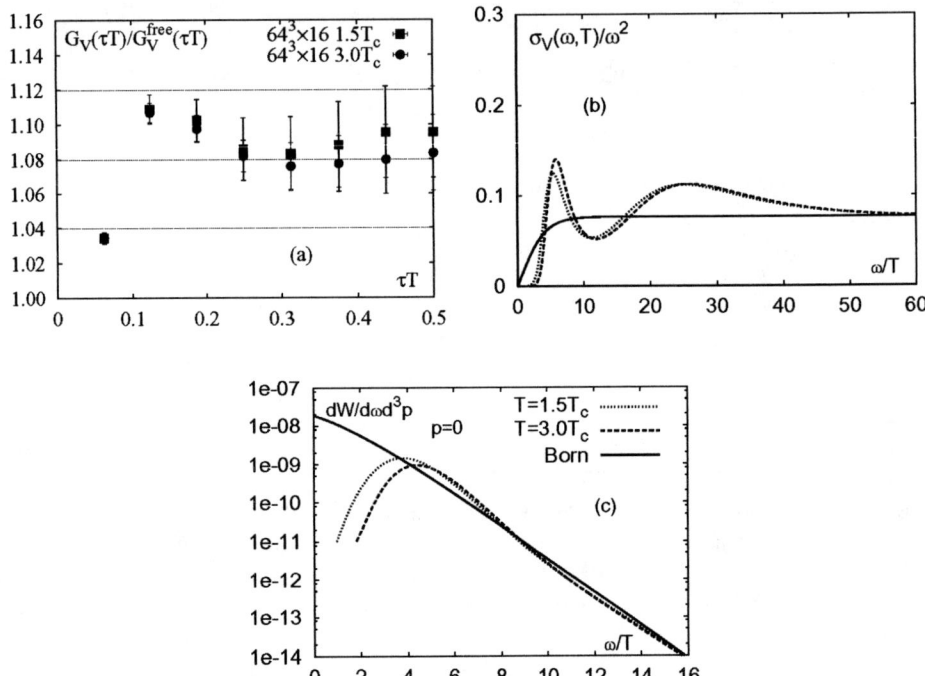

FIGURE 15. The ratio of the vector correlation function $G_V(\tau T)$ and the corresponding free correlation function for massless quark anti-quark pairs, $G_V^{free}(\tau T)$ versus Euclidean time in units of the temperature (a), reconstructed vector spectral functions σ_V in units of ω^2 at zero momentum (b) and the resulting zero momentum differential dilepton rate (c) at $T/T_c = 1.5$ (dots) and 3 (dashes). Also shown with solid lines is the free spectral function in (b) and the resulting Born rate for thermal dilepton production in (c).

Dilepton rates: Of particular interest is the influence of a rising temperature on the spectral properties in the vector meson channel. This is directly related to the thermal dilepton production which can be analyzed in heavy ion experiments. In fact, an enhanced production of low mass dileptons has been observed [55]. For two massless

---

[8] Below T_c calculations have been performed with non-zero quark masses. For this reason, the pseudo-scalar is not massless.

quark flavours the thermal rate resulting from quark anti-quark annihilation in the QCD plasma phase is related to the vector spectral function σ_V through

$$\frac{dW}{d\omega d^3p} = \frac{5\alpha^2}{27\pi^2} \frac{1}{\omega^2(e^{\omega/T}-1)} \sigma_V(\omega, \vec{p}, T) \quad . \tag{44}$$

While in the (pseudo-)scalar channel the correlation functions above T_c still differ substantially from that of the free case the discrepancy is much smaller in the vector channel. Already for $T \simeq 1.5T_c$ the vector correlator $G_V(\tau T)$ differs by less than 15% from the limiting form expected at infinite temperature, *i.e.* the correlation function of freely propagating quark anti-quark pairs. This is shown in Fig. 15a. As a consequence also the spectral functions reconstructed from the correlators (Fig. 15b) and in turn the the dilepton rates (Fig. 15c) stay close to the leading order perturbative results [56].

CONCLUSIONS

We have given a brief introduction into the lattice formulation of QCD thermodynamics and presented a few of the basic results on the equation of state, the critical parameters for the transition to the QCD plasma phase and the fate of hadrons in this new phase of matter. We now have reached a first quantitative understanding of the quark mass dependence of QCD thermodynamics which justifies a first attempt to extrapolate the results to the physical quark spectrum. At present calculations with light quarks still correspond to a world in which the pion would have a mass of about (300-500) MeV. This mass still is too large to become sensitive to details of the physics of chiral symmetry breaking. Nonetheless, lattice calculations performed with different lattice fermion formulations start to produce a consistent picture for the quark mass dependence of the equation of state as well as the influence of the number of light flavours on the phase transition and they yield compatible results for the transition temperature. These calculations suggest that over a wide range of quark masses the transition from the hadronic phase to the plasma phase occurs in a narrow temperature interval. When varying the quark masses the transition parameters (T_c, ε_c) change little compared to the change in the lightest hadron masses. This suggests that at least for intermediate values of the quark masses the transition is not controlled by symmetries of QCD. It rather seems to be density driven and; the quark mass dependence of T_c and ε_c reflects the importance of heavier resonances for the onset of critical behaviour.

The strong modifications of the thermal environment that occur during the transition to the plasma phase influence the properties of hadrons. Thermal hadron correlation functions clearly show traces of chiral symmetry restoration and suggest that basic hadron properties, e.g. their masses and widths, are modified in a thermal heat bath. A new, promising tool to analyze the structure of thermal correlation functions is the maximum entropy method [53, 54]. First applications of this method are encouraging and led to a first calculation of the vector spectral function at high temperature and a determination of thermal dilepton rates [56]. It is to be expected that this approach will lead to further insight into the in-medium properties of hadrons in the near future.

ACKNOWLEDGMENTS

I would like to thank the organizers of the Pan American Studies Institute on New States of Matter in Hadronic Interactions (PASI 2002) for inviting me to contribute to this stimulating school. Furthermore, I would like to thank Andre' Ribeiro Taurines for his help in preparing this contribution. The work presented here has partly been supported by the DFG under grant FOR 339/1-2.

REFERENCES

1. K. G. Wilson, Phys. Rev. D10, 2445 (1974).
2. M. Creutz, Phys. Rev. D21, 2308 (1980).
3. L. D. McLerran and B. Svetitsky, Phys. Lett. B98, 195 (1981) and Phys. Rev. D24, 450 (1981).
4. J. Kuti, J. Polonyi and K. Szlachanyi, Phys. Lett. B98, 199 (1981).
5. J. Engels, F. Karsch, I. Montvay and H. Satz, Phys. Lett. B101, 89 (1981).
6. Y. Iwasaki, Nucl. Phys. B258, 141 (1985) and Univ. of Tsukuba report UTHEP-118 (1983).
7. J. Kogut and L. Susskind, Phys. Rev. D11, 395 (1975).
8. D. B. Kaplan, Phys. Lett. B288, 342 (1992).
9. P. Chen et al., Phys. Rev. D64 (2001) 014503.
10. U. M. Heller, F. Karsch and B. Sturm, Phys. Rev. D60 (1999) 114502.
11. B. Svetitsky and L. G. Yaffe, Nucl. Phys. B210 [FS6], 423 (1982).
12. R. D. Pisarski and F. Wilczek, Phys. Rev. D29, 338 (1984).
13. S. Gavin, A. Gocksch, and R. D. Pisarski, Phys. Rev. D49, 3079 (1994).
14. E. Shuryak and T. Schaefer, Phys. Rev. Lett. 75, 1707 (1995).
15. Y. Iwasaki, K. Kanaya, S. Sakai and T. Yoshie, Nucl. Phys. (Proc. Suppl.) B42, 261 (1995).
16. F. Karsch, E. Laermann and C. Schmidt, Phys. Lett. B520, 41 (2001).
17. A. D. Linde, Phys. Lett. B96, 289 (1980).
18. P. Arnold and C.-X. Zhai, Phys. Rev. D50, 7603 (1994).
19. B. Kastening and C. Zhai, Phys. Rev. D52, 7232 (1995).
20. J.-P. Blaizot, E. Iancu and A. Rebhan, Phys. Rev. Lett. 83, 2906 (1999).
21. J.-P. Blaizot and E. Iancu, Phys. Rep. 359 (2002) 355.
22. K. Kajantie, M. Laine, K. Rummukainen and Y. Schröder, Phys. Rev. Lett. 86, 10 (2001).
23. Naik, S., Nucl. Phys. B316 (1989) 238.
24. G. Boyd, et al., Phys. Rev. Let. 75, 4169 (1995) and Nucl. Phys. B469, 419 (1996).
25. A. Papa, Nucl. Phys. B478, 335 (1996).
26. M. Okamoto, et al. (CP-PACS). Phys. Rev. D60, 094510 (1999).
27. B. Beinlich, et al., Eur. Phys. J. C6, 133 (1999).
28. J.-P. Blaizot, E. Iancu and A. Rebhan, Phys. Lett. B470, 181 (1999).
29. A. Peshier, B. Kämpfer, O. P. Pavlenko and G. Soff, Phys. Rev. D54,2399 (1996);
    P. Lévai and U. Heinz,. Phys. Rev. C57,1879 (1998).
30. B. Beinlich, F. Karsch and A. Peikert, Phys. Lett. B390, 268 (1997).
31. F. Karsch, E. Laermann and A. Peikert, Phys. Lett. B478, 447 (2000).
32. C. Bernard, et al., Phys. Rev. D55, 6861 (1997).
33. A. Ali Khan, et al., Phys. Rev. D64 (2001) 074510.
34. H. Wittig, Int. J. Mod. Phys. A12, 4477 (1997).
35. K. M. Bitar, et al., Phys. Rev. D43, 2396 (1991).
36. C. Bernard, et al., Phys. Rev. D56, 5584 (1997) and references therein.
37. A. Ali Khan, et al.(CP-PACS), Phys. Rev. D63, 034502 (2001).
38. R. G. Edwards and U. H. Heller, Phys. Lett. B462, 132 (1999.
39. F. Karsch, E. Laermann and A. Peikert, Nucl. Phys. B605, 579 (2001).
40. F. Karsch, Nucl. Phys. A698 (2002) 199.
41. H. Meyer-Ortmanns and B.-J. Schaefer, Phys. Rev. D53, 6586 (1996).
42. J. Berges, D. U. Jungnickel and C. Wetterich, Phys. Rev. D59, 034010 (1999).

43. T. Matsui and H. Satz, Phys. Lett. B178, 416 (1986).
44. F. Zantow, O. Kaczmarek, F. Karsch and P. Petreczky, Nucl. Phys. (Proc.Suppl) B106 (2002) 519 and hep-lat/0110106.
45. F. Karsch, T. Mehr and H. Satz, Z. Phys. C 37 (1988) 617.
46. S. Digal, P. Petreczky and H. Satz, Phys. Lett. 514 (2001) 57 and Phys. Rev. D 64 (2001) 094015.
47. M. C. Abreu, et al. (NA50 Collaboration), Phys. Lett. B450 (1999) 4R56.
48. M. Le Bellac, Thermal Field Theory, Cambridge University Press 1996.
49. W. Florkowski and B. L. Friman, Z. Phys. A 347, 271 (1994).
50. F. Karsch, M. G. Mustafa and M. H. Thoma, Nucl. Phys. B497, 249 (2001).
51. R. K. Bryan, Eur. Biophys. J. 18, 165 (1990)..
52. M. Asakawa, T. Hatsuda and Y. Nakahara, Prog. Part. Nucl. Phys. 46, 459 (2001).
53. Y. Nakahara, M. Asakawa and T. Hatsuda, Phys. Rev. D60, 091503 (1999).
54. I. Wetzorke and F. Karsch, Proceedings of the International Workshop on Strong and Electroweak Matter 2000 (SEWM 2000) (Edt. C.P. Korthals-Altes, World Scientific 2001), p.193, hep-lat/0008008.
55. G. Agakichiev, et al. (CERES Collaboration), Phys. Lett. B422 (1998) 405.
56. F. Karsch, E. Laermann, P. Petreczky, S. Stickan and I. Wetzorke, Phys. Lett. B530 (2002) 147.

Quark–Gluon Plasma, and Strangeness

Johann Rafelski* and Jean Letessier[†]

*Department of Physics, University of Arizona, Tucson, AZ, 85721
[†]Laboratoire de Physique Théorique et Hautes Energies
Université Paris 7, 2 place Jussieu, F–75251 Cedex 05

Abstract. In order to recognize the new form of matter created at RHIC and SPS as the deconfined quark–gluon plasma state (QGP), we need to understand the expected properties of this phase near to the conditions of its formation and disintegration. Thus, we first develop a model of QGP considering the constrains arising from QCD properties and lattice results, and explore its properties. In the second part, we describe the kinetic theory of strangeness production in the QGP phase. We show that gluon fusion dominate and evaluate the degree of equilibration expected at RHIC.

1. INTRODUCTION

If a new state of matter has been formed in relativistic nuclear collisions at RHIC(BNL) and SPS(CERN), the question is, if this is the long aspired discovery of the hot deconfined quark–gluon plasma state of matter. To answer this, we must understand well the properties of quark matter, a topic explored for some time. Quark matter presence in collapsed stars has been postulated soon after the quark model of hadrons was developed [1]. These seminal ideas were deepened by the development of the quantum many body theory of quark matter [2, 3], which lead on to the formal recognition within the framework of asymptotically free quantum-chromodynamics (QCD) that a very high temperature perturbative quark matter state must exist [4].

Other reasoning based on a study of the 'boiling state' of dense hadron gas within the scheme of Hagedorn's statistical bootstrap of hadronic matter has independently lead from a different direction to the consideration of the transition to a hadron substructure phase [5]. Considering the present day lattice-QCD numerical simulations [6], we understand today well the properties of the baryon number free quark–gluon plasma (QGP), as we call the hot quark matter phase.

As the understanding of the possible QGP formation in high energy nuclear collision matured, an unexpected challenge emerged 20 years ago: how can the locally deconfined state, which exists a mere 10^{-23}s, be distinguished from the gas of confined hadrons? This is also a question of principle, considering that both quark and hadron pictures of the reaction could be equivalent, for it has been argued that a quark–gluon based description is merely a change of Hilbert space expansion basis, considering the rules of quantum mechanics with pure states. In the Galilean tradition, such a difficult question about existence and observability of a new phase of elementary matter, the QGP, must be settled by an experiment. This requires that a probe of QGP operational on the collision time scale of $2 \cdot 10^{-23}$ s is developed, which is sensitive to the local color charge deconfinement, and that it depends on the gluon degree of freedom, which is the characteristic new ingredient of the quark matter phase.

One of us (JR) proposed strangeness as the signature of QGP noticing more than 20 years ago that, when color bonds are broken, the chemically equilibrated deconfined state has an unusually high number of strange quark pairs [7, 8, 9, 10]. A study of the dynamics of strangeness (chemical) equilibration process has shown that only the gluon component in the QGP is capable to produce strangeness rapidly [11, 12], allowing the formation of (nearly) chemically equilibrated dense phase of deconfined, hot, strangeness-rich quark matter in relativistic nuclear collisions. Therefore, strangeness enhancement is related directly to the presence of gluons in QGP.

The high density of strangeness available in the deconfined fireball favors during hadronization the formation of multi-strange hadrons [13, 14], which are produced quite rarely if only individual hadrons collide [15, 16]. A large enhancement of multi strange (anti)baryons has been therefore proposed and studied in depth over the past 20 years as a characteristic signature of the QGP. A systematic enhancement has been reported at SPS, rising with strangeness content [17, 18, 19]. At RHIC a large yield of multistrange hyperons is also observed but further experimental results are required for a complete picture to emerge. The enhancement of strange antibaryons at SPS is reported with a base being the expectation derived from the scaled nucleon-nucleon and nucleon-nucleus reactions. These experimental results are consistent with particle production occurring from a strangeness dense source in which the strange quarks are already made, are freely moving, being readily available to form final state hadrons.

Strangeness enhancement has turned out to be a very practical signature of quark–gluon plasma physics. The absolute yield provides information about initial conditions of the interaction. The large abundance of strange hadrons which follows is allowing a precise study of the chemical freeze-out conditions in the dense matter fireball. Strange antibaryons are produced in a fashion expected for the deconfined source. This article summarizes foundations of this research area, i.e., the properties of quark–gluon plasma and production of strangeness, for many further details the reader may consult our recent book [20]. Also, the use of strangeness as a diagnostic tool of hadronic fireball hadronization is discussed in our other contribution in this volume [21].

In the following section 2, we will describe the properties of QGP at time of its formation and its sudden breakup [22]. The importance of this is that the initial conditions are important in determining the overall strangeness yield, while the final state conditions are responsible for the formation of otherwise rarely produced strange antibaryons. The evolution between the initial and final conditions is assumed to proceed by entropy conserving (hydrodynamic) flow. However, the high pressures available in the initial state lead to the development of rather rapid transverse outflow of matter. This ultimately leads to rather sudden break up of the hadronic fireball due to the development of an undercooled QGP phase [23].

Having established the properties and evolution of the plasma phase in section 2, we study, in section 3, the mechanism of production of strangeness in QGP. In historical perspective, it was an unexpected surprise: strangeness was not believed to be produced sufficiently rapidly in the short lived deconfined phase [24], to allow for near chemical equilibration of this semi-heavy flavor. It was shown that only the multitude of independent collisions between two gluons can lead to this making strangeness sensitive to the properties of the QGP phase [11].

2. PROPERTIES OF QUARK–GLUON PLASMA

2.1. Formulation of the QGP-liquid model

Using the latest lattice-QCD results [25], we have been able to better model the properties of the QGP [22], compared to our earlier considerations [26]. The key ingredients of our approach are:

1. We relate the QCD scale to the temperature $T = 1/\beta$, and we use for the scale the Matsubara frequency [27] (here μ without a subscript is *not* a chemical potential, it is the QCD scale):

$$\mu = 2\pi\beta^{-1}\sqrt{1 + \frac{1}{\pi^2}\ln^2\lambda_q} = 2\sqrt{(\pi T)^2 + \mu_q^2}. \tag{1}$$

This extension to finite chemical potential μ_q, or equivalently quark fugacity $\lambda_q = \exp(\mu_q/T)$, is motivated by the form of plasma frequency entering the computation of the vacuum polarization function [28].

2. In order to describe the effect of QCD interactions in the thermal quark–gluon plasma phase, it is necessary to know precisely the strength of scale dependent $\alpha_s(\mu)$ [29]. We integrate numerically the renormalization group equations,

$$\mu\frac{\partial \alpha_s}{\partial \mu} = -b_0\alpha_s^2 - b_1\alpha_s^3 + \cdots. \tag{2}$$

In the two loop approximation renormalization scheme independent result is:

$$b_0 = \frac{11 - 2n_f/3}{2\pi}, \quad b_1 = \frac{51 - 19n_f/3}{4\pi^2}.$$

Solutions of Eq. (2) differ from higher order (three, and four 'loop') renormalization scheme dependent results generally by less than the error introduced by the experimental uncertainty in the measured value of $\alpha_s(\mu = M_Z) = 0.118 + 0.001 - 0.0016$. To obtain the solid line for $\alpha_s^{(4)}$ in Fig. 1 (left) the four-loop β-function (upper index '4') was used, derived in the $\overline{MS}$ modified minimum subtraction scheme. The series of dotted lines demonstrate considerable sensitivity to the initial value $\alpha_s(M_Z)$. In Fig. 1 (right), the solid line corresponds to an exactly computed α_s with physical quark thresholds, evaluated at the scale defined by Eq. 1 for $\mu_q = 0$, and expressed in terms of T/T_c for $T_c = 0.16$ GeV. A precise approximant is obtained fitting $\alpha_s(T)$ with a logarithmic form, valid for $T < 5T_c$,

$$\alpha_s(T) \simeq \frac{\alpha_s(T_c)}{1 + C\ln(T/T_c)}, \quad \alpha_s(T_c) = 0.50^{-0.05}_{+0.03}, \quad C = 0.760 \pm 0.002. \tag{3}$$

The dotted line, in Fig. 1, corresponds to the high temperature approximation applicable for $\mu > 2m_b \simeq 9$ GeV (five active fermions), but is extended to near the critical temperature. This yields $\alpha_s(T)$ which is a factor two too small near T_c, and consequently to a significant underestimation of the effect of QCD interactions.

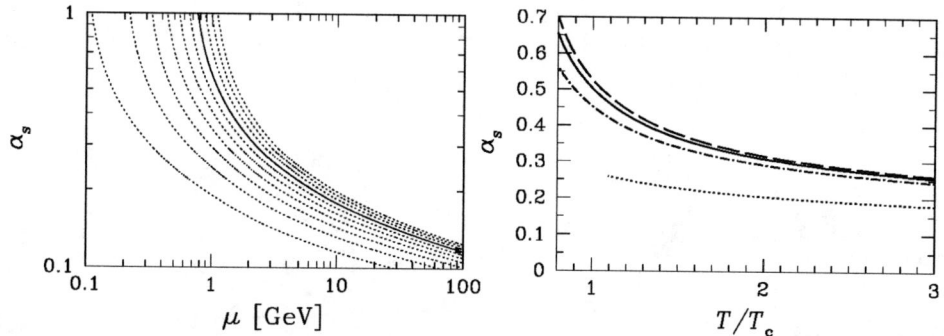

FIGURE 1. Left: $\alpha_s^{(4)}(\mu)$ as function of energy scale μ for a variety of initial conditions. Solid line: $\alpha_s(M_Z) = 0.1182$ (see the experimental point, which includes the error bar at $\mu = M_Z$); dotted lines: sensitivity to variation of the initial condition. Right: $\alpha_s(2\pi T)$ for $T_c = 0.160\,\text{GeV}$. Dashed line: $\alpha_s(M_Z) = 0.119$; solid line = 0.1181; dot-dashed line = 0.1156. Dotted line: 2-loop solution with 5 active fermions.

3. We introduce, in the domain of freely mobile quarks and gluons, a finite vacuum energy density:

$$\mathcal{B} = 0.19 \frac{\text{GeV}}{\text{fm}^3}. \quad (4)$$

This also implies, by virtue of relativistic invariance, that there must be a (negative) associated pressure acting on the surface of this volume, aiming to reduce the size of the deconfined region. These two properties of the vacuum follow consistently from the vacuum partition function:

$$\ln \mathcal{Z}_{\text{vac}} \equiv -\mathcal{B} V \beta. \quad (5)$$

$\mathcal{B}$ must be seen at present as a parameter with value given in Eq. (4) chosen to be well reproducing the lattice results.

4. The partition function of the quark–gluon liquid comprises interacting gluons, n_q flavors of light quarks [30], and the vacuum $\mathcal{B}$-term. We incorporate further the strange quarks by assuming that their mass in effect reduces their effective number $n_s < 1$,

$$\frac{T}{V} \ln \mathcal{Z}_{\text{QGP}} \equiv P_{\text{QGP}} = -\mathcal{B} + \frac{8}{45\pi^2} c_1 (\pi T)^4$$
$$+ \frac{n_q}{15\pi^2} \left[\frac{7}{4} c_2 (\pi T)^4 + \frac{15}{2} c_3 \left(\mu_q^2 (\pi T)^2 + \frac{1}{2} \mu_q^4 \right) \right]$$
$$+ \frac{n_s}{15\pi^2} \left[\frac{7}{4} c_2 (\pi T)^4 + \frac{15}{2} c_3 \left(\mu_s^2 (\pi T)^2 + \frac{1}{2} \mu_s^4 \right) \right], \quad (6)$$

where:

$$c_1 = 1 - \frac{15\alpha_s(\mu)}{4\pi} + \cdots, \quad c_2 = 1 - \frac{50\alpha_s(\mu)}{21\pi} + \cdots, \quad c_3 = 1 - \frac{2\alpha_s(\mu)}{\pi} + \cdots. \quad (7)$$

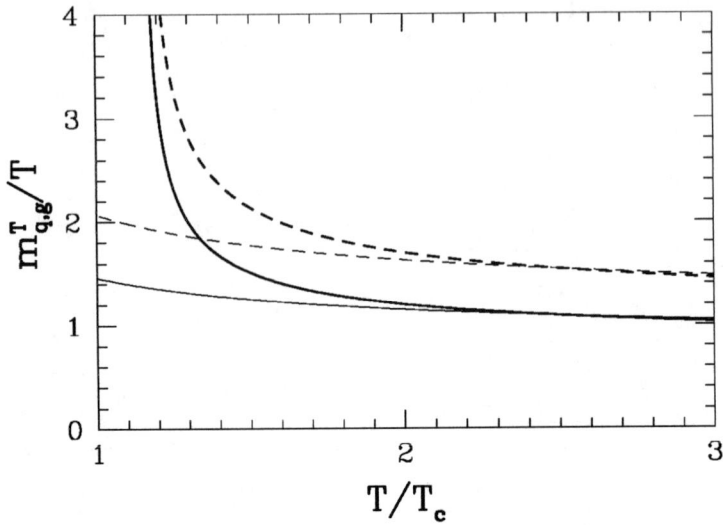

FIGURE 2. Thermal masses fitted to reproduce Lattice-QCD results [27], thick solid line for quarks, and thick dashed line for gluons. Thin lines, perturbative QCD masses for $\alpha_s(\mu = 2\pi T)$.

The perturbative QCD corrections in higher order are available. However, the use of the lowest order expression, with the scale defined by Eq. (1), allows to describe the lattice results very well, and thus this choice of the scale is adapted to the lowest order expansion shown in Eq. (7).

We have described a model of QGP-liquid where, in comparison to a free particle gas, we have introduced vacuum pressure $\mathscr{B}$ and perturbative first order interactions evaluated with thermal $\alpha_s(T)$. It has been shown that it is also possible to reproduce the lattice results using only the fine tuned thermal masses (see table I in [27]). In such an approach without $\mathscr{B}$ one obviously side-steps the appearance of a phase transition completely.

In Fig. 2, we show the light quark (solid thick line) and gluon (dashed thick line) thermal masses which were fitted to the lattice QGP data. The perturbative thermal masses, definitions slightly modified compared to [31, 32], are for quarks and gluons,

$$(m_q^T)^2 = \frac{4\pi}{3}\alpha_s T^2, \qquad (m_g^T)^2 = 2\pi\alpha_s T^2\left(1 + \frac{n_f}{6}\right), \qquad (8)$$

and are also shown in Fig. 2, thin lines (dashed for gluons) obtained using $\alpha_s(\mu = 2\pi T)$, see Fig. 1 (right).

We conclude that the thermal masses required to describe the reduction of the number of degrees of freedom for $T > 2T_c$ are corresponding outside of the range of the vacuum structure influence ($\mathscr{B}$), to the perturbative QCD result. Importantly, this means that thermal masses express, in a different way, the effect of perturbative QCD, and thus for $T > 2T_c$, we have the option to use Eq. 6, or the more complex thermal mass approach, which does not allow for the possibility of a phase transition. While there is no evidence

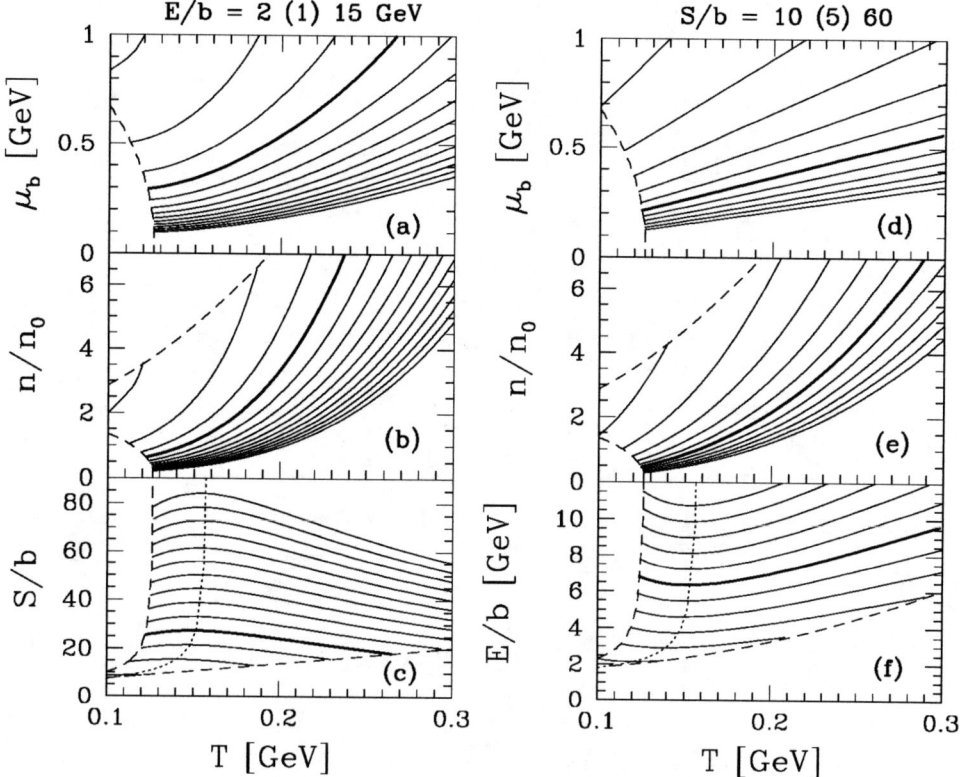

FIGURE 3. Left: lines corresponding to fixed energy per baryon $E/b = 2$ to 15 GeV in steps of 1 GeV, with $E/b = 5$ highlighted. Right: E/b lines corresponding to fixed entropy per baryon $S/b = 10$ to 60 in steps of 5, with $S/b = 35$ highlighted. (a) and (d) (top section), baryo-chemical potential μ_b, (b) and (e) (middle section), baryon density n/n_0 in units of equilibrium nuclear density, and (c) bottom portion: on left S/b, the entropy per baryon (highest E/b at the top), and (f) on right E/b, (highest S/b at the bottom).

for the phase transition in current lattice simulations, this feature is highly sensitive to the presence of quark masses, which are relatively high in the lattice simulations available today [25].

2.2. Properties of QGP-liquid

We consider the properties of QGP-liquid, defined by Eq. 6, in the domains applicable to SPS and RHIC. In Fig. 3, we show the SPS range. In the left-hand panels, we see the physical properties at fixed energy per baryon, in the range $2 \text{ GeV} \leq E/b \leq 15 \text{ GeV}$, as functions of temperature, while in the right-hand panels we study the behavior at fixed value of the (dimensionless) entropy per baryon $10 \leq S/b \leq 60$. In panels (a) and (b) as we step from line to line from left to right, the energy per baryon is incremented by 1

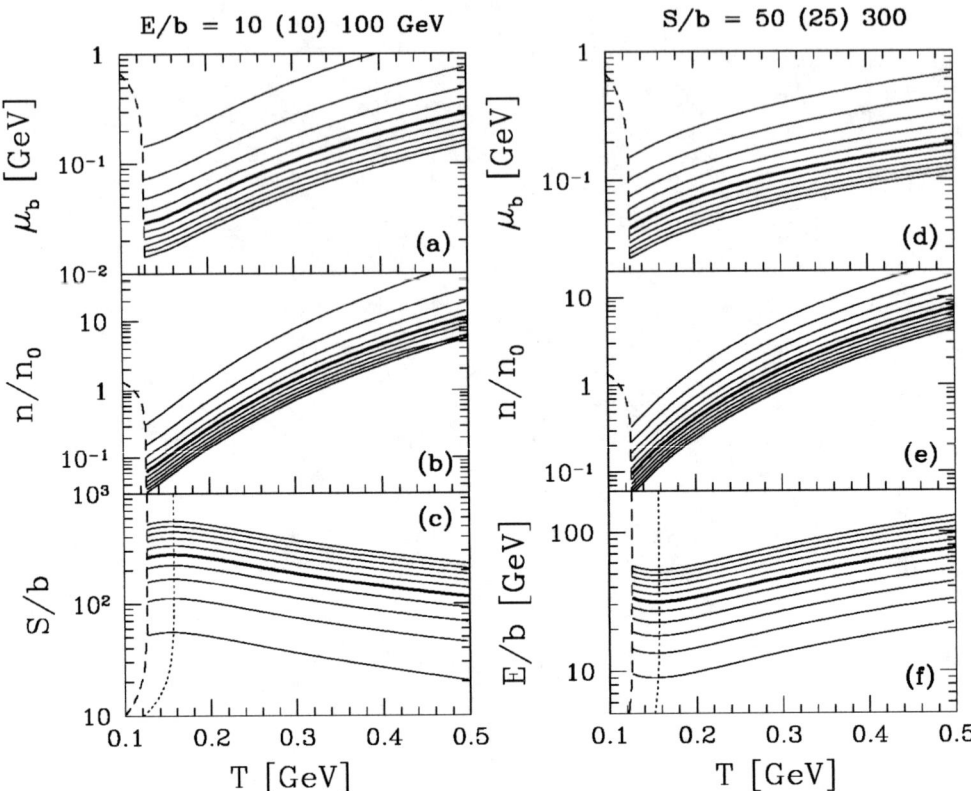

FIGURE 4. The analog of Fig. 3 for the RHIC energy domain. Left, lines at fixed energy per baryon $E/b = 10$ to $100\,\text{GeV}$ in steps of $10\,\text{GeV}$, with $E/b = 50\,\text{GeV}$ highlighted. Right, lines at fixed entropy per baryon $S/b = 50$ to 300 in steps of 25, with $S/b = 175$ hi lighted.

GeV; in panels (d) and (e) the entropy per baryon is incremented by 5 units; in panel (c) we step from bottom to top incrementing by 1 GeV; and in panel (f) from bottom to top by 5 entropy units. We highlight the result for $E/b = 5\,\text{GeV}$ by using a thick line on the left in Fig. 3, and that for $S/b = 35$ by using a thick solid line on the right. The dotted nearly vertical line in panel (f) show where the pressure balance between the particles (quarks, gluons) and the external vacuum $\mathcal{B}$ occurs.

The same set of results for RHIC range of physical properties are shown in Fig. 4, where we use logarithmic scales. On the left-hand side, the energy range is now 10 GeV $\leq E/b \leq 100$ GeV, and lines are in steps of 10 GeV. The thick lines are for $E/b = 50$ GeV. On the right-hand side, the specific entropy range is $50 \leq S/b \leq 300$, in step of 25 units. The thick lines are for $S/b = 175$. The lines become denser toward higher energy, or entropy. The dotted lines in panels (c) and (f) indicate where the particle pressure is balanced by the vacuum pressure.

Since little entropy is produced during the (nearly adiabatic) evolution of the QGP fireball [33, 34, 35], it is interesting to explore the relation of entropy content with

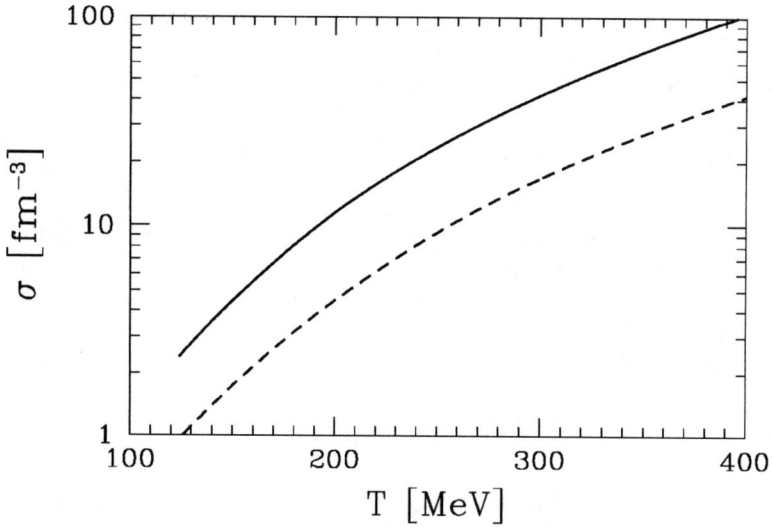

FIGURE 5. Entropy density in chemically equilibrated QGP at $\lambda_q = 1$ as function of temperature. Solid line $n_f = 2$, long dashed line $n_f = 2.5$, short dashed line 'pure glue' $n_f = 0$.

temperature further. First, we note that the lines in the lower-left panel (c) of figures 3 and 4 characterize the approximate trajectory in time of the fireball, ignoring changes in chemical composition. After initial drop in energy per baryon due to transfer of energy to the accelerating expansion of the fireball, during the supercooling process, see next subsection 2.3, the motion is slightly slowed and thus the local thermal energy per baryon increases.

How the entropy content depends on equilibration of quark degrees of freedom is seen in Fig. 5. The solid line is for the case of equilibrated quark-light glue system, in the limit of vanishing chemical potential. The 'pure glue' case (short-dashed line) contains as expected about half of the entropy when comparing at equal temperature, the addition of strangeness adds at the level of 10% to the entropy content, at a given temperature. However, as strangeness is produced temperature of the isolated fireball cools, and thus the chemical entropy production is smaller [33, 35].

The entropy decreases faster than the asymptotic T^3 behavior, since the QCD interactions strengthen as temperature decreases, leading to a decrease in the effective number of acting quark and gluon degrees of freedom. On the other hand, it can be expected that quark chemical equilibration occurs during the cooling of the QGP and a physical system 'moves' from a point on the dashed line to a point on the solid line in Fig. 5 as the temperature decreases, with an increasing effective number of degrees of freedom.

We will use the model here developed in next subsection 2.3 in order to study the QGP pressure and to establish the phase boundary between the hadron gas phase, and the deconfined phase. The discussion of other properties of QGP-liquid is a wide subject with many further developments possible, which are left at this point to the interested reader.

2.3. Supercooling and freeze-out condition

In presence of a phase transition of first order, both the QGP and the HG phases have metastable phase branches, and there can be supercooling when the QGP fireball expands rapidly, responding to the high initial pressure generated in the collision. The fluid-flow motion of colored quarks and gluons expands the domain of deconfinement by exercising against the vacuum a force originating not only from the thermal motion of quark and gluons but also in their flow velocity $\vec{v}_c$. The magnitude of this effect is well known and we will now explore its consequence.

Let P and ε be the pressure and energy density of the deconfined phase in the local restframe, subject to flow velocity $\vec{v}_c = (v_1, v_2, v_3)$. The pressure tensor component in the energy–momentum tensor is:

$$T^{ij} = P\delta_{ij} + (P+\varepsilon)\frac{v_i v_j}{1-\vec{v}^2}. \tag{9}$$

The rate-of-momentum flow vector $\vec{\mathscr{P}}$ at the surface of the fireball is obtained from the energy-stress tensor T_{kl}:

$$\vec{n} \cdot \widehat{\mathscr{T}} \cdot \vec{n} = P + (P+\varepsilon)\frac{(\vec{v}_c \cdot \vec{n})^2}{1-\vec{v}_c^2}. \tag{10}$$

The pressure and energy comprise particles (subscript p) and the vacuum properties, as we discussed above:

$$P = P_p - \mathscr{B}, \quad \varepsilon = \varepsilon_p + \mathscr{B}. \tag{11}$$

Eq. (10) for the condition $0 = \vec{n} \cdot \widehat{\mathscr{T}} \cdot \vec{n}$ thus reads,

$$\mathscr{B} = P_p + (P_p + \varepsilon_p)\frac{\kappa v_c^2}{1-v_c^2}, \tag{12}$$

where we introduced the geometric factor κ:

$$\kappa = \frac{(\vec{v}_c \cdot \vec{n})^2}{v_c^2}. \tag{13}$$

Eq. (12) describes the (equilibrium) condition under which the pressure of the expanding quark–gluon fluid is just balanced by the external vacuum pressure. κ characterizes the angular relation between the surface-normal vector and the direction of flow. Under the condition Eq. (12) of local dynamical balance between the force of the vacuum and that of the quark–gluon matter in motion, the local QGP-phase pressure $P = P_p - \mathscr{B}$, Eq. (11), is *always* negative, the collective flow of particles against the vacuum will in general lead to the formation of a supercooled quark–gluon phase.

Homogeneous expansion beyond condition Eq. (12) is in general not possible. A surface region of a fireball that continues to flow outward must be torn apart. This is a collective instability and the ensuing disintegration of the fireball matter should be very rapid [23]. We conclude that a rapidly evolving fireball that deeply supercools leads to a

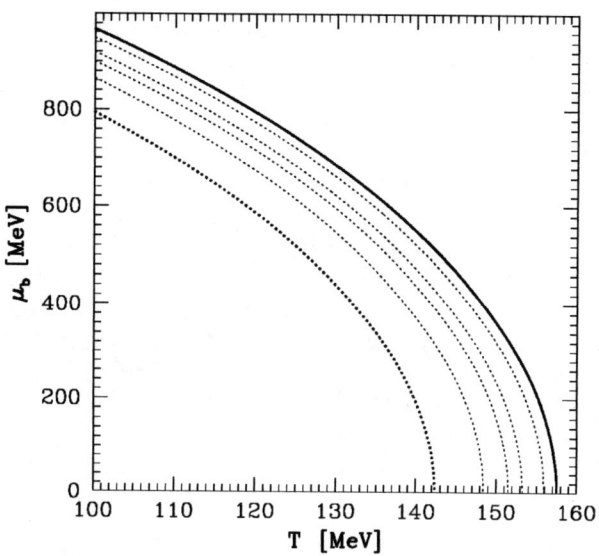

FIGURE 6. Solid line: equilibrium phase transition from hadron gas to QGP liquid accounting for the excluded volume correction. Dotted: breakup condition at shape parameter $\kappa = 1$ (exact spherical expansion), for expansion velocity $v_c^2 = 1/10, 1/6, 1/5, 1/4$ and $1/3$, from right to left.

sudden transformation (hadronization) into confined matter. The situation here described could arise only since the vacuum-pressure term is not subject to flow and always keeps the same value. In that regard the expansion of quark–gluon particle fluid is akin to the dynamics of a gas bubble in a liquid (here vacuum term $\mathscr{B}$).

We briefly survey the expected chemical freeze-out conditions. For hadron gas phase, we consider all known hadronic resonances and allow for the finite volume correction [36]. Equating this hadron and QGP pressure leads to the solid line to the right in the T, μ_b plane in Fig. 6. For vanishing baryo-chemical potential, we note in Fig. 6 that the equilibrium phase transition temperature is $T_c \simeq 157$ MeV. The scale in temperature we obtained is result of comparison with lattice gauge results and pressure balance with physical hadrons. Within the lattice calculations [6, 25], a scale of energy arises from the comparison with, e.g., the string tension.

The dotted lines, in Fig. 6, show where the flowing QGP liquid has a pressure which balances the hadron pressure, and thus these lines correspond to sudden break up at a velocity for (from right to left) $v_c^2 = 1/10, 1/6, 1/5, 1/4$ and $1/3$ using the shape parameter $\kappa = 1$, Eq. (13). Thus these lines maximize the flow effect of flow. For $\kappa = 0$, e.g., for longitudinal flow without transverse expansion of a infinite cylindrical structure, we would revert to the solid line as the transition condition. We see in the figure that the effect of flow velocity and geometry of the expanding system dominates the slight reduction in temperature when the baryochemical potential μ_b varies between 0 and 250 MeV. Freeze-out temperature $T = 150$–155 MeV seen in chemical nonequilibrium studies at both SPS and RHIC conditions [37], are consistent with the conditions expected at the sudden breakup of the QGP phase for velocity near to $v^2 \simeq 1/3$ and $\kappa = 0.1$–0.6.

But at lower heavy-ion interaction energies, i.e., at lowest CERN energies corresponding to top AGS energy collisions, near to $\mu_b \simeq 500$ MeV, the increase of $\kappa \to 1$ reduces strongly the expected chemical freeze-out condition temperature at fixed μ_b. We thus expect a faster decrease of chemical freeze-out temperature with increasing chemical potential than otherwise would be expected without supercooling.

2.4. Consistency with QGP properties

We consider now in how far the chemical freeze-out conditions and other features of hadronization process point to the mechanical instability and a sudden QGP breakup. We seek a way to determine if indeed supercooling developed. The interest in this resides, of course, in the question if first order phase transition separates the QGP and HG phases. To this end, we would like so observe 'experimentally' the condition $P < 0$. We employ the Gibbs-Duham relation for a unit volume,

$$P = T\sigma + \mu_b v_b - \varepsilon, \qquad (14)$$

which relates the pressure, to entropy density $\sigma = S/V$, energy density $\varepsilon = E/V$, and baryon density $v_b = b/V$, V is the volume, T is the temperature, and μ_b the baryochemical potential. Dividing by ε we obtain:

$$\frac{P}{\varepsilon} = \frac{T_h}{E/S} + \frac{\mu_b}{E/b} - 1. \qquad (15)$$

The quantities entering the right hand side of Eq. (15) can only be observed in the final hadronic phase. The microscopic processes governing the fireball breakup determine how these are related to properties of the deconfined phase. Certainly, the energy E and baryon content b are conserved. Entropy S is (nearly) conserved when the gluon content of a QGP fireball is transformed into quark pairs in the nearly entropy conserving process $G + G \to q + \bar{q}$. Similarly, when quarks and antiquarks recombine into hadrons, entropy remains nearly conserved in the range of parameters of interest here. Thus aside of E/b also S/b is conserved across hadronization condition. The sudden hadronization process also maintains the temperature T and baryo-chemical potential μ_b across the phase boundary. What changes in QGP breakup are the chemical phase space occupancy parameters, which describe the number of quark pairs, as this number has to change due to gluon processes we just mentioned: as gluons convert into quark pairs and hadrons, $\gamma_g \to 0$ and light valance quark pair number and thus phase space occupancy increases $\gamma_q > \gamma_{q_0} \simeq 1$, along with the number occupancy of strange quark pairs $\gamma_s > \gamma_{s_0} \simeq 1$, for a precise meaning of these parameters see [37] in this volume.

Evaluating Eq. (15) using the results of the data analysis presented in [21, 37], we indeed obtain $P/\varepsilon|_f \simeq \frac{-1}{22\pm1.5}$ for the RHIC freeze-out condition with chemical nonequilibrium. On the other hand the parameters of the chemical equilibrium statistical hadronization lead to $P/\varepsilon|_f \simeq 1/6.5$ which is appropriate for the known behavior of this phase. The nonequilibrium model is favored by the consistently much smaller χ^2/dof with the chemical freeze-out analysis, as well as by the consistency with the low temperature expected for the phase transformation, based on lattice calculations.

There is an interesting relation between the depth of supercooling and the geometry of the hadronization front arising from the mechanical instability condition: Eq. (10) for $0 = \vec{n} \cdot \widehat{\mathscr{T}} \cdot \vec{n}$ can be solved to yield P/ε:

$$-\frac{P}{\varepsilon} = \kappa \frac{v_c^2}{1-(1-\kappa)v_c^2} \tag{16}$$

We see that the value $P/\varepsilon|_{\rm f} \simeq -1/24$ implies $\kappa \simeq 0.1$ for $v_c^2 \simeq 1/3$, as is appropriate when the longitudinal flow dominates the transverse flow at RHIC.

2.5. Remarks about initial conditions in QGP formation

The formation of the QGP occurs well before chemical equilibration of quarks. We model this situation varying n_f. For $n_f = 1$ (roughly speaking 40% of accessible fermionic degrees of freedom) we show, in Fig. 7, lines of fixed energy per baryon $E/B = 3, 4, 5, 6, 8, 10, 20, 50$ and 100 GeV, The horizontal solid line is where the equilibrated hadronic gas phase has the same pressure as QGP-liquid with semi-equilibrated quark abundance. The free energy of the QGP liquid must be lower (pressure higher) in order for hadrons to dissolve into the plasma phase. The dotted lines in Fig. 7, from bottom to top, show where the pressure of the semi-equilibrated QGP phase is equal to $\eta = 20\%, 40\%, 60\%, 80\%$ and 100%, η being the 'stopping' fraction of the dynamical collisional pressure [26]:

$$P_{\rm col} = \eta n_0 \frac{P_{\rm CM}^2}{E_{\rm CM}}. \tag{17}$$

Here, the nuclear matter density is as before n_0, and the momentum and energy of each colliding nucleon in the nucleus seen in the CM (center of momentum) frame of reference is $P_{\rm CM}$ and $E_{\rm CM}$, respectively.

The rationale to study, in Fig. 7, lines at fixed E/b is that, during the nuclear collision which in the CM frame lasts about $2R_N/\gamma_L 2c \simeq 1$ fm/c, where γ_L is the Lorentz factor between the lab and CM frame and R_N is the nuclear radius, parton collisions lead to a partial (assumed here to be about 40%) chemical equilibration of u, d, s quarks. At that time, the pressure exercised on the parton matter corresponds to collisional pressure $P_{\rm col}$. This stopping fraction, seen in the transverse energy produced, is about 30–40% for S–S collisions at $200A$ GeV and 50–60% for Pb–Pb collisions at $158A$ GeV.

If the momentum-energy and baryon number stopping are similar, as we see in the experimental data at SPS, then the 160–200A GeV reactions are found in the highlighted area left of center of the figure. In the middle of upper boundary of this area, we would expect the beginning evolution of the thermal but not yet chemically equilibrated Pb–Pb fireball, and in the lower left corner of the S–S fireball. We note that the temperature reached in S–S case is seen to be about 25 MeV lower than in the Pb–Pb case. The lowest dotted line (20% stopping) nearly coincides with the non-equilibrium phase boundary (solid horizontal line, in Fig. 7) and thus we conclude that this is, for the condition $n_f = 1$, the lowest stopping that can lead to formation of a deconfined QGP phase, since for yet smaller η the lower collisional pressure can be balanced by the lower confined

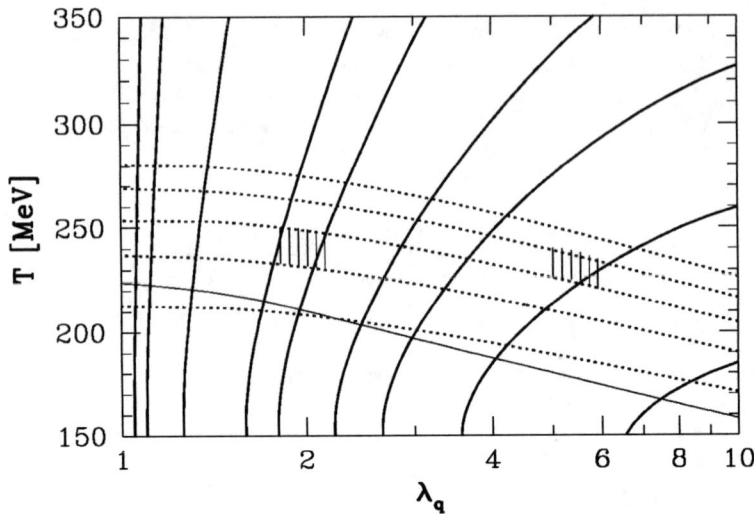

FIGURE 7. Solid lines: contours of energy per baryon in QGP in the T–λ_q plane for $n_f = 1$: From right to left E/b=3, 4, 5, 6, 8, 10, 20, 50 and 100 GeV. Thin and nearly horizontal line: hadronic gas phase has the same pressure as the QGP-liquid with semi-equilibrated quark flavor. Dotted lines from bottom to top: pressure in QGP liquid equals 20%, 40%, 60%, 80%, and 100%, of the dynamical collisional pressure.

HG matter pressure. Such a very low stopping would be encountered in small collision systems or/and at more peripheral, large impact parameter interactions of larger nuclei.

The highlighted area, right of center of the Fig. 7, corresponds to the expected conditions in Pb–Pb collisions at $40A$ GeV. If we assume that the stopping here is near 80%, then the initial conditions for fireball evolution would be found toward the upper right corner of this highlighted area. We recognize that the higher stopping nearly completely compensates the effect of reduced available energy in the collision and indeed, we expect that we form QGP also at these collision energies. It is important to realize that we are entering a domain of parameters, in particular λ_q, for which the equation of state extrapolation of the lattice QCD results is not necessarily reliable.

We have not indicated the domain of initial conditions at RHIC as there is still considerable uncertainty about this and the assumption of similar stopping of energy-momentum and baryon number cannot be maintained. Recent particle production results at $\sqrt{s_{NN}} = 130$ GeV, which we have discussed elsewhere in this volume [21], yield $S/b \simeq 180$ and $E/b \simeq 29$ GeV for $T \simeq 153$ MeV with $\mu_b = 32$ MeV as the chemical freeze-out condition. We see, in figure 4, that these values are consistent with the computed properties of the supercooled QGP-liquid model.

The final energy per baryon of $E/b = 29$ GeV implies that prior to the development of transverse flow the energy content per baryon must have been at (see Fig. 4):

$$\left.\frac{E}{b}\right|_{\text{initial}}^{\text{RHIC130}} \simeq 36\,\text{GeV},$$

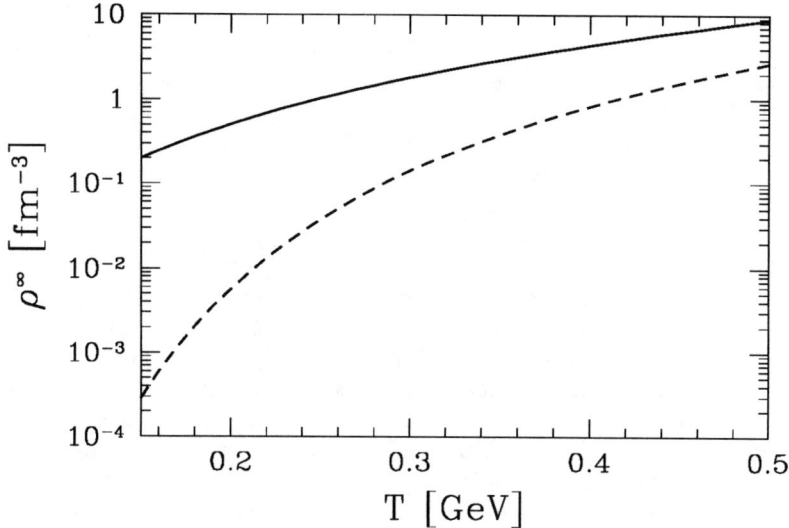

FIGURE 8. The statistical equilibrium density of strange or antistrange quarks with $m_s = 160$ MeV (solid line) and charmed or anticharmed quarks with $m_c = 1500$ MeV (dashed line), as function of temperature T.

(for the surface flow velocity $v = 1/\sqrt{3} = 0.58$ the Lorentz factor is 1.225). This is 55% of the available energy per baryon and thus the energy stopping is at RHIC noticeably less effective than baryon stopping.

We can use these results to estimate the RHIC-130 initial conditions assuming that the flow was entropy conserving. In panel (c), we see that, at $T \simeq 300$ MeV, we have nearly the expected energy content associated with $S/b \simeq 180$. Thus, a fully equilibrated QGP with $n_f = 2, n_s \simeq 1/2$ would imply an initial QGP matter with $T \simeq 280$ MeV which develops flow and thus cools by factor two down. Actually, the temperature reached was certainly higher since full chemical equilibrium of both light quarks and strangeness is established during the temporal evolution of the system. Assuming like in the SPS study that aside of gluons also about 40% of the light quark degrees of freedom are present, the initial temperature at RHIC-130 would be about 10% higher, slightly above 300 MeV.

3. KINETIC THEORY AND FLAVOR PRODUCTION

3.1. Approach to chemical equilibrium

The thermal equilibrium densities of strange and charm quarks in a quark gas is shown in Fig. 8 as function of temperature. The high strangeness density can only be reached in multitude of individual thermal collisions of QGP constituents and this type of production will be discussed in greater depth in what follows.

At QGP hadronization temperature around $T = 150$ MeV, the equilibrium charm density seen in figure 8 is relatively low, $n_c = 3 \cdot 10^{-4}/\text{fm}^3$. This yield can originate from initial 'direct' hard scattering processes of the primary partons at SPS energies. Indeed, there is no soft, in QGP, production of charm of significance at SPS or RHIC. Since the temperatures expected to be reached at LHC are well below the production threshold $T < 2m_c$, this remark also applies to LHC.

The direct yield of charm quark pairs in central SPS interactions, expected at 0.3 $c\bar{c}$-pairs in central Pb–Pb collisions at 158A GeV, corresponds to the thermal equilibrium yield in a $V = 1000\,\text{fm}^3$ fireball at hadronization temperature of 150 MeV. Since the direct charm production at RHIC energies is expected to be 30–50 times greater than at SPS, at the level of 10–15 $c\bar{c}$-pairs, and the freeze-out volume grows perhaps only by factor 4–5, there is potential that charm yield available at hadronization will by factor 5–10 exceed the chemical freeze-out equilibrium yield at RHIC. Thus in QGP at hadronization, we have the remarkable situation that strangeness may be approaching chemical equilibrium from below, while charm at least at RHIC could be well above chemical equilibrium abundance. This would favor in an unexpected way the formation of multi-charmed hadrons, including charmonium [38, 39]. This qualitative result amplifies the importance of, and need for, the kinetic theory studies of flavor evolution in QGP.

The approach to chemical equilibrium is, in comparison with the little understood kinetic (thermal, temperature forming) equilibration, much better understood theoretically. We will address, in the following, the case of particular interest in the physics of quark–gluon plasma: the approach to saturation ('absolute' chemical equilibration) of the phase space of strange particles. This process requires, as we shall see, just the lifespan of the QGP. The mechanism of production will be identified to be gluon fusion process, $GG \to s\bar{s}$ and thus in order to be able to evolve in time the population of (strange) quarks, we need to understand the population of gluons, i.e., γ_g.

Several workers have considered the glue approach to equilibrium in perturbative processes such as gluon splitting, e.g., $gg \to ggg$ [40, 41, 42, 43]. They find that in the thermal environment this process is not fast enough for gluons to reach chemical equilibrium. Were this true, there would be too few gluons to facilitate the approach to chemical equilibrium of light quarks, and, even more so, of strange quarks. We believe that multi gluon processes $gg \to ng$, $n > 3$ can equilibrate the abundance of gluons much faster [44]. Therefore, we assume that gluon chemical equilibrium is reached rapidly, $\mathcal{O}(1\,\text{fm})$, on a scale not much longer than the time required to reach thermal equilibrium. Our choice of an initial condition for consideration of strange flavor equilibration is $\gamma_g = 1$. In the study of evolution of strange quarks, we also take $\gamma_q = 1$. A large error, in this last assumption, is without significance for what follows, since gluons dominate the in QGP production of strange quarks.

In the quark–gluon plasma phase, there is no need to redistribute strange quarks among different carriers and relative chemical equilibrium is automatically established. The population master equation, can be easily cast into the general form

$$\frac{2\tau_{\text{chem}}^s}{\rho_i^\infty} \frac{d\rho_i}{dt} = 1 - \frac{\rho_s \rho_{\bar{s}}}{\rho_s^\infty \rho_{\bar{s}}^\infty}, \quad i = s, \bar{s}. \tag{18}$$

The last term describes the annihilation of strange quark pairs. The relative magnitude to the first production term is dictated by detailed balance, which assures that chemical equilibrium density ρ_i^∞ is reached. Assuming that the density of strange and antistrange quarks is equal, $\rho_s = \rho_{\bar{s}}$, equation

$$\frac{2\tau_{\text{chem}}^s}{\rho_s^\infty} \frac{d\rho_s}{dt} = 1 - \left(\frac{\rho_s}{\rho_s^\infty}\right)^2, \tag{19}$$

characterizes the population evolution of strangeness, and similarly of charm, but with much different initial conditions, within the scattering theory [11]. τ_{chem}^s is the time constant for chemical relaxation. The solution of Eq. (19) approaches equilibrium exponentially for $t \to \infty$:

$$\rho_s = \rho_s^\infty \tanh[t/(2\tau_{\text{chem}}^s)] \to (1 - e^{-t/\tau_{\text{chem}}^s})\rho_s^\infty. \tag{20}$$

The chemical equilibration (relaxation) time constant τ_{chem}^s, is computed as an inverse of the invariant reaction rate per unit volume R_s:

$$\tau_{\text{chem}}^s = \frac{\rho_s^\infty}{2R_s}. \tag{21}$$

R_s is the rate at which the system 'chases' the equilibrium density ρ_s^∞; the ratio is a characteristic time defining when the chase is over. The factor 2 in Eq. (21) is introduced to assure that the approach to equilibrium due to two-body reactions is governed by an exponential function with the time-decay parameter τ_{chem}^s, as is seen on the right-hand side of Eq. (20).

We consider the inelastic production process $a + b \to s + X$ with the reaction cross section, $\bar{\sigma}_{ab \to sX}(\sqrt{s})$. The invariant reaction rate per unit of time and volume is obtained from

$$R_s(x) = \sum_{a,b,X} \int_{(m_s+m_X)^2}^\infty 2\lambda_2(s)\,ds \int \frac{d^3k_a}{(2\pi)^3 2E_a} \int \frac{d^3k_b}{(2\pi)^3 2E_b}$$

$$\times f_a(k_a,x) f_b(k_b,x)\, \bar{\sigma}_{ab \to iX}(\sqrt{s})\, \delta[s - (k_a+k_b)^2], \tag{22}$$

where $\lambda_2(s) = [s - (m_a+m_b)^2][s - (m_a-m_b)^2]$, $f_a(k_a,x)$ and $f_b(k_b,x)$ are the phase-space distributions of the colliding particles. In Eq. (22), we are neglecting Pauli or Bose quantum effects (suppression or stimulated emission factors) in the initial and final states. The 'bar' over the reaction cross section indicates that the dependence on transfer of momentum (scattering angle) is averaged over. For the simplest case of $f_a(k_a,x)$ and $f_b(k_b,x)$ being relativistic Boltzmann momentum distribution

$$R_s(x) = \frac{\sum_{a,b,X} \int_{w_0}^\infty dw\, \lambda_2 \bar{\sigma}_{ab \to sX}(w) K_1(w/T)}{4 T m_a^2 m_b^2 K_2(m_a/T) K_2(m_b/T)}, \tag{23}$$

where $w = \sqrt{s}$ is the CM energy and $w_0 = m_i + m_X$ is the reaction threshold. This formula is presented in this form in [16], Eq. (5.7); it is stated there for the special case in which the reacting particles a and b are identical bosons, which, to avoid double counting of indistinguishable pairs of particles, requires an extra factor $\frac{1}{2}$, which is not included in Eq. (23). The interesting $m_{a,b} \to 0$ limit is implemented with a replacement of each factor $m^2 K_2(m/T)$ by $2T^2$ in Eq. (23), and $\lambda_2 \to s$, which reduces Eq. (23) to the result presented in [11], Eq. (2); [24] lacks the factor $1/T$.

We see explicitly, in Eq. (23), the mass threshold in the s-integration occurring for inelastic (particle-producing) rates. A high threshold combines with the exponentially small K_1-Bessel function, to reduce the strength of inelastic hadronic particle-production rates which are usually much smaller than the total rates of reaction (scattering). For this reason, the time scale of chemical equilibration is, in general, considerably longer than the time needed to equipartition the energy among light partons in the thermal equilibration process.

A particularly important property of the QGP matter evolution is that the volume grows as T decreases such that $VT^3 \simeq$ Const. with T^{-3} in order to preserve the entropy content. It turns out that this can be used to simplify the kinetic evolution Eq. (19). We introduce local phase-space occupancy γ_s of strange quarks thus $\rho_s = \gamma_s(t)\rho_s^\infty$. We use the Boltzmann equilibrium abundance,

$$\delta N_s = \delta V \, \gamma_s \rho_s^\infty = [T^3 \, \delta V] \gamma_s \frac{3}{\pi^2} z^2 K_2(z), \quad z = \frac{m_s}{T}. \tag{24}$$

In an entropy-conserving evolution, e.g., subject to (ideal) hydrodynamic flow, the first factor on the right-hand side in Eq. (24) (in square brackets) is a constant in time, $\delta V T^3 = \delta V_0 T_0^3 = $ constant. The change in the number of strange quarks is:

$$\frac{1}{\delta V_1} \frac{d \delta N_s}{dt} = \frac{d\rho_s}{dt} + \rho_s \frac{1}{\delta V_1} \frac{d \delta V_1}{dt}. \tag{25}$$

We now obtain, using the recursion relation for the K-functions in Eq. (25),

$$\frac{1}{\delta V_1} \frac{d \delta N_s}{dt} = \dot{T} \rho_s^\infty \left(\frac{d\gamma_s}{dT} + \frac{\gamma_s}{T} z \frac{K_1(z)}{K_2(z)} \right), \tag{26}$$

where $\dot{T} = dT/dt$. Only a part of the usual flow-dilution term is left, since we implemented the adiabatic volume expansion, and study the evolution of the phase-space occupancy in lieu of the particle density.

Generalizing slightly Eq. (19), stating explicitly the two production channels, the fusion of gluons and light-quark–antiquark fusion into a pair of strange quarks:

$$\dot{T} \rho_s^\infty \left(\frac{d\gamma_s}{dT} + \frac{\gamma_s}{T} z \frac{K_1(z)}{K_2(z)} \right) = \gamma_g^2(\tau) R^{gg \to s\bar{s}} \left(1 - \frac{\gamma_s^2(\tau)}{\gamma_g^2(\tau)} \right)$$
$$+ \gamma_q^2(\tau) R^{q\bar{q} \to s\bar{s}} \left(1 - \frac{\gamma_s^2(\tau)}{\gamma_q^2(\tau)} \right). \tag{27}$$

When all $\gamma_i \to 1$, the right-hand side vanishes, chemical equilibrium is established. In presence of the chemical equilibrium of gluons and light quarks, the dynamic equation describing the evolution of the local phase-space occupancy of strangeness is (and, in analogy, charm, with $s \to c$),

$$2\tau_s \dot{T}\left(\frac{d\gamma_s}{dT} + \frac{\gamma_s}{T}z\frac{K_1(z)}{K_2(z)}\right) = 1 - \gamma_s^2. \tag{28}$$

For a static, non-expanding system $\gamma_s \to 1$ monotonically as a function of time. However, Eq. (28) allows the range $\gamma_s > 1$, for it incorporates the possibility of a rapidly expanding high yield of strangeness created in the early stage at high T. Initially at a high background temperature, the approach of $\gamma_s(t) \to 1$ may produce a high particle yield, which at the lower temperature established after expansion of the system, implies $\gamma_s > 1$, as long as the re-annihilation of strange flavor cannot keep up with the rapid evolution of a fireball of quark–gluon plasma. Annihilation is slow, since the density of strange and antistrange quarks is about four times smaller than the density of gluons (an effect of color and mass) and thus the rate of annihilation for strange quarks is 16 times slower. With a relaxation time for the production of strangeness of 1.5–10 fm (depending on temperature), see Fig. 12 below, the relaxation time for $s\bar{s}$ annihilation is 20–150 fm, so practically all strangeness is preserved on the time scale of 5 fm of a quark–gluon plasma fireball [12]. For charm this argument is much stronger, once charm has been produced, this heavy flavor has no time to annihilate and chemically re-equilibrate in rapidly evolving QGP. This non-annihilation of heavy flavor is a very important feature which makes the total yield of strangeness along with charm a 'deep' probe of the deconfined phase.

3.2. Thermal Production Rates of Strangeness

The production processes involving quark and gluon degrees of freedom in QGP are

$$u + \bar{u} \to s + \bar{s}, \quad d + \bar{d} \to s + \bar{s}, \quad g + g \to s + \bar{s}. \tag{29}$$

These three processes describing perturbative production of pairs of strange quarks are represented to lowest order in Fig. 9, and have to be summed incoherently. These lowest-order diagrams were studied in the early eighties, for the quark process [45] and for the gluon process [11], employing fixed values of $\alpha_s = 0.6$ and $m_s = 160$–200 MeV.

The evaluation of the lowest order Feynman diagrams shown in Fig. 10 yields the cross sections [46],

$$\sigma_{q\bar{q}\to f\bar{f}}(s) = \frac{8\pi\alpha_s^2}{27s}\left(1 + \frac{2m_f^2}{s}\right)w(s), \quad w(s) = \sqrt{1 - \frac{4m_f^2}{s}}, \tag{30}$$

$$\sigma_{gg\to f\bar{f}}(s) = \frac{\pi\alpha_s^2}{3s}\left[\left(1 + \frac{4m_f^2}{s} + \frac{m_f^4}{s^2}\right)\ln\left(\frac{1+w(s)}{1-w(s)}\right) - \left(\frac{7}{4} + \frac{31m_f^2}{4s}\right)w(s)\right]. \tag{31}$$

Inspecting Fig. 10(a), we see that the magnitudes (up to 0.4 mb) of both types of reactions considered, quark fusion and gluon fusion, are similar. At this stage, it is not immediately apparent that gluons dominate the production of flavor.

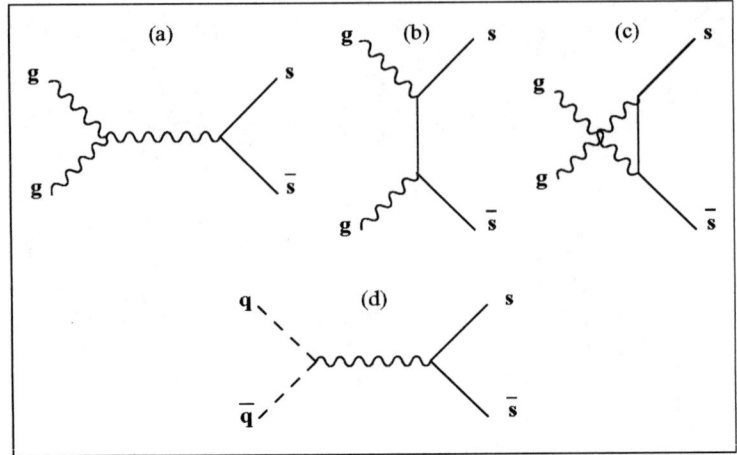

FIGURE 9. Leading order Feynman diagrams for production of $s\bar{s}$ by fusion of gluons and pairs of quarks (q = u, d).

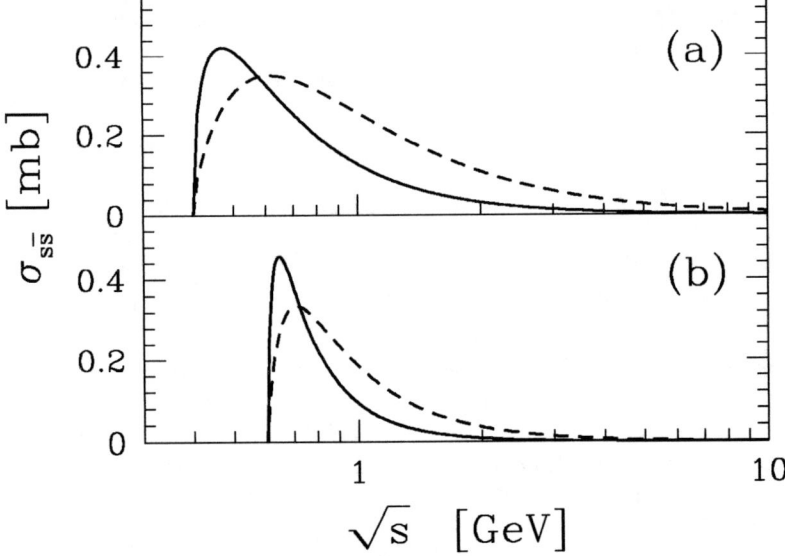

FIGURE 10. Production cross sections for strangeness in leading order: (a) for $\alpha_s = 0.6$ and $m_s = 200$ MeV; (b) for running $\alpha_s(\sqrt{s})$ and $m_s(\sqrt{s})$, with $\alpha_s = 0.118$. Solid lines, $q\bar{q} \to s\bar{s}$; dashed lines, $gg \to s\bar{s}$.

The magnitude of the cross section of interest is normalized by α_s. To obtain Fig. 10(a), we took $\alpha_s = 0.6$. While the value seems reasonable, a value of $\alpha_s = 0.3$ would lengthen the relaxation time of strangeness, $\tau_s \propto \alpha_s^{-2}$, by a factor four, nearly beyond the expected lifespan of the QGP fireball. Thus, we must improve the determination of α_s.

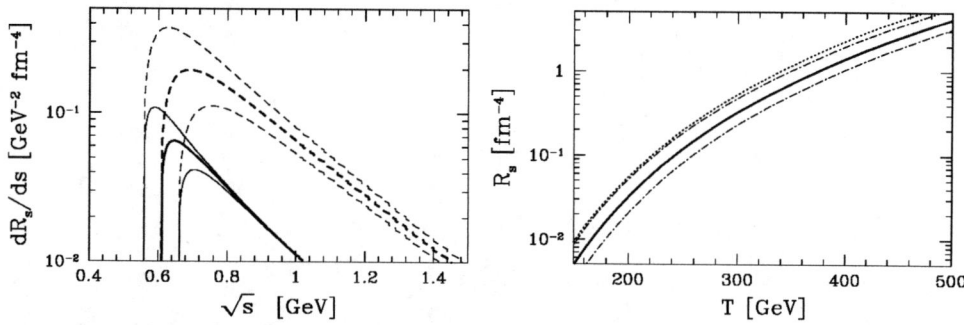

FIGURE 11. Left: The differential thermal production rate for strangeness dR_s/ds, with $T = 250$, $\lambda_q = 1.5$ for gluons (dashed line), and $q\bar{q} \to s\bar{s}$ (solid line, includes two interacting flavors), for the running $\alpha_s(M_Z) = 0.118$ and running mass of the strange quark $m_s(M_Z) = 90$ MeV $\pm$ 20% (thin lines). Right: Thermal production rates for strangeness R_s in QGP (thick solid line), calculated for $\lambda_q = 1.5$, $\alpha_s(M_Z) = 0.118$, and $m_s(M_Z) = 90$ MeV, as a function of temperature. Chain lines show the effect of variation of the mass of the strange quark by 20%. The dotted line shows for comparison the result for fixed $\alpha_s = 0.6$ and $m_s = 200$ MeV.

There are two natural ways to do this; the easier one is to adopt the functional $\alpha_s(T)$ seen in Fig. 1 (right). However, within such an approach, two-body collisions occurring in a thermal bath at the same temperature T but at a very different $\sqrt{s}$ are evaluated with the same value of α_s. Only for the thermal production of charm does this approach turn out to yield the same result as does the more complex, but more precise, consideration of an appropriate value of α_s for each individual collision, governed by the applicable $\alpha_s(\mu)$, Fig. 1 (left), with $\mu \simeq \sqrt{s}$. This second method, in which for each collision in the thermal bath an appropriate coupling strength is selected, is necessary for studying the production of strangeness in order to account for the growth of the cross section for soft scattering. We show this cross section in Fig. 10(b).

The effect of the increase of interaction strength in soft collisions is, however, largely balanced by the concurrent suppression of the cross section due to the increase in the mass of the strange quark m_s at the soft momentum scale, as is seen by the increased threshold comparing Fig. 10(a) and (b). These competing effects seem to be at the origin of the phenomenon that we find in next to leading order calculations a result for the cross section very similar to the lowest order result obtained with running mass.

The contemporary results for the invariant production rates of strangeness seen Fig. 11 agree very well with the early results [11] (see the dotted line in left panel of Fig. 11) within the uncertainty in mass of the strange quark (a smaller mass leads to a bigger value of R). Insertion of the rates R_s into Eq. (21) allows us to obtain the time constants for chemical relaxation τ_s, see Fig. 12. We see τ_s evaluated with $m_s(1\,\text{GeV}) = 200$ MeV. The range of the assumed 20% uncertainty in $m_s(M_Z)$ is indicated by the hatched areas. The initial predictions obtained 20 years ago [11] are shown here at fixed values $\alpha_s = 0.6$ for $m_s = 200$ MeV (see thick dotted line in Fig. 12). This result is overlapping with the band of τ_s seen in Fig. 11. The thin dotted line shows a significant reduction in the relaxation time when the strange quark mass is reduced by a factor 5/9.

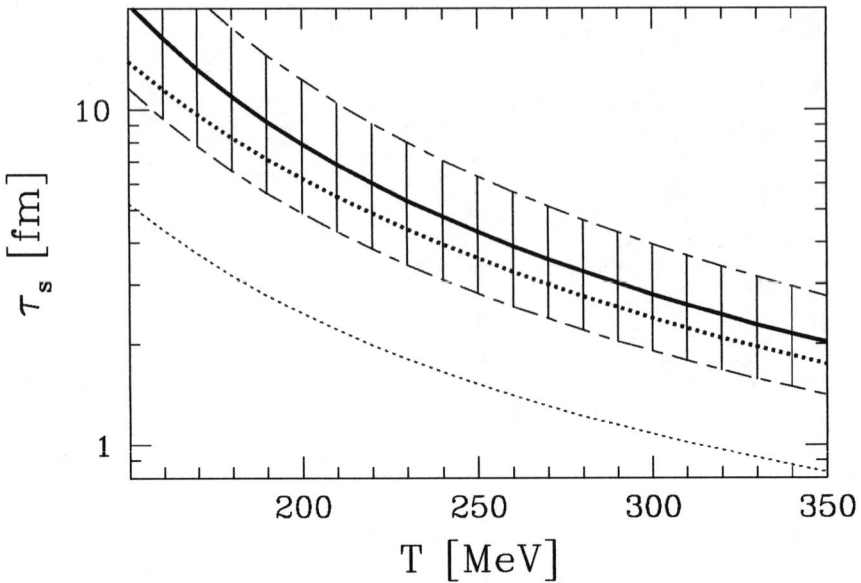

FIGURE 12. The QGP chemical relaxation time for strangeness τ_s, for $\alpha_s(M_Z) = 0.118$ with $m_s(M_Z) = 90$ MeV and $\rho_s^{\infty}(m_s \simeq 200 \text{ MeV})$ (thick line). Hatched areas show the effect of the variation mass of the strange quark by 20%. The fat-dotted shows comparison results for fixed $\alpha_s = 0.6$ and $m_s = 200$ MeV. The thin-dotted line shows the result for $m_s(M_Z) = 50$ MeV.

The approximate formula obtained in [11],

$$\tau_f = \frac{1.6}{\alpha_s^2 \gamma_g^2 T} \frac{m_f/T \, e^{m_f/T}}{[1 + (99/56)T/m_f + \cdots]}, \qquad (32)$$

allows a quick estimate of the expected relaxation time in all the environments discussed in this subsection. We have added the pre-factor γ_g^{-2} relevant in case the dominant source of heavy flavor, gluons, is not in chemical equilibrium. We see that the equilibration time lengthens accordingly.

Thermal nonperturbative effects on the relaxation of strangeness were studied by introducing thermal, temperature-dependent, particle masses [47]. After the new production rates, including the now possible gluon decay, were added up, the total rate of production of strangeness was found to be little changed compared with the free-space rate. This finding was challenged [48], but a further reevaluation [49] confirmed that the rates obtained with perturbative glue-fusion processes are describing precisely the rates of production of strangeness in QGP, for the relevant temperature range $T > 200$ MeV. We can thus assume today that the 'prototype' strangeness-production processes seen in Fig. 9, re-summed using the renormalization-group method, are dominating the rates of production of strangeness in QGP.

With these methods, we can evaluate the chemical charm relaxation time due to thermal processes, shown in Fig. 13. We see that the time constants are too long to

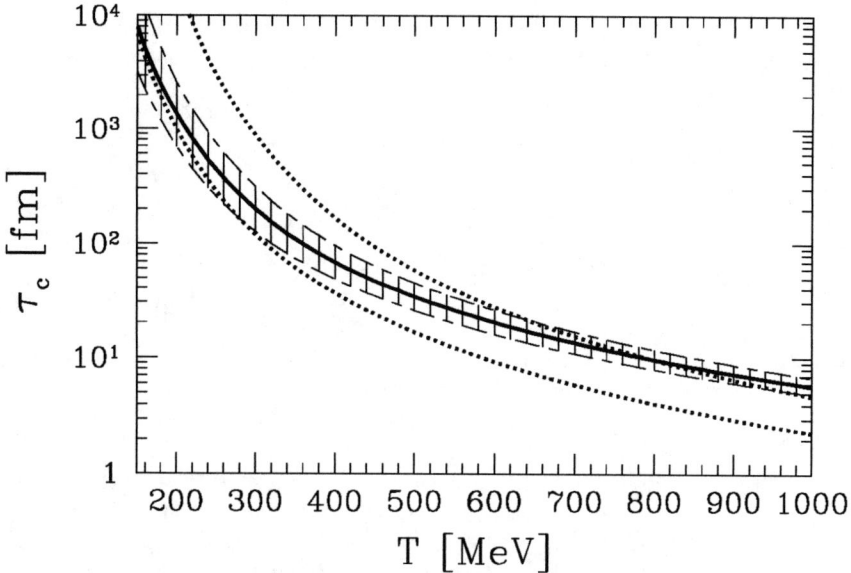

FIGURE 13. Solid lines show the thermal relaxation constant for charm in QGP, calculated for running $\alpha_s(M_Z) = 0.118$, $m_c(M_Z) = 0.7\,\mathrm{GeV}$ and $\rho_c^\infty(m_c \simeq 1.5\,\mathrm{GeV})$. Lower dotted line, for fixed $m_c = 0.9\,\mathrm{GeV}$ and $\alpha_s = 0.35$; upper dotted line, for fixed $m_c = 1.5\,\mathrm{GeV}$ and $\alpha_s = 0.4$. The hatched area shows the effect of variation $m_c(M_Z) = 0.7\,\mathrm{GeV} \pm 7\%$.

allow charm production or annihilation in the QGP phases produced at present, and this remark probably extends to LHC where one expects today a temperature $T < 600$ MeV.

Using strangeness relaxation time seen in Fig. 12, we now qualitatively explore the time evolution of the strange quark flavor in rapidly evolving QGP. For this we must model the time dependence of the temperature. For a purely longitudinal expansion, the local entropy density scales according to $S \propto T^3 \propto 1/\tau$. The transverse flow of matter accelerates the drop in entropy density. To model this behavior without too great a numerical effort, considering the other uncertainties, the following temporal-evolution function of the temperature was proposed [50]:

$$T(\tau) = T_0 \left(\frac{1}{(1 + 2\tau c/d)(1 + \tau v_\perp/R_\perp)^2} \right)^{1/3}. \tag{33}$$

Considering various values of T_0, the temperature at which the gluon equilibrium is reached, the longitudinal dimension is scaled according to:

$$d(T_0) = (0.5\,\mathrm{GeV}/T_0)^3 \times 1.5\,\mathrm{fm}. \tag{34}$$

This adjustment of the initial volume V_0 assures that the different evolution cases refer to a fireball with a similar entropy content. The following results are thus a study of one and the same collision system, and the curves reflect the uncertainty associated with unknown initial conditions of a fireball of quark–gluon plasma with in each case same, but large by current standards, entropy content.

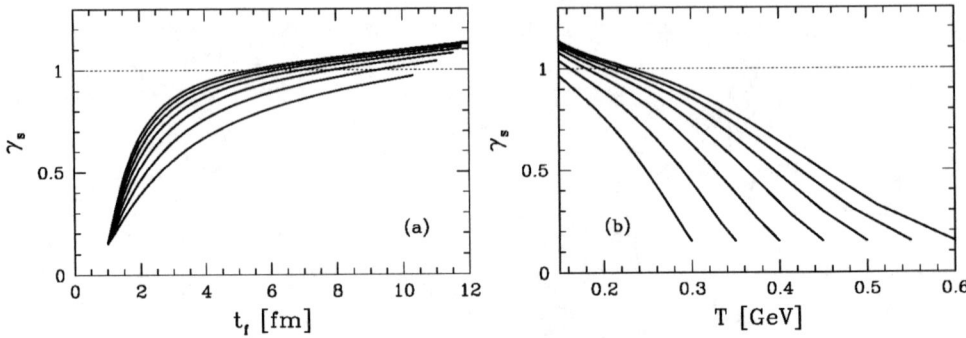

FIGURE 14. The evolution of QGP phase-space occupancy for strangeness γ_s: (a) as a function of time and (b) as a function of temperature for $m_s(1\,\text{GeV}) = 200\,\text{MeV}$; see text for details.

The numerical integration of Eq. (28) is started at $\tau_0 = 1$ fm/c, the time at which thermal initial conditions are reached. A range of initial temperatures $300\,\text{MeV} \leq T_0 \leq 600\,\text{MeV}$, varying in steps of 50 MeV, is considered. Since the initial p–p collisions also produce strangeness, to estimate the initial abundance a common initial value $\gamma_s(T_0) = 0.15$ is used. For $T_0 = 0.5\,\text{GeV}$, the thickness of the initial collision region is $d(T_0 = 0.5)/2 = 0.75$ fm. The initial transverse dimension in nearly central Au–Au collisions is taken to be $R_\perp = 4.5$ fm. The initial volume of quark–gluon plasma is 190 fm^3, which, at the temperature of $T_0 = 0.5\,\text{GeV}$, implies, according to results seen in Fig. 5, a total entropy content of $S = 38\,000$. We divide this by the specific entropy content per hadron in the final state, $S/N = 4$. We see that the primary final-state hadron multiplicity has implicitly been assumed to be 9500. This is somewhat above results seen even during the RHIC run at $\sqrt{s_N} = 200$ GeV, for which we estimate, for the 3% most central events, a total hadron multiplicity, after resonance cascading, of 8000. However, these model calculations originate in pre-RHIC era [51].

The evolution with time in the plasma phase is followed up to the breakup of the QGP at a temperature $T_f^{\text{RHIC}} \simeq 150 \pm 5$ MeV. The numerical solution of Eq. (28) for γ_s is shown as a function both of time t, in Fig. 14(a), and of temperature T, in Fig. 14(b). This evolution is physically meaningful until it reaches the QGP breakup condition. Since the results for higher temperatures are also displayed, the reader who prefers a higher hadronization temperature can easily draw his own conclusions.

We see, in Fig. 14, the following phenomena:

- A steep rise at early times, showing actual production of strangeness, which is followed by a dilution-driven increase of γ_s near the breakup temperature.
- Widely different initial conditions (but with similar initial entropy contents) lead to rather similar chemical conditions at chemical freeze-out of strangeness.
- Despite the use of a high mass of $m_s = 200$ MeV, we find that strangeness nearly equilibrates chemically, and that the dilution effect allows in certain cases a small over-population of the strange-quark phase-space even in the strangeness-dense quark–gluon plasma.

- For a wide range of initial conditions and final freeze-out temperature a narrow band of final result is seen, $1.10 > \gamma_s(T_f) > 0.9$.

Since strangeness is more easily made in the 'hot' era in glue–glue interactions, we can estimate that, if the abundance of glue had been at the time 70% chemically equilibrated, then a final value $\gamma_s \simeq 0.5$ would result. This high sensitivity to the glue chemical equilibration is at the origin of the claim that measuring the yield of strangeness probes the presence and abundance of glue, which is a specific property of quark–gluon plasma.

In the schematic model presented on Eq. (33), the fireball begins to expand in the transverse direction instantly at the full velocity. For this reason, the initial drop in temperature is very rapid. This also makes the transverse size at the end of the expansion too large, $R_\perp \simeq R_0 + t_f/\sqrt{3} \simeq 9$ fm, compared with the results of HBT analysis. This can easily be fixed by introducing a more refined model of the transverse velocity, which needs time to build up. The yield of strangeness may slightly increase in such a refinement, since the fireball will spend more time near to the high initial temperature.

The results presented are typical for all RHIC collision systems. In the top SPS energy range, the initial temperature reached is certainly less (by 10%–20%) than that in the RHIC 130-GeV run, and the baryon number in the fireball is considerably greater; however, the latter difference matters little for production of strangeness, which is driven by gluons. A model similar to the above yields $\gamma_s^{QGP} \simeq 0.6$–$0.7$, the upper index reminds us that in this section the strangeness occupancy factor γ_s refers to the property of the deconfined phase. The final state γ_s^{HG} is about a factor 3 larger [21]. The experimental observable directly related to γ_s^{QGP} is the total yield of strangeness per participating baryon.

4. SUMMARY AND CONCLUSIONS

We have presented two lectures addressing physics of quark–gluon plasma in section 2, and inside quark–gluon plasma strangeness production, section 3. We refer the reader to the other closely related contributions in this volume [21, 52] where further applications of the developments here presented can be found.

The first part, section 2 on QGP, provides the theoretical foundation for study of QGP hadronization. We have shown that the properties of QGP at hadronization are those expected at the condition of sudden breakup and we have shown the mechanism which can lead to this, the supercooling originating in fast expansion. The theoretical hypothesis of sudden QGP break up and hadronization is strongly supported by the experimental observation that strange baryons and antibaryons have very similar $m_\perp$-spectra [53] and are thus produced by the same mechanism [54]. Moreover, these spectra imply that the thermal freeze-out of these rarely produced particles occurs together with the chemical (abundance) freeze-out [52, 55]. This occurs for all collision centralities measured by the experiment WA97.

The second part, section 3, provides the foundation for the understanding of strangeness as a diagnostic tool allowing the study of QGP properties. We have in some detail demonstrated, in subsection 3.2, that strangeness can nearly saturate the phase-space of QGP, being produced in gluon fusion processes predominantly. These results agree in qualitative manner with the expectations arising from the analysis of the chemical freeze-out, see [21].

The results we have presented explain how the coincidence of the magnitudes of the critical temperature T_c of the confined to deconfined phase transformation, with strange quark mass and the fact that we reach in SPS and RHIC reactions the initial conditions where threshold for strangeness production and the initial temperature T_0 are nearly equal,

$$T_c \simeq m_s, \qquad T_0 \simeq 2m_s,$$

render strangeness as a unique observable of the physics of relativistic heavy-ion collisions.

ACKNOWLEDGMENTS

Work supported in part by a grant from the U.S. Department of Energy, DE-FG03-95ER40937 and by NSF grant INT-0003184. Laboratoire de Physique Théorique et Hautes Energies, LPTHE, at University Paris 6 and 7 is supported by CNRS as Unité Mixte de Recherche, UMR7589.

REFERENCES

1. N. Itoh. *Prog. Theor. Phys.*, **44**, 291, 1970.
2. P. Carruthers. *Collective Phenomena*, **1**, 147, 1973.
3. F. Iachello, W. D. Langer, and W. D. Lande. *Nucl. Phys.* A, **219**, 612, 1974.
4. J. C. Collins and M. J. Perry. *Phys. Rev. Lett.*, **34**, 1353, 1975.
5. R. Hagedorn. How we got to QCD matter from the hadron side by trial and error. In *Quark Matter '84*, page 53, 1984.
6. F. Karsch, E. Laermann, and A. Peikert. *Phys. Lett.* B, **478**, 447, 2000.
7. J. Rafelski. Extreme states of nuclear matter. In R. Bock and R. Stock, editors, *Workshop on Future Relativistic Heavy Ion Experiment*, page 282. GSI-Yellow Report 81-6, Darmstadt, 1981.
8. J. Rafelski and R. Hagedorn. From hadron gas to quark matter II. In H. Satz, editor, *Statistical Mechanics of Quarks and Hadrons*, page 253. North Holland, Amsterdam, 1981.
9. J. Rafelski. Hot hadronic matter. In J. Tran Thanh Van, editor, *New Flavors and Hadron Spectroscopy*, page 619. Editions Frontières, 1981.
10. J. Rafelski. *Nucl. Phys.* A, **374**, 489c, 1982.
11. J. Rafelski and B. Müller. *Phys. Rev. Lett.*, **48**, 1066, 1982. See also *Phys. Rev. Lett.*, **56**, 2334E (1986).
12. P. Koch, B. Müller, and J. Rafelski. *Z. Phys.* A, **324**, 453, 1986.
13. J. Rafelski. *Phys. Rep.*, **88**, 331, 1982.
14. J. Rafelski and M. Danos. Perspectives in high energy nuclear collisions. Updated version appeared in *Nuclear Matter under Extreme Conditions*, D. Heiss, Ed., Springer Lecture Notes in Physics **231**, 362 (1985), 1983.
15. P. Koch and J. Rafelski. *Nucl. Phys.* A, **444**, 678, 1985.
16. P. Koch, B. Müller, and J. Rafelski. *Phys. Rep.*, **142**, 167, 1986.
17. F. Antinori *et al.*, WA97 collaboration. *Nucl. Phys.* A, **663**, 717, 2000.
18. E. Andersen *et al. Phys. Lett.* B, **433**, 209, 1998.
19. E. Andersen *et al.*, WA97 collaboration. *Phys. Lett.* B, **449**, 401, 1999.
20. J. Letessier and J. Rafelski, 2002. *Hadrons and Quark Gluon Plasma*. Cambridge University Press, New York.

21. J. Rafelski and J. Letessier. Nonequilibrium hadrochemistry in qgp hadronization. In this volume, 2002.
22. S. Hamieh, J. Letessier, and J. Rafelski. *Phys. Rev. C*, **62**, 64901, 2000.
23. J. Rafelski and J. Letessier. *Phys. Rev. Lett.*, **85**, 4695, 2000.
24. T. Biró and J. Zimányi. *Phys. Lett. B*, **113**, 6, 1982.
25. F. Karch. Lattice QCD at finite temperature. In this volume, 2002.
26. J. Rafelski, J. Letessier, and A. Tounsi. *Acta. Phys. Pol. B*, **27**, 1037, 1996.
27. A. Peshier, B. Kämpfer, and G. Soff. *Phys. Rev. C*, **61**, 45 203, 2000.
28. H. Vija and M. H. Thoma. *Phys. Lett. B*, **342**, 212, 1995.
29. G. Altarelli. A QCD primer. In this volume, 2002.
30. S. A. Chin. *Phys. Lett. B*, **78**, 552, 1978.
31. H. A. Weldon. *Phys. Rev. D*, **26**, 1394, 1982.
32. H. A. Weldon. *Phys. Rev. D*, **26**, 2789, 1982.
33. J. Letessier, J. Rafelski, and A. Tounsi. *Phys. Rev. C*, **50**, 406, 1994.
34. J. Rafelski, J. Letessier, and A. Tounsi. *Acta. Phys. Pol. A*, **85**, 699, 1994.
35. H.-Th. Elze, J. Rafelski, and L. Turko. *Phys. Lett. B*, **506**, 123, 2001.
36. R. Hagedorn and J. Rafelski. *Phys. Lett. B*, **97**, 136, 1980.
37. J. Rafelski and J. Letesier. Quark-gluon plasma and strangeness. In this volume, 2002.
38. R. L. Thews, M. Schroedter, and J. Rafelski. *Phys. Rev. C*, **63**, 54905, 2001.
39. R. L. Thews. Nonlinear behavior of quarkonium formation and deconfinement signals. In this volume, 2002.
40. T. S. Biró, E. van Doorn, B. Müller, M. H. Thoma, and X.-N. Wang. *Phys. Rev. D*, **48**, 1275, 1993.
41. S. M. H. Wong. *Phys. Rev. C*, **54**, 2588, 1996.
42. D. K. Srivastava, M. G. Mustafa, and B. Müller. *Phys. Lett. B*, **45**, 396, 1997.
43. D. K. Srivastava, M. G. Mustafa, and B. Müller. *Phys. Rev. C*, **56**, 1064, 1997.
44. L. Xiong and E. V. Shuryak. *Phys. Rev. C*, **49**, 2203, 1994.
45. T. Biró and J. Zimányi. *Nucl. Phys. A*, **395**, 525, 1983.
46. B. Combridge. *Nucl. Phys. B*, **151**, 429, 1979.
47. T. S. Biró, P. Lévai, and B. Müller. *Phys. Rev. D*, **42**, 3078, 1990.
48. T. Altherr and D. Seibert. *Phys. Lett. B*, **313**, 149, 1993.
49. N. Bilić, J. Cleymans, I. Dadić, and D. Hislop. *Phys. Rev. C*, **52**, 401, 1995.
50. J. Rafelski and J. Letessier. *Phys. Lett. B*, **469**, 12, 1999.
51. J. Letessier and J. Rafelski. *Int. J. Mod. Phys. E*, **9**, 107, 2000.
52. C. Markert, G. Torrieri, and J. Rafelski. Strange hadron resonances: Freeze-out probes in heavy-ion collisions. In this volume, 2002.
53. F. Antinori *et al.*, WA97 collaboration. *Eur. Phys. J. C*, **14**, 633, 2000.
54. J. Rafelski and M. Danos. *Phys. Lett. B*, **192**, 432, 1987.
55. G. Torrieri and J. Rafelski. *New J. Phys.*, **3**, 12, 2001.

QCD Vacuum and Confinement

Adriano Di Giacomo

University of Pisa and INFN

Abstract. This course consists of two lectures. In the first lecture I discuss why a non perturbative formulation of QCD is needed ,and I show that lattice formulation copes with this need, even if it mainly produces numerical results. In the second lecture I discuss how lattice can help to understand the deconfinement transition.Such understanding is also important to predict parameters that can help in the interpertation of heavy ions high energy experiments.

INTRODUCTION

Vacuum is by definition the ground state of a field system:it is stable against quantum fluctuations.

An exact knowledge of the ground state provides complete information on the system.This property goes under the name of reconstruction theorem,the exact statement being that from the field correlators

$$\langle 0|T(\varphi_1(x_1)\ldots\varphi_n(x_n)|0\rangle \quad (1)$$

the Hilbert space and the matrix elements of physical observables can be constructed.

Textbook quantization is perturbative .The Lagrangean is split in a free term $\mathscr{L}_0$ and a perturbation $\mathscr{L}_I$

$$\mathscr{L} = \mathscr{L}_0 + \mathscr{L}_I \quad (2)$$

$\mathscr{L}_0$ is exactly solvable ,and defines as vacuum the ground state of Fock space,i.e. empty space. $\mathscr{L}_I$ is a small perturbation producing scattering between fundamental particles, and small changes in the wave function of the ground state.

In QCD $\mathscr{L}_0$ describes free gluons and quarks and their interactions. For sure this is not a good starting point ,since quarks and gluons are confined.Fock vacuum is not the ground state,and therefore it is not stable against perturbations.

This is most probably the reason why the perturbative expansion is not convergent,not even as an asymptotic series,in spite of the fact that it seems to work at small distances [1].

The knowledge of the true vacuum is,in particular,relevant to the understanding of one of the most intriguing properties of QCD, namely confinement of color.

In spite of the well established evidence that quarks and gluons are the fundamental constituents of hadronic matter, they have never been observed in Nature as free particles.

All particles produced in particle reactions are color singlets.This property is known as confinement of color.

Since the pioneering papers of Gellmann in which quarks were introduced as fundamental constituents of hadrons ,quarks have been searched for in particle reactions and in Nature.None has been found,so that experiments establish upper limits,which can be found in Particle Data Group [2]. The cross section σ_q for the inclusive reaction

$$p+p \to q(\bar{q}) + X \tag{3}$$

in which a q, $(\bar{q})$ is detected as a fractionally charged particle has an upper limit

$$\sigma_q < 10^{-40}\,\text{cm}^2 \tag{4}$$

to be compared to the total cross section σ_T at the same energy $\sigma_T \sim 10^{-25}\,\text{cm}^2$. In the absence of confinement σ_q/σ_T would be of the order of unity,and is instead $\sigma_q/\sigma_T \simeq 10^{-15}$.

Similarily the abundance n_q of free quarks in Nature is measured by looking for particles of fractional charge $q = \pm 1/3, \pm 2/3$ in Millikan like experiments.

No quarks have been found,and the resulting upper limit is $n_q/n_p < 10^{-27}$ corresponding to the analysis of $\sim 1g$ of matter.

In the absence of confinement the Standard Cosmological Model (SCM) predicts $n_q/n_p \sim 10^{-12}$ [3]. Again a factor 10^{-15} between upper limit and expectation.

Such a small factor cannot have a natural theoretical explanation ,except if the actual value of the above quantities is exactly zero.Confinement is an absolute property,and this can only be explained in terms of a symmetry property of the vacuum.The knowledge of the true vacuum is necessary to understand confinement .A quantization procedure that is not based on perturbation theory is needed.

FEYNMAN PATH INTEGRAL [4] AND LATTICE FORMULATION OF QCD [5].

A quantum system is defined by the canonical variables ,q, p and the Hamiltonian H. Solving the system means to construct a Hilbert Space on which q, p act as operators obeying the equations of motion and the canonical equal-time commutation relations.

$$[q_i, p_j] = i\delta_{ij} \tag{5}$$

A ground state must exist.

Field theory is a special case. The fields $\varphi_i(\vec{x}, t)$ play the role of the q's ,their conjugate momenta $\Pi_i(\vec{x}, t) = \partial \mathcal{L}/\partial(\partial_0 \varphi_i(\vec{x}, t))$ the role of the p's. The system has infinitely many degrees of freedom.To simplify the notation we shall refer in what follows to a system with one q and one p ,since our arguments will apply to any number of degrees of freedom. Problems with divergencies ,arising from infinite number of degrees of freedom are technical in nature and do not affect the arguments below.The same is for problems in defining the conjugate momenta, which are a consequence of gauge invariance.

Since the $|q\rangle$'s are a complete set of states ,the knowledge of the amplitudes

$$A(q',t',q,t) = \langle q'|e^{-iH(t'-t)}|q\rangle \tag{6}$$

contains all physical information.

Let us divide the time interval $t' - t$ into $n+1$ intervals of equal length δ,

$$\delta = \frac{t' - t}{n+1}$$

δ is a small quantity in the limit of large n. We can write

$$e^{-iH(t'-t)} = e^{-iH\delta} e^{-iH\delta} \ldots e^{-iH\delta} \tag{7}$$

the product of $n+1$ factors.

Again for the sake of notational simplicity we shall assume that

$$H = \frac{p^2}{2m} + V(q) \tag{8}$$

The generic case with linear terms in p, or q-dependent coefficients of p^2, p only present technical complications, but do not alter the results which follow.

The amplitude eq (6) becomes, by inserting eq (7) and a number of projectors on a complete set $\int dq |q\rangle\langle q| = 1$

$$A = \int dq_1 \ldots dq_n \langle q'|e^{-iH\delta}|q_n\rangle\langle q_n|e^{-iH\delta}|q_{n-1}\rangle \ldots \langle q_1|e^{-iH\delta}|q\rangle \tag{9}$$

On the other hand, by use of the well known Baker-Haussdorf formula

$$e^{-iH\delta} \simeq e^{-i\frac{p^2}{2m}\delta} e^{-iV(q)\delta} [1 + \mathcal{O}(\delta^2)] \tag{10}$$

so that

$$\langle q_{i+1}|e^{-iH\delta}|q_i\rangle = e^{-iV(q_i)\delta} \langle q_{i+1}|e^{-i\frac{p^2\delta}{2m}}|q_i\rangle [1 + \mathcal{O}(\delta^2)]$$

and since

$$\langle q_{i+1}|e^{-i\frac{p^2\delta}{2m}}|q_i\rangle = \int \frac{dp}{2\pi} \langle q_{i+1}|p\rangle\langle p|q_i\rangle e^{-i\frac{p^2\delta}{2m}} = \int \frac{dp}{2\pi} e^{ip(q_{i+1}-q_i) - i\frac{p^2\delta}{2m}} = e^{i\frac{m}{2\delta}(q_{i+1}-q_i)^2}$$

finally

$$\langle q_{i+1}|e^{-iH\delta}|q_i\rangle \simeq e^{i\left[\frac{m}{2}\frac{(q_{i+1}-q_i)^2}{\delta^2} - V(q_i)\right]\delta} = e^{i\delta \mathcal{L}[q_i]} \tag{11}$$

Inserting eq(11) into eq (9)

$$A = \lim_{n \to \infty} \int \left[\prod_1^n dq_i\right] e^{i\int_t^{t'} \mathcal{L} dt} \tag{12}$$

In the limit $n \to \infty$, $\delta \to 0$ and eq(11) becomes exact. Eq(12) is usually written as

$$A = \int [\mathcal{D}q] e^{iS[q',t',q,t]} \tag{13}$$

The limit in eq(12) defines an integral over an infinite number of variables (a functional integral) if it exists. The amplitude eq(6) is an integral over all paths from (q,t) to (q',t') and is defined as a limit of a lattice in time, as the points of the lattice fill densely the time interval (t,t').

Exponentials in eq(11) are analytic functions: the construction allows an analytic continuation of the amplitude to euclidean time $-t \to \tau = -it$. If $J(t)$ is an arbitrary function in the time interval (t,t') we define

$$S_J \equiv S(q,t,t') + \int_t^{t'} dt\, q(t) J(t)$$

$$A_J(q',t',q,t) = \int [\mathcal{D}q] \exp\left[iS_J(q',t',q,t)\right] \tag{14}$$

In the discretized version,

$$\int_t^{t'} dt\, q(t) J(t) = \sum_{i=1}^{n} q(t_i) J(t_i)$$

We define the amplitude

$$A(t',t_1\ldots t_k,t) = \langle q',t'|T(q(t_1)\ldots q(t_k))|q,t\rangle \qquad t \leq t_i \leq t' \tag{15}$$

by the same construction which led to eq(13) as a limit of a discretized integral

$$A(t',t_1\ldots t_k,t) = \int [\mathcal{D}q]\, q(t_1)\ldots q(t_k) e^{-iS[t',t]}$$

It follows from the definition of A_J, eq.(14), that

$$A(t',t_1\ldots t_k,t) = i^n \frac{\delta^n}{\delta J(t_1)\ldots \delta J(t_k)} A_J(q',t',q,t)\big|_{J=0} \tag{16}$$

We shall now compute the amplitude for immaginary times $A_J(q',-iT;q,iT)$, in the limit $T \to \infty$. We shall assume that the external current J is different from zero only in the time interval $-T/2 \leq t \leq T/2$. Then, calling $|E_n\rangle$ a complete set of states with definite energy,

$$A_J(q',-iT;q,iT) = \sum_{E_n,E_{n'}} \langle q'|e^{-HT/2}|E_n\rangle\langle E_n,-i\frac{T}{2}|E_{n'},i\frac{T}{2}\rangle_J \langle E_{n'}|e^{-HT/2}|q\rangle \tag{17}$$

since by assumption from $-T$ to $-T/2$ and from $T/2$ to T the evolution is governed by H. In the limit of large T

$$A_J(q',-iT;q,iT) \underset{T\to\infty}{\simeq} e^{-E_0 T} \psi_0(q') \psi_0^*(q) \langle 0| \int [\mathcal{D}q] e^{-S[J]} |0\rangle \tag{18}$$

$|0\rangle$ is the ground state and E_0 its energy. Excited states are exponentially depressed in the sum eq(17). Modulo a multiplicative constant which will be shown to be unimportant, the amplitude of eq(16) is then equal to

$$Z[J] = \langle 0| \int [\mathscr{D}q] e^{-S[J]} |0\rangle \tag{19}$$

and, as in eq's (15), (16)

$$\langle 0|T(q(t_1)\ldots q(t_k))|0\rangle = \left.\frac{\delta^n Z[J]}{\delta J(t_1)\ldots \delta J(t_k)}\right|_{J=0} \tag{20}$$

$Z[J]$ is known as the functional generator of the field correlators (20).

Since the field correlators fully describe the theory, knowledge of $Z[J]$ means solution of the system. In particular the euclidean Feynman integral uniquely identifies the true ground state of the theory.

Feynman integral quantization is more fundamental than perturbative quantization.

For QCD the euclidean functional integral will be

$$Z[J] = \int [\mathscr{D}A_\mu][\mathscr{D}\psi][\mathscr{D}\bar\psi] e^{-S_J[A,\bar\psi,\psi]} \tag{21}$$

where the role of q is plaied by the gluon fields A_μ, the quark field ψ and its hermitian conjugate $\bar\psi$. The integral is defined by discretizing a finite volume V of space time to a lattice of spacing a and taking first the limit in which the lattice spacing $a \to 0$, so that the lattice densely fills the volume V, and finally sending $V \to \infty$.

In QCD both limits exist. By renormalization group arguments the lattice spacing a, at a given value of the coupling constant g, is related to the physical momentum scale Λ_L as

$$a \underset{g^2 \to 0}{\simeq} \frac{1}{\Lambda_L} \exp(-\frac{b}{g^2}) \tag{22}$$

where b is minus the inverse of the first coefficient of the beta function of the theory and is positive, because of asymptotic freedom. Eq.(22) will be valid at sufficiently small values of g^2.

At sufficiently small g^2 a is small in physical units and the granularity of the lattice becomes irrelevant.

Moreover a mass gap exists in the theory (a non zero minimum mass), which determines a finite correlation length ξ. If the lattice is sufficiently large with respect to ξ the infinite volume limit is reached. Lattice will be a good approximant to the continuum if

$$a \ll \xi \ll La$$

L being the linear size of the lattice. Lattice calculations are in that case an element of the sequence eq(12) near to the limit. Lattice will also uniquely identify the ground state.

The symmetry of the ground state, which is at the basis of the confinement mechanism can safely be studied on the lattice.

FEYNMAN INTEGRAL AND PERTURBATION THEORY.

The only known computable functional integral is the gaussian integral, with a lagrangean quadratic in the fields, which corresponds to free fields.

The action can be written as

$$S_0 = \frac{1}{2}\int dx dy \varphi_i(x)\mathscr{D}_{ij}^{-1}(x-y)\varphi_j(y) \qquad (23)$$

The resulting equations of motion are $\delta S/\delta \varphi(x) = 0$ or $\int \mathscr{D}_{ij}^{-1}(x-y)\varphi_j(y) = 0$. The matrix $\mathscr{D}_{ij}^{-1}(x-y)$ is the inverse of the propagator. For the scalar field

$$\mathscr{D}_{ij}^{-1}(x-y) = (\partial^2 + m^2)\delta_{ij}\delta(x-y)$$

The discretized version for S_0 is

$$S_L^0 = \frac{1}{2}a^4 \sum_{ab} \varphi_a \mathscr{D}_{ab}^{-1} \varphi_b$$

$\mathscr{D}_{ab}^{-1}$ is a symmetric matrix which can be diagonalized by an orthogonal transformation

$$O_T \mathscr{D}^{-1} O = \mathscr{D}_{diag}^{-1} \qquad (24)$$

The jacobian of the transformation is equal to 1 so that

$$Z = \int \prod d\varphi_i e^{-\frac{1}{2}a^4 \sum \varphi_i^2 (\mathscr{D}_{diag}^{-1})_{ii}} = c(\det \mathscr{D})^{1/2} \qquad (25)$$

The correlation functions are given by

$$\langle 0|T(\varphi_{i_1}(x_1)\ldots \varphi_{i_n}(x_n)|0\rangle = \frac{1}{Z}\int \prod d\varphi_i \varphi_{i_1}(x_1)\ldots \varphi_{i_n}(x_n) e^{-S_0} \qquad (26)$$

The integral is easy to compute, again by diagonalizing $\mathscr{D}_{ij}^{-1}$, giving

$$c(\det \mathscr{D})^{1/2} \sum_{perm} \prod \mathscr{D}_{i_a i_b}$$

The sum runs over all possible choices of the pairs i_a, i_b.

The only non zero correlators are the two point functions. The Hilbert space is Fock space. The generic correlator is the product of two point functions.

In the perturbative approach to quantization the action S is split in a quadratic part S_0 and terms containing higher powers of the fields

$$S = S_0 + S_I$$

The functional integral is then computed by expanding the weight in powers of S_I

$$Z = \int \mathscr{D}\varphi e^{-S_0 - S_I} = \int \mathscr{D}\varphi e^{-S_0} \sum_n \frac{(-1)^n}{n!} S_I^n \qquad (27)$$

S_I is a polynomial in the fields. The generic correlation function is then given by

$$\langle 0|T(\varphi_{i_1}(x_1)\ldots\varphi_{i_n}(x_n)|0\rangle = \frac{1}{Z}\int\prod d\varphi_i e^{-S_0}\varphi_{i_1}(x_1)\ldots\varphi_{i_n}(x_n)\sum_k \frac{(-1)^k}{k!}S_I^k \quad (28)$$

and again is a gaussian integral which can be computed.

The result, a sum of products of propagators, is nothing but the well known Wick's theorem, and each term corresponds to a Feynman diagram.

Eq(26) is the interchange of the order of two limits, which is not always legitimate.

The perturbative expansion can be seen as the functional version of the saddle point method for evaluating ordinary integrals. Consider a one variable integral of the form

$$I = \int_a^b dx \exp(if(x))g(x) \quad (29)$$

If f and g are analytic in a domain of the complex plane including the real interval (a,b) the path of integration can be deformed in that domain to include the points $\bar{x}$ where the phase $f(x)$ is stationary, i.e. $\frac{df}{dx} = 0$. The neighborhood of these points will give the main contribution to the integral: far from it the phase factor is rapidly oscillating, and the contribution to the integral negligible. The saddle point method consists in writing I in the form

$$I = \int_{a,C}^b d\delta\, e^{if(\bar{x})+\frac{i}{2}f''(\bar{x})\delta^2} g(\bar{x}+\delta) \sum_n \frac{i^n R^n(\delta)}{n!}) \quad (30)$$

where $R(\delta) = f - f(\bar{x}) - \frac{1}{2}f''(\bar{x})\delta^2$ (the linear term drops at $\bar{x}$), and the factor multiplying the exponential is intended as a power series in δ. The integral is then gaussian. If more saddle points $\bar{x}_i$ exist I will be the sum of analogous expansions around $\bar{x}_i$.

The method works in practice. No systematic control, however, exists of the approximations involved.

For the Feynman integral the idea is the same, except that now there are infinitely many integration variables. The phase is the action S, the integration variables are the fields, and a saddle point

$$\frac{\delta S}{\delta \varphi(x)} = 0 \quad (31)$$

is nothing but a classical solution of the equations of motion.

The point $\bar{\varphi} = 0$ is certainly a saddle point. Around it one can expand the action as

$$S[\varphi] = S[\bar{\varphi}] + \frac{1}{2}\frac{\delta^2 S}{\delta\varphi_i\delta\varphi_j}\bigg|_{\varphi=\bar{\varphi}}\delta_i\delta_j + S_I \quad (32)$$

For $\bar{\varphi} = 0$ $S[\bar{\varphi}] = 0$, $\frac{1}{2}\frac{\delta^2 S}{\delta\varphi_i\delta\varphi_j}\delta_i\delta_j = S_0$ and the saddle point formula is identical to the perturbative expansion eq (27).

In general, if different saddle points exist, the expression for the partition function becomes

$$Z = \sum_{\bar{\varphi}_i} e^{-S[\bar{\varphi}_i]} \int [d\delta] e^{-\frac{1}{2}\frac{\delta^2 S}{\delta\varphi_i\delta\varphi_j}\delta_i\delta_j} \sum_n \frac{(-1)^n}{n!} S_I^n$$

and all the classical solutions with finite action contribute.

Such solutions (if non trivial) are called instantons. In a non abelian gauge theory finiteness of the euclidean action means

$$\left| \int d^4x Tr \left[G_{\mu\nu}(x) G_{\mu\nu}(x) \right] \right| < \infty$$

or that the field $G_{\mu\nu}$ decreases more rapidly than $1/r^2$ as $r \to \infty$. The fields A_μ at large distance are a pure gauge: $A_\mu = i\partial_\mu U U^\dagger$ with U an element of the gauge group. A topological number exists which counts how many times in this mapping the group is covered when the point at infinity sweeps the sphere S_3. The topological charge is

$$Q = \frac{1}{16\pi^2} \int d^4x Tr \left[G_{\mu\nu}(x) G^*_{\mu\nu}(x) \right]$$

and is an integer. Explicit instanton solutions were found [6] which are self dual or antiselfdual

$$G_{\mu\nu} = \pm G^*_{\mu\nu}$$

A non zero value of Q implies then a non zero value of $G^2 = G_{\mu\nu} G_{\mu\nu}$, or that $\langle 0 | G^2 | 0 \rangle \neq 0$, a phenomenon which is known as gluon condensation. $\langle G^2(x) \rangle$, which must be x independent by translation invariance, is known as gluon condensate. A non zero value of $\langle G^2 \rangle$ is a highly non-perturbative result. Indeed renormalization group dictates for $\langle G^2 \rangle$ the form of an operator of dimension 4,

$$\langle G^2 \rangle \simeq e^{-4/b_0 g^2} \tag{33}$$

where $b_0 > 0$ is proportional to the first non zero coefficient of the beta function. An expression like eq(33) is non analytic at $g^2 = 0$, and therefore non expandable in a power series of g^2. A complete classification of the instantonic solutions does not exist.Models have been developed based on quasi-solutions,consisting of ensembles of single instantons.If their average distance is large compared to their size the ensemble is called an instanton gas [7], if the overlapping is important the ensemble is called an instanton liquid [8]. Attempts to improve perturbative expansion by adding to the perturbative saddle point the saddle points corresponding to instantons have been done during the years,starting from the pioneering approach of ref[7]. Although useful to describe chiral properties, instantons fail in accounting for confinement of color.

Instantons were at the basis of the SVZ [9] sum rules of QCD: that approach, which will be described in next section, is based on the existence of condensates, the original idea being that instantons provide a slowly varying background field for quantum fluctuations.

CONDENSATES.

The time ordered product of two currents $j_\mu(x), j_\nu(0)$, e.g. electromagnetic currents,can be written at short distances as a sum of local operators times c-number coefficients [10].

$$T(j_\mu(x) j_\nu(0)) \simeq C^0_{\mu\nu}(x) I + C^A_{\mu\nu}(x) G^a_{\mu\nu} G^a_{\mu\nu} + C^\psi_{\mu\nu}(x) m \bar\psi \psi + \ldots \tag{34}$$

The operators in eq(34) are ordered by increasing order of dimension in mass. The expansion is rigorously valid in perturbation theory, but is assumed to hold also when perturbation theory is not expected to work. Take the vev of eq.(34) and Fourier transform. The left side gives

$$\Pi_{\mu\nu}(q) \equiv \int e^{iqx} \langle 0|T(j_\mu(x)j_\nu(0))|0\rangle d^4x = (q_\mu q_\nu - q^2 g_{\mu\nu})\Pi(q^2) \qquad (35)$$

the tensor structure being dictated by gauge invariance.

The spectral representation reads

$$\Pi(q^2) = \Pi(0) - q^2 \int_{4m_\pi^2}^{\infty} \frac{ds}{s} \frac{R(s)}{s - q^2 + i\varepsilon} \qquad (36)$$

with

$$R(s) = \frac{\sigma_{e^+e^- \to hadrons}}{\sigma_{e^+e^- \to \mu^+\mu^-}}$$

The right side of eq.(34) gives

$$(q_\mu q_\nu - q^2 g_{\mu\nu})\left\{C_0(q^2) + C_g(q^2)G_2 + C_\psi(q^2)\langle m\bar{\psi}\psi\rangle + \ldots\right\} \qquad (37)$$

The term $C_0(q)$ corresponds to the usual perturbative expansion and is constant modulo log's. The other terms, known as "higher twists", are non perturbative in nature.

The approximate equality

$$-q^2 \int_{4m_\pi^2}^{\infty} \frac{ds}{s} \frac{R(s)}{s - q^2 + i\varepsilon} \simeq \left\{C_0(q^2) + C_g(q^2)G_2 + C_\psi(q^2)\langle m\bar{\psi}\psi\rangle + \ldots\right\} \qquad (38)$$

can be exploited by appropriate weighting which emphasises the region where the equality is a good approximation on the average: the procedure is known as sum rules [9].

The dispersion integral in eq (38) is evaluated by inserting for $R(s)$ the resonances at low s and a constant above some threshold. A nice phenomenology results, relating resonance parameters to the condensates, which can be determined consistently giving [9]

$$\begin{aligned} G_2 &= 0.012 \,\text{GeV}^4 \\ \langle \bar{\psi}\psi \rangle &= -0.13 \,\text{GeV}^3 \end{aligned}$$

A more recent determination for G_2 is [11]

$$G_2 = (0.024 \pm 0.011)\,\text{GeV}^4$$

LATTICE DETERMINATION OF THE CONDENSATES.

Consider the gauge invariant correlator

$$\mathscr{D}^C_{\mu\nu,\rho\sigma}(x) = \langle 0|Tr\left\{G_{\mu\nu}(x)S_C(x,0)G_{\rho\sigma}(0)S^\dagger_C(x,0)\right\}|0\rangle \qquad (39)$$

where $G_{\mu\nu} = \sum_a \frac{\lambda^a}{2} G^a_{\mu\nu}$, $A_\mu = \sum_a \frac{\lambda^a}{2} A^a_\mu$ and λ^a are the generators of the gauge group in the fundamental representation.

$$S_C(x,0) = P\left\{\exp(i\int_{0,C}^x A_\mu(x)dx^\mu)\right\}$$

is the parallel transport from 0 to x along the path C. Under a gauge transformation $U(y)$, $S_C(x,0) \to U(x) S_C U^\dagger(0)$, which makes $\mathscr{D}^C_{\mu\nu,\rho\sigma}$ gauge invariant. In what follows we will chose for C the straight line connecting 0 and x. $\mathscr{D}^C_{\mu\nu,\rho\sigma}$ can be considered as the split point regulator a la Schwinger of the gluon condensate G_2. The Wilson operator product expansion gives indeed

$$\mathscr{D}^C_{\mu\nu,\rho\sigma}(x) \underset{x\to 0}{\simeq} \frac{c_1}{x^4}\langle I \rangle + c_4 G_2 \tag{40}$$

whence G_2 can be extracted. Poincare' invariance implies [12]

$$\mathscr{D}_{\mu\nu,\rho\sigma}(x) = (g_{\mu\rho}g_{\nu\sigma} - g_{\mu\sigma}g_{\nu\rho})[D(x) + D_1(x)] + \tag{41}$$
$$+ (x_\mu x_\rho g_{\nu\sigma} - x_\nu x_\rho g_{\mu\sigma} - x_\mu x_\sigma g_{\nu\rho} + x_\nu x_\sigma g_{\mu\rho})\frac{\partial D_1}{\partial x^2}$$

or, choosing the x^0 axis along x^μ one can define [13]

$$\mathscr{D}_\parallel = \frac{1}{3}\sum_{i=1}^3 \mathscr{D}_{0i,0i} = D + D_1 + x^2 \frac{\partial D_1}{\partial x^2}$$
$$\mathscr{D}_\perp = \frac{1}{3}\sum_{i,j} \mathscr{D}_{ij,ij} = D + D_1$$

On the lattice, which is a formulation in terms of parallel transports, the correlators are easy to implement: the field strength is the open plaquette, the parallel transport the product of elementary links. The correlator is measured as the average on the ensemble of the configurations produced by MonteCarlo simulations of the operator:

$$\mathscr{D}^L_{\mu\nu\rho\sigma} = \left\langle \underset{P_{\mu\nu}}{\boxed{}} \underset{P_{\rho\sigma}}{\boxed{}} - \frac{1}{N_c} \underset{P_{\mu\nu}}{\boxed{}} \underset{P_{\rho\sigma}}{\boxed{}} \right\rangle$$

The figure represents the product of elementary links. The correlators D and D_1 are plotted versus the distance in fm in fig. 1.

The lines represent a best fit of the form

$$D = Ae^{-x/\lambda} + \frac{a}{|x|^4} \tag{42}$$
$$D_1 = Be^{-x/\lambda} + \frac{b}{|x|^4}$$

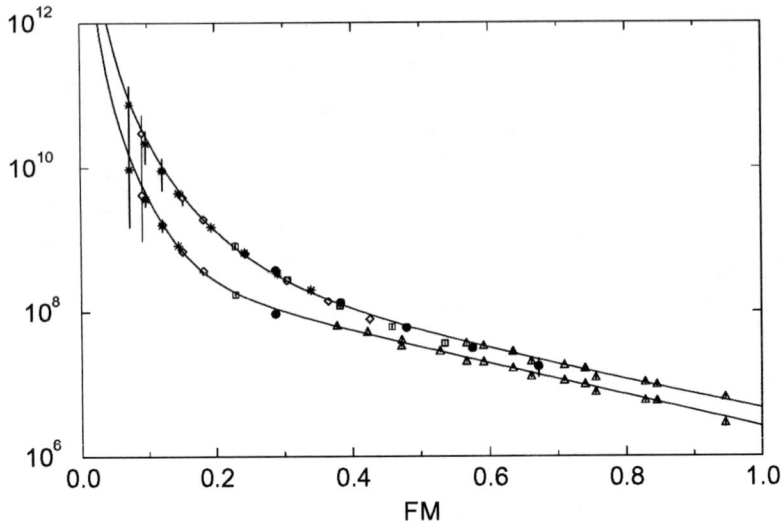

FIGURE 1. D (upper curve) and D_1 vs distance in fm, as determined on lattice [13, 14].

which corresponds to the OPE eq.(40). One can extract A, B and the correlation length λ. G_2 is proportional, with known coefficients to $A+B$. The result is

Quenched [15] $SU(2)$ $G_2 = (.33 \pm .01)\,\text{GeV}^4$ $\lambda = .16 \pm .02\,\text{fm}$
Quenched [13, 14] $SU(3)$ $G_2 = (.15 \pm .03)\,\text{GeV}^4$ $\lambda = .22 \pm .02\,\text{fm}$
Full QCD, extrapolated $G_2 = (.022 \pm .005)\,\text{GeV}^4$ $\lambda = .34 \pm .03\,\text{fm}$
to physical q masses [16]

The value of G_2 for full QCD agrees with recent phenomenological determinations [11]. The correlation length is definitely too small in quenched theories, both SU(2) and SU(3) to consider the background field as slowly varying, it is slightly larger in full QCD.

By similar techniques the chiral condensate can be extracted from the correlator [17]

$$\mathcal{D}_\psi(x) = m\langle \bar\psi(x) S_C(x,0) \psi(0) \rangle$$

which allows the OPE

$$\mathcal{D}_\psi(x) \simeq \frac{C_1}{|x|^2} + C_2 \langle m\bar\psi\psi \rangle$$

THE STOCHASTIC VACUUM.

In the Schwinger gauge $(x-z_0)^\mu A_\mu(x) = 0$, the gauge field A_μ can be expressed in terms of $G_{\mu\nu}$

$$A_\mu^a(x) = \int \alpha d\alpha (x-z_0) G_{\mu\nu}^a(z_0 + \alpha(x-z_0)) \tag{43}$$

and any physical quantity can be expressed in terms of correlators of field strengths in the vacuum. A cluster expansion is made, and the assumption that higher correlators are products of two point correlators, higher clusters being negligible (stochastic vacuum model). The two point correlators determined on the lattice are used as an input, and several physical quantities can be computed: the heavy $q\bar{q}$ bound states or high energy cross sections. A successfull phenomenology results [18].

We have shown that non perturbative effects, like gluon condensate, are important in QCD and can be studied on the lattice. In the next lecture we shall investigate what symmetry of the vacuum is at the origin of color confinement.

CONFINEMENT OF COLOR.

A deconfinement transition is observed in lattice simulations of QCD.

The static thermodynamics of a field system at temperature T is described by the partition function

$$Z = tr\left\{e^{-H/T}\right\}$$

It is a known theorem that Z is given by a Feynman euclidean integral, extending on the 3-dimensional physical space and on the time interval $(0, 1/T)$, with periodic boundary conditions for bosons, antiperiodic for fermions.

On the lattice Z is obtained by simulating the theory on a lattice $N_S^3 \times N_T$, with $N_S \gg N_T$. In terms of the lattice spacing a the temperature is then given by $T = 1/N_T a$. The only free parameter in the simulation is the gauge coupling constant g^2 or better the parameter $\beta = 2N_c/g^2$. In terms of β the lattice spacing can be written as [see eq(22)]

$$a = \frac{1}{\Lambda}\exp(-\frac{\beta}{b})$$

or

$$T = \frac{1}{N_T a} = \frac{\Lambda}{N_T}\exp(\frac{\beta}{b}) \tag{44}$$

Low temperature corresponds to small β (strong coupling), high temperature to large β (weak coupling).

This is the opposite to what happens in ordinary spin systems, where the temperature plays the role of coupling constant, and is due to asymptotic freedom. Confinement, for pure gauge theories, is related to the Polyakov line, which is the parallel transport along the time (temperature) axis from 0 to N, closed by periodic boundary conditions. The vev of the Polyakov line $\langle L \rangle$ can be interpreted as $\exp(-\mu_q)$, μ_q being the chemical potential of an isolated quark. In the confined phase $\mu_q = \infty$, $\langle L \rangle = 0$, while one can have $\langle L \rangle \neq 0$ in the deconfined phase. $\langle L \rangle$ can be called an order parameter for confinement, the symmetry being Z_N. What is usually determined on the lattice [19] is the correlator of two Polyakov lines, which by cluster property behaves at large distances as

$$G(d) = \langle L(\vec{x})\bar{L}(\vec{y})\rangle \simeq A\exp(-\sigma d a N_T) + |\langle L \rangle|^2 \tag{45}$$

with $d = |\vec{x} - \vec{y}|$.

The energy of two static quarks at distance d is related to $G(d)$ by the equation

$$V(d) = -\frac{1}{aN_T}\log(G(d)) \tag{46}$$

What is found on the lattice is that a temperature T_c exists such that

$$\begin{array}{llll} T < T_c & |\langle L \rangle| = 0 & V(d) \simeq_{d \to \infty} \sigma d & \text{(Confinement)} \\ T > T_c & |\langle L \rangle| \neq 0 & V(d) \simeq_{d \to \infty} \text{const.} & \text{(Deconfinement)} \end{array}$$

For quenched $SU(2)$ one finds $T_c/\sqrt{\sigma} \simeq 0.7$.

A finite size scaling analysis of G at different spatial sizes N_S, allows to extract the critical index ν The result is $\nu \simeq 0.62$. The phase transition belongs to the same class of universality as the 3d ising model,as expected [20]. For quenched $SU(3)$ the transition is found to be weak first order, $\nu = 1/3$ and $T_c/\sqrt{\sigma} = 0.65$ which means $T_c = 270$ Mev if the conventional value $\sqrt{\sigma} = 425$ MeV is assumed for the string tension.

In the presence of quarks $\langle L \rangle$ cannot be an order parameter, since the symmetry Z_{N_c} is explicitely broken by the very presence of the quarks. At $m_q = 0$, the chiral limit, an order parameter is the chiral condensate $\langle \bar{\psi}\psi \rangle$: however in reality the chiral symmetry is explicitely broken by quark masses. This variety of order parameters contrasts with the ideas of the limit $N_c \to \infty$.

The idea is that the number of colors N_c can be considered as a parameter of the theory. In the limit $N_c \to \infty$ with $g^2 N_c = \lambda$ fixed, a theory is defined , which differs little from the theory at finite N_c, say $N_c = 3$.

In this spirit the mechanism of confinement should be fixed by the limiting theory ,and be essentially the same at all values of N_c.

Quark loops being non leading in the expansion ,the mechanism of confinement should be the same in full QCD and in quenched theory.

The symmetry of the vacuum which is behind confinement, as explained in sect (1) should be N_c independent. In this respect the existing situation for the order parameters looks rather confusing.

Fig.2 shows schematically the phase dyagram for 2-flavor QCD as a function of the quark masses. The border between confined and deconfined phases is defined as the simultaneous maximum of various susceptibilities [21]. >From $m_q = \infty$ to $m_q = 3$ GeV the transition is first order, and the Polyakov line works as an order parameter as in quenched QCD. At $m_q = 0$ there are arguments that the transition is second order [22]. In the central region none of the susceptibilities which have been determined ($\bar{\psi}\psi$, susceptibility, derivative of the specific heat) goes large at the border between the two regions as the volume goes to infinity, and the conclusion is that maybe there is no transition, but only a crossover.

Still, if the deconfining transition is a change of symmetry,as argued in sect.1, there must exist an order(or disorder) parameter which distinguishes confined from deconfined, possibly the same as in quenched theory,in the line of $N_c \to \infty$ ideas.

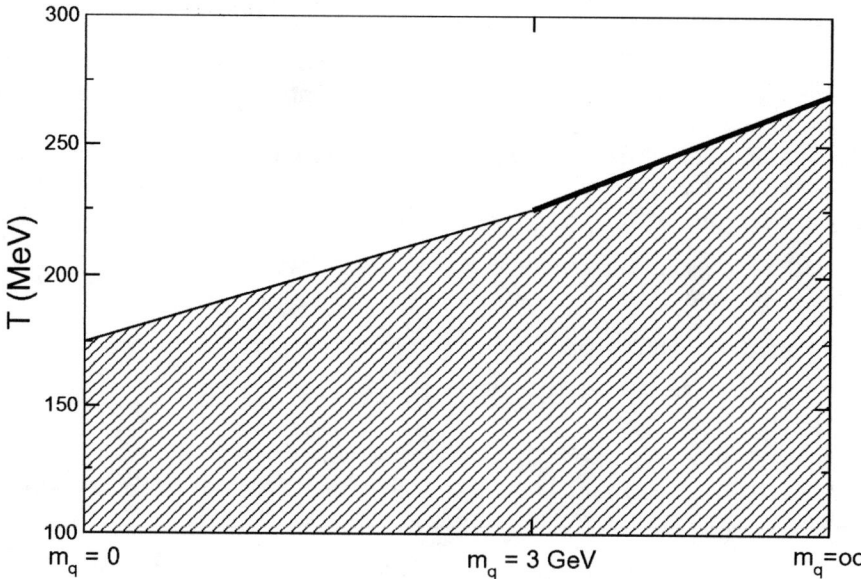

FIGURE 2. A schematic phase dyagram for 2 flavour QCD. Shaded area is confined, upper area deconfined.

DUALITY

The symmetry responsible for confinement has to be a symmetry of the (strong coupling) confining phase, which, in the langauge of statistical mechanics, is disordered. The natural question is: what can be the symmetry of a disordered phase? The answer is in a known concept in statistical mechanics, namely duality [23]. d dimensional systems admitting non local configurations with non trivial topology in $d-1$ dimensions, can have two complementary descriptions.

One description in terms of the ordinary local fields ϕ, in which topological excitations μ are non local, and $\langle\phi\rangle$ are the order parameters; this description is convenient in the weak coupling regime $g \ll 1$ (ordered phase).

The other one, dual description, in which the excitations μ become local fields, their vev's $\langle\mu\rangle$ are the order parameters: the coupling constant g_D in this description is related to the original g as $g_D \sim 1/g$. Duality maps the strong coupling regime of the direct description into the weak coupling regime of the dual.

A number of systems admitting dual descriptions are known in the literature. The prototype system is the 2d Ising model [24], in which the dual excitations are kinks, and the dual description is again a 2d ising model with $g_D \sim 1/g$; the N=2 SUSY QCD of ref [25], in which dual excitations are monopoles; the 3d XY model [26] where the dual excitations are vortices, and the Heisenberg magnet in 3d, where the dual excitations are 2d Weiss domains [27]. In QCD the dual excitations are not exactly known, but important information on their properties exists as we shall see below.

Two main candidates were originally proposed:

a) monopoles [28, 29]. Their condensation in the confining vacuum generates magnetic superconductivity, in the same way as the condensation of Cooper pairs generates ordinary superconductivity.
The chromoelectric field between a $q\bar{q}$ pair is channeled into electric Abrikosov flux tubes, so that energy is proportional to the distance.

b) Z_N vortices [30]

In what follows we shall analyze the option a). As for vortices we refer to [31, 32] and to references therein.

BASIC SUPERCONDUCTIVITY [33].

Before looking for dual (magnetic) superconductivity, we recall the basic concepts of ordinary superconductivity. The Landau-Ginzburg [34] density of free energy (effective lagrangean) is determined by arguments of symmetry and of scale.

$$\mathscr{L} = -\frac{1}{4}F_{\mu\nu}F_{\mu\nu} + \frac{1}{2}(D_\mu\phi)^*D_\mu\phi + \frac{m^2}{2}\phi^*\phi - \frac{\lambda}{4}(\phi^*\phi)^2 + \text{irrelevant terms} \qquad (47)$$

ϕ is a charged scalar field describing Cooper pairs, $D_\mu = \partial_\mu - iqA_\mu$ is the covariant derivative and the irrelevant terms are those of dimension higher than 4. $m^2 = m^2(T)$ and $\lambda = \lambda(T)$ depend on the temperature.

For $T < T_c$ $m^2 > 0$ and the potential has a mexican hat shape as a function of $R(\phi)$ an of $Im(\phi)$.

For $T > T_c$ $m^2 < 0$ and the minimum of the potential is at $\phi = 0$. If we parametrize $\phi = \rho e^{iq\theta}$, then $\tilde{A}_\mu = A_\mu - \partial_\mu \theta$ is a gauge invariant quantity: indeed under a gauge transformation $A_\mu \to A_\mu + \partial_\mu \Lambda$, $\theta \to \theta + \Lambda$. In terms of the new variables

$$\mathscr{L} = -\frac{1}{4}F_{\mu\nu}F_{\mu\nu} + \frac{q^2\rho^2}{2}\tilde{A}_\mu\tilde{A}_\mu + \frac{1}{2}(\partial_\mu\rho)^2 + \frac{m^2}{2}\rho^2 - \frac{\lambda}{4}\rho^4 \qquad (48)$$

A static homogeneous solution is $\rho = \bar{\rho}$, a constant. The effective lagrangean for the photon is in that case

$$\mathscr{L} = -\frac{1}{4}F_{\mu\nu}F_{\mu\nu} + q^2\bar{\rho}^2\tilde{A}_\mu\tilde{A}_\mu$$

The equation of motion is

$$\vec{\nabla}\wedge\vec{H} + q^2\rho^2\vec{\tilde{A}} = 0 \qquad (49)$$

Taking the curl of both sides of eq(49)

$$\nabla^2\vec{H} - q^2\rho^2\vec{H} = 0 \qquad (50)$$

The second term in eq (49) is a gauge invariant current, and is known as London current. A non zero current with zero electric field means zero resistivity or superconductivity.

Eq(50) means that the magnetic field has a finite penetration depth in the superconductor, $1/q\rho$, and this is nothing but Meissner effect. If $m \gg q\bar{\rho}$ the superconductor is named Type II. In that case, when trying to introduce a magnetic field in the bulk, a penetration through separated Abrikosov flux tubes is energetically favoured.

Outside the tube $\tilde{\vec{A}} = 0$ and $\oint_C \tilde{\vec{A}} d\vec{x} = 0$, C being a path which encercles the flux tube. Since $\tilde{\vec{A}} = \vec{A} - \vec{\nabla}\phi$, this means

$$\int_C \vec{A} d\vec{x} = \int_C \vec{\nabla} d\vec{x} = \frac{2\pi n}{q} \tag{51}$$

The magnetic flux in the tube is quantized (Dirac quantization condition). A flux tube is to all effects a monopole antimonopole pair sitting at the ends with energy proportional to the length.

The order parameter of the system is $\Phi = q\bar{\rho}$, the vev of a charged field.

MONOPOLES IN QCD.

Monopoles are always abelian [35]: the magnetic monopole term in the multipole expansion of the field produced at large distances by any hadronic matter distribution, obeys abelian field equations, and can always be reduced by a gauge transformation to the form $(0,0,a/r)$ in polar coordinates, with

$$a_\varphi = Q(1 - \cos\theta) \tag{52}$$

and the Dirac string along the north pole. The $N_c \times N_c$ matrix Q obeys the condition $e^{ig4\pi Q} = 1$, implying that, in the representation in which Q is diagonal, it has integer or half integer eigenvalues in units $1/g$. A monopole is identified by a constant diagonal matrix of the algebra with integer or half integer matrix elements: there exist $N_c - 1$ independent monopole charges.

The same result follows from the procedure known as abelian projection [28]. We shall illustrate it for the case of SU(2) gauge group: generalization to arbitrary N_c is straightforward. Let be any field in the adjoint representation. Define $\hat{\phi} = \vec{\phi}(x)/|\vec{\phi}(x)|$, a direction in color space, and

$$F_{\mu\nu} = \hat{\phi} \cdot \vec{G}_{\mu\nu} - \frac{1}{g}\hat{\phi}\left(D_\mu\hat{\phi} \wedge D_\nu\hat{\phi}\right) \tag{53}$$

Both terms in the expression (53) are color singlets and gauge invariant. The combination is chosen in such a way that bilinear terms $A_\mu a_\nu$ and $A_\mu \partial_\nu \phi$ cancel. Indeed, by explicit computation

$$F_{\mu\nu} = \hat{\phi}(\partial_\mu \vec{A}_\nu - \partial_\nu \vec{A}_\mu) - \frac{1}{g}\hat{\phi}(\partial_\mu\hat{\phi} \wedge \partial_\nu\hat{\phi}) \tag{54}$$

A gauge transformation which brings $\hat{\phi}$ to a constant, e.g. to $(0,0,1)$ is called an abelian projection on $\vec{\phi}(x)$. In that gauge indeed

$$F_{\mu\nu} = \partial_\mu(\hat{\phi}\vec{A}_\nu) - \partial_\nu(\hat{\phi}\vec{A}_\mu) \tag{55}$$

is an abelian field. If, in the usual notation $F^*_{\mu\nu} = \varepsilon_{\mu\nu\rho\sigma}F_{\rho\sigma}$, a magnetic current can be defined as

$$j_\nu = \partial_\mu F^*_{\mu\nu} \tag{56}$$

This current is identically conserved, because of the antisymmetry of $F^*_{\mu\nu}$

$$\partial_\nu j_\nu = 0 \tag{57}$$

and identifies a $U(1)$ magnetic symmetry of the system.

In the usual continuum formulation j_ν is zero (Bianchi identities).

In a compact formulation, like lattice, j_ν can be non zero.

The U(1) magnetic symmetry can be either realized à la Wigner, and then the Hilbert space consists of superselected sectors with definite magnetic charge, or can be Higgs broken. In the first case the vev $\langle\mu\rangle = \langle 0|\mu|0\rangle$ of any operator μ carrying magnetic charge is strictly zero. In the second case there exists some μ such that $\langle\mu\rangle \neq 0$.

The effective action for $\langle\mu\rangle$ will then be of the form eq(47) and the system behaves as a dual superconductor. The operator μ is a disorder operator

$\langle\mu\rangle \neq 0$ in the disordered, confining phase
$\langle\mu\rangle = 0$ in the ordered, deconfined phase

A magnetically charged operator μ can be constructed, which is magnetically charged in any given abelian projection [36, 37]. The basic principle to construct μ is the well known formula for translations

$$e^{ipa}|x\rangle = |x+a\rangle \tag{58}$$

In field theory the field $\phi(y)$ is the analog of x. Its conjugate momentum $\Pi(y)$ the analog of p, and the operator

$$\mu(\vec{x},t) = \exp\left[i\int d^3\vec{y}\,\Pi(\vec{y},t)\tilde{\phi}(\vec{x}-\vec{y})\right] \tag{59}$$

gives

$$\mu(\vec{x},t)|\phi(\vec{y},t)\rangle = |\phi(\vec{y},t)+\tilde{\phi}(\vec{x}-\vec{y})\rangle \tag{60}$$

i.e. it adds the classical configuration $\tilde{\phi}(\vec{x}-\vec{y})$ to any field configuration.

For monopoles the field is $\vec{A}(\vec{x},t)$, the conjugate momentum $\partial_0\vec{A}(\vec{x},t) = \vec{E}(\vec{x},t)$, and the classical configuration to be added is the vector potential generated at the point $\vec{y}$ by a monopole sitting at the point $\vec{x}$, e.g.

$$\vec{b}(\vec{x}-\vec{y}) = -\frac{m}{2e}\frac{\vec{n}\wedge\vec{r}}{r(r-\vec{n}\vec{r})}$$

where the Dirac string has been put along the $\vec{n}$ axis. The definition (60) has to be adapted to compact formulation. The details are published in ref [37]. The result is of the form

$$\langle\mu(\vec{x},t)\rangle = \frac{\tilde{Z}(\beta,\vec{x},t)}{Z(\beta)} \qquad \beta = \frac{2N_c}{g^2} \tag{61}$$

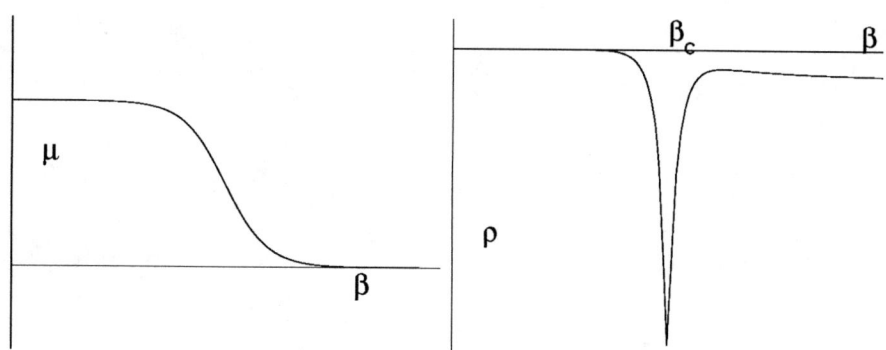

FIGURE 3. Typical dependence on β of μ and ρ.

where $Z(\beta)$ is the partition function

$$Z(\beta) = \int [\mathscr{D}A][\mathscr{D}\psi][\mathscr{D}\bar{\psi}] \exp(-\beta S_E)$$

and

$$\tilde{Z}(\beta) = \int [\mathscr{D}A][\mathscr{D}\psi][\mathscr{D}\bar{\psi}] \exp(-\beta \tilde{S}_E)$$

$\tilde{S}_E$ differs from S_E by the introduction of a dislocation on the slice $x^0 = t$. ΔS is proportional to the spatial volume, and therefore fluctuates as $N_S^{3/2}$, so that $\tilde{Z}$ fluctuates as $\exp(N_S^{3/2})$. It proves convenient to define instead

$$\rho(\beta) \equiv \frac{d}{d\beta} \log\langle \mu \rangle = \langle S \rangle_S - \langle S + \Delta S \rangle_{S+\Delta S} \qquad (62)$$

which is easier to measure. In terms of ρ

$$\langle \mu \rangle = \exp \int_0^\beta \rho(\tau) d\tau \qquad (63)$$

Fig.3 shows schematically the shape of $\langle \mu \rangle$ as a function of β, and that of ρ.

For $\beta < \beta_c$ (confinement) $\langle \mu \rangle \neq 0$, for $\beta > \beta_c$ $\quad \langle \mu \rangle = 0$, at $\beta \sim \beta_c$

$$\langle \mu \rangle \sim (\beta - \beta_c)^\delta \equiv (1 - \frac{T}{T_c})^\delta$$

For finite lattices $\langle \mu \rangle$ cannot be strictly zero above β_c, because it is analytic in β and therefore if it were zero on a line it would be identically zero. Only in the infinite volume limit singularities can develop [39] and can be identically zero above β_c. The limit $N_s \to \infty$ can be studied by finite size scaling techniques.

I In the region of low β, at N_s larger than any physical length scale ρ becomes N_s independent and finite, and from eq(63) one concludes that $\langle \mu \rangle \neq 0$ (fig.4).

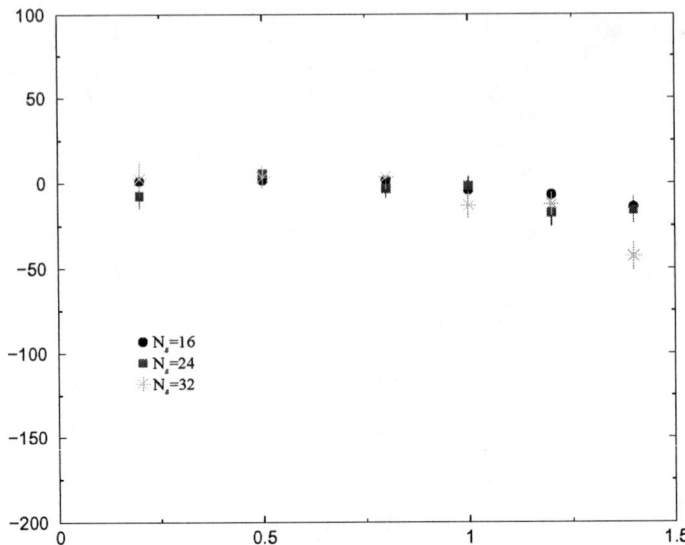

FIGURE 4. Limit at large N_s of ρ for $\beta < \beta_c$, $SU(2)$ gauge group.

II. In the region $\beta > \beta_c$ the dependence of ρ on N_s can be studied. It is found (see fig.5)
$$\rho = -kN_s + k' \qquad k > 0 \tag{64}$$
In the limit $N_s \to \infty$, $\langle \mu \rangle$ is strictly zero. A direct determination of $\langle \mu \rangle$ would give zero within large errors.

III In the vicinity of β_c by dimensional arguments
$$\langle \mu \rangle = \tau^\delta \Phi(\frac{a}{\xi}, \frac{N_s}{\xi}) \tag{65}$$
where ξ is the correlation length in units of lattice spacings. ξ diverges at the critical point with an index ν
$$\xi \underset{\tau \to 0}{\sim} \tau^{-\nu} \qquad \tau = 1 - \frac{T}{T_c} \tag{66}$$
If $\xi \gg 1$ then $1/\xi \sim 0$ and
$$\langle \mu \rangle \sim \tau^\delta \Phi(0, \frac{N_s}{\xi}) \xi \underset{\tau \to 0}{\sim} \tau^{-\nu} \tag{67}$$
One can trade N_s/ξ with $N_s^{1/\nu} \tau$ and
$$\rho = -\frac{\delta}{\tau} + N_s^{1/\nu} \Phi'(0, N_s^{1/\nu} \tau) \tag{68}$$
or
$$\rho/N_s^{1/\nu} = -\frac{\delta}{\tau N_s^{1/\nu}} + \Phi'(0, N_s^{1/\nu} \tau) \tag{69}$$

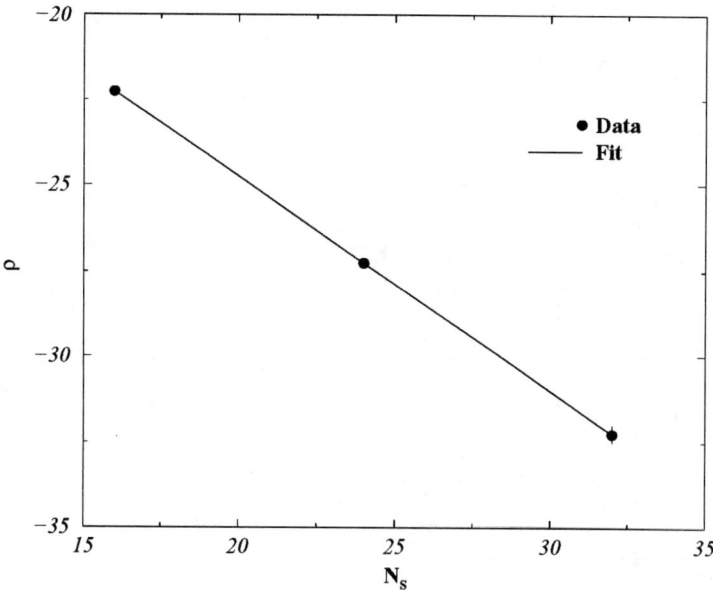

FIGURE 5. ρ vs N_s for $\beta > \beta_c$.

A scaling law results: $\rho/N_s^{1/\nu}$ is a universal function of $N_s^{1/\nu}\tau$, independent of N_s. Fig.6 shows a typical dependence of ρ on N_s. Fig.7 shows the quality of scaling for SU(2). A best square fit to the data allows to determine β_c, ν, δ.

For pure gauge theory one obtains [35]

$$SU(2) \quad \nu = .62(1) \quad \delta = 0.20(3)$$
$$SU(3) \quad \nu = .33(1) \quad \delta = .50(3)$$

The result is independent on the choice of the field used to perform the abelian projection. Confining vacuum is a dual superconductor in all abelian projections [38].

In all projections magnetic symmetry is Wigner in the deconfined phase.

Whatever the dual excitations of QCD are, they must be magnetically charged in all the abelian projections.

In QCD with dynamical quarks the same operator $\langle\mu\rangle$ can be defined ,and the corresponding $\rho = \frac{d}{d\beta}\log(\langle\mu\rangle)$ can be studied [40]. Results already exist that $\langle\mu\rangle \neq 0$ in the confining phase, and $\langle\mu\rangle = 0$ in the infinite volume limit. . The finite size scaling in the vicinity of the transition is on the way.

Now another scale, the quark mass enters, which was absent in the quenched case. Eq(65) -(67) become

$$\langle\mu\rangle \simeq \tau^\delta \Phi(\frac{1}{\xi}, \frac{\xi}{N_s}, m_q\xi) \underset{\xi\to\infty}{\simeq} \tau^\delta \Phi(0, N - s^{1/\nu}\tau, m_q N_s^{1/\gamma}) \tag{70}$$

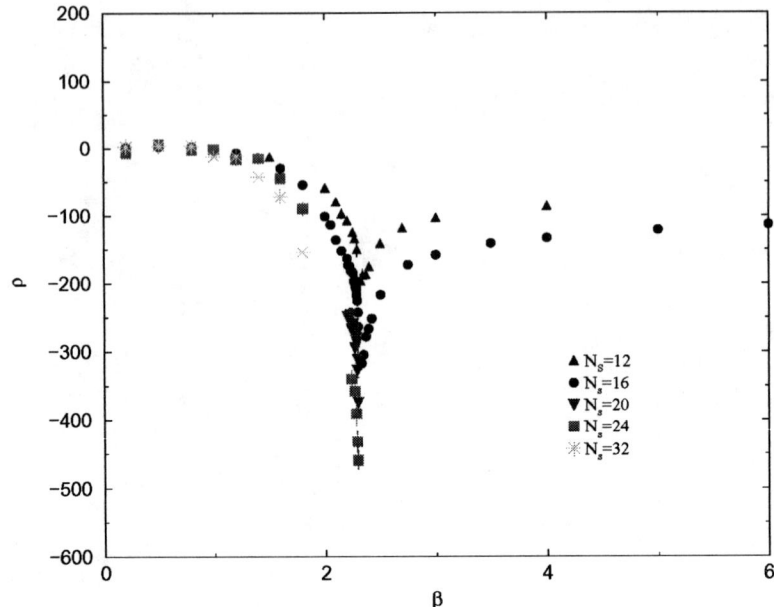

FIGURE 6. Volume dependence of ρ, pure gauge $SU(2)$, deconfinement transition.

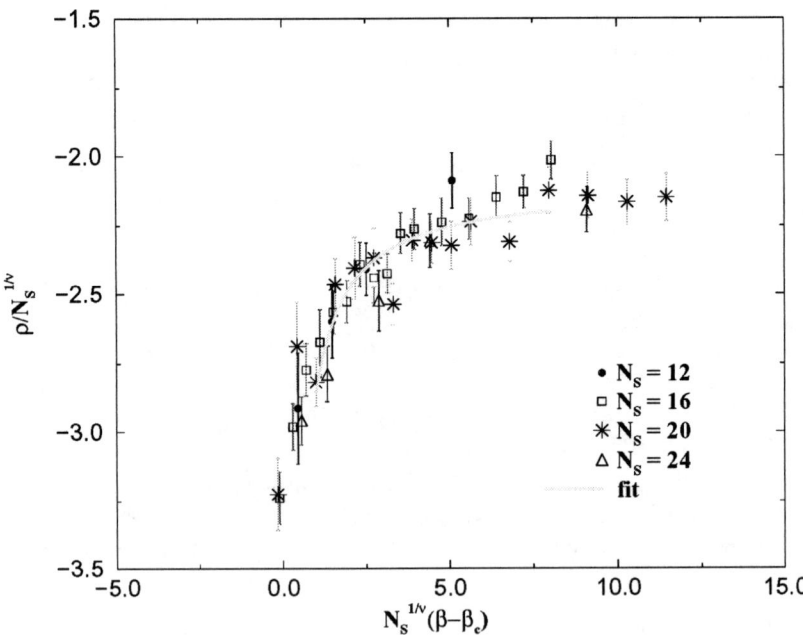

FIGURE 7. Scaling, eq.(69). $SU(2)$ pure gauge.

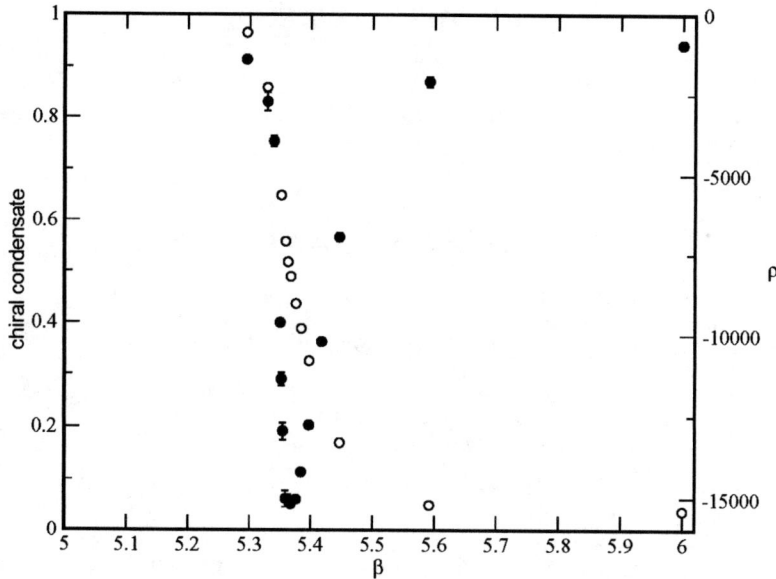

FIGURE 8. 2 flavour QCD. Open circles represent the chiral condensate, with the scale axis on the left. Full circles represent ρ, with the scale on the right. The peak of ρ coincides with the drop of $\langle\bar{\psi}\psi\rangle$.

The index γ is known. In order to determine ν one can choose values of N_s such that $m_q N_s^{1/\gamma}$=const., and test the scaling law

$$\rho/N_s^{1/\nu} \sim \Phi(N_s^{1/\nu}\tau) \qquad (71)$$

to determine ν and to investigate the nature of the transition. This will complete the analysis. The existing data already show, however, that dual superconductivity is the mechanism of confinement also in the presence of dynamical quarks, in agreement with the ideas of $N_c \to \infty$. Fig.8 shows the negative peak of ρ at the same temperature where the chiral condensate drops to zero [40].

CONCLUDING REMARKS.

Large distance QCD is an intrinsecally non perturbative system. Lattice provides a tool to investigate this regime.

Much progress has been done towards the understanding of confinement. The mechanism of confinement is definitely dual superconductivity both for pure gauge and full QCD. The dual excitations which condense to produce confinement are not yet identified: what is known is that they are magnetically charged in all abelian projections.

ACKNOWLEDGMENTS

I am indebted to L Del Debbio, M.D'Elia, B.Lucini, E Meggiolaro, G.Paffuti, who have collaborated to the results presented in these lectures. I am also grateful to Sergio Lerma for help in preparing these notes

REFERENCES

1. A.H.Mueller *Nucl.Phys.***B250**(1985)327;
2. Review of Particle Physics *E.P.J.* **15** (2000)
3. L. B. Okun *Leptons and Quarks*, North Holland (1982);
4. R.P.Feynman *Rev.Mod.Phys.* **20** (1947)267;
5. K.G. Wilson *Phys.Rev.***D10**(1974) 2445;
6. A. Belavin, A. Polyakov, A. Schwartz, Yu. Tyupkin *Phys.Lett.***B205** (1988) 339;
7. C.G.Callan, R.F. Dashen, D.J. Gross *Phys. Rev.***D19** (1979) 1826;
8. E.V. Shuryak *Rev. Mod. Phys***65**(1993) 1
9. M.A. Shifman, A.I. Vainshtein, V.I. Zakharov *Nucl.Phys.***B147**(1979) 385,448,519;
10. K.G. Wilson *Phys.Rev.***D3**(1971) 1818;
11. S. Narison *Phys.Lett.***B387**(1996) 121;
12. H.G. Dosch *Phys.Lett.***B190**(1987)555,Yu.A, Simonov *Phys. Lett.***B205**(1988)339;
13. A. Di Giacomo, H. Panagopoulos *Phys.Lett.***B285**(1992)133;
14. A. Di Giacomo, E. Meggiolaro, H. Panagopoulos *Nucl.Phys.***B483**(1997)371;
15. M. Campostrini, A. Di Giacomo, G. Mussardo *Zeits.Phys.***C25**(1984)173;
16. M. D'Elia, A. Di Giacomo, E. Meggiolaro *Phys.Lett.***B408**(1997)315;
17. M. D'Elia, A. Di Giacomo, E. Meggiolaro *Phys.Rev* **D59**(1999)54503;
18. For a recent review and for references see: H.G. Dosch, V.I. Shevchenko, Yu.A. Simonov /hep-ph/0007223;
19. J. Engels, F. Karsch, H. Satz, I. Montvay *Nucl.Phys.* **B205**(1982) 239;
20. B. Svetitsky, L.G. Yaffe *Nucl. Phys***B210** (1982)423;
21. F. Karsch, E. Laerman *Phys.Rev>***D50**(1994)6954;
22. R.D. Pisarski, F. Wilczek *Phys. Rev.* **D29** (1984)338
23. H.A. Kramers, G.H. Wannier *Phys.Rev.***60** (1941)252;
24. L.P. Kadanoff, H. Ceva *Phys. Rev.***B3** (1971)3918;
25. N. Seiberg, E. Witten *Nucl. Phys.***B341**(1994)484;
26. G. Di Cecio, A. Di Giacomo, G. Paffuti, M. Triggiante *Nucl. Phys.***B489**(1997)739;
27. A. Di Giacomo, D. Martelli, G. Paffuti*Phys. Rev.* **D60**(1999)094511;
28. G. 'tHooft *Nucl. Phys.***B190**(1981)455;
29. G. 'tHooft *High Energy Physics EPS International Conference* Palermo 1975, A.Zichichi ed., S. Mandelstam *Phys. Rep.* **23C**(1976)245;
30. G. 'tHooft *Nucl. Phys.***B138**(1978)1
31. L. Del Debbio, A. Di Giacomo, B. Lucini *Nucl. Phys.* **B594**(2001)287;
32. L. Del Debbio, A. Di Giacomo, B. Lucini *Phys. Lett.* **B500**(2001)326;
33. S. Weinberg *Progr. Theor. Phys.***Suppl.86** (1986)43;
34. V.L. Ginzburg, L. Landau *Zh.Exp.Teor.Fiz.***20** (1950)1064;
35. S. Coleman *Erice Summer School 1981* Plenum Press, A. Zichichi ed.;
36. L. Del Debbio, A. Di Giacomo, G. Paffuti, P. Pieri *Phys. Lett.***B355**(1995)255;
37. A. Di Giacomo, B. Lucini, L. Montesi, G. Paffuti *Phys. Rev.***D61**(2000)034503,034504;
38. J.M. Carmona, M. D'Elia, A. Di Giacomo, B. Lucini,G. Paffuti *Phys. Rev.***D64**(2001)114507;
39. T.D. Lee, C.N. Yang *Phys.Rev>***87**(1952)404;
40. J.M. Carmona, M. D'Elia, L. Del Debbio, A. Di Giacomo, B. Lucini, G. Paffuti *Nucl. Phys* **B106** Proc.Suppl. (2002) 607 and *in preparation*

Introduction to Effective Field Theories in QCD[1]

U. van Kolck*†, L.J. Abu-Raddad** and D.M. Cardamone*

*Department of Physics, University of Arizona, Tucson, AZ 85721
†RIKEN-BNL Research Center, Brookhaven National Laboratory, Upton, NY 11973
**Research Center for Nuclear Physics, Osaka University, 10-1 Mihogaoka, Ibaraki, Osaka 567-0047, Japan

Abstract. We present a simple introduction to the techniques of effective field theory (EFT) and their application to QCD. For problems with more than one energy scale, the EFT approach is a useful alternative to more traditional model-building strategies. The most relevant such problem for this discussion is that of making contact between QCD and the hadronic phase of matter. As a simple example, an EFT calculation of the bound states of hydrogen within QED is sketched. A more significant demonstration of the power of EFTs, the construction of the chiral Lagrangian and chiral perturbation theory, is also included. The results provide us with the road map to a complete QCD-based theory of nuclear matter at nonzero temperatures and densities, a vital component to a quantitative understanding of the phase transition from hadron gas to quark-gluon plasma.

INTRODUCTION

Effective field theory (EFT) is a theoretical prescription for constructing theories spanning multiple energy scales. The physics of a system may appear radically different at different energy scales, due to low-energy restrictions on available degrees of freedom and symmetries. When trying to construct a theory which spans energy scales, traditional methods of physics can therefore be difficult to apply. Rather than stumbling on this obstacle, however, EFT provides a method to use the physical difference between energy regimes to advantage. We present an introduction to the basic ideas of this useful technique, from the perspective of its application to nuclear physics. The emphasis is pedagogical. Comprehensive and up-to-date reviews can be found in Ref. [1]. For more extensive lectures with a similar perspective, the reader should refer to Ref. [2].

In a typical experiment designed to produce a quark-gluon plasma (QGP), two heavy nuclei collide at relativistic speeds. The experimental signal that results is necessarily dependent not just on the nature of the plasma, but on the nature of the hadronic phase as well as the physics of the phase transformation. Trying to understand such an experiment with only an understanding of the plasma phase, therefore, is as doomed to failure as attempting to understand the latent heat of a water-to-steam transformation using only the kinetic theory of gases.

To remedy this problem, we must seek a true quantitative understanding of the physics of the hadronic phase. The theory we seek should furthermore be based on the micro-

---

[1] Notes taken by LJA and DMC on lectures delivered by UvK.

scopic theory of QCD. The development of theoretically-sound hadron interactions has been the major problem of nuclear physics since its inception. This theory must be valid over the significant temperatures and densities of the phase transformations typical in experiment. However, even a more limited approach that is restricted to the hadronic phase can be useful. One would like, for example, to estimate the position of the QGP transition line. Moreover, it is believed that the phase diagram of QCD presents other interesting features such a liquid-gas transition at lower temperatures and at densities close to that of equilibrium for cold nuclear matter.

At zero baryon density, an expansion in temperature T of the free energy density of a pion gas can be derived in EFT and reveals the presence of a transition temperature T_c at the point where the expansion diverges [3]. A more precise estimate can then be obtained by considering the effects of higher-mass mesons [4], in good agreement with lattice results [5].

At non-zero baryon density ρ, on the other hand, the situation is much more complicated. Lattice QCD simulations are hampered by the infamous sign problem. EFT might therefore be the only way to study the problem, but it requires a conceptual leap from the work with mesons alone, because nucleons bind. Perturbative expansions for the energy density of dilute boson and fermion gases are well known: see, e.g., Ref. [6] for $T = 0$. These are expansions in ρa^3, where a is the scattering length, which is essentially the scattering amplitude at zero energy. The problem is that the nucleon-nucleon (NN) scattering length is large, so that the corresponding expansion fails way before the QGP phase transition. Indeed, the interesting physics of bound states such as nuclei is associated with $\rho^{1/3} \gtrsim 1/a$. The failure of a perturbative expansion in density implies that we must effectively resum such an expansion.

The reader may be somewhat surprised by the apparent failure in this regime of the perturbative approach, upon which so much of particle physics is based. This circumstance may be attributed to the complexities of the nonzero temperature and density requirements, as well as the striking disparities of the two energy scales we would like to connect. On the one hand, the consensus of the majority of the nuclear-physics community holds that in nuclei

- nucleons are non-relativistic;
- they interact via essentially two-body forces, with smaller contributions from many-body forces;
- the two-nucleon interaction generally possesses a high degree of isospin symmetry;
- external probes usually interact with mainly one nucleon at a time.

By contrast, in QCD

- the u and d quarks are relativistic;
- the interaction is manifestly multi-body, involving exchange of multiple gluons;
- there is no obvious isospin symmetry;
- external probes can, and often do, interact with many quarks at once.

It should not be surprising, then, that some new ideas are required to merge these two extraordinarily different bodies of theory. Of course, we expect that QCD encompasses the physics of hadronic interactions. The root of the problem must therefore lie in the difference of energy scales.

In fact, constructing a QCD-based theory of the hadronic phase is a problem which involves three separate energy scales spanning three orders of magnitude. The first and most obvious to a reader well versed in high-energy physics is the typical energy scale of QCD,

$$M_{QCD} \sim 1 \text{ GeV}. \tag{1}$$

The masses of all hadrons except the pion fall within this scale [2], and the scale of chiral symmetry breaking is thought to be $M_\chi = 4\pi f_\pi$, where $f_\pi \simeq 93$ MeV is the pion decay constant. The second scale,

$$M_{nuc} \sim 100 \text{ MeV}, \tag{2}$$

represents the typical momentum of nucleons in a nucleus: the inverse root-mean-square charge radius of light nuclei, or the Fermi momentum of equilibrium nuclear matter. It contains also the pion decay constant itself, the mass difference between the delta isobar and the nucleon, and the mass of the pion. The final energy scale is the typical energy scale of a nucleus,

$$\frac{M_{nuc}^2}{M_{QCD}} \sim 10 \text{ MeV}. \tag{3}$$

The binding energy per nucleon of a nucleus is typically a few MeV. For example, the binding energy of ^{4}He is 28.296 MeV, and the binding energy per nucleon of infinite nuclear matter is 16 MeV.

Since the goal is to construct a theory valid over an energy range of three orders of magnitude, and spanning regimes in which the physics is quite disparate, the problem is somewhat daunting. We might hope to seek inspiration from an analogous problem in atomic physics, the simple case of a hydrogen atom in the ground state. A qualitative argument suffices for this discussion. We know from experiment that the mass of an electron is

$$m_e \simeq 0.511 \text{ MeV}. \tag{4}$$

The Hamiltonian is, to a good approximation, $H = p^2/2m_e - \alpha/r$, where $\alpha = e^2/4\pi \simeq 1/137$ is the fine-structure constant. From the uncertainty principle we know that the electron momentum is $p \sim 1/R$, if the size of the atom is R. Minimizing $E(R) = 1/2m_e R^2 - \alpha/R$ yields $R = 1/\alpha m_e$, or

$$p \sim \alpha m_e = 3.6 \text{ keV}. \tag{5}$$

Finally, the binding energy $B = -E(1/\alpha m_e)$ is

$$B \sim \frac{1}{2}\alpha^2 m_e = 13.6 \text{ eV}. \tag{6}$$

This problem, like ours, possesses three distinct energy scales, on the order of m_e, p, and B. It is easy to see from the above, however, that in the atomic problem all three scales

---

[2] Here and throughout the paper we use units where $\hbar = c = 1$.

are coupled by the small fine-structure constant α. It is not so clear what the analogous coupling constant should be in the nuclear problem. In fact, the lack of a clear small coupling constant is one of the main difficulties of nuclear physics. Even though our problem involves QCD, the strong coupling strength α_s is not useful to us, for at the highest energy scale in the problem $\alpha_s(1\text{GeV}) \approx 1$ and perturbative QCD, valid at larger energies, breaks down.

If all obvious coupling parameters are of order one in the regime of our problem, we have no choice but to seek an alternative formulation. What is required is a theoretical framework that has the ability to sensibly and efficiently deal with multiple energy scales, as the perturbative approach does, but one that also does not rely *ab initio* on the existence of a small coupling constant. The framework which possesses both of these characteristics is EFT, as we shall see.

EFFECTIVE FIELD THEORY

EFT is a technique for developing theories of problems with multiple energy scales. It is applicable in situations where we wish to understand the physics at some low-energy scale as the limiting case of a more general problem whose full features are apparent only at some higher energy.

For simplicity, consider a generic problem with two energy scales. The complete physics of the problem throughout the full spectrum of energies can be said to be described by some Lagrangian density $\mathscr{L}(\varphi)$ in terms of some degrees of freedom φ. At a characteristic energy scale E_{under}, the full properties of $\mathscr{L}$ are vital to an understanding of the physics. In general, we may or may not know $\mathscr{L}$. Even if we know $\mathscr{L}$, we may or may not be able to solve it for the dynamics of the underlying theory. In our problem of the hadronic phase, for example, $\mathscr{L}$ is the QCD Lagrangian, and the underlying, full physics would be QCD. Processes explicitly involving quarks and gluons are seen to be important at energies of order $E_{under} \approx M_{QCD}$. At significantly lower energies, however, these processes "freeze out" and only hadronic degrees of freedom are available to the system. In the EFT, we therefore set some energy cutoff Λ which divides energies of order E_{under} from energies at which some of the freedoms of the full Lagrangian can be neglected.

The S-matrix contains all of the information relating initial states to final states in a many-body problem. The elements of the matrix can be calculated from the path integral

$$Z = \int \mathscr{D}\varphi\, e^{i\int d^D x\, \mathscr{L}(\varphi)}, \qquad (7)$$

where D is the number of spacetime dimensions in the problem. This expression is based in the full physics of the elementary, microscopic theory contained in $\mathscr{L}$.

We can begin to exploit the differences in physics between the two energy scales by using Λ to relabel the fields according to momentum,

$$\varphi = \begin{cases} \varphi_H & (p > \Lambda) \\ \varphi_L & (p < \Lambda), \end{cases} \qquad (8)$$

and integrating over the degrees of freedom in φ_H. This gives us

$$Z = \int \mathscr{D}\varphi_L \int \mathscr{D}\varphi_H e^{i\int d^D x \mathscr{L}(\varphi_L, \varphi_H)} = \int \mathscr{D}\varphi_L e^{i\int d^D x \mathscr{L}_{eff}(\varphi_L)}, \tag{9}$$

where $\mathscr{L}_{eff}$ is defined by

$$\int d^D x \mathscr{L}_{eff}(\varphi_L) = -i \ln \int \mathscr{D}\varphi_H e^{i\int d^D x \mathscr{L}(\varphi_L, \varphi_H)}. \tag{10}$$

$\mathscr{L}_{eff}$ is obviously a function only of φ_L, since we have integrated over φ_H. We can create a series representation for Eq. (10),

$$\int d^D x \mathscr{L}_{eff}(\varphi_L) \equiv \int d^D x \sum_i g_i O_i(\varphi_L). \tag{11}$$

In general, Eq. (11) is a pure mathematical statement without any real physical content, but it becomes more meaningful when we make some statements about the properties of the operators O_i and coefficients g_i.

The operators O_i can in general be quite complicated. We can see, however, that they must possess two important physical properties. The first is that, although they contain an arbitrary number of derivatives, the O_i's are in fact local operators in the sense that they involve fields at the same spacetime point. Nevertheless, their necessary dependence on derivatives follows from the fact that they depend only on fields with momentum components below Λ. As such, the uncertainty principle tells us that the operators can probe length scales only down to a minimum distance $\sim 1/\Lambda$, and so high-order derivatives will be necessary to perform the averaging over smaller length scales. The idea is the same as that of a multipole expansion in electrodynamics. The second observation we can make regarding the O_i's is that, for an appropriate decomposition (8), they must possess all of the symmetries and transformation properties of the underlying high-energy theory. Even if a particular symmetry is broken, it will manifest itself in the same way in the effective Lagrangian.

The coefficients g_i can likewise be put in perspective with some physical arguments. It is clear that they reflect the "freeze out" of the high-energy degrees of freedom below the cutoff Λ. Therefore, it is true that their particular form is dependent on the nature of the underlying theory and the structure of $\mathscr{L}(\varphi_L, \varphi_H)$. Furthermore, they are obviously a function of the cutoff parameter Λ. Note, however, that the specific dependence of the g_i on Λ is constrained by the familiar principle of renormalization-group invariance: physical observables cannot depend on the cutoff because the choice of the latter is arbitrary.

The physical motivation for EFT should now be apparent. It is an extremely common circumstance in physics that we wish to treat some low-energy limit of a fundamentally higher-energy problem. In such a case, we expect that the low-energy physics will prove to be some reflection of the full problem, but with the phase space restricted such that high-energy excitations or degrees of freedom are not accessible. Indeed, this is arguably the case in any physical problem short of a theory of everything: there exists some smaller-length and higher-energy scale at which new physics occurs, but whose details are unimportant to a description of a particular system. Were this not the case, physics as we know it would be impossible. The power of EFT, then, is that not only does it provide a framework to unify a problem with the energy scale above it, but it turns the energy difference from an obstacle into part of the solution: it recognizes this principle of limiting the system's freedom at low energy and incorporates it directly into the theoretical strategy.

The EFT is useful even when we know and can solve the underlying theory at higher energies. In this case, it is possible to derive the coefficients g_i from our knowledge of the underlying Lagrangian $\mathscr{L}(\varphi)$. The advantage lies in the reorganization of the dynamics in the EFT, as it is often the case that the effective degrees of freedom are "collective" excitations of the underlying degrees of freedom. This is the case whenever there is spontaneous breaking of a continuous symmetry and Goldstone bosons appear. An example is magnons, which are a better way to describe low-energy excitations in a system with spontaneous magnetization than individual magnetic moments. Another example are pions in QCD.

But EFT is a necessity when either the underlying theory is not known or, as in the case of QCD, it is not currently solvable. In this case, we can rely on Weinberg's "theorem": "if one writes down the most general possible Lagrangian, including *all* terms consistent with assumed symmetry principles, and then calculates S-matrix elements with this Lagrangian to any order in perturbation theory, the result will simply be the most general possible S-matrix consistent with analyticity, perturbative unitarity, cluster decomposition and the assumed symmetry principles" [7]. There is no known general proof of this "theorem", although it has been proven for a scalar field with Z_2 symmetry in Euclidean space [8]. Nevertheless, it is certainly plausible, as it is really stating that the most general quantum field theory is a direct consequence only of analyticity, unitarity, cluster decomposition, and symmetry. This, combined with the fact that there are no known counterexamples, suggests that we would do well to consider the ramifications of the "theorem" regardless of its current lack of rigor.

Weinberg's "theorem" allows us to formulate a plan to solve problems with an unknown or insoluble Lagrangian by using EFT.

1. Identify the relevant degrees of freedom and symmetries of the problem.
2. Construct the most general Lagrangian consistent with these limitations.
3. Do standard quantum field theory with this Lagrangian.

"Standard quantum field theory" consists of computing the contributions from all diagrams with momenta $Q < \Lambda$, and then renormalizing the result to relate the coefficients g_i to the physical observables of the problem. After renormalization, observables should be independent of Λ. According to Weinberg's "theorem", these steps give the dynamics of the general system, and the process of building an EFT will have introduced no spurious information.

An important issue is that of ordering the infinite number of contributions to any observable, known as "power counting". The minimal assumption is that, barring a suppression by some symmetry, the coefficients in the expansion will be roughly of $\mathscr{O}(1)$ once expressed in units of E_{under} according to dimensional analysis. This is called the assumption of "naturalness", and its validity is a consequence of our having chosen an appropriate cutoff Λ. The perturbative result of this process is a controlled expansion in energy, E/E_{under}. However, since observables are cutoff independent, the same expansion is valid for any cutoff. As we vary the cutoff, strength moves from one contribution to another that appears at the same order in E/E_{under}. So, regardless of the cutoff, to any given order in E/E_{under} only a finite number of g_i's need to be considered. One can use a finite number of experimental data as input to determine these g_i's, and then use the known g_i's to predict everything else, with an error given by the estimated size of higher-order terms. Thus, the most important ingredient to the final stage of building the EFT is power counting.

The controlled, natural expansion resulting from an EFT can be contrasted favorably with the characteristics of a theory constructed with traditional model-building strategies. A successful model takes a complex problem and reduces it reasonably to only a few degrees of freedom and types of interactions. This is a valid way to build a physical understanding of the problem, in that it allows the identification of the important symmetries and degrees of freedom. It falls short in quantitative predictive power, however, because there is no way to place a bound on the error arising from effects not included in the model. EFT, on the other hand, provides a controlled expansion of the most general dynamics, so errors are well bounded and predictable. The value of such an advantage in comparing theoretical predictions to experiment cannot be overstated.

A classical example

At this point, some simple examples of EFT calculations will prove fruitful. We begin with a problem drawn from classical physics, a light object interacting via gravity with a much larger object, for example an apple near the surface of the earth. We can easily identify the relevant degree of freedom and symmetries of this simple problem. By experimenting with various objects close to the surface of the earth, we find that, to a good approximation, the degree of freedom is the mass m, while the symmetries involve translations parallel to the earth's surface and rotations about an axis normal to it. According to our recipe for an effective theory, we write down the most general potential reflecting these properties which is a power series in the height h:

$$V_{eff}(h) = m \sum_{i=0}^{\infty} g_i h^i. \tag{12}$$

Each of the g_i's are parameters that can be fit to experimental data. (Here we are of course neglecting any quantum corrections, so there are no terms which are non-analytic in h and V_{eff} is directly observable.) The first term is an irrelevant constant that depends on the arbitrary choice of the zero in energy. The second term is the familiar mgh, where g is the acceleration of a free body due to gravity as measured at $h = 0$. The linear form of the gravitational potential, however, is simply an approximation: higher-order terms are corrections that could in principle be extracted from careful measurements. This expansion will breakdown at some energy E_{under}, which can be used to define a distance scale R through $E_{under} \equiv mgR$. This is a controlled expansion in h suitable for the low-energy regime $E \ll E_{under}$. The assumption of naturalness means

$$\frac{mg_{i+1}h^{i+1}}{mg_i h^i} = c_i \frac{E}{E_{under}} = c_i \frac{h}{R}, \qquad (13)$$

where the c_i's are coefficients of $\mathcal{O}(1)$. This reduces to

$$g_{i+1} = \mathcal{O}\left(\frac{g}{R^i}\right). \qquad (14)$$

Thanks to Newton's consideration of large distances, in this case we know the underlying theory. A more accurate expression for the gravitational potential follows from Newton's law of universal gravitation,

$$V(h) = -GMm \frac{1}{R+h}, \qquad (15)$$

where G is the gravitational constant, M is the mass of the earth, and R is its radius. For small h, we can expand Eq. (15) as

$$V(h) = mgR \sum_{i=0}^{\infty} (-1)^{i-1} \left(\frac{h}{R}\right)^i, \qquad (16)$$

where we have simplified the notation with the relation $g \equiv GM/R^2$. The underlying potential (16) indeed gives the effective gravitational potential (12) for $h \ll R$. Matching the two expressions, the coefficients of the effective potential are easily seen to be

$$g_{i+1} = (-1)^i \frac{g}{R^i}. \qquad (17)$$

The naturalness assumption (14) is verified —all of the c_i's are ± 1— with the added understanding of the scale R as the earth's radius.

We have seen that the familiar linear form of gravitational potential energy can be treated as an effective theory of the more general theory given by Newton's law of gravitation. It is worth noting in passing, however, that Newton's law could also be viewed as an effective theory of a general relativistic formulation of gravitation whose effects are only visible at even larger energies. Thus, we can begin to develop a picture of nature as an "onion", with each successively smaller energy scale being described by an effective theory of the last.

Non-relativistic QED

While a simple example from classical mechanics is instructive, we are now ready to increase our insight into the nature of EFT by examining a quantum-mechanical problem that bring us closer to our nuclear-physics interest. Let us attempt to find the effective Lagrangian for QED at low energies.

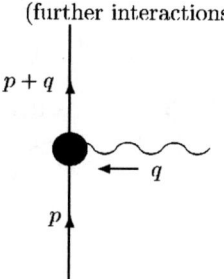

FIGURE 1. Schematic representation of a class of processes involving a low-momentum fermion. The fermion first interacts with a real or virtual photon and then propagates until it interacts further.

Consider fermions described by a field ψ of mass m and charge e, and assume the underlying theory to be given by the QED Lagrangian

$$\mathscr{L}_{QED} = \overline{\psi}[i(\slashed{\partial} - ie\slashed{A}) - m]\psi + \frac{1}{4}F_{\mu\nu}F^{\mu\nu}, \qquad (18)$$

where A_μ ($F_{\mu\nu}$) is the photon field (field strength).

What happens in processes where all particles have momenta of the same order $Q \ll m$? Consider the part of a diagram found in Fig. 1, where the fermion interacts with a low-energy photon, either real or virtual. In this non-relativistic regime we can write

$$|\vec{p}| \sim |\vec{q}| \sim Q \ll m, \qquad (19)$$

$$p^0 = \sqrt{\vec{p}^2 + m^2} \sim m + \mathcal{O}\left(\frac{Q^2}{m}\right), \qquad (20)$$

$$q^0 \sim |\vec{q}| \sim Q. \qquad (21)$$

The internal fermion line (with momentum $p+q$) contributes a factor to these diagrams

$$\begin{aligned}
\frac{i}{\slashed{p}+\slashed{q}-m+i\varepsilon} &= \frac{i(\slashed{p}+\slashed{q}+m)}{(p^0+q^0)^2-(\vec{p}+\vec{q})^2-m^2+i\varepsilon} \\
&= \frac{i(\slashed{p}+\slashed{q}+m)}{p^{0^2}+2q^0p^0+q^{0^2}-(\vec{p}+\vec{q})^2-m^2+i\varepsilon} \\
&= \frac{i(\slashed{p}+m)}{2p^0q^0+i\varepsilon}+\ldots = \frac{i}{q^0+i\varepsilon}\frac{1+\gamma^0}{2}+\ldots
\end{aligned} \qquad (22)$$

Here "…" stand for terms which are suppressed by powers of Q/m.

The last equality in Eq. (22) contains the operator we will call P_+, which can be defined together with the complementary P_-,

$$P_\pm \equiv \frac{1 \pm \gamma^0}{2}. \qquad (23)$$

It is easy to verify that these operators possess the properties of idempotent projection operators:

$$P_\pm^2 = P_\pm, \quad P_\pm P_\mp = 0. \tag{24}$$

The presence of P_+ in Eq. (22) indicates that the fermion propagation is given by the particle propagating forward in time. This suggests that we separate the nearly-inert mass m and the antiparticle component from the field, by rewriting the ψ in a "heavy-fermion formalism" [9]

$$e^{-imt}\underset{\sim}{\psi}_\pm \equiv P_\pm \psi. \tag{25}$$

The Lagrangian then becomes

$$\mathscr{L}_{QED} = \overline{\underset{\sim}{\psi}}_+ i\partial_0 \underset{\sim}{\psi}_+ - \overline{\underset{\sim}{\psi}}_- i\vec{\gamma}\cdot\vec{\nabla}\underset{\sim}{\psi}_+ + \overline{(i\vec{\gamma}\cdot\vec{\nabla}\underset{\sim}{\psi}_-)}\underset{\sim}{\psi}_+ + \overline{\underset{\sim}{\psi}}_-(-i\partial_0 - 2m)\underset{\sim}{\psi}_- + \ldots \tag{26}$$

We calculate the low-energy Lagrangian from Eq. (10). Completing the square, doing the gaussian integral over $\underset{\sim}{\psi}_-$, and renaming $\underset{\sim}{\psi}_+$ to ψ yields

$$\mathscr{L}_{eff}(\psi) = \overline{\psi} i(\partial_0 - ieA_0)\psi + \frac{1}{2m}\overline{\psi}(\vec{\nabla} - ie\vec{A})^2\psi + \ldots + \frac{1}{4}F_{\mu\nu}F^{\mu\nu} + \ldots \tag{27}$$

Of course, in the real world, there exist more than one particle coupling to photons, and other gauge bosons coupling to these particles. Because the photon is massless while weak bosons are not, the latter can be integrated out at low energies. Likewise, we can integrate out other, heavier fermions. This results in additional terms in the effective Lagrangian, such as a Pauli term that gives rise to an anomalous magnetic moment through a parameter κ:

$$\mathscr{L}_{eff}^{(\text{higher})}(\psi) = \frac{e}{2m}\kappa \varepsilon_{ijk}\overline{\psi}\sigma_k\psi F_{ij} + \ldots \tag{28}$$

The new parameters can be calculated from the underlying theory. All interactions in $\mathscr{L}_{eff}^{(\text{higher})}$ are suppressed by powers of a heavy mass. The Pauli term, for example, is $\propto 1/m$, and κ is of $\mathcal{O}(1)$ unless the particle represented by ψ has only weak interactions.

Now, according to Weinberg's "theorem", we do not need to go through this whole song and dance. If we directly construct the most general effective Lagrangian involving the fermion ψ and photon A that is invariant under gauge transformations, parity, time reversal, and (non-relativistic) Lorentz boosts, we find the $\mathscr{L}_{eff}$ above.

Bound states in QED

With the above effective Lagrangian, we can return to the problem of the hydrogen atom, considered in the Introduction. We have already seen that this problem possesses multiple energy scales, much like the nuclear-physics problem we wish to solve. Discussing non-relativistic bound states in QED is a good warm-up for tackling nuclear bound states.

For simplicity, let us consider the interaction of two non-relativistic fermions of the same type. We denote the initial (final) center-of-mass momentum p (p'), with

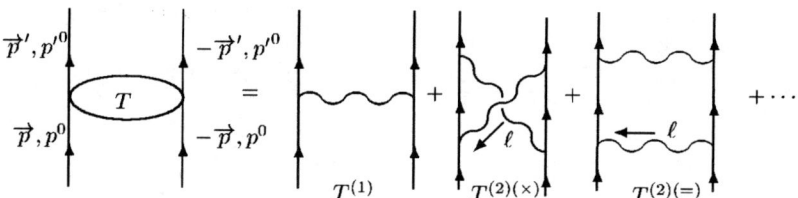

FIGURE 2. Diagrams representing all photon-exchange interactions between two heavy fermions in the center of momentum frame. The interaction T is shorthand for all interactions which can occur. Note that ℓ is a dummy 4-momentum, and as such should be integrated over in the two loop diagrams.

$|\vec{p}| \sim |\vec{p}'| \sim Q$ and $p^0 \sim p'^0 \sim Q^2/2m$. Since heavy-fermion number is conserved, all diagrams that contribute to the scattering amplitude T have two fermion lines that go through. Let us consider first the diagrams made of photon exchange only, that is, of the type in Fig. 2, where the photon-fermion vertex comes from the first term in the Lagrangian (27). We want to estimate each to leading order in Q.

The first-order diagram contributes a term

$$T^{(1)} = \frac{e^2}{(p-p')^2 + i\varepsilon} = \frac{e^2}{(p^0-p'^0)^2 - (\vec{p}-\vec{p}')^2 + i\varepsilon} = -\frac{e^2}{(\vec{p}-\vec{p}')^2 - i\varepsilon} + \ldots$$
$$\sim \frac{e^2}{Q^2}. \qquad (29)$$

The second-order term in which the photon lines cross is

$$T^{(2)(\times)} = -ie^4 \int \frac{d^4\ell}{(2\pi)^4} \frac{1}{p^0 + \ell^0 - \frac{(\vec{\ell}+\vec{p})^2}{2m} + i\varepsilon} \times \frac{1}{p'^0 + \ell^0 - \frac{(\vec{\ell}-\vec{p}')^2}{2m} + i\varepsilon}$$
$$\times \frac{1}{(p^0 - p'^0 + \ell^0)^2 - (\vec{p}-\vec{p}'+\vec{\ell})^2 + i\varepsilon} \times \frac{1}{\ell^{0^2} - \vec{\ell}^2 + i\varepsilon}. \qquad (30)$$

The integral in ℓ^0 can be evaluated as usual with a contour integral. It is clear that the first and second factors possess poles below the real axis, while the third and fourth terms each produce a pole above and below the real axis. We therefore minimize the algebra by closing in an infinite semicircle above the real axis: this way we avoid the two shallow poles from the nucleon propagators, and

$$T^{(2)(\times)} = e^4 \int \frac{d^3\ell}{(2\pi)^3} \frac{1}{p^0 - |\vec{\ell}| - \frac{(\vec{\ell}+\vec{p})^2}{2m} + i\varepsilon} \times \frac{1}{p'^0 - |\vec{\ell}| - \frac{(\vec{\ell}-\vec{p}')^2}{2m} + i\varepsilon}$$
$$\times \frac{1}{(p^0 - p'^0 - |\vec{\ell}|)^2 - (\vec{p}-\vec{p}'+\vec{\ell})^2 + i\varepsilon} \times \frac{1}{2|\vec{\ell}| - i\varepsilon} + \ldots$$
$$\sim e^4 \left(\frac{Q^3}{4\pi}\right) \left(\frac{1}{Q}\right) \left(\frac{1}{Q}\right) \left(\frac{1}{Q^2}\right) \left(\frac{1}{Q}\right) \sim \alpha \frac{e^2}{Q^2}. \qquad (31)$$

There is no surprise here: these contributions are, as ordinarily expected from loop diagrams, down by a factor of α compared to Eq. (29).

Life is not so boring, fortunately. In the other second-order term, the photon lines do not cross. This contributes

$$T^{(2)(=)} = -ie^4 \int \frac{d^4\ell}{(2\pi)^4} \frac{1}{p^0 + \ell^0 - \frac{(\vec{\ell}+\vec{p})^2}{2m} + i\varepsilon} \times \frac{1}{p^0 - \ell^0 - \frac{(\vec{\ell}+\vec{p})^2}{2m} + i\varepsilon}$$
$$\times \frac{1}{(p^0 - p'^0 + \ell^0)^2 - (\vec{p} - \vec{p}' + \vec{\ell})^2 + i\varepsilon} \times \frac{1}{\ell^{0\,2} - \vec{\ell}^2 + i\varepsilon}. \quad (32)$$

The pole structure of Eq. (32) is the same as that of Eq. (30), except that the pole of the second factor lies in the upper half complex plane. In addition to residues from poles from photon propagators that contribute terms similar to the ones in Eq. (31), we cannot avoid a contribution from a shallow pole,

$$T^{(2)(=)} = e^4 \int \frac{d^3\ell}{(2\pi)^3} \frac{1}{2p^0 - \frac{(\vec{\ell}+\vec{p})^2}{m} + i\varepsilon} \times \frac{1}{\left[2p^0 - p'^0 - \frac{(\vec{\ell}+\vec{p})^2}{2m}\right]^2 - (\vec{p} - \vec{p}' + \vec{\ell})^2 + i\varepsilon}$$
$$\times \frac{1}{\left[p^0 - \frac{(\vec{p}+\vec{\ell})^2}{2m}\right]^2 - \vec{\ell}^2 + i\varepsilon} + \ldots$$
$$\sim e^4 \left(\frac{Q^3}{4\pi}\right)\left(\frac{m}{Q^2}\right)\left(\frac{1}{Q^2}\right)\left(\frac{1}{Q^2}\right) + \alpha \frac{e^2}{Q^2} \sim \alpha \frac{e^2}{Q^2}\left(\frac{m}{Q} + \ldots\right). \quad (33)$$

We note that

$$\frac{m}{Q} \gg 1. \quad (34)$$

Thus, the term in $T^{(2)(=)}$ from the shallow pole is enhanced over the other terms. Having done the time-component integrals in these diagrams, we can now think in terms of *time-ordered* perturbation theory. The reason time-ordered diagrams are useful here is that we are considering the interactions of non-relativistic particles. In time-ordered perturbation theory loops correspond to three-dimensional integrals. Each diagram in covariant perturbation theory unfolds into various time-ordered diagrams. The time-ordering of the four vertices in each second-order diagram is significant. Second-order terms, then, can be said to fall into two groups, those for which both photons are created and then both are destroyed, and those for which the first photon is destroyed before the second is created. These second diagrams are really iterations of first-order diagrams, and it is these that lead to the enhancement. It is clear that they can only occur as part of $T^{(2)(=)}$, since $T^{(2)(\times)}$ diagrams contain a crossing of photon lines by definition.

Infrared enhancements will appear at all orders in perturbation theory, always from "two-fermion reducible" diagrams. The leading terms at each order form a series that goes roughly as

$$T \sim \frac{e^2}{Q^2}\left[1 + \mathcal{O}\left(\alpha\frac{m}{Q}\right) + \cdots + \mathcal{O}(\alpha)\right] \sim \frac{e^2}{Q^2}\left[\frac{1}{1 - \mathcal{O}\left(\alpha\frac{m}{Q}\right)} + \cdots + \mathcal{O}(\alpha)\right]. \quad (35)$$

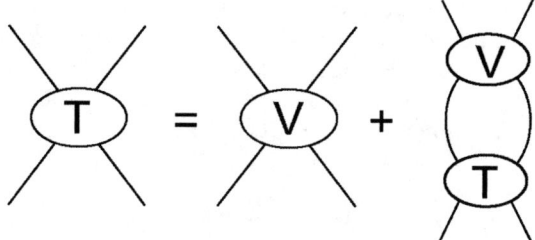

FIGURE 3. The Lippmann-Schwinger equation with potential V.

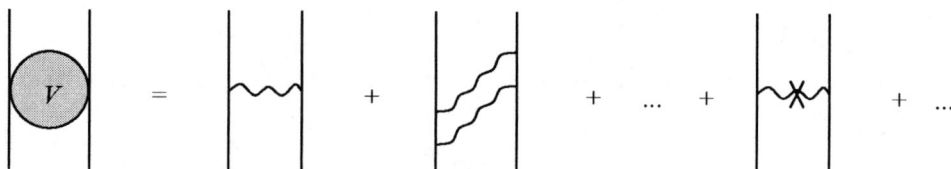

FIGURE 4. Some time-ordered diagrams contributing to the two-particle potential V in non-relativistic QED. All orderings with at least one photon in intermediate states are included. A cross denotes energy corrections.

It is clear from Eq. (35) that the series expansion breaks down at sufficiently small momenta such that m/Q compensates for α, that is, at

$$Q \sim \alpha m, \qquad (36)$$

indicating the existence of a bound state with binding energy

$$B \sim \frac{Q^2}{m} \sim \alpha^2 m. \qquad (37)$$

Hence we reproduce the estimates (5) and (6) arrived at earlier.

This is *not* just a more complicated way of generating known results: it is a more *fundamental* way. Of course, the actual resummation of this series can only be carried out by explicitly solving the Lippmann-Schwinger (or, equivalently, the Schrödinger) equation with a potential V, as represented in Fig. 3. The potential is given in leading order by the Coulomb potential given in Eq. (29). But from quantum mechanics alone we are at a loss about how to improve on the Coulomb potential. With non-relativistic QED, the prescription is clear, and it is illustrated in Fig. 4. The Coulomb potential of $\mathcal{O}(e^2/Q^2)$ is the result from the lowest-order, static photon exchange. The corrections in the potential come from all other two-fermion irreducible time-ordered diagrams. As we have seem above, the irreducible second-order diagrams generate corrections of $\mathcal{O}(\alpha e^2/Q^2)$. The recoil part of the one-photon exchange (29) is an $\mathcal{O}(Q/m \times e^2/Q^2)$ correction, which for the bound state (where $Q/m \sim \alpha$) is comparable to two-photon exchange. And so on.

A remarkable feature of the discussion above is that all the photon-exchange diagrams considered scaled with negative powers of Q. This is rooted on our temporary restriction to diagrams built from the simplest interaction in the Lagrangian (27). It means that no new ultraviolet cutoff is needed in loop integrals, once the bare fermion charge and mass are adjusted to remove cutoff dependence in diagrams involving a single fermion. At some point in the past great importance was attributed to this "renormalizability". Nowadays, we realize that its simple form in QED is limited. Interactions with more derivatives, which are not forbidden by symmetry and thus exist, spoil this simplification. Piling up sufficiently many higher-derivative interactions from Eq. (28), a two-fermion diagram will scale as a positive power of Q, and diverge in the absence of a cutoff. However, because four-fermion interactions can be constructed that are gauge invariant, the new cutoff dependence can still be eliminated by a shift in four-fermion parameters. The EFT is still renormalizable, in the sense that at any given power of Q a finite number of parameters will remove cutoff dependence from observables (renormalization-group invariance). QED can be deceivingly simple because the symmetries allow couplings of dimension four, which dominate. As we are going to see next, the same does *not* happen in the EFT of QCD.

BUILDING A QCD-BASED EFT

Standard model and QCD

We now focus on the energy region of interest. The EFT corresponding to the standard model at an energy scale of a few GeV involves only the lightest leptons and quarks, gluons, and the photon; weak gauge bosons, and heavy leptons and quarks can be integrated out. For simplicity, here we focus on interactions involving the lightest u and d quarks. They can be arranged in a flavor doublet $q = \binom{u}{d}$. The relevant pieces of the effective Lagrangian at this scale are

$$\begin{aligned}\mathscr{L} = & -\sum_{f=1}^{3} \bar{l}_f(\slashed{\partial} - ie\slashed{A} + m_f)l_f - \frac{1}{4}F_{\mu\nu}F^{\mu\nu} \\ & -\bar{q}(\slashed{\partial} - ig_s\slashed{G} - ieQ\slashed{A})q - \frac{1}{2}(m_u + m_d)\bar{q}q + \frac{1}{2}(m_d - m_u)\bar{q}\tau_3 q \\ & -\frac{1}{2}Tr[G_{\mu\nu}G^{\mu\nu}] + \frac{\bar{\theta}g_s^2}{32\pi^2}\varepsilon_{\mu\nu\rho\sigma}Tr[G^{\mu\nu}G^{\rho\sigma}] + \ldots \end{aligned} \quad (38)$$

Here the l_f's are the lepton fields with mass m_f, G_μ (A_μ) is the gluon (photon) field of strength $G_{\mu\nu}$ ($F_{\mu\nu}$), τ_3 is the usual Pauli matrix, $Q = 1/6 + \tau_3/2$ is the quark charge matrix, and "..." denote higher-dimension terms. Higher-dimension interactions are suppressed by powers of the masses of the heavy particles that have been integrated out. We will neglect them in a (very good) first approximation. We will also neglect the theta term, since the strong CP parameter $\bar{\theta}$ is found to be unnaturally small (the so-called strong CP problem). The effect of these terms can be easily incorporated in a more comprehensive analysis.

The leading Lagrangian (QCD + QED) becomes then invariant under parity and time-reversal transformations. The quark/gluon sector has 4 remaining parameters: the gauge couplings for strong (g_s) and electromagnetic (e) interactions, and the masses of the up (m_u) and the down (m_d) quarks.

In the chiral limit, that is, when $m_u = m_d = 0$ and $e = 0$, the action of the QCD Lagrangian becomes invariant under scale transformations: $x \to \lambda^{-1} x$, $q \to \lambda^{3/2} q$, and $G_\mu \to \lambda G_\mu$. Thus there is no dimensionful parameter in the Lagrangian; there is only one dimensionless parameter which is the strong-interaction coupling constant g_s. However, the scale symmetry is explicitly broken due to quantum corrections that probe high energies. A regulator is necessary, thus introducing a dimensionful parameter in the problem. As a result, g_s becomes a "running coupling" that is a function of a momentum scale. We observe in fact that g_s grows as the scale is lowered, and reaches $g_s \sim 1$ near 1 GeV. Perturbation theory in g_s, a useful tool at high energies, fails miserably at nuclear energies.

Basic assumptions about a QCD-based EFT

Thus in our pursuit to build an EFT, we have to take the properties of the nonperturbative QCD into account, even though we cannot at present deduce them. The first assumption we make is that QCD "confines": only colorless states (hadrons) are asymptotic states. In the low-energy EFT, only hadrons need to be incorporated as fields. Our second assumption is that QCD is natural. Since almost all hadron masses $\gtrsim 1$ GeV, we conclude that there is a characteristic QCD mass scale $M_{QCD} \sim 1$ GeV. This sets the limit of validity of the EFT.

Because we want to discuss nuclei, we certainly want to include the nucleon in the EFT. Because the delta-nucleon mass difference $\delta \equiv m_\Delta - m_N \simeq 300$ MeV $\ll M_{QCD}$, if we want to explore the full region of energies accessible to EFT, we incorporate the delta isobar as well. Nucleon and delta fields form isospin multiplets as following:

$$N = \begin{pmatrix} p \\ n \end{pmatrix} \quad \text{and} \quad \Delta = \begin{pmatrix} \Delta^{++} \\ \Delta^{+} \\ \Delta^{0} \\ \Delta^{-} \end{pmatrix}. \tag{39}$$

Below we denote these fermions by ψ. Other baryon states such as N^* are not included, as their mass differences to the nucleon approach M_{QCD} ($m_{N^*} - m_N \simeq 500$ MeV, etc.). Mesons such as the ρ and the ω are not included either because they have masses $\mathcal{O}(M_{QCD})$. (Moreover, they interact with the pions and nucleons via dimension-four interactions that are not weak, which is at present an insurmountable obstacle for a systematic approach.)

Does the pion mass, $m_\pi \simeq 140$ MeV $\ll M_{QCD}$, imply a breakdown of naturalness? No! We notice that in the chiral limit the leading-order Lagrangian (38) has an $SU_L(2) \times SU_R(2) \sim SO(4)$ chiral symmetry, since it is invariant under independent $SU(2)$ rotations of left- and right-handed quarks,

$$q_{R(L)} \equiv \frac{1 + (-)\gamma_5}{2} q \to \exp(i\vec{\alpha}_{R(L)} \cdot \vec{\tau}) q_{R(L)}. \tag{40}$$

This symmetry is not manifest in the hadron spectrum. There is, for example, no scalar particle degenerate with the three pions. We thus make our third assumption, that chiral symmetry is spontaneously broken down to its diagonal subgroup, the $SU_{L+R}(2) \sim SO(3)$ of isospin. Although we cannot at present calculate it, the effective potential of QCD, when plotted as function of quark bilinears ($\bar{q}\tau i\gamma_5 q, \bar{q}q$), has to have a "Mexican-hat" shape. The symmetry of the potential is $SO(4)$. The degenerate minima form thus a (four-dimensional) "chiral circle", the radius of which we call f_π. (Later, we find out that f_π so defined coincides with the pion decay constant.) The symmetry is broken because the world actually sits on one particular point of this circle (which is selected by the quark-mass perturbation). Excitations orthogonal to the circle should have mass $m_\sigma \sim M_{QCD}$, since this is the scale that sets the curvature of the effective potential. Excitations along the circle, on the other hand, are massless; the three Goldstone bosons can be associated with the light pions.

Interactions of pions in the EFT have to reproduce these symmetry properties. There are several possible parametrizations of the chiral circle which correspond to different choices of pion fields $\boldsymbol{\pi}$, such as the σ-model-like, the Callan-Coleman-Wess-Zumino construction, and the stereographic projection. Since a chiral transformation rotates the chiral circle, the infinitesimal transformation of the pions is $\boldsymbol{\pi} \to \boldsymbol{\pi} + \boldsymbol{\varepsilon}$. The EFT Lagrangian has to have the same symmetry, which implies that the pion fields can be chosen to couple derivatively. This is very important because it means that pion interactions involve the small momentum Q. Moreover, because the derivative is on the chiral circle, non-linear terms

$$D^{-1} \equiv 1 - \boldsymbol{\pi}^2/4f_\pi^2 + \cdots \qquad (41)$$

appear. There is a well-defined procedure to construct the Lagrangian, the theory of non-linear realizations of a symmetry [10]. We first define covariant objects, such as covariant derivatives that involve the associated non-linear terms: the pion covariant derivative $\boldsymbol{D}_\mu$ and the fermion covariant derivative $\mathscr{D}_\mu$. Then, the invariant Lagrangian can be constructed as the most general $SO(3)$-invariant Lagrangian made out of covariant objects.

We can now relax the restriction to the chiral limit and consider the effect of non-vanishing quark masses. The important point is that the mass terms break $SO(4)$ in a specific way. The common-mass term is the fourth component of an $SO(4)$ vector, and the mass-difference term is the third component of another $SO(4)$ vector. The common-mass term, for example, tilts the Mexican hat in the $\bar{q}q$ direction. The chiral circle is no longer degenerate, and the pions get a common mass $m_\pi^2 \sim (m_u + m_d)M_{QCD}$. In the EFT, we construct all operators that break chiral symmetry in the same way. Their coefficients will be proportional to powers of $(m_u + m_d)$ and $(m_d - m_u) \equiv \varepsilon(m_u + m_d)$.

Finally, electromagnetic interactions via "soft" photons can be added as well through all gauge-invariant terms. The one complication is that the integrating out of "hard" photons generates operators that do not necessarily involve soft photons. Such "indirect" electromagnetic operators can be constructed by looking at the chiral transformation properties of the (non-local) four-quark operators they produce. One can show that these operators break chiral symmetry as 34 and 34-34 components of $SO(4)$ antisymmetric tensors. The corresponding operators in the EFT will break isospin with strengths proportional to powers of e^2.

To sum it up, the EFT Lagrangian has two classes of interactions. One class is chiral invariant; they involve powers of the momentum and no dimension-four interactions are permissible. Another class is chiral-symmetry breaking; dimension-four operators are among the ones allowed, but they are suppressed by powers of the small quark masses or charges. As a consequence, all pion interactions are weak at low energies.

The chiral Lagrangian

Based on the above, the most general Lagrangian takes the schematic form:

$$\mathcal{L} = \sum_{mnpqf}^{\infty} C_{mnpqf} \left(\frac{\boldsymbol{D}_\mu}{M_{QCD}}\right)^n \left(\frac{\mathcal{D}_\nu}{M_{QCD}}\right)^m \left(\frac{\bar{\psi}\psi}{f_\pi^2 M_{QCD}}\right)^{\frac{f}{2}} \left(\frac{\delta}{M_{QCD}}\right)^q \left(\frac{m_\pi^2}{M_{QCD}^2}\frac{\boldsymbol{\pi}^2}{f_\pi^2}\right)^p$$
$$\times f_\pi^2 M_{QCD}^2, \qquad (42)$$

where the C_{mnpqf} are the unknown constants in the EFT. By naturalness these are of $\mathcal{O}(1)$ if isospin conserving and $\mathcal{O}(\varepsilon)$ and $\mathcal{O}(e^2)$ if isospin breaking.

It is convenient to introduce the "chiral index" Δ that counts the inverse powers of the large mass scale M_{QCD},

$$\Delta = n + m + q + 2p + \frac{f}{2} - 2 \equiv d + \frac{f}{2} - 2. \qquad (43)$$

Because of chiral symmetry, there are no interactions with negative chiral index. We can then split the chiral Lagrangian in pieces $\mathcal{L}^{(\Delta)}$:

$$\sum_{\Delta=0}^{\infty} \mathcal{L}^{(\Delta)} = \mathcal{L}^{(0)} + \mathcal{L}^{(1)} + \mathcal{L}^{(2)} + \ldots \qquad (44)$$

The form of the Lagrangian depends on the parametrization of the pion fields. For example, using stereographic coordinates, the lower-order Lagrangians are [11, 12, 13]

$$\begin{aligned}\mathcal{L}^{(0)} =\ & -2f_\pi^2 \boldsymbol{D}_\mu^2 - \frac{m_\pi^2}{2}D^{-1}\boldsymbol{\pi}^2 + N^\dagger i\mathcal{D}_0 N - 2g_A \vec{\boldsymbol{D}} \cdot N^\dagger \boldsymbol{t}\vec{\sigma}N \\ & + C_0^{(S)} N^\dagger N\, N^\dagger N + C_0^{(T)} N^\dagger \vec{\sigma} N \cdot N^\dagger \vec{\sigma} N \\ & + \Delta^\dagger (i\mathcal{D}_0 - \delta)\Delta - 2h_A \vec{\boldsymbol{D}} \cdot \left[N^\dagger \boldsymbol{T}\vec{S}\Delta + h.c.\right] + \ldots, \end{aligned} \qquad (45)$$

$$\begin{aligned}\mathscr{L}^{(1)} = &\frac{1}{2m_N}N^\dagger \vec{\mathscr{D}}^2 N - \frac{g_A}{2m_N f_\pi}D_0 \cdot \left[iN^\dagger t\vec{\sigma}\cdot\vec{\mathscr{D}}N + h.c.\right]\\&-B_1 D_\mu^2 N^\dagger N - B_2(\vec{D}\times\vec{D})\cdot N^\dagger t\vec{\sigma} N - \frac{B_3 m_\pi^2}{4f_\pi^2}D^{-1}\pi^2 N^\dagger N - B_4 D_0^2 N^\dagger N\\&-D_1\vec{D}\cdot N^\dagger t\vec{\sigma} N\, N^\dagger N - D_2\vec{D}\cdot\left(N^\dagger t\vec{\sigma} N \times N^\dagger t\vec{\sigma} N\right)\\&-\frac{E_1}{2}N^\dagger N\, N^\dagger t N\cdot N^\dagger t N - \frac{E_2}{2}N^\dagger N\, N^\dagger t\vec{\sigma} N\cdot N^\dagger t\vec{\sigma} N\\&-\frac{E_3}{2}N^\dagger t\vec{\sigma} N\cdot\left(N^\dagger t\vec{\sigma} N\times N^\dagger t\vec{\sigma} N\right)\\&+\frac{1}{2m_N}\Delta^\dagger\vec{\mathscr{D}}^2\Delta - \frac{h_A}{2m_N f_\pi}D_0\cdot\left[iN^\dagger T\vec{S}\cdot\vec{\mathscr{D}}\Delta + h.c.\right] + \ldots\end{aligned} \quad (46)$$

Here g_A, $h_A = \mathcal{O}(1)$ and $B_i = \mathcal{O}(1/M_{QCD})$ are undetermined constants, to be obtained either by solving QCD or by fitting data; "..." stand for other terms involving the delta isobar. Higher-index interactions can be constructed similarly.

Why is nuclear physics so interesting?

Processes that involve at most one nucleon ($A \leq 1$) can be easily described in the EFT. If all external momenta are of the same order $Q \sim M_{nuc}$, there are only two scales Q and M_{QCD}. A generic contribution to an amplitude can be written as

$$T \sim \mathcal{N}\left(\frac{Q}{M_{QCD}}\right)^\nu \mathcal{F}\left(\frac{Q}{m_\pi}\right), \quad (47)$$

where $\mathcal{F}$ is a dimensionless non-analytic function and $\mathcal{N}$ is a normalization factor. Counting powers of Q in a particular diagram can be done as for the superficial degree of divergence. For a diagram with L loops and V_Δ vertices of index Δ we find [7]

$$\nu = 2 - A + 2L + \sum_\Delta V_\Delta \Delta. \quad (48)$$

This formula is important because chiral symmetry places a lower bound on the interaction index $\Delta \geq 0$. Since L is bounded from below ($L \geq 0$), $\nu \geq \nu_{min} = 2 - A$ for strong interactions. An expansion in Q/M_{QCD} results. It starts at $\nu = \nu_{min}$ with tree ($L = 0$) diagrams built out of vertices of index 0 ($\sum V_\Delta \Delta = 0$), then proceeds at $\nu = \nu_{min} + 1$ with further tree diagrams, now with one vertex of index 1, the remaining having index 0 ($\sum V_\Delta \Delta = 1$). These first two orders are equivalent to the current algebra of the 1960's, but now unitarity corrections can be accounted for systematically. At $\nu = \nu_{min} + 2$, for example, besides tree diagrams with one index-2 interaction or two index-1 interactions ($\sum V_\Delta \Delta = 2$), there are also one-loop ($L = 1$) diagrams built out of index-0 vertices ($\sum V_\Delta \Delta = 0$). This is generalized to higher orders in obvious fashion. An estimate of the expansion parameter is $\sim m_\pi/m_\rho \approx 0.2$. In this context the EFT is called Chiral Perturbation Theory (ChPT). See Ref. [14] for a review.

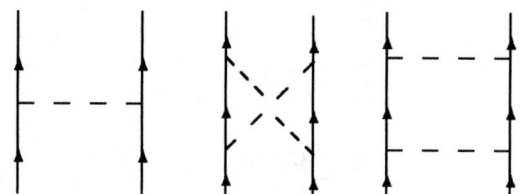

FIGURE 5. Simplest pion-exchange diagrams in the *NN* amplitude.

Processes with two or more nucleons ($A \geq 2$) are much more interesting. One can see this by studying the simplest Feynman diagrams contributing to the *NN* amplitude T_{NN}, shown in Fig. 5. The first diagram results in the term:

$$T_{NN}^{(1)} = \frac{g_A^2}{f_\pi^2} \frac{\vec{\sigma}_1 \cdot \vec{q} \, \vec{\sigma}_2 \cdot \vec{q}}{q^2 + m_\pi^2} \sim \frac{g_A^2}{f_\pi^2}. \tag{49}$$

This is analogous to Coulomb-photon exchange in QED. By following the same steps as in that example, we see that the crossed-box diagram is

$$T_{NN}^{(2)(\times)} \sim \frac{g_A^4}{f_\pi^4}\left(\frac{Q^3}{4\pi}\right)\left(\frac{1}{Q}\right)\left(\frac{1}{Q}\right)\left(\frac{Q^2}{Q^2}\right)\left(\frac{Q^2}{Q}\right) \sim \frac{g_A^2 Q^2}{4\pi f_\pi^2} \frac{g_A^2}{f_\pi^2}, \tag{50}$$

which is small, given that $4\pi f_\pi \sim M_{QCD}$.

However, the box diagram is

$$T_{NN}^{(2)(x)} \sim \frac{g_A^4}{f_\pi^4}\left(\frac{Q^3}{4\pi}\right)\left(\frac{m_N}{Q^2}\right)\left(\frac{Q^2}{Q^2}\right)\left(\frac{Q^2}{Q^2}\right) + \ldots \sim \frac{g_A^2 m_N Q}{4\pi f_\pi^2} \frac{g_A^2}{f_\pi^2} + \ldots \tag{51}$$

The scale that sets the relative size of this diagram is not M_{QCD} but

$$M_{NN} \equiv \frac{4\pi f_\pi^2}{g_A^2 m_N}, \tag{52}$$

which numerically is $\sim f_\pi \sim M_{nuc} \ll M_{QCD}$.

As in the QED case, the infrared enhancement of m_N/Q over irreducible states appears in all reducible intermediate states. Contrary to ordinary ChPT, we need to resum the leading terms; in a schematic way,

$$\begin{aligned} T_{NN} &\sim \frac{g_A^2}{f_\pi^2}\left[1 + O\left(\frac{Q}{M_{NN}}\right) + \cdots + O\left(\frac{Q}{M_{QCD}}\right)\right] \\ &\sim \frac{g_A^2}{f_\pi^2}\left[\frac{1}{1 - O\left(\frac{Q}{M_{NN}}\right)} + \cdots + O\left(\frac{Q}{M_{QCD}}\right)\right]. \end{aligned} \tag{53}$$

Nuclear bound states thus appear at

$$Q \sim M_{NN}, \tag{54}$$

with binding energy

$$B \sim \frac{Q^2}{m_N} \sim \frac{M_{NN}^2}{m_N}. \tag{55}$$

Numerically this is $\sim M_{QCD}/(4\pi)^2 \approx 10$ MeV, which explains why nuclei are so shallow when compared to the characteristic QCD scale.

Deriving a potential for nuclear physics

We can now carry out systematic calculations by finding a power counting for the potential, defined as the sum of all irreducible diagrams, and then solving the Lippman-Schwinger equation (Fig. 3) order by order.

Weinberg [15] was the first to suggest a simple power counting. He reasoned that, because the enhanced states are by construction removed from the potential, powers of Q could be counted in the potential as in ChPT. This results in an expansion of the form:

$$V = \sum_{\nu=0}^{\infty} c_\nu Q^\nu, \tag{56}$$

where

$$\nu = 4 - A + 2(L - C) + \sum_\Delta V_\Delta \Delta. \tag{57}$$

The only difference compared to Eq. (48) is the number of separately connected pieces $C \geq 1$. They arise because the potential is just part of the full amplitude and does not need to be fully connected.

According to this power counting, in leading order we have, besides one-pion exchange (OPE) between two nucleons, also non-derivative two-nucleon contact interactions (the $C_0^{(S)}$ and $C_0^{(T)}$ terms in Eq. (45).) A calculation of all (isospin-conserving) contributions to the two-nucleon potential up to $\nu = \nu_{min} + 3$ was carried out in Refs. [11, 12] using time-ordered perturbation theory. Some diagrams are shown in Fig. 6. To this order, the potential has all the spin-isospin structure of the phenomenological models, but its profile is determined by explicit degrees of freedom, symmetries, and power counting. The power counting suggests a hierarchy of short-range effects: S waves should depend strongly on the short-range parameters $C_0^{(S,T)}$; contact interactions affect P-wave phase shifts only in subleading order, so their effect should be smaller and approximately linear; D waves are affected by contact interactions only via mixing, while higher waves should be essentially determined by pion exchange. Chiral symmetry is particularly influential in the two-pion exchange (TPE) piece. The latter includes a particular form of terms previously considered [16], plus a few new terms. Those terms involving the B_i's and the deltas provide the only form of correlated TPE to this order, as graphs where pions interact in flight appear only in next order and should thus be

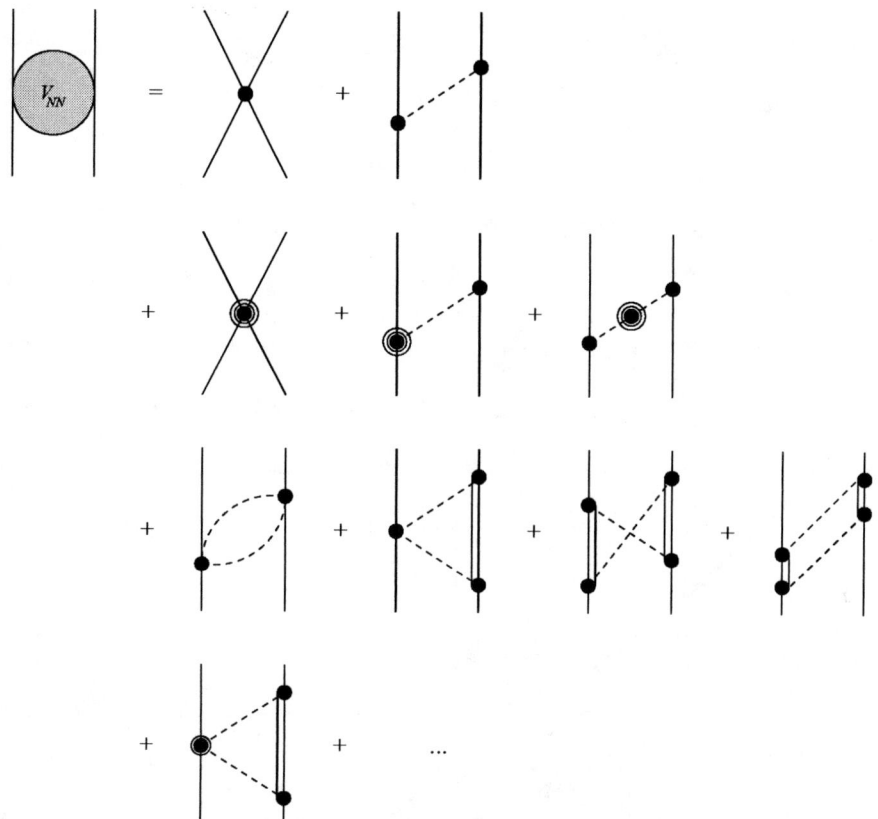

FIGURE 6. Some time-ordered diagrams contributing to the two-nucleon potential V_{NN} in the EFT. (Double) solid lines represent nucleons (and/or deltas), dashed lines pions, a heavy dot an interaction in $\mathscr{L}^{(0)}$, a dot within a circle an interaction in $\mathscr{L}^{(1)}$, and a dot within two circles an interaction in $\mathscr{L}^{(2)}$. First line corresponds to $v = v_{min}$, second and third lines to $v = v_{min} + 2$, fourth line to $v = v_{min} + 3$, and "..." denote $v \geq v_{min} + 4$. All orderings with at least one pion or delta in intermediate states are included. Not shown are diagrams contributing only to renormalization of parameters.

relatively small. The sum of the B_1 term and the corresponding delta term (related to the nucleon axial polarizability), is particularly important in providing an isoscalar central force. Not surprisingly, in the chiral limit these potentials behave at large separations as van der Waals forces. Isospin-violating pieces of the potential can be calculated as well [19].

Using this potential, one can then solve the Schrödinger equation. For each cutoff the parameters are fitted to data at selected energies. A sample of the results [12] for the lowest, most important partial waves is presented in Fig. 7, together with phases from the Nijmegen phase-shift analysis [17]. Quality of the fits is typical of other waves. Waves higher than F-waves were found to be mostly described well by pion exchange alone, as expected. Deuteron quantities are shown in Table 1. Electromagnetic quantities refer to

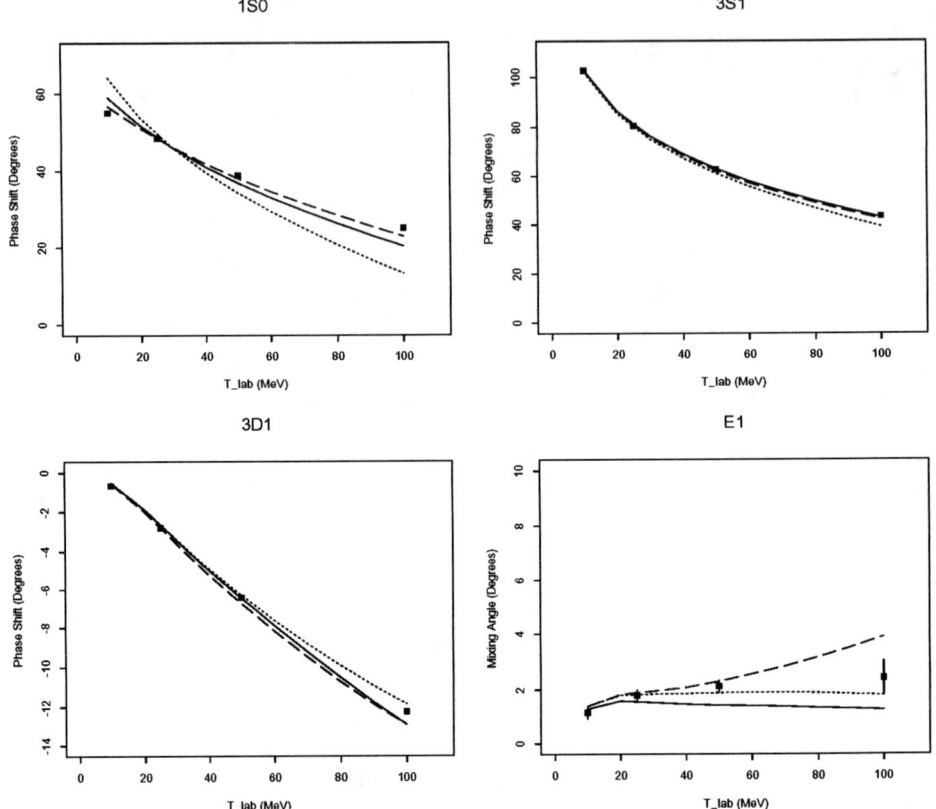

FIGURE 7. $^1S_0$, $^3S_1$, and $^3D_1$ NN phase shifts and ε_1 mixing angle in degrees as functions of the laboratory energy in MeV: EFT up to $\nu = 3$ for cutoffs of 500 (dotted), 780 (dashed), and 1000 MeV (solid line); and Nijmegen PSA (squares).

the contributions from lowest-order γNN couplings only, not to a consistent calculation which would include sub-leading one- and two-nucleon effects. The predicted S-wave scattering lengths (not used to constrain the fit) were found to be $a_2^{(^1S_0)} \simeq -15.0$ fm and $a_2^{(^3S_1)} \simeq 5.46$ fm. Important central attraction comes from the B_i's and deltas, and indeed the central potential does resemble those from models that include σ and ω meson exchange explicitly [18] (see Fig. 8). Thus the properties of the mesons extracted from phenomenological models have limited meaning. Values for the parameters are listed in Ref. [12]. Reasonable values were found for quantities known at the time, for example $g_A = 1.33$ (in agreement with the Goldberger-Treiman relation) and $h_A = 2.03$ (smaller but not too far off the large-$\mathcal{N}_c$ value). However, the values for the B_i's came out somewhat different than those found later in πN scattering. For a cutoff of $\Lambda = 780$ MeV, the coefficients C_{2n} of the contact interactions were found to scale approximately with $M_{QCD} \sim 500$ MeV.

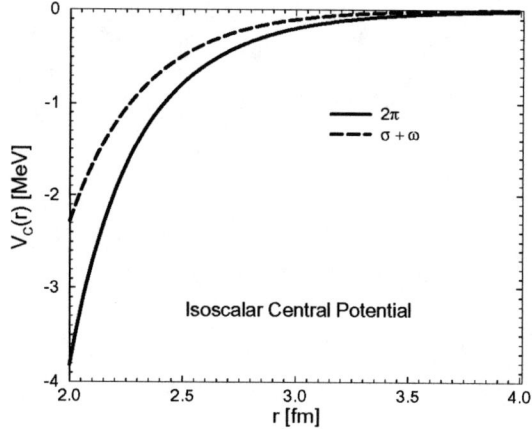

FIGURE 8. The isoscalar central potential generated by two-pion exchange compared with phenomenological $\sigma + \omega$ contributions. (Reprinted from Ref. [18], Copyright 1997, with permission from Elsevier Science.

The EFT also offers some insight into few-nucleon forces. Weinberg's power counting embodied in Eq. (57) suggests a hierarchy of few-nucleon forces. In leading order ($\nu = \nu_{min} = 6 - 3A$), C is maximum, so we have only pairwise interactions via the leading two-nucleon potential. We can easily verify that, if the delta is kept as an explicit degree of freedom, a 3N potential will arise at $\nu = \nu_{min} + 2$, a 4N potential at $\nu = \nu_{min} + 4$, and so on. It is (approximate) chiral symmetry therefore that implies that n-nucleon forces V_{nN} obey a hierarchy of the type

$$\frac{\langle V_{(n+1)N} \rangle}{\langle V_{nN} \rangle} \sim \mathscr{O}\left(\frac{Q}{M_{QCD}}\right)^2, \tag{58}$$

TABLE 1. Results from EFT fits at $\nu = 3$ for various cut-offs Λ and experimental values for the deuteron binding energy (B), magnetic dipole moment (μ_d), electric quadrupole moment (Q_E), asymptotic D/S ratio (η), and D-state probability (P_D).

Deuteron	Λ (MeV)			
quantities	500	780	1000	Experiment
B (MeV)	2.15	2.24	2.18	2.224579(9)
μ_d (μ_N)	0.863	0.863	0.866	0.857406(1)
Q_E (fm^2)	0.246	0.249	0.237	0.2859(3)
η	0.0229	0.0244	0.0230	0.0271(4)
P_D (%)	2.98	2.86	2.40	

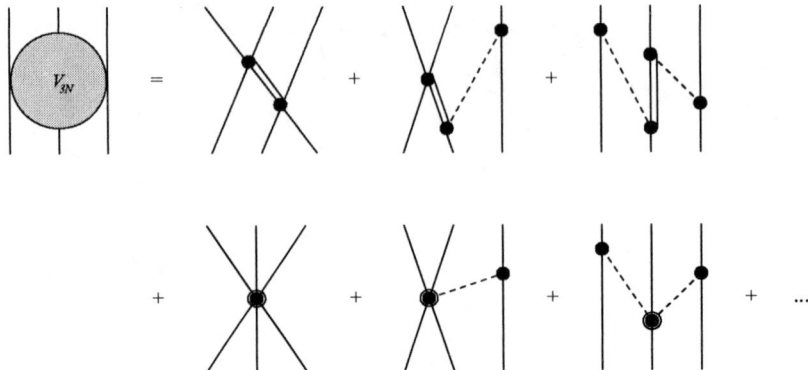

FIGURE 9. Some time-ordered diagrams contributing to the 3N potential in the pionful EFT. (Double) solid lines represent nucleons (and/or deltas), dashed lines pions, a heavy dot an interaction in $\mathscr{L}^{(0)}$, and a dot within a circle an interaction in $\mathscr{L}^{(1)}$. First line corresponds to $v = v_{min} + 2$, second line to $v = v_{min} + 3$, and "..." denote $v \geq v_{min} + 4$. All nucleon permutations and orderings with at least one pion or delta in intermediate states are included.

with $\langle V_{nN} \rangle$ denoting the contribution per n-plet. If $|\langle V_{2N} \rangle| \sim M_{QCD}/(4\pi)^2 \approx 10$ MeV, we can expect $|\langle V_{3N} \rangle| \sim 0.5$ MeV, $|\langle V_{4N} \rangle| \sim 0.02$ MeV, and so on. This is in accord with detailed few-nucleon phenomenology based on potentials that include small 3N and no 4N forces. This is shown in Table 2 in the case of the AV18/IL2 potential [20].

The new forces that appear in systems with more than two nucleons have been derived in Refs. [11, 13]. The relevant terms up to $v_{min} + 3$ are shown in Fig. 9. The leading 3N potential has components with three different ranges: TPE; OPE/short; and purely short. The TPE part of the potential is determined in terms of πN scattering observables, and is similar to existing two-pion-exchange 3N potentials [21]. The novel OPE/short-range components involve $\pi(N^\dagger N)^2$ interactions of strengths that are not fixed by chiral symmetry alone but can be determined from reactions involving only two nucleons, such as $NN \to NN\pi$ [22]. The purely short-range components of the potential can only be determined from few-nucleon systems.

The nuclear potential from EFT has been further elaborated in the last couple of years. Calculations are being pushed to next order [23], better fits (with more extensive input

TABLE 2. Contributions of the two-, three- and four-nucleon potentials (per doublet, per triplet, and per quadruplet, respectively): Weinberg power counting (W pc) and calculations with the AV18/IL2 potential for the ground states of various light nuclei (^{2}H, ^{3}H, *etc.*).

(MeV)	W pc	^{2}H	^{3}H	^{4}He	^{6}He	^{7}Li	^{8}Be	^{9}Be	^{10}B
$|\langle V_{2N} \rangle|$	~ 10	22	20	23	13	11	11	9.4	8.9
$|\langle V_{3N} \rangle|$	~ 0.5	–	1.5	2.1	0.55	0.43	0.38	0.29	0.30
$|\langle V_{4N} \rangle|$	~ 0.02	–	–	?	?	?	?	?	?
$|\langle V_{3N} \rangle|/|\langle V_{2N} \rangle|$	~ 0.05	–	0.075	0.091	0.042	0.039	0.035	0.031	0.034
$|\langle V_{4N} \rangle|/|\langle V_{3N} \rangle|$	~ 0.05	–	–	?	?	?	?	?	?

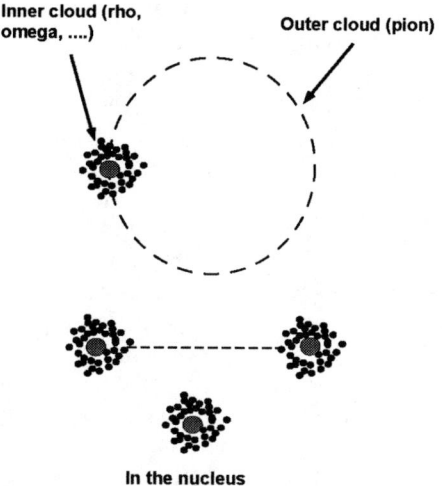

FIGURE 10. The short-ranged (ρ, etc.) cloud around the nucleon and the long-ranged (π) cloud surrounding it.

from πN scattering) to NN data have been achieved [24], and the first results for $3N$ and $4N$ systems have appeared [25]. These developments are all reviewed in Ref. [1].

Weinberg's power counting thus provides a rationale to understand much of the phenomenology of light nuclei. At low energies, the nucleon can be visualized as in Fig. 10: the light pion states form a sparse outer cloud, leading to a loop expansion in $Q/4\pi F_\pi$, while the high-energy states form a dense inner cloud, amenable to a multipole expansion in terms of Q/m_ρ. Since most of the time there is only a single pion "in the air", in the nucleus the interaction among nucleons is mostly pairwise, resulting also in a cluster expansion for the potential.

Renormalization

Despite all similarities, there is an important difference between the nuclear and atomic potentials. Because of chiral symmetry, the potential from one-pion exchange goes as Q^0, from two-pion exchange as Q^2, and so on. The loops generated by the iteration of these potentials are then sensitive to the cutoff. In coordinate-space parlance, these potentials are singular, that is, they behave for small radial distances r as $1/r^3$, $1/r^5$, and so on. The Schrödinger equation for such potentials is not well defined. Clearly, one cannot consider pion exchange without including enough short-distance operators to erase the cutoff dependence from observables.

Now, Weinberg's power counting automatically predicts which contact operators have to be taken at each order. Following this prescription, it does seem that the fits are of the same quality for different cutoffs. However, the numerical character of the calculations makes it hard to offer a proof of consistency. It is not apparent, for example, how a momentum-independent short-range interaction can renormalize the OPE potential (49).

Much effort has been devoted to this issue in the last five years [1]. A search for more analytical approaches led to extensive studies of the EFT at momenta $Q \ll m_\pi$. In this regime, pions can be treated as heavy particles. All interactions are of contact type, and form an expansion in Q/M_{nuc}. This "pionless" EFT can be solved by hand in the NN sector, and by very simple numerical means in the $3N$ and $4N$ systems. Power counting and renormalization are not without surprises, but they can be done consistently. Of course, the pionless EFT does not address the problem of pion exchange directly, but it goes a long way in elucidating how EFTs work in a non-perturbative context. Moreover, it is useful for many low-energy reactions, to which it has by now been applied.

As Q is increased, the approximation of zero-range propagation for pions becomes less and less reliable. At momenta $Q \sim m_\pi$, pions have to be included explicitly in the theory. Once pions have been reinstated, we can still consider the low-Q region. Because of chiral symmetry, it is reasonable to suppose that pion interactions are perturbative at sufficiently low Q. Based on the power counting of the pionless EFT, a new power counting was formulated that led to a manifestly consistent EFT where pion exchange was treated in perturbation theory [26]. Unfortunately, it has been shown that the range of validity of this power counting is $Q \lesssim 100$ MeV [27]. In this region the simplest pionless EFT is more useful.

Recently, the non-perturbative renormalization of pion exchange was analyzed [28]. It was found that OPE can be renormalized à la Weinberg, provided one expands in the pion mass even the pion propagators. If one does not do expand, one makes numerically small but conceptually large errors. The full implication of this conclusion to previous results remains to be explored.

Nuclear matter on a lattice

The nuclear EFT that has been sketched above can be used to study nuclear matter at finite temperature. This provides a link with RHIC physics that is the subject of this meeting.

The general concept of a nuclear matter calculation consists of nucleons interacting via a variety of components of the nuclear potential. While the ultimate goal is to use EFT interactions, let us concentrate for simplicity on few parts of the NN potential, namely central, spin- and isospin-exchange. The Hamiltonian,

$$\hat{\mathcal{H}} = \hat{\mathcal{K}} + \hat{\mathcal{V}}, \qquad (59)$$

can be expressed in second quantization and contains kinetic and potential operators. The kinetic term is written as

$$\hat{\mathcal{K}} = -\psi^\dagger \frac{\nabla^2}{2m_N} \psi, \qquad (60)$$

while the potential is taken as

$$\hat{\mathcal{V}} = C_0 \left(\psi^\dagger \psi\right)^2 + C_2 \left(\psi^\dagger \nabla \psi\right)^2. \qquad (61)$$

The fermion operator $\psi^\dagger$ creates a nucleon of spin and isospin (σ, τ) at location $\vec{x}$, while its adjoint ψ destroys it. This Hamiltonian describes merely a toy model for developing the formalism. More complicated Hamiltonians arising from full-fledged nuclear EFT can be later used for more realistic calculations.

How can we deal with the complication of many nucleons? A natural approach, patterned after similar attempts in QCD, is to investigate nuclear matter on a three-dimensional cubic lattice of spacing a and periodic boundary conditions. We describe the nuclear-matter Monte Carlo method [29], which consists of the thermal formalism to express the grand-canonical partition function as an integral over single-body evolution operators. At its center stands the Hubbard-Stratonovitch transformation, which is used to reduce the many-body problem to an effective one-body problem.

In order to study thermal properties of nuclear matter, the grand-canonical partition function at a given temperature $T = \beta^{-1}$ needs to be determined:

$$Z = \hat{\mathrm{Tr}}\left[\exp\left(-\beta\left(\mathcal{H} - \psi^\dagger \mu \psi\right)\right)\right] \equiv \hat{\mathrm{Tr}}\left[\hat{U}\right], \qquad (62)$$

with μ as the chemical potential. $\hat{U}$ is called the imaginary-time evolution operator of the system and is a many-body operator; the trace is taken over all many-body states. The partition function Z is an exponential over all one- and two-body operators (and therefore interactions) present in the system. It is impossible to deal with Z in this form, because the number of many-body correlations that have to be kept track of grows rapidly with system size. We therefore find an expression for Z that is based on a single-particle representation, replacing the many-body problem with that of non-interacting nucleons that are coupled to a heat bath of auxiliary fields. A thermal observable $\langle \hat{O} \rangle$ is then expressed as [30]

$$\langle \hat{O} \rangle = \frac{1}{Z}\hat{\mathrm{Tr}}\left[\hat{O}\exp\left(-\beta\left(\mathcal{H} - \psi^\dagger \mu \psi\right)\right)\right] = \frac{\int \mathscr{D}[\chi] G(\chi) \langle \hat{O}(\chi) \rangle \xi(\chi)}{\int \mathscr{D}[\chi] G(\chi) \xi(\chi)}, \qquad (63)$$

where the Gaussian factor G is given by

$$G(\chi) = \prod_{m=1}^{n_t} \prod_{\vec{x}_n} \prod_i \exp\left(-|\alpha_i| \chi^2_{m,\vec{x}_n,i}\right). \qquad (64)$$

The integrals in this equation can then be evaluated using the Metropolis algorithm [31] (Monte Carlo simulations).

Fig. 11 [29] shows that such calculations are feasible. It displays the energy per particle E/A for symmetric nuclear matter and for pure neutron matter as a function of density ρ and for different temperatures. The two interaction parameters were adjusted so that the known *qualitative* features of the symmetric nuclear matter are reproduced. With decreasing temperature, symmetric nuclear matter develops a minimum at $\rho = 0.32$ fm^{-3} first, which is most pronounced between $10 - 14$ MeV, before it shifts to lower densities. At $T = 3.3$ MeV and $T = 5.9$ MeV the minimum is very broad, making matter softer. For high temperatures and/or high density, the simulation suffers from the fact that it runs out of model space. At $T = 50$ MeV the system behaves almost like a Fermi gas and the energy per particle should behave like $\sim \rho^{2/3}$. Yet, the curve

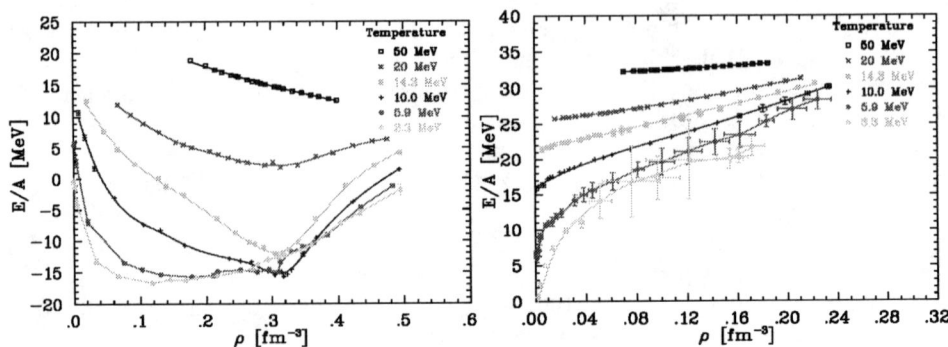

FIGURE 11. E/A for symmetric nuclear (left) and pure neutron (right) matter as a function of density ρ and for different temperatures. The purpose of the lines is to guide the eye.

bends down. Also, for all other temperatures, the curves converge to the energy of the full lattice state, $E/A = 5.96$ MeV, as density increases. For sub-saturation densities the model gives more binding if compared to other calculations (see, for example, Ref. [32]), and the energy is not as high for densities beyond saturation. At $\rho = 0.32$ fm^{-3}, E/A as a function of temperature has a minimum at $T \approx 10$ MeV which means that at even lower temperatures E/A increases again. We see some evidence for a liquid-gas phase transition. The uncertainties for pure neutron matter are much larger than for symmetrical nuclear matter. As a potential, we used the parameters obtained from the fit to symmetric nuclear matter, even though we could have fitted the potential parameters for this case anew. Therefore we view the results for pure neutron matter more as a test to see how well the given potential already reproduces the energy. Note that the slopes of the curves at high temperatures are not negative as they are for symmetric nuclear matter. But clearly, we cannot conclude that the energies at $T = 3.3$ MeV have converged to that of the ground state because the curve differs quite a bit from that of $T = 5.9$ MeV. At the lowest temperature they are $4-5$ MeV higher than those of the ground state as calculated in Ref. [32], but the general shape of the curve is very similar. This is no surprise, since pure neutron matter is like a Fermi gas, with attractive forces between neutrons lowering the energies with respect to the non-interacting system. The search for any kind of phase transition in the range of $5-50$ MeV was to no avail.

CONCLUSION

We have presented an introduction to EFT and some of the mileposts along the sinuous road from QCD to nuclear physics. We discussed some of the essential features of the interactions among nucleons and some of their consequences to few-nucleon systems. We can already see how to approach the solution of the important problem of finite-temperature nuclear matter. Yet, there is no denying that much is still to be done in straightening the path already traveled, and in breaking new ground. It is a good time to be on the road!

ACKNOWLEDGMENTS

The authors acknowledge the support and dedication of all PASI sponsors and organizers, in particular Jan Rafelski, Bob Thews, Thomas Elze, Erasmo Ferreira, and Takeshi Kodama. UvK thanks the Nuclear Theory Group at the University of Washington for hospitality while part of this work was carried out, and RIKEN, Brookhaven National Laboratory and to the U.S. Department of Energy [DE-AC02-98CH10886] for providing the facilities essential for the completion of this work. This research was supported in part by a DOE Outstanding Junior Investigator Award (UvK).

REFERENCES

1. Bedaque, P.F., and van Kolck, U., nucl-th/0203055, to appear in *Ann. Rev. Nucl. Part. Sci.*; van Kolck, U., *Nucl. Phys.* **A699**, 33 (2002).
2. Phillips, D., nucl-th/0203040, to appear in *Czech. J. Phys.*
3. Gerber, P., and Leutwyler, H., *Nucl. Phys.* **B321**, 387 (1989).
4. Hagedorn, R., and Rafelski, J., *Phys. Lett.* **B97**, 136 (1980).
5. Karsch, F., *AIP Conf. Proc.* **602**, 323 (2001).
6. Braaten, E., and Nieto, A., *Eur. Phys. J.* **B11**, 143 (1999); Hammer, H.-W., and Furnstahl, R.J., *Nucl. Phys.* **A678**, 277 (2000).
7. Weinberg, S., *Physica* **A96**, 327 (1979).
8. Ball, R.D., and Thorne, R.S., *Ann. Phys.* **236**, 117 (1994).
9. Georgi, H., *Phys. Lett.* **B240**, 447 (1990).
10. Coleman, S., Wess, J., and Zumino, B., *Phys. Rev.* **177**, 2239 (1969); Callan, C.G., Coleman, S., Wess, J., and Zumino, B., *Phys. Rev.* **177**, 2247 (1969).
11. Ordóñez, C., and van Kolck, U., *Phys. Lett.* **B291**, 459 (1992).
12. Ordóñez, C., Ray, L., and van Kolck, U., *Phys. Rev.* **C53**, 2086 (1996).
13. U. van Kolck, *Phys. Rev.* **C49**, 2932 (1994).
14. Bernard, V, Kaiser, N, and Meißner, U-G., *Int. J. Mod. Phys.* **E4**, 193 (1995).
15. Weinberg, S., *Phys. Lett.* **B251**, 288 (1990); *Nucl. Phys.* **B363**, 3 (1991).
16. Brueckner, K., and Watson, K., *Phys. Rev.* **92**, 1023 (1953); Sugawara, H., and Okubo, S., *Phys. Rev.* **117**, 605 (1960); Sugawara, H., and von Hippel, F., *Phys. Rev.* **172**, 1764 (1968).
17. Stoks, V.G.J., Klomp, R.A.M., Rentmeester, M.C.M., and de Swart, J.J., *Phys. Rev.* **C48**, 792 (1993).
18. Kaiser, N., Brockmann, R., and Weise, W., *Nucl. Phys.* **A625**, 758 (1997).
19. van Kolck, U., *Few-Body Syst. Suppl.* **9** 444 (1995); van Kolck, U., Friar, J.L., and Goldman, T., *Phys. Lett.* **B371**, 169 (1996); van Kolck, U., Rentmeester, M.C.M., Friar, J.L., Goldman, T., and de Swart, J.J., *Phys. Rev. Lett.* **80**, 4386 (1998); Friar, J.L., and van Kolck, U., *Phys. Rev.* **C60**, 034006 (1999); Niskanen, J.A., *Phys. Rev.* **C65**, 037001 (2002).
20. Pieper, S.C., and Wiringa R.B., *Ann. Rev. Nucl. Part. Sci.* **51**, 53 (2001).
21. Friar, J.L., Hüber, D., and van Kolck, U., *Phys. Rev.* **C59**, 53 (1999).
22. Hüber, D., Friar, J.L., Nogga, A., Witała, H., and van Kolck, U., *Few-Body Syst.* **30**, 95 (2001); Hanhart, C., van Kolck, U., and Miller, G.A., *Phys. Rev. Lett.* **85**, 2905 (2000).
23. Kaiser, N., *Phys. Rev.* **C61**, 014003 (2000); **C62**, 024001 (2000); **C63**, 044010 (2001); **C64**, 057001 (2001); **C65**, 017001 (2002).
24. Epelbaum, E., Glöckle, W., and Meißner, U.-G., *Nucl. Phys.* **A671**, 295 (2000); Entem, D.R., and Machleidt, R., *Phys. Lett.* **B524**, 93 (2002); Walzl, M., Meißner, U.-G., and Epelbaum, E., *Nucl. Phys.* **A693**, 663 (2001).
25. Epelbaum, E., Kamada, H., Nogga, A., Witała, H., Glöckle, W., and Meißner, U.-G., *Phys. Rev. Lett.* **86**, 4787 (2001).
26. Kaplan, D.B., Savage, M.J., and Wise, M.B., *Nucl. Phys.* **B534**, 329 (1998).
27. Fleming, S., Mehen, T., and Stewart, I.W., *Nucl. Phys.* **A677**, 313 (2000).
28. Beane, S.R., Bedaque, P.F., Savage, M.J., and van Kolck, U., *Nucl. Phys.* **A700**, 377 (2002).

29. Müller, H.-M., Koonin, S.E., Seki, R., and van Kolck, U., *Phys. Rev.* **C61**, 044320 (2000).
30. Loh Jr., E.Y., and Gubernatis, J., in *Electronic Phase Transitions*, ed. by W. Hanke and Yu.V. Kopaev Elsevier, New York, 1992; Lang, G., Johnson, C., Koonin, S., and Ormand, W., *Phys. Rev.* **C48**, 1518 (1993).
31. Metropolis, N., Rosenbluth, A., Rosenbluth, M., Teller, A., and Teller, E., *J. Chem. Phys.* **21**, 1087 (1953).
32. Wiringa, R.B., Fiks, V., and Fabrocini, A., *Phys. Rev.* **C38**, 1010 (1988); Akmal, A., and Pandharipande, V.R., *Phys. Rev.* **C56**, 2261 (1997).

The Nucleon and the Nuclear Force: Effective Theory and Path-Integral Methods

L. J. Abu-Raddad[1]

Theory Group, Research Center for Nuclear Physics, Osaka University, 10-1 Mihogaoka, Ibaraki City, Osaka 567-0047, Japan

Abstract. The nucleon structure and the nuclear force are investigated in the context of the non-perturbative path-integral method of hadronization. Starting from a microscopic quark-diquark model, the nucleon is generated as a relativistic bound state and an effective chiral meson-nucleon Lagrangian is derived. Many of the nucleon physical properties are studied using a theory of at most two free parameters.

INTRODUCTION

The central challenge in nuclear physics remains to understand the origin and nature of the nuclear force due to our inability to solve quantum chromodynamics (QCD), the fundamental theory for the strong interactions. The basic problem of QCD is that its fundamental degrees of freedom, quarks and gluons, are not the observable baryon and meson states. Thus bridging the missing link between the fundamental and observable degrees of freedom stands as one of the stark challenges of nuclear/elementary particle physics today. Although we do have an ab initio approach to solve this problem, that is lattice QCD, this endeavor is still miles away from achieving such a goal. For the time being, we have no alternative but to resort to effective non-perturbative approaches of which this study is one.

This presentation describes our work [1] in addressing this missing link by deriving a chiral meson-nucleon Lagrangian from a microscopic model of quarks and diquarks using the path-integral method of hadronization. Chiral symmetry and its spontaneous breaking have proven to be key concepts in understanding meson and baryon structure and many features of the nuclear force. Therefore, we start from a QCD-based chiral effective field theory, the Nambu Jona-Lasinio (NJL) model that accommodates most of QCD symmetries [2]. Then, the nucleon is described as quark-diquark correlations. This assertion is vindicated by a mounting experimental evidence that diquarks play a dynamical role in hadrons [3]. Using the path integral hadronization, we calculate nucleon properties and derive an effective chiral meson-nucleon Lagrangian of the quantum hadrodynamics (QHD) type [4] that describes the rich meson-nucleon interactions in a fully covariant and chirally symmetric formalism.

While this program is applied to the case of nucleons and mesons, it is certainly

---

[1] Email: laith@rcnp.osaka-u.ac.jp

of general nature and can possibly be applied prolifically to yield other baryons and their interactions. Moreover, the idea of using path-integral techniques to transform a Lagrangian from its fundamental to its composite degrees of freedom is a powerful concept in physics of immense impact and utility. As a matter of fact, the authors of Ref. [5] have recently invoked such techniques in their study of high-temperature superconductivity. They succeeded in doing so by converting a model of strongly-correlated electrons into an effective U(1) gauge field theory in terms of composite fields.

The use of hadronization has been introduced in Ref. [6, 7]. Consequently, the authors of Ref. [8] attempted to construct an effective Lagrangian for the nucleon using only scalar diquarks. In the present work, we extend their work by deriving the structure using both axial-vector and scalar diquarks and we employ, as opposed to Ref. [8], a gauge-invariant regularization scheme throughout our analysis. Furthermore, we verify the Ward identity and the Goldberger-Treiman relation and present a full numerical study of various nucleon observables for the case of scalar diquarks drawing special attention to the role of an intrinsic diquark form factor. Hence, this work is the first calculation of an extensive set of nucleon observables using path-integral hadronization since the introduction of the idea more than ten years ago.

FORMALISM

We start from an NJL Lagrangian satisfying $SU(2)_L \times SU(2)_R$ chiral symmetry:

$$\mathscr{L}_{NJL} = \bar{q}(i\partial\!\!\!/ - m_0)q + \frac{G}{2}\left[(\bar{q}q)^2 + (\bar{q}i\gamma_5\vec{\tau}q)^2\right], \qquad (1)$$

where q is the current quark field, $\vec{\tau}$ are the Pauli matrices, G is the NJL coupling constant, and m_0 is the current quark mass which explicitly breaks chiral symmetry. The color and flavor indices are suppressed for brevity.

By introducing composite scalar ($\sigma \sim \bar{q}q$) and pseudoscalar ($\vec{\pi} \sim \bar{q}i\gamma_5\vec{\tau}q$) fields through the Hubbard-Stratonovich transformation [9], we can rewrite the NJL lagrangian into a semi-bosonized Lagrangian $\mathscr{L}'_{NJL} = \left[\bar{q}(i\partial\!\!\!/ - \sigma - i\gamma_5\vec{\tau}\cdot\vec{\pi})q - \frac{1}{2G}(\sigma^2 + \vec{\pi}^2)\right]$, where we have absorbed the bare quark mass m_0 into the sigma field σ. Further, we transform the meson fields according to the non-linear parameterization $[\sigma, \pi] \to [\sigma', \Phi]$: $\sigma + i\gamma_5\vec{\tau}\cdot\vec{\pi} = (m_q + \sigma')e^{-\frac{i}{F_\pi}\gamma_5\vec{\tau}\cdot\vec{\Phi}}$. Here, $F_\pi = 93$ MeV is the pion decay constant and $m_q \equiv \langle\sigma\rangle_0$ is the constituent quark mass which is fixed through a gap equation in the meson sector [2].

As a consequence of considering Lorentz structure, there are five types of possible diquark qq correlations. These are the scalar, pseudo-scalar, vector, axial-vector, and tensor diquarks. For each of these Lorentz qq formations, we have also an isoscalar and isovector diquarks. Using permutation symmetry and Fierz transformation, we have verified an earlier assertion [10] that only two diquark formations are independent for the nucleon if its field is to be written as a local operator of three quarks. Hence, we introduce $\vec{D}^\mu$ as an axial-vector isovector diquark field and D as a scalar isoscalar one.

Next we introduce a quark-diquark interaction term in such a manner to generate the nucleon as a linear combination of axial-vector and scalar diquarks. It is convenient here, considering chiral symmetry, to work with the chirally rotated χ "constituent" quark field defined by $\chi \equiv e^{\frac{-i}{F_\pi}\gamma^5 \frac{\vec{\tau}}{2}\cdot\vec{\Phi}} q$. By introducing electromagnetic interactions and batching the semi-bosonized NJL Lagrangian, the diquark contributions and the quark-diquark interaction term, we obtain the following Lagrangian as our microscopic model:

$$\mathscr{L} = \bar{\chi} S^{-1} \chi - \frac{1}{2G}(\sigma' + m_q)^2 + \delta\mathscr{L}_{sb} + D^\dagger \Delta^{-1} D + \vec{D}^{\dagger\mu} \tilde{\Delta}^{-1}_{\mu\nu} \vec{D}^\nu +$$
$$\tilde{G}\left(\sin\theta\, \bar{\chi}\gamma^\mu\gamma^5 \vec{\tau}\cdot\vec{D}^\dagger_\mu + \cos\theta\, \bar{\chi} D^\dagger\right)\left(\sin\theta\, \vec{D}_\nu \cdot \vec{\tau}\gamma^\nu\gamma^5\chi + \cos\theta\, D\chi\right), \quad (2)$$

where

$$S^{-1} = S_0^{-1} + \mathscr{M}, \quad (3a)$$

$$\mathscr{M} = -\left[\gamma^\mu \frac{\vec{\tau}}{2}\cdot\vec{\mathscr{V}}^\pi_\mu + \gamma^\mu\gamma^5 \frac{\vec{\tau}}{2}\cdot\vec{\mathscr{A}}^\pi_\mu + \sigma' + \gamma^\mu Q_q A^{EM}_\mu\right], \quad (3b)$$

$$\Delta^{-1} = \Delta_0^{-1} + iQ_S A^{EM}_\mu(\overrightarrow{\partial^\mu} - \overleftarrow{\partial^\mu}), \quad (3c)$$

$$\tilde{\Delta}^{-1}_{\mu\nu} = \tilde{\Delta}_0{}^{-1}_{\mu\nu} + iQ_A\left[(A^{EM}_\mu \overleftarrow{\partial_\nu} - A^{EM}_\nu \overrightarrow{\partial_\mu}) - g_{\mu\nu} A^{EM}_\alpha(\overleftarrow{\partial^\alpha} - \overrightarrow{\partial^\alpha})\right]. \quad (3d)$$

Here θ is a mixing angle for the two diquark contributions, $\tilde{G}$ is the quark-diquark coupling constant and Q_q, Q_S and Q_A are the quark and diquark charges while S_0, Δ_0 and $\tilde{\Delta}_{0\,\mu\nu}$ are the free quark and diquark propagators (Notice that $\mathscr{O}(Q^2_{S,A})$ terms are discarded in Eq. (3)). The $\mathscr{M}$ matrix contains all interaction vertices of the quark field with meson and electroweak fields (weak part is not shown), where the vector $\vec{\mathscr{V}}^\pi_\mu$ and the axial vector $\vec{\mathscr{A}}^\pi_\mu$ fields are defined through the Cartan decomposition ($\vec{\xi} \equiv \frac{\vec{\Phi}}{F_\pi}$):
$\exp\left(-\frac{i}{2}\gamma^5 \vec{\tau}\cdot\vec{\xi}\right)\partial_\mu \exp\left(\frac{i}{2}\gamma^5\vec{\tau}\cdot\vec{\xi}\right) = \frac{i}{2}\gamma^5\vec{\tau}\cdot\vec{\mathscr{A}}^\pi_\mu(\xi) + \frac{i}{2}\vec{\tau}\cdot\vec{\mathscr{V}}^\pi_\mu(\xi)$.

Subsequently, we introduce collective nucleon fields ($B \sim \sin\theta \vec{D}_\nu \cdot \vec{\tau}\gamma^\nu\gamma^5\chi + \cos\theta D\chi$) through another Hubbard-Stratonovich transformation. At this point only quarks and diquarks are dynamical fields with kinetic terms while the meson and nucleon fields are merely auxiliary ones. By integrating over the quark and then over the diquark fields, we obtain a meson-nucleon effective Lagrangian. Accordingly, we have "hadronized" the microscopic theory by producing the dynamical meson (bosonization) and nucleon (fermionization) fields.

Thereupon, we arrive at a compact Lagrangian given by

$$\mathscr{L}_{eff} = \delta\mathscr{L}_{sb} - \frac{1}{2G}(\sigma' + m_q)^2 - i\,\mathrm{tr}\ln S^{-1} - \frac{1}{\tilde{G}}\bar{B}B + i\,\mathrm{tr}\ln(1 - \Box) +$$
$$i\,\mathrm{tr}\ln(1 - \Delta_0\,\mathrm{EM\,Int}) + i\,\mathrm{tr}\ln(1 - \tilde{\Delta}_0\,\mathrm{EM\,Int}). \quad (4)$$

Here the trace is over color, flavor and Lorentz indices while the "EM Int" label stands for the diquark electromagnetic interaction terms. Furthermore,

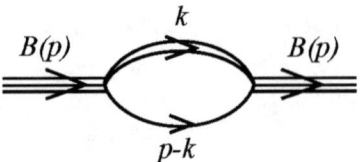

FIGURE 1. The self-energy diagram which generates the nucleon kinetic and mass terms and produces the mass equation that determines the nucleon mass.

$$\Box = \begin{pmatrix} \mathscr{A} & \mathscr{F}^{\dagger\,\nu j} \\ \mathscr{F}^{\mu i} & \mathscr{S} \end{pmatrix}, \tag{5a}$$

$$\mathscr{A} = \sin^2\theta\,\bar{B}\,\gamma^\mu\gamma^5\,\tau_i\,\tilde{\Delta}^{ij}_{\mu\nu}\,S\,\tau_j\,\gamma^\nu\gamma^5\,B, \tag{5b}$$

$$\mathscr{S} = \cos^2\theta\,\bar{B}\Delta S B, \quad \text{and} \tag{5c}$$

$$\mathscr{F}^{\mu i} = \sin\theta\cos\theta\,\bar{B}\,\gamma^\mu\gamma^5\,\tau^i S B. \tag{5d}$$

This effective Lagrangian contains plenty of rich physics: kinetic and mass terms for nucleons and mesons together with a multitude of interaction terms of mesons, nucleons, and electroweak gauge bosons. Nonetheless, the most desired part of the Lagrangian is the prized chiral meson-nucleon interaction and nucleon-nucleon vertices which delineates the nuclear force.

In order to explore the physics of the nucleon sector, we take the leading term in the loop and derivative expansion of $i\,\mathrm{tr}\ln(1-\Box) - \frac{1}{\tilde{G}}\bar{B}B \to -\int d^4x d^4y \bar{B}(x)\left[\Sigma(x,y) + \frac{1}{\tilde{G}}\delta(x-y)\right]B(y)$, which is nothing but the nucleon self-energy (see Fig. 1). The Fourier transform of Σ is then decomposed as $\Sigma(p) = \Sigma_s(p^2) + \not{p}\,\Sigma_v(p^2)$. This leading term generates dynamically the nucleon mass M_B which is extracted as the pole of the propagator:

$$\frac{1}{\tilde{G}} + \Sigma_s(M_B^2) + M_B\,\Sigma_v(M_B^2) = 0. \tag{6}$$

Thus, near the mass shell the inverse nucleon propagator takes the form:

$$\left[\Sigma(p^2) + \frac{1}{\tilde{G}}\right] \sim (\not{p} - M_B)\,Z^{-1}, \tag{7}$$

where Z is the wave-function renormalization constant ($B = \sqrt{Z}\,B_{\mathrm{ren}}$). Evidently, the nucleon has finally acquired the desired status as a dynamical degree of freedom in the problem.

In computing the various Feynman diagrams in the problem, we encounter divergent integrals that must be regularized. Several regularization schemes were attempted. We found that the most suitable scheme is the PV technique which we have adopted as the standard method in this work. Accordingly, we verified the Ward-Takahashi identity by computing the electromagnetic vertex shown in Fig. 2. As a matter of principle, the PV mass in the nucleon sector can be different from the NJL cut-off arising in the meson

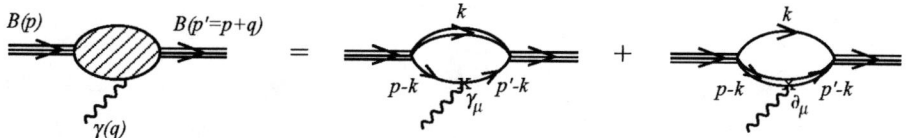

FIGURE 2. The Feynman diagrams for the electromagnetic coupling which generate the electromagnetic vertex of the nucleon and subsequently are used to test the validity of the Ward identity.

sector [2]. Nonetheless, to minimize the number of free parameters, we elected to equate them. It is noteworthy here that all observables were found to be very insensitive to the value of the PV mass upholding the futility of using it as a free parameter.

We have also verified one of the chiral symmetry relations, the Goldberger-Treiman relation, by computing both the axial-vector coupling constant g_A and the pion nucleon coupling constant $g_{\pi NN}$. In the hadronization formalism, this relation emerges naturally intact as opposed to large violations in the Bethe-Salpeter equation approach [11]

NUMERICAL STUDY

Having derived the structure of the Lagrangian, we proceed to generate numerical results using only scalar diquarks ($\theta = 0$ in Eq. (2) and (5)), thereby admitting the possibility of an intrinsic diquark form factor (IDFF). Including only scalar diquarks is not out of place as many recent studies using such diquarks have reported good results for most of the nucleon observables [12, 13, 11]. Moreover, there are strong indications of large scalar diquark dominance in the nucleon [14].

The parameters G and the cut-off Λ are fixed to yield the constituent quark mass and the pion decay constant [2]. The diquark masses are also determined in the NJL model [15]. This leaves us with only one free parameter in our model: the quark-diquark coupling constant $\tilde{G}$ which is fixed to determine the nucleon mass of 0.94 GeV through the mass equation (6). Thus the basic quantities in our model are the constituent quark mass $m_q = .390$ GeV, the scalar diquark mass $M_D = .600$ GeV, the quark-diquark coupling constant $\tilde{G} = 159.1$ GeV^{-1}, and the Pauli-Villars mass $\Lambda = .600$ GeV. We obtain a binding energy of $\Delta E_{\text{bin}} \equiv m_q + M_D - M_B = 50$ MeV, suggesting a loosely bound state for the nucleon.

Tab. 1 displays our predictions for some of the static properties of the nucleon. For

TABLE 1. Some of the nucleon static properties as predicted in the present calculation using the intrinsic diquark form factor (IDFF) or without it. Experimental values are taken from Ref. [16, 17].

	μ_p	μ_n	g_A	$<r^2>_E^p$ (fm^2)	$<r^2>_E^n$ (fm^2)	$<r^2>_M^p$ (fm^2)	$<r^2>_M^n$ (fm^2)
Theory with IDFF	1.57	-.75	.87	.77	-.11	.82	.84
Theory without IDFF	1.57	-.75	.87	.68	-.19	.82	.85
Experiment	2.79	-1.91	1.26	.74	-.12	.74	.77

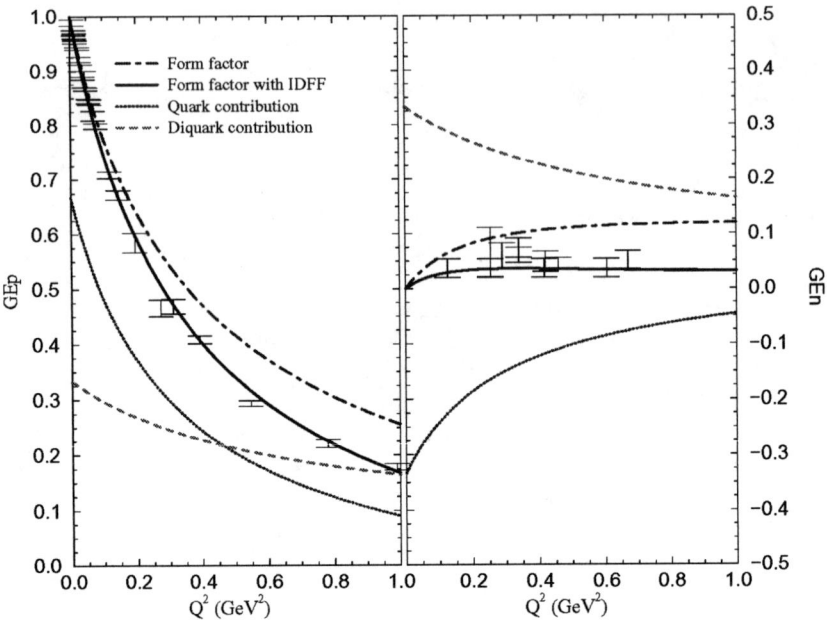

FIGURE 3. The nucleon electric form factors with and without the intrinsic diquark form factor (IDFF) along with its quark and diquark contribution. The left panel shows the proton results while the right panel displays those for the neutron. Experimental data can be found in Ref. [19, 20].

the nucleon magnetic moments, our treatment predicts a number that is two-third of the experimental value for the proton and one-half of that for the neutron. This is not a surprising result as we have not included the axial-vector diquark in the present calculation. The predicted value for the axial-vector coupling g_A of 0.87 is less than the experimental one of 1.26 indicating here also the importance of the axial-vector diquark.

The nucleon size is nicely well-produced in our model as the electric and magnetic radii for the nucleon are very close to the experimental measurements. The negative charge radius of the neutron has been suggested as an indication of a scalar diquark clustering in the nucleon [18] and our treatment dynamically manifests this assertion. These numbers point to a physical picture of a "heavy" diquark at the center with a quark rotating around it. The extended size of the diquark contributes a positive value of about 0.10 fm^2 for the nucleon electric radii. As expected, the IDFF has virtually no effect on the magnetic radii as the scalar diquark has a negligible contribution to the magnetic form factors.

Next we calculate the nucleon form factors. The left panel of Fig. 3 displays the proton form factor with and without the IDFF along with its quark and diquark contribution. Our treatment produces beautifully this observable. It is evident here that the IDFF [21] plays an important role specially at large values of momentum transfer ($Q^2 \equiv -q^2$ where q^μ is the momentum transfer). The neutron electric form factor tells a similar story (right panel). Clearly, the quark contribution is negative in value (d-quark) and thus cancels

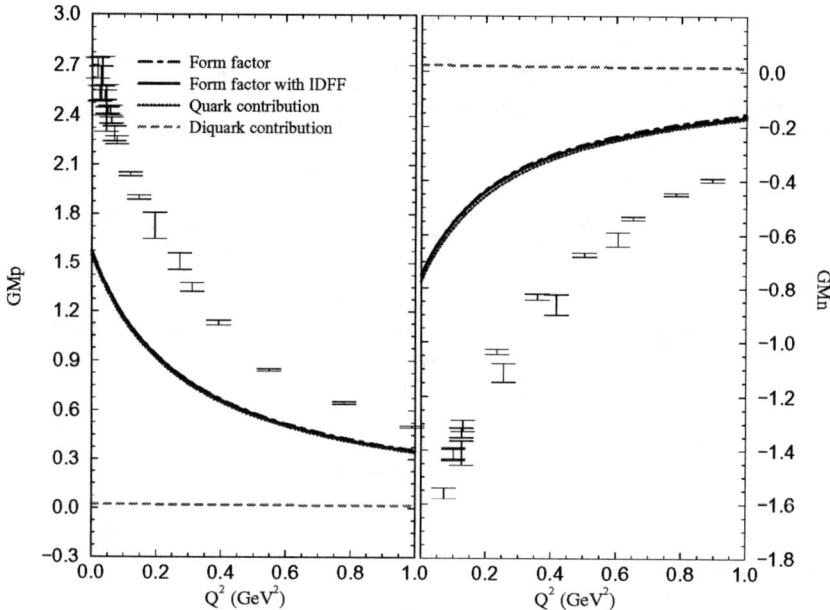

FIGURE 4. The nucleon magnetic form factors with and without the intrinsic diquark form factor (IDFF) along with its quark and diquark contribution. The left panel shows the proton results while the right panel displays those for the neutron. Experimental data can be found in Ref. [19, 20].

much of the diquark contribution leading to a small form factor. It is noteworthy here that the neutron form factor is a potent test of any treatment as it is a delicate cancellation of two large contributions [13]. Saliently, the cancellation is naturally produced in our study.

In Fig. 4 we present the nucleon magnetic form factor as calculated with or without the intrinsic diquark form factor. Unmistakably, the scalar diquark contribution is virtually vanishing due to the lack of an intrinsic spin. Nevertheless, there is a very small contribution due to a small orbital angular-momentum effect in the bound quark-diquark system. A comparison with experimental data suggests the need for the axial-vector diquark, which does have an intrinsic spin, to supplement the quark contributions and to provide the missing strengths for the magnetic form factors.

CONCLUSIONS

In conclusion, we have tackled the nucleon structure and the challenging problem of understanding the origin and nature of the nuclear force by deriving a meson-nucleon Lagrangian using the path-integral method of hadronization. The treatment produced a remarkable agreement with experimental data for the nucleon size and its electric form factors, while our calculations show missing strengths for the magnetic properties and

g_A. The discrepancy is likely due to the absence of the axial-vector diquark in the present numerical study. This presentation describes the first work in our program of using path-integral hadronization to study baryon structure and the nuclear force. Deriving this force provides nuclear physics with a solution to the stigma of no fundamental foundation that has tarnished its image for decades.

ACKNOWLEDGMENTS

I would like to acknowledge the support of a joint fellowship from the Japan Society for the Promotion of Science and the United States National Science Foundation as well as the support of all of the sponsors of the "New States of Matter in Hadronic Interactions" conference.

REFERENCES

1. Abu-Raddad, L. J., Hosaka, A., Toki, H., and Ebert, D., Nucleon structure and interactions in the context of path-integral methods and the nambu-jona-lasinio model (2002), to be submitted shortly for publication.
2. Hatsuda, T., and Kunihiro, T., *Phys. Rept.*, **247**, 221–367 (1994), and references therein.
3. Anselmino, M., Predazzi, E., Ekelin, S., Fredriksson, S., and Lichtenberg, D. B., *Rev. Mod. Phys.*, **65**, 1199–1234 (1993).
4. Serot, B. D., and Walecka, J. D., *Adv. Nucl. Phys.*, **16**, 1–327 (1986).
5. Ichinose, I., Matsui, T., and Onoda, M., *Phys. Rev.*, **B64**, 104516 (2001).
6. Cahill, R., *Aust. J. Phys.*, **42**, 171–86 (1989).
7. Reinhardt, H., *Phys. Lett.*, **B244**, 316 (1990).
8. Ebert, D., and Jurke, T., *Phys. Rev.*, **D58**, 034001 (1998).
9. Ebert, D., Reinhardt, H., and Volkov, M. K., *Prog. Part. Nucl. Phys.*, **33**, 1–120 (1994), and references therein.
10. Espriu, D., Pascual, P., and Tarrach, R., *Nucl. Phys.*, **B214**, 285–298 (1983).
11. Hellstern, G., Alkofer, R., Oettel, M., and Reinhardt, H., *Nucl. Phys.*, **A627**, 679–709 (1997).
12. Hellstern, G., and Weiss, C., *Phys. Lett.*, **B351**, 64–69 (1995).
13. Keiner, V., *Z. Phys.*, **A354**, 87 (1996).
14. Mineo, H., Bentz, W., Ishii, N., and Yazaki, K. (2002).
15. Vogl, U., and Weise, W., *Prog. Part. Nucl. Phys.*, **27**, 195–272 (1991).
16. Dumbrajs, O., et al., *Nucl. Phys.*, **B216**, 277–335 (1983).
17. Groom, D. E., et al., *Eur. Phys. J.*, **C15**, 1–878 (2000).
18. Dziembowski, Z., Metzger, W. J., and Van de Walle, R. T., *Zeit. Phys.*, **C10**, 231 (1981).
19. Hohler, G., et al., *Nucl. Phys.*, **B114**, 505 (1976).
20. Lomon, E. L., *Phys. Rev.*, **C64**, 035204 (2001).
21. Weiss, C., Buck, A., Alkofer, R., and Reinhardt, H., *Phys. Lett.*, **B312**, 6–12 (1993).

Relativistic Quantum Transport Theory

Hans-Thomas Elze

Universidade Federal do Rio de Janeiro, Instituto de Física
Caixa Postal 68.528, 21945-970 Rio de Janeiro, RJ, Brazil

Abstract. Relativistic quantum transport theory has begun to play an important role in the space-time description of matter under extreme conditions of high energy density in out-of-equilibrium situations. The following introductory lectures to some of its basic concepts and methods comprise the sections: 1. Introduction; 2. Aims of transport theory (classical); 3. Quantum mechanical distribution functions - the density matrix and the Wigner function; 4. Transport theory for quantum fields; 5. Particle production by classical fields; 6. Fluid dynamics of relativistic quantum dust.

1. INTRODUCTION

The study of the behavior of matter under more and more extreme conditions has a long tradition motivated by the quest for understanding the forces among its constituents on smaller and smaller scales. Not only the attempts to understand such spectacular phenomena as the stellar supernova explosions or theories of even the primordial stages of cosmological evolution, but also the ever increasing collision energy of high-energy particle accelerators and the heavy-ion programs at CERN and RHIC, in particular, witness the most recent stages of this scientific development.

In the latter context, often the complicated space-time dependence of what is really a quantum many-body system or what are highly dynamical interacting quantum fields is described in terms of a perfect fluid model. Since the seminal work by Fermi and Landau this approach has been applied successfully, in order to study global features, such as multiplicity distributions and apparently thermal transverse momentum spectra of produced particles, in high-energy collisions of strongly interacting matter [1, 2, 3, 4]. Similarly, the hydrodynamic approximation is often invoked in astrophysical applications and cosmological studies of the early universe [5]

The limitations of and likely necessary corrections to the fluid picture, however, have rarely been explored in the microscopic or high energy density domain. Difficulties reside in the derivation of consistent transport equations and in the amount of computational work required to find realistic solutions; see Refs. [6, 7, 8], for example, for a review and recent progress concerning selfinteracting scalar particles and the quark-gluon plasma, respectively. More understanding of related hydrodynamic behavior, if any, seems highly desirable.

For example, it has recently been shown that a free scalar field indeed behaves like a perfect fluid in the semiclassical (WKB) regime [9]. More generally, the mechanisms of quantum decoherence and thermalization in such systems which can be described hydrodynamically, i.e. the emergence of classical deterministic evolution from an underlying quantum field theory, are of fundamental interest [10, 11, 12, 13].

Having said this, it becomes obvious that not only particular applications of relativistic quantum transport theory motivated by experiments or observations – such as a suitable quark-gluon plasma transport theory to be applied to the phenomenology of high-energy heavy-ion collisions, or a transport theory for the electro-weak interactions of the intense neutrino flux from a supernova core with its electron-positron plasma sphere – are of interest, but that many interesting conceptual problems can be found in this field. It is the aim of the present introduction to describe some of its basic concepts and methods.

It seems worth while to emphasize here that transport theory by its very nature aims to describe highly dynamical systems where the time dependence of the phenomena to be studied cannot be neglected. Therefore, one necessarily has to go beyond (thermal) equilibrium field theory, for example. A partial exception consists in linear response theory, where standard field theory methods are employed to calculate the short-time response of the system to necessarily small perturbations.

The plan of these lectures is as follows. In Section 2 we present the motivation for transport theory by taking a cursory look at classical relativistic transport theory and its relation to relativistic hydrodynamics. In Section 3 we introduce quantum mechanical distribution functions. Especially, the need for the density matrix formalism is reviewed in basic terms, and the Wigner function is introduced. In Section 4, as an example, we develop the transport theory for the particular model of interacting quantum fields with a global O(4) symmetry, i.e. the linear sigma model, in the Hartree approximation. In Section 5 particle production by classical fields is described, presumably an important effect during early stages of heavy-ion collisions, solving fermion quantum transport equations perturbatively. Finally, in Section 6, we investigate the fluid dynamical behavior of relativistic quantum dust, solving the free quantum transport equations for arbitrary initial conditions exactly.

2. AIMS OF TRANSPORT THEORY

Concerning the historical development as well as the systems under study, the subject matter of transport theory is most frequently associated with nonequilibrium plasmas of all sorts:

- The quark-gluon plasma with QCD interactions formed shortly after the Big Bang initiating the observable Universe and possibly recreated during high-energy collisions, in particular with heavy nuclear projectiles/targets (**R+Q**).
- The $\nu\bar{\nu}e^+e^-$ plasma with electro-weak interactions created during supernova explosions between the proto-neutron star core and the leftover outer layers of the collapsing star (**R**).

- H, He, ... fusion plasmas in burning stars.
- Electrodynamic plasma phenomena in the Earth's ionosphere leading to polar lights and thunderstorms with lightnings.
- Discharge plasmas used in neon lighting and plasma welding.
- The e^- plasma in metals or semiconductors, the study of which has been advanced in solid state physics with particular attention to quantum effects (**Q**).

Here we marked by **R** and/or **Q** the systems where relativistic and/or quantum effects are known or expected to play an essential role. – Abstracting from these examples, we notice two common qualitative features among these systems:

- microscopic distance scales (average interparticle distance $n^{-1/3}$, mean free path λ_{mfp}, etc.)
  $\approx$ homogeneity scale L;
- microscopic time scales (average lifetime τ_l, relaxation time(s) τ_r, etc.)
  $\approx$ hydrodynamic time scale $\tau_h = L/c_s$, where c_s denotes the sound velocity.

Clearly, there will be exceptions to these qualitative statements and more precise characterizations of the plasma state can be given in the individual cases. However, this may suffice here, and we embark on the more formal description in the following. – We refer the reader to the monograph on relativistic kinetic theory of Ref. [14], which presents an excellent detailed exposition of the more traditional material in this context.

2.1 Classical phase space description of many-body dynamics

For a classical many-particle system, consider the probability to find a particle in the 8-dimensional phase space volume element at the (four-vector) position x with (four-vector) momentum p,

$$dP(x,p) \equiv f(x,p) d^4x d^4p , \qquad (1)$$

where $f(x,p)$ denotes the corresponding Lorentz scalar phase space density. Generally, in a classical system, the four-momentum is on-shell, such that not all components of p are independent. In particular, we assume that the constraint expressing the energy in terms of the three-momentum and particle mass, $p^0 = +\sqrt{\vec{p}^2 + m^2}$, is incorporated in f. – Except when explicitly stated, our units are such that $\hbar = c = k_B = 1$, and we use the Minkowski metric $g_{\mu\nu} = \text{diag}(1,-1,-1,-1)$.

A remark is in order here. Clearly, in order to learn not only about the behavior of a typical particle but also about its correlations with others, one principally should study the one-body density f along with with a two-body distribution $f(x_1, x_2, p_1, p_2)$, three-body distribution ..., etc. This is quite complicated in general and we restrict our attention to the one-body density here. Sometimes it is useful to visualize this function alternatively as describing a collection of (test) particles or a single particle with an ensemble of initial conditions.

Now, how does the one-body density f evolve, e.g. from one time-like hypersurface to another? We recall two ingredients of Liouville's theorem, which will provide the basic tool to answer this question [15]: **i)** Consider a phase-space volume element which is defined by 'tracer' particles forming its surface; then, due the uniqueness of the Newtonian motion, or its relativistic generalization, the number of particles inside is constant (in the absence of scattering interactions). **ii)** The size of the volume element, being associated with one of Poincaré's integral invariants (under contact transformations), is also a constant of motion. Then, Liouville's theorem follows: *Iff there are only conservative forces, then the phase space density is a constant of motion.*

Beginning with this theorem, we derive the evolution equation for f in terms of the relativistic proper time τ as follows:

$$0 = \frac{d}{d\tau}f(x,p) \equiv \frac{d}{d\tau}\sum_i \delta^4[x_i(\tau)-x]\delta^4[p_i(\tau)-p] , \tag{2}$$

$$= \sum_i \left\{\frac{dx_i^\mu}{d\tau}\frac{\partial}{\partial x^\mu} + \frac{dp_i^\mu}{d\tau}\frac{\partial}{\partial p^\mu}\right\}\delta^4(x_i(\tau)-x)\delta^4(p_i(\tau)-p)$$

$$= \left\{\frac{1}{m}p^\mu\frac{\partial}{\partial x^\mu} + \mathcal{F}^\mu(x)\frac{\partial}{\partial p^\mu}\right\}f(x,p) , \tag{3}$$

where we used the constancy of the density which is represented in terms of the particle (index i) trajectories $\{x_i(\tau), p_i(\tau)\}$ in the first line, carried out the differentiation in the second, and employed the definition of the four-velocity and the equation of motion in the last, respectively; here $\mathcal{F}^\mu(x)$ denotes the external or selfconsistent internal four-force(s). This is the relativistic Vlasov equation.

More generally, allowing for scattering of particles into and out of phase space volume elements, i.e. $(d/d\tau)f(x,p) \neq 0$, the Vlasov equation is replaced by the Boltzmann equation:

$$\left\{\frac{1}{m}p\cdot\partial_x + \mathcal{F}(x)\cdot\partial_p\right\}f(x,p) = \mathcal{C}[f](x,p) , \tag{4}$$

where $\mathcal{C}[f]$ denotes the collision term. It can be derived from an analysis of the equations including the two-body densities and generally turns out to be a nonlinear functional of the one-body densities.

A popular set of assumptions for this derivation is the following: **i)** Only binary or two-body collisions contribute (in a sufficiently dilute system); **ii)** Boltzmann's "Stosszahlansatz" according to which the number of collisions at x is proportional to $f(x,p)f(x,p')$; **iii)** the distribution function varies slowly on the scale of the mean free path, $\lambda_{mfp}|\nabla \log f(x,p)| \ll 1$.

We remark that while the Vlasov equation is appropriate for systems with conservative forces only and, thus, describes nondissipative phenomena, the Boltzmann equation incorporates dissipative scattering processes, which lead to entropy production.

In order to illustrate this, we introduce the simple relaxation time collision term, which could be derived rigorously as an approximation of a two-body scattering term but can also be seen as a phenomenological ansatz taking the dissipation into account:

$$\mathcal{C}[f] \equiv -\frac{1}{\tau_r}(f - f_0) ,\qquad(5)$$

where τ_r denotes the relaxation time parameter and f_0 the equilibrium one-body density, towards which the system will relax.

A particular equilibrium solution of the Boltzmann equation with the above collision term can be obtained from the Jüttner distribution,

$$f_0(x,p) \equiv \exp[-\beta(U \cdot p + \mu)] ,\qquad(6)$$

where the local parameters $\{\beta \equiv 1/T, U^\mu, \mu\}(x)$ denote the inverse temperature, flow four-velocity, and chemical potential, respectively. Instead of the exponential 'Boltzmann factor' one may also use a Fermi-Dirac or Bose-Einstein distribution, depending on the nature of the considered particles.

In the example of a plasma of charged particles, including a static homogeneous neutralizing background, the Lorentz force is $\mathcal{F}^\mu(x) \equiv F^{\mu\nu} j_\nu = (e/m) F^{\mu\nu} p_\nu$, in terms of the field strength tensor $F^{\mu\nu}$ of external and/or selfconsistently generated internal electromagnetic fields and the particle electric current j^μ. Inserting this together with the Jüttner distribution, i.e. $f = f_0$, into the Boltzmann Eq. (4) together with Eq. (5), we obtain the equation:

$$\frac{\partial}{\partial x_\mu}[\beta(U \cdot p + \mu)] + e\beta F^{\mu\nu} U_\nu = 0 ,\qquad(7)$$

which constrains the parameters of Jüttner distribution. Thus, the simplest solution indeed is the global equilibrium distribution with a constant temperature, a global rest frame, and where the gradient of the chemical potential compensates the electric field $F^{\mu 0}$, $-\partial_x^\mu \mu = eF^{\mu 0}$.

2.2 Relation to relativistic hydrodynamics

Turning to the observables to be described by the phase space density f introduced in Eq. (1), we define the particle (mass, charge) four-current,

$$N^\mu(x) \equiv \int d^4p\, p^\mu f(x,p) ,\qquad(8)$$

and the energy-momentum tensor,

$$T^{\mu\nu}(x) \equiv \int d^4p\, p^\mu p^\nu f(x,p) ,\qquad(9)$$

as the first and second moment of the distribution function, respectively; it is important to recall here that we assume f to implicitly contain the usual on-shell constraint, i.e. the factors $\Theta(p^0)\delta(p^2 - m^2)$, as we discussed.

These are the main quantities of interest in the hydrodynamic description of matter, where one integrates out the momentum space information contained in f [16].

Indeed, it is straightforward to show that N^μ and $T^{\mu\nu}$ obey the appropriate continuity equations related to mass (or charge) and four-momentum conservation. To begin with, using the Boltzmann Eq. (4) together with Eq. (5), we obtain:

$$\begin{aligned}\partial_\mu N^\mu &= \int d^4 p\, p \cdot \partial_x f = -m \int d^4 p \left(\mathcal{F} \cdot \partial_p f + \frac{f - f_0}{\tau_r} \right) \\ &= -\frac{m}{\tau_r} \int d^4 p\, (f - f_0) \equiv -\partial_\mu \delta N^\mu \, ,\end{aligned} \qquad (10)$$

where we find a dissipative contribution δN^μ on the right-hand side, which vanishes only if the ordinary density equals the equilibrium density determined by f_0. Similarly, one obtains:

$$\begin{aligned}\partial_\mu T^{\mu\nu} &= -m \int d^4 p\, p^\nu \left(\mathcal{F} \cdot \partial_p f + \frac{f - f_0}{\tau_r} \right) \\ &= m\mathcal{F}^\nu (\int d^4 p\, f) - \frac{m}{\tau_r}(N^\nu - N_0^\nu) \, ,\end{aligned} \qquad (11)$$

where the second term on the right-hand side is related to a dissipative contribution $-\partial_\mu \delta T^{\mu\nu}$ to the energy-momentum tensor, while the first term presents the external or selfconsistent force density acting on the system; N_0^μ is defined like N^μ, however, with f replaced by f_0.

More generally, based on the Chapman-Enskog method, standard forms of the dissipative terms can be constructed incorporating the transport coefficients of shear and bulk viscosity, heat conductivity, particle production/annihilation, and diffusion [16, 17].

Finally, we consider the entropy. It is an important quantity not only in equilibrium thermodynamics, but can be used, for example, to characterize the bulk properties of matter produced in high-energy collisions. In particular, it can be related more or less directly to the observed particle multiplicities [3, 4, 16]. Here we define the entropy four-current:

$$S^\mu(x) \equiv -\int d^4 p\, p^\mu f(x,p) \ln[f(x,p)/f_0(x,p)] \, . \qquad (12)$$

Calculating as before, we obtain the entropy production formula:

$$\partial_x \cdot S = \frac{m}{\tau_r} \int d^4 p\, (\ln[f/f_0] + 1)(f - f_0) \qquad (13)$$

Since $(\ln x + 1)(x - 1) \geq 0$, for all $x \geq 0$, we recover Boltzmann's "H-theorem":

$$\partial_x \cdot S \geq 0 \, , \qquad (14)$$

expressing a positive entropy production which vanishes only in equilibrium, when $f = f_0$.

This completes our overview of classical relativistic transport theory and some of its ramifications and we turn to quantum mechanics next.

3. QUANTUM MECHANICAL DISTRIBUTION FUNCTIONS - THE DENSITY MATRIX AND THE WIGNER FUNCTION

In order to motivate the necessity for a density matrix formulation in quantum mechanics, we recall Feynman's famous division [18, 19],

$$\mathcal{U}niverse \equiv \mathcal{S}ystem + \mathcal{R}est \,, \tag{15}$$

which says that doing physics and paying attention to the system requires precisely to separate off the rest of the Universe, i.e. the 'environment' of the system in modern terminology.

Following the presentation in Ref. [18], we introduce generic coordinates x and y for the description of $\mathcal{S}$ and $\mathcal{R}$, respectively. Assuming the existence of corresponding complete sets of states or wave functions, $\{\phi_i(x)\}$ and $\{\psi_j(y)\}$, the most general state vector and normalized wave function of $\mathcal{U}$ can be expanded, respectively, as:

$$|\Psi\rangle = \sum_{i,j} c_{ij}|\phi_i\rangle|\psi_j\rangle \,, \quad \Psi(x,y) = \sum_{i,j} c_{ij}\langle y|\phi_i\rangle\langle x|\psi_j\rangle \equiv \sum_i c_i(y)\phi_i(x) \,, \tag{16}$$

with complex expansion coefficients c_{ij} and functions $c_i(y)$.

Considering an operator $\hat{A}$ which acts on $\mathcal{S}$ only, it is given in terms of its matrix elements by:

$$\hat{A} \equiv \sum_{ii'j} A_{ii'}|\phi_i\rangle|\psi_j\rangle\langle\psi_j|\langle\phi_{i'}| \,, \tag{17}$$

i.e., it acts as a projector on $\mathcal{R}$. Following the rules, we obtain the expectation value of $\hat{A}$ in the state $|\Psi\rangle$ of Eq. (16):

$$\begin{aligned}\langle \hat{A} \rangle &\equiv \langle\Psi|\hat{A}|\Psi\rangle = \sum_{i'ij'} A_{i'i} C_{ij'} C^*_{i'j'} \\ &\equiv \sum_{i'i} A_{i'i}\rho_{ii'} = \text{Tr}\,\hat{A}\hat{\rho} = \text{Tr}\,\hat{\rho}\hat{A} \,.\end{aligned} \tag{18}$$

It follows from its definition here that the density operator $\hat{\rho}$, with the density matrix elements identified as $\rho_{ii'} \equiv \langle\phi_i|\hat{\rho}|\phi_{i'}\rangle$, is a hermitean operator.

Therefore, we may introduce a complete orthonormal basis $\{|i\rangle\}$ diagonalizing $\hat{\rho}$: $\hat{\rho} = \sum_i w_i |i\rangle\langle i|$, with real eigenvalues w_i. Furthermore, choosing $\hat{A} = 1_\mathcal{S} \otimes 1_\mathcal{R}$, i.e. the identity operator, one finds with the help of Eq. (18) the sum rule: $1 = \langle\hat{A}\rangle = \text{Tr}\,\hat{\rho} = \sum_i w_i$. Finally, choosing instead $\hat{A} = |i'\rangle\langle i'| \otimes 1_\mathcal{R}$, a similar calculation yields: $w_{i'} = \langle\hat{A}\rangle = \langle\Psi|\hat{A}|\Psi\rangle = \sum_j |\langle\psi_j|\langle i'|\Psi\rangle|^2 \geq 0$.

Abstracting from the present example, it is postulated that *any quantum mechanical system* is to be described by a hermitean density operator,

$$\hat{\rho} = \sum_i w_i |i\rangle\langle i| , \qquad (19)$$

where $\{|i\rangle\}$ forms a complete orthonormal set, the real expansion coefficients are non-negative, $w_i \geq 0$, and fullfill a normalization condition, $\sum_i w_i = 1$. – We remark that the density matrix was first introduced by von Neumann in 1932 in his by now famous book [20]. – Observables in particular and expectation values of operators in general are always to be calculated by:

$$\langle \hat{A} \rangle = \mathrm{Tr}\, \hat{\rho}\hat{A} = \sum_i w_i \langle i|\hat{A}|i\rangle . \qquad (20)$$

Consequently, the coefficients w_i are interpreted as describing the probability to find the system in the state $|i\rangle$.

As a matter of nomenclature, one distinguishes pure and mixed states of a system, the former being defined by a density operator which is a projector, $\hat{\rho} = \hat{\rho}^2 \iff (w_{i^*} = 1$, all other $w_i = 0)$, and the latter comprising all other cases. – We recall the case of a system in contact with a heat bath, which leads to the mixed state density operator of thermal equilibrium:

$$\hat{\rho}(\beta) = Z^{-1}(\beta) \sum_n e^{-\beta E_n} |E_n\rangle\langle E_n| , \quad Z(\beta) \equiv \sum_n e^{-\beta E_n} , \qquad (21)$$

where $\{|E_n\rangle\}$ denotes the complete set of normalized energy eigenstates of the system and Z is its partition function necessary to normalize the exponential 'Boltzmann factor'. All canonical thermodynamical relations can be easily derived from this density operator, see, for example, [18].

Finally, we may address also in the present context the question how the system, i.e. its density operator, evolves in time. This is easily answered by expanding the eigenstates $\{|i\rangle\}$ of $\hat{\rho}$ in terms of the energy eigenstates,

$$|i(0)\rangle \equiv |i\rangle = \sum_n |E_n\rangle\langle E_n|i(0)\rangle , \qquad (22)$$

which implies according to the Schrödinger equation: $|i(t)\rangle = \exp(-i\hat{H}t)|i(0)\rangle$, given the Hamiltonian $\hat{H}$. Then, the evolution of $\hat{\rho}$ follows:

$$\hat{\rho}(t) = e^{-i\hat{H}t} \hat{\rho}(0) e^{+i\hat{H}t} . \qquad (23)$$

Equivalently, we obtain:

$$\frac{\mathrm{d}}{\mathrm{d}t} \hat{\rho}(t) = i[\hat{\rho}(t), \hat{H}] , \qquad (24)$$

which differs from the usual operator evolution equation in quantum mechanics by an additional minus sign on the right-hand side. Both signs, of course, are consistent, since we have:

$$\langle \hat{A} \rangle_{(t)} \equiv \mathrm{Tr}\, \hat{\rho}(t)\hat{A} = \mathrm{Tr}\, \hat{\rho}(0) e^{+i\hat{H}t} \hat{A} e^{-i\hat{H}t} \equiv \langle \hat{A}(t) \rangle , \qquad (25)$$

using the cyclicity of the trace.

The equations (24) and (25) present the problem of quantum transport theory in its most compact form. In particular, Sections 4–6 are devoted to explorations of various more detailed forms of these abstract results and their applications.

In order to connect the density matrix formalism with the classical transport theory, we turn to the density matrix in coordinate and momentum representation.

It helps to visualize the following by imagining a single particle to be described quantum mechanically. Then, in the coordinate representation, using corresponding single-particle wave functions, we have the density matrix elements:

$$\rho(x',x) = \sum_i w_i \langle x'|i\rangle\langle i|x\rangle = \sum_i w_i \phi_i(x')\phi_i^*(x) , \qquad (26)$$

see Eq. (19). Then, we calculate immediately the probability to find the particle at x:

$$P(x) \equiv \rho(x,x) = \sum_i w_i |\phi_i(x)|^2 , \qquad (27)$$

in agreement with the standard rules of quantum mechanics. Similarly, in momentum representation,

$$\rho(p',p) = \sum_i w_i \langle p'|i\rangle\langle i|p\rangle = \sum_i w_i \phi_i(p')\phi_i^*(p) , \qquad (28)$$

which yields:

$$P(p) \equiv \rho(p,p) = \sum_i w_i |\phi_i(p)|^2 , \qquad (29)$$

i.e. the probability to find the particle with momentum p.

Now, considering a function O – representing some observable, for example – which is defined over phase space, we obtain its average value by calculating:

$$\overline{O} \equiv \int \mathrm{d}x \mathrm{d}p\, O(x,p) f(x,p) , \qquad (30)$$

involving the one-body probability density function f. The question then arises, whether there exists a corresponding quantum mechanical density function which yields the expectation value of operators in the form of phase space integrals, generalizing Eq. (30). One possible answer is provided by the Wigner function [18, 19, 21]:

$$W(x,p) \equiv \int \mathrm{d}y\, e^{ipy/\hbar} \rho(x+y/2, x-y/2) , \qquad (31)$$

i.e. a particular Fourier transform of the density matrix in coordinate space. We remark that the momentum appearing as an argument of W is the conjugate variable to the relative coordinate separating the pair of wave functions which enter ρ, cf. Eq. (26); we also indicate here explicitly the $\hbar$-dependence arising in the phase factor.

Furthermore, even though the Wigner function presents a real distribution, due to $W(x,p) = W^*(x,p)$, it may oscillate and indeed does so in most cases.

Nevertheless, we easily obtain the following results which match their classical counterparts:

$$\int \frac{\mathrm{d}p}{2\pi\hbar} W(x,p) = \rho(x,x) = P(x) , \tag{32}$$

$$\int \mathrm{d}x\, W(x,p) = \rho(p,p) = P(p) . \tag{33}$$

This implies for the expectation values of functions of operators:

$$\langle O(\hat{x})\rangle = \mathrm{Tr}\,\hat{\rho} O(\hat{x}) = \int \frac{\mathrm{d}x \mathrm{d}p}{2\pi\hbar} O(x) W(x,p) , \tag{34}$$

$$\langle O(\hat{p})\rangle = \mathrm{Tr}\,\hat{\rho} O(\hat{p}) = \int \frac{\mathrm{d}x \mathrm{d}p}{2\pi\hbar} O(p) W(x,p) , \tag{35}$$

which should be compared to Eq. (30).

However, besides the fact that the Wigner function generally is not positive definite, the expected limitation of the classical/quantum correspondence of various (probability) density functions also shows up, when one considers operators of the form $O(\hat{x},\hat{p})$. In this case, operator ordering becomes an issue and the above formulae have to be generalized with care. On the other hand, as we shall see shortly in Section 4, the appropriate semiclassical expansion of the Wigner function transport equation does yield the transport equation for the classical probability density function, which we discussed in Section 2.

4. TRANSPORT THEORY FOR QUANTUM FIELDS

As an example of a field theory we choose the O(4) *linear σ-model*, which has a long history of applications in various phenomenological contexts. Recently it has been argued by Wilczek that it represents the QCD chiral order parameter for $n_f = 2$ massless quark flavors [22]. Furthermore, it has been demonstrated by nonperturbative calculation that this model possesses a second order finite temperature phase transition between the spontaneously broken and symmetry restored phases [23]. Most interestingly, the effective mass and coupling go to zero at the phase transition with well-determined critical exponents. This may influence the hydrodynamic behavior of matter described by this model in interesting ways, which we presently study.

Here, our aim is to derive the quantum transport equations in the Hartree approximation. In particular, we will illustrate how an appropriate Wigner function can be introduced and how field theory aspects are related to our earlier considerations of classical transport theory.

To begin with, the O(4)-invariant σ-model action is defined by:

$$S[\vec{\phi}] \equiv \int \mathrm{d}^4 x \left(\frac{1}{2}(\partial\vec{\phi})^2 - \frac{1}{2}\mu^2 \vec{\phi}^2 - \frac{1}{4!}\lambda(\vec{\phi}^2)^2 \right) , \tag{36}$$

with $\vec{\phi} \equiv (\phi_1, \phi_2, \phi_3, \phi_4)$ and where the mass parameter is chosen with the 'wrong' sign, i.e. $\mu^2 < 0$. The corresponding potential, $V(\vec{\phi}) \equiv \frac{1}{2}\mu^2\vec{\phi}^2 + \frac{1}{4!}\lambda(\vec{\phi}^2)^2$, is of the so-called 'Mexican hat' form which leads to spontaneous symmetry breaking at the tree level.

We parametrize the four-component vector $\vec{\phi}$ in terms of three-component 'pion' and one-component 'sigma' fields, $\vec{\phi} \equiv (\vec{\pi}, \sigma')$, at each space-time point. Shifting the σ'-field by the vacuum expectation value σ_0 of $\vec{\phi}$, which is determined by the minimum of the potential,

$$\frac{dV(\vec{\phi})}{d\vec{\phi}} = (\mu^2 + \frac{1}{3!}\lambda\vec{\phi}^2)\vec{\phi} = 0 \implies |\sigma_0| = \sqrt{-6\mu^2/\lambda} \, , \tag{37}$$

we define the fluctuation field $\sigma \equiv \sigma' - \sigma_0$.

The effect of this parametrization combined with the shift of one component by the vacuum expectation value is easily seen in the corresponding <u>Heisenberg operator</u> equations of motion,

$$\partial^2 \vec{\pi} + m_\pi^2 \vec{\pi} + \frac{\lambda}{3!}\vec{\pi}^2\vec{\pi} = 0 \, , \tag{38}$$

$$\partial^2 \sigma + m_\sigma^2 \sigma + \frac{\lambda}{2}\sigma_0\sigma^2 + \frac{\lambda}{3!}\sigma^3 = 0 \, , \tag{39}$$

which are obtained by varying the action $S[\vec{\phi}]$ with respect to $\vec{\phi}$, introducing parametrization and shift, and considering the fields as quantum field operators. Here we introduced the effective masses:

$$m_\pi^2 \equiv \frac{\lambda}{3!}(\sigma^2 + 2\sigma_0\sigma) \, , \quad m_\sigma^2 \equiv \frac{\lambda}{3!}(\vec{\pi}^2 + 2\sigma_0^2) \, . \tag{40}$$

. We observe that for $\sigma \to 0$ we have $m_\pi^2 \to 0$, while $m_\sigma^2 \to \lambda\sigma_0^2/3$, yielding one massive 'radial' mode together with three massless 'Goldstone modes', in accordance with the Goldstone theorem.

Taking the expectation value Eqs. (38) and (39), we observe that one-point functions, such as the <u>mean field</u> $\bar{\sigma}(x) \equiv \langle\sigma(x)\rangle$, generally are coupled to two-point functions, such as $\lim_{x' \to x}\langle\sigma(x)\sigma(x')\rangle$, and higher n-point functions. The coincidence limit of these <u>Wightman functions</u> produces divergences necessitating a renormalization procedure, which we will discuss at the end of this section.

In order to solve the resulting equations, one has to specify the density operator which enters the expectation values, e.g. $\langle\sigma\rangle \equiv \text{Tr}\,\rho\sigma$, omitting the operator signs used previously, cf. Section 3. This can be done, for example, on a fixed time-like hypersurface, which will be demonstrated for fermions in Section 6. Furthermore, we need to derive equations of motion for the higher n-point functions, since taking expectation values of nonlinear operator equations automatically generates an infinite hierarchy of equations, similarly as the Schwinger-Dyson equations for propagators or the BBGKY hierarchy in classical transport theory.

The simplest nonperturbative truncation of the hierarchy of Wightman function equations is produced by the <u>Hartree approximation</u>. It consists in factorizing the

n-point functions into products of one- or two-point functions, properly taking into account all possible factorizations. For example, we obtain:

$$\langle \sigma_1 \sigma_2 \sigma_3 \rangle = \bar{\sigma}_1 \langle \underline{\sigma}_2 \underline{\sigma}_3 \rangle + \bar{\sigma}_2 \langle \underline{\sigma}_1 \underline{\sigma}_3 \rangle + \bar{\sigma}_3 \langle \underline{\sigma}_1 \underline{\sigma}_2 \rangle + \bar{\sigma}_1 \bar{\sigma}_2 \bar{\sigma}_3 \;, \quad (41)$$

where the subscripts refer to different space-time points. Note that the expectation value of any odd power of the proper quantum field vanishes, since we define $\underline{\sigma} \equiv \sigma - \bar{\sigma}$. – This approximation is known to be equivalent to summing all iterated bubbles ('superdaisies') in the Feynman diagram calculation of the vacuum effective action in a ϕ^4-model [24].

Here we will furthermore assume that cross terms vanish, e.g. $\langle \vec{\pi} \underline{\sigma} \rangle = 0$. It turns out to be consistent with this assumption to set $\langle \vec{\pi} \rangle = 0$, since the classical pion field obtains no source term. In distinction, the Eq. (39) yields the Klein-Gordon type <u>mean field equation</u>:

$$\left(\partial^2 + m_\sigma^2 + \frac{\lambda}{2} (\frac{1}{3} \bar{\sigma}^2 + \sigma_0 \bar{\sigma} + \langle \underline{\sigma}^2 \rangle) \right) \bar{\sigma} + \frac{\lambda}{2} \sigma_0 \langle \underline{\sigma}^2 \rangle \equiv \left(\partial^2 + m_\sigma^2 + \delta m^2 \right) \bar{\sigma} + J = 0 \;, \quad (42)$$

where the two-point function adds to the nonlinear classical force term shifting the effective σ-mass, which contains $\langle \vec{\pi}^2 \rangle$, and to a source term, δm^2 and J, respectively. For a homogeneous system, for $\delta m^2 \gg m_\sigma^2$, and for sufficiently large $\langle \underline{\sigma}^2 \rangle$, the mean field is $\langle \sigma \rangle = -\sigma_0$, indicating symmetry restoration with $\langle \sigma' \rangle = 0$.

Multiplying Eq. (39) by one more power of the field operator from the left, before applying the Hartree approximation as before, we obtain the <u>two-point function equation</u>:

$$\left(\partial_2^2 + m_{\sigma,2}^2 + \lambda \sigma_0 \bar{\sigma}_2 + \frac{\lambda}{2} (\bar{\sigma}_2^2 + \langle \underline{\sigma}_2^2 \rangle) \right) \langle \underline{\sigma}_1 \underline{\sigma}_2 \rangle = 0 \;, \quad (43)$$

where subscripts "1,2" refer to two different space-time points. Here we also used Eq. (42) at point "2", multiplied by $\bar{\sigma}_1$, in order to simplify Eq. (43) considerably. A similar equation follows from Eq. (38) for the 'pion' modes.

The next step consists in defining a suitable <u>Wigner operator</u>,

$$W_{ab}(x,p) \equiv \int \frac{\mathrm{d}^4 y}{(2\pi)^4} e^{-ip \cdot y} \underline{\Phi}_a(x + \frac{1}{2}y) \underline{\Phi}_b(x - \frac{1}{2}y) \;, \quad (44)$$

where $\vec{\Phi} \equiv (\vec{\pi}, \sigma)$ and $\underline{\Phi} \equiv \Phi - \bar{\Phi}$. This should be compared to the quantum mechanical Wigner function introduced previously, Eq. (31). We remark that it might be useful for some applications not involving vacuum properties to normal-order the field operators in the definition of W, see, for example, Section 6.

Illustrating the usefulness of W, we write down the energy-momentum tensor for the O(4)-model:

$$T_{\mu\nu} = \left\langle (\partial_\mu \vec{\Phi}) \cdot (\partial_\nu \vec{\Phi}) - g_{\mu\nu} [\frac{1}{2}(\partial \vec{\Phi})^2 - \frac{1}{2}\mu^2 \vec{\Phi}^2 - \frac{1}{4!}\lambda(\vec{\Phi}^2)^2] \right\rangle = \quad (45)$$

$$= \int d^4p \left(p_\mu p_\nu + \frac{1}{4}\partial_{x^\mu}\partial_{x^\nu} - \frac{1}{2}g_{\mu\nu}(p^2 + \frac{1}{4}\partial_x^2) \right) \langle W_{aa}(x,p) \rangle$$

$$+ g_{\mu\nu} \int d^4p \left(\frac{1}{2}\mu^2 \langle W_{aa}(x,p) \rangle + \frac{1}{4!}\lambda \int d^4p' \, \langle W_{aa}(x,p) W_{bb}(x,p') \rangle \right) , \quad (46)$$

to which must be added the purely classical terms plus mean field dependent interaction terms ($\propto \lambda$) involving W, which can all be further evaluated in Hartree approximation.

Heading for the transport theory, we will express Eq. (43) in terms of the Wigner operator. We introduce the abbreviation:

$$\mathcal{M}^2(x) \equiv m_\sigma^2(x) + \lambda \sigma_0 \bar{\sigma}(x) + \frac{\lambda}{2}\left(\bar{\sigma}^2(x) + \langle \underline{\sigma}^2(x) \rangle \right) . \quad (47)$$

The two-point functions contained here, $\langle \underline{\vec{\pi}}^2 \rangle$ in m_σ^2 and $\langle \underline{\sigma}^2 \rangle$, respectively, can be rewritten using the Wigner operator, e.g.: $\langle \underline{\vec{\sigma}}^2(x) \rangle = \int d^4p \langle W_{\sigma\sigma}(x,p) \rangle$. Then we obtain instead of Eq. (43):

$$\left(\frac{1}{4}\partial_x^2 - p^2 + ip \cdot \partial_x + \exp(-\frac{i}{2}\partial_x \cdot \partial_p) \mathcal{M}^2(x) \right) \langle W_{\sigma\sigma}(x,p) \rangle = 0 , \quad (48)$$

where the x-derivative in the exponential acts only on $\mathcal{M}^2$. In the derivation of this result, and similarly for the 'pions', one frequently makes use of suitable partial integrations under the momentum integral from W, as well as expressing the shifted argument of $\mathcal{M}^2$ by a translation operator giving rise to the exponential [6, 25].

The final step consists in adding to/subtracting from the complex Eq. (48) its adjoint. This yields the <u>transport equation</u>:

$$\left(p \cdot \partial_x - \frac{1}{\hbar}\sin(\frac{\hbar}{2}\partial_x \cdot \partial_p) \mathcal{M}^2(x) \right) \langle W_{\sigma\sigma}(x,p) \rangle = 0 , \quad (49)$$

together with a generalized <u>mass-shell constraint</u>:

$$\left(p^2 - \frac{\hbar^2}{4}\partial_x^2 - \cos(\frac{\hbar}{2}\partial_x \cdot \partial_p) \mathcal{M}^2(x) \right) \langle W_{\sigma\sigma}(x,p) \rangle = 0 , \quad (50)$$

where the appropriate powers of $\hbar$ are reinserted. We stress that the covariant transport equation alone does not suffice to determine the dynamics.

A systematic semiclassical expansion of Eqs. (49)-(50) in powers of $\hbar$ becomes feasible now. To leading order we obtain:

$$\left(p \cdot \partial_x - \frac{1}{2}\partial_x \cdot \partial_p \mathcal{M}^2(x) \right) \langle W_{\sigma\sigma}(x,p) \rangle = 0 , \quad (51)$$

$$\left(p^2 - \mathcal{M}^2(x) \right) \langle W_{\sigma\sigma}(x,p) \rangle = 0 , \quad (52)$$

which are indeed of the form of a classical Vlasov and mass-shell equation, respectively, discussed in Section 2.1. Note, however, that the effective mass or effective potential, appearing here contains contributions from the mean field, which have to be determined selfconsistently from Eq. (42), and 'selfenergy' terms, which we will discuss shortly.

We remark that effects of the higher-order $\hbar$-corrections to Eq. (51) and (52) are largely unexplored, since numerically the corresponding higher derivatives lead to strong instabilities. This provides part of the motivation to develop an analytical approach solving the exact quantum transport equations, the first step of which is described in Section 6.

Furthermore, we observe that the equations derived here do not yield any collision terms. This was to be expected, since the Hartree approximation, which is essentially equivalent to the Gaussian approximation in the Schrödinger functional approach, leads to an effective 'quantum' Hamiltonian evolution which is nondissipative [11]. Only an improved treatment of correlation terms, such as the last term in Eq. (46), will go beyond the collisionless Vlasov dynamics obtained here, see, for example, Refs. [7, 8].

Finally, we turn to the 'selfenergy' terms mentioned above and to the divergences caused by them, in particular. These terms involve integrals of the kind $\int d^4p \langle W_{\sigma\sigma}(x,p) \rangle$. Let us consider the simplest case of a density operator projecting on the vacuum state, $\rho_{vac} = |0\rangle\langle 0|$. Then, for a generic free scalar field, the vacuum Wigner function is:

$$W_{vac}(p) \equiv \text{Tr } \rho_{vac} W(x,p) = \langle 0|W(x,p)|0\rangle = (2\pi)^{-3}\delta(p^2 - m^2) \ , \qquad (53)$$

where m denotes its *physical* mass, and the calculation proceeds, for example, by expanding the field operators appearing in the definition of the Wigner operator in terms of creation and annihilation operators. In this case, we obtain:

$$\int d^4p \, W_{vac}(p) = \int \frac{d^3p}{(2\pi)^3 \omega_p} \equiv I_m \ , \qquad (54)$$

with $\omega_p \equiv (\vec{p}^2 + m^2)^{1/2}$. This 'selfenergy' integral is quadratically divergent.

Since divergences arise from field operators at coinciding points and, as a first approximation, independently of which point it is, we rewrite Eq. (50) accordingly, using Eqs. (40) and (47):

$$\begin{aligned} 0 &= (-p^2 + m_{bare}^2 + \frac{\lambda}{2} I_m \\ &\quad + \frac{\lambda}{2}\int d^4k\, \{W(x,k) - W_{vac}(k)\})(W(x,p) - W_{vac}(p) + W_{vac}(p)) \quad (55) \\ &\equiv \left(-p^2 + m_r^2 + \frac{\lambda}{2}\int d^4k\, W_r(x,k)\right)(W_r(x,p) + W_{vac}(p)) \ , \quad (56) \end{aligned}$$

with $W(x,p) \equiv \langle W_{\sigma\sigma}(x,p)\rangle$ and the Lagrangian 'bare' mass $m_{bare}^2 \equiv \frac{\lambda}{3}\sigma_0^2$, cf. Eq. (40). Here we left out all 'pion' and mean field contributions to keep things transparent. These can be added following the same strategy, while the gradient corrections would force us to keep $W(x,p)$ unrenormalized at first sight.

Note that we added and subtracted suitably the divergent term in Eq. (55), thus introducing in Eq. (56) the renormalized mass and Wigner function, m_r and W_r, respectively. We complete the renormalization by identifying the physical mass:

$$m^2 \equiv m_r^2 + \frac{\lambda}{2} \int d^4k \, W_r(x,k) \ . \tag{57}$$

With this, the Eq. (56) assumes a simple form, $(p^2 - m^2)W_r(x,p) = 0$, indicating that W_r is (in our first approximation) an on-shell distribution:

$$W_r(x,p) = \delta(p^2 - m^2) f(x,p) \ , \tag{58}$$

such as a Bose-Einstein distribution with (weakly) x-dependent parameters, cf. Section 2.1. Neglecting this dependence altogether and assuming a simple thermal distribution turns Eq. (57) into a mass-gap equation with m_r^2 as input parameter and a resulting temperature dependent mass $m^2(T)$.

A mass renormalization as performed here is sufficient to render the mean field Eq. (42) and the operator of two-point function Eq. (43) finite. The gradient corrections in the transport and generalized mass-shell constraint, however, indicate that the renormalized Wigner (or equivalently the two-point) function has to be be renormalized separately, subtracting as above, and thus closing the set of equations. Generally, the situation gets more complicated, when the x-dependence of W_r has to be taken serious and, thus, the physical mass becomes space-time dependent, e.g. in the case of a system which evolves from strongly inhomogeneous initial conditions. We will not address these issues here, but turn to a related physical effect in the next section.

5. PARTICLE PRODUCTION BY CLASSICAL FIELDS

The color string, rope, or flux tube models are widely used in phenomenological descriptions of particle production, in high-energy nuclear collisions in particular [4]. The basis here is Schwinger's nonperturbative calculation of vacuum decay due to charged particle production in *constant* and *homogeneous* external electric fields [26, 27]. Considering the inhomogeneous and rapid evolution of the system during a heavy-ion collison, the basic assumptions of this picture are questionable.

Since an appropriate generalization of Schwinger's result is still not available, one may as well consider the perturbative evaluation of particle production, however, for arbitrarily varying fields. This was originally discussed in Ref. [28] for QCD in abelian dominance approximation. There, also the ensuing modification of semiclassical transport equations, cf. Section 4, and the related vacuum polarization current were obtained. Here we briefly recall part of this calculation which is based on an $O(g^2)$ solution of the quark transport equation in external color fields.

Simplifying the notation, we consider electrically charged Dirac fermions (mass m) in arbitrary electromagnetic fields. We define the Fourier transformed vacuum Wigner function:

$$f_{vac}(q,p) \equiv \int d^4x d^4y \, e^{iq \cdot x} e^{-ip \cdot y} \langle 0|\bar{\psi}(x+\frac{1}{2}y)\psi(x-\frac{1}{2}y)|0\rangle \quad (59)$$

$$= -(2\pi)^5 \delta^4(q)\delta(p^2-m^2)\theta(-p^0)(\gamma \cdot p + m) \;, \quad (60)$$

which is a 4x4 spinor matrix. The final form here follows similarly as in the scalar case, Eq. (53), using a standard expansion of the field operators [27]. In distinction to the fermion Wigner function introduced in the following Section 6, Eq. (70), where the notation is explained in more detail, we presently do not normal-order the field operators, since we are interested to study the response of the vacuum, i.e. the modification of f_{vac}, due to external fields.

Our task is to somehow solve the full quantum transport equation which determines the fermionic Wigner function. For the present study the transport equation based on the linear form of the Dirac equation is most useful [6, 25]. – The quadratic form yields transport equations which are closer to the usual form, such as in the scalar field case studied in Section 4. – Presently, following Fourier transformation, we have:

$$\left(\gamma \cdot (p+\frac{1}{2}q) - m\right) f(q,p) = g \int \frac{d^4q'}{(2\pi)^4} \, \gamma \cdot A(q') f(q-q', p-\frac{1}{2}q') \;, \quad (61)$$

where A^μ denotes the vector potential of the external field. This equation has to be solved together with the constraint:

$$f^\dagger(q,p) = \gamma^0 f(-q,p) \gamma^0 \;, \quad (62)$$

which follows from the definition of the fermion Wigner function; equivalently, one may solve Eq. (61) simultaneously with its adjoint.

The perturbative solution in powers of the coupling constant g of Eq. (61) can be found interatively, starting with the zeroth order solution f_{vac} as input on the right-hand side.

Thus, at first order one finds the induced vacuum current, $J^\mu_{(1)}(q) = \int d^4 p \, \text{Tr} \gamma^\mu f^\mu_{(1)}(q,p)$, which can be written explicitly involving the first order (one-loop) QED vacuum polarization tensor and reproduces the known result [27, 28]. Furthermore, since there is a gauge invariant version of Eq. (61) based on a modified definition of the Wigner function [6, 25, 28], this type of calculation might allow to calculate the polarization tensor in a manifestly gauge invariant way.

For our present purposes, the second order solution of the equations is most interesting. The main algebraic complication arises from Eq. (62), which has to be satisfied order by order. We do not give the lengthy expressions here, but turn to the calculation of the spectrum of the produced particles. We have to use reduction formulae, in order to relate the 'in' field operators defining the Wigner function to asymptotic on-shell 'out' particle states [27]. In this way, we obtain:

$$2\omega_p \frac{dN}{d^3 p} \equiv \mathcal{N}_{(2)}(p) = \quad (63)$$

$$= \frac{1}{2(2\pi)^3} \sum_r \bar{u}^r(p)(\gamma \cdot p - m) f_{(2)}(0,p)(\gamma \cdot p - m) u^r(p) \qquad (64)$$

$$= \frac{g^2}{4\pi^2} \int d^4q \, \theta(q^0 - p^0) \delta\left((q-p)^2 - m^2\right)$$
$$\cdot \left(\vec{E}(q) \cdot \vec{E}(-q) - \vec{B}(q) \cdot \vec{B}(-q) - |A(q) \cdot (q-2p)|^2\right) , \qquad (65)$$

where the electromagnetic fields and the vector potential enter; here $\omega_p \equiv \sqrt{\vec{p}^2 + m^2}$ and u^r ($\bar{u}^r$) denotes the single-particle (adjoint) spinor wave function, respectively, in the normalization of Ref. [27]. Also the last term of this result is gauge invariant, as can be demonstrated using the constraints of the integral and the fact that the particles are on-shell. Interestingly, with the external fields left completely general, still there is no spatial dependence of this spectrum.

Finally, we calculate the vacuum decay rate $\mathcal{R}$ per unit four-volume, using a more suitable intermediate form of the spectrum result, Eq. (65):

$$\mathcal{R} = 2 \int \frac{d^3 p}{2\omega_p} \mathcal{N}_{(2)}(p) \qquad (66)$$

$$= \frac{g^2}{\pi^2} \int \frac{d^3 p \, d^3 p'}{\omega_p \, \omega'_p} d^4 q \, \delta^4(q - p - p')$$
$$\cdot \left(-\frac{1}{2} q^2 g_{\mu\nu} + p_\mu p'_\nu + p'_\mu p_\nu\right) A^\mu(q) A^\nu(-q) \qquad (67)$$

$$= \frac{g^2}{12\pi} \int d^4 q \, \theta(q^2 - 4m^2)$$
$$\cdot \left(1 - \frac{4m^2}{q^2}\right)^{1/2} \left(1 + \frac{2m^2}{q^2}\right) \left(|\vec{E}(q)|^2 - |\vec{B}(q)|^2\right) , \qquad (68)$$

which confirms the result obtained by other methods in Ref. [27]. Clearly, particle production is an electric field effect.

For completeness, let us quote Schwinger's nonperturbative result [26]:

$$\mathcal{R} = \frac{g^2 \vec{E}^2}{4\pi^3} \sum_{n=1}^{\infty} \frac{1}{n^2} \exp\left(-\frac{n\pi m^2}{|gE|}\right) , \qquad (69)$$

which forms the basis to date of the phenomenological models mentioned in the beginning of this section. For reasons which we discussed, a future calculation bridging the gap between our perturbative calculation for arbitrary fields and the nonperturbative result for constant electric fields would be extremely useful.

6. FLUID DYNAMICS OF RELATIVISTIC QUANTUM DUST

In this section, we will study the relation between relativistic hydrodynamics and the full quantum evolution of a free matter field [29]. In particular, we try to answer how a free fermion field and its energy-momentum tensor will evolve, given arbitrary initial conditions and especially those of the Landau and Bjorken models.

In the absence of interactions, decoherence or thermalization may be present in the initial state, corresponding to an impure density matrix, but is followed by unitary evolution. We consider this as a "quantum dust" model of the expansion of matter originating from a high energy density preparation phase, which the Landau and Bjorken models describe classically [2, 3].

Our approach is independent of the nature of the field, as long as it obeys a standard wave equation. To be definite, we choose to work with Dirac fermions and comment about neutrinos later. We introduce the spinor Wigner function, i.e., a (4x4)-matrix depending on space-time and four-momentum coordinates:

$$W_{\alpha\beta}(x;p) \equiv \int \frac{d^4y}{(2\pi)^4} e^{-ip\cdot y} \langle :\bar{\psi}_\beta(x+y/2)\psi_\alpha(x-y/2): \rangle \ , \qquad (70)$$

where the expectation value refers to the (mixed) state of the system; without interactions, the vacuum plays only a passive role and, therefore, is eliminated by normal-ordering the field operators. Note that the normalization of the Wigner function is a matter of convention, and we have chosen the most convenient one for us here.

All observables can be expressed in terms of the Wigner function here. In particular, the (unsymmetrized) energy-momentum tensor:

$$\langle :T_{\mu\nu}(x): \rangle \equiv i\langle :\bar{\psi}(x)\gamma_\mu \overleftrightarrow{\partial}_\nu \psi(x): \rangle = \mathrm{tr}\, \gamma_\mu \int d^4p\, p_\nu W(x;p) \ , \qquad (71)$$

where $\overleftrightarrow{\partial} \equiv \frac{1}{2}(\overrightarrow{\partial} - \overleftarrow{\partial})$ and with a trace over spinor indices (conventions as in [25]). Furthermore, the dynamics of W reduces to the usual phase space description, as in Section 2, in the classical limit [25].

The following study is based on the simple fact that the propagation of the free fields entering in Eq. (70) from one time-like hypersurface to another is described by the Schwinger function. It is the solution of the homogeneous Dirac equation, $[i\gamma\cdot\partial_x - m]S(x,x') = 0$, for the initial condition $S(\vec{x},\vec{x}',x^0 = x'^0) = -i\gamma^0\delta^3(\vec{x}-\vec{x}')$. Thus, $\psi(x) = i\int d^3x' S(x,x')\gamma^0\psi(x')$, and similarly for the adjoint. An explicit form is:

$$iS(x,x') = iS(x-x' \equiv \Delta) = (i\gamma\cdot\partial_\Delta + m)\int \frac{d^3k}{(2\pi)^3 2\omega_k}\left(e^{-ik_+\cdot\Delta} - e^{-ik_-\cdot\Delta}\right) \ , \qquad (72)$$

where $k_\pm \equiv (\pm\omega_k, \vec{k})$ and $\omega_k \equiv (\vec{k}^2 + m^2)^{1/2}$.

Making use of Eqs. (70) and (72), we relate the Wigner function at different times, $t = x^0, x'^0$:

$$W(x;p) = \int \frac{d^4k}{(2\pi)^3} e^{-ik\cdot x} \delta^{\pm}(p_k^+) \delta^{\pm}(p_k^-)$$

$$\cdot \int d^3x' e^{ik\cdot x'} \int dp'^0 \Lambda(p_k^+) \gamma^0 W(x';p') \gamma^0 \Lambda(p_k^-) , \qquad (73)$$

where $p_k^{\pm} \equiv p \pm \frac{k}{2}$, $\delta^{\pm}(q) \equiv \pm \delta(q^2 - m^2)$ (for $q^0/|q^0| = \pm 1$), $\Lambda(q) \equiv \gamma \cdot q + m$, and $p'^{\mu} \equiv (p'^0, \vec{p})$.

The Eq. (73) implies that the Wigner function obeys a generalized <u>mass-shell constraint</u> and a proper <u>free-streaming transport equation</u>:

$$[p^2 - m^2 - \frac{\hbar^2}{4} \partial_x^2] W(x;p) = 0 , \qquad (74)$$

$$p \cdot \partial_x W(x;p) = 0 , \qquad (75)$$

separately for each matrix element. The reinserted $\hbar$ indicates the important quantum term in the equations, which otherwise have the familiar classical appearance.

Thus Eq. (73) presents an <u>integral solution</u> of the microscopic transport equations for a given initial Wigner function. Furthermore, a semiclassical approximation of the Schwinger function may be used to generate an integral solution of the corresponding classical transport problem.

Next, we decompose the Wigner function with respect to the standard basis of the <u>Clifford algebra</u>, $W = \mathcal{F} + i\gamma^5 \mathcal{P} + \gamma^{\mu} \mathcal{V}_{\mu} + \gamma^{\mu}\gamma^5 \mathcal{A}_{\mu} + \frac{1}{2}\sigma^{\mu\nu} \mathcal{S}_{\mu\nu}$, i.e., in terms of scalar, pseudoscalar, vector, axial vector, and antisymmetric tensor components [27]. The functions, $\mathcal{F} \equiv \frac{1}{4} \text{tr} W$, $\mathcal{P} \equiv -\frac{1}{4} i \text{tr} \gamma^5 W$, $\mathcal{V}_{\mu} \equiv \frac{1}{4} \text{tr} \gamma_{\mu} W$, $\mathcal{A}_{\mu} \equiv \frac{1}{4} \text{tr} \gamma_5 \gamma_{\mu} W$, and $\mathcal{S}_{\mu\nu} \equiv \frac{1}{4} \text{tr} \sigma_{\mu\nu} W$, which represent physical current densities, are real, due to $W^{\dagger} = \gamma^0 W \gamma^0$ [25]. They individually obey Eqs. (74) and (75).

We assume $\mathcal{P} = 0 = \mathcal{A}_{\mu}$, i.e., we consider a spin saturated system for simplicity. – From this point on, a corresponding study of (approximately) massless Standard Model $\nu_L \bar{\nu}_R$ neutrinos differs, and is simpler, since $\mathcal{V}_{\mu} = \mathcal{A}_{\mu}$, while all other densities vanish identically; see, e.g., Ref. [30]. – Then, using the 'transport equation' which follows directly from the Dirac equation applied to W, $[\gamma \cdot (p + \frac{i}{2} \partial_x) - m] W(x;p) = 0$, and decomposing it accordingly, the following additional relations among the remaining densities are obtained:

$$\mathcal{V}^{\mu}(x;p) = \frac{mp^{\mu}}{p^2} \mathcal{F}(x;p) , \qquad (76)$$

$$\mathcal{S}^{\mu\nu}(x;p) = \frac{1}{2p^2}(p^{\nu} \partial_x^{\mu} - p^{\mu} \partial_x^{\nu}) \mathcal{F}(x;p) . \qquad (77)$$

Note that $\mathcal{S}^{\mu\nu}$ is intrinsically by one order in $\hbar$ smaller than the other two densities.

We conclude that presently the dynamics of the system is represented completely by the <u>scalar</u> phase space <u>density</u> $\mathcal{F}$, which obeys the same transport equations as the full W itself. Using Eqs. (71) and (76), we obtain in particular:

$$\langle :T^{\mu\nu}(x):\rangle = 4m \int d^4p \, \frac{p^\mu p^\nu}{p^2} \mathcal{F}(x;p) \;, \qquad (78)$$

which is symmetric and conserved, $\partial_\mu T^{\mu\nu}(x) = 0$, on account of Eq. (75). Furthermore, this implies the 'equation of state':

$$\langle :T^{00}(x):\rangle - \sum_{i=1}^{3} \langle :T^{ii}:\rangle = 4m \int d^4p \, \mathcal{F}(x;p) \;, \qquad (79)$$

which relates energy density and pressure(s). However, applying Eq. (74), we find that this relationship evolves in a wavelike manner, driven by off-shell contributions to the evolving $\mathcal{F}$:

$$\partial_x^2 \langle :T^\mu_\mu(x):\rangle = 16m \int d^4p \, (p^2 - m^2)\mathcal{F}(x;p) \;. \qquad (80)$$

This differs from classical hydrodynamics with a fixed functional form of the equation of state. Eqs. (78)-(80) hold independently of the initial state, of course, if it evolves without further interaction.

Making use of Eq. (73) in Eq. (78), we now calculate the energy-momentum tensor at any time in terms of the initial scalar density. Employing the decomposition of the Wigner function and commutation and trace relations for the γ matrices, as well as Eqs. (74)-(77), we obtain:

$$\langle :T^{\mu\nu}(x):\rangle = 8m \int \ldots \int \frac{p^\mu p^\nu}{p^2} \left(p^2 + \frac{m^2}{p'^2} \left(p^0 p'^0 + \vec{p}^2\right) - \frac{1}{4p'^2}\left((p^0 k^0)^2 - \vec{p}^2 \vec{k}^2 \right.\right.$$
$$\left.\left. + p^0 p'^0 k^2 + \frac{p^0}{p'^0}(k^0)^2 p^2 \right) \right) \mathcal{F}(x';\vec{p},p'^0) \;, \qquad (81)$$

where $\int\ldots\int \equiv (2\pi)^{-3} \int d^4p \int d^4k \, e^{-ik\cdot x} \delta^\pm(p_k^+) \delta^\pm(p_k^-) \int d^3x' e^{ik\cdot x'} \int dp'^0$; we also made use of partial integrations and the δ-function constraints. The three terms on the right-hand side stem from the scalar, vector, and antisymmetric tensor components of the initial Wigner function, respectively.

If the initial distribution is an isotropic function of the three-momentum, then $T^{\mu\nu}$ is diagonal at all times, implying that the absence of flow in the initial state will be preserved.

Indeed, we expect the (non-)flow features of the initial distribution to be preserved during the evolution, due to the absence of interactions. Kinetic energy from microscopic particle degrees of freedom will not be converted into collective motion. An interesting question is, how the classical hydrodynamic acceleration of fluid cells due to pressure gradients arises in our present model after coarse graining [10, 11, 12, 13]. This has not been studied yet. Recalling earlier work on the hydrodynamic representation of quantum mechanics, e.g. Refs. [31], and recently deduced classical fluid behavior of quantum fields in WKB approximation [9], however, we study the full quantum evolution here.

We are particularly interested in the exact evolution of $T^{\mu\nu}$, assuming a particle-antiparticle symmetric initial state. This is believed to hold, for example, close to midrapidity in the center-of-mass frame of central high-energy collisions [3, 4]. It implies that the initial $\mathcal{F}$ is an even function of the energy variable, $\mathcal{F}(x';\vec{p},p'^0) = \mathcal{F}(x';\vec{p},-p'^0)$. While Eq. (81) allows general initial conditions, we follow the implicit on-shell assumption in classical hydrodynamic models:

$$\mathcal{F}(x';\vec{p},p'^0) = (2\pi)^{-3} m\,\delta(p'^2 - m^2)\left(\Theta(p'^0)F(x';\vec{p},p'^0) + \Theta(-p'^0)F(x';\vec{p},-p'^0)\right). \tag{82}$$

Fermion blackbody radiation is described by $F(x';\vec{p},p'^0) \equiv f(p'^0/T(x'))$, where T denotes the local temperature, and with $f(s) \equiv (e^s + 1)^{-1}$; this is easily illustrated with the help of Eqs. (78) and (82).

Implementing Eq. (82), we obtain the simpler result:

$$\langle :T^{\mu\nu}(x): \rangle = \int \frac{d^3 x' d^3 p}{(2\pi)^3} \frac{d^3 k}{(2\pi)^3} \frac{p^\mu p^\nu \cos[\vec{k}\cdot(\vec{x}-\vec{x}')]}{\omega_p \omega_+ \omega_-} F(x';\vec{p},\omega_p) \tag{83}$$

$$\cdot\{((\omega_+ + \omega_-)^2 - \vec{k}^2)\cos[(\omega_+ - \omega_-)t] - ((\omega_+ - \omega_-)^2 - \vec{k}^2)\cos[(\omega_+ + \omega_-)t]\},$$

where $t \equiv x^0 - x'^0$, $\omega_p \equiv (\vec{p}^2 + m^2)^{1/2}$, $\omega_\pm \equiv ((\vec{p}\pm\vec{k}/2)^2 + m^2)^{1/2}$; furthermore, $p^0 \equiv \frac{1}{2}|\omega_+ \pm \omega_-|$, with "+" when multiplying the first and "−" when multiplying the second term of the difference, respectively. Depending on geometry and initial state, further integrations can be done analytically.

Consider a (1+1)-dimensional system for illustration, assuming that the particles are approximately massless, i.e. $\omega_p \approx |p|$, and that F is even in p (no flow). Specializing to a kind of Landau initial condition, the distribution is prepared on a fixed timelike hypersurface at $t = 0$ [2]. We find the ultrarelativistic equation of state for the only nonvanishing components of $T^{\mu\nu}$, $\epsilon \equiv T^{00} = T^{11} \equiv P$ ($d = 1+1$), which are calculated as a momentum integral following Eq. (83):

$$T^{00}(x,t) = 2\int \frac{dp}{2\pi}|p|\bigl(F(x-t;|p|) + F(x+t;|p|)\bigr) \tag{84}$$

$$= \frac{1}{2}\bigl(T^{00}(x-t,t=0) + T^{00}(x+t,t=0)\bigr), \tag{85}$$

i.e., a superposition of wavelike propagating momentum contributions in accordance with Eq. (80).

Similarly, a kind of Bjorken initial condition can be specified on a surface of constant proper time [3]. A transformation of Eq. (84) to space-time rapidity and proper time coordinates yields:

$$T^{00}(y,\tau) = 2\int \frac{dp}{2\pi}|p|\bigl(F(-\tau_0 e^{-y/2+\ln\tau/\tau_0};|p|) + F(\tau_0 e^{y/2+\ln\tau/\tau_0};|p|)\bigr), \tag{86}$$

since $x \equiv \tau\sinh y/2$ and $t \equiv \tau\cosh y/2$ ($\tau \geq \tau_0 > 0$).

Our results for free fermions show the free-streaming behavior of classical dust, associated with the independent propagation and linear superposition of the momentum contributions to the scalar component F of the Wigner function here. In particular, the shape function of each mode is preserved and translated lightlike (with dispersion for massive particles). Due to the assumed momentum symmetry, the initial distribution will separate into two components after a finite time, travelling into the forward and backward direction, respectively, with a corresponding dilution at the center. For example, an initial distribution of Gaussian shape will separate into two corresponding humps.

We recall that $T^{\mu\nu}$ being diagonal implies the absence of ideal hydrodynamic flow, given $\epsilon = (d-1)P$. This does not depend on whether the initial state is on- or off-shell, see Eq.(81). Therefore, any hydrodynamic behavior must be the effect of a pecularity of the semiclassical limit [9], of coarse graining [11, 12, 13], or of interactions [32], or a combination of these.

Despite the apparently classical evolution, however, all initial state quantum effects are incorporated and preserved. If the initial dimensionless distribution F has a dependence on products of momentum and space-time variables, which is characteristic for matter waves, such terms invoke a factor $1/\hbar$. Similarly, if it is thermal (T) but includes the finite size (L) shell effects or global constraints, then there are quantum corrections involving $LT/\hbar$ ($k_B = c = 1$) [33]. For typical values of $LT/\hbar \approx 1$ these latter corrections are known to lead to corrections on the order of 30% in the thermodynamical quantities. They have not been included in semiclassical transport or classical hydrodynamic models of high-energy (nuclear) collisions, but may be large. Here the quantum dust model provides a testing ground to assess the importance of these quantum effects.

Let us summarize briefly the results and perspective of this section:

- Based on the Schwinger function, we obtain the solution of the free quantum transport problem for arbitrary on- or off-shell initial conditions.
- In 1+1 dimensions free fermions, i.e. their observables embodied in the energy-momentum tensor, collectively behave like classical dust showing no flow effects, if there is no flow present in the initial condition. Corresponding analytical results in three dimensions can be obtained.
- The method presented here may lead to an efficient way of treating interacting particles. Especially when a low-order perturbative expansion is meaningful, interactions could be incorporated in a multiple scattering expansion. With free propagation in between scattering events, a consistent and conceptually simple space-time description of transport phenomena seems feasible.

Thus we conclude our introductory lectures on relativistic quantum transport theory, which should convey some of its basic concepts and hopefully will lead to some of the interesting topics for further study.

ACKNOWLEDGMENTS

I thank A.G. Grunfeld and A.M.S. Santo for careful reading and their helpful suggestions improving the manuscript.

REFERENCES

1. E. Fermi, Progr. Theor. Phys. **5**, 570 (1950); Phys. Rev. **81**, 683 (1951).
2. L. D. Landau, Izv. Akad. Nauk SSSR, Ser. fiz., **17**, 51 (1953); S. Z. Belenkij and L. D. Landau, N. Cim., Suppl., **3**, 15 (1956).
3. J. D. Bjorken, Phys. Rev. **D27**, 140 (1983).
4. See the series of proceedings of the International Conferences on Ultra-Relativistic Nucleus-Nucleus Collisions; e.g., *Quark Matter '99*, eds. L. Riccati, M. Masera and E. Vercellini (Elsevier, Amsterdam, 1999).
5. S. Weinberg, *Gravitation and Cosmology* (Wiley, New York, 1972).
6. H.-Th. Elze and U. Heinz, Phys. Rep. **183**, 81 (1989).
7. D. Boyanovsky, F. Cooper, H. J. de Vega and P. Sodano, Phys. Rev. **D58**, 025007 (1998).
8. J.-P. Blaizot and E. Iancu, Nucl. Phys. **B557**, 183 (1999); *The Quark-Gluon Plasma: Collective Dynamics and Hard Thermal Loops*, submitted to Physics Reports, [hep-ph/0101103].
9. G. Domenech and M. L. Levinas, Physica **A278**, 440 (2000).
10. W. Zurek (ed.), *Complexity, Entropy and the Physics of Information* (Addison-Wesley, Reading, Mass., 1990).
11. H.-Th. Elze, Nucl. Phys. **B436**, 213 (1995); Phys. Lett. **B369**, 295 (1996); [quant-ph/9710063].
12. A. M. Lisewski, *On the classical hydrodynamic limit of quantum field theories*, [quant-ph/9905014].
13. T. A. Brun and J. B. Hartle, Phys. Rev. **D60**, 123503 (1999).
14. S. R. de Groot, W. A. van Leeuwen and Ch. G. van Weert, *Relativistic Kinetic Theory* (North-Holland, Amsterdam, 1980).
15. H. Goldstein, *Classical Mechanics* (Addison-Wesley, Cambridge, Mass., 1953), Ch. 8-8.
16. L. P. Csernai *Introduction to Relativistic Heavy Ion Collisions* (John Wiley & Sons, New York, 1994);
  see the lecture notes by L. P. Csernai in these proceedings.
17. H.-Th. Elze, J. Rafelski and L. Turko, Phys. Lett. **B506** (2001) 123.
18. R. P. Feynman, *Statistical Mechanics* (Benjamin/Cummings, Reading, Mass., 1972).
19. D. Han, Y. S. Kim and M. E. Noz, Am. J. Phys. **67** 1999 61.
20. J. von Neumann, *Die mathematischen Grundlagen der Quantenmechanik* Springer, Berlin, 1932; translated in: *Mathematical Foundation of Quantum Mechanics* (Princeton University, Princeton, 1955).
21. P. Carruthers and F. Zachariasen, Rev. Mod. Phys. **55** (1983) 245.
22. K. Rajagopal and F. Wilczek, Nucl. Phys. **B399** (1993) 395.
23. M. Reuter, N. Tetradis and C. Wetterich, Nucl. Phys. **B401** (1993) 567;
  H.-Th. Elze, Nucl. Phys. **A566** (1994) 571c.
24. J. M. Cornwall, R. Jackiw and E. Tomboulis, Phys. Rev. **D10** (1974) 2428.
25. D. Vasak, M. Gyulassy and H.-Th. Elze, Ann. Phys. (N.Y.) **173**, 462 (1987).
26. J. Schwinger, Phys. Rev. **82** (1951) 664.
27. C. Itzykson and J.-B. Zuber, *Quantum Field Theory* (McGraw-Hill, New York, 1980).
28. H.-Th. Elze, Z. Phys. **C38** (1988) 211.
29. H.-Th. Elze, *Fluid dynamics of relativistic quantum dust*, J. Phys. **G** (2002), in press; [hep-ph/0201007].
30. H.-Th. Elze, T. Kodama and R. Opher, Phys. Rev. **D63** (2001) 013008.
31. E. Madelung, Z. Phys. **40** (1926) 322; G. Holzwarth and D. Schütte, Phys. Lett. **B73** (1978) 255; S. K. Ghosh and B. M. Deb, Phys. Rep. **92** (1982) 1.
32. L. M. A. Bettencourt, F. Cooper and K. Pao, *Hydrodynamic scaling from the dynamics of relativistic quantum field theory*, [hep-ph/0109108].
33. H.-Th. Elze and W. Greiner, Phys. Lett. **B179**, 385 (1986); Phys. Rev. **A33**, 1879 (1986).

EXPERIMENTS

RHIC Physics Overview

J. W. Harris*, C. Beckmann†, J. Gans* and K. H. Gulbrandsen**

*Physics Deparment, Yale University, 272 Whitney Ave, New Haven, CT 06520-8124, USA
†Institut für Theoretische Physik, J. W. Goethe-Universität, Frankfurt am Main, Germany
**Laboratory for Nuclear Science, MIT, Cambridge, MA 02139-4307, USA

Abstract. The field of relativistic heavy ion physics has entered a new regime with the commissioning of the Relativistic Heavy Ion Collider (RHIC). This article strives to present in a pedagogical manner a brief introduction to the field, an overview of RHIC and its experiments, and the first exciting physics results from RHIC.

INTRODUCTION

The motivation for studying relativistic heavy ion collisions is to understand at a fundamental level the behavior of matter at extremely high temperature and/or density. This is the study of Quantum Chromodynamics (QCD) at high energy density, which has implications for nuclear physics, astrophysics, cosmology and particle physics. The behavior of matter as a function of temperature and density (or pressure), depicted in Fig. 1, is governed by its equation of state. The lower left portion of the diagram at low temperatures and near normal nuclear matter density is of interest in nuclear physics.

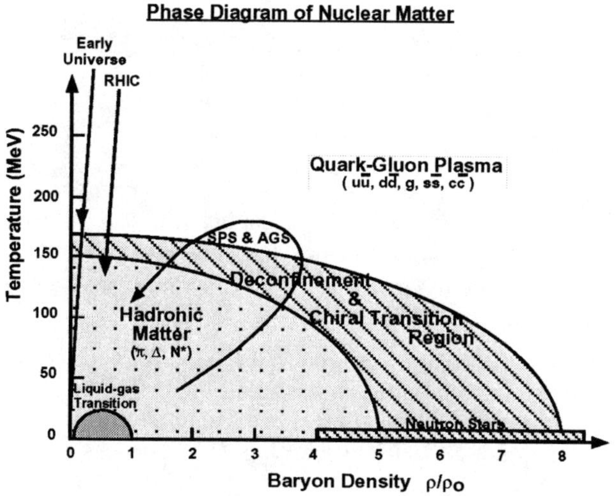

FIGURE 1. Schematic phase diagram of subatomic matter.

CP631, *New States of Matter in Hadronic Interactions,* edited by H.-T. Elze et al.

Here normal nuclei exist and at low excitation a liquid-gas phase transition is expected to occur. This is the focus of experimental studies using low energy heavy ions. At higher excitation, nucleons are excited into baryonic resonance states, accompanied by particle production and hadronic resonance formation. This region of hadronic resonance matter has been accessible in heavy ion studies at the AGS at the Brookhaven National Laboratory and at the SPS at CERN. These collisions may partially traverse the transition region into the quark-gluon plasma regime. Formation of a quark-gluon plasma, a deconfined state of quarks and gluons, [1] is a major focus of relativistic heavy ion experiments.

The study of the equation of state of nuclear, hadronic and partonic matter is not limited to just heavy ion physics, but is of cross-disciplinary interest. The physics of the equation of state affects low energy nuclear structure, and includes collective nuclear phenomena and effects in a nuclear medium. Astrophysicists are interested in the precise form of the nuclear matter equation of state to explain neutron star stability and supernova expansion dynamics. The physics of the phase transition affects cosmologists' theories of the evolution of the early universe. Heavy ion physics also has direct ties to the study of traditional high energy physics topics such as symmetry breaking mechanisms, high temperature and matter density QCD and may lend insight into the particle masses themselves.

In order to answer these questions the Relativistic Heavy Ion Collider (RHIC) [2] and its experiments have accumulated data in collisions of gold on gold (Au + Au) at center of mass energy per nucleon pair $\sqrt{s}$ = 130 GeV and 200 GeV. Proton on proton (p + p) data at $\sqrt{s_{NN}}$ = 200 GeV has also been taken. This article will present a brief introduction to the field of relativistic heavy ion physics, an overview of RHIC and its experiments, and initial results from RHIC including what we have learned after the first year of results. We shall attempt to emphasize what is understood and what is still puzzling, with comments on what to expect. We shall report on the initial Au + Au collisions at $\sqrt{s_{NN}}$ = 130 GeV that are just becoming available.

MELTING THE QCD VACUUM TO PRODUCE THE QGP AND POTENTIAL SIGNATURES

According to T.D. Lee, [3]

> "In high-energy physics we have concentrated on experiments in which we distribute a higher and higher amount of energy into a region with smaller and smaller dimensions. In order to study the question of 'vacuum', we must turn to a different direction; we should investigate some bulk phenomena by distributing high energy over a relatively large volume."

This is what happens in relativistic heavy ion collisions. In such studies, a better understanding of symmetry breaking mechanisms and the origin of particle masses may be found. The QCD vacuum is predicted to have a complex internal structure, that is made up of a quark-antiquark condensate sea. This sea possesses energy and mass, and can undergo fluctuations about its zero point. At low energy densities, quarks and gluons are confined in hadrons (colorless baryons and mesons). Here the vacuum acts as a color dielectric, and isolated quarks (which have color) are not expected. At the high temperatures anticipated in high energy collisions of heavy ions, the quark-antiquark condensate vacuum is expected to melt. The hadrons dissolve into freely propagating quarks and gluons, and the vacuum becomes a color conductor, the true perturbative QCD vacuum. This is the quark-gluon plasma (QGP) phase. The melting of the vacuum and production of a QGP is predicted to have various experimental effects that can be used to determine whether the QGP has actually been produced. We divide these into four categories and introduce them next.

Deconfinement Signatures

A system of deconfined quarks and gluons will have various effects on particle production and particle propagation. One such effect is that heavy quarkonium states, such as $c\bar{c}$ and $b\bar{b}$ states (e.g. J/Ψ and Υ, respectively), will be suppressed. [4] To form these states, $q\bar{q}$ pairs are created by the fusion of two gluons. In a deconfined medium (QGP) Debye screening of a $q\bar{q}$ pair will suppress binding into these states relative to purely hadronic production. [5] Less tightly bound excited states of the ($c\bar{c}$) system, such as ψ' and χ_c, are more easily dissociated and will be suppressed even more than the J/ψ. Similar arguments can be made for the heavier $\Upsilon(b\bar{b})$ system [6]. However the temperature at which the Υ ground state "melts" is predicted to be around 2.5 times the transition temperature (T_c), and that of the larger Υ' state only slightly above T_c.

Furthermore, production of strange hadrons will be suppressed [7] relative to non-strange hadrons in hadronic reactions. This suppression should increase with increasing strangeness content of the hadron. In a QGP the strange quark content is rapidly saturated by $s\bar{s}$ pair production in gluon-gluon fusion reactions, resulting in an enhancement in the production of strange hadrons relative to production in hadronic interactions [8]. Thus, multi-strange baryons and strange antibaryons are predicted to be strongly enhanced when a QGP is formed. [9, 10] In addition, it has been shown [10, 11] that an enhanced strangeness content is not expected to be destroyed nor generated by interactions during expansion and freezeout.

Kinematic Signatures

Since the QGP is a separate phase of matter, thermodynamic quantities such as temperature T, energy density ε, pressure p, and entropy density s will change due to formation of the plasma. Experimental observables can be associated with these quantities and their relative behavior determined from measurements. The quantities T, ε, p, and s can be identified with the average transverse momentum, transverse energy density, transverse flow, and hadron rapidity density, respectively. [12] A potential signature of a phase change would be a rapid rise in the effective number of degrees of freedom, expressed by the ratios ε/T^4 or s/T^3 as shown in Fig. 2.

The plasma is expected to have high pressure that will lead to collective outward flow during the expansion. Measurements to determine the amount of flow in the system

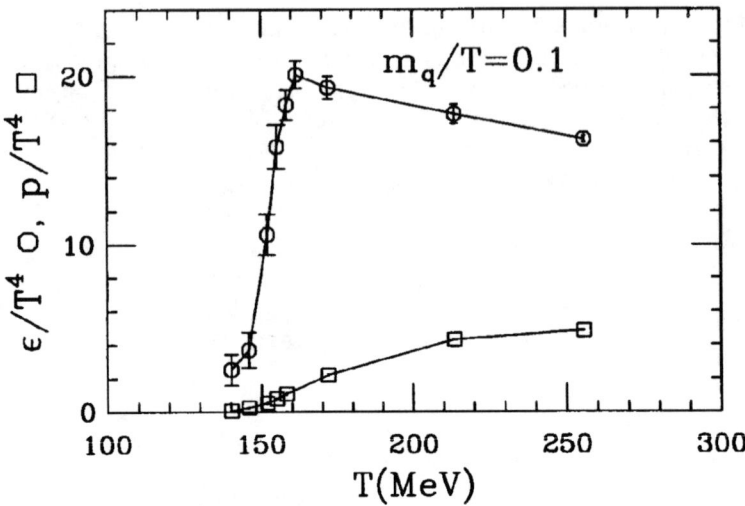

FIGURE 2. Energy density ε (upper curve) and pressure p (lower curve) from a numerical evaluation of QCD "on the lattice" with two light flavors of quarks. ε and p are divided by T^4 and exhibit a sudden rise in the number of degrees of freedom at the critical temperature $T_c \approx 150$ MeV due to liberation of color. [13]

can provide information on pressure gradients in the system. The flow signal would be further enhanced by the formation of a detonation wave during the hadronization transition. [14, 15, 16, 17]

Identical particle interferometry, eg. $\pi\pi$, KK, or NN correlations, allows for the measurement of the source size, lifetime, and flow patterns of the fireball at thermal freeze-out, when interactions sieze. Theoretical work has shown the importance of the finite lifetime of the fireball [18], of flow patterns [19] and of shadowing effects [20] on particle correlations.

Chiral Transition Signatures

A temporary restoration of chiral symmetry during the evolution of these collisions may result in the formation of domains of disoriented chiral condensates, which are identified with coherent excitation of the pion field. During the collision process, the eventual hadronization forces these domains to decay, producing neutral and charged pions that may reflect the chirally restored phase. This is predicted to result in pion ratios $N_{\pi^o} / (N_{\pi^+} + N_{\pi^-})$ substantially different from the ratio $1/3$ found in our everyday "chirally broken" world.

Another possible set of signatures is a shifting of the masses and changes in the widths of resonances. This results from the sensitivity of the widths and positions of the mass peaks of resonances to medium-induced changes of the hadronic mass spectrum, as seen in the lepton-pair spectrum from the decays of the ρ, ω, and ϕ. Modification of the

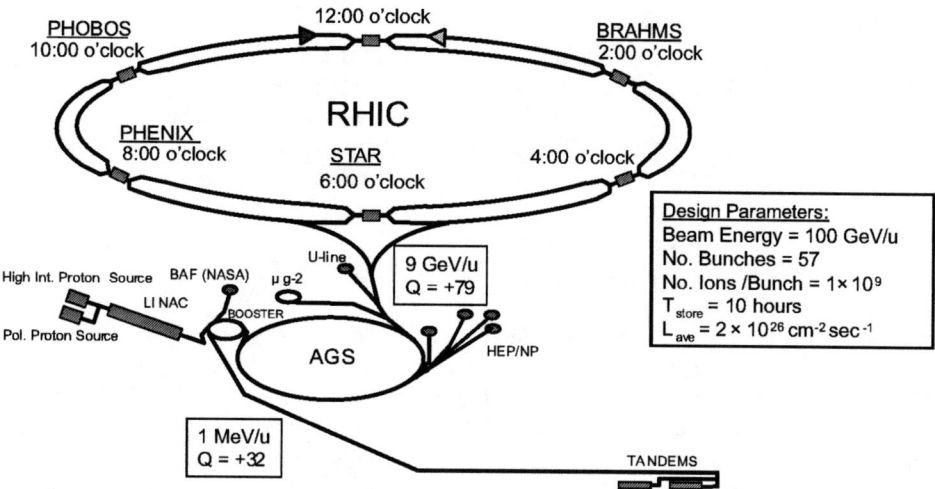

FIGURE 3. The Relativistic Heavy Ion Collider (RHIC) accelerator complex at Brookhaven National Laboratory. Nuclear beams are accelerated from the tandem Van de Graaff, through the transfer line into the AGS Booster and AGS prior to injection into RHIC. Some details of the characteristics of the Au beams are indicated (in the boxes) after acceleration in each phase.

these mass peaks in the lepton pair spectrum may be indicative of a long-lived mixed phase.[21]

Electromagnetic Probes

Since photons are unaffected by final state interactions, they can be used as probes of the interior of the quark-gluon plasma during the earliest and hottest phase of evolution of the fireball. The dominant source of photons from quarks and gluons is the reaction $gq \rightarrow \gamma q$. [22] Photons from the thermal hadronic phase form predominantly from the $\pi \rho \rightarrow \gamma \rho$ reaction. [23] The intensity and spectral shapes of these photons from the various different processes are similar. However, if a very hot plasma is formed initially, a signal of the QGP is predicted to be visible in the transverse momentum range 2-5 GeV/c. [24, 25, 26]

RELATIVISTIC HEAVY ION COLLIDER OVERVIEW

A schematic diagram of the RHIC accelerator complex at Brookhaven is displayed in Fig. 3. Nuclear beams are accelerated from the tandem Van de Graaff accelerator through a transfer line into the AGS Booster synchrotron and then into the AGS, that serves as an injector for RHIC. The beams collide in the six interaction regions, although only 4 of these regions currently house detectors. RHIC can accelerate and collide ions from protons (polarized and unpolarized) up to the heaviest nuclei. RHIC was designed to provide Au+Au collisions up to $\sqrt{s_{NN}}$ = 200 GeV with a luminosity of 2×10^{26} cm^{-2} s^{-1}. This energy was obtained during the 2001-2002 run at a luminosity sometimes higher than design. The RHIC design also allows for p+p collisions at $\sqrt{s}$ up to 500 GeV. However, only 200 GeV was utilized in the 2002 proton run in order to provide a comparison for the 200 GeV Au+Au data. In 2002, RHIC also initiated its spin physics program [27] by colliding polarized protons.

THE DETECTORS AT RHIC

Near head-on collisions of heavy nuclei at RHIC produce around 1000 particles per unit pseudorapidity (at $|\eta| < 1$). This presents a formidable environment in which to detect the products of these reactions. The four experiments constructed at RHIC take various approaches to search for the QGP. The STAR experiment concentrates on measurements of hadron production over a large solid angle in order to measure single- and multi-particle spectra, and global observables on an event-by-event basis. The PHENIX experiment focuses on precision measurements of lepton and photon production with the capability of measuring hadrons in a limited range of pseudorapidity. Both STAR and PHENIX have active spin physics programs to study the spin content of the proton using collisions of polarized protons at RHIC. In addition to STAR and PHENIX, there are two smaller experiments BRAHMS (a forward and midrapidity hadron spectrometer) and PHOBOS (a compact multiparticle spectrometer). The four collaborations, in total, consist of approximately 1000 scientists from over 80 institutions internationally.

The STAR Experiment

The STAR (Solenoidal Tracker At RHIC) experiment [28] investigates the behavior of strongly interacting matter at high energy density and searches for signatures of QGP formation in central collisions of relativistic nuclei. STAR simultaneously measures many experimental observables in order to study signatures of the QGP phase transition as well as the space-time evolution of the collision process over a variety of colliding nuclear systems. [28] In addition, STAR investigates very peripheral collisions of relativistic nuclei to study photon and pomeron interactions resulting from the intense electromagnetic fields of colliding ions and colorless strong interactions, respectively. [29]

The layout of STAR is shown in Fig. 4. The year 2000 configuration of STAR consisted of a large time projection chamber (TPC) covering $|\eta| < 2$, trigger detectors, a ring imaging Cherenkov detector [30] covering $|\eta| < 0.3$ and $\Delta\phi = 0.1\pi$ (not shown in Fig. 4), and an initial partial coverage of electromagnetic calorimetry of $0 < \eta < 1$ and $\Delta\phi = 0.14\pi$ inside a room temperature solenoidal magnet with 0.25 T magnetic field. The solenoid provides a uniform magnetic field for tracking, momentum analysis and particle identification via ionization energy loss measurements in the TPC. Measurements in the TPC and trigger detectors are carried out at mid-rapidity with full azimuthal coverage ($\Delta\phi = 2\pi$) and azimuthal symmetry. During the 2001 run a silicon

STAR Detector

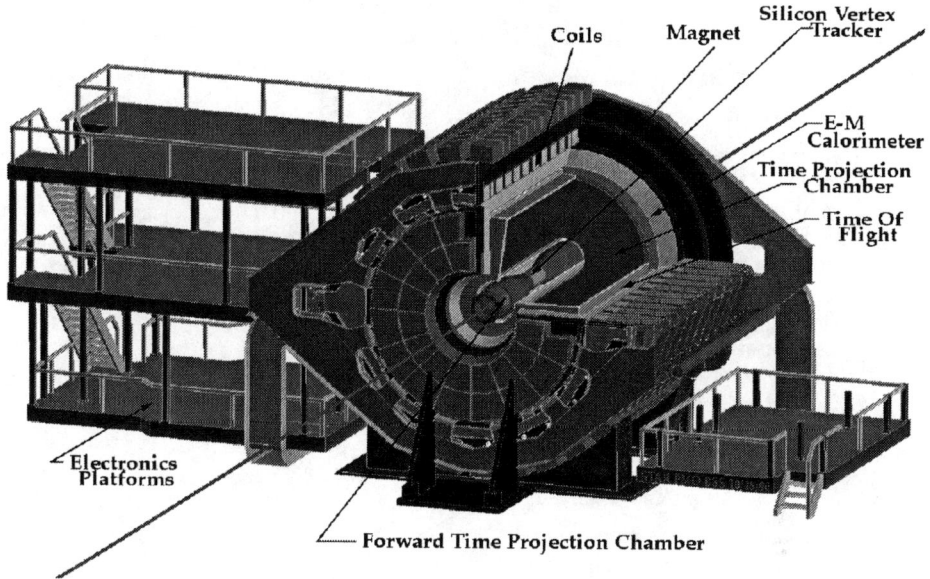

FIGURE 4. View of the STAR Detector.

vertex tracker (SVT) covering $|\eta|<1$, and two forward radial-drift TPC's (FTPC's) covering $2.5<|\eta|<4$ were installed, as seen in Fig. 4. A time-of-flight (TOF) patch covering $0<\eta<1$ and $\Delta\phi=0.04\pi$ is also used to extend the particle identification in STAR. For the 2002-2003 run, the solid angle of the electromagnetic calorimeter (EMC) will be increased to almost 50% of its eventual $|\eta|<1$ and $\Delta\phi=2\pi$ coverage and will measure the transverse energy of events, as well as trigger on and measure high transverse momentum photons, particles and jets. The remainder of the EMC will be completed and installed over the next 2 years.

The PHENIX Experiment

The physics goals of PHENIX (Pioneering High Energy Nuclear Interaction eXperiment) [31, 32] are to measure as many potential signatures of the QGP as possible as a function of a well-defined common variable such as impact parameter or pseudorapidity density. PHENIX measures lepton pairs (di-electrons and di-muons), photons and hadrons. The experiment is sensitive to very small cross section processes such as the production of the J/ψ, ψ', Υ and particle spectra at high p_T. It also has the capability for high rates with p+p and p+A collisions.

A diagram of the PHENIX detector system is displayed in Fig. 5. The magnet has an axial field along the beam direction with tracking chambers and detectors for the iden-

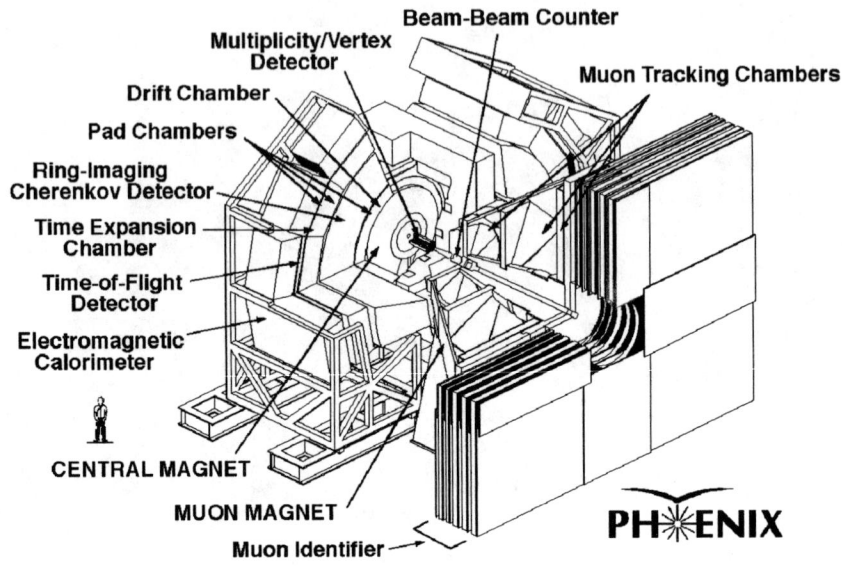

FIGURE 5. A diagram of the PHENIX experiment at RHIC in 2001. A second muon arm (on upstream-side), not shown in this diagram, will be completed in 2002. Detector components are defined in the text.

tification of electrons, muons, photons and hadrons installed outside the field. There are two arms for dielectron measurements, each with 1 steradian acceptance at midrapidity. Each arm is equipped with a ring-imaging Cerenkov detector (RICH), a time-expansion chamber (TEC) for dE/dx measurements, time-of-flight detectors (TOF), and an electromagnetic calorimeter (EM Cal). Photons and hadrons are measured at midrapidity with 2 and 0.36 steradian acceptances, respectively. Separate muon spectrometers with 1 steradian acceptance each are located at forward rapidities (only one is shown in Fig. 5). The magnetic field in each muon arm is transverse to the beam direction. A silicon multiplicity-vertex detector (MVD) and beam-beam coincidence counters are located near the interaction vertex. The MVD covers $|\eta| < 2.7$ about midrapidity for event selection via charged-particle multiplicity.

The PHOBOS Experiment

The physics goals of the PHOBOS experiment [33, 34] are to measure single particle spectra and correlations between particles with low transverse momenta and to characterize events using a multiplicity detector. Charged particles are measured and identified in the range $0 < \eta < 1.5$ and 15 MeV/c $< p_T <$ 600 MeV/c for pions and 45 MeV/c $< p_T <$ 1200 MeV/c for protons. The range of particles studied in-

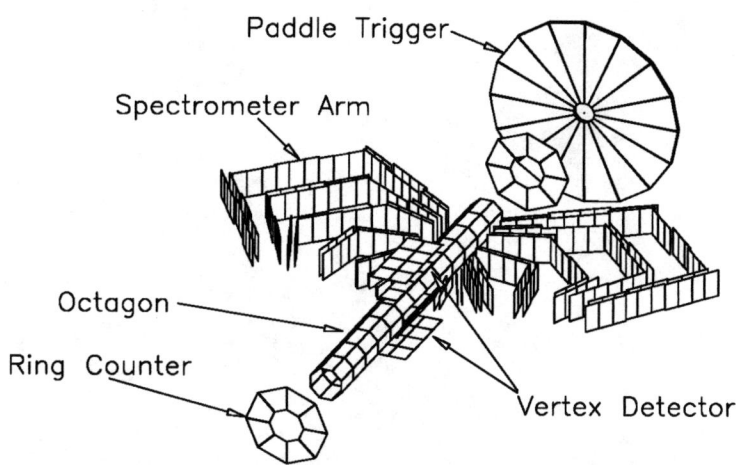

FIGURE 6. A diagram of the PHOBOS two-arm multiparticle spectrometer. The two arms are located on opposite sides of the beam pipe. Each arm has a 2 Tesla magnet (not shown), with silicon detector planes for tracking. (A time-of-flight array located behind one of the spectrometer arms is not shown.) The beams collide in the middle of the octagon multiplicity array.

clude $\gamma, \pi, K, \rho, \bar{\rho}, \phi, \Lambda, \bar{\Lambda}, d$, and $\bar{d}$. Particle ratios, p_T spectra, strangeness production ($K, \phi, \Lambda, \bar{\Lambda}$) and particle correlations are being studied. An illustration of the experiment is shown in Fig. 6. Two dipole magnets (not shown) with field strength 2 Tesla surround the spectrometer arms. The multiplicity detector consists of silicon strip detectors and silicon pad detectors. Eleven layers of silicon, 5 layers of pads and 6 layers of strips, are installed in each magnetic spectrometer for tracking and momentum measurements.

The BRAHMS Experiment

The physics goals of the BRAHMS (BRoad RAnge Hadron Magnetic Spectrometers) experiment [35] are to achieve a basic understanding of relativistic heavy ion collisions at RHIC through a systematic study of particle production in AA collisions from the peripheral to the most central in impact parameter. Measurements are performed using two high resolution magnetic spectrometers at various angles to cover both the baryon-rich fragmentation regions and the high temperature, baryon-depleted midrapidity region. A diagram of the BRAHMS forward and midrapidity spectrometers is shown in Fig. 7. The spectrometers measure and identify inclusive and semi-inclusive π, K and p, and their momenta with high statistics over a small solid angle, and over a wide range of pseudorapidity ($0 < \eta < 4$) and transverse momentum. Particle identification is performed using various combinations of time-of-flight arrays, and threshold and ring-imaging Cherenkov counters. BRAHMS measures inclusive and semi-inclusive particle spectra, and extracts the net baryon densities and temperatures from spectral slopes as a function of rapidity to determine whether thermal and chemical equilibrium are reached in these collisions. It studies both high and low transverse momentum processes. Cen-

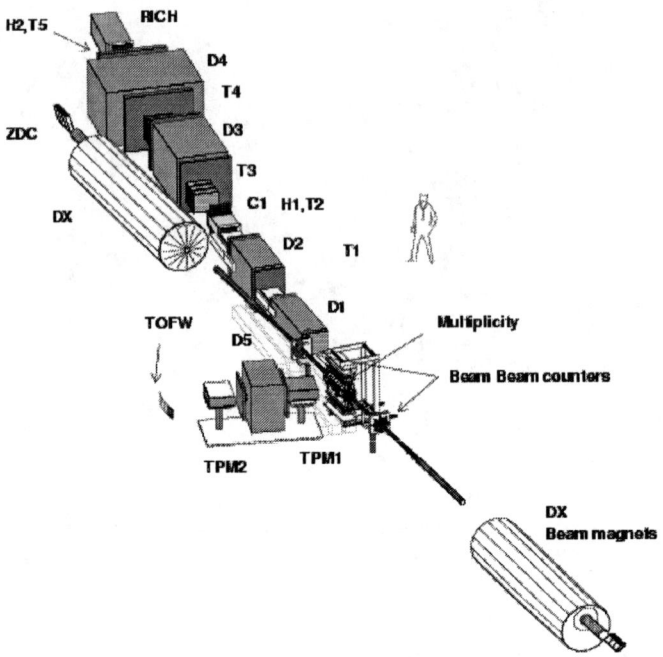

FIGURE 7. A diagram of the BRAHMS forward and midrapidity spectrometers at RHIC. For reference, the abbreviations labeling components of the spectrometer arms in the figure are defined as D = dipole magnet, TP = TPC, T and H = scintillator timing detectors and hodoscopes.

trality is measured using a global multiplicity detector.

PARTICLE IDENTIFICATION TECHNIQUES

Details of the particle production and spectra are best understood when experiments provide results sorted by particle type. In order to provide particle identification experiments must distinguish particles with different masses. This can be done in various ways, and an overview of the general techniques used at RHIC will be presented. To start off, all experiments at RHIC employ magnetic analysis to determine the momentum to charge ratio of each particle. From classical physics we know that a charged particle moving in a magnetic field suffers a force that deflects its trajectory along a helical path. The radius of curvature of the particle trajectory through the magnetic field is used to determine the momentum per charge of the track transverse to the field:

$$mv_\perp^2/r = qv_\perp \times B \Rightarrow mv_\perp/r = qB \Rightarrow p_\perp = qBr \qquad (1)$$

and

$$p_T = p_\perp/q = Br \qquad (2)$$

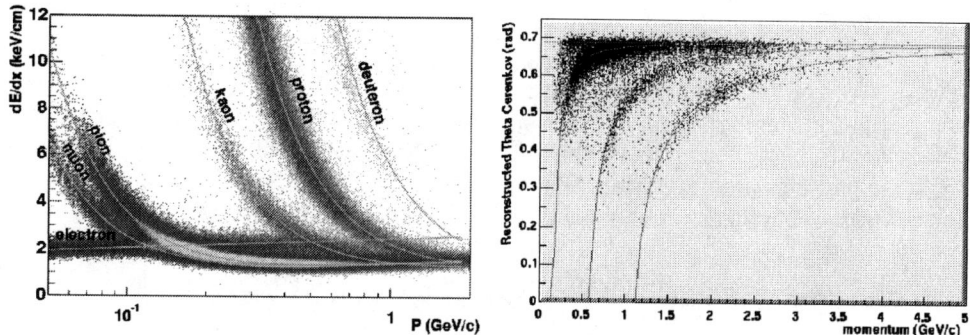

FIGURE 8. Left: dE/dx of tracks versus momentum in the STAR TPC. The e, μ, π, k, p, and d bands can be seen (Reprinted from Ref. [38], Copyright 2002, with permission from Elsevier Science.) Right: Cherenkov angle (θ) of tracks versus momentum in the STAR RICH. The e, π, k, and p bands can be seen from top to bottom.

The z component of the momentum per charge in the direction of the magnetic field can be determined by plotting the path length of the track relative to its z position. The angle λ that this line makes with the z axis is called the dip angle of the track. From this angle, p_z can be determined:

$$p_z = p_T \tan \lambda \tag{3}$$

Thus, the 3-momentum $\vec{p}$ per charge can be determined by adding p_T and p_z in quadrature. If the magnetic field is parallel to the beam line, then these components become the transverse and longitudinal momentum (per charge) of the particle relative to the beam direction. Otherwise, the mathematical formulation to determine the curvature is slightly more complicated, but similar and feasible. Since most of the particles produced in these reactions are singly-charged, the equations lead directly to the momentum of the particle. However, although we determine the momentum, we still do not know the particle mass. Relativistically, $p = \gamma m v$, thus the velocity must be determined to know the mass.

One method to determine the particle velocity, is to use ionization energy loss in a material or gas and measure the energy loss per unit length ($\frac{dE}{dx}$) in the known material. This $\frac{dE}{dx}$ measures the energy loss of the particle versus the distance of material traversed. This energy loss is purely a function of velocity β and other constants:

$$-\frac{dE}{dx} = K z^2 \frac{Z}{A} \frac{1}{\beta^2} [\frac{1}{2} \ln \frac{2 m_e c^2 \beta^2 \gamma^2 T_{max}}{I^2} - \beta^2 - \frac{\delta}{2}] \tag{4}$$

[36]. Thus, when energy loss is plotted versus momentum the various particles in the plot separate out into bands as shown in the left panel of Fig. 8. Notice the bands converge around 1 GeV/c. Therefore, another method must be used for particle identification above this limit because here all particles, no matter their mass, have the same $\frac{dE}{dx}$ value.

One such method employed by the RHIC experiments is to use a Ring Imaging Cherenkov Detector (RICH). In analogy to a sonic shock wave, when a particle travels faster than the speed of light in a material, it will emit photons in a cone at a specific angle given by

$$\theta = \arccos\frac{c}{nv} = \arccos\frac{1}{n\beta} \quad (5)$$

where c is the speed of light, n is the index of refraction of the medium, and v is the velocity of the particle. Once again, as this is purely a function of velocity, bands appear when the particles are plotted versus momentum. Notice the bands do not merge until about 4 or 5 GeV/c in the right panel of Fig. 8. [37] Remember, relativistically

$$\beta = \frac{pc}{E} = \frac{\sqrt{E^2 + m^2c^2}}{E} \to 1 \quad as\ E >> mc^2 \quad (6)$$

Therefore high momentum tracks, despite their different masses, will produce a ring with the same angle theta.

Closely related to the RICH detector is a Threshold Cherenkov Detector. As can be seen by the above equation, if c/nv is not greater than unity, θ will be undetermined and no light will be produced. Typically, several planes of Threshold Cherenkov Detectors are used in succession, all with increasing indices of refraction. By establishing from which plane(s) the particle radiates Cherenkov light, the particle velocity can be determined to be between the threshold velocity values of the previous and radiating plane (assuming e, π, k, and p).[39]

Another method to determine the velocity of the track is to measure its velocity directly with a time-of-flight (TOF) detector.[40] These detectors usually consist of two scintillating planes, one that starts a high precision clock and another that stops it. In a magnetic field, if the track has a very high momentum, it will have little curvature so the velocity is the distance between the scintillators divided by the elapsed time of traversal. If the track has curvature, its path length must be used instead of the distance between the scintillators. The one draw-back of this method is that it has tended to be very expensive at RHIC, due to the large number of produced particles. In order to work in such a high multiplicity environment as RHIC, the scintillators must have small pixel size to reduce the probability of multiple particles per pixel. Moreover, the photo-tubes must be magnetically shielded because they must be close to the experiment in order to register the

fast signals provided by the start and stop scintillators.

Once particle identification is accomplished, short-lived resonances (with $c\tau <$ tracking distance in detector) can be found. There are two distinct ways to do this, through topology or through a combinatorial technique. One can use the Λ-decay vertex (V_o) in Fig. 9 as an illustrative example to introduce the topology method. After tracking all particles in an event, an identified proton and pion are found to intersect forming a secondary vertex (V_o in Fig. 9) that is displaced from the primary interaction vertex in the event. One assumes that the secondary vertex (V_o) was produced from the decay of a Λ as in Fig. 9. The invariant mass of the particle pair can be computed and plotted along with all other invariant masses of pairs summing over all events. The resultant invariant mass spectrum is shown in the left panel of Fig. 10. Finding a secondary vertex in the detector, especially for short-lived particles, is difficult and typically has low but accurately calculable efficiency.

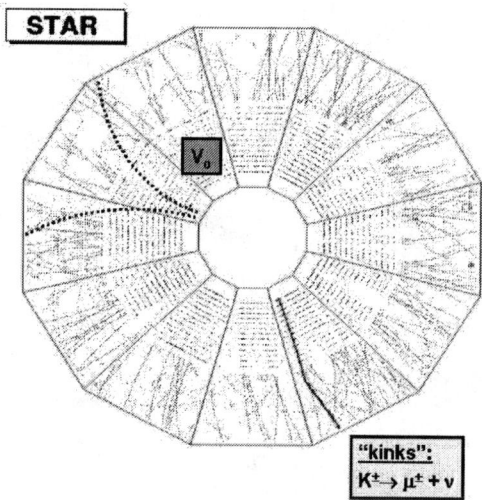

FIGURE 9. Cross-section of track hits in the STAR-TPC. A V_o decay and a kink are highlighted.

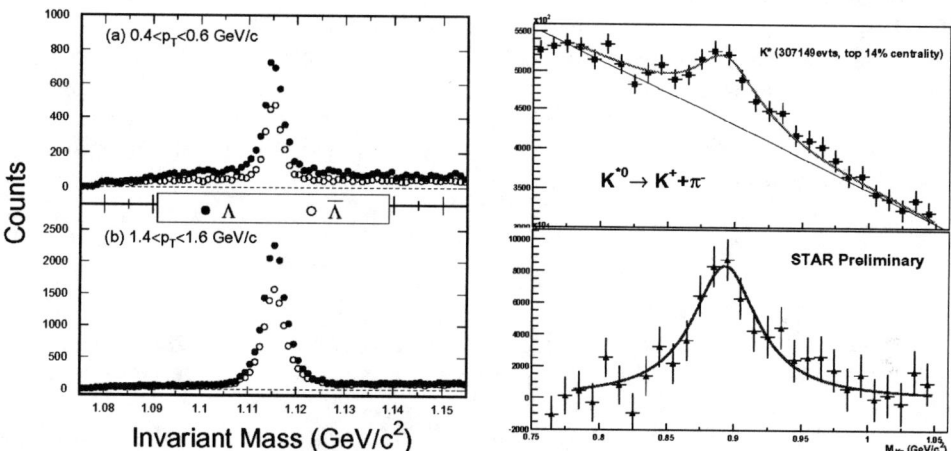

FIGURE 10. Left: Invariant mass distributions of Λ (solid circles) and $\overline{\Lambda}$ (open circles) for two regions of p_T noted in figure. Right: Invariant mass distributions of K^{0*}. See text for details.

To bypass the difficulty and inefficiency of secondary vertex identification, especially in a high multiplicity environment, the combinatorial technique can be employed. Here one uses all particles, regardless of topology, and constructs the invariant mass of all possible combinations of particle-pairs in each event. This method inherently results in a higher background that must be removed. To form a background distribution with the appropriate phase-space, all like- and unlike-sign particle pairs are combined. For example, the K^{0*} can be reconstructed from its K^+ and π^- daughters. The raw signal

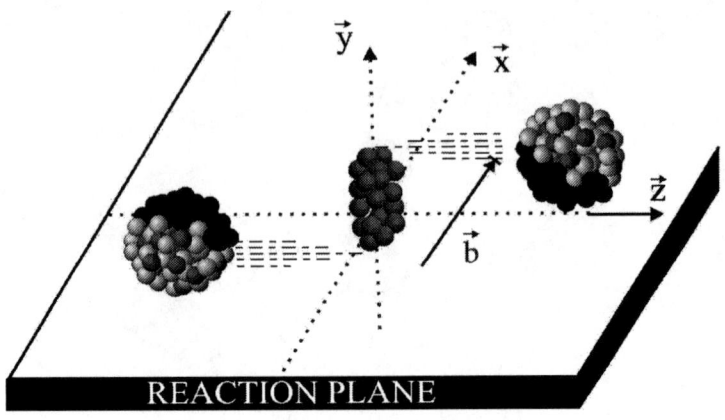

FIGURE 11. Schematic diagram of a collision between two nuclei, depicting the impact parameter (b) and the reaction plane. The beams collide along the z-axis.

distribution is formed from all K^+ and π^- pairs. The background can be formed from all K^+ and π^+ or K^- and π^-, which cannot be decays of the K^{0*} and thus model the background. Then, the background distribution can be subtracted from the raw signal leaving only the background-subtracted signal. The right-upper panel of Fig. 10 shows the raw K^{0*} invariant mass calculations with a straight line fit to the background. The right-lower panel is the background-subtracted distribution. Along with the aforementioned resonances, STAR is studying decays of the Ξ, Ω, $\Sigma(1385)$, $\Lambda(1520)$, and others.

COLLISION GEOMETRY

Collisions between nuclei possess added complexity compared to single nucleon collisions. At high energies, collisions of nuclei can be viewed in many respects as simultaneous collisions of groups of nucleons bound together to form the nucleus. Thus, a nucleus-nucleus collision can be represented by multiple nucleon-nucleon collisions. However, to understand nucleus-nucleus collisions it is necessary to know the collision geometry and specifically the fraction of nucleons that interact for each incoming collision trajectory. Fig. 11 shows a schematic diagram of a collision between two nuclei. The region of overlap between the incoming nuclei, is usually characterized by the impact parameter (b) of the collision as depicted in Fig. 11. The impact parameter is the lateral distance between the centers of the incident nuclei. Central collisions are events with small impact parameter, while peripheral collisions have a large impact parameter. There is a continuum of degrees of centrality between these two limits. Nucleons in the overlap region that interact are called participants and non-interacting nucleons are called spectators. Clearly, the numbers of interacting nucleons and non-interacting spectators depend on the overlap between the incident nuclei, and thus the impact parameter of the collision.

The production of particles depends directly on the number of nucleons that participate in a collision. Particle production peaks in the center-of-mass (c.m.) of the colliding system, mid-way between the rapidity of the incident beams - thus, the term mid-rapidity to describe the c.m. of the system. As the nuclear overlap increases, particle production increases, and progressively more particles and energy are transported away from the beam rapidity towards mid-rapidity. Meanwhile, spectator nucleons continue along at beam rapidity. The beam-remnant nuclei, composed of spectators, become excited as a consequence of the nuclear interaction, and subsequently break-up. These excited nuclei typically eject excess neutrons along the beam direction. The beam fragments with a charge-to-mass ratio slightly different from the incident nuclei also continue along the beam direction, but are not captured in stable machine orbit nor re-accelerated in RHIC. Furthermore, neutrons continue forward, unaffected by experiment or beam magnets. These neutrons are detected in the RHIC Zero Degree Calorimeters (ZDC) [41] that are present in each RHIC experiment. The commonality of identical ZDC's in each experiment provides the ability for comparison of centrality between experiments (in principle) without invoking models for experiment-to-experiment comparison of collision centrality.

However, the Zero Degree Calorimeters alone are insufficient to unambiguously characterize the centrality of a collision. In principle, a smaller signal is expected in the ZDC's in a central collision than in a peripheral collision, simply because more nucleons participate in a central collision due to the larger geometric overlap. This leaves fewer nucleons, and beam fragments with low charge-to-mass ratio and fewer neutrons to continue forward into the ZDC. In extremely peripheral collisions a small signal will also register in the ZDC since the primary beam fragments, that contain most of the nucleons, will continue to travel at rigidities near that of the beam. These are then magnetically deflected and miss the ZDC's. The ZDC's only detect particles and fragments traveling straight-through, primarily neutrons undeflected at zero degrees. It is therefore important for each experiment to incorporate an additional detector system away from beam rapidity to resolve the ambiguity in centrality discrimination for small ZDC signals. Fig. 12 exhibits the correlation between signals in the ZDC and in a typical off-beam-rapidity detector.

The characteristic shape is an inverse relationship between the off-beam-rapidity counter signal and that of the ZDC. This comes simply from the fact that the number of participant nucleons (that produce particles away from beam-rapidity and thus the signal in an off-beam-rapidity detector) is the difference between the number of incoming nucleons and the number of spectator nucleons (registering a proportional signal in the ZDC). This correlation is valid until collisions become so peripheral that fragments start missing the ZDC and the measurement of the number of spectators becomes poor. The correlation between the two signals can be utilized to provide a measure of and trigger for centrality in nuclear collisions at RHIC.

Determination of the number of nucleons that actually participate in the reaction or the inelastic cross-section is model dependent. This is a result of the increasing inefficiency of the ZDC's to detect the most peripheral of collisions, where little if any signal remains outside of the beam-pipe. Models are invoked to determine the fraction of peripheral events, from among the total cross section, that are missed by the detector triggers. Fig. 13 shows a multiplicity distribution for negative hadrons as measured in the STAR

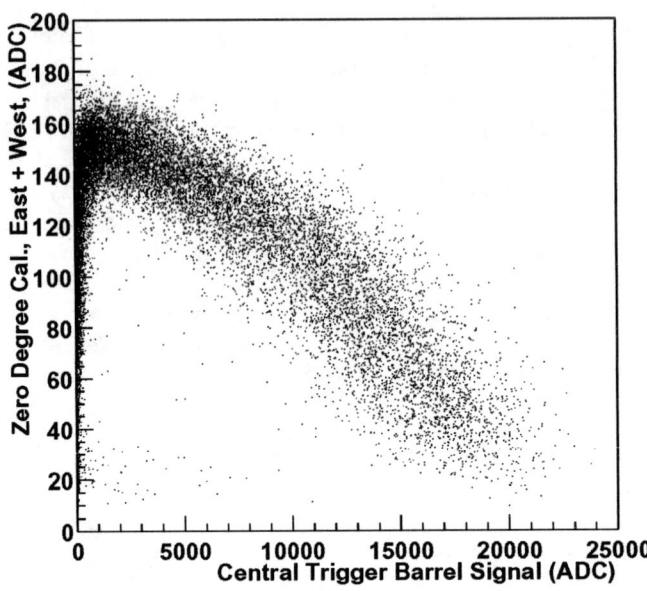

FIGURE 12. Typical response of ZDC signal vs response of off-beam-rapidity detector signal at RHIC. This specific plot shows the pulse height in the ZDC versus that in the Central Trigger Barrel at midrapidity in STAR. (Reprinted from Ref. [38], Copyright 2002, with permission from Elsevier Science.)

TABLE 1. $dN_{ch}/d\eta$ at midrapidity measured by RHIC experiments.

Experiment	$dN_{ch}/d\eta$	Centrality	Midrapidity definition
PHENIX	$633 \pm 1(\text{stat}) \pm 41(\text{sys})$	5% most central	$|\eta| < 0.35$
PHOBOS	$555 \pm 12(\text{stat}) \pm 35(\text{sys})$	6% most central	$|\eta| < 1$
STAR	$567 \pm 1(\text{stat}) \pm 34(\text{sys})$	5% most central	$\eta = 0$

detector. [42] Detection efficiency is good down to very low N_{h_-} where the missing part of the cross-section must be modeled. Shown in Fig. 14 is the multiplicity distribution generated from a geometrical Glauber approach [43] for comparison. One sees the behavior of the charged particle multiplicity as a function of the impact parameter from the model calculations. From this comparison, it is easy to see how experiments use charged particle multiplicity away from beam-rapidity to trigger on centrality.

Various observables, such as the produced particle multiplicity and transverse energy, are expected to scale with the number of participants. One of the first measurements performed by RHIC experiments was $dN_{ch}/d\eta$ at midrapidity for central collisions. Table 1 lists these values for central collisions of Au + Au at $\sqrt{s_{NN}} = 130$ GeV. Although a variety of techniques and instruments were used for these measurements, the measured values are consistent between experiments and minor discrepancies are due primarily to differing centrality and midrapidity definitions.

The measurement of $dN_{ch}/d\eta$ or $dN_{h_-}/d\eta$ can be performed as a function of the centrality selection to ascertain the scaling of the number of produced particles with

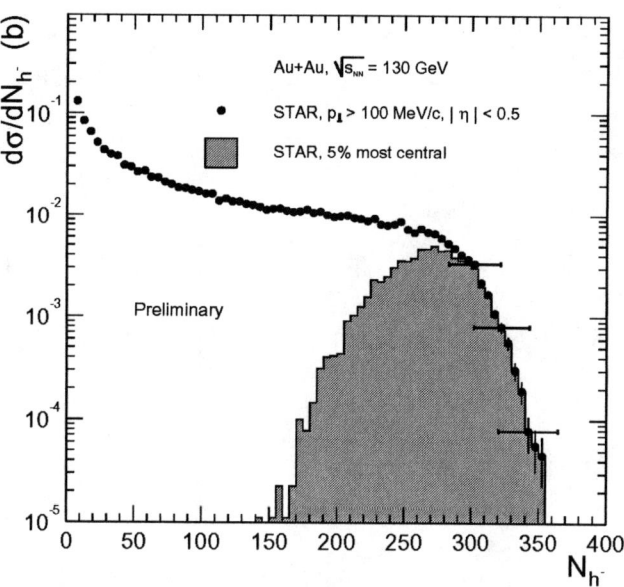

FIGURE 13. Characteristic charged particle multiplicity distribution for Au + Au events at 130 GeV as measured in STAR around mid-rapidity. (Reprinted from Ref. [38], Copyright 2002, with permission from Elsevier Science.)

collision geometry. Fig. 15 provides this information [44]. The number of produced particles at $\eta = 0$ is observed to increase with centrality. This suggests that a scaling exists between the number of participants and $dN_{ch}/d\eta$, and that the ratio of $dN_{ch}/d\eta$ with the number of participants should be constant independent of centrality. However, this is not the case. [32, 34, 45] An increase is observed, and is strongest for more peripheral collisions, and only gradual for more central collisions. Therefore, effects other than collision geometry must play a role in the final determination of the number of produced particles.

The measurement of $dN_{ch}/d\eta/(0.5N_{part})$ is therefore of interest in that the quantity contains information other than just collision geometry that still must be understood. The ratio $dN_{ch}/d\eta/(0.5N_{part})$ can be measured as a function of center of mass collision energy to determine more precisely the role that energetics plays in particle production. Fig. 16 shows the dependence of particle production on the available c.m. energy. In principle, an increase in particle production is expected as the available c.m. energy is increased. However, the increased number of particles produced at higher energies have a progressively larger phase space available due to the larger beam rapidity. Taking this into consideration, one would just as likely expect a broader and broader rapidity distribution rather than an increase at mid-rapidity per participant. This might even be preferred by the argument that the rapidity difference between beam and c.m. continues to increase as the collision energy is increased, thus resulting in a larger rapidity difference between the produced particles at mid-rapidity and the initial beam rapidity. From the measurements it appears that there is a complicated combination of these effects re-

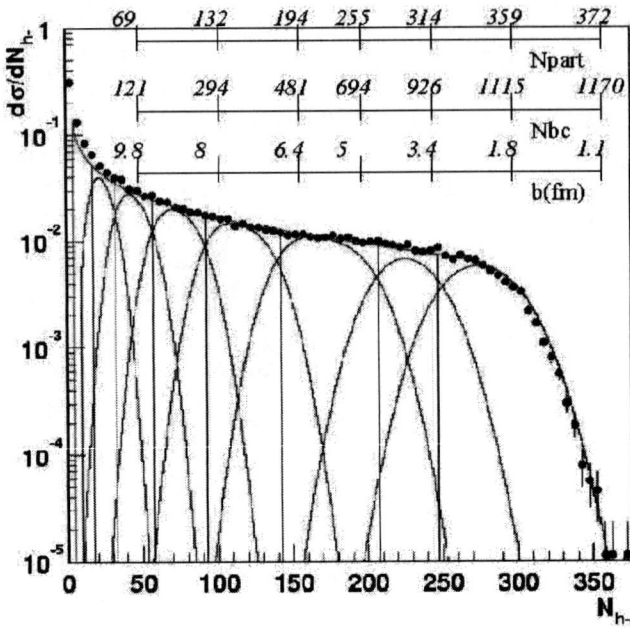

FIGURE 14. Glauber model calculations [43] for the data shown in Fig. 13. Curves represent impact parameter selections in the calculation. Table at the top provides values of number of participants, number of binary collisions, and impact parameter corresponding to the N_{h_-} on the abscissa.

sulting in perhaps scaling with the participant number for the most central collisions, but with other kinematic effects being more influential for more peripheral collisions.

INITIAL HARD SCATTERING AND PARTON ENERGY LOSS

RHIC collisions offer the ability to measure hard scattering for the first time in relativistic nuclear collisions, and to potentially use the hard-scattered partons to probe the medium through which they propagate. [46] Hard scattering processes are calculable in perturbative QCD and scale with the number of binary collisions (N_{binary}) between individual nucleons, rather than the number of nucleon participants. Scaling by N_{binary} is a faster growing function than N_{part} scaling because any participant must have had one binary collision (i.e. $N_{binary} > N_{part}$). At RHIC energies, approximately 30 - 50% of the particle production is expected to result from hard scattering (i.e. large momentum transfer processes).

When a hard scattering or large momentum transfer process occurs, particles with high transverse momentum ($p_T > 2$ GeV/c) are often produced. In elementary collisions in the absence of a medium, hard-scattered partons fragment into hadrons according to the fragmentation function, giving rise to high momentum, leading hadrons that carry a large part of the momentum of the original parton. In nucleus-nucleus collisions at

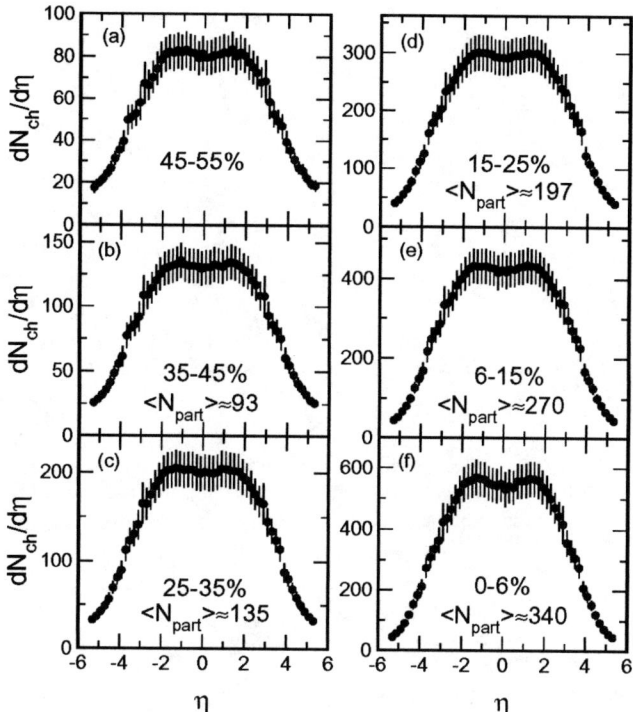

FIGURE 15. $dN_{ch}/d\eta$ versus η and centrality from PHOBOS. An obvious increase in production at midrapidity exists along with a slight change in shape.

RHIC, the hard-scattered parton must traverse the medium, which is colored if a QGP is formed, and subsequently radiates and loses energy as it interacts with constituents of the medium (whether quarks and gluons, or hadrons). As a consequence of the energy loss of a parton traversing a medium, the fragmentation function is also modified [47, 48] and the momentum distribution of leading hadrons will be suppressed relative to that measured in elementary processes, as described by the normal fragmentation function.

The distribution of produced hadrons from nucleus-nucleus collisions at the CERN-SPS is compared to that from elementary p + p collisions in Fig. 17. Plotted is the ratio of charged hadrons in Pb + Pb (normalized by the number of binary collisions) relative to p + p as a function of transverse momentum (p_T). Without modification due to the medium, pQCD predicts a value of 1 for this ratio. The ratio in Fig. 17 indicates that particles from nucleus-nucleus collisions with p_T less than about 1.5 GeV/c have a ratio less than 1, possibly because they are produced in softer collisions. The ratio rises as a function of p_T and exceeds 1 up to the highest p_T shown (4 GeV/c) for SPS energies. This excess has been described in terms of an anomalous nuclear enhancement, an initial state effect, known as the Cronin effect. [49] However, no energy loss or quenching effects due to the medium are observed in the hadron spectra from the SPS.

The first results from RHIC have recently been presented. [50] Shown in Fig. 18 are

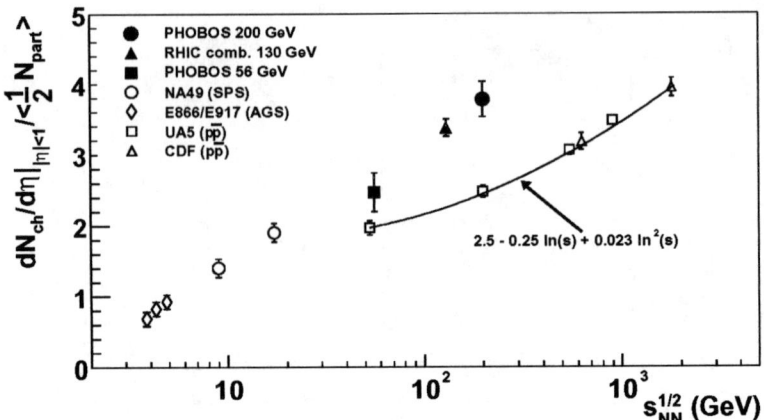

FIGURE 16. $dN_{ch}/d\eta/(0.5N_{part})$ is observed to increase as a function of $\sqrt{s_{NN}}$, even in $p+\bar{p}$ data. The fact that the Au + Au curve sits above the $p+\bar{p}$ curve indicates that particle production also depends on quantities other than the number of participants and energy.

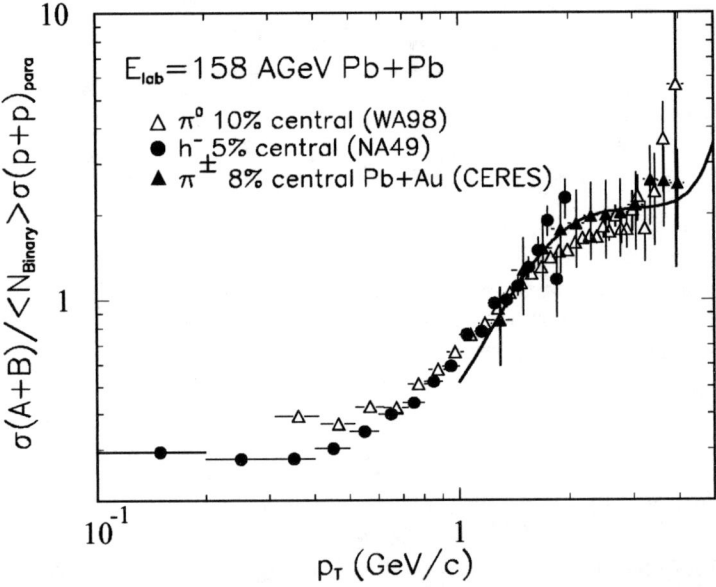

FIGURE 17. Ratio of SPS hadron p_T distributions (scaled by the number of binary collisions) in AA relative to p + p.

negatively charged hadron p_T distributions measured for central collisions of $\sqrt{s_{NN}} = 130$ GeV Au + Au in the STAR experiment at RHIC [51] and $\sqrt{s_{NN}} = 17.2$ GeV Pb + Pb in the NA49 experiment at the CERN-SPS [52], and for minimum-bias $\sqrt{s} = 200$ GeV $p + \bar{p}$ collisions in UA1 [53] at the $Sp\bar{p}S$ collider at CERN. Although exponentials are often fitted to particle spectra to extract inverse slope (temperature) parameters, a power law fit ($A(1 + p_t/p_0)^{-n}$) has been found to best describe particle spectra over a wide range of p_T and especially at high p_T. The free parameters in these fits are A, n, and p_o. As seen in Fig. 18 the power law form describes the data well. The spectra in Fig. 18 have progressively higher $\langle p_T \rangle$ as one goes from UA1 data to NA49 data and then to STAR data. This suggests a higher contribution to hard processes as $\sqrt{s_{NN}}$ increases and in going from p + p collisions to A + A collisions (i.e. increasing the number of binary collisions).

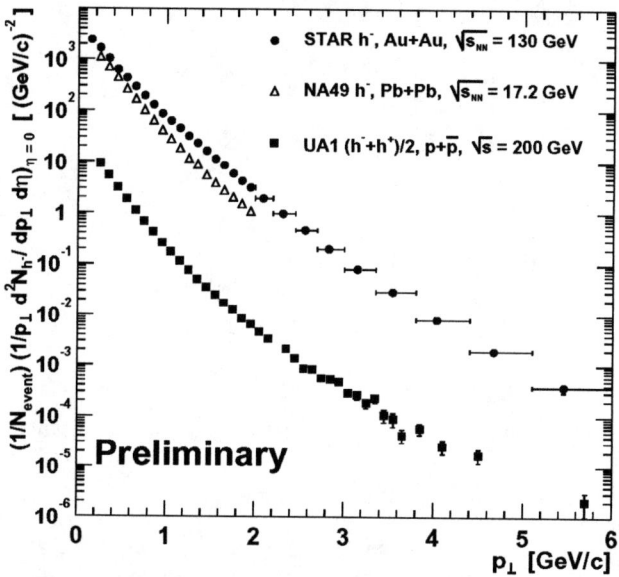

FIGURE 18. Negative hadron p_T distributions from STAR, NA49, and UA1 experiments. (Reprinted from Ref. [38], Copyright 2002, with permission from Elsevier Science.)

In order to compare Au + Au spectra to p + p spectra at $\sqrt{s}$ = 130 GeV, a few extra steps must be taken. Since there is no elementary collision data for comparison at the same energy, the $p + \bar{p}$ data over a large energy range encompassing this energy are utilized to make a best approximation to the 130 GeV p + p distribution. Fig. 19 shows p_T distributions at other energies for $p + \bar{p}$ collisions from the ISR, UA1, and CDF. Each distribution is fit to a power law and A, p_0, and n are interpolated to give the 130 GeV $p + \bar{p}$ spectrum. Then, the number of binary collisions must be determined. T_{AA} is the nuclear overlap integral [42]. It is calculated from the geometrical overlap of the colliding nuclei and is used to determine the average number of overlapping nucleons at a given impact parameter for normalization to p + p collisions. This requires an assumption about the nuclear profile of the nucleus under study. This information (with the $p + \bar{p}$ cross-section) provides an estimate for the average number of binary collisions. The ratio of the hadron spectra for Au + Au with appropriate T_{AA} weighting is divided by the $p + \bar{p}$ reference distribution to obtain the ratio of interest, shown in Fig. 20. Fig. 20 also contains two lines, one for each type of scaling discussed. The lower line at $\sim$ 0.16 represents what is expected to be seen when (soft) scaling due to N_{part} applies,

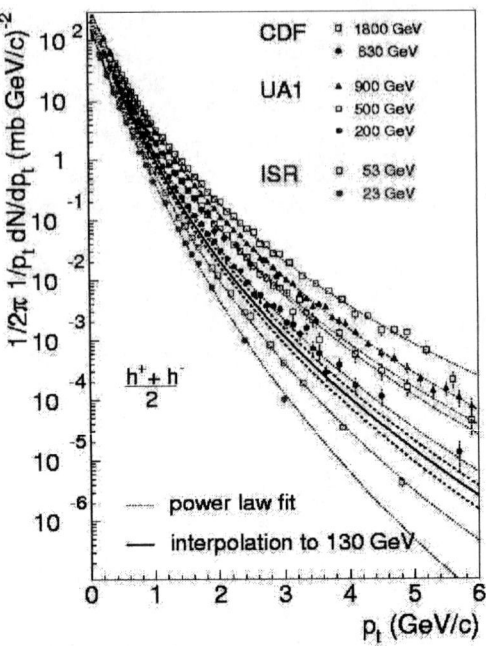

FIGURE 19. Measured hadron p_T distributions over a range of energies for elementary $p + p$ and $p + \bar{p}$ collisions from the ISR, UA1, and CDF.

or according to the number of "wounded nucleons". The line at the ratio 1 is what is expected for (hard collision) scaling according to the number of binary collisions. A striking difference is seen in the ratio at RHIC energy as compared to that observed for the SPS data in Fig. 17. As a function of the p_T of the measured hadron, the ratio rises from the "wounded nucleon" (soft) limit towards the binary scaling limit, without reaching it, and decreases again above $p_T \sim 2$ GeV/c. There is clearly no enhancement with the ratio above 1, as seen in the SPS data. The number of high p_T particles is suppressed. This suppression supplies clear evidence for parton energy loss at RHIC in Au + Au collisions at $\sqrt{s_{NN}} = 130$ GeV. The PHENIX Collaboration has also reported similar results. [54]

New experimental results at higher transverse momentum are eagerly anticipated. At sufficiently high p_T the ratio should reach the pQCD prediction of unity. The experiments are also expected to map the suppression as a function of pathlength through the medium by incorporating measurements of central collisions of lighter nuclei and will eventually utilize coincidence measurements detecting both sides of dijet fragmentation to gain better control of the pathlength in the medium.

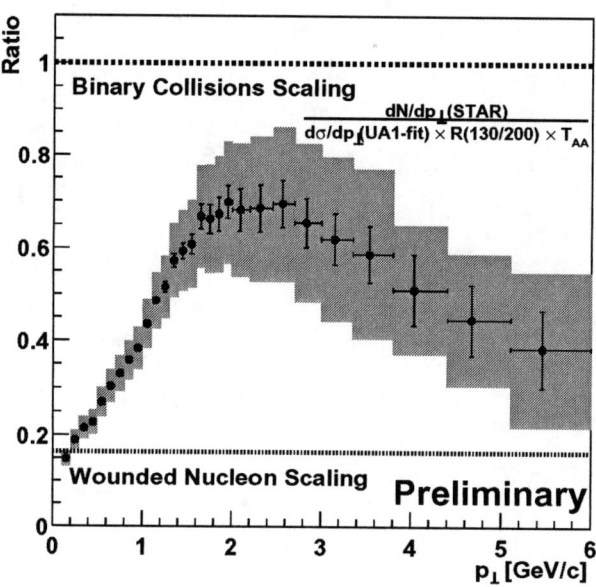

FIGURE 20. Ratio of 130 GeV Au + Au spectrum as measured in STAR scaled by number of binary collisions relative to the 130 GeV p + p reference distribution. See text for details. (Reprinted from Ref. [38], Copyright 2002, with permission from Elsevier Science.)

ELLIPTIC FLOW

Elliptic flow is a measure of an azimuthal asymmetry in the number of particles emerging from a collision in the plane transverse to the beam direction. It is sensitive to the pressure felt by particles in the early stages of the collision. Fig. 11 shows a schematic diagram of the reaction for a non-central collision of two nuclei. The reaction plane is defined by the plane containing the impact parameter and the beam axis. Non-zero impact parameter collisions have an initial spatial asymmetry, as depicted in Fig. 11, that may manifest itself in an azimuthal asymmetry in the emission of particles from the impact zone. If high pressure is present and the medium is highly interacting, the initial spatial anisotropy will be transferred into a momentum anisotropy. The momentum anisotropy can be viewed by measuring the angles that tracks make relative to the reaction plane ($\phi = arctan(py/px)$) after they emerge from the collision.

Mathematically, the elliptic flow quantity is defined as $v_2 = \langle cos2\phi \rangle$ (averaging over all particles measured in a collision). This is the second order Fourier coefficient of the expansion of the azimuthal distribution of particles. In principle, the reaction plane must be determined. To accomplish this, the ϕ distribution of produced particles is measured and is fit by $A(1 + v_2 cos2(\phi - \phi_0))$. The measured ϕ_0 determines the reaction plane, assumed to be the orientation of the azimuthal anisotropy of particle emission for each event.

Effects such as production of back-to-back jets, particle decays, and other types of particle correlations can also create a measurable value of the elliptic flow v_2. If a finite

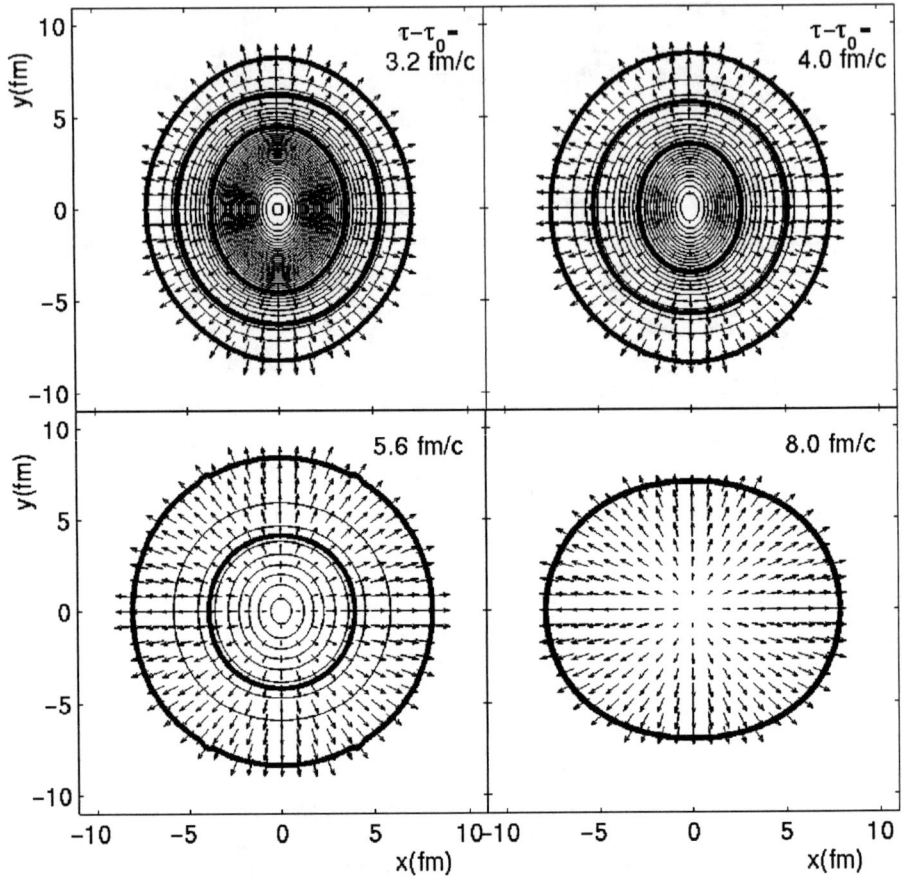

FIGURE 21. Evolution of spatial dimensions of the overlap region of the colliding nuclei. The spatial distribution becomes fairly isotropic, but the velocity distributions (represented by the arrows) are noticeably azimuthally asymmetric.

value of elliptic flow is measured and it is larger than these "background" effects, then the elliptic flow can be used to study the early stage of a collision, when pressure gradients are present, and how much and how quickly the medium responds (interacts).

Hydrodynamical models have been used to predict the evolution of the particle distributions during a collision. Local thermal equilibrium and an initial pressure gradient are assumed. The evolution as a function of time, as shown in Fig. 21, indicates that the initial spatial anisotropy is converted into a final momentum space anisotropy and that this results in a large, measurable value of elliptic flow. [55]

Experiments have measured v_2 versus centrality. Fig. 22 shows experimental results from STAR along with the hydrodynamical prediction. Estimates of the effects of jets and particle decays result in much smaller values of v_2. The data and the hydrodynamical model agree for central collisions. This provides experimental evidence consistent with

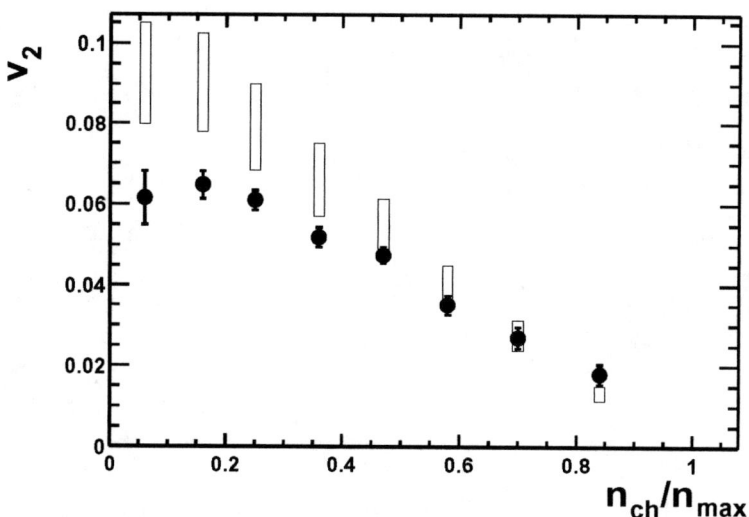

FIGURE 22. v_2 versus centrality as measured by STAR in Au + Au at 130 GeV, along with the hydrodynamic model prediction (open bars). v_2 is observed to decrease with increasing centrality (to the right).

the hydrodynamic prediction that the initial spatial asymmetry evolves into a measurable momentum asymmetry during the early stage of the collision. The dependence of the elliptic flow on c.m. energy is shown in Fig. 23. The v_2 is observed to increase with increasing energy of the colliding heavy system from the AGS (Au + Au) to the SPS (Pb + Pb) to RHIC (Au + Au).

Fig. 24 shows the value of v_2 vs p_T and centrality. The centrality dependence in Figs. 22 and 24 indicates that in central collisions particles emanate from a source that is more spatially symmetric and therefore has lower v_2 values associated with it. There is a linear increase of v_2 with p_T, similar to the case for hydrodynamics, until a saturation sets in at $p_T > 2$ GeV/c. Whether the v_2 values decrease at higher p_T must still be determined by experiment.

In order to understand the saturation of v_2 at higher p_T, a model must be employed. Pure hydrodynamical models predict the initial linear dependence up to $p_T \sim 1.5$ GeV/c. However, they predict even larger values of v_2 for higher p_T and do not level off as observed in the data. The observed saturation could be a result of the medium-induced energy loss of high p_T particles, i.e. consistent with the high p_T energy loss (jet quenching) above around 2 GeV/c. Displayed in Fig. 25 is a model that incorporates hydrodynamics and energy loss of high p_T particles in the medium. [56, 57] The energy loss is an increasing function of the gluon density. The v_2 is observed to increase approximately linearly in the low p_T region, while at higher p_T the v_2 saturates and even turns over for larger gluon densities. These results suggest that a combination of hydrodynamic behavior, which manifests itself in the low p_T region, and parton energy loss, which affects the energy loss and elliptic flow of high p_T particles, can be used to fit and describe the data. However, it has been pointed out that transverse expansion of

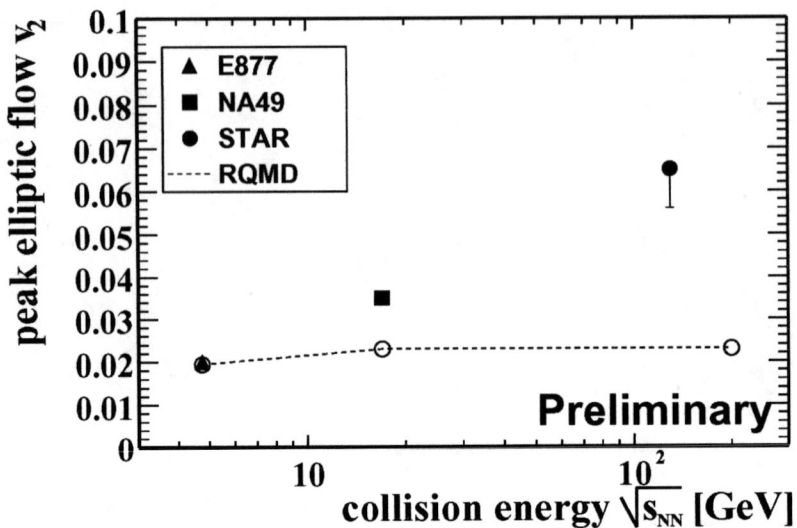

FIGURE 23. Energy dependence of the elliptic flow as measured in minimum bias Au + Au at the AGS, Pb + Pb at the SPS and Au + Au at RHIC.

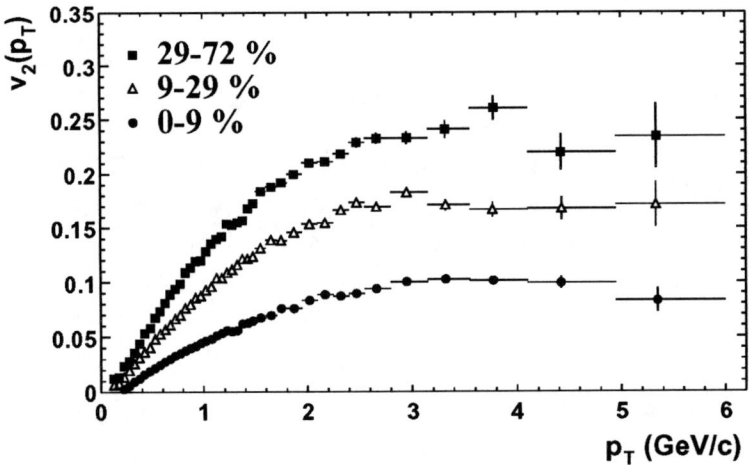

FIGURE 24. v_2 versus p_T for three different collision centralities of Au + Au at 130 GeV as measured by STAR. v_2 increases with increasing p_T of the particles, with saturation observed at high p_T.

the system will further decrease the v_2 at high p_T and must be taken into consideration. [57]

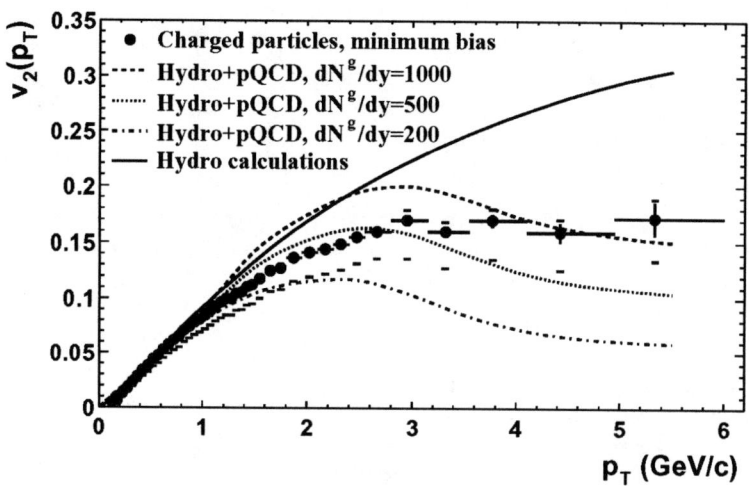

FIGURE 25. Overlay of hydrodynamic model predictions incorporating a hard scattering energy loss mechanism for v_2 versus p_T for Au + Au at 130 GeV data (solid points) as measured in STAR.

BARYON STOPPING AND TRANSPORT

A large number of baryons and anti-baryons are observed in collisions of heavy ions at high energy. The two primary sources for baryons and anti-baryons in these collisions are the incident nuclei and particle-pair production. Since the incoming nuclei contain no anti-baryons, the main source for anti-baryons is pair production, whereas baryons can either originate from the incident nuclei or from pair production.

Simply from energetics and the available energy, pair production is expected to increase with the center-of-mass (c.m.) energy of the system. Thus, production of particles at mid-rapidity reflects the transfer of energy from the incoming nuclei into pair production. The larger the pair production, the more energy was transformed away from the incident motion and the larger the stopping of energy. However, from a purely kinematic perspective the stopping of matter and baryon number (in the form of valence quarks) should decrease with increasing collision or c.m. energy. As the collision energy increases individual or successive collisions at the quark or hadron level become less and less efficient at slowing down the valence quarks (which carry baryon number) over the larger and larger difference in rapidity between the incident beams and center-of-mass frames.

Therefore, it is expected that increased pair production, which provides information on the energy density, will accompany increased stopping in the collision. A ratio $\bar{B}/B = 1$ indicates that most or all baryons result from pair production, so a high degree of energy stopping and a high energy density are expected. A ratio $\bar{B}/B \to 0$ signifies that there is little pair production or energy transport, and little stopping. Various $\bar{B}/B$ ratios have been measured in heavy ion collisions over three incident energies. The $\bar{p}/p$ and $\bar{\Lambda}/\Lambda$ for collisions of heavy nuclei ($A \sim 200$) is shown as a function of c.m.

energy in Fig. 26 [38]. The $\overline{B}/B$ ratios are observed to increase with increasing c.m. energy and approach a value near unity at the RHIC energy of $\sqrt{s}$ = 130 GeV. This indicates an increase in pair production with energy, approaching low net baryon density $(B - \overline{B} \to 0)$ and larger c.m. energy densities at the higher c.m. energies. $\overline{B}/B$ ratios near unity resemble those that existed in the early Universe.

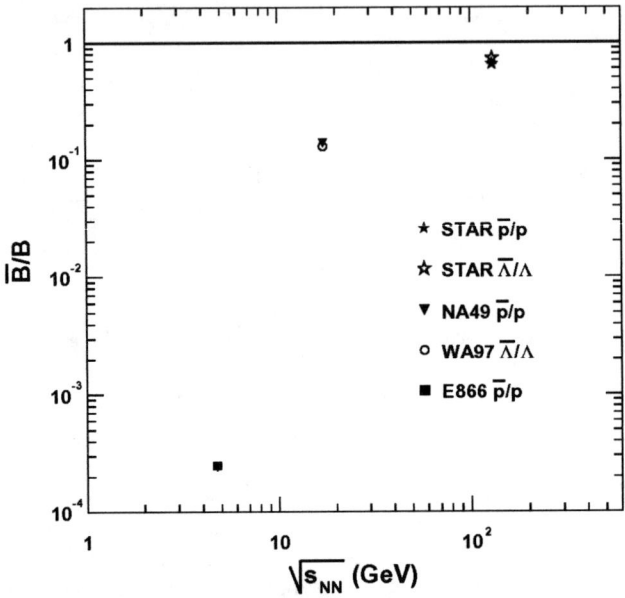

FIGURE 26. Measured anti-baryon to baryon ratios as a function of c.m. energy. (Reprinted from Ref. [38], Copyright 2002, with permission from Elsevier Science.)

However, particle production at mid-rapidity at RHIC is still not baryon-free. The measured $\overline{p}/p$ ratio is approximately 0.65. Taking into account that the $\overline{p}/p$ ratios can be directly associated with the fraction of protons from pair production relative to all protons (considering only pair production and baryon transport) gives the simple relation

$$\overline{p}/p = \frac{yield(pair\ production)}{yield(pair\ production) + yield(transport)} = 0.65. \qquad (7)$$

This results in

$$\frac{yield(pair\ production)}{yield(transport)} \approx 2. \qquad (8)$$

One way to view this is that two thirds of the measured protons originate from pair production and one third is transported over five units of rapidity from beam to c.m. rapidity.

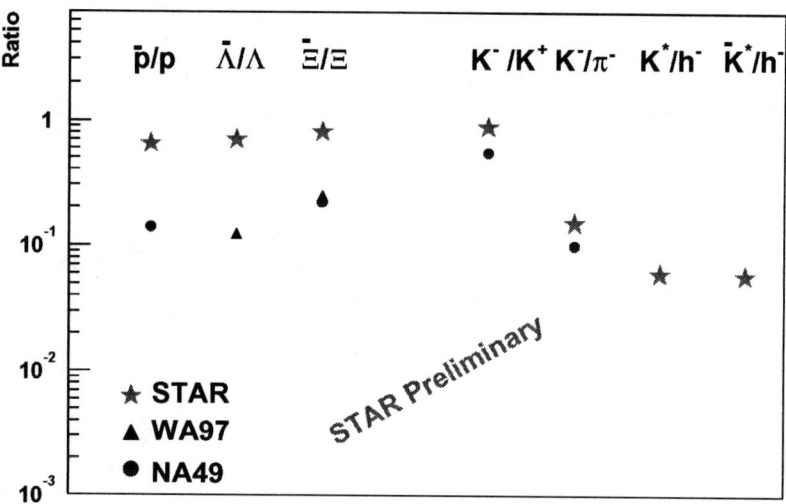

FIGURE 27. Ratios of anti-particles to particles (four left-most ratios) and dissimilar particle ratios measured by STAR in central collisions of Au + Au at RHIC. (Reprinted from Ref. [38], Copyright 2002, with permission from Elsevier Science.)

CHEMICAL AND THERMAL EQUILIBRIUM

An important question in relativistic heavy ion physics is whether the matter reaches equilibrium during the collision process. If the system is in equilibrium, then thermodynamics can be used to describe the system. In this case, the particle yields and spectra will reflect the equilibrium conditions. It is important to note that there are two different types of equilibrium that are commonly discussed - chemical and thermal. Chemical equilibrium occurs when the rates for competing processes, e.g. creation and annihilation, reach their equilibrium values and the resulting particle abundances no longer change on average. Thermal equilibrium occurs when all parts of the system reach the same temperature, after which the particle spectra no longer change. Once the system starts expanding, the mean distance between particles and the time between interactions increase. This leads to freeze-out when interactions cease, because the probability of another interaction is small. The inelastic cross section determines when chemical freeze-out occurs and the total cross section for thermal freeze-out. Since inelastic cross sections are always smaller than the total cross section, inelastic reactions that change the particle species cease (chemical freeze-out) at an earlier time that those that change kinematics and temperature (thermal freeze-out).

To determine whether a system has reached equilibrium, particle yields and spectra are compared with models that assume equilibrium. The particle ratios and particle spectra provide a test for chemical and thermal equilibrium, respectively, in the system. Note these are necessary conditions for equilibrium and act as consistency tests, but are not sufficient conditions to prove the existence of equilibrium in the system.

Chemical Equilibrium: Particle Ratios

Anti-particle to particle ratios $\bar{x}/x$ are often the first measurements from new experiments. In such measurements many of the systematic errors of a given experiment cancel, allowing rapid analysis and early results, as compared to ratios of different particles x/y. First results on ratios of strange anti-baryons and baryons are seen on the left-side of Fig. 27 [38].

With the help of some quark algebra as presented in [58] it is possible to extract information about the quark flavor content and the ratios of various particles produced in the collision. For example the ratios of $\bar{\Lambda}/\Lambda$, $\bar{\Xi}/\Xi$ and K^+/K^- are connected as follows in this quark coalescence model:

$$\frac{\bar{\Lambda}}{\Lambda}\left[\frac{\bar{u}\bar{d}\bar{s}}{uds}\right] = \left(\frac{\bar{u}}{u}\right) \times \left(\frac{\bar{s}}{s}\right) \times \frac{\bar{p}}{p}\left[\frac{\overline{uud}}{uud}\right] = D \times \frac{\bar{p}}{p} \qquad (9)$$

and

$$\frac{\bar{\Xi}}{\Xi}\left[\frac{\overline{uss}}{uss}\right] = \left(\frac{\bar{u}}{u}\right) \times \left(\frac{\bar{s}}{s}\right) \times \frac{\bar{\Lambda}}{\Lambda}\left[\frac{\bar{u}\bar{d}\bar{s}}{uds}\right] = D \times \frac{\bar{\Lambda}}{\Lambda} \qquad (10)$$

where

$$D = \left(\frac{\bar{u}}{u}\right) \times \left(\frac{\bar{s}}{s}\right) = \frac{K^+}{K^-}\left[\frac{u\bar{s}}{\bar{u}s}\right] \qquad (11)$$

Therefore, if the physics assumed in this quark coalescence model is correct, then the measured ratios of particles should follow those predicted by the above algebra. This is a simple exercise and one finds that the measured values of $\bar{p}/p$, $\bar{\Lambda}/\Lambda$, and $\bar{\Xi}/\Xi$ can be reproduced using the measured K^+/K^- ratio and the algebra above.

In heavy ion physics it is not usually possible to measure thermodynamic variables directly. Normally they must be extracted indirectly using a theoretical model to connect the desired variables, e.g. temperature or entropy, with measurable quantities like the particle spectra. One model for this purpose is the statistical thermal model (see for e.g. [59]). This model incorporates a thermally and chemically equilibrated fireball at the hadro-chemical freeze-out and assumes the law of mass action. The partition function can be defined in a grand canonical ensemble to calculate the density of particles of each species. The constraints in this model are the volume, the strangeness chemical potential, and the isospin. If the particle ratios from the model are fit to the experimental data, the temperature and baryo-chemical potential can be determined. Another model is the chemical freeze-out model [60]. Here an ideal gas of hadron resonances is assumed and the particle density of each particle is determined from

$$\rho_i = \gamma_s^{|s_i|} \frac{g_i}{2\pi^2} T_{ch}^3 \left(\frac{m_i}{T_{ch}}\right)^2 K_2(m_i/T_{ch}) \lambda_q^{Q_i} \lambda_s^{s_i} \qquad (12)$$

with

$$\lambda_q = \exp\left(\frac{\mu_q}{T_{ch}}\right)$$

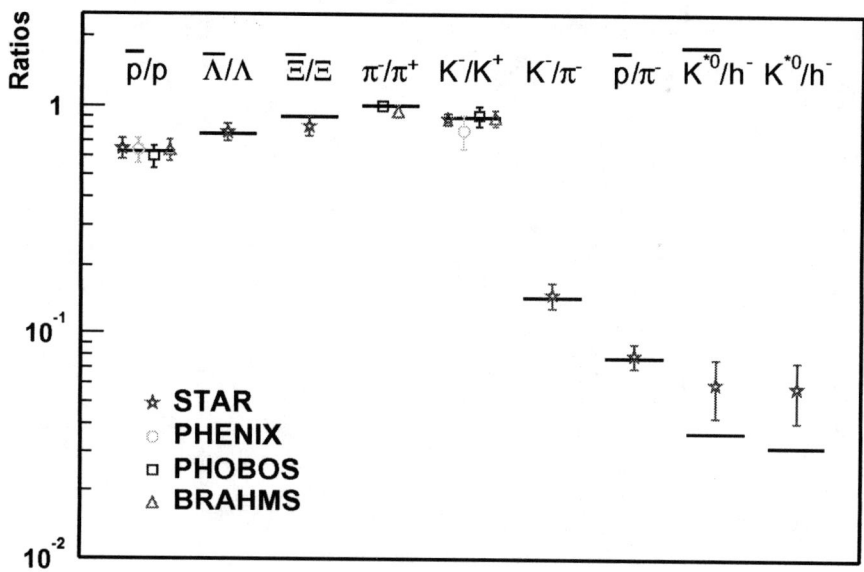

FIGURE 28. Ratios of various particles for 130 GeV Au + Au collisions at RHIC, with thermal model [59] ratio predictions from best fit to data.

$$\lambda_s = \exp\left(\frac{\mu_s}{T_{ch}}\right) . \qquad (13)$$

The variables are:

Q_i : 1 for u and d, -1 for $\bar{u}$ and $\bar{d}$

s_i : 1 for s and -1 for $\bar{s}$

g_i : spin-isospin freedom

m_i : particle mass

T_{ch} : chemical freeze-out temperature

μ_q : light-quark chemical potential

μ_s : strangeness chemical potential

g_s : strangeness saturation factor.

The parameters of this model can be adjusted to reproduce reasonably well the measured particle ratios. This gives the chemical freeze-out temperature and the chemical potentials under the assumption that the hadrons behave as an ideal gas.

A comparison of the data to the best global fit to the particle ratios using the statistical thermal model is shown in Fig. 28. The horizontal bars are the ratios from the model calculation. There is reasonable agreement. The best fit parameters of this model yield $\mu_B(\text{RHIC}) = 51 \pm 6 \text{MeV}$ and $T_{ch}(\text{RHIC}) = 175 \pm 7 \text{MeV}$. This is very close to the value of T_c from lattice QCD. Calculations assuming QCD with two flavors on a lattice predict a critical transition temperature of $T_c = 173 \pm 8 \text{MeV}$ at $\mu_B = 0$ [61].

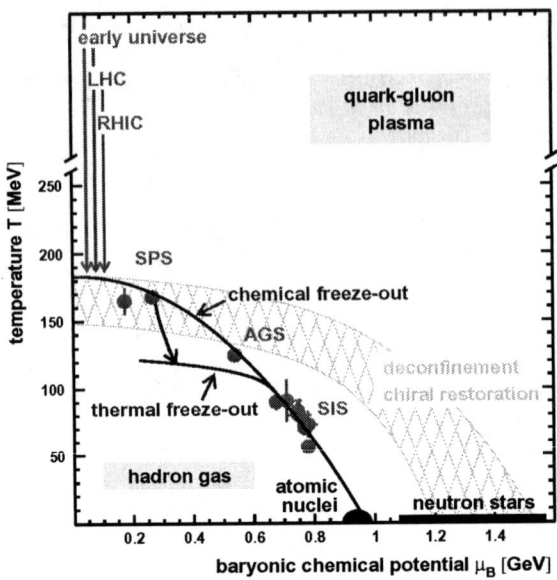

FIGURE 29. Phase diagram of temperature versus baryo-chemical potential. (Reprinted from Ref. [62], Copyright 2001, with permission from Elsevier Science.) Displayed are data points from heavy ion measurements along with descriptions of various anticipated regions of the diagram.

For illustration, Fig. 29 shows a phase diagram of the temperature versus baryo-chemical potential plane [62]. The hatched region represents the phase boundary and transition region between a hadron gas and the quark-gluon plasma. The data points indicate measurements in heavy ion collisions at SIS at the GSI, the SPS at CERN, and the AGS at BNL. The temperature increases while the baryo-chemical potential decreases with increasing beam energy. This follows from the fact that the fraction of net baryons (originally the excess of initial nucleons) gets smaller relative to the produced pairs of baryons and anti-baryons. The solid black curve corresponds to the freezeout curves for chemical and thermal freezeout, as labeled. RHIC reaches the highest temperature of heavy ion machines thus far, and may yield a temperature sufficiently high to create the quark gluon plasma. The Large Hadron Collider (LHC) will extend to even higher temperatures as indicated in the diagram.

Thermal Equilibrium, Expansion and Freeze-out: Particle Spectra and Flow

Particle spectra are often studied and compared to thermal model predictions, to check for consistency with thermal equilibrium. A simple thermal distribution plotted on a semi-logarithmic scale yields a spectrum with a slope ($1/T$) that is given by

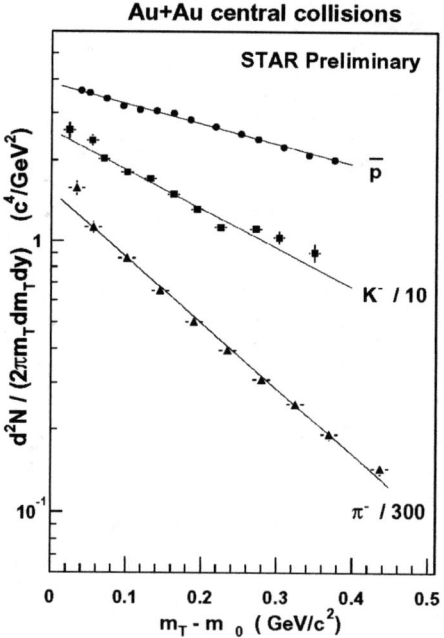

FIGURE 30. Transverse mass distributions of negatively charged pions, kaons, and anti-protons as measured in STAR. (Reprinted from Ref. [38], Copyright 2002, with permission from Elsevier Science.)

$$\frac{1}{m_T}\frac{dN}{dm_T} \propto A\exp\left(-\frac{m_T}{T}\right) \quad (14)$$

which is the number of particles per unit transverse mass as a function of the transverse mass $m_T = \sqrt{p_T^2 + m^2}$.

A typical result from the RHIC data is shown in Fig. 30. Notice that the inverse slope (T) of the curves can be associated with the thermal freeze-out temperature, if the system is thermalized. Thus, a steeper slope corresponds to a lower temperature. The spectra in Fig. 30 appear to have a mass dependence, that is unexpected and inconsistent with purely thermal spectra. This mass dependence can be seen in Fig. 31 for results from RHIC and from the CERN SPS. However, this approach of comparing the slopes of spectra does not take into consideration the effects of collective flow.

If hot and dense matter is produced, the matter is expected to expand due to the pressure long before thermal freeze-out. This collective expansion can leave the particles with a common boost velocity at the time of freeze-out. This boost contributes to the particle spectra, and thus the measured slopes. In a naive hydrodynamical approach, where nuclear matter is assumed to be a fluid, a simple relationship between the freeze-out temperature, the measured temperature, and the boost velocity of the collective expansion (flow) can be derived. This is

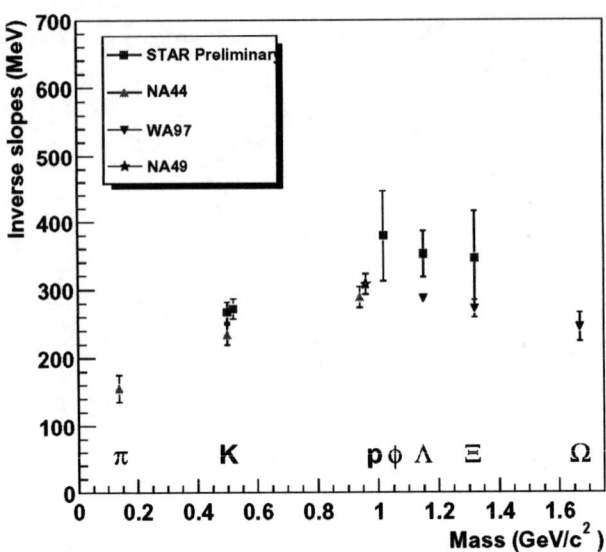

FIGURE 31. Inverse slopes of transverse mass distributions of various particles plotted as a function of the particle mass as measured in STAR at RHIC, and in NA44, WA97 and NA49 from the CERN SPS.

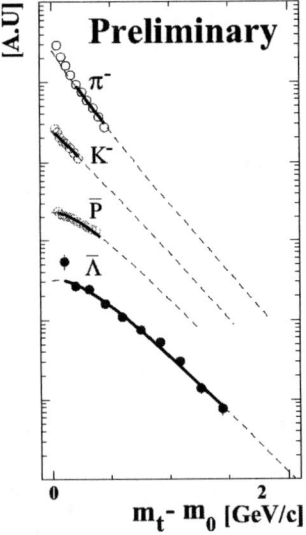

FIGURE 32. STAR particle spectra with simple hydrodynamical model fits.

$$T = T_{\text{freeze-out}} + \frac{1}{2}m\langle \beta_t \rangle^2 \tag{15}$$

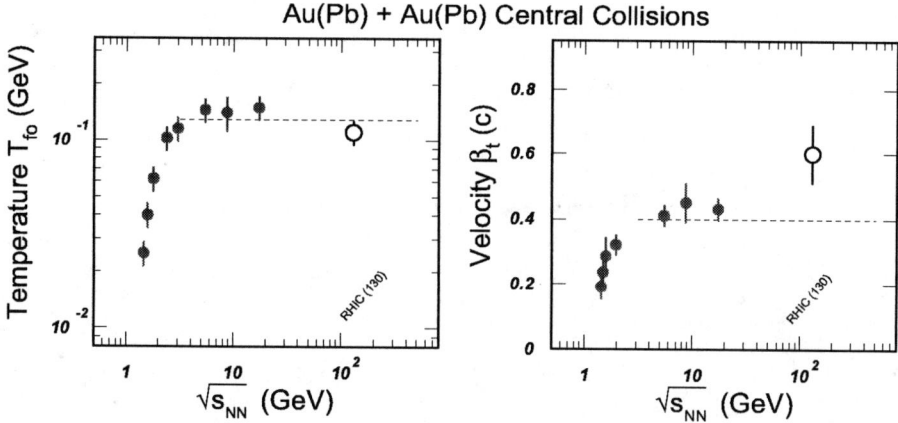

FIGURE 33. Systematics of the thermal freezeout temperature parameter and the average transverse flow velocity extracted from hadron spectra as a function of energy. (Reprinted from Ref. [63], Copyright 2002, with permission from Elsevier Science.)

with $\langle \beta_t \rangle$ the average transverse flow velocity. Thus, there is a particle mass-dependent relationship between the temperature and the freezeout temperature. The mass-dependent term depends upon the average transverse flow velocity. The difference in the spectra for the various particles can be attributed to collective flow. Fig. 32 shows the m_T spectra for different particles measured in STAR. The temperature resulting from this fit via the hydrodynamical model [63] is 120+50-25 MeV with $\langle \beta_t \rangle$=0.52+0.12-0.08 c. The temperature and average flow velocity can be extracted under these simple assumptions, and is presented in Fig. 33 for data from the AGS, SPS and RHIC. [63] The thermal freeze-out temperature parameter increases over the energy range of the AGS and SPS, but is rather similar between the SPS and RHIC energies. The average transverse flow velocity appears to increase strongly at lower energies and continues to increase between the SPS and RHIC energies, suggesting an increasing amount of transverse flow. The results point to strong collective transverse expansion at RHIC and thus to high pressure and a large rescattering rate in a scenario such that thermalization appears likely in these events.

TWO-PARTICLE CORRELATIONS: SIZE OF THE EMITTING SOURCE

Two-particle interferometry can be utilized too obtain information on the space-time geometry of the fireball at thermal freeze-out. When applied to identical bosons this is often referred to as Hanbury-Brown Twiss (HBT) interferometry. This method was originally developed using photons for determination of the size of stars in astronomy. In high energy and heavy ion physics identical pions or kaons are often used. Pions can be emitted from anywhere in the source. To perform an HBT interferometry measurement,

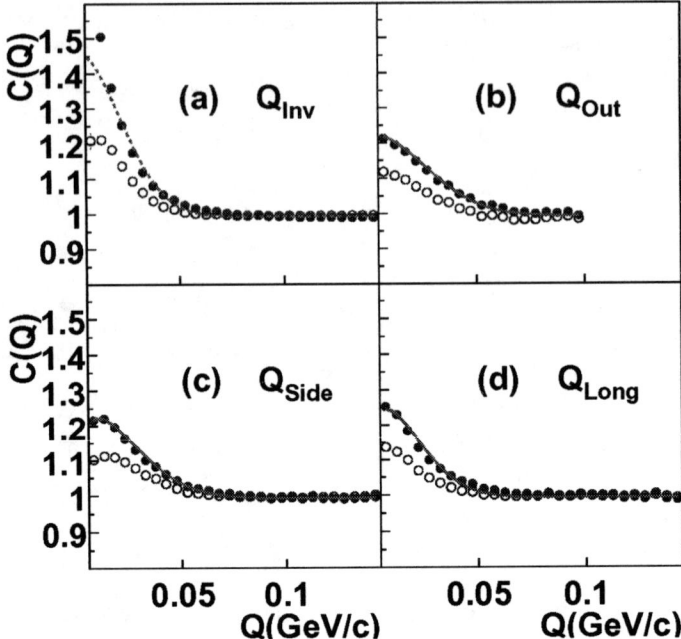

FIGURE 34. STAR two-pion correlation functions of negative pions for $q_{\text{invariant}}, q_{\text{out}}, q_{\text{side}}, q_{\text{long}}$ for 130 GeV Au + Au central collisions. Solid points are Coulomb-corrected and open circles are uncorrected. Curves are fits to the Coulomb-corrected data.

the pions measured in each event are grouped in pairs. From these the relative probability of finding a particle pair is constructed as a function of the three-dimensional momentum space separation. Due to the Bose-Einstein statistics for identical bosons, the probability is higher to find two particles at zero relative momentum q. The width of the resultant Bose-Einstein enhancement peak at $q = 0$ is inversely related to the size of the emitting source. The correlation function can now be determined from all identical particles in an event after integrating over many events. Normally the Bertsch-Pratt parameterization is used where $C(q_{\text{out}}, q_{\text{side}}, q_{\text{long}}) = 1 + \lambda \exp(\sum q_i^2 R_i^2)$.

The one-dimensional projection of the three-dimensional Bertsch-Pratt parameterization for STAR central collision events is shown in Fig. 34. The solid points are corrected for Coulomb effects, and all results are restricted to particles with $|y| < 1$ and $0.125 < p_T < 0.225$ GeV/c. The enhancement at low values of q is apparent. The radii extracted from these measurements are:

$$R_{\text{out}} = (5.86 \pm 0.11 \pm 0.23)\,\text{fm}$$
$$R_{\text{side}} = (5.47 \pm 0.09 \pm 0.16)\,\text{fm}$$
$$R_{\text{long}} = (7.07 \pm 0.12 \pm 0.21)\,\text{fm}$$

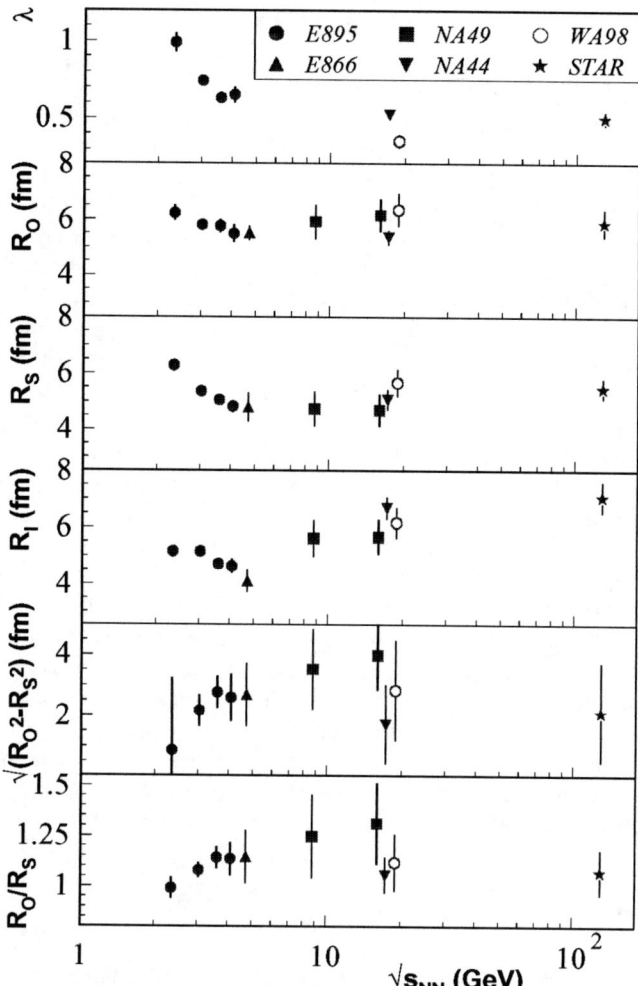

FIGURE 35. World's compilation of mid-rapidity, negative-pion correlation source parameters as a function of c.m. energy for Au + Au or Pb + Pb central collision data at $p_t \approx 0.17$ GeV/c.

with $\lambda = 0.50 \pm 0.01 \pm 0.03$.

A compilation of the world's 3D $\pi\pi$-HBT results (parameters) as a function of $\sqrt{s}$ is given in Fig. 35 [64]. Surprisingly, the source sizes are similar at RHIC, AGS and SPS, namely smaller than 10 fm. Furthermore, the radii increase with centrality and decrease with increasing pair transverse momentum ($k_\perp$) (not shown). Unexpectedly, one observes that $R_{out}/R_{side} \sim 1$. All models to date are unable to predict this ratio and are unable to predict its decrease with increasing $k_\perp$. Thus far, speculation points to an explosive source with short emission times during freezeout.

CONCLUSIONS

The initial round of results from RHIC experiments are just emerging. They provide us with our first view of heavy ion collisions at ultra-relativistic energies. By comparison with models, some still-speculative conclusions can be made. In the early stage of the collision there appears to be rapid thermalization, on very short timescales (less than to a few fm/c). This allows for a large elliptic flow signal due to early pressure gradients in the system, that is well reproduced by hydrodynamics as a function of transverse momentum up to $p_T \sim 2$ GeV/c. Beyond $p_T \sim 2$ GeV/c the elliptic flow signal departs from the hydrodynamical prediction and saturates or starts to decrease. At similar values of p_T the particle spectra exhibit a departure from scaling of the elementary spectra with geometry and number of binary collisions, suggesting a large amount of energy loss for hard-scattered partons traversing the medium. The particle ratios are consistent with chemical equilibrium at temperatures and baryo-chemical potentials near the quark-hadron phase transition predicted in lattice QCD. The hadron spectra confirm that there is significant flow in the system and that the expansion cools the system somewhat prior to final freeze-out. Two-particle interferometry measurements are consistent with rapid hadronization, but no dynamical models to date are successful at reproducing the momentum dependence of the source parameters.

There are many measurements eagerly awaited from RHIC. These include further, more detailed studies and systematics of those mentioned above. Yields, spectra and collision systematics on multiply-strange baryons are anticipated in the near future, as are correlations among various observables. Further studies of hard scattering should provide more detailed evidence on parton energy loss and the medium through which the partons propagate. Results on charmonium suppression will require larger integral luminosities and new detectors, presently being installed. We look forward to more exciting and definitive results from RHIC.

REFERENCES

1. J.C. Collins and M.J. Perry, Phys. Rev. Lett. 34, 1353 (1975); G. Chapline and L. Susskind, Phys. Rev. D 20, 2610 (1979).
2. Conceptual Design of the Relativistic Heavy Ion Collider (Report BNL 52195, 1989).
3. T.D. Lee, Rev. Mod. Phys. 47 (1975) 267.
4. T. Matsui and H. Satz, *Phys. Lett.* B. 178, 416 (1986).
5. F. Karsch, M.T. Mehr and H. Satz, *Z. Phys.* C 37, 617 (1988).
6. F. Karsch and H. Satz, *Z. Phys.* C 51, 209 (1991).
7. J.P. Bailly et al, *Phys. Lett.* B 195, 609 (1987).
8. J. Rafelski and B. Müller, *Phys. Rev. Lett.* 48, 1066 (1982). [Erratum: *ibid.* 56, 2334 (1986).]
9. J. Rafelski, *Phys. Rep.* 88, 331 (1982).
10. P. Koch, B. Müller and J. Rafelski, *Phys. Rep.* 142, 167 (1986).
11. H.W. Barz, G.L. Friman, J. Knoll and H. Schulz, *Nucl. Phys.* A 484, 661 (1988); *Nucl. Phys.* A 519, 831 (1990); *Phys. Lett.* B 254, 315 (1991).
12. L. van Hove, Phys. Lett. B118, 138 (1982); Z. Phys. C21, 93 (1983).
13. T. Blum, L. Kärkkäinen, D. Toussaint, S. Gottlieb, *Phys. Rev.* D51, 5153 (1995)
14. M. Gyulassy, K. Kajantie, H. Kurki-Suonio, L. McLerran, *Nucl. Phys.* B237, 477 (1984).
15. H.W. Barz, L..P Csernai, B. Kämpfer, B. Lukács, *Phys. Rev.* D32, 115 (1985).
16. D. Seibert, *Phys. Rev.* D32, 2812 (1987); *Phys. Rev.* D35, 2013 (1987).

17. N. Bilić, J. Cleymans, E. Suhonen, D.W. van Oertzen, *Phys. Lett.* B311, 266 (1993).
18. T. Csörgö, *Phys. Lett.* B347, 354 (1994).
19. T. Csörgö, B. Lörstad, *Nucl. Phys.* A590, 465 (1995).
20. M.C. Chu, S. Gardner, T. Matsui, R. Seki, *Phys. Rev.* C50, 3079 (1994).
21. M. Asakawa, C.M. Ko, P. Lévai and W.J. Qiu, *Phys. Rev.* C 46, 1159 (1992).
22. L. Xiong, E. Shuryak, G.E. Brown, *Phys. Rev.* D46, 3798 (1992).
23. J. Kapusta, P. Lichard, D. Seibert, *Phys. Rev.* D44, 2774(1991); [Erratum: ibid. D47:4171 (1993).
24. M.T. Strickland, *Phys. Lett.* B 331, 245 (1994).
25. D.K. Srivastava, B. Sinha, M. Gyulassy and X.N. Wang, *Phys. Lett.* B 276, 285 (1992).
26. S. Chakrabarty, J. Alam, S. Raha, B. Sinha and K. Srivastava, *Phys. Rev.* D 46, 3802 (1992).
27. Proposal on Spin Physics Using the RHIC Polarized Collider, RHIC Spin Collaboration (1992).
28. Conceptual Design Report for the Solenoidal Tracker At RHIC, The STAR Collaboration, PUB-5347 (1992); J.W. Harris et al, Nucl. Phys. A 566, 277c (1994).
29. S. Klein and E. Scannapieco in Proceedings on Intersections Between Particle and Nuclear Physics: 6th Conference, ed. T.W. Donnelly, LBNL Report LBNL-40495, (AIP Press, 1997); J. Nystrand and S. Klein in Proceedings of Hadron '97, (LBNL Report LBNL-41111, BNL, 1997).
30. A Ring Imaging Cherenkov Detector for STAR, STARnote 349, STAR/ALICE RICH Collaboration (1998); ALICE Collaboration, Technical Design and Report, Detector for High Momentum PID, CERN/LHCC 98-19.
31. PHENIX Experiment at RHIC - Preliminary Conceptual Design Report, PHENIX Collaboration Report (1992).
32. see also talk of W. Zajc, Proceedings of this Institute (2002).
33. RHIC Letter of Intent to Study Very Low pt Phenomena at RHIC, PHOBOS Collaboration (1991).
34. see also talks of W. Busza and C. Roland, Proceedings of this Institute (2002).
35. Interim Design Report for the BRAHMS Experiment at RHIC, BNL Report, (1994).
36. D.E. Groom et al., *The European Physical Journal* C15, 1(2000).
37. M. Horsley et al., *The International Nuclear Physics Conference* CP610, 542 (2002).
38. C. Adler et al.(STAR Collaboration) *Nucl. Phys.* A698, 64 (2002).
39. C.W. Fabjan et al., *Nucl. Instrum. Meth. Phys. Res.* A367, 240 (1995).
40. P. Grabmayr et al., *Nucl. Instr. Methods* A402, 85 (1998).
41. C. Adler et al., Nucl. Instrum. Meth. A470, 488 (2001).
42. C. Adler et al. Phys. Rev. Lett. 87, 112303 (2001).
43. M. Miller, private communication (2002).
44. B.B. Back et al., Phys. Rev. Lett. 87, 102303 (2001).
45. B.B. Back et al., Phys. Rev. Lett. 88, 22302 (2002).
46. M. Gyulassy and M. Pluemmer, Phys. Lett. B243, 432 (1990); X.N. Wang and M. Gyulassy, Phys. Rev. Lett. 68, 1480 (1992).
47. X.N. Wang and Z. Huang, Phys. Rev. Lett. 77, 231 (1996); Phys. Rev. C55, 3047 (1997).
48. X.N. Wang and X.F. Guo, Phys. Rev. Lett. 85, 3591 (2000); Nucl. Phys. A696, 788 (2001).
49. J.W. Cronin et al., Phys. Rev. D11, 3105 (1975); L. Kluberg et al., Phys. Rev. Lett. 38, 670 (1997).
50. see Proceedings of Quark Matter 2001, Nucl. Phys. A698 (2002).
51. J.C. Dunlop (STAR and STAR-RICH Collaboration), Nuclear Physics A698 (2002) 515c-518c.
52. H. Appelshauser et al, Phys. Rev. Lett. 82, 2471 (1999).
53. C. Albajar et al., Nucl. Phys. B335, 261 (1990).
54. K. Adcox et al, Phys. Rev. Lett. 88, 022301 (2002).
55. P.F. Kolb, J. Sollfrank, and U. Heinz, Phys. Rev. C 62, 054909 (2000).
56. P. Huovinen et al., Phys. Lett. B503, 58 (2001).
57. M. Gyulassy et al., Phys. Lett. B526, 301 (2002).
58. T.S. Biro, P. Levai and J. Zimanyi, Phys. Lett. B347, 6 (1995).
59. P. Braun-Munzinger et al. Phys. Lett. B518, 41 (2001); ibid. Phys. Lett. B465, 15 (1999).
60. J. Sollfrank et al., Phys. Rev. C59, 1637 (1999).
61. F. Karsch, Nucl. Phys. A698, 199c (2002).
62. P. Braun-Munzinger, Nucl.Phys. A681, 119 (2001).
63. N. Xu and M. Kaneta, Nucl. Phys. AA698, 306c (2002).
64. C. Adler et al., STAR Collaboration, Phys. Rev. Lett. 87, 082301 (2001).

Introduction to Ultrarelativistic Heavy-Ion Physics at the CERN-SPS

F. Antinori*, A. Billmeier[†] and J. Zaranek**

*INFN Padova, Italy
[†]Wayne State University, Detroit, USA, and Brookhaven National Laboratory, New York, USA
**IKF Frankfurt, Germany

Abstract. This material is meant as an introduction to the physics of ultrarelativistic heavy-ion collisions at the CERN-SPS from an experimental viewpoint. The basic concepts, quantities and methods are introduced and a selection of experimental results from the CERN lead beam programme is presented and discussed.

This material is meant as an introduction to the physics of ultrarelativistic heavy-ion collisions at the CERN-SPS from an experimental viewpoint.

Within the available space, we have attempted to offer to the student an organic view of the highlights from the SPS while at the same time providing enough details so that the newcomer is equipped with some minimal basic tools of the trade. Obviously, it was possible to present only a selection of results from the CERN Pb programme. Such a selection inevitably entails a subjective judgment on the part of the authors.

The paper is organized as follows: Section 1 contains a brief description of the CERN PS-SPS accelerator complex. In section 2 we discuss the basic features of Pb+Pb collisions at the SPS, such as collision geometry, energy density, transverse mass spectra and freeze-out conditions. We then turn to a more detailed discussion of the main results obtained on the study of specific Quark-Gluon Plasma signatures in section 3. Section 4 contains the summary and conclusions.

Throughout the paper, we have inserted tutorial discussions aimed at introducing the basic concepts, quantities and methods for the benefit of the younger students.

1. THE SPS LEAD BEAM PROGRAMME

1.1. The accelerator complex

The CERN accelerator complex is sketched in figure 1. A beam of Pb^{27+} ions with a current of 80 mA originating from the Electron Cyclotron Resonance (ECR) source goes through a Radio Frequency Quadrupole (RFQ) from which it emerges with an energy of 250 AkeV (250 keV per nucleon). The ions are then injected into a linac where they are accelerated to 4.2 AMeV before passing the first electron stripper foil. The stripper produces a Pb^{53+} beam which is then accelerated first to 94 AMeV in the

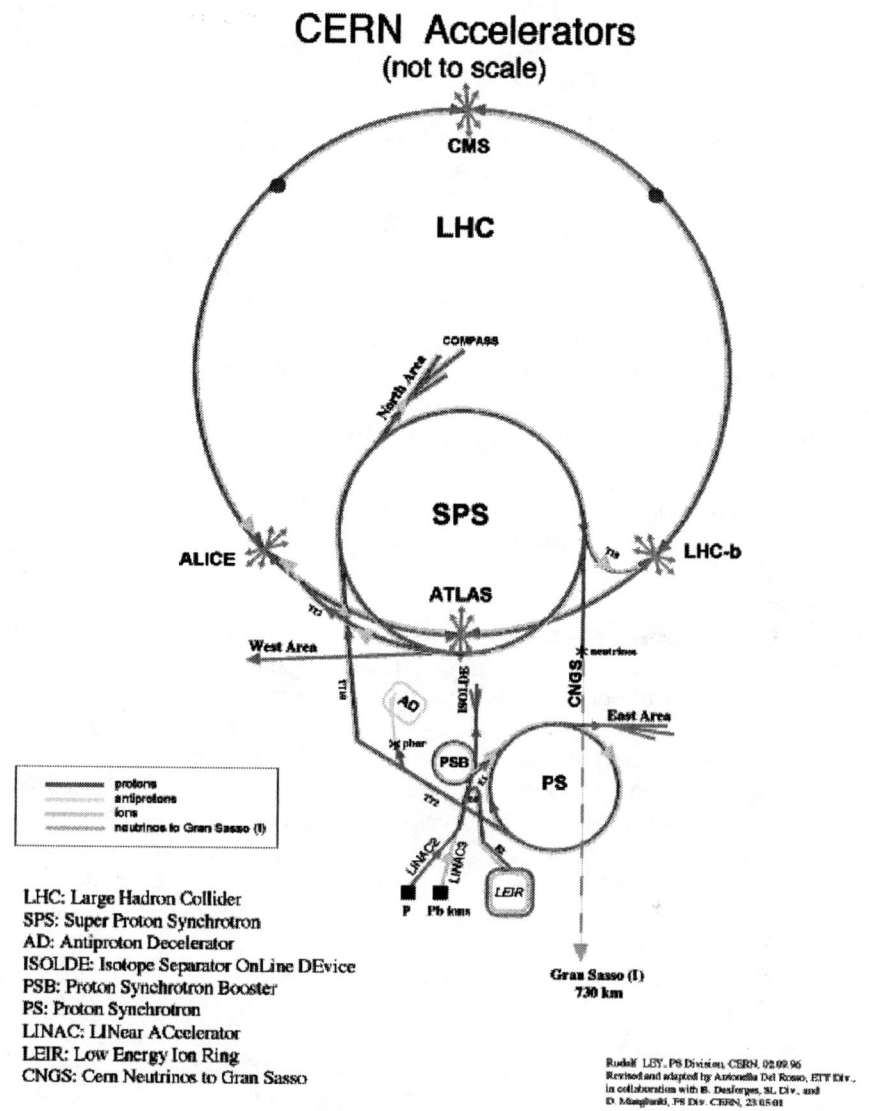

FIGURE 1. A sketch of the CERN accelerator complex.

Proton Synchrotron Booster (PSB) and then to 4.25 AGeV in the Proton Syncrotron. At this point the ions pass through the second and final stripper from which they emerge fully ionised. The resulting Pb^{82+} beam is injected in the SPS where it can be accelerated up to the top energy of 158 AGeV. The SPS beam is finally extracted to seven beam lines to supply the fixed target experiments located in the CERN North and West experimental areas.

1.2. The CERN Pb ion experiments

Lead ions have been accelerated at CERN since 1994. The beam is typically made available for four to six weeks per year. A Pb+Pb event reconstructed in the Time Projection Chamber detectors of experiment NA49 is shown in figure 2. Many hundreds of particles emerge from such collisions. In order to make the pattern recognition possible in such a high track density environment, high granularity detectors such as Time Projection Chambers and Silicon Pixel Detectors are employed. In this section, we briefly describe the seven main experiments that have collected data at the SPS. Sketches of the layouts of a few of them are included to illustrate the variety of the experimental techniques employed.

- **NA44**: a single arm spectrometer designed for the measurement of particle spectra and correlations.
- **NA45**: an e^+e^- spectrometer designed to study the production of low mass lepton pairs.
- **NA49**: the experimental setup is sketched in figure 3. Four large Time Projection Chambers (TPC) allow the measurement and correlation of a wide range of observables, such as particle spectra and yields, two-particle correlations and event-by-event fluctuations.
- **NA50**: a dimuon spectrometer designed for the study of the production of J/ψ and high mass lepton pairs. The experimental setup is sketched in figure 4. The target is followed by the multiplicity detector and the transverse energy calorimeter, providing information on the centrality of the events, and by an absorber where all the particles except the muons are stopped. A "zero degree" hadron calorimeter for the measurement of the beam spectators is embedded in the absorber. The muons emerging from the absorber pass through a spectrometer that allows the measurement of their momentum. The spectrometer consists of a series of Multi-Wire Proportional Chambers (MWPC) placed before and after a magnet. Scintillator hodoscopes placed between the spectrometer chambers and behind an iron wall at the end of the layout provide the information used to define the dimuon trigger.
- **NA52**: a spectrometer designed to analyse the charge-to-mass ratio of the collision products in order, in particular, to search for strangelets. Strangelets are hypothetic particles: chunks of quark matter that may be stable or metastable and may be formed in heavy-ion collisions as a result of deconfinement. They would be recognisable by an anomalously low value of their Z/A ratio.
- **WA97, NA57**: these experiments are based on the use of a telescope of silicon pixel detectors to reconstruct the weak decays of strange and multi-strange hadrons produced at central rapidity. A series of multiplicity detectors sampling the charged multiplicity around midrapidity provides information on the centrality of the events. A sketch of the WA97 apparatus is shown in figure 5.
- **WA98** is a multi-purpose apparatus (figure 6) equipped for a wide range of measurements such as particle tracking and identification, hadron and electromagnetic calorimetry, nuclear fragment detection.

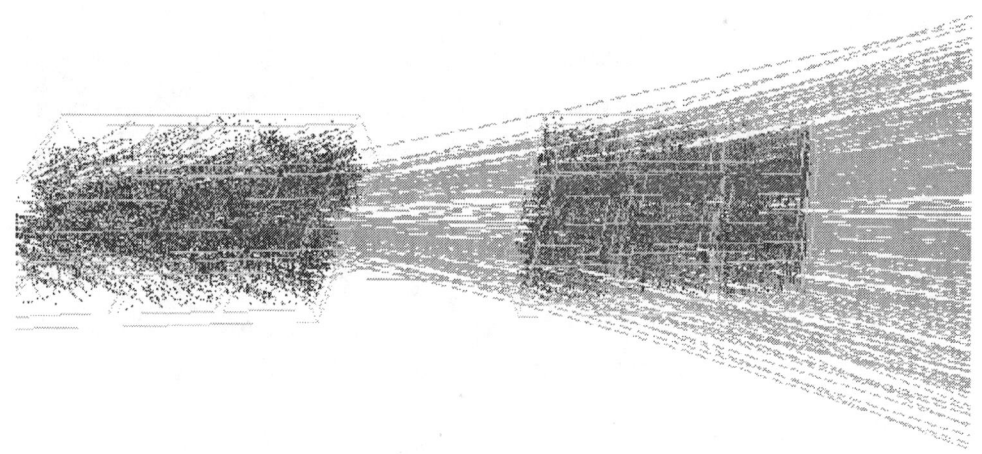

FIGURE 2. Display of a Pb+Pb collision event as detected in the NA49 TPCs.

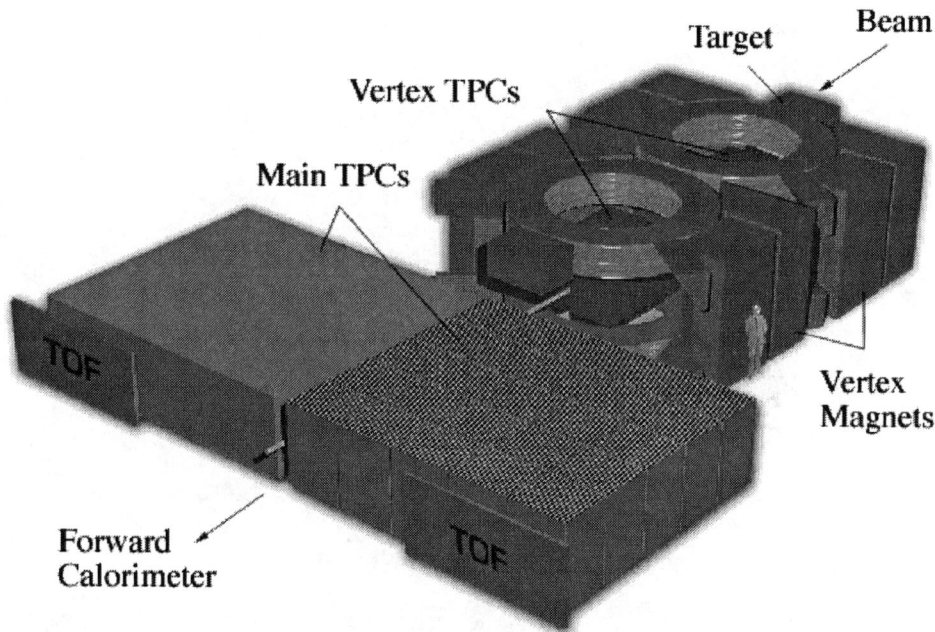

FIGURE 3. The NA49 experimental setup.

The variety of the experimental tools employed is a distinctive feature of the rich SPS lead beam programme. It allowed physicists to collect information on a wide spectrum of observables and ensured a high level of cross-check for many of the results.

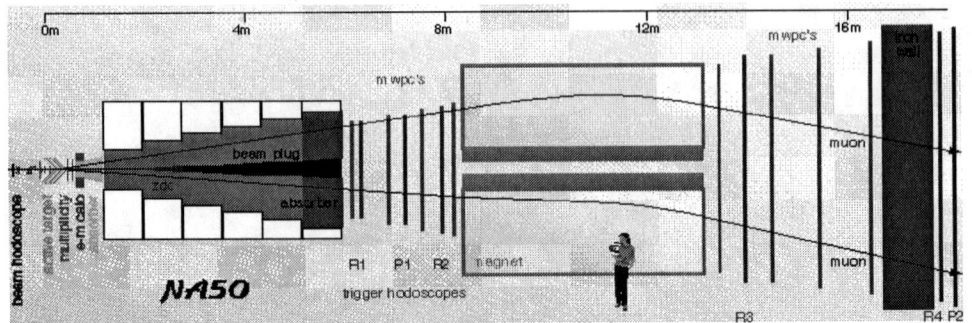

FIGURE 4. The NA50 experimental setup.

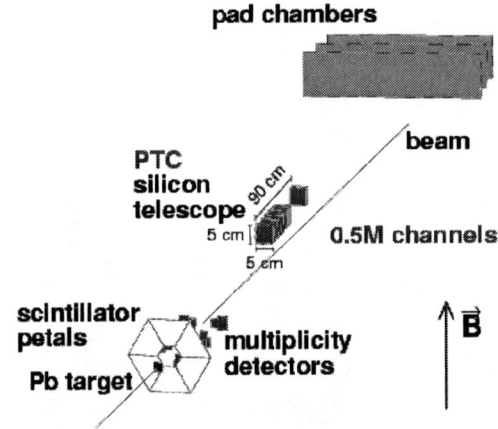

FIGURE 5. The WA97 experimental setup.

2. BASIC FEATURES OF PB+PB COLLISIONS AT THE SPS

2.1. Kinematic variables

In this section we introduce some of the variables commonly employed to describe the particles produced in high energy collisions. We take $c = 1$, and the z coordinate along the collision axis. The 4-momentum of a particle is: $p^\mu = (p^0, p^1, p^2, p^3) = (E, \vec{p}) = (E, \vec{p_t}, p_z = p_{//})$ where $\vec{p_t}$ is the transverse momentum.

The relativistic sum of two velocities along the z-axis:

$$\beta = \frac{\beta_1 + \beta_2}{1 + \beta_1 \cdot \beta_2} \tag{1}$$

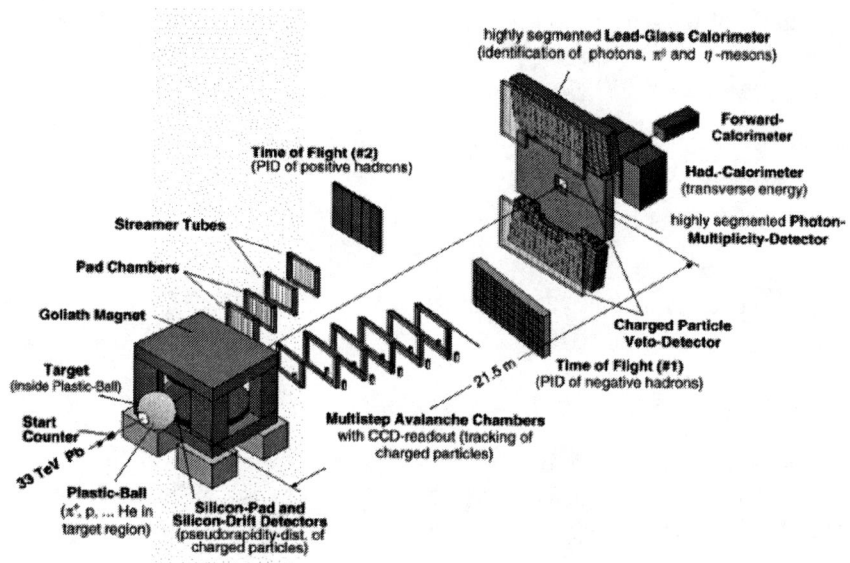

FIGURE 6. The WA98 experimental setup.

has the same form as the expression for the sum of hyperbolic tangents:

$$\tanh(y_1 + y_2) = \frac{\tanh(y_1) + \tanh(y_2)}{1 + \tanh(y_1) \cdot \tanh(y_2)}. \tag{2}$$

A convenient relativistic extension of the velocity can then be obtained by introducing a new variable y, called "rapidity", such that $\beta = \tanh(y)$. Then:

$$y(\beta) = \tanh^{-1}\beta = \frac{1}{2}\log\frac{1+\beta}{1-\beta}. \tag{3}$$

The rapidity, which reduces to the velocity β in the non-relativistic limit, has a simple additive behaviour under a Lorentz transformation along z with a velocity β:

$$y \to y' = y - y(\beta). \tag{4}$$

Rapidity distributions are therefore invariant for a boost along z: $dN/dy' = dN/dy$. It can be easily verified that:

$$y = \frac{1}{2}\log\frac{E + p_z}{E - p_z}. \tag{5}$$

In a Pb+Pb collision at a beam momentum of 158 AGeV per nucleon (i.e. at the top SPS energy for Pb beams) a particle at rest in the center-of-mass frame has a rapidity $y \simeq 3$ (this is called "mid-rapidity").

The transverse momentum p_T of a particle is defined as:

$$p_T = \sqrt{p_x^2 + p_y^2}. \tag{6}$$

When the mass of the particle is known, an equivalent variable is the transverse mass, defined as:

$$m_T = \sqrt{m^2 + p_T^2}. \tag{7}$$

Both variables are invariant under Lorentz transformations along the z-axis.

The energy E of a particle can be expressed as:

$$E = \sqrt{m_T^2 + p_z^2}. \tag{8}$$

The following useful relations can be shown to hold:

$$p_z = m_T \sinh(y) \tag{9}$$

$$E = m_T \cosh(y). \tag{10}$$

Sometimes the energy or the momentum of a particle are not known, but its emission angle θ with respect to the beam axis is measured. In this case, for high energy particles, it is possible to approximate the rapidity by the pseudorapidity η:

$$\eta = -\log\tan(\theta/2) = \frac{1}{2}\log\frac{p+p_z}{p-p_z}. \tag{11}$$

In the ultrarelativistic limit $p \approx E$ and $\eta \approx y$.

p_z and p can be expressed in terms of η and p_T as:

$$p_z = p_T \sinh(\eta) \tag{12}$$

$$p = p_T \cosh(\eta). \tag{13}$$

2.2. Collision Centrality

A generic nucleus-nucleus collision is sketched in figure 7. The nucleons outside of the overlap of the two colliding nuclei continue to move essentially undisturbed along their original trajectories. They are referred to as the "spectators". The nucleons within the collision overlap, called "participants", take part in the violent reaction that gives rise to the strongly interacting fireball that is the subject of our study. Collision centrality is a measure of how "frontal" the collision has been. Maximum centrality corresponds to maximum overlap of the two colliding nuclei. The distance of closest approach between the centers of the two colliding nuclei is called impact parameter (b, see figure 7). The collision centrality is usually expressed by indicating the value of the impact parameter in fm or the number of participants N_{part}. The integrated cross section corresponding to collisions below a certain value of the impact parameter

$$\sigma(b) = \int_0^b \frac{d\sigma}{db'} db' \tag{14}$$

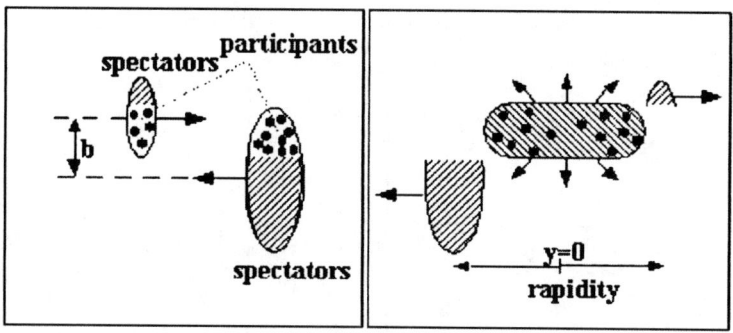

FIGURE 7. Sketch of the geometry of a nucleus-nucleus collision.

or its ratio to the total inelastic cross section $\sigma(b)/\sigma_{TOT}$, can also be used to indicate a centrality range. For instance, one can speak of "the most central 5% of collisions" referring to all the collisions with impact parameter between 0 and b such that $\sigma(b)/\sigma_{TOT}$ = 0.05. The use of the absolute value of the cross section expressed in barn is preferable to the use of the ratio, since in order to express the cross section as a fraction of the total one introduces the additional uncertainty on the value of σ_{TOT}. An often unnecessary complication.

Experimentally, the centrality is evaluated by measuring at least one of the following variables:

- N_{ch}: the number of charged particles produced in a given rapidity interval (near mid-rapidity). It increases almost linearly with N_{part}.
- the transverse energy $E_T = \sum E_i \cdot \sin\theta_i$. It is the sum over all the particles of the energy associated with the motion transversal to the collision axis. Is is also found to be practically proportional to N_{part}.
- E_{ZDC}: the energy collected in a "zero degree" hadron calorimeter. It is a measure of how many spectator nucleons kept moving along their original trajectory after the collision. It decreases linearly as N_{part} increases.

An example of centrality measurement, from experiment NA57, is shown in figure 8. The distribution of the charged particle multiplicity measured by the multiplicity detectors in the pseudorapidity range $2 \leq \eta \leq 4$ (left) is used to divide the event sample into five centrality classes. The value of the cross section corresponding to the five classes, expressed in barn, is also indicated. The distribution of the number of participants in the five bins (right) is obtained with the help of a Monte Carlo simulation.

2.3. Bjorken's formula, initial energy density

The energy deposited by the participants in the collision region manifests itself in the form of produced particles. Figure 9, from NA49, shows the rapidity distribution of

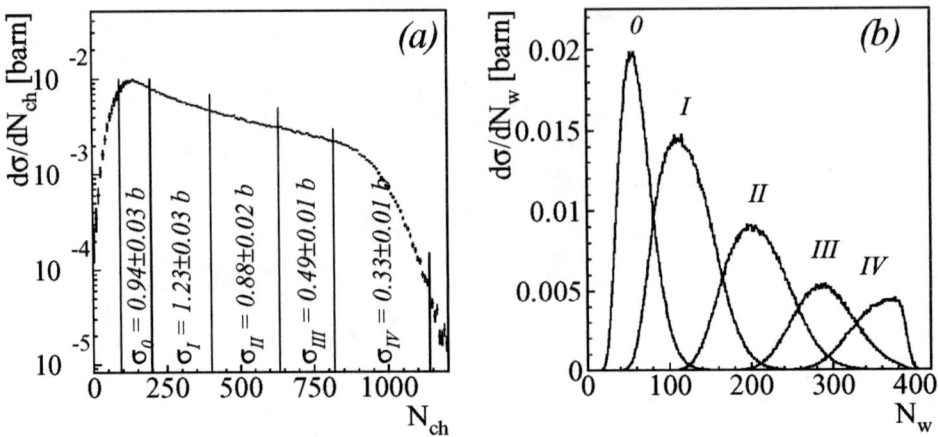

FIGURE 8. Distribution of the multiplicity in the multiplicity detector for a sample of NA57 data (*a*), subdivided into five centrality classes. Distribution of the number of participants within each class (*b*).

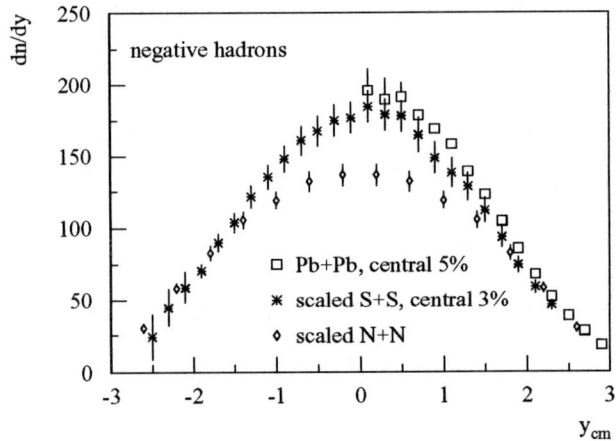

FIGURE 9. Rapidity distribution of negative particles for central Pb+Pb collisions (from NA49). Also shown are the distributions for central S+S and N+N collisions, scaled with the number of participants.

negative hadrons, for the most part pions, for a sample of 158 *A*GeV Pb+Pb collisions (open squares) corresponding to the most central 5% or the inelastic cross section. Approximately 2500 particles (of all charges) are produced in each event. The energy associated with the motion of these particles can be "projected back" in time in order to obtain an estimate of the energy density reached in the early stages of the collision.

Let's consider a thin cylindrical slab of free-streaming expanding matter contained within a thickness $\pm dz$, centered in the center-of-mass of the collision and of transverse dimension S (see figure 10). The velocity in the center-of-mass frame at the center of the slab is $v = 0$.

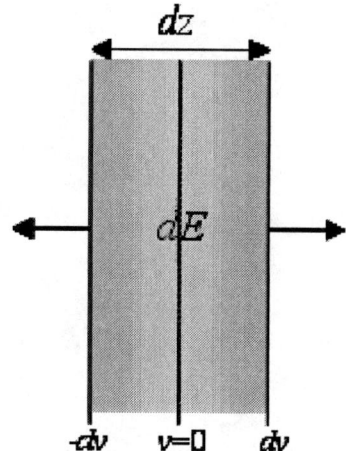

FIGURE 10. A cylindrical slab of expanding matter (see text).

At an arbitrary time τ, the matter at the edge of the slab will be moving with a velocity along the z-axis given by
$$d\beta = dz/c\tau. \tag{15}$$
The density of energy in the slab will be given by:
$$\varepsilon = dE/dV \tag{16}$$
where the volume of the slab dV can be expressed as:
$$dV = S dz = S c \tau d\beta \tag{17}$$
or, since dealing with infinitesimal velocities we are surely in the non-relativistic limit, by:
$$dV = S c \tau dy. \tag{18}$$
One can therefore write the expression:
$$\varepsilon = \frac{1}{Sc\tau} \frac{dE}{dy}\Big|_{y=0}. \tag{19}$$
known as "Bjorken's formula" [1].
Bjorken's formula can be employed to obtain an estimate of the initial energy density for central Pb+Pb collisions. Experimentally:
$$\frac{dE}{dy}\Big|_{y=0} = \frac{dE_T}{dy}\Big|_{y=0} = \langle m_T \rangle \cdot \frac{dN}{dy}\Big|_{y=0} \approx 400 \text{ GeV}. \tag{20}$$

The initial time τ_0 is usually taken to be 1fm. This is the time it takes for the energy initially stored in the field to materialize into hadrons (the so-called "formation time").

The transverse dimension S of a Pb-ion is:

$$S = \pi \cdot R_{Pb}^2 = \pi \cdot (1.2 \text{ fm} \cdot A^{1/3})^2 = \pi \cdot (1.2 \text{ fm} \cdot 208^{1/3})^2 \approx 160 \text{ fm}^2. \qquad (21)$$

In this way, one obtains an estimated value of the initial energy density of:

$$\varepsilon \approx 2.5 \text{ GeV}/\text{fm}^3. \qquad (22)$$

A more refined value, published by NA49 [2] is:

$$\varepsilon = 3.2 \pm 0.3 \text{ GeV}/\text{fm}^3. \qquad (23)$$

In principle, such values of the energy density should be sufficient for the deconfinement phase transition to occur.

2.4. Invariant cross sections

The expression for the differential cross section for particle production

$$\frac{d^3\sigma}{dp_x dp_y dp_z} \qquad (24)$$

is not Lorentz-invariant, since while dp_x and dp_y are invariant for a Lorentz boost along the z-axis, dp_z is not.

A Lorentz-invariant expression can be obtained dividing d^3p by the energy: $p_z = m_T \sinh(y)$ and $dp_z = m_T \cosh(y) dy = E dy$, hence $\frac{dp_z}{E} = dy$ which, as discussed earlier, is Lorentz-invariant.

The "invariant cross section" is then defined as:

$$E \cdot \frac{d^3\sigma}{d^3p} = \frac{d^3\sigma}{dp_x dp_y dy}. \qquad (25)$$

Using the substitution

$$p_t = \sqrt{p_x^2 + p_y^2} \qquad (26)$$

and

$$\phi = \tan^{-1} \frac{p_y}{p_x}, \qquad (27)$$

the invariant cross section can be expressed in terms of p_T, y and the polar angle ϕ:

$$E \cdot \frac{d^3\sigma}{d^3p} = \frac{d^3\sigma}{p_t dp_t d\phi dy}. \qquad (28)$$

Normally, when the distribution does not depend on ϕ, the polar angle is integrated over, and one of the following equivalent expressions is used:

$$\frac{1}{2\pi p_T} \frac{d^2\sigma}{dp_T dy} = \frac{1}{\pi} \frac{d^2\sigma}{d(p_T^2) dy} = \frac{1}{\pi} \frac{d^2\sigma}{d(m_T^2) dy} = \frac{1}{2\pi m_T} \frac{d^2\sigma}{dm_T dy}. \qquad (29)$$

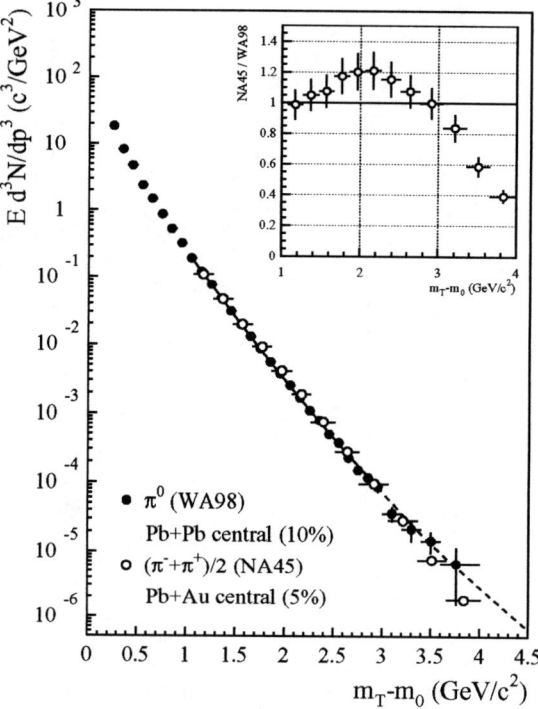

FIGURE 11. Pion transverse mass distributions (from NA45 and WA98).

2.5. Transverse Mass Distributions

The transverse mass distributions dN/dm_T are usually well fitted by thermal distributions of the form:

$$\frac{1}{m_T}\frac{dN}{dm_T} \propto \exp(-\frac{m_T}{T}) \qquad (30)$$

T is called "apparent temperature" or "inverse slope" or "m_T slope". Figure 11 shows a thermal fit to high statistics WA98 and NA45 pion m_T data.
Thermal fits for hyperons from WA97 are shown in figure 12.
The values of the m_T slopes T measured at the SPS for various species of particles are plotted in figure 13. The slopes appear to increase linearly with the particle mass, at least for pions, kaons and protons (the hyperons' behaviour seems to deviate from this trend; this is an important point which we shall discuss later).
The fireball formed in the collision expands and cools until the point where the system is so dilute that the particles stop interacting among themselves. From this moment on the momenta of the particles "freeze out" and the particles just stream out towards the detector. The expansion causes a collective motion of the escaping particles. This collective motion is called "transverse flow". It adds to the thermal motion and modifies the m_T spectra. The apparent temperature is a function of the freeze-out temperature T_F

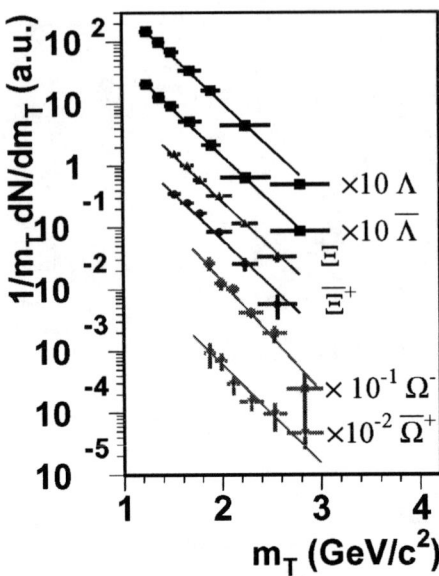

FIGURE 12. Hyperon transverse mass distributions (from WA97).

and the mean transverse flow velocity $\langle \beta_T \rangle$. For particles with $m_T < 2m$ the apparent temperature can be written as:

$$T \approx T_F + \frac{1}{2}m\langle \beta_T \rangle^2 \tag{31}$$

which indeed predicts a linear dependence of the inverse slope on the particle mass (although the pions are perhaps a bit too light for the formula to be valid).

In principle, such measurements should allow us to extract information on the freeze-out temperature and the flow velocity. In practice, however, it is hard to disentangle the thermal and flow contributions relying only on the transverse slope information. Extra information can however be obtained from HBT interferometry.

2.6. HBT interferometry

Hanbury-Brown and Twiss (HBT) interferometry is a vast and rich subject both theoretically and experimentally. Here, we just give a brief introduction concentrating on the information relative to the freeze-out conditions. In 1956 R. Hanbury-Brown and R.Q. Twiss measured the size of a star using the Bose-Einstein correlation between pairs of photons [3]. A similar method was first applied in particle physics in 1960 by G. Goldhaber, S. Goldhaber, W. Lee and A. Pais. They studied the correlations of pairs of pions in proton-antiproton collisions [4]. The correlation function for two particles with four-momenta k_1 and k_2 is defined as the ratio between the conditional probability $p(k_1, k_2)$ of detecting the pair in coincidence and the product of the individual

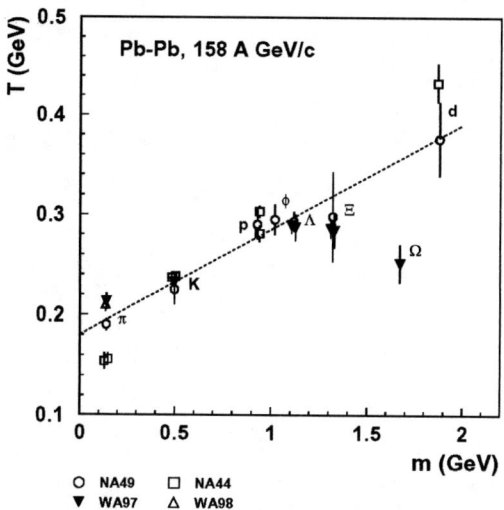

FIGURE 13. Dependence of the m_T slope on the particle mass. The line indicates a linear dependence (see text).

probabilities $p(k_1)$ and $p(k_2)$. It can be shown that the correlation function of a pair of identical bosons emitted from an extended source depends on the four-momentum difference $q = k_1 - k_2$ and on the source size four-vector R as follows:

$$C(q) = \frac{p(k_1, k_2)}{p(k_1)p(k_2)} = 1 + \lambda \exp\{-(q \cdot R)^2\} \quad (32)$$

where λ is a chaoticity parameter ranging from 0 to 1 (the effect disappears for completely coherent sources). A detailed derivation can be found in [17].

The experimental information on the HBT source size provides a relation between the transverse flow velocity β_T and the freeze-out temperature T_F. This information can be combined with that coming from the particle spectra to obtain an estimate of the freeze-out parameters. This is shown in figure 14: the band indicated by h^- contains the set of (T_F, β_T) values compatible with the measured m_T distribution of negative particles (mostly π^-), while the $2\pi - BE$ band indicates the constraint coming from the two-pion HBT measurement [5]. The intersection identifies an allowed region for the freeze-out parameters where:

$$T_F \approx 120 \text{ MeV} \quad \beta_T \approx 0.55. \quad (33)$$

2.7. Summary

In Pb+Pb collisions at the SPS we initially create a fireball with an energy density $\varepsilon \approx 3$ GeV/fm^3. In principle this is enough for deconfinement to occur.

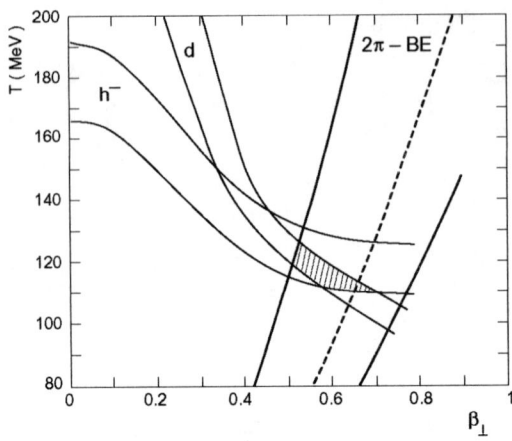

FIGURE 14. Constraints on the relation between freeze-out temperature and transverse flow velocity from the HBT radii ($2\pi - BE$) and the inverse slope for negative particles (h^-). An additional constraint coming from the deuteron data is also shown (d). (Data from NA49)

The fireball expands and cools down until it finally "freezes out" at a temperature of about 120 MeV. At this point it is "exploding" at about one-half of the speed of light. This indicates that a large pressure has been generated inside the fireball - by which mechanism? In order to explore what happens during the lifetime of the fireball, we shall now turn to two more specific observables: charmonium and strangeness.

3. QUARK GLUON PLASMA SIGNATURES

3.1. Charmonium suppression

The bound system of a $Q\bar{Q}$ heavy quark-antiquark pair can essentially be treated as a non-relativistic fermion-antifermion bound system. Such systems are known as "quarkonia". The bound system formed by a $c\bar{c}$ pair is called "charmonium", while the bound system formed by a $b\bar{b}$ pair is known as "bottonium". Quarkonia display a rich spectrum of narrow excited states, as shown in figures 15 and 16 [13].

The suppression of the production of charmonium states as a signature of deconfinement was suggested by T. Matsui and H. Satz in 1986 ([6]).

In strong interactions, heavy quarks are produced in pairs: a c quark must be produced together with a $\bar{c}$ antiquark, due to flavour conservation. Usually, the quark and the antiquark fly out in different directions and eventually end up in particles containing a single c or $\bar{c}$ quark, such as D mesons. In a small fraction of cases however, before the heavy quark and antiquark separate colour interaction causes them to bind into a quarkonium state such as the $c\bar{c}$ J/ψ hadron.

In the plasma phase, where colour charges are liberated, the colour interaction potential between two quarks is expected to be screened for distances beyond a screening

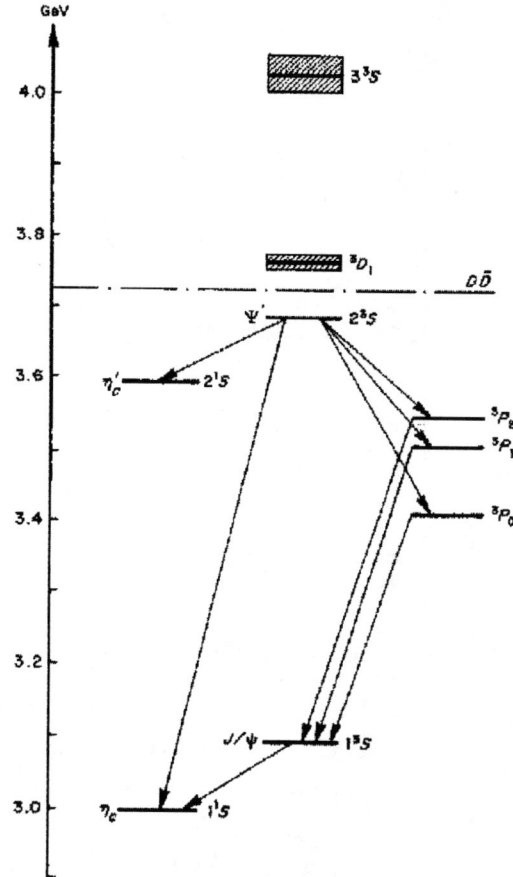

FIGURE 15. The $c\bar{c}$ spectrum.

length λ_D ("Debye length"), similarly to what happens in the case of the Debye screening in an electromagnetic plasma. Hadrons with a radius $r > \lambda_D$ will therefore not bind in the QGP phase. In particular, charmonium ($c\bar{c}$) and bottomium ($b\bar{b}$) states with a radius $r > \lambda_d$ cannot be produced at this stage: the Q and $\overline{Q}$ quarks lose their correlation and follow independent trajectories. What makes quarkonium states particularly interesting is that heavy quarks can only be produced at the onset of the collision, in a high energy interaction between two energetic partons. Later, when the energy is redistributed among the many hundreds of produced particles, no collision in the system can involve enough energy to allow for heavy quark production any more. Therefore, as long as the $Q\overline{Q}$ yield per event is so low that the probability of later combining an uncorrelated $Q\overline{Q}$ pair at the hadronization stage is negligible, as it is at SPS energies, the only chance of producing a bound quarkonium state would be shortly after the pair is produced. Debye screening prevents this, leading to a suppression in the yield of quarkonia in the final state.

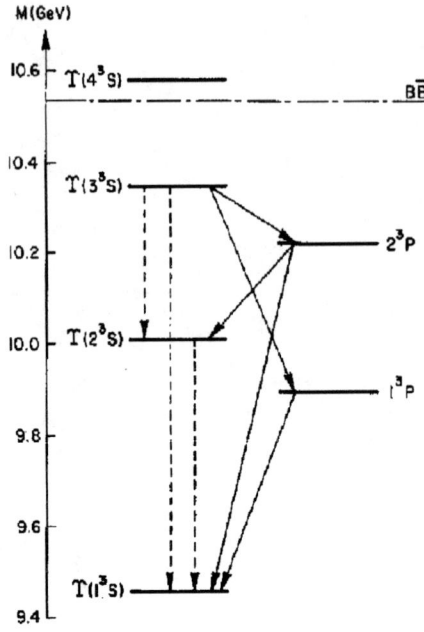

FIGURE 16. The $b\bar{b}$ spectrum.

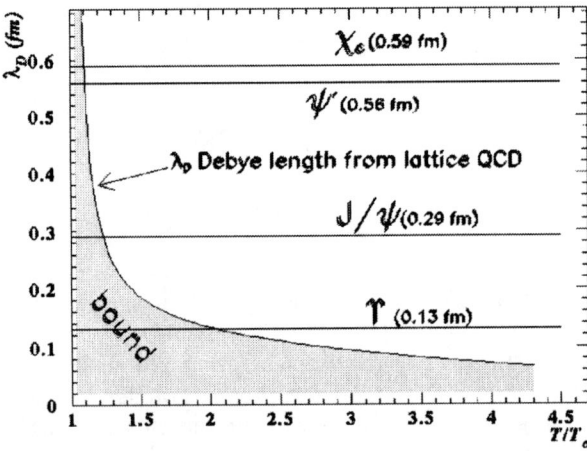

FIGURE 17. Dependence of the Debye length on the temperature (expressed in units of the critical temperature T_c). The values of the radii of a few quarkonium states are also shown.

The value of the screening length λ_D, and therefore which quarkonium states will be suppressed, depends on the value of the temperature. This dependence is shown in figure 17, together with the values of the radii of some of the most frequent quarkonium states.

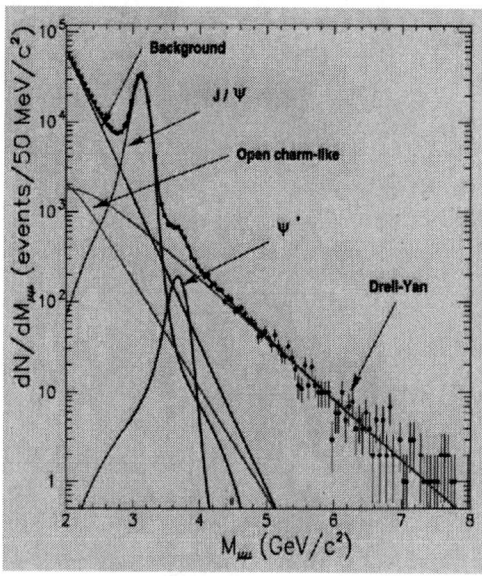

FIGURE 18. The dimuon mass spectrum (NA50).

Due to their large mass, bottonium states are very difficult to produce at SPS energies. The experimental efforts have therefore so far been concentrated on the study of charmonium production. Experiment NA50 was designed to study the production of J/ψ particles at the SPS. The apparatus, sketched in figure 4, is concisely described in section 1. The experimental technique consists in absorbing all the charged particles produced in the collision except for the muons. Dimuon combinations emerging from the absorber can then be very cleanly reconstructed. J/ψ particles can be detected by reconstructing the decays $J/\psi \longrightarrow \mu^+\mu^-$ which have a branching ratio of about 5.9 %. The total four-momentum p^μ of a system of two particles, such as a dilepton, is constructed as $p^\mu = p_1^\mu + p_2^\mu$. The invariant mass M of the combination is obtained by calculating the norm:

$$p^\mu p_\mu = E_1 E_2 - \vec{p_1} \cdot \vec{p_2} = M^2. \tag{34}$$

For a dimuon pair coming from a J/ψ decay: $M \simeq M(J/\psi) \simeq 3.1$ GeV.

Charmonium production is normalised to the production of dimuon pairs by quark - antiquark annihilation to a virtual photon (the Drell-Yan process). The production of Drell-Yan pairs with an invariant mass in the J/ψ mass range is another process which can only take place in the early parton-parton collisions. Unlike J/ψ production, however, Drell-Yan dilepton production is an electromagnetic process not influenced by strong interaction. It therefore provides a suitable reference. Like-sign ($\mu^+\mu^+$ and $\mu^-\mu^-$) dimuon combinations are used to estimate and subtract the combinatorial background. The remaining unlike-sign ($\mu^+\mu^-$)dimuon invariant mass spectrum is fitted to a cocktail of sources in order to extract the amount of J/ψ, ψ' and Drell-Yan contributions. Such a spectrum is shown in figure 18.

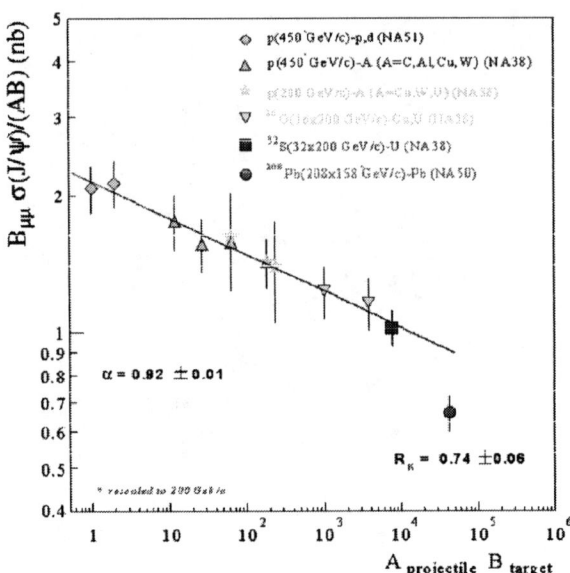

FIGURE 19. Data on J/ψ suppression for different collision systems (see text).

In the earlier rounds of experiments with lighter ions at the SPS it was understood that a "normal" suppression of the production of J/ψ is present already in proton-nucleus and light ion-ion collisions. This effect is attributed to the absorption of J/ψ particles in the nuclear matter they traverse to emerge from the collision. This is shown in figure 19, where the J/ψ cross section multiplied by the branching ratio into $\mu^+\mu^-$ and divided by the product AB of the mass numbers of the colliding nuclei is plotted versus AB. The nuclear absorption curve is a straight line in the log-log plot. While the results for light ion collisions follow this curve, the Pb+Pb point falls significantly below. This effect is referred to as "anomalous J/ψ suppression". The effect can be studied as a function of the event centrality within the sample of Pb+Pb collisions. This is shown in figure 20, where the centrality is gauged by the value of the transverse energy measured in the NA50 electromagnetic calorimeter and the J/ψ dimuon cross section is normalized to the cross section for Drell-Yan dimuon production. The data from three NA50 data sets are shown. The curve indicates the nuclear absorption prediction. While for peripheral collisions the data are close to the nuclear absorption curve, they clearly deviate from it for more central collisions, the deviation increasing with increasing collision centrality.

The results are summarized in figure 21, where the ratio of the measured J/ψ suppression to the suppression expected by nuclear absorption is plotted versus an estimate of the energy density reached in the collision for various collision systems. The anomalous suppression sets in at a energy density around 2.3 GeV/fm^3 (corresponding to an impact parameter $b \sim 8$ fm) and seems to accelerate at an energy density around 3 GeV/fm^3 ($b \sim 3.6$ fm). This pattern has been interpreted as due to the successive melting, as the energy density is increased, first of the χ_c and then of the J/ψ itself. (The χ_c is itself a charmonium state. A non-negligible fraction of the J/ψ are produced in the decay of a

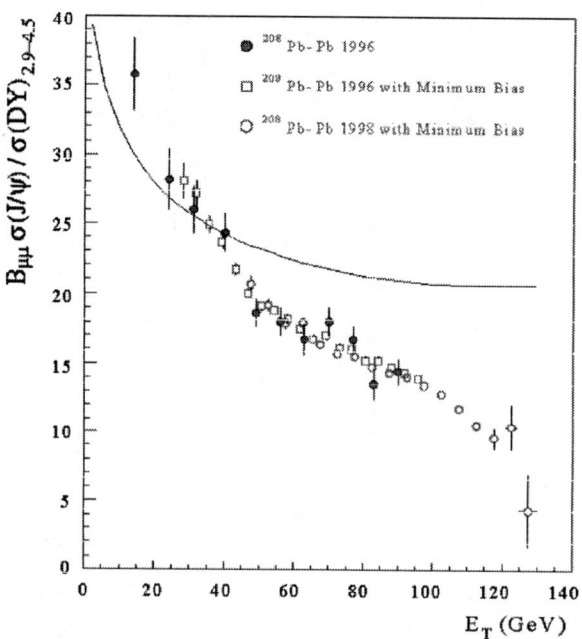

FIGURE 20. Centrality dependence of the J/ψ suppression within the Pb+Pb system (NA50, see text).

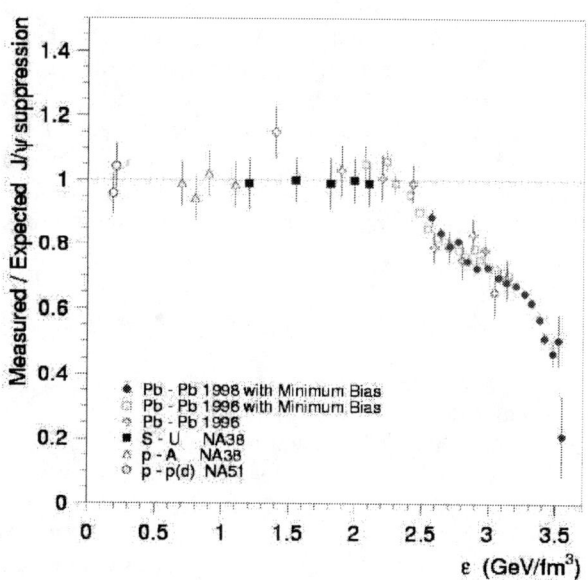

FIGURE 21. Measured/expected (from nuclear absorption) J/ψ suppression for different collision systems as a function of the estimated energy density of the collisions.

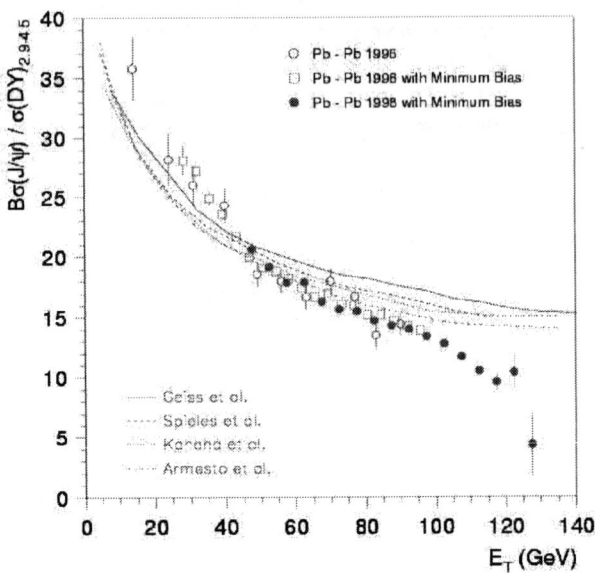

FIGURE 22. Centrality dependence of the J/ψ suppression within the Pb+Pb system: comparison with models (see text).

parent χ_c particle to $J/\psi + \gamma$. The χ_c is expected to melt at a lower value of the energy density than the J/ψ, due to its larger radius, see figure 17.)

Many theoretical efforts have been made to try and reproduce the observed J/ψ suppression without invoking deconfinement. The basic idea is that additional dissociation of J/ψ particles might occur due to their final state interactions with produced hadrons emerging from the collision together with the J/ψ (the so-called "comovers"). Some such attempts are shown in figure 22. The models are in better agreement with the data than the nuclear absorption model, but it has so far not been possible to find a purely hadronic explanation for the J/ψ suppression in the most central collisions.

3.2. Strangeness Enhancement

The enhancement of the production of strange particles as a QGP signature has been proposed by J. Rafelski and B. Müller in 1982 ([7]).

When confined inside hadrons, quarks acquire an additional contribution to the mass due to the confining effects of the strong interaction. The mass of a confined quark is called its "constituent" mass. In the deconfined phase the masses of the quarks are expected to recover their bare values (the values which appear in the Lagrangean). This effect is often referred to as "partial restoration of the chiral symmetry" (due to chiral symmetry fermions and antifermions have opposite helicity; the symmetry is exact only for massless particles, therefore its restoration here is only partial). The mass of the strange quark in the QGP is therefore expected to be of the order of 150 MeV (the bare value), to be compared with a constituent value of the strange quark mass of about 500 MeV. With an estimated value of the critical deconfinement temperature T_c in the 100 - 200 MeV range, the strange quark is expected to be particularly sensitive to chiral symmetry restoration. If the system goes through a deconfined phase, where the value of the effective mass of the s quark is comparable with the temperature, we expect $s\bar{s}$ pairs to be copiously produced, mainly via gluon-gluon fusion. If in addition the system is baryon-rich, i.e. if there is a sizeable excess of quarks over antiquarks in the central rapidity region (as is the case at the SPS), the production of strange quarks and antiquarks can be further enhanced due to the Pauli principle. This effect is called "Pauli blocking of light quark production". In such a situation, the lowest-lying energy levels for u and d quarks are filled. The production of an $s\bar{s}$ pair may be favoured over the production of a $u\bar{u}$ or $d\bar{d}$ pair, if the energy of the lowest lying free level for a light quark is comparable with the energy required for the production of the $s\bar{s}$ pair (for instance: for a system with a baryon density of the order of that of the nuclei, the Fermi momentum for a light quark is estimated to be in the 200 MeV range).

As the QGP cools down, eventually the quarks recombine into hadrons. The abundance of strange hadrons in the final state should also be enhanced. The enhancement effect is expected to be larger for particles of higher strangeness content. For instance, the production of the Ω^- (sss, $|S|=3$) should be more enhanced than the production of the Ξ^- (ssd, $|S|=2$) which should in turn be more enhanced than that of the Λ (sud, $|S|=1$).

If a relatively long-lived strongly interacting hadronic fireball is formed in the collision, a certain amount of enhancement of the abundance of strange particles could be expected even in the absence of QGP. Particles such as kaons and lambdas should be relatively easy to produce in such a system via reactions like: $\pi + \pi \longrightarrow K + \overline{K}$; $\pi + N \longrightarrow \Lambda + K$. Such processes are indeed expected to be relatively easy (i.e. fast on the collision timescale) for $|S|=1$ particles. They are however estimated to be progressively harder (i.e. slower on the collision timescale) for particles of higher and higher strangeness. In this case, one would expect to see a larger enhancement effect for the $|S|=1$ particles and a progressively smaller effect for higher strangeness states: an opposite pattern to that expected from QGP enhancement. The production of hadrons with multiple strangeness such as Ξ^- and Ω^- is therefore expected to be particularly sensitive to deconfinement.

Experiments WA97 and NA57 have been designed to study the production of strange and multi-strange particles in Pb+Pb collisions at the SPS. The apparatus of WA97 is sketched in figure 5 and was concisely described in section 1. The experimental technique consists in employing a high granularity silicon pixel tracker to detect strange particles produced around central rapidity by reconstructing their weak decay topologies.

The main detection channels for strange particles via reconstruction of weak decays are (B.R. = Branching Ratio):

$K_S^0 \to \pi^+\pi^-$ ($c\tau \simeq 2.68$ cm, B.R. $\simeq 69\%$)
$\Lambda \to p\pi^-$ ($c\tau \simeq 7.89$ cm, B.R. $\simeq 64\%$)
$\Xi^- \to \Lambda\pi^-$ ($c\tau \simeq 4.9$ cm, B.R. $\simeq 100\%$)
$\Omega^- \to \Lambda K^-$ ($c\tau \simeq 2.46$ cm, B.R. $\simeq 68\%$)

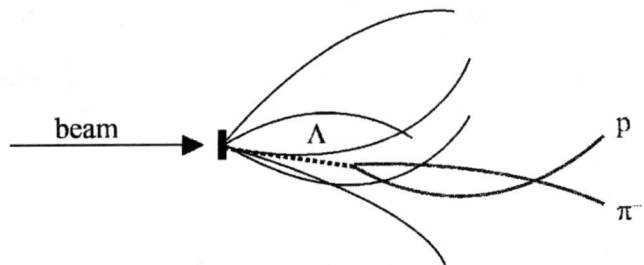

FIGURE 23. Sketch of a $\Lambda \longrightarrow p\pi^-$ decay.

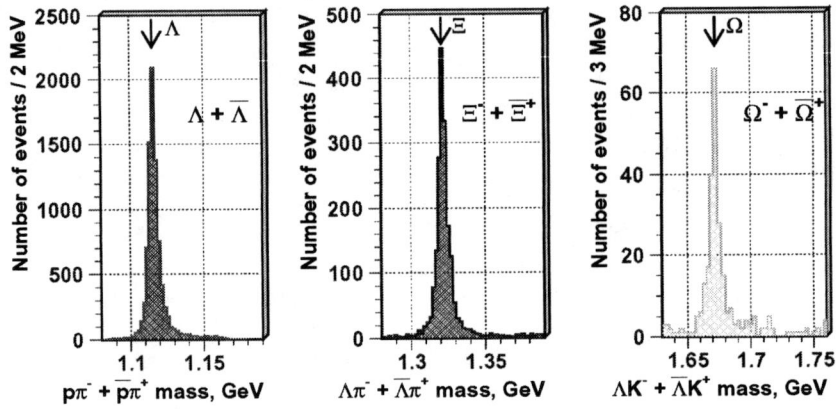

FIGURE 24. Sample Λ, Ξ^- and Ω^- invariant mass peaks (WA97).

The decay of a Λ baryon is schematically shown in figure 23. Weakly decaying hyperons produced at central rapidity at the top SPS energy have decay distances of tens of centimeters. This allows to separate the secondary decay products from the much more numerous primary collision products, drastically reducing the combinatorial backgrounds.

After suitable geometrical and kinematical cuts are applied, the hyperon signals emerge very clearly in the invariant mass plots, with very little background (see figure 24).
WA97 collected data from p+Be, p+Pb and Pb+Pb collisions at a beam momentum of 158 AGeV at the SPS. The Pb+Pb data have been divided into four centrality classes according to the multiplicity of charged particles (N_{ch}) measured in a Silicon Microstrip centrality detector. The multiplicity distribution is shown in figure 25 (left), together with the definition of the four classes and the result of a model fit. The data is in good agreement with the fit down to a multiplicity in the centrality detector of about 300, below which the events are suppressed by the experiment's centrality trigger. The model allows to estimate the distribution of the number of nucleons participating in the initial Pb+Pb collision within each class (N_{part}, figure 25, right). The average number

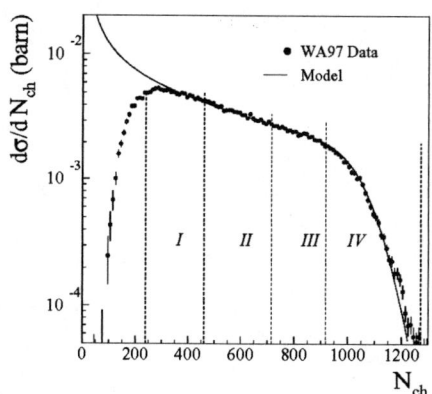 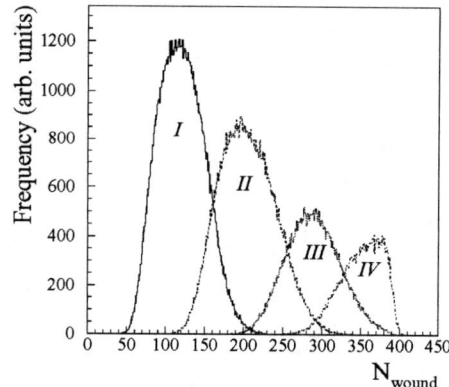

FIGURE 25. Distribution of the multiplicity in the multiplicity detector for a sample of WA97 data (left), subdivided into four centrality classes. Distribution of the number of participants within each class (right) (see text).

of participants in the four Pb+Pb centrality classes are respectively 120, 205, 289 and 351. The average number of participants in p+Be and p+Pb collisions are estimated to be respectively 2.50 and 4.75.

The yield (multiplicity per event) per unit rapidity of the various particles is determined by integrating the invariant cross section over the full p_T range and over one unit of central rapidity centered at mid-rapidity ($y_{CM} = 0$):

$$N = \frac{1}{\sigma_{TOT}} \int_{y_{CM}=-0.5}^{y_{CM}=+0.5} dy \int_0^\infty dp_T^2 \frac{d^3\sigma}{E d^3 p}. \quad (35)$$

The result is shown in figure 26, as a function of the number of participants for p+Be, p+Pb and the four centrality classes in Pb+Pb collisions. The plot on the left contains particles which can have a valence quark in common with the nucleon, while the plot on the right contains particles which have no valence quark in common with the nucleons. The two sets are kept separate since it is known that they may have different behaviour in their rapidity distributions. The relative abundances measured in Pb+Pb collisions are close to the the thermodynamical (chemical) equilibrium (i.e. maximum entropy) values corresponding to a chemical freeze-out temperature of about 170 MeV. (The chemical freeze-out temperature should not be confused with the kinetic freeze-out temperature discussed in section 1. At chemical freeze-out inelastic interactions stop and the final particle composition is frozen. Elastic collision among particles, however, can still affect the particles' momentum distributions until the final, kinetic, freeze-out when all interactions cease). In other words, the system behaves as if the particle abundances at hadronization were determined by the statistical recombination of a soup of uncorrelated quarks. This is displayed in figure 27 which shows a collection of particle ratios measured at the SPS, including those involving multi-strange hyperons, together with "thermal model" predictions from [8]).

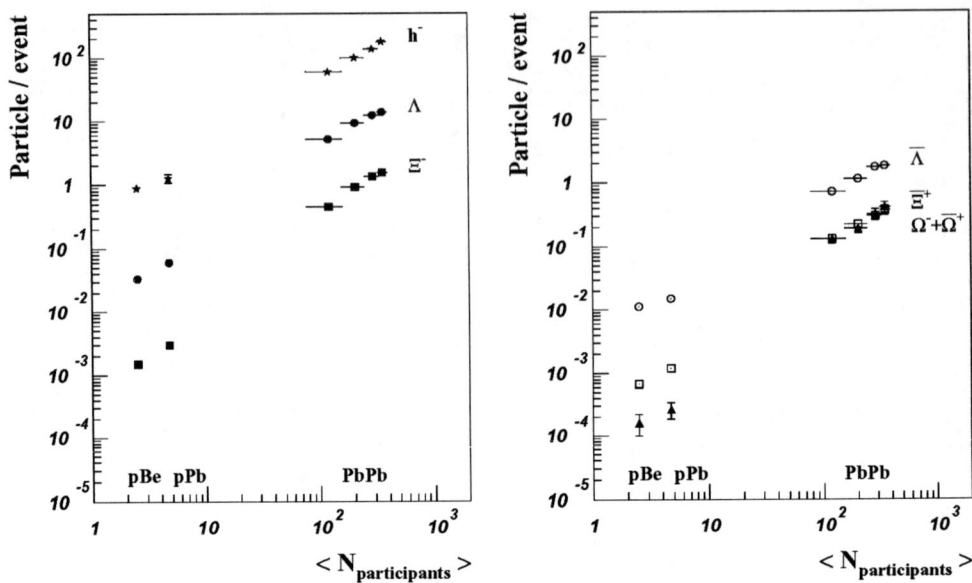

FIGURE 26. Yields per unit of rapidity at central rapidity as a function of the number of participants for negative particles, Λ and Ξ^- (left) and for $\overline{\Lambda}$, $\overline{\Xi}^+$ and $\Omega^- + \overline{\Omega}^+$ (right) (WA97).

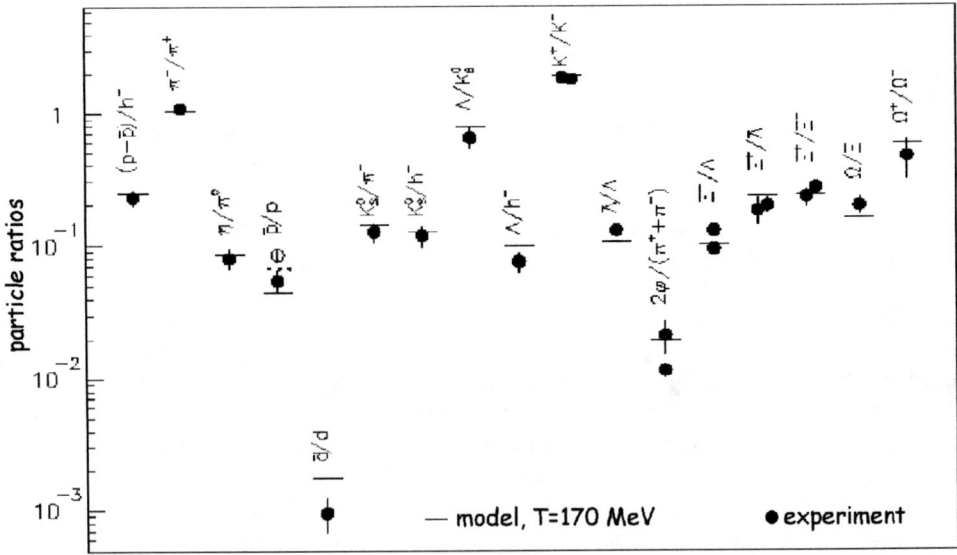

FIGURE 27. Compilation of SPS Pb-Pb particle ratios (data points) compared with thermal model predictions (horizontal lines) for a temperature of 170 MeV (see: [8]).

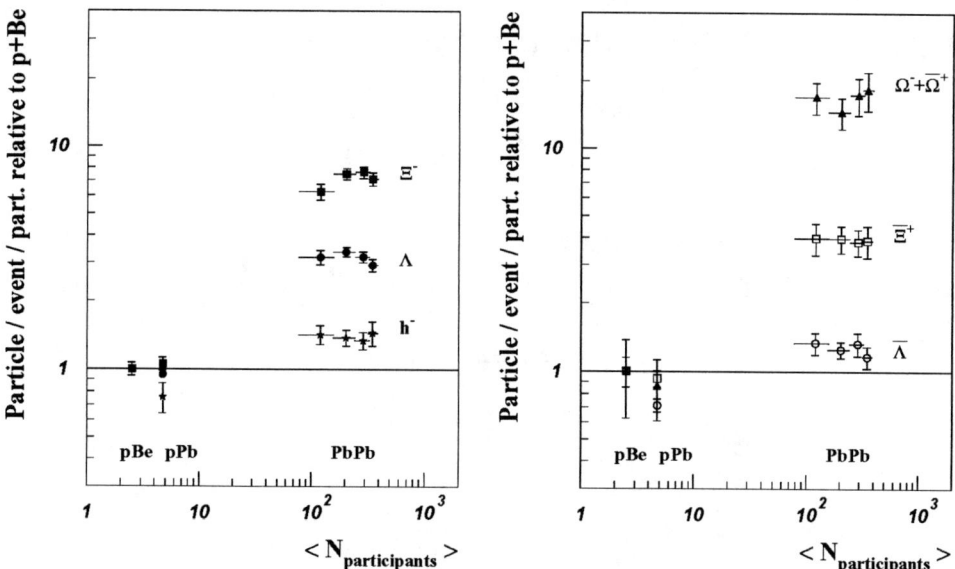

FIGURE 28. Yields per unit rapidity (same as in figure 26) per participant relative to the p+Be yields (WA97).

Figure 28 shows the yields per participant measured by WA97, rescaled so that the p+Be yields are equal to one, plotted versus the number of participants. If the yields per participant were to remain the same from p+Be collisions all the way to central Pb+Pb collisions, they would lie on the solid line. As can be seen, all yields, including those of negative particles (h^-, mostly π^-) are enhanced. The enhancement increases with the strangeness content of the particles up to an enhancement of about a factor 15 for the $|S| = 3$ ($\Omega^- + \overline{\Omega}^+$). As discussed above, this is the pattern expected in case of QGP formation.

As we saw in section 2, multi-strange hyperons, particularly the Ω^-, deviate from the linear increase of the apparent temperature with particle mass attributed to the collective transverse expansion of the system. This anomalous behaviour has been reproduced by transport calculations as due to the low interaction cross section of the Ω^- particles in the hadronic fireball, due to the absence of resonances of the Ω^- with the pion: after hadronization, the Ω^- particles decouple very early from the rest of the hadronic soup, and are therefore less affected by the collective flow. Obviously, such a picture yields further support to the conclusion that the large observed Ω abundance is determined very early in the history of the collision and cannot be the result of hadronic reinteractions.

Some novel hadronic mechanisms have been proposed, such as string merging (into so-called "string ropes"), that, in conjunction with large values of the baryon stopping (i.e. with a large transfer of the participants' baryon number into the mid-rapidity region), can yield to an enhanced production of multistrange hyperons [9]. No such model has however succeeded in reproducing the full set of WA97 observations, in particular the large value of the Ω^- enhancement.

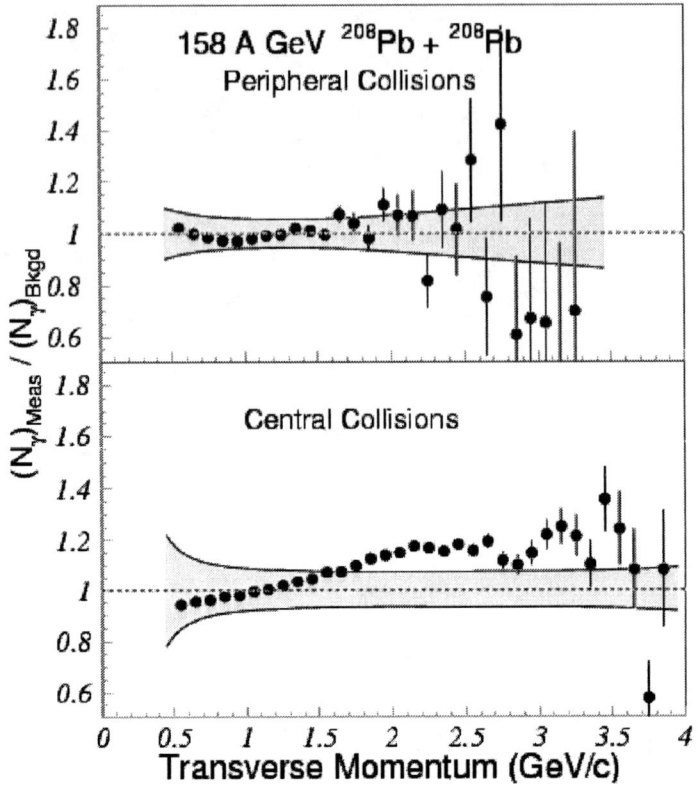

FIGURE 29. Excess of photons over the estimated background from hadron decays for peripheral (top) and central (bottom) collisons (WA98).

3.3. Electromagnetic Signatures

A direct signal from the QGP phase would be provided by the detection of electromagnetic radiation from the hot initial phases of the collision. This radiation should manifest itself as real photons (in the form γ rays) or virtual photons (in the form of dilepton l^+l^- pairs) emitted with a thermal spectrum of the form $\exp(-E_\gamma/T)$. The search for the γ radiation is experimentally very hard, due to the huge background of photons from hadron (mainly π^0 and η) decays. Experiment WA98 is equipped with high quality electromagnetic calorimetry which enabled the collaboration to undertake the daunting task of reconstructing the original π^0 and η spectra, estimating the single photon spectrum from the decay of these particles and subtracting it from the measured photon spectrum in order to search for an excess of "prompt" photons from the electromagnetic radiation. The result is shown in figure 29 for peripheral (top) and central (bottom) collisions as a function of the photon transverse momentum. The grey bands indicate the uncertainty on the γ signal expected from hadron decays. The data indicate a possible excess in central events for transverse momenta above 1.5 - 2 GeV.

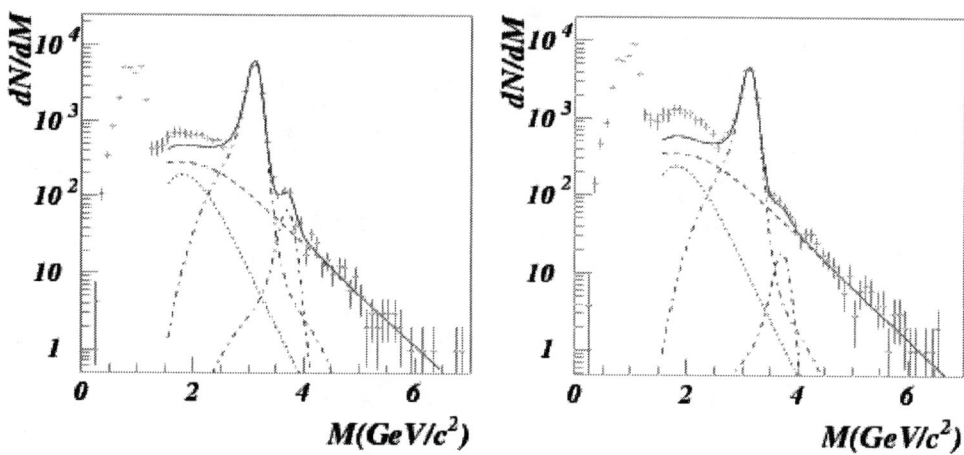

FIGURE 30. Dimuon invariant mass spectra from NA50. The solid line indicates the extrapolation of known dimuon sources from p-A collisions for peripheral (left) and central (right) collisions. An excess is visible in the 2 GeV mass region.

NA50 has observed an excess of dimuons over the expected sources in the dimuon mass region below the J/ψ. The effect, shown in figure 30, is especially pronounced for central events. Various explanations have been proposed, including thermal radiation of virtual photons materializing as thermal dileptons and excess production of pairs of charmed hadrons which are a major source of dimuons in this mass region due to events in which both charm hadrons undergo a semileptonic decay with emission of a muon. In order to investigate the latter possibility, a new collaboration, NA60, is planning to upgrade the NA50 dimuon spectrometer with the addition of a Silicon Pixel vertex detector. Such a detector would measure the tracks close to the primary interaction vertex with enough precision to allow for the separation from the primary vertex of the tracks emerging from the weak decays of charm particles and would therefore allow to test the hypothesis of an excess production of charm hadrons.

The dielectron experiment NA45 has measured the dilepton invariant mass spectrum at lower values of the mass, compared to NA50. The dielectron spectrum measured in p+Au collisions is shown in figure 31. The data are well described by a superposition of expected sources. The Pb+Au data shown in figure 32, however, show an excess of dielectrons over the expected sources in the region between the low mass peak (due to e^+e^- pairs emerging from the "Dalitz decay" $\pi^0 \longrightarrow \gamma e^+e^-$) and the peak from the ρ^0 and ω decays to e^+e^- pairs which in fact has practically disappeared. This effect has been interpreted as being due to a smearing of the spectral function of the ρ meson, an effect connected with the partial restoration of chiral symmetry expected in the QGP phase (discussed above in the strangeness enhancement section). Alternative explanations are also being discussed. The NA45 apparatus has been upgraded with a TPC. This addition should improve the experiment's invariant mass resolution and provide more precise data points for comparison with models.

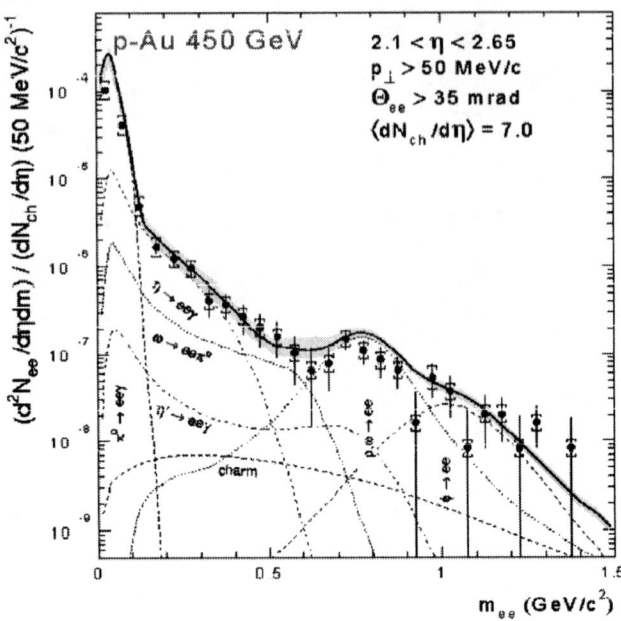

FIGURE 31. p+Au dielectron mass spectrum from NA45. The expectation from the superposition of known sources is indicated.

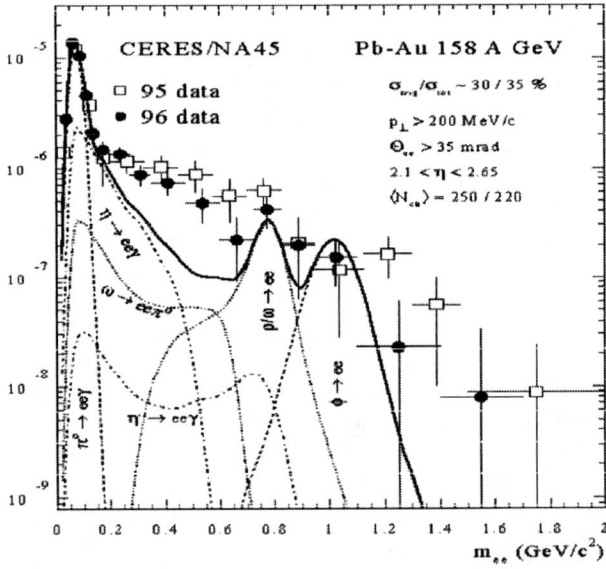

FIGURE 32. Pb+Au dielectron mass spectrum from two NA45 data sets. The expectation from the superposition of known sources is indicated. An excess is visible in the region between the π^0 Dalitz and the $\rho - \omega$ peaks.

4. SUMMARY AND CONCLUSIONS

In Pb+Pb collisions at the SPS energy we create a dense strongly interacting fireball with an estimated initial energy density of the order of 3 GeV/fm^3. Theory predicts that such a system should be in a deconfined state. From thermal analyses of the particle abundances we see that the particle composition is close to that expected from maximum entropy at a temperature $T \simeq 170$ MeV. This suggests statistical hadronization followed very closely by chemical freeze-out when the expanding system has cooled down to a temperature around that value, corresponding to an energy density of about 1 GeV/fm^3. At this point, the chemical composition of the system is frozen; the production of strange particles is enhanced - the enhancement increasing with the strangeness content - and the J/ψ production has been suppressed. The two "historic" deconfinement signatures are there. The information extracted from transverse mass spectra and HBT correlations suggests that elastic interactions cease when the system has cooled down to about 120 MeV (kinetic freeze-out) [5, 10]. An alternative analysis suggests that the chemical and kinetic freeze-out may even coincide at an intermediate temperature of about 140 MeV [11], indicating a sudden breakup of the fireball at hadronization, without much further elastic rescattering either. At freeze-out, the system is expanding at over one-half of the speed of light. This indicates that a large buildup of pressure has occurred, as could be expected if the partonic degrees of freedom had become active as a result of a deconfinement phase transition. The mass spectrum of dilepton pairs emitted by the system also shows a few anomalous features.

What can we conclude about the state of the system before chemical freeze-out? The system is in a state of very large energy density, at which the very concept of individual, separated hadrons is not too meaningful. The properties of this state are not explained in terms of those of a conventional system of strongly interacting hadrons: we seem to have found a new regime for strongly interacting systems which certainly has to be further investigated. The state exhibits many of the predicted features of QGP, in particular the expected effects of colour deconfinement (J/ψ suppression) and chiral symmetry restoration (strangeness enhancement) are there. The evidence for deconfinement, however, is still rather indirect. In particular, there is no clear evidence of thermal QGP radiation. It has been suggested that exotica such as free quarks or strangelets may be produced following a deconfinement phase transition. None of these have been observed. The results from the seven SPS collaborations where summarized in a special seminar which was held at CERN on February 10, 2000 [20].

There are still a number of open points. How do the observed effects depend on the beam energy? Is the dependence the same for heavier and lighter colliding systems? How do the various hyperon enhancements depend on the number of participating nucleons below the centrality range accessible to WA97? Is it possible to pinpoint the transition? What is the origin of the deviations of the dilepton mass spectra from what we would expect by extrapolating the known sources from p+A?

Some of these questions are still being addressed at the SPS, with data already on tape or to be soon collected. Others will be answered as we increase the interaction energy and approach the expected "deep deconfinement" region at RHIC and LHC. What will be the hadronization temperature at higher energies? How will the values of the hyperon enhancements change at higher energy? Will we finally be able to observe thermal photon radiation? Will the J/ψ still be suppressed? How will the J/ψ suppression pattern look like? A new window will open as heavy quarks will enter the stage. What surprises wait for us there?

Certainly we have some exciting times ahead to look forward to. A lot remains to be learned in the study of condensed strongly interacting matter. The field is much richer than just the search for the Quark-Gluon Plasma, and as we learn more, the Quark-Gluon Plasma itself may well turn out to be too crude a description of reality.

ACKNOWLEDGMENTS

The authors would like to thank the organizers of the PASI school for the excellent scientific organization and for providing a unique opportunity for a broad and deep exchange of ideas. F. A. would like to thank N. Carrer, E. Quercigh, J. Rafelski and K. Safarik for many stimulating discussions and suggestions.

REFERENCES

1. Bjorken, J.D., Phys. Rev. D **27** (1983) 140.
2. Margetis, S. et al.(NA49 Collaboration), Phys. Rev. Lett. **75** (1995) 3814.
3. Hanbury-Brown, R., and Twiss, R.Q., Nature **178** (1956) 1046.
4. Goldhaber, G. et al., Phys. Rev. **120** (1960) 300.
5. Appelshäuser, H. et al. (NA49 Collaboration), Eur. Phys. J. **C2** (1998) 661.
6. Matsui, T. and Satz, H., Phys. Lett. **B178** (1986) 416.
7. Rafelski, J., and Müller, B., Phys. Rev. Lett. **48** (1982) 1066 and Phys. Rev. Lett. **56** (1986) 2334.
8. Braun-Munzinger, P., Heppe, I. and Stachel, J., Phys. Lett. **B465** (1999) 15.
9. see, for instance: Antinori, F. et al., Nucl. Phys. **A661** (1999) 130. and references therein.
10. Heinz, U., Nucl. Phys. **A685** (2001) 414.
11. Letessier, J. and Rafelski, J., J. Phys. **G25** (1999) 295.
    In preparation of the lectures and proceedings, we have found the following particularly useful:
12. Fritzsch, H., *Quarks*, Basic Books 1983 (English Edition).
13. Gottfried, K. and Weisskopf, V.F., *Concepts of Particle Physics*, Oxford University Press 1984 (2 vol.).
14. Groom, D.E. et al., *Review of Particle Physics*, Eur. Phys. J. **C15** (2000) 1.
15. Jackson, J.D., *Classical Electrodynamics* (Second Edition), John Wiley and Sons 1975.
16. Perkins, D.H., *Introduction to High Energy Physics* (Third Edition), Addison Wesley 1987.
17. Wong, C., *Introduction to High-Energy Heavy-Ion Collisions*, World Scientific 1994.
18. An excellent reference for all the SPS experimental results are the contributions of the SPS collaborations to the proceedings of the last two Quark Matter conferences and the bibliography therein: Riccati, L., Masera, M. and Vercellin, E. (eds.) *Quark Matter '99* Proceedings of the 14th International Conference on Ultra-Relativistic Nucleus-Nucleus Collisions, Torino, Italy, 10-15 May, 1999, Nucl. Phys. **A661** (1999).
19. Hallman, T.J., Karzheev, D.E., Mitchell, J.T. and Ullrich, E.T. (eds.) *Quark Matter 2001*, Proceedings of the 15th International Conference on Ultra-Relativistic Nucleus-Nucleus Collisions, Long Island, New York, USA, 15-20 January, 2001, Nucl. Phys. **A698** (2002).
20. The results from the seven SPS collaborations were summarized in a special seminar which was held at CERN on February 10, 2000. The material from that event can be accessed on the web starting at: http://cern.web.cern.ch/CERN/Announcements/2000/NewStateMatter/

PHENIX for Beginners

W. A. Zajc (for the PHENIX Collaboration)* and P. Fachini†

*Physics Department, Columbia University, New York 10027,NY
†Brookhaven National Laboratory, Department of Physics, Bldg. 510 A, Upton, NY 11973-5000

Abstract. An introduction to the PHENIX detector and to the PHENIX physics program is presented. The PHENIX physics results presented here are those from the the first RHIC (Relativistic Heavy Ion Collider) run with Au+Au collisions at $\sqrt{s_{NN}} = 130$ GeV. A brief overview of the PHENIX detector is provided. The systematic variation with centrality of charged particle multiplicity, transverse energy, identified particle spectra and yield ratios, production of charged hadrons and π^0's at high transverse momenta are reported, together with first results on charm production at RHIC.

INTRODUCTION

The primary motivation for studying relativistic heavy ion collisions is to understand of the equation of state of nuclear matter. A phase transition from ordinary hadronic matter to a new state of deconfined partons (quarks and gluons) is predicted by non-perturbative Quantum Chromo-Dynamics (QCD) under extreme conditions of high density and high temperature that can be reached in relativistic heavy ion collisions [1]. This new form of matter, called the Quark Gluon Plasma (QGP), is somewhat analogous to the plasma phase of ordinary atomic matter. However, unlike such ordinary plasmas, the deconfined quanta of a quantum chromodynamics are not directly observable because of the confining property of the physical QCD vacuum. This fundamental difference may be traced back to the non-Abelian nature of QCD, which allows (in fact requires) interactions between the gauge quanta (gluons), as opposed to QED, where there is no direct interaction between photons. As a result, the observables are hadronic and leptonic residues of the transient QGP state, and there is a large variety of such individual probes [2]. The leptonic probes γ, e^+e^- and $\mu^+\mu^-$ carry information about the spectrum of electromagnetic current fluctuations in the QGP state. The abundance of quarkonia Ψ, Ψ', Υ and Υ' (also observed via l^+l^-) are sensitive to the the chromo-electric field fluctuations in the QGP; in particular the systematics of their production is expected to characterize the role of Debye screening in the plasma. The hadronic probes $\pi, K, p, \bar{p}, \Lambda, \Xi, \Omega, \rho$... provide information on the quark flavor chemistry and baryon number transport. Theory also suggests that with decays such as $\rho \to e^+e^-$ the properties of the hadronization and chiral symmetry breaking can be indirectly studied. Quantum statistical interference patterns in $\pi\pi, KK, pp, \Lambda\Lambda$... correlations measures the space-time geometry of hadronic probes from the QGP. The detailed rapidity and transverse momentum spectra of hadrons provide information on the dynamical evolution of the system.

The outstanding problem with all the proposed probes is their indirect nature. It would be 'trivial' to verify the QCD prediction of the QGP state if we could see free quarks and gluons (as in ordinary plasmas); however, quarks and gluons are confined within color neutral hadrons. Further complicating matters is the extremely short duration $10^{-(23-22)}$ s of the QCP state; even the briefest atomic plasma studied in the laboratory persists for times many orders-of-magnitude longer. The QGP state formed in nuclear collisions is a transient rearrangement of the correlations among quarks and gluons contained in the incident baryons into a larger but globally color neutral system with remarkable theoretical properties. The task of heavy ion collisions is to provide experimental information on this fundamental prediction of the Standard Model. This is the goal of the PHENIX experiment at the Relativistic Heavy Ion Collider (RHIC) at the Brookhaven National Laboratory in New York, USA.

RHIC is a versatile collider that can produce Au+Au collisions up to $\sqrt{s_{NN}}$ = 200 GeV, polarized proton-proton collisions up to 500 GeV and various ion combinations at intermediate values of these maximum energies. During the first RHIC run in the summer of 2000 (referred to as 'Run-1'), gold ions were collided at $\sqrt{s_{NN}}$ = 130 GeV and the PHENIX detector recorded approximately 5 million events. In this contribution we describe the PHENIX detector and report on a subset of PHENIX's results from the first year of data taking.

THE PHENIX DETECTOR

The PHENIX experiment [3] has been designed to measure a broad variety of signals from both heavy ion and polarized proton-proton collisions at RHIC. The pursuit of penetrating probes generated in the early stages of the collision, combined with a program of hadron measurements, provides a detector with unparalleled capabilities to address observables sensitive to all stages of the collision process. This same detector is also very well-suited to the study of gluon and anti-quark contributions to the proton spin [4, 5].

The PHENIX detector consists of three spectrometers: two muon spectrometers covering the full azimuth for $1.1 < |\eta| < 2.4$ and a central spectrometer consisting of two arms each subtending 90^0 in azimuth and with $|\eta| < 0.35$. A central magnet provides an axial field, while each muon spectrometer contains a magnet that produces a roughly radial field. The central arms contain three tracking sub-systems: pad chambers (PC), drift chambers (DC) and time-expansion chambers (TEC); two forms of electromagnetic calorimetry (PbSc and PbGl); a time-of-flight hodoscope (TOF) and ring imaging Cerenkov counter (RICH). These sub-systems, together with a set of beam-beam counters (BBC) located in the region $3.0 < |\eta| < 3.9$, provide superb hadron and electron identification over a broad range of transverse momentum [6]. The muon spectrometers use cathode strip chambers in three stations for tracking (muTr) and five layers of Iarocci tubes interleaved with iron absorber for muon identification (muID). Global event characterization is achieved via a multiplicity and vertex detector (MVD) consisting of silicon strips and pads covering $|\eta| < 2.5$, and the RHIC-standard Zero-Degree Calorimeters (ZDCs) that detect neutral particles emitted along the beam directions [7, 8]. The front end electronics for all sub-systems are clocked synchronously with the beam cross-

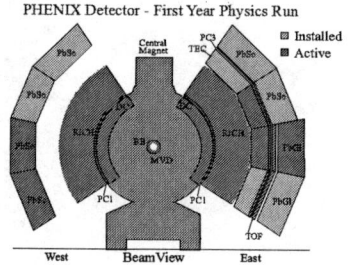
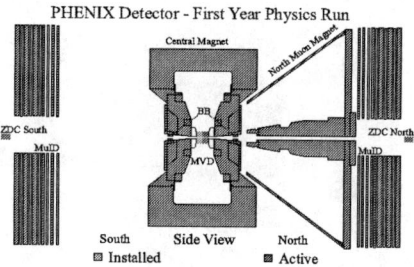

FIGURE 1. Installed and active detectors for the RHIC Run-1 configuration of the PHENIX experiment.

ing frequency of 9.4 MHz. A set of Level-1 triggers derived from various sub-systems is used to initiate readout of the entire detector through a pipelined high bandwidth data acquisition system capable of archiving 20 MB/s.

For the first physics run of RHIC in the summer of 2000, the portions of the PHENIX detector shown in Figure 1 were instrumented. Elements of all subsystems, with the exception of the muTr, were in place and were read out. Small subsets of the MVD and muID front end electronics were connected and tested as part of an engineering run. All other subsystems were instrumented in fractions ranging from 25% to 100% of their ultimate aperture and were used in the physics results presented here. Independent minimum bias triggers were formed using coincidences between the BBC counters and between the ZDC counters. A total of approximately 5 million events was recorded at $\sqrt{s_{NN}}=130$ GeV. The primary trigger used for most of the results presented in this contribution is based on the BBC coincidence with an additional offline requirement that restricts the collision vertex to $|z| < 20$ cm.

For the second physics run of RHIC('Run-2', in 2001-2), the portions of the PHENIX detector shown in Figure 2 were instrumented. The complete aperture of the central arms was available (compare to Figure 1), much of the MVD was instrumented and an entirely new spectrometer to measure muons was deployed. These additions in aperture and in capability, coupled with significant upgrades to the data acquisition and triggering system, resulted in approximately a hundredfold increase in the event sample obtained from Run-1. Since the analysis of this new data is still in progress, in this contribution we will not report on the PHENIX physics results from the second year run of RHIC.

PHENIX PHYSICS RESULTS

Centrality

In order to understand whether particle production in nuclear collisions is fundamentally different than in pp or pA collisions, it is important to understand how the various observables scale with N_p, the number of nucleons that interact inelastically (i.e. participate) in the collision. In fixed target experiments, this quantity is directly available (at

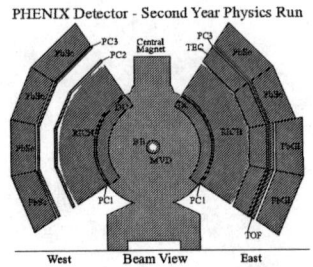 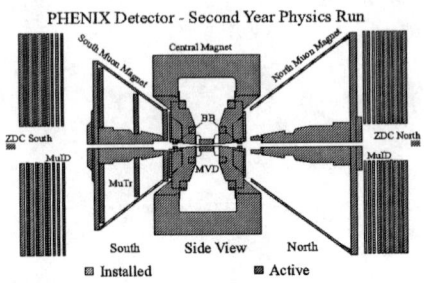

FIGURE 2. RHIC Run-2 configuration of the PHENIX experiment.

least for symmetric collisions) through measurement of the forward energy E_F, since this is proportional to the number of *non*-participants, i.e., the number of spectators N_S: $E_F/E_{NN}^{BEAM} \approx \frac{1}{2}N_S = A - \frac{1}{2}N_P$. Determining N_P specifies in turn the impact parameter in the collision (in the limit of straight line geometry, which is a good approximation at relativistic energies).

In colliders, a different approach is required, since the charged forward-going fragments detected in fixed-target experiments are deflected by the accelerator magnets to experimentally inaccessible regions. Nonetheless, it is possible to infer the collision centrality by mapping the production of some observable (multiplicity, transverse energy, ...) that is assumed to be monotonically increasing with increasing collision centrality (that is, decreasing impact parameter). The mapping is done via a geometric model that provides the fraction of events expected to lie in some impact parameter range, computing the number of participants for that impact parameter, identifying the corresponding fraction in the observable distribution, then mapping that range of observable with a class of events with the associated number of participants.

PHENIX has chosen to apply this procedure to a two-dimensional distribution formed from the ZDC and BBC signals. At RHIC, the ZDCs are placed behind dipole magnets which sweep away charged particles and nuclear fragments, leaving primarily spectator neutrons. Figure 3 shows the relationship between the energy deposition in the ZDCs and the charge deposition in the BBCs. It is clear from this figure that a given value of ZDC energy could arise from either a low or a high multiplicity event in the BBC's; i.e., the ZDC's do not uniquely specify the overall multiplicity. On the other hand, there is a wide variation of ZDC energy at low and high values of the BBC distribution, suggesting that the BBC's by themselves may not provide ideal event selection in this regions. By calculating what fraction of events occur along the 2-D distribution, event classes corresponding to 0-5% most central, the next 5-10%, etc. are defined.

Although one cannot directly measure the number of participants in a given collision, one can use the (assumed) monotonic relationship of N_P with respect to charged particle multiplicity (N_{ch}) or transverse energy (E_T) to relate a fraction of the cross section to a range in N_p. More accurately, the range of impact parameters that contributes to some range in centrality (say 10-20%) is used to compute the corresponding number of participants. This is done by means of the Glauber model of nuclear collisions [9, 10, 11] that allows the calculation of N_p as well as the number of binary collisions

(N_c) experienced by the colliding system as a function of the impact parameter. Since PHENIX has made a systematic effort to analyze many of its results for their scaling properties with respect to N_p and/or N_c, we present here some of the primary methods used to calculate these quantities:

The underlying assumption of the Glauber modelling used in relativistic heavy ion collisions is that of straight-line trajectories. Then, for a given nuclear density distribution $\rho_A(\mathbf{r})$, the *profile function*

$$T_A(\mathbf{s}) \equiv \int_{-\infty}^{+\infty} \rho_A(\mathbf{s},z)\, dz \qquad (1)$$

measures the integrated nucleon density along a flight path at a distance **s** from the z-axis. In computing the profile function, we will express all results in terms of standard density distributions normalized to the atomic number of the nucleus, that is, $\int_{-\infty}^{+\infty} \rho_A(\mathbf{r})\, d\mathbf{r} = A$. The probability of a nucleon with cross section σ_{NN} *not* interacting when traversing a nucleus A is then given by

$$P_0(\mathbf{s}) = \left[1 - \frac{\sigma_{NN} T_A(\mathbf{s})}{A} \right]^A . \qquad (2)$$

In the case of two nuclei A and B colliding at an impact parameter **b**, the density of *particpants* with coordinate **s** in the transverse plane is then readily found to be

$$\begin{aligned}\frac{dN_p}{d^2 s} &= T_B(\mathbf{s} - \tfrac{1}{2}\mathbf{b}) \left\{ 1 - \left[1 - \frac{\sigma_{NN} T_A(\mathbf{s} + \tfrac{1}{2}\mathbf{b})}{A} \right]^A \right\} \\ &+ T_A(\mathbf{s} + \tfrac{1}{2}\mathbf{b}) \left\{ 1 - \left[1 - \frac{\sigma_{NN} T_B(\mathbf{s} - \tfrac{1}{2}\mathbf{b})}{B} \right]^B \right\} \end{aligned} \qquad (3)$$

which can then be integrated over **s** to find the number of participants $N_p(\mathbf{b})$ at impact parameter **b**.

To calculate the number of *collisions*, it is convenient to define the *overlap function* $T_{AB}(\mathbf{b})$ given by

$$T_{AB}(\mathbf{b}) \equiv \int_{-\infty}^{+\infty} T_A(\mathbf{s} + \tfrac{1}{2}\mathbf{b}) T_B(\mathbf{s} - \tfrac{1}{2}\mathbf{b})\, d\mathbf{s} \quad , \qquad (4)$$

in terms of which one can express the total cross-section as

$$\frac{d\sigma_{AB}}{d^2 \mathbf{b}} = 1 - \left[1 - \frac{\sigma_{NN} T_{AB}(\mathbf{b})}{AB} \right]^{AB} \qquad (5)$$

The mean number of nucleon-nucleon collisions as a function of impact parameter is then given by the simple expression

$$N_C(\mathbf{b}) = \sigma_{NN} T_{AB}(\mathbf{b}) \quad , \qquad (6)$$

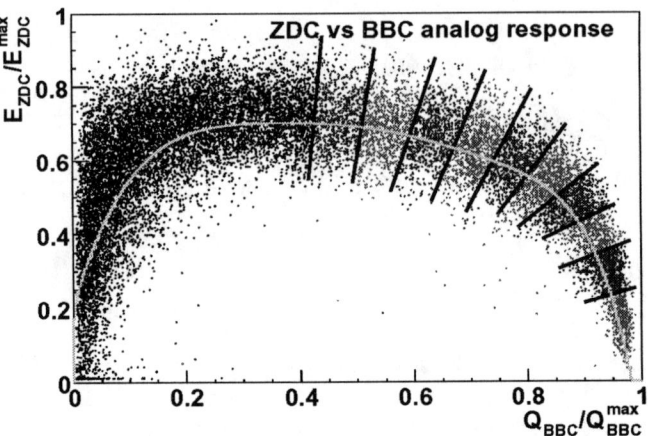

FIGURE 3. BBC vs. ZDC analog response related to the centrality classes of the collisions. The rightmost interval is the 0-5% centrality class, the next is 5-10% and so on.

which simply counts the total number of potential collisions ($\sim \sigma_{NN}$) between two slabs of nuclear matter ($\sim T_{AB}(\mathbf{b})$)[1].

Note that we can apply the formalism of Equation 5 to any sub-process (for example, the cross section for hard scattering σ^H). For those processes in which the cross-section is 'small' (such that expanding the product is a valid approximation), one can then easily show that the total cross section in and A-B collision is given by

$$\sigma_{AB}^i = \int d^2\mathbf{b} \left\{ 1 - \left[1 - \frac{\sigma_{NN}^i T_{AB}(\mathbf{b})}{AB} \right]^{AB} \right\} \approx \int d^2\mathbf{b}\, \sigma_{NN}^i T_{AB}(\mathbf{b}) = AB\, \sigma_{NN}^i \quad . \quad (7)$$

In fact, the integration over impact parameter is not required; and our baseline hypothesis for any rare process with cross section σ^i in a nucleon-nucleon is that it should have a differential yield in a nucleus-nucleus collision given by

$$\frac{d\sigma_{AB}^i}{d^2\mathbf{b}} = \sigma_{NN}^i T_{AB}(\mathbf{b}) \quad . \quad (8)$$

This is often referred to as *point-like* or *binary* scaling. Note in particular that Equations 6 and 8 together imply that such processes should scale as N_C, that is,

$$\frac{d\sigma_{AB}^i}{d^2\mathbf{b}} = \frac{\sigma_{NN}^i}{\sigma_{NN}} N_C(\mathbf{b}) \quad . \quad (9)$$

---

[1] There is an implicit assumption of transparency here– the formalism would fail spectacularly if applied to the collision of two neutron stars!

Global Observables

The systematic variation of particle multiplicities and the produced transverse energy with the number of participants reflects the underlying reaction mechanisms. For example, such a study can discriminate between cascade models and models which incorporate gluon saturation effects [12]. Gluons can begin to fuse with high enough gluon density, limiting the parton production. If the gluon density is indeed high enough to saturate, it is plausible that the system would thermalize quickly. This would be important as it has implications on the initial energy density achieved. If the parton production is limited, we expect that the final state charged particle yields to also be limited.

Charged Particle Multiplicity

The multiplicity of charged particles produced in heavy ion collisions arises from a variety of physics processes. In addition to the expected soft processes seen at lower energies, hard processes, nuclear shadowing and hadronic re-scattering all play a role [12]. Each of these has an effect on the number of degrees of freedom available to the colliding system (i.e. the entropy). Although a value of N_{ch} in isolation does not provide insight into the relative contributions of the various processes, we can attempt to disentangle them by systematically varying the initial conditions of the collision and comparing the results to pp and pA data as well as theoretical models. For this, it is useful to study the charged particle multiplicity per participant pair $dN_{ch}/d\eta/\frac{1}{2}N_p$.

The PHENIX measurement of $dN_{ch}/d\eta_{|\eta=0}$ = 622 ± 1(stat.) ± 41(sys.) (5% most central collisions) [13] shows that ∼ 80% more particles are produced at RHIC than at SPS at $\sqrt{s_{NN}}$ = 17.2 GeV [13] and ∼ 40% more than the extrapolation from $\bar{p}p$ would predict at $\sqrt{s_{NN}}$ = 130 GeV [14]. This is a strong evidence that particle production is not simply due to independent N+N interactions. Instead, whatever process amplifies the production at SPS energies relative to p+p collisions is even stronger at RHIC. Both the PHOBOS [15] and STAR [16] measurements are consistent with the PHENIX measurement within systematic errors.

The centrality dependence of $dN_{ch}/d\eta$ allows the discrimination between various models of particle production [12]. Figure 4 shows the PHENIX results for $dN_{ch}/d\eta$ per participant pair as a function of N_p [13]. The results from PHOBOS [17] are also incorporated in Figure 4, which agree very well with the PHENIX measurements within systematic errors. Figure 4 also shows the value from $\bar{p}$ + p at the same $\sqrt{s}$ taken from UA5 analysis [18]. It is interesting to note that the extrapolation of the data points to low multiplicities approaches the $\bar{p}$ + p value.

Theoretical models of particle production in RHIC collisions broadly fall into two classes [19]. The first class is based on modifying the wounded nucleon model to include hard processes (pQCD jets). In these, one assumes that hard and soft processes scale with binary collisions and wounded nucleons, respectively. Following this approach, the data in Figure 4 is fitted with the function

$$\frac{dN_{ch}}{d\eta} = A \times N_p + B \times N_c \qquad (10)$$

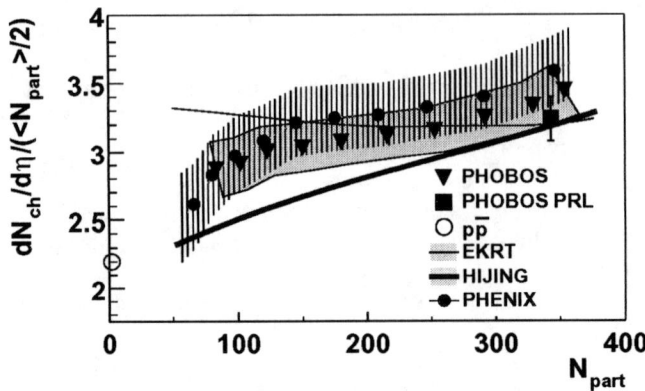

FIGURE 4. Charged particle pseudorapidity density per participant pair as a function of the number of participants measured by PHENIX [13] and PHOBOS [17]. The $\bar{p}p$ value taken from UA5 analysis [22] is also presented. Predictions from HIJING [12, 20] and EKRT [24, 19] are also shown. Both the shaded area and the doted area represent the systematic errors of $dN_{ch}/d\eta$ and N_p.

and we obtain $A = 0.88 \pm 0.28$ and $B = 0.34 \mp 0.12$ (note that the errors in A and B are anticorrelated) [13]. The relative values of A and B, taken at face value, would imply a large contribution of hard processes[2]. The HIJING model [12, 20] follows this approach and includes additional physics effects such as jet quenching and nuclear shadowing, which leads to a linear rise in $dN_{ch}/d\eta/\frac{1}{2}N_p$ vs. N_p. HIJING predicts the same trend as observed in the data although the calculated values are lower by $\sim 15\%$. On the other hand, at the AGS energies, where hard processes do not occur, the particle production per participant was also observed to increase with centrality [21]. The eikonal model calculation [11] also assumes that there is a component from soft interactions that scales linearly wit N_p and a second component from hard processes that scales with N_c. However, the eikonal model only uses as input the fraction of hard processes and the PHOBOS result [15]. This leads to a dependence of $dN_{ch}/d\eta$ similar to that measured by WA98 [22] and WA97 [23] which is well described by a simple power-law form CN_p^α.

The other class of calculations, based on parton saturation, predict a very different dependence on N_p. EKRT [19, 24] predict that at RHIC energies the large production of semi-hard gluons in a small volume may saturate the gluon density. The resulting gluon fusion limits the total entropy production and thus lowers the final particle production per participant. EKRT found that a geometry-dependent saturation scale predicts a nearly-constant dependence of $dN_{ch}/d\eta/\frac{1}{2}N_p$ as a function of N_p. The prediction of the EKRT model is shown in Figure 4. No such saturation effect is observed within errors for Au + Au collisions at $\sqrt{s_{NN}} = 130$ GeV, instead the particle production per participant pair rises steadily. In [11] a calculation based on parton saturation is also performed

---

[2] It must be noted that this parameterization is not sufficient to establish the presence of hard scattering. For example, simple cascading can lead to multiplicities increasing faster than N_P.

including the running of the strong coupling constant with saturation scale and finds $dN_{ch}/d\eta$ per participant to scale as $\ln(Q_s^2/\Lambda_{QCD}^2)$, where Q_s^2 is the impact parameter dependent saturation momentum scale and Λ_{QCD}^2 is the energy scale parameter of QCD. This result agrees well with their eikonal calculation.

Transverse Energy

Lattice calculations predict a phase transition to a quark-gluon plasma to occur for zero baryon density at energy densities on the order of a few GeV/fm^3 [26]. Assuming a boost invariant expanding cylinder of dense nuclear matter in which a thermalized system reaches after a formation time τ_0, Bjorken [27] derived a formula in which the measured transverse energy density in rapidity (dE_T/dy) can be related to the energy density ε_{Bj} as

$$\varepsilon_{Bj} = \frac{dE_T}{dy} \frac{1}{\tau_0 \pi R^2} \tag{11}$$

where the formation time τ_0 is usually taken as 1 fm/c and πR^2 is the effective area of the collision. The transverse energy (E_T) is a multiparticle variable defined as

$$E_T = \sum_i E_i \sin \theta_i \tag{12}$$

which is intended as an experimental approximation to the explicitly Lorentz-invariant quantity

$$E_T^{theory} = \sum_i m_{Ti} \sin \theta_i = \sum_i \sqrt{m_i^2 + p_i^2 \sin^2 \theta_i} \tag{13}$$

It is clear that the experimental definition is equivalent to the theoretical quantity only in the limit of $\langle p_i \rangle \gg m_i$, which is appropriate for pions but not for kaons and nucleons. Further complicating this is the convention[25] (which stems from the intrinsic response of calorimeters) that E_i is taken as the kinetic energy for nucleons and the total energy for other particles. (This somewhat odd definition results from the intrinsic response of calorimeters, in that the mesons and anti-baryons may be totally absorbed, while baryons always leave a nucleon remnant whose rest mass does not appear in the calorimeter signal. The alert reader will note that in principle anti-baryons may 'extract' energy from the calorimeter material via annihilation; this is discussed further below.)

Assuming that these difficulties in measurement and interpretation may be overcome, transverse energy measurements give an excellent measure of the degree to which ordered longitudinal momentum is converted to excitations in the transverse direction. This information is of course also available from the convolution of the charged multiplicity measurements with the charged particle momentum spectrum, but calorimetry also provides a measurement of the neutral energy produced in the event, as well as automatically summing over any correlations present that would complicate the simple-minded convolution of multiplicity and momenta spectra. Even in limited apertures at mid-rapidity E_T measurements provide excellent characterization of the nuclear geometry of a reaction on an event-by-event basis and are sensitive to the underlying reaction dynamics [26].

The PHENIX measurement of E_T is performed using the PbSc calorimeter [29]. A careful treatment of the contributions from produced energy in the aperture, the in-flux from scattering sources, and both the in-flux and out-flux from decays is performed to convert the deposited energy seen in the calorimeter to the equivalent transverse energy. The total transverse energy density $dE_T/d\eta|_{\eta=0}$ per participant pair is then calculated and the trend is essentially identical to that found for charged particle multiplicity, as shown in Figure 5. The WA98 measurements at mid-rapidity $(dE_T/d\eta|_{mid})$ [22] from Pb + Pb collisions at $\sqrt{s_{NN}}$ = 17.2 GeV are also shown in Figure 5. We observe that $dE_T/d\eta|_{\eta=0}$ for central Au + Au collisions at $\sqrt{s_{NN}}$ = 130 GeV is about 40% higher than the WA98 measurement. A straightforward derivation of ε_{Bj} from the measured $dE_T/d\eta|_{\eta=0} = 578^{+26}_{-39}$ GeV for the most central 2% of the inelastic cross section, using $\pi R^2 = 148$ fm^2 with $R = A^{1/3} \times 1.18$ fm gives ε_{Bj} = 4.6 GeV/fm^3 [29]. This represents an increase of 60% over ε_{Bj} = 2.9 GeV/fm^3 reported by WA98 [22] and NA49 [30] at the same centrality cut.

As shown in Figure 5, the dependence on the number of participants for both the PHENIX and WA98 measurements of $dE_T/d\eta$ is well described by a simple power-law form CN_p^α. Figure 6 shows that for both $\sqrt{s_{NN}}$ energies the ratio between $dE_T/d\eta$ and $dN_{ch}/d\eta$ remains constant as a function of centrality at roughly the same value of ~ 0.8 GeV. This result suggests that no dramatic modification of the particle spectra between peripheral and central collisions. In addition, the transverse energy at mid-rapidity should scale as the charged particle multiplicity, which is confirmed in Figure 5. This is observed, but leads to the somewhat surprising result that the transverse energy per charged particle (dE_T/dN_{ch}) as a function of $\sqrt{s_{NN}}$ measured at RHIC (also shown in Figure 6) appears to be nearly the same as that measured at the AGS by E877/E814 [28], and at the SPS by WA98 [22] and NA49 [30]. One possible source of this striking lack of variation may be the differing experimental conventions used to define E_T. For instance, WA98 uses the convention that the contribution to E_T of nucleons is $E - m$, of anti-nucleons $E + m$, and all other particles is simply E. If applied to the PHENIX data, this convention would increase all E_T values by a few percent, due to the much larger anti-baryon to baryon ration found at RHIC as compared to the SPS[31]. However, any such modifications to definition of E_T do not drastically change the result that while the mid-rapidity transverse energy density at $\sqrt{s_{NN}}$ = 130 GeV is 60% larger than at $\sqrt{s_{NN}}$ = 17.2 GeV, the transverse energy per charged particle increases only incrementally, indicating that (despite the increased role of hard processes at RHIC) the additional energy density at RHIC is achieved mainly by an increase in particle production rather than by an increase in transverse energy per particle.

Identified Hadrons

The spectra of various identified hadrons provides additional information into the reaction dynamics beyond that from global event characterization. The yields of hadrons

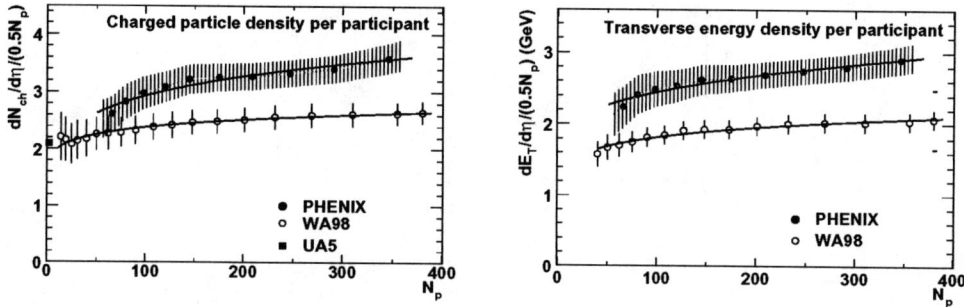

FIGURE 5. PHENIX $dE_T/d\eta|_{\eta=0}/\frac{1}{2}N_p$ (right) and $dN_{ch}/d\eta|_{\eta=0}/\frac{1}{2}N_p$ (left) as a function of N_p compared to WA98 [22]. The $\bar{p}p$ value taken from UA5 analysis [22] is also shown in the left plot. The solid line is the best fit to CN_p^α and the shaded area represent the systematic errors of $dN_{ch}/d\eta$, $dE_T/d\eta$ and N_p. The WA98 data has an additional ±20% overall systematic error that is not shown.

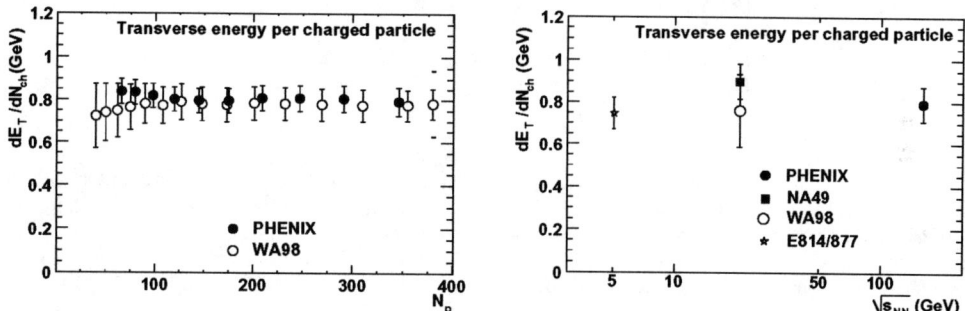

FIGURE 6. Left: PHENIX dE_T/dN_{ch} as a function of N_p compared to WA98 [22]. The PHENIX data includes all systematic errors. The WA98 data has an additional ±20% overall systematic error that is not shown. Right: dE_T/dN_{ch} as a function of $\sqrt{s_{NN}}$ for the most central 2% of the inelastic cross section.

reflect the particle production mechanism, while the spectral shapes are sensitive to the dynamical evolution of the system. The mass and centrality dependence of the spectra can help to differentiate between competing theoretical descriptions such as collective hydrodynamical expansion [32, 33, 34] or transverse momentum (p_T) broadening in the partonic stage of the reaction [35]. In addition, the relative yields of baryons and mesons at high p_T (> 2 GeV/c) may give insight into the baryon number transport [36] and the interplay between soft and hard processes.

The PHENIX central arm detectors have been designed to provide particle identification over the broadest possible range of transverse momentum[6]. The primary tool for charged hadron identification is the time-of-flight difference between the BBC and the highly-segmented TOF hodoscope, which spans $\Delta\Phi = 45^0$ in the East spectrometer arm. The overall time resolution $\sigma \sim 115$ ps permits unambiguous π/K separation to at least $p_T = 1.5$ GeV/c.

Figure 7 depicts the invariant yield as a function of p_T for both positive and negative identified hadrons for three centrality selections. We observe that in peripheral events

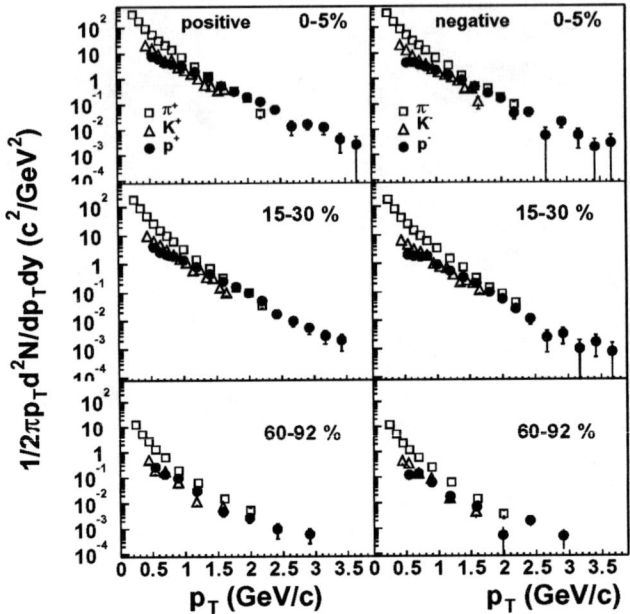

FIGURE 7. Transverse momentum spectra measured at mid-rapidity for π^+, K^+, p (left) and $\pi^-, K^-, \bar{p}$ (right) at three different centrality selections indicated in each panel. The symbols indicated in the top panels apply for all centrality selections. The error bars are composed of statistical errors and the systematic errors associated with acceptance and decay corrections. There is an additional $\sim 11\%$ systematic error associated with the overall normalization.

the pion spectra exhibit a concave shape, well described by a modified power-law parameterization

$$\frac{1}{p_T}\frac{dN}{dp_T} = \frac{C}{(p_T + p_0)^n} \qquad (14)$$

as observed in hadron-hadron collisions [37]. With increasing centrality the curvature in the spectra decreases, leading to an almost exponential dependence on p_T for the most central events. Over the measured p_T range, the kaon spectra can be described by an exponential distribution either in p_T on in $m_T = \sqrt{p_T^2 + m^2}$, while the p and $\bar{p}$ spectra can be described either as a Boltzmann ($dN/m_T dm_T \sim m_T e^{-m_T/T}$) or as a exponential ($dN/m_T dm_T \sim e^{-m_T/T}$) distribution in m_T. The slopes of the m_T spectra flatten and the mean transverse momentum $<p_T>$ increases with particle mass and with centrality [38]. This behavior has been previously observed in lower energy heavy ion collisions at the AGS [39] and at the SPS [40, 41] and has been attributed to collective radial motion (flow) in the rapidly expanding system.

At lower energies, it is not uncommon for the proton yields to equal or exceed the π^+ yields, since most of the protons observed at mid-rapidity are those contained in the initial nuclei. Said differently, the very limited production of baryons and anti-baryons means that nearly all baryons are those found in the initial state. The ratio of anti-protons to protons at mid-rapidity is thus of particular interest, since it is a direct measure of the net baryon content in the central region. The $\bar{p}/p$ ratio has been studied as a function of both transverse momentum and centrality [42], and found to be only weakly dependent on p_T and independent of centrality within systematic errors (the K^+/K^- ratio was found to behave similarly [42]). For minimum bias collisions, the $\bar{p}/p$ ratio in the interval 0.8 GeV/c < p_T < 3.0 GeV/c is 0.64 ± 0.01(stat.) ± 0.07(sys.) which is consistent with that observed by BRAHMS [43], PHOBOS [44] and STAR [45]. This result together with the measured yields shown in Figure 7 means that the central region in heavy ion collisions at RHIC is dominated by mesons, which is very distinct from the SPS where the central region is dominated by baryons and the $\bar{p}/p$ ratio is never greater than ~ 0.1 [46, 47].

A new feature observed for the first time at RHIC is that in central collisions at $p_T \simeq$ 2 GeV/c the anti-proton yields are comparable to the negative pion yields. Positive and negative hadrons behave in a similar way. In the central events the proton yields become comparable to the pion yields for $p_T \simeq 1.6$ GeV/c. The relative yield of protons compared to pions is highest in central events; in peripheral events, both the proton and the anti-proton spectra are below the pion spectra in the whole measured p_T range. Since anti-protons are not as numerous as the protons, $\bar{p}$ and π^- yields become comparable only at the high end of the measured pion p_T range in the most central collisions. The high p_T (anti)proton/pion ratios observed at RHIC are of the order of 1 which is much larger than in hadron-hadron collisions [48, 49].

Different theoretical explanations have been proposed to describe such behavior. It has been suggested that hydrodynamic expansion alone or combined with hadronic re-scattering is responsible for the baryon dominance at high p_T [32, 33, 34, 50]. Protons/anti-protons produced via a baryon junction mechanism combined with jet-quenching in the pion channel are shown to exhibit the same effect [36]. Intrinsic p_T broadening in the partonic phase caused by gluon saturation expected in high density QCD [35] gives yet another alternative explanation.

The identified particle yields per participant as a function of N_p are shown in Figure 8. Unlike in the centrality dependence of $<p_T>$, where the average p_T rises with centrality similarly for all particle species [38], the total increase in the yields per participant differs among particle species when going from peripheral to central events. The kaon, proton and anti-proton yields per participant behave similarly and increase faster than the pion yields [38].

Hard Probes

In relativistic heavy ion collisions it is interesting to study the products of parton scattering with large momentum transfer, i.e. 'hard scattering'. In p + p collisions hard scattered partons fragment into jets of hadrons, whose fragments are the primary source of hadrons at high transverse momentum (p_T), typically above ~ 2 GeV/c [51]. In a high-energy nuclear collision hard scattering will occur at the earliest time during the collision, well before the QGP is expected to form, and thus the scattered partons will

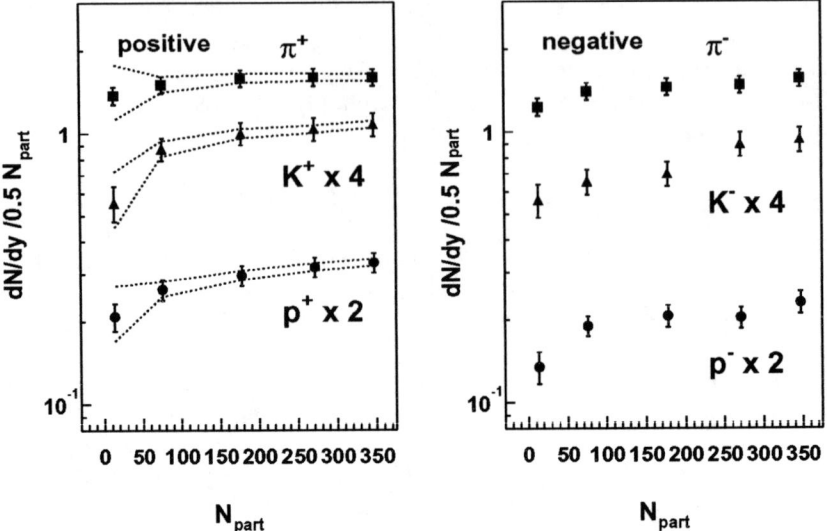

FIGURE 8. $dN/dy|_{y=0}$ per participant for π^+, K^+, p (left) and $\pi^-, K^-, \bar{p}$ (right) as a function of N_p. The error bars include statistical and systematic errors in dN/dy. The dashed lines around the positive hadrons show the effect of the systematic error on N_p and affects all curves in the same way.

subsequently experience the strongly interacting medium created in the collision. These partons are expected to lose energy [52, 53] in the hot and dense matter through gluon bremsstrahlung, effectively quenching jet production. This would have many observable consequences, such as a depletion in the yield of high p_T hadrons [54, 55]. It has been suggested that the energy loss is larger in a medium of deconfined color charges than in hadronic matter [56], making jet quenching a potential signature for the formation of the QGP.

In order to quantify such modifications, we need a baseline expectation for spectra from nuclear collisions (A + A) collisions in the absence of nuclear medium effects. Since hard parton scatterings have small cross sections compared to the total nucleon-nucleon inelastic cross section σ_{NN}, we can apply the results of Equation 8 and 9, that is, we expect the yield of hard processes to scale as the number of underlying nucleon-nucleon collisions. Using the same methods of event characterization described previously, the mean number of collisions $\langle N_c \rangle$ can be determined for a given centrality range, and then used to compute *the nuclear modification factor* $R_{AA}(p_T)$, defined as

$$R_{AA}(p_T) = \frac{\frac{1}{\langle N_C \rangle} \frac{d^2 N^{AA}}{dp_T d\eta}}{\frac{1}{\sigma_{NN}} \frac{d^2 \sigma_{NN}}{dp_T d\eta}} \tag{15}$$

Under the plausible assumption of point-like scattering for hard scattering processes, the ratio $R_{AA}(p_T)$ will be unity. It is important to realize that this applies only for a transverse momenta in the regime where hard scattering becomes important. For example, previous

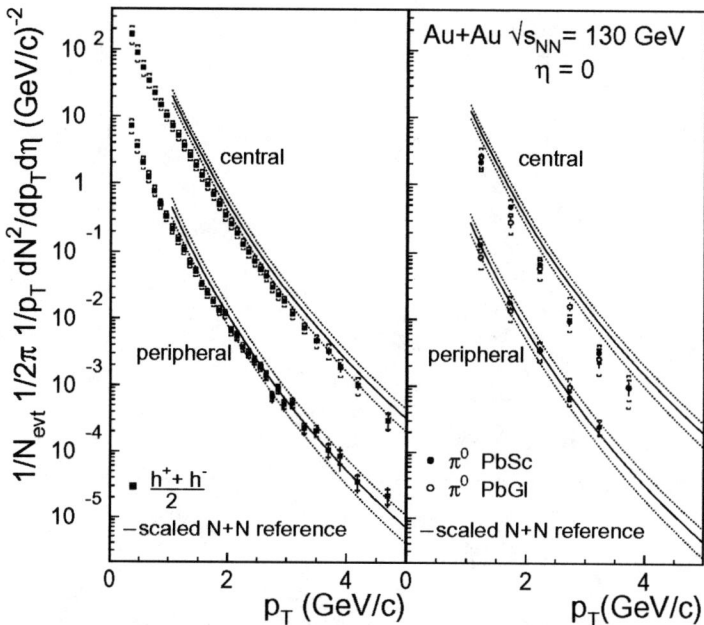

FIGURE 9. The yields per event at mid-rapidity for charged hadrons (left) and neutral pions (right) are shown as a function of p_T for peripheral and central events. The error bars indicate the statistical errors on the yields and the surrounding brackets indicate the systematic errors. Also shown are the N + N references for charged hadrons and neutral pions, each scaled up by $<N_c>$. The dotted lines indicate the uncertainty in the N + N reference and in the $<N_c>$.

measurements indicate that for p_T below 2 GeV/c, R_{AA} is smaller than one since the bulk of particle production scales with the number of the nucleons participating in the reaction [13, 16][3]. For p_T above 2 Gev/c particle production in p+A collisions is enhanced compared to binary scaling, commonly referred as the Cronin effect [57]. Parton shadowing as measured in lepton + A collisions [58, 59] is also expected to modify the hadron spectra in p + A and A + A compared to binary scaling.

The high-p_T spectra of both charged hadrons and neutral pions is measured by the PHENIX experiment in a central and a peripheral class of Au + Au collisions at $\sqrt{s_{NN}}$ = 130 GeV [60]. The charged particles are reconstructed using a drift chamber (DC) and two layers of pad chambers (PC) in the east arm. The neutral pions are measured via their $\pi^0 \to \gamma\gamma$ decay on a statistical basis (that is, not on an event-by-event basis) by computing the background for each p_T bin using a mixed-event technique [61, 62]. In the case of the neutral pions, two analyses were performed. The first uses the lead-scintillator (PbSc) sampling calorimeter in the west arm, and the second analysis uses the lead-glass Cerenkov (PbGl) calorimeter in the east arm aperture. In both cases, pairs of

---

[3] This remark assumes that $N_P < N_C$, which is the case in nucleus-nucleus collisions for all but the very most peripheral (and usually not measured) collisions.

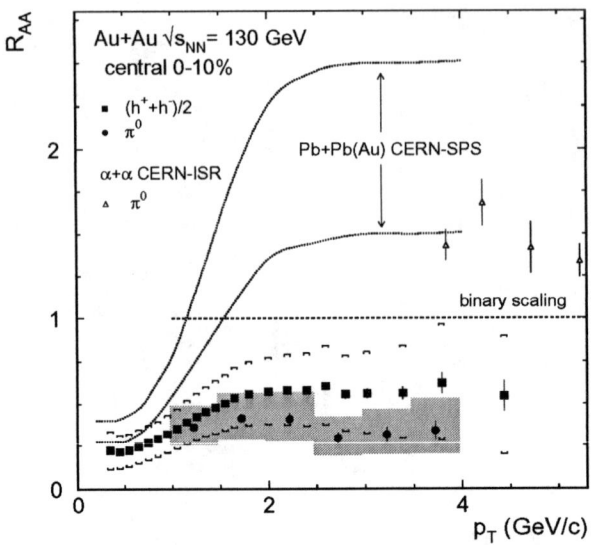

FIGURE 10. The ratio R_{AA} for charged hadrons and neutral pions (weighted average of PbSc and PbGl results) in central Au+Au collisions. The error bars indicate the statistical errors on the measurement. The brackets (for $(h^+ + h^-)/2$) and the shaded area (for π^0's) are the quadrature sums of the systematic errors on the measurement, the uncertainty in the N+N reference and the uncertainty in $\langle N_c \rangle$. Also shown are the ratio of inclusive cross sections in $\alpha + \alpha$ compared to p+p at $\sqrt{s_{NN}} = 31$ GeV [63] (open triangles), and spectra from central Pb + Pb(Au) compared to p+p collisions at $\sqrt{s_{NN}} = 17$ GeV [64] depicted as a band indicating the range of uncertainty.

calorimeter showers (photon 'clusters') are combined and the invariant mass $m_{\gamma\gamma}$ given by

$$m_{\gamma\gamma} = \sqrt{(E_{\gamma_1} + E_{\gamma_2})^2 - (\vec{p}_{\gamma_1} + \vec{p}_{\gamma_2})^2} = \sqrt{E_{\gamma_1} E_{\gamma_2}(1 - \cos\theta_{12})} \qquad (16)$$

is calculated. The invariant mass distribution obtained from the combination of all clusters corresponds to the true π^0 distribution and a combinatorial pair background. The combinatorial pair background is then estimated using the mixed-event technique, in which clusters from different events with similar centrality are mixed and the invariant mass distribution for this mixed distribution is then calculated. After normalizing the mixed $m_{\gamma\gamma}$ distribution in a region outside the π^0 mass peak, the excess in the region of the π^0 mass is extracted either by direct subtraction or by fitting the peak (the two different methods provide one of many systematic checks applied to the data). The procedure is applied for each p_T bin, and thereby produces the π^0 spectrum as a function of transverse momentum.

Figure 9 depicts both π^0 and charged hadron $((h^+ + h^-)/2)$ p_T distributions. The data presented are from two event classes, central and peripheral. The central events cover the 0-10% most central fraction of the geometrical Au + Au cross section, while the peripheral sample contains events in the 60-80% selection. The number of binary collisions are estimated to be $<N_c> = 905 \pm 96$ for central events, $<N_c> = 20 \pm 6$

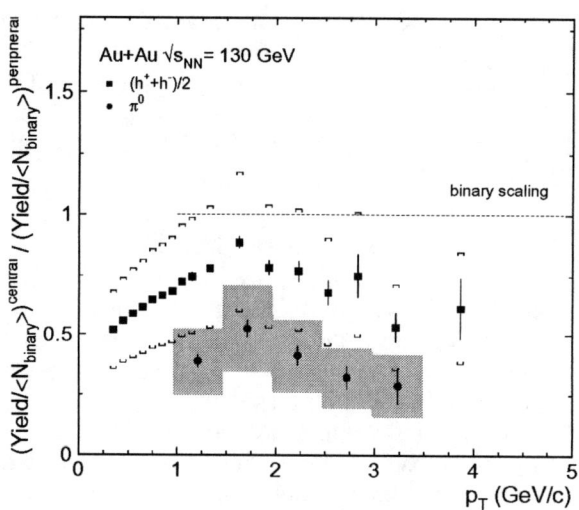

FIGURE 11. Ratio of central and peripheral yields per event, each divided by $\langle N_c \rangle$ for the corresponding centrality class. For π^0's the weighted average of PbSc and PbGl results are shown. The error bars indicate the statistical errors on the spectra measurement. The brackets (for $(h^+ + h^-)/2$) and the shaded area (for π^0's) are the quadrature sums of the uncertainty in $\langle N_c \rangle$ and the systematic errors on the spectra that do not cancel in the ratio.

for peripheral events and 45 ± 13 for the ratio between them [60]. It is apparent from this figure that the yields in peripheral events for $p_T > 2$ GeV/c are consistent with the assumption of binary scaling in $\langle N_C \rangle$, while in central events charged hadrons fall below this expectation, with π^0's showing an even larger suppression.

This difference can be better examined by looking at the ratio R_{AA} for central collisions that is shown in Figure 10. The ratio R_{AA} rises up to 2 GeV/c for the charged spectrum, as expected. However, above 2 GeV/c R_{AA} remains significantly below unity for both spectra. The depletion is quite striking, since the production of high-p_T hadrons in p+A collisions at fixed target energies is known to be enhanced compared to binary-scaling expectation for $p_T > 2$ GeV/c due to the Cronin effect [57]. A similar enhancement has also been observed in heavy ion collisions at lower energies [63, 64] as shown in Figure 10. Phenomenological calculations[64] including shadowing and the Cronin effect predict that for central Au + Au collisions at $\sqrt{s_{NN}} = 130$ GeV $R_{AA} > 1$ for hadron spectra in the p_T range 3-9 GeV/c.

Figure 10 shows that R_{AA} is lower for pions than for charged hadrons, which implies that the π/h ratio is smaller in central RHIC Au + Au collisions than in ISR p + p collisions. This effect is consistent with the trends observed in the identified charged hadron spectra measured by PHENIX (Figure 7) in which the yield of protons and anti-protons becomes comparable to that of pions for $p_T \sim 2$ GeV/c, unlike the corresponding ratio observed in p+p collisions.

While the p+p reference spectrum used in these comparisons may be determined with some confidence by interpolating between ISR (63 GeV) and Sp$\bar{p}$S (200 GeV) energies, the suppression may also be checked directly by comparing Au+Au spectra from central events to those from peripheral events (again after scaling by the appropriate number of collisions for the two centrality classes). as shown in Figure 11, the trends observed in R_{AA} are also present in this comparison. indicating a suppression of the yield per N+N collision in central collisions relative to peripheral. Again, the difference between the two ratios implies that the π/h ratio is smaller in central than in peripheral collisions.

The Charm of PHENIX

There are many motivations for studying charm production in RHIC collisions. Since charm is produced dominantly through the gluon fusion process $g+g \to c+\bar{c}$, the yield of charmed mesons is a sensitive probe of the the gluon distribution. Proton-proton collisions provide the fundamental calibration of our ability to calculate charm yields via next-to-leading-order (NLO) pQCD using measured gluon structure functions. Comparison of the proton-proton yields to those in proton-nucleus collisions determine the role of *shadowing*, that is, the suppression of yields compared to the factor of A enhancement expected in the absence of nuclear effects. Finally, charm production in nucleus-nucleus collisions is sensitive to the time evolution of the gluon density[65, 66, 67], explores the limits of applicability of thermal models[68], provides a probe which should have different energy loss characteristics in a dense medium[69], and (if nothing else) must be measured as an unavoidable background to di-lepton measurements[70]. Perhaps the most succinct statement is to simply note that the charm quark, with a mass $m_c \sim 1.3$ GeV, is a truly heavy quark as compared to u, d and s quarks, and since $m_c \gg \Lambda_{QCD}$, may be expected to have qualitatively different properties in a medium characterized by a temperature scale $T \sim \Lambda_{QCD}$. This, together with the flavor independence of QCD, should make it clear that much of this also applies to the production of bottom quarks, once one accounts for their even larger mass $m_B \sim 4.3$ GeV/c.

Most of the charm produced in hadronic reactions appears as *open charm*, that is, in mesons containing one charm or anti-charm quark (and thus containing one unit of charm). These are generically referred to as *D-mesons*, with the lowest lying members $D^0 = (c\bar{u})$, $D^+ = (c\bar{d})$, together with the $\bar{D}^0$ and D^-. Roughly 1% of the produced charm appears as *charmonium*, that is, as bound $c\bar{c}$ pairs. The lowest lying member of the vector (spin-1) charmonium family is the famed J/Ψ resonance. The 1986 prediction by Matsui and Satz[71] of strong suppression of charmonium in a quark-gluon plasma, together with the observation of suppression by NA38[72] and NA50[73] in collisions at the SPS, has stimulated tremendous theoretical activity directed towards understanding the origin of the suppression. PHENIX is ideally suited to measuring charmonium production at RHIC via the leptonic decay channels J/$\Psi \to e^+e^-$ in the central arms and J/$\Psi \to \mu^+\mu^-$ in the two muon arms. However, the much smaller production cross-section for the onium states requires weeks of running at or above RHIC design luminosity, and hence is not accessible in the Run-1 data set.

In principle, open charm may be detected via exclusive decay channels such as $D^+ \to \bar{K}^0 \pi^+$ and $D^0 \to \bar{K}^- \pi^+$. This approach, coupled with detection of the corresponding displaced vertices ($c\tau(D^\pm) = 317\ \mu$m, $c\tau(D^0) = 124\ \mu$m), is indeed the primary method used in particle physics to amass the tremendous variety of D-decay modes found in the

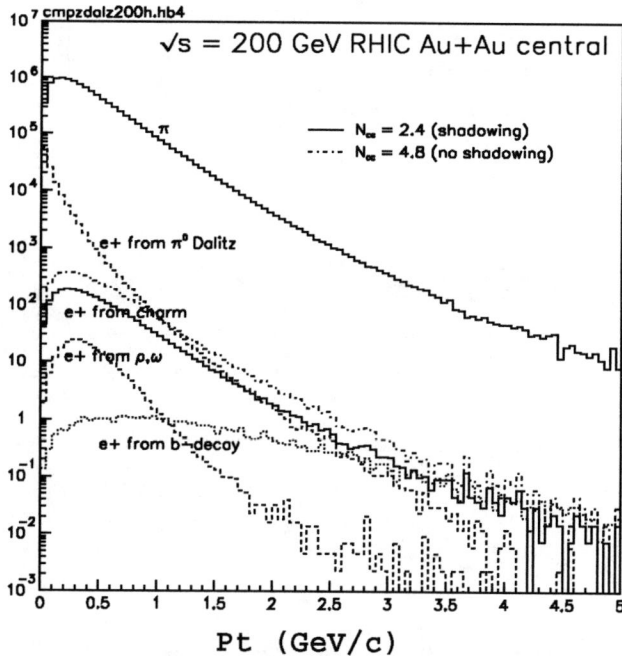

FIGURE 12. Calculated yields of pions and electrons from various sources as a function of transverse momentum for Au+Au collisions at $\sqrt{s_{NN}} = 200$ GeV[75].

Particle Data Book[74]. However, it presents extraordinary challenges in the heavy ion environment, since the the very large multiplicities not only present huge combinatoric backgrounds to reconstruction of invariant mass distributions but also make detection of secondary vertices displaced by ~ 100 μm extremely difficult[4].

The technique used by PHENIX to detect open charm is also borrowed from high energy physics, and in fact dates to the earliest evidence for charm production via the observation of *prompt single leptons*. We now know that roughly 10% of the D-decays occur semi-leptonically, e.g., the branching ration to $D^+ \rightarrow e^+X$ is 17%. This was not the case when Hinchliffe and Llewellwyn Smith[76] and Bourquin and Gaillard[77] made the suggestion that the excess electron yield observed at the CERN ISR was the result of the (then conjectured) charm meson. Nonetheless, subsequent events have verified the validity of this hypothesis, thereby providing a detection technique immune to the combinatoric backgrounds present in more conventional approaches[78].

This should not be interpreted as a statement that the measurement is background free; nothing could be further from the truth. As shown in Figure 12, the pion yields are roughly 10^3 times larger than the yields of electrons from charm decays over a

---

[4] Difficult, but not impossible. A major component of the PHENIX upgrade plan is a vertex detector capable of resolving the displaced vertices from open charm.

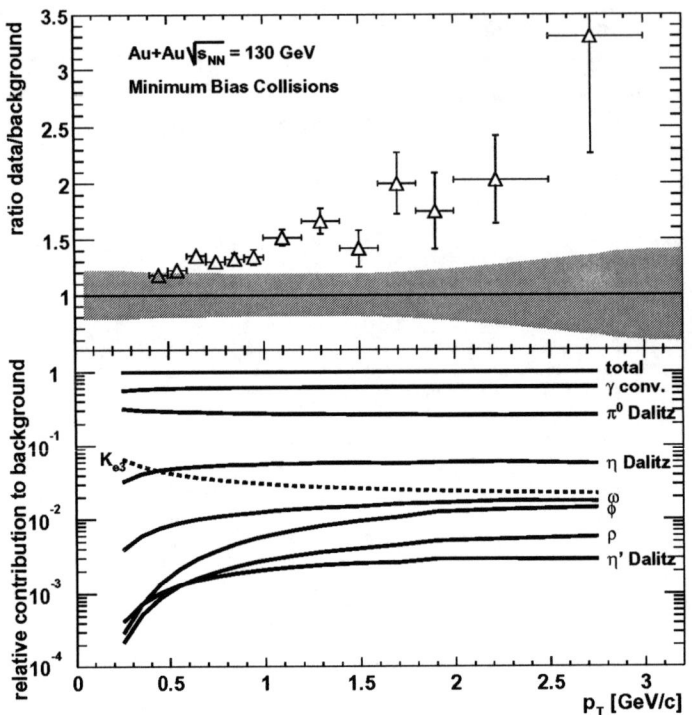

FIGURE 13. The upper panel shows the ratio of the extracted electron signal to the background as a function of p_T for minimum bias collisions. The lower panel shows the relative contributions as a function of p_T to the background.

broad range in transverse momentum. This presents two experimental difficulties: 1) One must reject charged pions at a level well in excess of 1000:1, lest misidentified pions be confused with signal electrons 2) The intrinsic backgrounds from internal conversions (Dalitz decays) $\pi^0 \to \gamma e^+ e^-$ and from external conversions $\pi^0 \to \gamma\gamma \to \gamma e^+ e^-$ must be very well known, in that these are present at the level of a few percent of the pion yields, and therefore also exceed the signal from semi-leptonic charm decay[5]

PHENIX has been designed with these issues in mind[80]. Electron rejection well beyond the required 1000:1 is provided by the combination of the RICH, energy-momentum matches in the EmCal, and (in the future) energy-loss measurements in the TEC. Neutral pions are measured in two separate calorimeter systems, and the comparison to the charged pion spectrum (measured by PHENIX in the same aperture) provides excellent control of the experimental systematics. The material in this aperture has been minimized to reduce the contribution from external conversions to a tractable

---

[5] It is important to note that *all* sources of photons and conversions must be considered. Roughly 80% of the background electrons result from the decays of pions, with η mesons being the next-most important contribution. See Figure 13 and Ref. [79] for details.

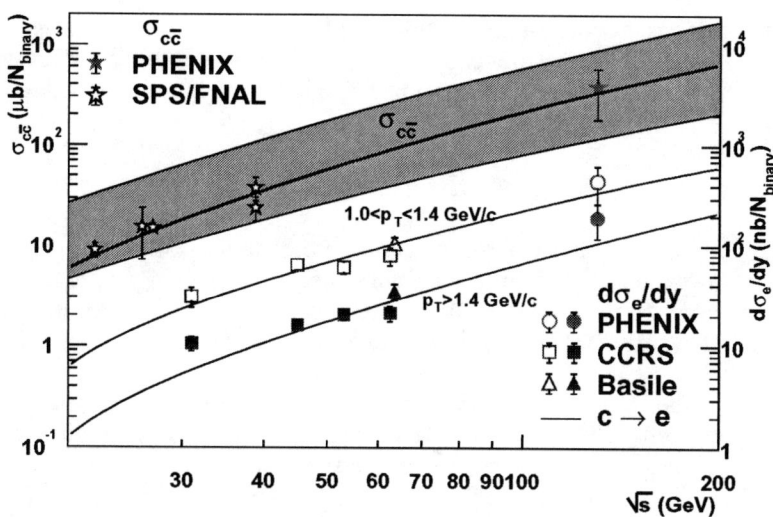

FIGURE 14. The single electron cross sections $d\sigma_e/dy|_{y=0}$ in Au+Au collisions at $\sqrt{s_{NN}} = 130$ GeV extracted by PHENIX[79] assuming scaling with the number of binary collisions are compared to the world's data as a function of $\sqrt{s}$. The right-hand scale applies to the lower two curves; the left-hand scale applies to the upper curve for the total charm production cross-section $\sigma_{c\bar{c}}$.

level. Using as input measurements of the charged and neutral spectra (Figures 7 and 9), along with plausible assumptions regarding various other background sources (Figure 13), a clear excess of electrons above the background is detected[79]. In Figure 14, the yields extracted by PHENIX are compared to previous measurements of charm hadro-production. The basis of comparison here is the presumed scaling with binary collisions discussed in the previous section. The extracted cross-sections $\sigma_{c\bar{c}}$ for charm production of $380 \pm 60(stat.) \pm 200(sys.)$ μb in minimum bias (0-92%) collisions, and $420 \pm 33(stat.) \pm 250(sys.)$ μb in central (0-10%) collisions are in good agreement with the value of $\sigma_{c\bar{c}} \approx 330$ μb obtained by PYTHIA[81] calculations at these energies.

It is intriguing to compare these results on charm production to those for mesons formed from lighter quarks. The charm yields, both for minimum bias *and* for central events, are equal to or slightly in excess of those predicted by binary scaling. This is in sharp contrast to the yields of π^0's, which show substantial suppression (Figure 10) out to the highest measured transverse momenta in Run-1. It is tempting to speculate that this is the result of enhanced energy loss for the light quarks[56], combined with the decreased energy loss expected for heavy quarks[69]. However, the relatively large errors in the charm measurement prevent any strong conclusion from the Run-1 data set. This situation should be dramatically improved in Run-2, with the more than twenty-fold increase in integrated luminosity, and with the substantially increased PHENIX aperture (compare Figures 1 and 2). These increases, together with special systematic studies made during Run-2, will significantly decrease both the statistical and systemic errors in the charm measurements, permitting direct comparisons between charm production and π^0 yields at transverse momenta up to at least 4 GeV/c. It is also worth noting that in light of recent predictions of significant JΨ formation through coalescence of open charm[82, 83], extending the precision of open charm detection at RHIC will be an essential component in understanding the fate of charmonia in the quark-gluon plasma.

CONCLUSIONS

We have reported on the systematic variation with centrality of charged particle multiplicity, transverse energy, identified particle spectra and yield ratios, production of charged hadrons and π^0's at high transverse momenta, and the production of open charm. The PHENIX physics results presented correspond to the first RHIC run with Au+Au collisions at $\sqrt{s_{NN}} = 130$ GeV.

The results of the charged particle multiplicity distribution at mid-rapidity analyzed as a function of centrality show a steady rise of the particle density per participant nucleon with centrality. The mid-rapidity transverse energy density for central Au + Au collisions is at least 1.6 times larger at $\sqrt{s_{NN}} = 130$ GeV (RHIC) than at $\sqrt{s_{NN}} = 17.2$ GeV (CERN-SPS). The E_T density per participant as a function of centrality is very similar to the charged multiplicity density per participant, and the constancy of dE_T/dN_{ch} at a value ~ 0.8 GeV indicate that the additional energy density at RHIC energies is achieved mainly by an increase in particle production rather than by an increase in transverse energy per particle.

In central Au+Au collisions the anti-proton yield is comparable to the negative pion yield at high p_T, a behavior never observed before in elementary or in heavy ion collisions. The average p_T rises with centrality similarly for all particle species, while kaon, proton and anti-proton yields per participant increase faster than the pion yields.

The spectra for charged hadrons and neutral pions above $p_T \sim 2$ GeV/c for peripheral collisions appear to be consistent with a simple and incoherent sum of underlying N+N collisions. On the other hand, the spectra from central collisions are systematically below the scaled N+N expectation when compared to both the data from p+p collisions and the spectra from Au+Au peripheral collisions, in sharp contrast to previous observations at lower energies. The suppression in central collisions is in qualitative agreement with the predictions of energy loss by scattered partons traversing a dense medium. However, it is clear that other nuclear effects should be understood before a quantitative conclusion can be drawn. In particular, measurements in proton-nucleus collisions at RHIC can help quantify the contributions of both nuclear shadowing and Cronin enhancements in this new energy regime.

The yields for open charm also appear to be consistent with expectations derived from binary scaling. Refining this measurement, and extending it towards higher transverse momenta where it can directly confront the established suppression in light meson yields, is an important goal of the PHENIX Run-2 program.

This brief discussion covers only a fraction of the many results obtained by PHENIX from RHIC Run-1. Not covered here are significant findings on elliptic flow[84], HBT correlations[87], charged fluctuations[86], fluctuations in p_T and E_T per particle[85], and Λ and $\bar{\Lambda}$ production[88]. Nonetheless, this sample of the Run-1 results is sufficient to validate the original design concept of the PHENIX detector. The combination of superb photon and electron detection, together with excellent hadron identification, has provided unique insights into RHIC collisions. Supplementing these abilities with the powerful muon detection capabilities and physics triggers in Run-2 and beyond will only add to the discover potential of PHENIX.

REFERENCES

1. C. Bernard et al., Phys. Rev. Lett. D **54** 4585-4594 (1996).
2. S.A. Bass et al., J. Phys. G **25** R1-R57 (1999).
3. D.P. Morrison et al., Nucl. Phys. A **638**, 565-570 (1998).
4. N. Saito et al., Nucl. Phys. A **638**, 575-578 (1998).
5. G. Bunce et al., Ann. Rev. Nucl. Part. Sci. **50** 525-575 (2000).
6. H. Hamagaki et al., Nucl. Phys. A **698** 412-415 (2002).
7. S. White et al., Nucl. Phys. A **698** 420-423 (2002).
8. C. Adler et al., Nucl. Instrum. Meth. A **470** 488-499 (2001).
9. R.J. Glauber, Phys. Rev. **100**, 252, (1955).
10. R.J. Glauber, *Lectures in Theoretical Physics*, edited by W.E. Brittin and L.G. Dunham, Interscience, N.Y., 1959, Vol. 1, pp. 315.
11. D. Kharzeev and M. Nardi, Phys. Lett. B **507** 121-128 (2001).
12. X.N. Wang and M. Gyulassy, Phys. Rev. D **44** 3501-3516 (2001).
13. K. Adcox et al., Phys. Rev. Lett. **86** 3500-3505 (2001).
14. P.A. Steinberg, Nucl. Phys. A **698** 314-322 (2002).
15. B.B. Back et al., Phys. Rev. Lett. **85** 3100-3104 (2000).
16. C. Adler et al., Phys. Rev. Lett. **87** 112303 (2001).
17. B.B. Back et al., Phys. Rev. C **65** 031901 (2002).
18. G.J. Alner et al., Z. Phys. C **33** 1-6 (1986).
19. K. Eskola, Nucl. Phys. A **698** 78-87 (2002).
20. X.N. Wang and M. Gyulassy, Phys. Rev. Lett. **86** 3496-3499 (2001).
21. L. Ahle et al., Phys. Rev. C **59** 2173-2188 (1999).
22. M.M. Aggarwal et al., Eur. Phys. J. C **18** 651-663 (2001).
23. F. Antinori et al., Nucl. Phys. A **661** 357-361 (1999) and Eur. Phys. J. C **18** 57-63 (2000).
24. K. Eskola et al., Nucl. Phys. B **570** 379-389 (2000).
25. T. Åkesson et al., Nucl. Phys. **353**, 1 (1991).
26. H. Satz, H.J. Specht and R. Stock, Z. Phys. C **38** 1-370 (1988).
27. D.J. Bjorken, Phys. Rev. D **27** 140-151 (1983).
28. J. Barrette et al., Phys. Rev. Lett. **70** 2996-2999 (1993).
29. K. Adcox et al., Phys. Rev. Lett. **87** 052301 (2001).
30. T. Alber et al., Phys. Rev. Lett. **75** 3814-3817 (1995).
31. For a discussion of this and several other effects affecting the systematic comparison of data sets, see the talk of I. Tserruya at the 2001 RHIC/INT workshop *"Ultra-relativistic Heavy Ion Collisions in the RHIC era"*: http://www-nsdth.lbl.gov/rhic-int-2001/talks/Tserruya.pdf
32. D. Teaney et al., Phys. Rev. Lett. **86** 4783-4786 (2001), *nucl-th/0110037*.
33. P. Kolb et al., Nucl. Phys. A **696** 197-215 (2001).
34. P. Huovinen et al., Phys. Lett. B **503** 58-64 (2001).
35. J. Schaffner-Bielich et al., nucl-th/0108048.
36. I. Vitev and M. Gyulassy, nucl-th/0104066.
37. C. Albajar et al., Nucl. Phys. B **335** 261-287 (1990).
38. K. Adcox et al., Phys. Rev. Lett. **88**, 242301 (2002)
39. L. Ahle et al., Nucl. Phys. A **638** 57-68 (1998).
40. I.G. Bearden et al., Phys. Rev. Lett. **78** 2080-2083 (1997).
41. J. Bachler et al., Nucl. Phys. A **661** 45-54 (1999).
42. H. Ohnishi et al., Nucl. Phys. A **698** 659-662 (2002).
43. I.G. Bearden et al., Phys. Rev. Lett. **87** 112305 (2001).

44. B.B. Back *et al.*, Phys. Rev. Lett. **87** 102301 (2001).
45. C. Adler *et al.*, Phys. Rev. Lett. **86** 4778-4782 (2001).
46. M. Kaneta *et al.*, J. Phys. G **23** 1865-1871 (1997).
47. H. Appelshäuser *et al.*, Phys. Rev. Lett. **82** 2471-2475 (1999).
48. B. Alpe *et al.*, Nucl. Phys. B **100** 237-290 (1975).
49. T. Alexopoulos *et al.*, Phys. Rev. D **48** 984-997(1993).
50. W. Broniowski and W. Florkowski, Phys. Rev. Lett. **87** 272302 (2001).
51. J.F. Owens *et al.*, Phys. Rev. D **18** 1501-1514 (1978).
52. M. Gyulassy and M. Plümer, Phys. Lett. B **243** 432-438 (1990).
53. R. Baier *et al.*, Phys. Lett. B **345** 277-286 (1995).
54. X.N. Wang and M. Gyulassy, Phys. Rev. Lett. **68** 1480-1483 (1992).
55. X.N. Wang, Phys. Rev. C **58** 2321-2330 (1998).
56. R. Baier, D. Schiff and B.G. Zakharov, Ann. Rev. Nucl. Part. Sci. **50** 37-69 (2000).
57. D. Antreasyan *et al.*, Phys. Rev. D **19** 764-778 (1979).
58. J.J. Aubert *et al.*, Phys. Lett. B **123** 275-278 (1983).
59. R.G. Arnold *et al.*, Phys. Rev. Lett. **52** 727-730 (1984).
60. K. Adcox *et al.*, Phys. Rev. Lett. **88** 022301 (2002).
61. D. Drijard *et al.*, Nucl. Instrum. Meth. A **225** 367-377 (1984).
62. D. L'Hôte, Nucl. Instrum. Meth. A **337** 544-556 (1994).
63. A.L.S. Angelis *et al.*, Phys. Lett. B **185** 213-226 (1987).
64. E. Wang and X.N. Wang, Phys. Rev. C **64** 034901 (2001).
65. B. Müller and X.N. Wang, Phys. Rev. Lett. **68**, 2437 (1992).
66. E. Shuryak, Phys. Rev. Lett. **68**, 3270 (1992).
67. Z. Lin and M. Gyulassy, Phys. Rev. C **51**, 2177 (1995).
68. P. Levai and R. Vogt, Phys. Rev. C **56**, 2707 (1997).
69. Y. Dokshitzer and D. Kharzeev, Phys. Lett. B **519**, 199 (2001).
70. R. Vogt *et al.*, Phys. Rev. D **49**, 3345 (1994).
71. T. Matsui and H. Satz, Phys. Lett. B **178**, 416 (1986).
72. M.C. Abreu *et al.*, Phys. Lett. B **438**, 35 (1998).
    M.C. Abreu *et al.*, Phys. Lett. B **449**, 128 (1999).
    M.C. Abreu *et al.*, Phys. Lett. B **466**, 408 (1999).
73. M.C. Abreu *et al.*, Phys. Lett. B **410**, 337 (1997).
    M.C. Abreu *et al.*, Phys. Lett. B **450**, 456 (1999).
    M.C. Abreu *et al.*, Phys. Lett. B **477**, 28 (2000).
74. D.E. Groom *et al.* [Particle Data Group Collaboration], Eur. Phys. J. C **15**, 1 (2000).
75. Y. Akiba, in *Physics with the Collider Detectors at RHIC and at the LHC*, proceedings of the pre-conference workshop, Quark Matter'95, Monterey, CA, eds. J. Thomas and T. Hallman, UCRL-ID-121571, pp. 131-142.
76. I. Hinchliffe and C.H. Llewellyn Smith, Phys. Lett. B **61**, 472 (1976).
    I. Hinchliffe and C.H. Llewellyn Smith, Nucl. Phys.. B **114**, 45 (1976).
77. M. Bourquin and J.M. Gaillard, Nucl. Phys. B **114**, 334 (1976).
78. For a complete discussion, see M.J. Tannenbaum, Heavy Ion Phys. **4**, 139 (1996).
79. K. Adcox *et al.*, Phys. Rev. Lett. **88**, 192303 (2002), .
80. *The PHENIX Conceptual Design Report*, Brookhaven National Laboratory document, January, 1993. Portions are available online at
    http://www.phenix.bnl.gov/phenix/WWW/CDR/03_physics/chapter2_1.html.
81. T. Sjostrand, Comp. Phys. Commun. **82**, 74 (1994).
82. P. Braun-Munzinger and J. Stachel, Phys. Lett. B **490**, 196 (2000).
83. R. L. Thews, M. Schroedter and J. Rafelski, Phys. Rev. C **63**, 054905 (2001).
84. K. Adcox *et al.*, submitted to Phys. Rev.Lett., nucl-ex/0204005.
85. K. Adcox *et al.*, submitted to Phys. Rev. C, nucl-ex/0203015.
86. K. Adcox *et al.*, to appear in Phys. Rev. Lett., nucl-ex/0203014.
87. K. Adcox *et al.*, Phys. Rev. Lett. **88**, 192302 (2002).
88. K. Adcox *et al.*, submitted to Phys. Rev.Lett., nucl-ex/0204007.

Survey of Recent Results from the PHOBOS Experiment at RHIC

Christof Roland for the PHOBOS Collaboration

B.B.Back[1], M.D.Baker[2], D.S.Barton[2], R.R.Betts[1,6], M.Ballintijn[4],
A.A.Bickley[7], R.Bindel[7], A.Budzanowski[3], W.Busza[4], A.Carroll[2],
M.P.Decowski[4], E.Garcia[7], N.George[1], K.Gulbrandsen[4], S.Gushue[2],
C.Halliwell[6], J.Hamblen[8], G.A.Heintzelman[2], C.Henderson[4],
D.J.Hofman[6], R.S.Hollis[6], R.Hołyński[3], B.Holzman[6], A.Iordanova[6],
E.Johnson[8], J.L.Kane[4], J.Katzy[4], N. Khan[8], W.Kucewicz[6], P.Kulinich[4],
C.M.Kuo[5], W.T.Lin[5], S.Manly[8], D.McLeod[6], J.Michałowski[3],
A.C.Mignerey[7], R.Nouicer[6], A.Olszewski[2,3], R.Pak[2], I.C.Park[8],
H.Pernegger[4], C.Reed[4], L.P.Remsberg[2], M.Reuter[6], C.Roland[4],
G.Roland[4], L.Rosenberg[4], P.Sarin[4], P.Sawicki[3], W.Skulski[8],
S.G.Steadman[4], P.Steinberg[2], G.S.F.Stephans[4], M.Stodulski[3],
A.Sukhanov[2], J.-L.Tang[5], R.Teng[8], A.Trzupek[3], C.Vale[4], G.J.van
Nieuwenhuizen[4], R.Verdier[4], B.Wadsworth[4], F.L.H.Wolfs[8], B.Wosiek[3],
K.Woźniak[3], A.H.Wuosmaa[1], B.Wysłouch[4]

[1] Physics Division, Argonne National Laboratory, Argonne, IL
[2] Chemistry and C-A Departments, Brookhaven National Laboratory, Upton, NY
[3] Institute of Nuclear Physics, Kraków, Poland
[4] Laboratory for Nuclear Science, Massachusetts Institute of Technology, Cambridge, MA
[5] Department of Physics, National Central University, Chung-Li, Taiwan
[6] Department of Physics, University of Illinois at Chicago, Chicago, IL
[7] Department of Chemistry and Biochemistry, University of Maryland, College Park, MD
[8] Department of Physics and Astronomy, University of Rochester, Rochester, NY

Abstract. We present an overview of the latest results for interactions of Au+Au ions at center-of-mass energies of $\sqrt{s_{NN}}$ of 56, 130 and 200 GeV obtained by the PHOBOS collaboration at the Relativistic Heavy Ion Collider (RHIC). These data have allowed us to perform an extensive study of the pseudorapidity density of primary charged particles as a function of incident energy, centrality and pseudorapidity. Our results show a non-trivial evolution of particle densities with both centrality and collision energy, reaching significantly higher values per participating nucleon than at lower energies or in nucleon-nucleon collisions. Further we present results on the azimuthal asymmetry of particle production observed in the $\sqrt{s_{NN}}$ of 130 GeV data set. The observed strong event anisotropy of $v_2^{max} > 0.06$, reaching beyond the value predicted in hadronic cascade models, indicates a closer approach to local thermal equilibrium than at lower collision energies. The measured antiparticle-particle ratios of production rates for pions kaons and protons in central Au+Au interactions at $\sqrt{s_{NN}}$ of 130 GeV are compatible with predictions from statistical models, showing an approach to a baryon free region in mid-rapidity with the increase in collision energy.

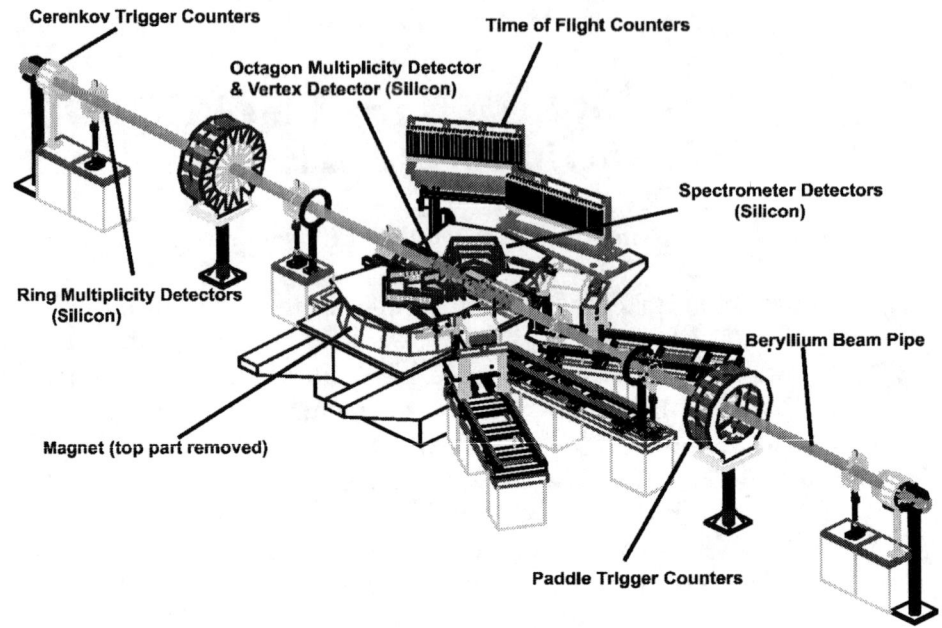

FIGURE 1. PHOBOS detector setup for the 2001 running period.

INTRODUCTION

In this paper we present data taken with the PHOBOS detector during the first two runs of the Relativistic Heavy-Ion Collider (RHIC) at Brookhaven National Laboratory (BNL). In the year 2000 run we recorded collisions of Au-nuclei at energies of $\sqrt{s_{NN}}$ = 56 and 130 GeV and in 2001 we took data at $\sqrt{s_{NN}}$ = 200 GeV.

The main goal of the RHIC research program is to study the behaviour of strongly interacting matter under conditions of extreme temperatures and energy densities satisfying the prerequisites predicted for the creation of the Quark Gluon Plasma (QGP).

In the measurements performed by the PHOBOS collaboration special emphasis has been placed on first characterizing the global conditions achieved in this new energy regime of nuclear collisions. In particular, we tried to obtain information on the overall density of particles produced in the collision, the azimuthal anisotropy of the final state particle distribution, and on particle ratios near mid-rapidity. As we will discuss later, these measurements are related to questions of entropy density, the approach to thermal equilibration and the hadro-chemical composition of the particle source.

EXPERIMENTAL SETUP AND EVENT SELECTION

The PHOBOS detector employs silicon pad detectors to perform tracking, vertex detection and multiplicity measurements. An overview of the apparatus as it was implemented for the year 2001 run is shown in Fig. 1. The main components are the single layer multiplicity detector which covers the pseudo-rapidity range $|\eta| < 5.4$ with nearly complete azimuthal coverage, a two-layer vertex detector (VTX) covering $|\eta| < 1.5$, and a two arm magnetic spectrometer providing momentum and particle ID information for particles near mid-rapidity. Each spectrometer arm consists of 16 layers of silicon, 6 of which are located in the field free region between the interaction point and the magnet. In this configuration, the detector has close to 140000 readout channels. Details of the experimental arrangement are described in [1]. In the year 2000 run only a fraction of the full setup was installed. A description of this setup can be found in [2]. The primary event trigger was provided by two sets of 16 scintillator paddle counters that detect charged particles in the pseudorapidity range $3 < |\eta| < 4.5$. The paddle counters were also used for offline event selection and determination of the collision centrality. Not shown in Fig. 1 are two zero-degree calorimeters at $z = \pm 18.5$ m, which provided additional information for event selection by measuring the energy deposited by spectator neutrons.

The online trigger selected events based on the coincidence of at least one hit in each set of paddle counters within a time window of 38 ns. This trigger was sensitive to 97% of the inelastic hadronic cross-section. By applying further timing cuts offline, on both the paddle counter and ZDC signals, we achieved an overall background rate of less than 1% of all events. Central events, which were characterized by large signals in both paddle counters, were essentially background-free.

To divide our data set into classes of events according to the number of participating nucleons, we used information from the paddle counters and the zero degree calorimeters. Every event was characterized by a centrality estimator based on the truncated mean of gain-normalized ADC values in the 16 scintillator elements in each paddle counter. We estimated the number of participants for the event classes using a Glauber calculation based on HIJING [3] and our GEANT simulation. Details of this procedure can be found in [7].

RESULTS

Energy dependence of $dN_{ch}/d\eta|_{|\eta|<1}$

Charged particle production, as characterized by $dN_{ch}/d\eta$, reflects the time integral of many processes in the evolution of a heavy-ion collision. This starts with the initial conditions of the collision, e.g. parton saturation, the interplay of hard and soft processes in the initial particle production extends to the effects of re-scattering close to the freeze-out stages. Clearly, a systematic study as a function of system size, collision energy and over a large range of η is needed to disentangle the various effects in comparison to collision models.

The first observable studied was the energy dependence of the charged particle pseudo-rapidity density near mid-rapidity for central collisions, $dN_{ch}/d\eta|_{|\eta|<1}$.

We selected the 6% most central collisions using the signal from the paddle counters. The measurement is described in detail in [2]. It was based on counting two hit-combinations, called tracklets, in consecutive layers of our silicon detectors, with the

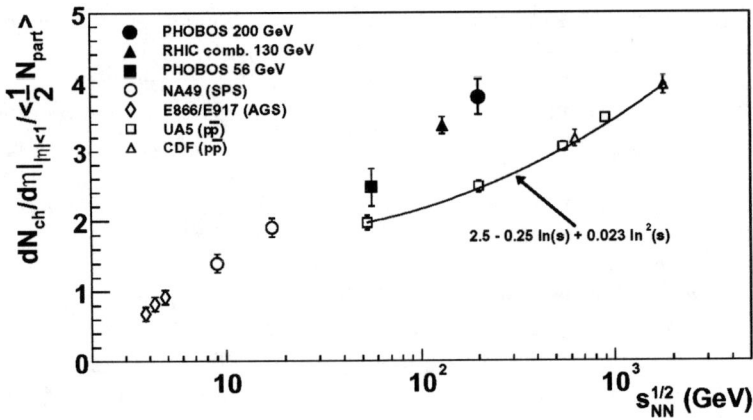

FIGURE 2. Measured pseudorapidity density normalized per pair of participating nucleons for central Au+Au collisions. The data are compared with $p\bar{p}$ data and nucleus-nucleus data from lower energies.

constraint that the two hits form a straight line pointing to the known primary collision vertex. The results are corrected for backgrounds from decay feed-down, secondary interactions, stopping particles in the beampipe and detector inefficiencies based on the MC simulation of our apparatus.

Fig. 2 shows a compilation of all results on the charged particle pseudo-rapidity density normalized yield per participant $dN_{ch}/d\eta|_{|\eta|<1}/\langle\frac{1}{2}N_{part}\rangle$ obtained for Au+Au collisions. It includes results from PHOBOS at $\sqrt{s_{NN}} = 56$ GeV [2], from all four RHIC experiments at $\sqrt{s_{NN}} = 130$ GeV [2, 4, 5, 6] and the latest PHOBOS result at $\sqrt{s_{NN}} = 200$ GeV [1]. For comparison also shown are proton-antiproton ($p\bar{p}$) collisions [11] and central Au+Au and Pb+Pb collisions at lower energies measured in fixed target experiments at the AGS and CERN-SPS [12, 13].

The energy evolution of $dN_{ch}/d\eta|_{|\eta|<1}/\langle\frac{1}{2}N_{part}\rangle$ in central heavy ion collisions follows an approximately logarithmic rise, within the precision of the existing data. This modest increase of particle production with collision energy provides important constraints on entropy production. It specifically limits the amount of particle production from hard scattering and gluon radiation (jet quenching) [14].

Centrality dependence of $dN_{ch}/d\eta|_{|\eta|<1}$

Model calculations suggest that important additional information on the mechanism responsible for the enhanced particle production observed in central Au+Au collisions can be obtained by studying the particle density as a function of system size [16]. With the present data set, a variation of system size can be achieved by studying $dN_{ch}/d\eta$ for events selected in different bins of fractional cross-section based on the paddle mean, corresponding to a variation in the average number of participants. A detailed discussion

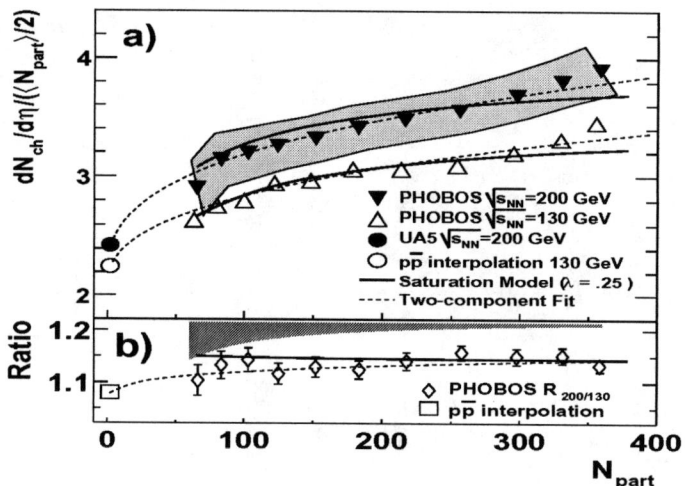

FIGURE 3. a) Normalized pseudorapidity density per pair of participating nucleons $dN_{ch}/d\eta|_{|\eta|<1}/\langle\frac{1}{2}N_{part}\rangle$ as a function of the number of participants. The error band around the 200 GeV data combines the error on $dN_{ch}/d\eta|_{|\eta|<1}$ and $\langle N_{part}\rangle$. b) Ratio $R_{200/130}$ of $dN_{ch}/d\eta|_{|\eta|<1}$ at $\sqrt{s_{NN}} = 130$ GeV and 200 GeV as a function of the number of participants. The grey band indicates the systematic error estimate.

of this measurement can be found in [7, 8]. The measured centrality dependence of $dN_{ch}/d\eta|_{|\eta|<1}/\langle\frac{1}{2}N_{part}\rangle$ at $\sqrt{s_{NN}} = 130$ GeV and $\sqrt{s_{NN}} = 200$ GeV is shown in Fig. 3. The grey band shows the systematic uncertainty in the multiplicity measurement and the determination of $\langle N_{part}\rangle$ based on the fractional cross-section [7, 9].

At $\sqrt{s_{NN}} = 200$ GeV we observe a steep initial rise from the $p\bar{p}$ value to our first data point at $\langle N_{part}\rangle = 83$ and then a slow increase up to our most central bin with $\langle N_{part}\rangle = 358$. For both incident energies we observe a similar evolution of the charged particle density per participant pair with the collision centrality. Panel b) of Fig. 3 shows the ratio $R_{200/130}$ of $dN_{ch}/d\eta|_{|\eta|<1}$ at $\sqrt{s_{NN}} = 130$ GeV and 200 GeV. This is found to be approximately constant at $1.14 \pm 0.01(stat.) \pm 0.05(syst.)$ over the range of the 50% most central collisions.

To put these results in a context, we also show two calculations. The first calculation shown is the prediction of the parton saturation model described in Ref. [16] for both energies, indicated by solid lines in Fig. 3. The second calculation, denoted by the dashed line, involves fits to the data using the two-component parametrization proposed in Ref. [15], $dN_{ch}/d\eta = n_{pp}((1-x)\langle N_{part}\rangle/2 + x\langle N_{coll}\rangle)$. The parameters have a simple interpretation as the fraction of production from hard processes (x) and the number of particles associated with a single pp interaction (n_{pp}) when $N_{part} = 2$ and $N_{coll} = 1$. Using the measured values for n_{pp} we find that values of $x = 0.09 \pm 0.02$ for $\sqrt{s_{NN}} = 130$ GeV and $x = 0.11 \pm 0.02$ for $\sqrt{s_{NN}} = 200$ GeV account well for the measured centrality dependence of $dN_{ch}/d\eta|_{|\eta|<1}/\langle\frac{1}{2}N_{part}\rangle$.

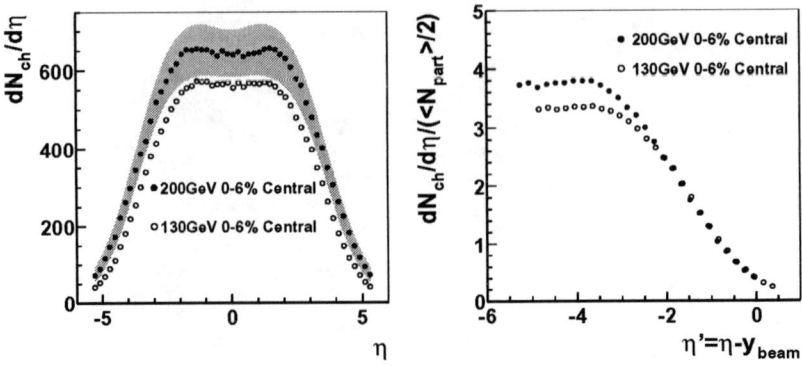

FIGURE 4. Left panel. Measured $dN_{ch}/d\eta$ for the 6% most central collision at $\sqrt{s_{NN}} = 200$ GeV (filled symbols) and at $\sqrt{s_{NN}} = 130$ GeV (open symbols) respectively. Right panel. Same data plotted as $dN_{ch}/d\eta'$ where $\eta' \equiv \eta - y_{beam}$.

Further we observe that, while the two calculations evolve in opposite directions in $R_{200/130}$ as N_{part} decreases, they remain sufficiently close down to $\langle N_{part} \rangle \sim 65$ such that our present data cannot resolve them definitely.

Charged particle density in $|\eta| < 5.4$

Our data on $dN_{ch}/d\eta|_{|\eta|<1}$ clearly show a non-trivial evolution of the mid-rapidity particle density going from the most peripheral to the most central collisions. Further information on the underlying physics can be obtained by studying the change in particle production over a wider range in η. We performed this measurement using the PHOBOS multiplicity detector, consisting of the Octagon barrel covering $|\eta| \leq 3.2$ and the Ring counters detecting particles with $3 \leq |\eta| < 5.4$.

The number of charged particles traversing the detector was estimated based on the number and energy of hits in the silicon detectors, i.e. Si pixels showing a charge deposition above a certain threshold. Details of the hit-counting procedure and the background correction can be found in [1, 10]. We performed this measurement for different centrality classes ranging from $\langle N_{part} \rangle = 83$ to $\langle N_{part} \rangle = 353$. In the left panel of Fig. 4 we show the $dN_{ch}/d\eta$ distributions for 6% most central collisions. Integrating these distributions we find an average number of charged particles within $|\eta| < 5.4$ of $4100 \pm 210(syst)$ for $\sqrt{s_{NN}} = 130$ GeV and $4960 \pm 250(syst)$ for 200 GeV.

To further compare these distributions away from mid-rapidity we can study particle production in the beam (or target) rest frame. The right panel of Fig. 4 shows the same data plotted as $dN_{ch}/d\eta'$ where $\eta' \equiv \eta - y_{beam}$. In $p+p$ and $p+\bar{p}$ collisions it was observed that particle production in the fragmentation region is independent of the collision energy, as predicted by the "limiting fragmentation" hypothesis [17, 18].

We find that the $dN_{ch}/d\eta'$ distributions at $\sqrt{s_{NN}} = 130$ GeV and 200 GeV overlap in

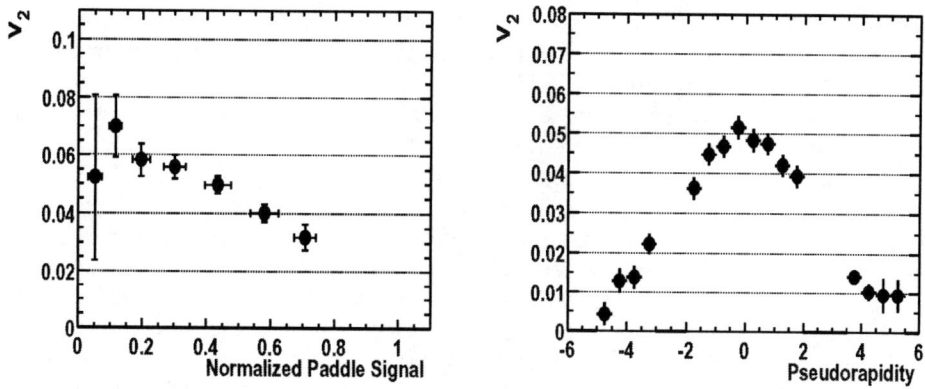

FIGURE 5. Left panel: Elliptic flow, v_2, in the region $-1.0 < \eta < 1.0$ as a function of centrality. Right panel: Elliptic flow, averaged over centrality, as a function of η. The error bars are statistical only. We estimate a systematic error of $\Delta v_2 = 0.007$.

the fragmentation region in broad range of more than 2 units of η'. Thus we observe a similar "limiting fragmentation" behavior in Au+Au collisions at the highest two RHIC energies as in nucleon-nucleon collisions. This universal scaling behavior limits the increased entropy production in high energy heavy ion interactions to be concentrated around mid-rapidity.

Event Anisotropy

Based on hydrodynamical model calculations, it has been argued that a systematic study of the final state momentum anisotropy can provide a sensitive probe of the approach to local thermal equilibrium in the particle source in heavy ion collisions and also of the equation of state in the various stages of the collision [19, 20].

The momentum space anisotropy, which is a reflection of the initial-state space anisotropy for non-central collisions, is commonly quantified by the first and second Fourier coefficients of the azimuthal particle distribution relative to the reaction plane determined for every event, $dN/d(\phi - \Psi_R) = N_0(1 + \sum_i 2v_i \cos[n(\phi - \Psi_R)])$, where Ψ_R is the reaction plane angle. Usually, the first Fourier component (v_1) is referred to as directed flow and the second one (v_2) as elliptic flow. Both components have been observed and studied in detail at lower collision energies. Early predictions of hadronic cascade models suggested that the elliptic flow should be of equal magnitude as that observed at collisions at the CERN SPS and, therefore, significantly below the limit expected for a system in local thermal equilibrium. For the directed flow, a trivial decrease of the observed strength is expected based on the larger longitudinal momenta at the higher RHIC collision energies.

In PHOBOS, we have studied elliptic flow based on the azimuthal distribution of hits in the Octagon detector. For every event the event plane angle was determined by a fit to the azimuthal hit distribution. The observed raw value for v_2 was corrected for the effects of detector granularity, the resolution of the event-by-event determination of the event plane relative to the true reaction plane, and the contribution of background hits. Details of the analysis can be found in [21].

We determined v_2 for events in seven classes of centrality, based on the paddle-mean. The resulting dependence on collision centrality is shown in the left panel of Fig. 5. We observed a strong centrality dependence of v_2 ranging from 0.03 for the most central events up to 0.07 for peripheral events. This follows the trend expected from the magnitude of the initial-state spatial anisotropy. The values observed by PHOBOS are in good agreement with those reported by the STAR collaboration [22]. The maximum value of $v_2 = 0.07$ is about twice as large as v_2 obtained in Pb+Pb collisions at the SPS. Contrary to cascade code predictions, we observe a strong increase between SPS to RHIC energies, indicating a much closer approach to local thermal equilibrium. It is a matter of intense theoretical discussion at which stage in the collision process this equilibrium is established [20]. We have also studied the pseudorapidity dependence of v_2 over a large range in centrality. Outside a central region of $|\eta| < 1$ v_2 drops very quickly (see right panel of Fig. 5), approximately following the shape of the charged particle distribution as shown in Fig. 4.

Particle ratios

The ratio of antibaryons to baryons plays a key role in understanding the properties of the hot and dense system formed in heavy-ion collisions at high energies. Using the mid-rapidity spectrometer, we have studied the ratios of multiplicities of particles and antiparticles for primary charged pions, kaons and protons at $\sqrt{s_{NN}} = 130$ GeV. The particle ratios were measured for the 12% most central events, for which we estimate $\langle N_{part} \rangle = 312 \pm 10$(syst). As the geometrical layout of the PHOBOS detector leads to an asymmetry in the acceptance and detection efficiency for positively and negatively charged particles in the same events, we used data taken with both polarities of the PHOBOS magnet. The most important systematic effects, such as geometrical acceptance and tracking efficiency, cancel in the ratio. The raw particle ratios are corrected for particle decays, secondary interactions, and the contribution from feed-down of strange hadrons. Further details can be found in [23]. Within our acceptance we find the following ratios:

$$\langle \pi^- \rangle / \langle \pi^+ \rangle = 1.00 \pm 0.01 (\text{stat}) \pm 0.02 (\text{syst})$$
$$\langle K^- \rangle / \langle K^+ \rangle = 0.91 \pm 0.07 (\text{stat}) \pm 0.06 (\text{syst})$$
$$\langle \bar{p} \rangle / \langle p \rangle = 0.60 \pm 0.04 (\text{stat}) \pm 0.06 (\text{syst})$$

We estimated a baryo-chemical potential μ_B from the $\langle K^- \rangle / \langle K^+ \rangle$ and $\langle \bar{p} \rangle / \langle p \rangle$ ratios, using a statistical model calculation [24]. For a realistic range of freeze-out temperatures of 160 to 170 MeV, both ratios are consistent with $\mu_B = 45 \pm 5$ MeV. This value is much lower than $\mu_B = 240 - 270$ MeV obtained in thermal fits to Pb+Pb data from the CERN SPS [25, 26], showing a closer, but not yet complete approach to a baryon-free regime at RHIC.

ACKNOWLEDGMENTS

This work was partially supported by U.S. DOE grants DE-AC02-98CH10886, DE-FG02-93ER40802, DE-FC02-94ER40818, DE-FG02-94ER40865, DE-FG02-99ER41099, and W-31-109-ENG-38 as well as NSF grants 9603486, 9722606 and 0072204. The Polish groups were partially supported by KBN grant 2 PO3B 04916. The NCU group was partially supported by NSC of Taiwan under contract NSC 89-2112-M-008-024.

REFERENCES

1. B.Back et al.,Phys. Rev. Lett. **88** (2002) 22302.
2. B.Back et al.,Phys. Rev. Lett. **85** (2000) 3100.
3. M.Gyulassy and X.N.Wang, Comp. Phys. Comm. **83** (1994) 307. We used HIJING V1.35 with standard parameter settings.
4. K. Adcox *et al.*, Phys. Rev. Lett. **86** (2001) 3500.
5. F. Videbaek *et al.*, Nucl. Phys. **A698** (2002) 29c.
6. C. Adler *et al.*,Phys. Rev. Lett. **87** (2001) 112303.
7. B.Back et al., Phys. Rev. **65** (2002) 31901R.
8. B.Back et al., nucl-ex/0201005, accepted by Phys. Rev. C - Rapid Communications.
9. A. Olszewski for the PHOBOS Collaboration, "Centrality measurements in the PHOBOS experiment", Epiphany 2002 Conference, Cracow, to be publlished in Acta Physica Polonica B.
10. B.Back et al., Phys. Rev. Lett. **87** (2001) 102303.
11. F.Abe et al., Phys. Rev. **D41** (1990) 2330.
12. L. Ahle et al., Phys. Lett. **B490** (2000) 53, Phys. Rev. **C60**, (1999) 064901, Phys. Rev. Lett. **81**, (1998) 2650 .
13. J. Bächler et al., Nucl. Phys. **A661** (1999) 45.
14. M. Gyulassy and X. N. Wang, Phys.Rev.Lett. **86** (2001) 3496-3499.
15. D. Kharzeev and M. Nardi, Phys. Lett.**B507** (2001) 121.
16. D. Kharzeev and E. Levin,Phys. Lett. **B523** (2001) 79-87
17. J. Benecke, T. T. Chou, C-N. Yang, E. Yen, Phys. Rev. **188** (1969) 2159.
18. UA5 Collaboraation, Z. Phys. **C33** (1986) 1.
19. J.-Y. Ollitrault, Phys. Rev. **D46**(1992) 229.
20. D. Teaney, J. Lauret, E.V. Shuryak, Phys.Rev.Lett. **86** (2001) 4783-4786.
21. I. C. Park et al., Nuclear Physics **A698** (2002) 564c.
22. K. Ackermann, STAR coll. Phys. Rev. Lett. **86** (2001) 402.
23. B. Back et al. Phys. Rev. Lett. **87** (2001) 102301.
24. K. Redlich, Nuclear Physics **A698** (2002) 94c.
25. F. Becattini, Z. Phys. **C69** (1996) 485.
26. P. Braun-Munzinger, I. Heppe, J. Stachel, Phys. Lett. **B465** (1999) 15.

$K^*(892)^0$ Production in Relativistic Heavy Ion Collisions

P. Fachini* and the STAR Collaboration[†]

*Upton 11973-5000, NY
[†]for a complete list see [1]

Abstract. Preliminary results on the first measurement of $K^*(892)^0 \to \pi K$ production using the mixed-event technique are presented. The measurements are performed at mid-rapidity by the STAR detector in $\sqrt{s_{NN}} =$ 130 GeV Au-Au collisions at RHIC. The K^{*0} to negative hadron, kaon and ϕ ratios are obtained and compared to the measurements in e^+e^-, pp and $\bar{p}p$ at various energies. Despite the K^{*0} short lifetime ($c\tau \simeq$ 4 fm) and the expected losses due to the re-scattering of the decay products in the dense matter, no dramatic reduction on this resonance production is observed.

INTRODUCTION

The main motivation for studying heavy ion collisions at high energy is to investigate the properties of the strongly interacting matter at high densities and temperatures. In particular, the masses and widths of resonances are expected to change in hadronic or nuclear matter compared to their vacuum values. Various theoretical models predict modification of resonance masses and widths in a dense and hot medium as a signal of a possible phase transition of nuclear matter in relativistic heavy ion collisions [2, 3]. In this context, the measurement of the properties of vector mesons whose lifetimes are of the order of the lifetime of the dense matter may be sensitive to the properties of the strongly interacting matter in which they are produced. For example, model calculations show that the K^{*0}/K ratio is sensitive to the mass modification of particles in-medium and the dynamic evolution of the source [4].

The resonances that decay into strongly interacting hadrons before thermal freeze-out may not be reconstructed due to the re-scattering of the daughter particles. However, resonances with higher transverse momentum (p_T) may have a larger probability of decaying outside the system and of being detected. Therefore, the measurement of the yields and the p_T distributions of resonances can provide information on the expansion time of the system between chemical and kinetic freeze-out. In addition, from a detailed comparison between the yield and the p_T distribution of resonances and stable hadrons, we may be able to distinguish between a sudden freeze-out [5, 6] or a smooth hadronic expansion [7, 8].

On the other hand, the elastic interactions $\pi K \to K^{*0} \to \pi K$ can increase the K^{*0} population due to the large population of π's and K's [9, 10, 11] after chemical freeze-out, when the inelastic interactions are too infrequent to change the particle species and the total number of particles. This regeneration mechanism may compensate for the K^{*0} resonances that decay during a possible long expansion time of the system between chemical and thermal (kinetic) freeze-out.

The leptonic decay modes of vector mesons have been studied extensively, even though the hadronic decay modes are dominant. This is largely due to the large background from other produced hadrons, the broad mass width and the final interactions of the hadronic decay products that destroy the early time information of these resonances. However, Monte Carlo calculations have shown that the mixed-event technique [12, 13] should allow a statistical measurement of both $K^*(892)^0$ and $\overline{K}^*(892)^0$ at the Relativistic Heavy Ion Collider (RHIC) energies because the significance of the signal increases with the square root of the number of events.

Preliminary results from the first observation of such short-lived resonance ($c\tau = 4$ fm) via its hadronic decay channel in Au-Au relativistic heavy ion collisions at $\sqrt{s_{NN}} = 130$ GeV using the STAR (Solenoidal Tracker At RHIC) detector at RHIC are presented. The ratios K^{*0}/h^-, K^{*0}/K and ϕ/K^{*0} in both central and minimum bias collisions are obtained and compared to the measurements in e^+e^-, pp and $\bar{p}p$ at various energies.

DATA ANALYSIS

The main STAR detector consists of a large Time Projection Chamber (TPC) [14] placed inside a uniform solenoidal magnetic field oriented parallel to the beam direction that provides the measurement of charged particle momenta. A Central Trigger Barrel (CTB) of scintillation counters surrounding the TPC is used to select the top 14% central events by measuring the charged particle multiplicity. Two hadronic calorimeters (ZDCs) located upstream along the beam axis intercept spectator neutrons from the collision and provide the minimum bias trigger.

The first collisions between Au nuclei at $\sqrt{s_{NN}} = 130$ GeV took place at RHIC in the summer of 2000. During this period, the solenoidal magnet operated with a magnetic field of 0.25 T. In this analysis we use about 440K central and 230K minimum bias events that survive the cuts mentioned below. The central events correspond to 14% of the hadronic cross-section. In order to assure uniform acceptance in the pseudo-rapidity range studied, only events with a vertex located within ±95 cm of the center of the TPC along the beam direction are selected in this analysis. The identification of the daughter particles from the K^{*0} decay is obtained from their energy loss (dE/dx) in the gas of the TPC. The tracks are required to cross the entire active area of the detector, to point to the event vertex within a certain accuracy, to have more than a designated minimum number of TPC space points along the reconstructed track and to have a transverse momentum between 0.2 GeV/c and 2.0 GeV/c.

For the measurements, the decay channels $K^{*0} \to \pi^- K^+$ and $\overline{K}^{*0} \to \pi^+ K^-$, each with a branching ratio of 2/3, are selected. Due to limited statistics, the term K^{*0} in this analysis refers to the average of K^{*0} and $\overline{K}^{*0}$ unless specified. The invariant mass of every opposite sign $K\pi$ pair is calculated for each event. The resulting invariant mass distribution consists of the resonance signal and the combinatorial background. The shape of the uncorrelated combinatorial background is determined by mixing events of comparable centrality and with primary vertex locations within 20 cm in the beam direction, which minimizes possible fluctuations and distortions in the uncorrelated background.

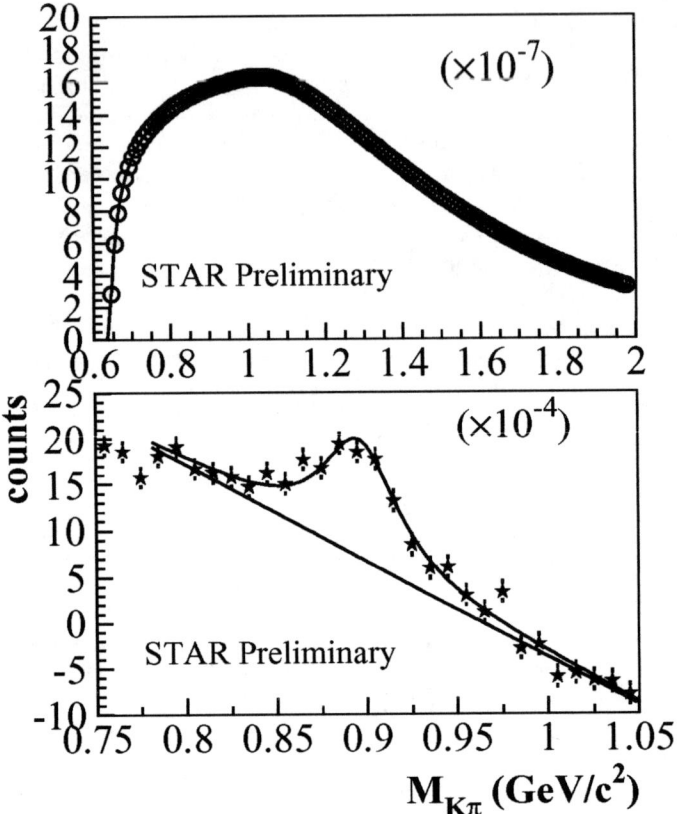

FIGURE 1. Top Panel: $K\pi$ invariant mass distribution from same event pairs (open circles) and mixed-event pairs (solid curve) for the 14% central collisions. The x-axis corresponds to 10 MeV per bin and the data are scaled down by a factor of 10^{-7}. Bottom Panel: K^{*0} invariant mass distribution after background subtraction for the 14% central collisions. The solid curve corresponds to a combination of a linear background and a simple Breit-Wigner function are used to fit the distribution with the resonance width and mass from the Particle Data Book. The x-axis corresponds to 10 MeV per bin and the data are scaled down by a factor of 10^{-4}.

RESULTS

The upper panel of Fig 1 depicts the invariant mass distributions from same-event $K\pi$ pairs and from mixed-event pairs for the 14% most central events. There are more than 14×10^9 pairs of selected kaons and pions from these central events. The signal to background ratio is about 1/1000 for central and 1/200 for minimum bias events, which is significantly lower than the value of 1/4 observed for pp at $\sqrt{s_{NN}} = 63$ GeV [15]. The $K\pi$ invariant mass distribution after mixed-event background subtraction for the 14% most central hadronic interactions is shown in the lower panel of Fig 1. A 15 standard deviation (σ) signal for the K^{*0} above the background is observed.

Only the signal and the correlated background are present in the distribution after background subtraction, since the mixed-event technique [12, 13] describes the shape of the uncorrelated background distribution. Besides the $K^*(892)^0$ resonance, in the $K\pi$ channel there is the $K\pi$ and S-wave correlation [16], which is not a resonance, as well as a long list of higher resonant states. All these contribute to the $K\pi$ correlation. In addition, particle misidentification of the decay products of ρ, ω, η, η' and K_S^0 produce correlations in the same-event distribution that are not present in the mixed-event distribution used to estimate the background. Studies using the HIJING model [17, 18] and an invariant mass distribution produced from like-sign $K\pi$ pairs from the data are consistent with the conclusion that the general features of the residual combinatorial background in the $K\pi$ invariant mass distribution after mixed-event background subtraction come from the correlations mentioned. Since a quantitative description requires the accurate knowledge of particle production and phase space distributions that are not yet measured at RHIC energies, the residual background was fit by both a linear and an exponential function. The differences between the results obtained from the two fits was about 20%. These differences are included in estimating the systematical uncertainties.

The m_T - m_0 spectra is obtained by fitting the K^{*0} signal to a Breit-Wigner resonant function and correcting for detector acceptance and efficiency. The acceptance and reconstruction efficiency is determined by embedding simulated tracks into real events at the raw data level, reconstructing the full events and comparing the simulated input to the reconstructed output. The acceptance and efficiency factor, ε, depends on the transverse momentum, the event centrality and the rapidity of the parent and daughter particles. The value of ε varies from under 15% for $p_T \simeq 0$ GeV/c to about 35% for $p_T \simeq 2.0$ GeV/c.

Fig 2 depicts the $d^2N/(2\pi m_T dm_T dy)$ distribution as a function of m_T - m_0 for central interactions at mid-rapidity ($|y| < 0.5$). An exponential fit is used to extract the K^{*0} yield per unit of rapidity around mid-rapidity and the inverse slope (T). We obtain dN/dy = 10.0 ± 0.8 (stat) and T = 0.41 ± 0.02 (stat) GeV. The K^{*0} and $\overline{K}^{*0}$ invariant mass distributions integrated in p_T are fit separately to a Breit-Wigner resonant function and a linear residual background. The ratio $\overline{K}^{*0}/K^{*0} = 0.92 \pm 0.14$ (stat) is obtained for central collisions. Therefore, the average value between K^{*0} and $\overline{K}^{*0}$ should represent the $K^*(892)^0$ production well within our statistics. The systematical uncertainty in dN/dy is estimated to be 25% due to both detector effects and the uncertainty in the background determination. The systematical uncertainty in the inverse slope is estimated to be 10% by varying the analysis cuts and studying the detector effects. In order to estimate the K^{*0} yield, we assume that the K^{*0} m_T distribution for both central and minimum bias events have the same shape. This is necessary due to limited statistics in the minimum bias data set. The K^{*0} yield per unit of rapidity around mid-rapidity obtained for the minimum bias data set is dN/dy = 4.5 ± 0.7 (stat) ± 1.4 (sys). The difference between [19] and the results presented here is that we are able to measure the slope of 410 ± 20 (stat) MeV instead of an assumed 300 MeV slope, and we use a linear function to describe the background instead of an exponential function. Each of these lowers the yield by about 20%.

The K^{*0}/h^- ratios measured for central and minimum bias events are compared to the measurements in pp [15] and e^+e^- [20, 21, 22]. The h^- yield corresponds to the

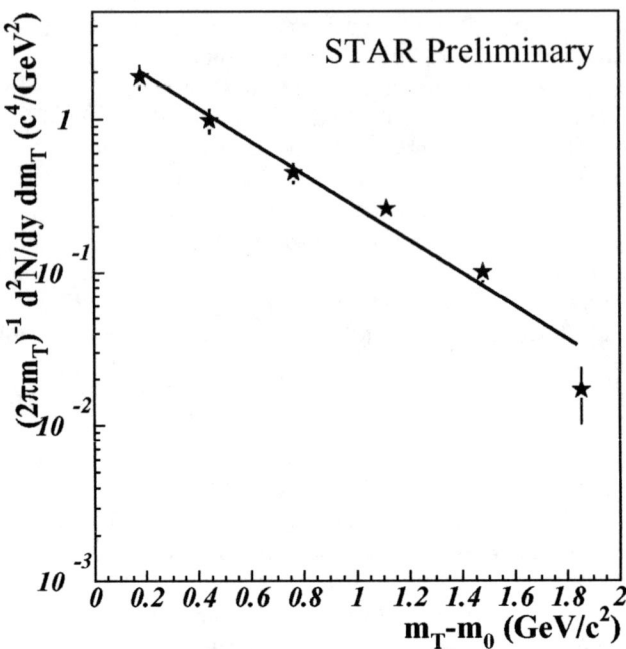

FIGURE 2. $(K^{*0} + \overline{K^{*0}})/2$ m_T spectrum at mid-rapidity ($|y| < 0.5$) for the 14% most central interactions in the p_T range from 0.4 GeV/c to 3.2 GeV/c. The errors shown are statistical only.

yield of primary negatively charged hadrons at $|\eta| < 0.5$ [23]. At $\sqrt{s_{NN}} = 130$ GeV, $K^{*0}/h^- = 0.042 \pm 0.004$ (stat) ± 0.01 (sys) for central collisions and $K^{*0}/h^- = 0.059 \pm 0.008$ (stat) ± 0.019 (sys) for minimum bias events. These results are compared to $K^{*0}/\pi = 0.057 \pm 0.012$ measured in pp at $\sqrt{s_{NN}} = 63$ GeV [15] and $K^{*0}/\pi = 0.044 \pm 0.003$ measured in e^+e^- at $\sqrt{s_{NN}} = 91$ GeV [20, 21, 22]. The error in the measurements in pp and e^+e^- corresponds to the quadratic sum of the systematic and statistical errors. We observe that the K^{*0}/h^- ratio is comparable to the K^{*0}/π ratios measured in pp and e^+e^-, noting that about 80% of h^- are pions at RHIC energies. We do not observe the enhancement in the K^{*0} production observed for other strange particles as function of increasing energies and colliding systems [24].

The K^{*0}/K and ϕ/K^{*0} ratios measured in different colliding systems at various energies are shown in Fig. 3. The K^{*0}/K ratio is interesting because K^{*0} and K have similar quark content and differ mainly in mass and spin. Fig 3 shows that the ratio $K^{*0}/K = 0.26 \pm 0.03$ (stat) ± 0.07 (sys) measured in central Au-Au collisions at 130 GeV beam energy is compatible or even lower than the measurements in $\bar{p}p$, pp and e^+e^- at lower energies.

The ϕ/K^{*0} ratio measures the strangeness suppression to good approximation since $\Delta S = 1$ with hidden strangeness in the ϕ, and there is only a small mass difference. Fig 3 shows that the ratio $\phi/K^{*0} = 0.49 \pm 0.05$ (stat) ± 0.12 (sys) measured in central Au-

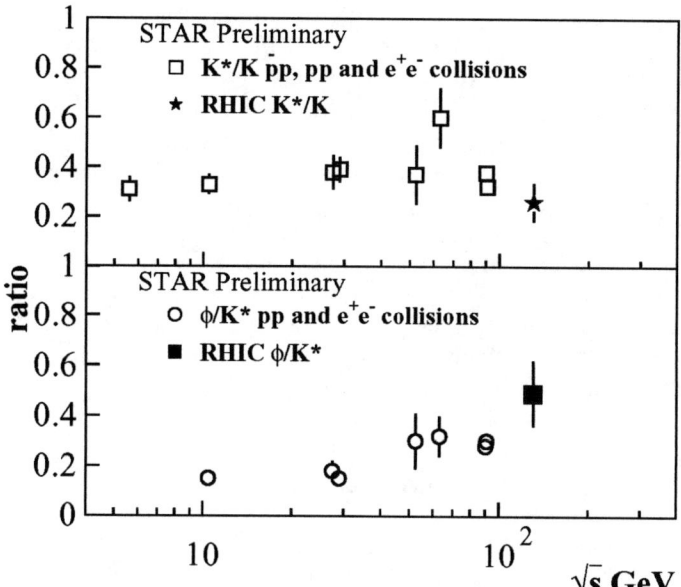

FIGURE 3. K^{*0}/K and ϕ/K^{*0} ratios measured in different colliding systems at various energies. The ratios are from measurements in e^+e^- collisions at 10.45 GeV [25], 29 GeV [26], 90 GeV [20] and 91 GeV [21, 22] beam energies, $\bar{p}p$ at 5.6 GeV [27] and pp at 27.5 GeV [28], 52.5 GeV [29] and 63 GeV [15]. The errors shown correspond to the quadratic sum of the statistical and systematical errors.

Au collisions at 130 GeV beam energy [30] increases compared to the measurements in pp and e^+e^- at lower energies. However, this increase may not be solely related to strangeness suppression due to additional effects (*e.g.* re-scattering) on short lived resonances in heavy ion collisions. It is also interesting to note that the ϕ/K^{*0} ratio already rises as a function of beam energy in pp and e^+e^-.

Since the K^{*0} lifetime is short ($c\tau = 4$ fm) and comparable to the time scale of the evolution of the system in relativistic heavy ion collisions, the K^{*0} survival probability needs to be taken into account. This survival probability depends on the expansion time between chemical and thermal freeze-out, the source size and the K^{*0} transverse momentum. If the K^{*0} decays between chemical and thermal freeze-out, the daughters from the decay may re-scatter and the K^{*0} will not be reconstructed. On the other hand, chemical freeze-out is only truly defined for non-resonant long-lived particles, and elastic interactions such as $\pi K \rightarrow K^{*0} \rightarrow \pi K$ are still effective and regenerate the K^{*0} until thermal freeze-out.

The K^{*0} measurement at RHIC may provide information on the expansion time of the system between chemical and thermal freeze-out due to these additional effects on short lived resonances in heavy ion collisions. However, final state interactions of the hadronic decay products destroy the early time information carried by these decays. Assuming that the difference in the K^{*0}/K ratio measured in Au-Au collisions at RHIC and the ratios measured in pp and e^+e^- is due to the K^{*0} survival probability, our measurement is consistent with an expansion time of only a few fm between chemical and thermal freeze-out (sudden freeze-out). In the scenario of a long expansion time between chemical and thermal freeze-out (~20 fm [7, 8]), without K^{*0} regeneration, not only is the measured rate of K^{*0} production reduced but there is also a low p_T suppression resulting in a larger effective inverse slope in m_T. However, the measured K^{*0} inverse slope is similar to the ϕ inverse slope [30] and higher than that of kaons [24] (effect of radial flow [31]), which is consistent with a short expansion time of the system. As a consequence, our measurement at RHIC is consistent with either a sudden freeze-out with no K^{*0} regeneration or a long expansion time scenario with significant K^{*0} regeneration. In the statistical model, the measured K^{*0} should be consistent with the condition at kinetic freeze-out instead of chemical freeze-out. Since the statistical model provides particle yields at chemical freeze-out, the fact that the statistical model [32] reproduces our measurement is consistent with a short expansion time between chemical and thermal freeze-out.

CONCLUSIONS

Preliminary results on $K^*(892)^0$ and $\overline{K}^*(892)^0$ production measured at mid-rapidity by the STAR detector in $\sqrt{s_{NN}}$= 130 GeV Au-Au collisions at RHIC were presented. The data show significant K^{*0} production despite the short lifetime ($c\tau$ = 4 fm) of this resonance. We do not observe an enhancement in the K^{*0} production as observed for other strange particles for increasing energies and differing colliding systems. Our measurement is consistent with a sudden freeze-out if the difference in the K^{*0}/K ratio measured in Au-Au collisions at RHIC and the ratios measured in pp and e^+e^- is due to the K^{*0} survival probability. The K^{*0} production at RHIC rules out a long expansion time between chemical and thermal freeze-out unless there is K^{*0} regeneration, due to elastic interactions after chemical freeze-out.

The study of resonances like the K^{*0} may provide important information on the collision process. The improvement of the uncertainties in the K^{*0} and ϕ measurements and the measurements of other resonances with different properties may help in understanding the development of the system after chemical freeze-out.

ACKNOWLEDGMENTS

We wish to thank the RHIC Operations Group and the RHIC Computing Facility at Brookhaven National Laboratory, and the National Energy Research Scientific Computing Center at Lawrence Berkeley Laboratory for their support. This work was supported by the Division of Nuclear Physics and the Division of High Energy Physics of the Office of Science of the U.S.Department of Energy, the United States National Science Foundation, the Bundesministerium fuer Bildung und Forschung of Germany, the Institut National de la Physique Nucleaire et de la Physique des Particules of France, the United Kingdom Engineering and Physical Sciences Research Council, Fundação de Amparo a Pesquisa do Estado de São Paulo, Brazil, and the Russian Ministry of Science and Technology.

REFERENCES

1. J. Harris, *These Contributions*.
2. R. Rapp and J. Wambach, *Adv. Nucl. Phys.* **25** 1-164 (2000).
3. R. Rapp, *Phys. Rev.* C **63** 054907 (2001).
4. J. Schaffner-Bielich, *Phys. Rev. Lett.* **84** 3261-3264 (2000).
5. J. Letessier et al., *Preprint nucl-th/0011048*.
6. G. Torrieri and J. Rafelski, *Phys. Lett.* B **509** 239-245 (2001).
7. L.V. Bravina et al., *Preprint hep-ph/0010172*.
8. D. Teaney et al., *Preprint nucl-th/0110037*.
9. H. Bebie et al., *Nucl. Phys.* B **378** 95-128 (1992).
10. R. Rapp and E.V. Shuryak, *Phys. Rev. Lett.* **86** 2980-2983 (2001).
11. C. Song and V. Koch, *Phys. Rev.* C **55** 3026-3037 (1997).
12. D. Drijard et al., *Nucl. Instrum. Meth.* A **225** 367-377 (1984).
13. D. L'Hôte, *Nucl. Instrum. Meth.* A **337** 544-556 (1994).
14. K.H. Ackermann et al., *Nucl. Phys.* A **661** 681 (1999).
15. T. Akesson et al., *Nucl. Phys.* B **203** 27-39 (1982).
16. M. Aguilar-Benitez et al., *Phys. Rev.* D **6** 11-28 (1972).
17. X.N. Wang and M. Gyulassy, *Phys. Rev.* D **44** 3501-3516 (1991).
18. X.N. Wang and M. Gyulassy, *Comp. Phys. Commu.* **83** 307-345 (1994).
19. Z. Xu et al., *Nucl. Phys.* A **698** 607-610 (2002).
20. K. Abe et al., 1999 *Phys. Rev.* D **59** 052001 (1999).
21. J.Y. Pei, *Z. Phys.* C **72** 39-46 (1996).
22. P.V. Chliapnikov, *Phys. Lett.* B **470** 263-267 (1999).
23. C. Adler et al., *Phys. Rev. Lett.* **87** 112303 (2001).
24. L. Barnby et al., *Proceedings of Strange Quarks in Matter 2001*, to be published in *J. Phys.* G.
25. H. Albrecht et al., *Z. Phys.* C **61** 1-17 (1994).
26. M. Derrick et al., *Phys. Lett.* B **158** 519-532 (1985).
27. J. Canter et al., *Phys. Rev.* D **20** 1029-1036 (1979).
28. M. Aguilar-Benitez et al., *Z. Phys.* C **50** 405-426 (1991).
29. D. Drijard et al., *Z. Phys.* C **9** 293-303 (1981).
30. F. Laue et al., *Proceedings of Strange Quarks in Matter 2001*, to be published in *J. Phys.* G.
31. N. Xu and M. Kaneta, *Preprint nucl-ex/0104021*.
32. D. Magestro et al., *Proceedings of Strange Quarks in Matter 2001*, to be published in *J. Phys.* G.

Azimuthal Anisotropy of K_S^0 and Λ Production at Mid-Rapidity from Au+Au Collisions at $\sqrt{s_{NN}} = 130$ GeV

Paul Sorensen for the STAR Collaboration

University of California, Los Angeles, California 90095

Abstract. We report STAR results on the azimuthal anisotropy parameter v_2 for strange particles K_S^0 and Λ at mid-rapidity in Au+Au collisions at $\sqrt{s_{NN}} = 130$ GeV at RHIC. The value of v_2 as a function of transverse momentum p_t and collision centrality is presented for both particles and compared with model calculations. A strong p_t dependence in v_2 is observed up to $p_t \sim 2.0$ GeV/c where v_2 begins to saturate.

INTRODUCTION

Measurements of azimuthal anisotropies in the transverse momentum distributions of particles probe early stages of ultra-relativistic heavy-ion collisions [1, 2, 3]. After an initial geometric anisotropy is established in an off-center or non-central collision, rescattering in the overlapping region of the colliding nuclei amongst collision participants transfers the spatial anisotropy into an anisotropy in momentum space. The extent of the transformation depends on the initial conditions and the dynamical evolution of the collision. As a result, anisotropy measurements for nucleus-nucleus collisions at RHIC energies may increase our understanding of the processes governing the evolution of the collision system and in particular, may provide information about an early partonic stage in the evolution of the system [1, 4, 5, 6, 7, 8].

For the purpose of studying azimuthal anisotropies it is advantageous to write the triple differential distribution of particles in the form of a Fourier series

$$E\frac{d^3N}{d^3p} = \frac{1}{2\pi}\frac{d^2N}{p_t dp_t dy}(1 + \sum_{n=1}^{\infty} 2v_n cos(n\phi)), \qquad (1)$$

where p_t is the transverse momentum of the particle, y is its rapidity and ϕ denotes its azimuthal angle of emmission with respect to the true reaction plane angle[1] [9, 10]. The harmonic coefficients v_n are anisotropy parameters and the second coefficient v_2 is called the *elliptic flow* parameter. Recent experimental results from RHIC [11, 12, 13, 14] include measurements of v_2 as a function of collision centrality and p_t for charged

---

[1] The reaction plane is the plane defined by the beam line and the line connecting the centers of the colliding nuclei.

particles up to a p_t of about 2.0 GeV/c, and for identified $\pi^\pm$, $K^\pm$ and $p(\overline{p})$ up to a p_t of about 0.8 GeV/c.

We report on the first measurement of the azimuthal anisotropy parameter v_2, as a function of p_t and collision centrality, for the strange particles K_S^0 and Λ from minimum bias Au + Au collisions at $\sqrt{s_{NN}} = 130$ GeV. Our measurements of v_2 for identified particles using the Solenoidal Tracker At RHIC (STAR) are the first to extend beyond the p_t range where particles are identified by their specific energy loss (dE/dx) in the gas of the Time Projection Chamber (TPC), and up to $p_t \sim 3.0$ GeV/c. Previously v_2 in this higher p_t range had only been measured for unidentified charged particles [15].

ANALYSIS

The STAR detector [16], due to its azimuthal symmetry and large acceptance, is ideally suited for measuring elliptic flow. For collisions in its center, the STAR TPC measures the tracks of charged particles in the pseudo-rapidity range $|\eta| < 1.8$ with 2π azimuthal coverage. A scintillator barrel, the Central Trigger Barrel (CTB), surrounding the TPC that measures the charged particle multiplicity within $|\eta| < 1$ was used for a central trigger. Two Zero-Degree Calorimeters [17] at both ends of the TPC in coincidence provided a minimum bias trigger.

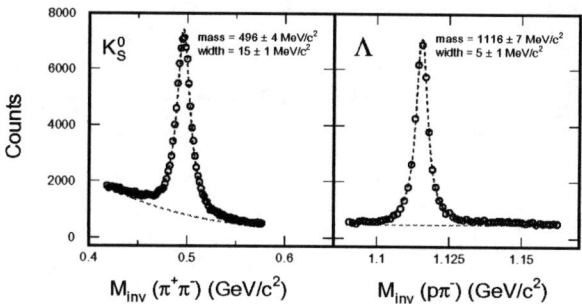

FIGURE 1. Invariant mass distributions for $\pi^+\pi^-$ showing an enhancement at the K_S^0 mass (left panel) and $p\pi^-$ ($\overline{p}\pi^+$) showing an enhancement at the $\Lambda(\overline{\Lambda})$ mass (right panel). Fitting results are shown as dashed lines in the figure.

The masses and kinematic properties of both $K_S^0 \to \pi^+ + \pi^-$ and $\Lambda(\overline{\Lambda}) \to p + \pi^-$ ($\overline{p} + \pi^+$), are reconstructed via their decay topologies in the TPC [18, 19, 20]. Figure 1 shows the invariant mass distributions for a $\pi^+\pi^-$ mass hypothesis and a $p\pi^-$ ($\overline{p}\pi^+$) hypothesis. The background is dominated by combinatoric counts and the observed masses, 496 ± 4 MeV/c^2 and 1116 ± 7 MeV/c^2, are consistent with values listed in the PDG [21] for K_S^0 and Λ respectively. The K_S^0 and Λ particles used in the v_2 analysis are from the kinematic region of $0.20 \leq p_t \leq 3$ GeV/c and $|y| \leq 1.0$. To compensate for limited statistics, Λ and $\overline{\Lambda}$ are summed together. To reduce the combinatoric background, for K_S^0, pion-like tracks are required to have a distance of closest approach dca > 1.0 cm, while for Λ, the pion-like tracks have a dca > 1.5 cm and the proton-like tracks have a dca > 0.8 cm. Tracks are determined to be either proton-like or pion-like based on their energy loss (dE/dx) in the TPC gas. The yield from the enhancement in the invariant mass peaks in each ϕ, p_t bin is used to evaluate $v_2 = \langle cos(2\phi) \rangle$ as a function of p_t. This method enables us to measure identified particle flow beyond the p_t range where dE/dx particle identification fails.

The event plane, an experimental estimator of the true reaction plane [10], is calculated from the azimuthal distribution of tracks using cuts similar to those used in reference [11]. To avoid auto-correlations, only tracks excluded from the neutral vertex reconstruction are used in the event plane calculation. The observed v_2 is corrected to account for the imperfect event plane resolution estimated using the method of subevents described previously [10]. The maximum resolution correction factor for the K_S^0 and Λ analysis is found to be 0.681 ± 0.004 and 0.582 ± 0.007 respectively and is reached in the centrality corresponding to 25 – 35% of the measured cross section, where the relative multiplicity distribution is used to estimate the event centrality as in reference [11].

For this analysis, three sources contribute to systematic errors in the measured anisotropy parameters: (1) particle identification; (2) background subtraction; (3) non-reaction plane related correlations contributing to v_2 such as resonance decays or Coulomb and Bose-Einstein correlations [22, 23]. The first two sources are estimated by examining the variation in v_2 after changing several track, event and neutral vertex cuts and are found to contribute an error of less than ± 0.005 to v_2. A previous study used the correlation of event plane angles from subevents to estimate the magnitude of non-reaction plane related correlations [12]. That analysis showed that these effects which always act to increase the measured value of v_2 above its true value typically contribute a systematic error to v_2 of -0.005, but that the magnitude is larger in the more peripheral events where the error increases to about -0.035 for the centrality corresponding to 58 – 85% of the measured cross section.

RESULTS

The centrality dependence of v_2 as a function of transverse momentum calculated from 201 thousand minimum bias and 180 thousand central events is shown in figure 2. The two particles show a similar p_t dependence in the respective centralities with more flow in the more peripheral collisions. This is similar to observations made previously at the same energy [12] where the agreement with hydrodynamic calculations in the lower p_t region was interpreted as evidence for early local thermal equilibrium in all but the most peripheral events (45 – 85% of the measured cross section).

In figure 3 (left) we plot $v_2(p_t)$ for K_S^0 and Λ from minimum bias collisions with results from hydrodynamic model calculations [5] and $v_2(p_t)$ for negatively charged particles [15]. The K_S^0 results are in agreement with the v_2 for $K^\pm$ in the p_t range they share ($300 \leq p_t \leq 700$ MeV/c) [12]. We observe that v_2 for both strange particles increases as a function of p_t similar to the hydrodynamic model prediction, up to about 1.5 GeV/c. In the higher p_t region however ($p_t \geq 2$ GeV/c), the values of v_2 seem to be saturated. It has been suggested [6] that the shape and height of v_2 above 2 – 3 GeV/c is related to energy loss in an early, high parton-density stage of the evolution.

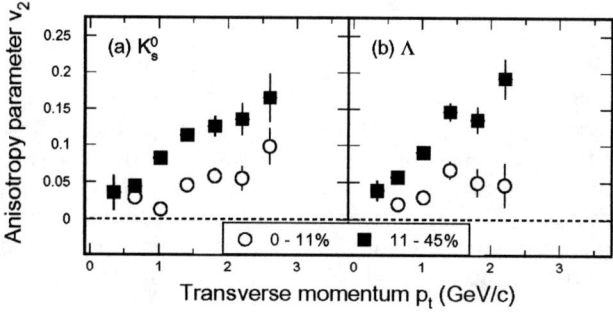

FIGURE 2. Elliptic flow, $v_2(p_t)$ for K_S^0 and Λ for central (0 – 11%) and mid-central (11 – 48%) collisions.

The p_t integrated v_2 from minimum bias collisions for K_S^0, Λ and negatively charged particles are shown in figure 3 (right). The integrated values of v_2 are calculated by parameterizing the yield with the inverse slope parameter of exponential fits to the K_S^0 or Λ transverse mass distributions and are dominated by the region near the particles mean p_t. The relatively larger v_2 of Λ reflects the higher mean p_t of the Λ compared to the K_S^0. Hydrodynamic model calculations [5], shown as a gray-band and central line, are, within errors, in agreement with this result. The width of the gray-band indicates the uncertainty of the model calculation, mostly due to the choice of the freeze-out conditions. The increase of v_2 with particle mass in figure 3 points to a significant commonality in velocities between particles of different masses that is perhaps, established early in the collision. The nature of the particles during this process however, whether parton or hadron, and the degree of thermalization remains unclear.

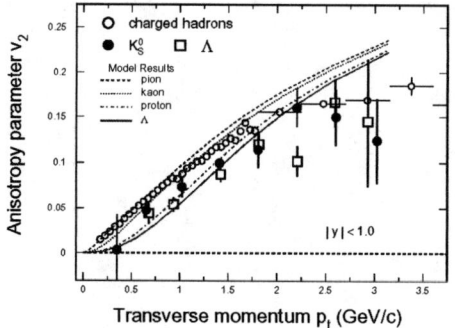 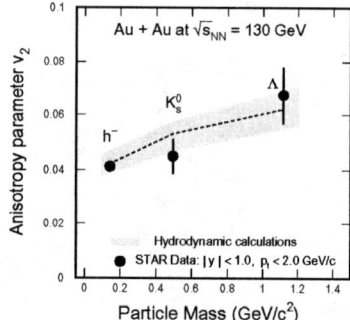

FIGURE 3. Elliptic flow, v_2 for K_S^0 and Λ as a function of p_t from minimum bias Au+Au collisions compared to results from hydrodynamic model calculations and v_2 of negatively charged particles [15] (left). Integrated azimuthal anisotropy parameters v_2 as a function of particle mass with a gray-band and central line indicating hydrodynamic model results [5] (right).

SUMMARY

We have reported the first measurement of v_2 for K_S^0 and Λ from Au + Au collisions at $\sqrt{s_{NN}} = 130$ GeV and the first measurement, at this energy, of v_2 above $p_t \sim 0.8$ GeV/c for any identified particle. For both particles more flow is seen in mid-central than in central collisions. The integrated values of v_2 show a mass dependence consistent with the development of a common velocity, a feature of hydrodynamic models, where local thermalization is assumed. In $v_2(p_t)$ however, we see a strong p_t dependence only up to $p_t \sim 2.0$ GeV/c where v_2 seems to saturate, suggesting that the hydrodynamic picture is incomplete for particle production above $p_t = 1.5 - 2.0$ GeV/c.

ACKNOWLEDGMENTS

We thank P. Huovinen for hydrodynamic model calculations and the RHIC Operations Group at Brookhaven National Laboratory for their support and for providing collisions for the experiment. This work was vsupported by the Division of Nuclear Physics and the Division of High Energy Physics of the Office Science of the U.S. Department of Energy, the United States National Science Foundation, the Bundesministerium für Bildung und Forschung of Germany, the Institut National de la Physique Nucleaire et de la Physique des Particules of France, the United Kingdom Engineering and Physical Sciences Research Council, and the Russian Ministry of Science and Technology.

REFERENCES

1. Sorge, H., *Phys. Rev. Lett.*, **82**, 2048–2051 (1999).
2. Sorge, H., *Phys. Lett.*, **B82**, 251–256 (1997).
3. Ollitrault, J.-Y., *Phys. Rev.*, **D46**, 229–245 (1992).
4. Zhang, B., Gyulassy, M., and Ko, C. M., *Phys. Lett.*, **B455**, 45–48 (1999).
5. Huovinen, P., Kolb, P. F., Heinz, U. W., Ruuskanen, P., and Voloshin, S. A., *Phys. Lett.*, **B503**, 58–64 (2001).
6. Gyulassy, M., Vitev, I., and Wang, X. N., *Phys. Rev. Lett.*, **86**, 2537–2540 (2001).
7. Teaney, D., Lauret, J., and Shuryak, E. V., *Phys. Rev. Lett.*, **86**, 4783–4786 (2001).
8. Lin, Z.-w., and Ko, C. M., *Phys. Rev.*, **C65**, 034904 (2002).
9. Voloshin, S., and Zhang, Y., *Z. Phys.*, **C70**, 665–672 (1996).
10. Poskanzer, A. M., and Voloshin, S. A., *Phys. Rev.*, **C58**, 1671–1678 (1998).
11. Ackermann, K. H., et al., *Phys. Rev. Lett.*, **86**, 402–407 (2001).
12. Adler, C., et al., *Phys. Rev. Lett.*, **87**, 182301 (2001).
13. Lacey, R. A., *Nucl. Phys.*, **A698**, 559–563 (2002).
14. Park, I. C., et al., *Nucl. Phys.*, **A698**, 564–567 (2002).
15. Snellings, R. J. M., *Nucl. Phys.*, **A698**, 193–198 (2002).
16. Ackermann, K. H., et al., *Nucl. Phys.*, **A661**, 681–685 (1999).
17. Adler, C., et al., *Nucl. Instrum. Meth.*, **A461**, 337–340 (2001).
18. Wieman, H., et al., *IEEE Trans. Nucl. Sci.*, **44**, 671–678 (1997).
19. Betts, W., et al., *IEEE Trans. Nucl. Sci.*, **44**, 592–597 (1997).
20. Klein, S. R., et al., *IEEE Trans. Nucl. Sci.*, **43**, 1768–1772 (1996).
21. Groom, D. E., et al., *Eur. Phys. J.*, **C15**, 1–878 (2000).
22. Dinh, P. M., Borghini, N., and Ollitrault, J.-Y., *Phys. Lett.*, **B477**, 51–58 (2000).
23. Borghini, N., Dinh, P. M., and Ollitrault, J.-Y., *Phys. Rev.*, **C62**, 034902 (2000).

Multi-strange baryon production at mid-rapidity in Au+Au collisions at $\sqrt{s_{NN}} = 130$ GeV

Javier Castillo for the STAR collaboration

SUBATECH, Nantes, France

Abstract. The STAR experiment is the most suited of four RHIC experiments for the measurements of strange hadrons production and especially multi-strange baryons and anti-baryons. Relative production of strange baryons and anti-baryons are important to learn about baryon densities achieved in heavy ions collisions and the validity of models for production yields. We present here preliminary results on multi-strange baryon production at mid-rapidity in $\sqrt{s_{NN}} = 130$ GeV Au+Au collisions from the STAR experiment.

INTRODUCTION

The formation of a quark gluon plasma (QGP) during ultra-relativistic heavy ion collisions [1] is predicted to exhibit an enhancement of strange $q\bar{q}$ pair production compared to that found from a hadron gas state [2]. The enhancement is thought to be more pronounced for multi-strange baryons and anti-baryons since their production is strongly suppressed in hadronic interactions due to their high mass thresholds [3]. Various models have been developed to address hadron production in heavy ion collisions. Statistical models have been very successful in describing hadron production in a wide variety of colliding systems and energies. Deviations from these models may be an indication of new physics.

The Relativistic Heavy Ion Collider (RHIC) allows for the study of nuclear matter in extreme conditions such as in the case of the theoretical QGP. We present here preliminary measurements of multi-strange baryon production and non-identical particle ratios at mid-rapidity in $\sqrt{s_{NN}} = 130$ GeV Au+Au collisions from the STAR experiment.

THE STAR EXPERIMENT

The data presented here were taken with the STAR detector during the year 2000 Au+Au at $\sqrt{s_{NN}} = 130$ GeV run. The experimental setup for this period consisted mainly of a large cylindrical Time Projection Chamber (TPC) used for charged particle tracking and identification. A Central Trigger Barrel (CTB) and two Zero Degree Calorimeters (ZDC) were used for triggering. More details of the experimental setup can be found in [4]. We present results from the 10% most central events which have been selected as described in [5].

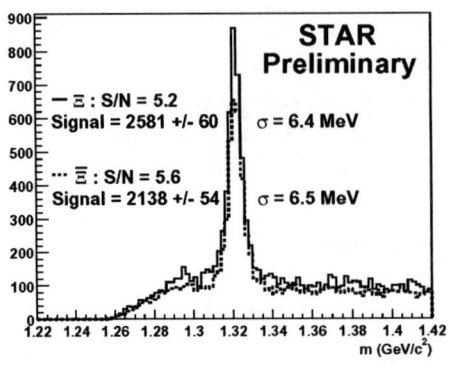 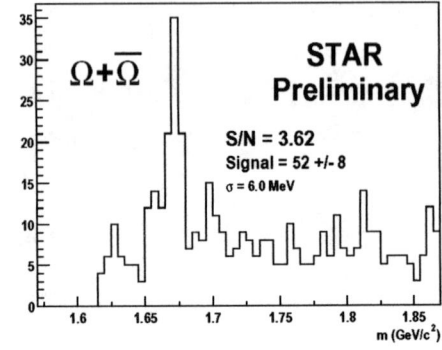

FIGURE 1. Left panel: Invariant mass distribution for Ξ^- (full) and $\overline{\Xi}^+$ (dashed) candidates. Right panel: Invariant mass distribution for $\Omega^- + \overline{\Omega}^+$ candidates

RECONSTRUCTION AND ANALYSIS OF MULTI-STRANGE BARYONS

The multi-strange Ξ^- particles are reconstructed via their decay topologies, $\Xi^- \to \Lambda\pi^-$ followed by $\Lambda \to p\pi^-$ with a branching ratio of 100% and 64% respectively. The reconstruction of Ξ^- particles is handled by first reconstructing Λ candidates. This is done by extrapolating backwards two oppositely charged tracks looking for a common origin point. Further details on Λ reconstruction can be found in [6]. This Λ candidates are then extrapolated further and combined with negatively charged tracks to determine Ξ^- candidates, and similarly for the charge conjugate $\overline{\Xi}^+$ decay. Particle identification from energy loss of charged tracks in the TPC as well as loose geometric and kinematic cuts are applied at both steps to reduce the large combinatorics background inherent to such a reconstruction process. Ω^- and $\overline{\Omega}^+$ candidates are also reconstructed with the same method described above using their decay channel, $\Omega^- \to \Lambda K^-$ followed by $\Lambda \to p\pi^-$ with a branching ratio of 67.9% and 64% respectively. The resultant invariant mass distributions are shown in, Figure 1 for Ξ^- and $\overline{\Xi}^+$, and $\Omega^- + \overline{\Omega}^+$.

For Ξ^- and $\overline{\Xi}^+$ the raw yields are extracted from the invariant mass distribution by counting the entries within ± 15 MeV/c^2 about the expected mass and then subtracting the background. The background under the peak is estimated by sampling two regions on either side of the peak. Due to very low statistics, a separate analysis of Ω^- or $\overline{\Omega}^+$ is more difficult to perform, and is still work in progress.

RESULTS AND DISCUSSION

The STAR detector being azimuthally symmetrical, the reconstruction efficiency is identical for positive and negative charged tracks, thus the Ξ^- and $\overline{\Xi}^+$ efficiencies are identical. Under this assumption $\overline{\Xi}^+/\Xi^-$ can be obtained directly from the raw

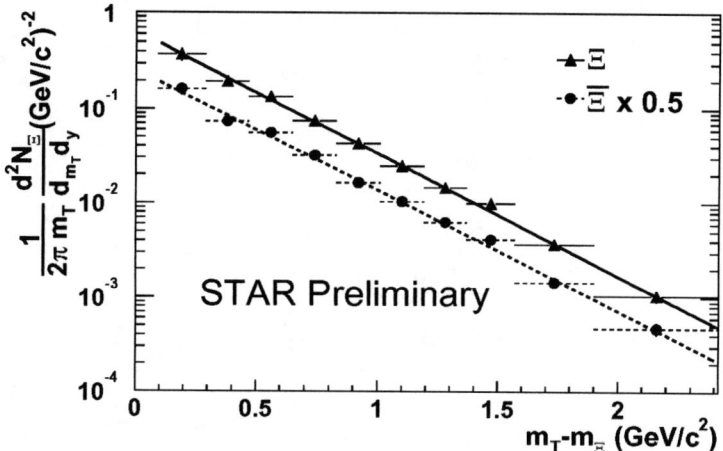

FIGURE 2. Transverse mass spectra of Ξ^- (triangles) and $\overline{\Xi}^+$ (circles) for central events. Lines are exponential fits to the data.

yields. The measured $\overline{\Xi}^+/\Xi^- = 0.83 \pm 0.03(stat.) \pm 0.05(syst.)$ is found to be in good agreement with a quark coalescence model [7].

The invariant mass distribution of $\Xi^-(\overline{\Xi}^+)$ candidates is then histogramed in transverse mass $m_T = \sqrt{p_T^2 + m_0^2}$ bins and the signal extracted in each bin as described above. Each m_T bin was then corrected for detector acceptance and reconstruction efficiency by the Monte Carlo technique where simulated Ξ^- particles were embedded into real events. The data cover $|y| < 0.75$, simulations have shown the acceptance and efficiency for Ξ's is constant, in y, over this range.

Figure 2 shows Ξ^- and $\overline{\Xi}^+$ preliminary m_T spectra for the 10% most central events. The spectra were fit to an exponential function ($\propto e^{-(m_T-m_0)/T}$) to determine the inverse slope, T, and the rapidity density, $\frac{dN}{dy}$, integrated over all m_T. For the yields we found $\frac{dN}{dy} = 2.30 \pm 0.09$ for Ξ^- and $\frac{dN}{dy} = 1.89 \pm 0.08$ for $\overline{\Xi}^+$ [1]. Systematic errors are estimated to be $\sim 20\%$. Note that the data have not been corrected for feed-down from weak decays of higher mass strange particles. The Ξ^- and $\overline{\Xi}^+$ particle yields in the measured m_T region correspond to $\sim 75\%$ of the total yield.

Non-identical particle ratios

Figure 3 shows the mid-rapidity ratios (a) Ξ/h^- and (b) Ξ/Λ from heavy ion collisions at AGS [8], SPS [9] energies, plus preliminary RHIC results. Our estimated systematic

---

[1] Between the conference and the submission of these proceedings, the analysis has evolved so that the values presented here are different from the ones presented in the talk.

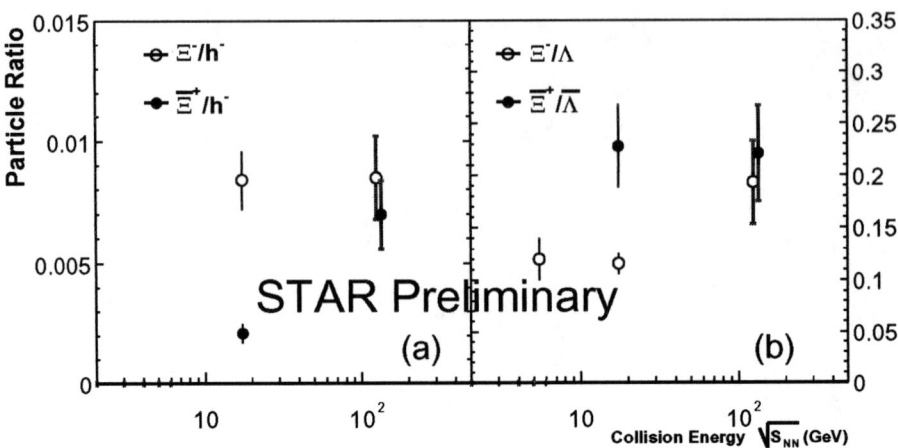

FIGURE 3. Mid-rapidity (a) Ξ/h^- ratios and (b) Ξ/Λ ratios as a function of beam energy. Statistical and systematic errors are added in quadrature to STAR points.

error on measured the Ξ/h^- and Ξ/Λ ratios are $\sim 20\%$. The $\overline{\Xi}^+/h^-$ ratio increase with beam energy indicating that the relative production of multi-strange anti-baryons with respect to negative hadrons is enhanced at RHIC energies compared to SPS energies. On the other hand, the Ξ^-/h^- ratio remains constant from SPS to RHIC energies indicating a reduced net-baryon density achieved at RHIC energies. Also, the Ξ^-/Λ ratio increase while the $\overline{\Xi}^+/\overline{\Lambda}$ is constant as a function of collision energy from SPS to RHIC. The bahaviour of $\overline{\Xi}^+/\overline{\Lambda}$ suggest a saturation of the $\bar{s}$ production with respect to $\bar{q}$ from SPS to RHIC. Note that Ξ^-/Λ and $\overline{\Xi}^+/\overline{\Lambda}$ ratios have been corrected for feeddown from weak decays of Ξ on Λ. For SPS values the quoted [10] upper limits of 5% for Λ and 10% for $\overline{\Lambda}$ were used.

Statistical models have shown to be very successful in reproducing particle ratios data at SPS energies, as well as anti-particle to particle ratios at RHIC energy. In statistical models frame-work these ratios are primarily driven by the relative difference in the abundance of quarks and anti-quarks. At RHIC energy, transported numbers of valence quarks to mid-rapidity are smaller than those from pair production, so this relative difference becomes smaller, diminishing statistical sensitivity to small differences. Thus all statistical models fit anti-particle to particle ratios rather well at these higher energies.

More powerful in distinguishing between models are ratios between highly dissimilar particle species such as Ξ/π. Combining the preliminary Ξ^- yields presented here with STAR preliminary π^- yield result in a $\Xi^-/\pi^- = 0.0088 \pm 0.0004$. Authors of references [11] and [12] assume local chemical equilibrium and strangeness neutrality to predict the Ξ^-/π^-, which is 10% lower than the measured ratio. ALCOR (a quark coalescence model), another statistical model but not an equilibrium one, over predicts the Ξ^-/π^- by almost a factor of two [13]. Also presented at this conference the model by Rafelski and collaborators [14] have performed two fits to particle ratios at RHIC energy. The authors need to assume chemical non-equilibrium to correctly address the Ξ/Λ ratios.

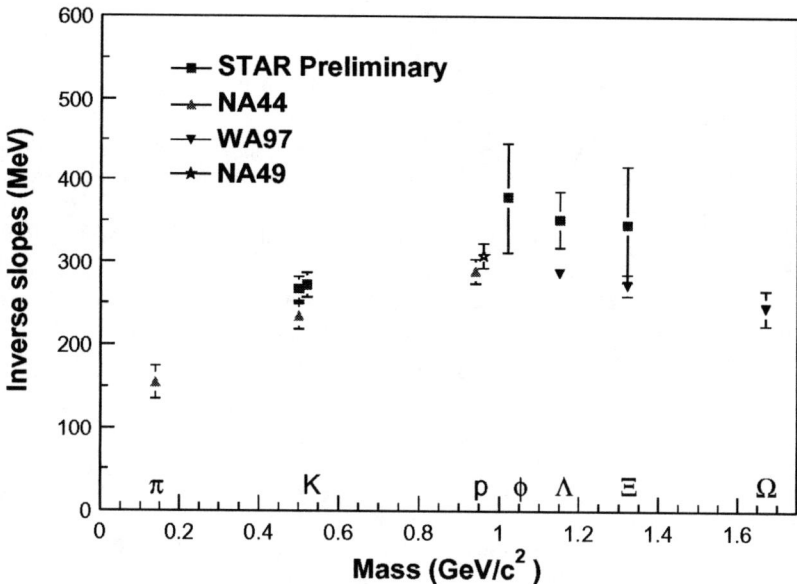

FIGURE 4. Inverse slope parameters as a function of particle mass for SPS and RHIC central collisions. Statistical and systematic errors are added in quadrature to STAR points.

Systematics of the inverse slope parameter

The inverse slope parameter, T, is found to be $T = 335 \pm 7(stat.)$ MeV and $T = 338 \pm 7(stat.)$ MeV for Ξ^- and $\overline{\Xi}^+$ respectively, the systematic errors are estimated to be $\sim 20\%$. As is the case for Λ and $\overline{\Lambda}$ [6], there is no significant difference between the slopes of Ξ^- and $\overline{\Xi}^+$. Figure 4 shows the mid-rapidity inverse slope parameters as a function of particle mass, for central Pb+Pb collisions at $\sqrt{s_{NN}} = 17.4$ GeV [10, 17, 18] and preliminary results from Au+Au collisions at $\sqrt{s_{NN}} = 130$ GeV [15, 6, 16]. It is clear that the overall inverse slope parameters are higher in RHIC collisions than in SPS collisions, also the same saturation trend is observed for both energies. This observation for strange hadrons has been suggested as an indication that strangeness freezes out earlier in the system's evolution [19].

CONCLUSION

In summary, we report the STAR measurements of multi-strange baryon (Ξ^- and $\overline{\Xi}^+$) mid-rapidity transverse momentum distributions from the Au+Au collisions at $\sqrt{s_{NN}} = 130$ GeV. The extracted inverse slope parameters of both Ξ^- and $\overline{\Xi}^+$ are found to be higher than that from collisions at the SPS energy. While the measured anti-baryon to baryon ratios are well described by all statistical models, the non-identical particle ratios allow us to discriminate between the different approaches of the statistical models.

ACKNOWLEDGMENTS

We wish to thank the RHIC Operations Group and the RHIC Computing Facility at Brookhaven National Laboratory, and the National Energy Research Scientific Computing Center at Lawrence Berkeley National Laboratory for their support. This work was supported by the Division of Nuclear Physics and the Division of High Energy Physics of the Office of Science of the U.S. Department of Energy, the United States National Science Foundation, the Bundesministerium fuer Bildung und Forschung of Germany, the Institut National de la Physique Nucléaire et de la Physique des Particules of France, the United Kingdom Engineering and Physical Sciences Research Council, Fundacao de Amparo a Pesquisa do Estado de Sao Paulo, Brazil, the Russian Ministry of Science and Technology and the Ministry of Education of China and the National Science Foundation of China.

REFERENCES

1. For reviews and recent developments see *"Quark Matter '2001"*, Nucl. Phys. **A663**, 1c(2001).
2. B. Müller, J. Rafelski, Phys. Rev. Lett. **48** (1982) 1066.
3. J. Rafelski, Phys. Lett. B **262** (1991) 333.
4. L.S. Barnby et al. (STAR Collaboration), J. Phys. G: Nucl. Part. Phys. **28** 1535-1542 (2002).
5. K.H. Ackermann et al. (STAR Collaboration), Phys. Rev. Lett. **86** (2001) 402.
6. M.A.C. Lamont et al. (STAR Collaboration), J. Phys. G: Nucl. Part. Phys. **28** 1721 (2002).
7. J. Zimányi, T.S. Biró, T. Csörgö and P. Lévai, Phys. Lett. **B472** (2000) 243.
8. S. E. Eiseman et al., Phys. Lett. **B325**, 322 (1994).
9. F. Antinori et al. (WA97 Collaboration), Nucl. Phys. **A661** (1999) 130c.-139c.
10. F. Antinori et al. (WA97 Collaboration), Eur. Phys. J. **C14** (2000) 633-641.
11. P. Braun-Munzinger, D. Magestro, K. Redlich, and J. Stachel, Phys. Lett. **B518**, 41(2001).
12. D. Magestro, J. Phys. G: Nucl. Part. Phys. **28** 1745-1752 (2002)
13. T. Biró, P. Lévai, and J. Zimányi, hep-ph/0112137.
14. J. Rafelski and J. Letessier, these proceedings, arXiv:hep-ph/0206145.
15. J. Harris et al. (STAR Collaboration), Nucl. Phys. **A698** (2002) 64c-77c.
16. F. Laue et al. (STAR Collaboration), J. Phys. G: Nucl. Part. Phys. **28** 2051 (2002).
17. I.G. Bearden et al. (NA44 Collaboration), Phys. Rev. Lett. **78**, (1997) 2080.
18. S. V. Afanasev et al. (NA49 Collaboration), Phys. Lett. **B491** (2000) 59.
19. N. Xu, J. Phys. G: Nucl. Part. Phys. **28** 2121-2126 (2002)

Particle Tracking

Karel Šafařík*, Ingrid Kraus†, Jason Newby** and Paul Sorensen‡

*CERN, 1211 Genève 23, Switzerland
†GSI, Planckstrasse 1, 64291 Darmstadt, Germany
**University of Tennessee, Knoxville, TN 37996-1200, USA
‡University of California, Los Angeles, CA 90095, USA

Abstract. In this lecture we will present a short historical overview of different tracking detectors. Then we will describe currently used gaseous and silicon detectors and their performance. In the second part we will discuss how to estimate tracking precision, how to design a tracker and how the track finding works. After a short description of the LHC the main attention is drawn to the ALICE experiment since it is dedicated to study new states in hadronic matter at the LHC. The ALICE tracking procedure is discussed in detail. A comparison to the tracking in ATLAS, CMS and LHCb is given.

TRACKING DETECTORS

History of particle tracking

Since the very beginning of particle physics, when J.J. Thomson realized that the cathode rays he was studying were not "rays" but streams of subatomic charged particles instead, our understanding of the subatomic realm and the mechanics that governs it has depended strongly on our ability to detect the tracks of charged particles. Thomson was able to surmise the existence of electrons, measure their charge to mass ratio and even measure their velocity as they were emitted by a hot filament because he could see their trajectory as they passed through crossed electric and magnetic fields. During the years since Thomson's experiments in 1897 many techniques have been developed to detect or visualize charged particle tracks—nuclear emulsions, cloud chambers, bubble chambers, spark chambers, streamer chambers, various gas and solid-state detectors. All of these techniques rely on high-energy charged particles ionizing atoms as they pass through matter. The ions left along the paths of the high-energy particles can then act as catalysts for a reaction that leaves an observable trace such as, a bubble, a spark, condensation, or a charge avalanche.

Many of these particle detecting techniques have—because of their inherent limitations—been for the most part abandoned in favor of gas or solid-state detectors. Experiments at the newest colliders like RHIC or the LHC rely almost exclusively on these two techniques because they lend themselves well to triggering, high-event rates and the digitization of huge amounts of data. Bubble chambers however, are of particular historic importance and produced a wealth of information from their inception in 1952 to well into the 1970s. Their importance to high-energy physics was acknowledged with a Nobel prize in 1960 awarded to Donald Glaser. Glaser struck upon the idea of a bubble chamber when he saw the tracks created by bubbles in beer.

Bubble chambers—three decades of physics

Bubble chambers initially used liquid in a super-heated state to detect the ions left along the tracks of high-energy charged particles passing through the liquid. In 1952 Glaser used diethyl-ether heated to $\sim 100°C$ above its boiling point to build the first bubble chamber. The super-heated liquid when struck by cosmic rays began boiling violently and a photograph made using a fast camera showed tracks left behind by the high-energy charged particles created by a cosmic ray. It is presumed that after a high-energy particle creates the initial ions along its path, heat generated by recombination is responsible for the boiling and bubble formation in the liquid.

Improvements to this technique including the use of pistons to create a sudden pressure drop to induce bubble formation, led to greater precision and larger chambers that could be placed in a magnetic field. Perhaps the most notable improvement to the bubble chamber came when Luis Alvarez substituted hydrogen for the ether used by Glaser. Alvarez's chamber produced much clearer tracks and this technological advance was deemed so important that he won the Nobel prize for physics in 1968 for his work, the second Nobel prize awarded for work related to the development of the bubble chamber.

In bubble chambers, the fluids in the chamber act both as the target and as the detector so different fluids—some cryogenic and some room temperature—were eventually used to suit the purpose of the experiment. Cryogenic liquids consisted of the simplest nuclei like, H_2, D_2, He, Ne, Ar and Xe while the room temperature "heavy liquids" like propane (C_3H_8) and Freon (CF_2Cl_2 or CF_3Br) offered short interaction lengths. The typical size of a bubble in a bubble chamber is ~ 10 μm and the bubble density can be used to determine $\beta \equiv v/c$ for the passing particle.

The advantages bubble chambers offered kept them in widespread use for three decades, from the 1950's to well into the 1970's. They had good spatial resolution (10 – 150 μm), a large sensitive volume, 4π geometrical acceptance, and allowed for the use of a variety of materials as targets. Eventually however, as physics began requiring more complex triggers and as large-volume high-precision detectors demanding electronic data recording came in use, the bubble chambers disadvantages rendered them obsolete. The analysis of photographs was a tedious task requiring expensive projectors for scanning the images and the whole process was only modestly scalable so that only limited statistics could be achieved. Bubble chambers were also complicated to operate, required cryogenics, and were a safety concern. In addition, bubble chambers weren't compatible with particle colliders—the now dominant high-energy accelerator, they provided no triggering for low cross-sections and they had a relatively long sensitive time (~ 1 ms) which necessitates a lower beam intensity.

Streamer chambers—a precursor to modern gas detectors

The streamer chamber developed by G.E. Chikovani in 1963—an improvement on the spark chamber—overcame some of the limitations of the bubble chamber and was the predecessor of the gaseous detectors of today. Like the bubble chamber, however, the streamer chamber also relied on photographic film to record the tracks of streamers, placing a limit on the statistics available for analysis.

The spark chamber uses a large potential across two parallel planes of electrodes to induce electrical breakdown—a spark—in gas between the electrodes. The ionization left from the passing of a high-energy particle acts as the catalyst for the spark. The spark chamber however, can only measure the position of the track in the direction parallel to the electric field to within the spacing of the electrodes. The streamer chamber overcomes this limitation by applying a high-voltage pulse for a short duration ($\sim$ 15 ns). The strong electric field ($\sim$ 20 kV/cm) from the high-voltage pulse induces an incomplete spark discharge. These electron avalanches or streamers form all along the particles path and the radiation of the gas in the streamer plasma can be recorded optically. Streamer chambers were built with sensitive volumes of several cubic meters that recorded particle tracks in any direction with equal efficiency. The density of the streamers can be used for particle identification up to particle momentums of $\sim$ 1 GeV/c.

The streamer chamber's two major advantages over the bubble chamber are its ability to be triggered by external devices and its very short sensitive time ($\sim$ 1 μs). Eventually however, its use of photographic film, its limited spatial resolution ($\geq$ 300 μm) and its relatively long dead time ($\sim$ 300 ms) turned out to favor the gaseous detectors that would rely on electronic, not optical, recording techniques.

Today's tracking detectors

Almost all tracking detectors, other than the modern solid state and gaseous detectors using electronic recording techniques have been abandoned. The modern-era detectors have shorter sensitive times and shorter dead times so that the beam intensity of the particle accelerator can be increased and greater statistics can be recorded. These newer detectors also tend to be easier to operate and have greater spatial resolution.

Gas detectors

Most gas detectors—multi-wire proportional chambers (MWPC), drift chambers, straw tubes, cathode strip or pad chambers, time projection chambers (TPC) and microstrip gas chambers (MSGC)—use the proportional counting mode of operation. In this mode, electrons produced by ionization are directed in an electrostatic field toward a very high field region (10 – 100 kV/cm) surrounding a very thin anode. In this region, the acceleration induced by the strong electric field gives the electrons enough energy to ionize other gas atoms as they undergo subsequent collisions. Each new electron produced by ionization in turn creates more electrons and ions in similar collisions. This chain of events creates an avalanche or cascade of electrons and ions and is called gas multiplication. Eventually the electrons freed in this gas multiplication reach the anode wires, however, the usual wisdom that they produce an analogue signal used for measurements is wrong. What happens is that when the electron and ion charges start to separate in the gas they are already seen by the anode due to the capacitive coupling. The electrons, carrying negative charge, induce at this time a positive signal. Later, when they themselves arrive, they deposit a negative charge on the anode. However, due their large mobility this process is so fast that these signals are integrated in the readout electronics with the zero net result. On the other hand, the mobility of ions is $10^3 - 10^4$ times slower and therefore they are inducing a negative signal (being themselves positive) on the anode for a long time. This analogue signal with a $1/t$ time behavior is the one which is detected and used for position and energy loss (dE/dx) measurements.

The mode of operation of a gaseous detector is determined by the response of the ions and gas to the field strength surrounding the anode. In the proportional mode of operation the field strength is great enough to induce gas amplification—typically up to $10^4 - 10^6$ times—but is not so great that it leads to non-negligible space charge effects caused by longer-lived positive ions or complete breakdown. In the proportional mode of operation the analogue signal on the anode is proportional to the number of primary ions which is in turn proportional to the energy loss of the traversing particle. The measured dE/dx, being a function of particle velocity, can then be used for particle identification (PID).

A multi-wire proportional chamber (MWPC) consists of planes of independent anode wires—typically spaced $1 - 2$ mm apart—set between two planes of cathodes held at a distance of $4 - 5$ times the wire spacing. A negative voltage is applied to the cathodes and the anode wires are held at ground. Each wire then acts as a proportional counter for primary ionization left in a track. The position of the track between the wires can be measured using the time delay of the pulses. The wire diameter should be about $20 - 50$ μm. The spatial resolution is given by $d/\sqrt{12} = 300 - 600$ μm, where d is the wire pitch or spacing. For this invention G. Charpak was awarded the 1992 Nobel prize in physics.

A drift chamber is a multi-wire proportional chamber with a large wire pitch—from several centimeters up to 50 cm but more typically 5 cm. Track position is determined by measuring the time electrons need to reach the nearest anode wires. The speed of the electron drift depends on the gas used and the pressure in the chamber and is typically ~ 5 cm/μs so that a timing resolution of 1 ns gives a spatial resolution of ~ 50 μm. For even better resolution the drift chambers with higher than atmospheric pressure are used. The disadvantage comparing to the MWPC is a lower rate capability. Different geometries and configurations can be used in order to create constant fields pointing toward the anode wires. A straw tube is a drift chamber composed of an individual straw shaped cathode (diameter of ~ 5 mm) with a single anode wire in the center. A continuous tracker can be constructed by packing many layers of straw tubes together. Straw tube detectors tolerate high loads because they don't use a common gas volume and can achieve a resolution of about 150 μm with coarse time measurements.

A time projection chamber (TPC) is a rather drastic variation on a simple drift chamber. A TPC consist of a large three-dimensional gas filled vessel with readout detectors on a wall at the end of a drift volume created by a high-voltage plane (cathode). The readout detectors are usually cathode pad chambers and the drift field is established by a strong electric field across the TPC. When a charged particle creates a track of electron-ion pairs within the drift volume the electric field prevents them from recombining and pulls the much faster electrons toward the readout detectors. The electric field is chosen so that no multiplication of the primary electrons occurs and is typically hundreds of volts per centimeter. The readout chamber is separated from the drift volume by a gating grid. The gating grid is a plane of wires that electrostatically separates the amplification region of readout chambers from the drift region. This prevents the ions created in the amplification region from getting back into the drift region and allows for triggering of the detector. When an interesting event occurs the gating grid wires are set to voltages that allow drifting electrons to pass through.

The TPC readout chambers use anode wires in front of a pad plane. The signal is typically read from the pad plane where image charges are created when ions are left after electron cascade nearby the anode wires. The pad plane is designed so that the signal will be seen on several pads. The position of the ion cloud in the vicinity of the anode wire along its direction can be determined precisely by fitting a modified Gaussian to the signals on several consecutive pads. This measurement gives two transverse coordinates and the drift time gives the third, making the TPC a fully three-dimensional detector. Unlike other historic three-dimensional detectors, such as the bubble or streamer chambers, the TPC is read completely electronically. The TPC also has the advantage that it has no pulsed very high-voltages and is fast compared to historic detectors; its speed is determined by the maximum drift time which, for large chambers is ~ 100 μm.

Inhomogeneities in the drift field and effects due to magnetic fields can distort the drift path of the electrons and further degrade the resolution. The electron clouds also diffuse at a rate of hundreds of μm/$\sqrt{cm}$ due to elastic rescattering in the gas as they drift toward the readout chamber. The TPC requires careful tuning of the drift field and a high degree of gas purity. Many parameters like drift length, track angle, or the number of primary ions affect the spatial resolution but a typical value is ~ 500 μm.

Solid state detectors

Solid state detectors—silicon micro-strip detectors, silicon pixel detectors and silicon drift detectors—offer very good resolutions of $10 - 100$ μm and are now in common use. Every detector planned at the Large Hadron Collider at CERN (LHC) will use trackers based on silicon devices. Silicon detectors require only 3.6 eV of energy from the traversing particle to create an electron-hole pair. That is roughly one order-of-magnitude less than gas detectors require (~ 30 eV). This, along with silicon's higher density, means that the number of electron-hole pairs created by a minimum ionizing particle (MIP) in silicon is much greater than the number of electron-ion pairs created over the same distance in gas. In 1 μm of silicon a MIP produces ~ 100 charge pairs. To produce that much charge in gas would require several centimeters. As a result, unlike gaseous detectors, silicon detectors don't require signal amplification inside the detector. In a typical silicon detector a MIP will produce $20 - 30$ thousand electrons.

Typically, silicon detectors are built using ~ 300 μm thick, high resistivity, n-doped silicon plates with a thin p-doped layer on one side. A reverse bias voltage—positive on the n-side and negative on the p-side—is applied to deplete the silicon of free charge carriers and to create an electric field that will cause the electrons and holes to drift to opposite surfaces where readout structures are organized. The highly developed state of silicon technology allows for the production of many different readout structures.

The readout structures for both silicon strip and silicon pixel detectors are layers of aluminum applied to the surface of the silicon. Silicon strip detectors use a solid layer of aluminum on the n-doped side of the silicon and a sequence of aluminum strips on the p-doped side. The strips typically have a pitch (d) of ~ 50 μm. The resolution for this pitch is ~ 15 μm or $d/\sqrt{12}$. The charge collected on the strip is electronically integrated and read out as an analogue or digital signal. If strips are placed on both sides of the silicon—a double sided silicon strip detector—two coordinates can be measured simultaneously.

The silicon pixel detector uses pixels instead of strips and has the advantage that it is a true two dimensional micro-detector. Amplifier and readout circuitry however, needs to be connected to each pixel which typically has a surface area of only $\sim 50 \times 400 \mu m^2$. This is done using specially designed multiplexing readout chips that are bump-bonded to the detector silicon. Silicon pixel detectors are fast, have very low noise (small capacitance) and have excellent pattern recognition for high particle densities. They are however, very fragile and offer many technological challenges in development.

Silicon drift detectors have two-dimensional capabilities but by their design avoid the large number of readout channels required by silicon pixel detectors. Silicon drift detectors use a silicon wafer, with an array of pixel anodes arranged at one edge and strips on both wafer surfaces. An electrostatic field valley, created by the diminishing toward the anode voltage on the strips, drifts the primary electrons—from the track of a passing particle—through the silicon, toward the array of anodes. A typical drift speed is ~ 1 cm/μs. The anode position along the edge of the wafer and the drift time give two coordinates for the position of the track. The third coordinate is given by the position of the wafer itself and, like a spark chamber, is only known to within the thickness of the wafer. Because of their high granularity they have an excellent high-occupancy capability. On the other hand, the time for the electrons to drift to the anodes (~ 5 μs), makes it a relatively slow detector. In addition because of the dependence of the drift time on temperature these detectors require very precise climate control.

The high resolution of all these solid state detectors however, makes them ideal for constructing vertex chambers that are particularly useful for detecting heavy-flavor decays nearby the interaction point.

TRACK FINDING AND FITTING

Momentum precision estimates

We will consider three contributions to the error in the reconstructed momentum made by measuring the track curvature. First, the precision of the reconstructed momentum can be estimated in part from the detector measurement error alone. The relative momentum error is proportional to the momentum p itself, the measurement error, and $1/(BL^2)$ where L is the distance from the track vertex to the farthest detector plane and B the magnitude of the magnetic field. Second, the contributions of multiple scattering to the relative momentum are proportional to the square root of the material thickness (expressed in radiation lengths) and $1/(BL)$. The third contribution to the relative momentum error is fluctuations in the ionization losses. An empirical approximation to this error demonstrates that it is proportional to the material thickness and $1/p$. The behavior of these contributions are summarized in Table 1.

Concerning the optimization of the positioning of the detectors along the particle

TABLE 1. Contributions to the relative momentum error

Measurement Error	p	$1/B$	$1/L^2$	1
Multiple Scattering Error	1	$1/B$	$1/L$	$\sqrt{X}$
Energy Loss Fluctuation	$1/p$	1	1	X

trajectory, there are different considerations to be taken into account. Let us try to design a detector measuring momentum inside the homogenous magnetic field using three measuring stations. It is easy to show that the relative momentum error due to the spatial measurement error itself behave as $\delta p/p \propto 1/x(1-x)$, where x is the fractional distance along the track from the beginning to the middle station. It has a minimum at $x = 1/2$, therefore from the point of view of the spatial measurement error the best choice would be to place the middle measuring station exactly in the center of our detector. On the other hand, the behavior of the multiple scattering contribution to the relative momentum error (and energy loss fluctuation as well) is completely different. The $\delta p/p$ is independent on the position of the middle station. Naively, one would now conclude that, anyway, the best solution would be to concentrate tracking detectors in the center of the magnet (in addition to the ones at the beginning and at the end of the tracking volume). However, considering the case of multiple scattering contribution to the $\delta p/p$ one can demonstrate the following peculiarity: the momentum estimates using the triplets of consequent measurements are uncorrelated, even if they share a common spatial point. This is because the momentum error is influenced only by (scattering in) the middle plane of the triplet. (Of course this is not true for the other limiting case—the spatial measurement error.) Having in the mind that the $\delta p/p$ goes as $1/L$, it is clear that in the case of four measuring stations the best solution would be to put two of them at beginning and another two at the end of the tracking volume. In this arrangement both triplets make use of the full magnet length (and the relative position of the middle station has no influence) minimizing thus momentum error. In conclusion, when the spatial resolution of the detectors dominates the momentum resolution, the arrangement with evenly spaced stations gives better precision. Contrary to that, when the momentum resolution is dictated by the multiple scattering, the grouping of measuring stations at the beginning and at the end of the tracking volume, and clearing the middle, will optimize the resolution.

The contributing errors from the spatial resolution and the multiple scattering can be added in quadrature. However, this has to be used with caution, since it is only valid for a uniform distribution of scattering material (i.e. for a bubble chamber). Nevertheless this receipt one can find practically everywhere, for example in Particle Data Booklet. In order to demonstrate how large the discrepancy could be let us imagine a detector inside the magnetic field made of two identical halves which have negligible amount of scattering material. These two parts are, however, separated by plate with infinite amount of scattering material. Because of that the two detector halves act as independent—there is no correlation between the track in the first part and the second part. Therefore the relative momentum resolution is $\delta p/p = (1/\sqrt{2})(\delta p/p)_{1/2}$, expressed in the resolution given by one half of the detector. The contribution caused by the spatial resolution we can estimate by going with thicknesses of scattering material to zero. In this case the two halves of detector are as one of them with doubled length L and twice the number of measured points. Knowing the dependence of the relative momentum error on the length (see Table 1) and on the number of the points we can write the spatial measurement contribution as $(\delta p/p)_s = (1/4\sqrt{2})(\delta p/p)_{12}$. The contribution from multiple scattering is $(\delta p/p)_m = 0$ because when we neglect the spatial measurement errors both halves measure with infinite precision (they have negligible material within them) and even being separated by infinitely thick plate the momentum error will be zero. In conclusion, the total relative momentum error is much larger than the contribution from spatial measurement error and from multiple scattering added in quadrature, it is even larger than the linear sum of the two (four times, to be precise). The detector we have discussed here is of course very badly designed. However, for modern detectors the assumption of uniform material distribution is usually not valid, and the discrepancy comparing with simple quadratic addition could very large even for higher momenta.

Track parametrization

In general a track is defined with five independent parameters. Mathematically, any choices for these five parameters is equally acceptable. However, one should use the inverse of the momentum instead of the momentum itself and in general avoid any variable which has discontinuities or highly correlated with the others. The choice of a definite set of track parameters is mainly driven by arguments for the numerical stability. For example choosing the momentum as one of the parameter could lead to problems during the track parameter fitting. One needs to go continuously form positive to negative charge track when approaching very high momentum (i.e. straight line). This is usually achieved using the inverse of the momentum with the sign corresponding to the track charge. The optimal choices depend on the exact situation such as whether it is used for collider or fix-target experiment.

Whatever the choice, these five parameters are estimated from point measurements in the detector. We have to know these measurements y_n and their covariance matrix C_{nm}. The diagonal terms of this matrix reflect the spatial measurement errors and the off-diagonal terms the multiple scattering contribution. To find the track parameters a_i, we usually minimize the quadratic form $S = \sum (Y(a_i) - y_n)^T (C_{nm})^{-1} (Y(a_i) - y_m)$. However, the covariance matrix C_{nm} is not uniquely defined. This sounds quite strange because one can always simulate many times the same track and construct the covariance matrix calculating the variances $(y_n - \overline{y_n})(y_m - \overline{y_m})$. The problem is what really means the same track: having the same parameters at the beginning, in the middle or at the end of the tracking volume. The structure of the covariance matrix depends what definition we adopt. The resulting track parameters will be optimal at the point which we choose as a reference. Therefore, the calculation of the optimal track parameters at a different location requires all the point measurements, and cannot be calculated from the track state at a different location by propagation.

Beware of approximations that seemingly linearize the minimization problem. Even the simplest case of a homogeneous magnetic field where the track is described by a helix cannot be linearized (it is possible only in non-optimal approximation). Without approximation there exists an iterative approach which capitalizes on the independence of the differences between the subsequent measurements y_n and y_{n-1}. By finding a linear transformation of y_n that results in a corresponding diagonal covariance matrix, a system of linear equations will have a nice iterative solution. The estimation of the track parameters in Kalman filter is based on this principle.

In a non-homogenous magnetic field the track trajectory is a solution to a second order partial differential equation. While this can solved using a straight forward numerical method like a Runge-Kutta integration, it can be quite time consuming when used in the fit. For a quasi-homogenous field a faster method is to build up the track from consecutive helices.

Track finding

The algorithm for track finding can be classified into two categories, local and global tracking methods. In the local method each measured point is considered one at a time and then after extrapolation the next point is considered etc. In the global method all points enter into the algorithm in the same way. The large advantage of the local method is that there is no need for a global track model of the track trajectory and therefore is easy to take into account stochastic processes like multiple scattering and energy loss fluctuation. But there are several disadvantages. First, the decision to associate a point to the track is made when only partial information about the track is available. Second, the result can depend on the order of the data. Finally, the local method usually needs seeding to start with. While the global method requires a precise global track model, it enables the decision to be made when all the information about the track is available. Sometimes the division between local and global method is not so clear. For example, the seeding for local method is often done using some variant of the global method. On the other hand, in some global method after the global transformation a local type algorithm is used.

Local methods

Different local methods are often employed depending on the detector configuration. In the track following method an initial track segment is formed by seeding and then the next point is predicted by extrapolation repeatedly until the candidate is projected through the whole detector. In the track road method points at the extremes of the detector are combined and then intermediate points are considered by interpolation. This can be much slower, but may be optimal for widely spaced detector layouts. The final is the track element method in which short track segments are formed through out some natural subdivisions of the detector and then connected by extrapolations or interpolations.

The most widely used local method is the track following method formulated as the Kalman filter. It was introduced in 1983 by P. Billoir [1] as a method for track fitting and only later recognized as an application of a more general method from the system evolution theory by R. Kalman. Usually this method is used as a backward propagation, we assume that we know the optimal estimation of the track parameters (state vector a_i)

and their covariance matrix at some point n and we want to get the optimal estimator for the state vector and its covariance matrix at the point $n-1$. The propagation of the state vector from the point n to the point $n-1$ can be formally written as the multiplication of the original state vector by some transport matrix. Doing this we get the prediction for the state vector at point $n-1$ but not yet the best estimate. Before we have to add the information from the measurement at that point. In order to make it in the optimal way we have to know the covariance matrix of the predicted state vector and the covariance matrix of measurement itself and to add them as (roughly speaking) weighted mean. The covariance matrix of the predicted state vector can be calculated from the matrix at the point n, first adding the noise corresponding to multiple scattering (basically increasing the matrix elements containing the angular errors of the track parameters) and other noise which we want to take into account, and second propagating the matrix by multiplying it from the right side by the transport matrix and from the left by the transpose of the transport matrix. Now we are ready to calculate the weighted mean for the state vector at the point $n-1$. If we have many candidates for the $n-1$ point we have to do filtering—to choose the one which adds more information to the state vector (i.e. lowers the track parameter errors). To calculate the weighted mean means that if the errors of the measured point at $n-1$ is large compared to the errors of the prediction the result will be close to the predicted track, and vice versa, if the point has small errors compared to the prediction errors (for example, due to a large multiple scattering at point n) the track will be tilted to go closer to the measured point. This way the track do not follow a global track model but rather simulate the stochastic behavior according the measured points.

This method has several advantages. It combines the processes of track finding and optimal track fitting. There are no inversions of large matrices, and it provides a natural way to connect to other detectors. As with all local methods, there is no global track model required, and it is easy to take into account other stochastic processes like fluctuations in energy loss. Some of the limitations of this local method can be minimized. By developing a hypothesis tree when filtering, one may avoid making the decision until more information is known. Furthermore, just as the state vector was propagated through the detector to provide predictions for filtering, once filtered the state vector can be smoothed by propagating the track in the reverse direction.

Global methods

The global methods usually used at the beginning some type of global transforation. One of the popular method is the conformal mapping. In the bending plane (perpendicular to the magnetic field) for homogenous magnetic field the track trajectory is a circle. Moreover, most of the tracks are coming from one common point (the primary vertex). We place in this point the origin of the complex plane u which we map to another complex plane $v = 1/u$. This conformal mapping between the two complex plane has few interesting geometrical features: the angles are conserved, the concentring circles with the center at the origin are mapped into the same set of circles in reverse order, the straight lines going through origin are mapped into the same set of straight lines but the orientation is changed, and the circles going through origin (most of our tracks) are mapped into straight lines. This last feature is used in the global tracking method. This way the task to find the circles is transformed to the task to find straight lines, which is easier (and done by other method). The distance of the line from the origin is proportional to the circle curvature in the original complex plane. The clear disadvantage of this method is that the transformation is non-linear. This means that even if we seemingly linearized the problem, now the non-linearity is back in the errors after the transformation.

The second method which we will discussed here is Hough-Radon transform. This is a more general method and, for example, it is used in the computer tomography. The easy way to explain the principle is the following toy example. Suppose we want to find straight lines (this is our track model) going through (within the error) measured points (x_n, y_n). The parametrization of our track model can be written as $y = ax + b$. The straight lines going through the measured point (x_n, y_n) have to have the parameters a and b satisfying the track equation $y_n = ax_n + b$ or in other way $b = -x_n a + y_n$. In other words the track parameters for a track going through the given point are constraint by this linear equation. Then we can defined a transformation from the normal space (x, y) to the parameter space (a, b) representing each measured point (x_n, y_n) by a straight line $b = -x_n a + y_n$. The point in the (a, b) plane in which many of these lines cross represents the parameters of the track we search for. The multiplicity of crossing lines gives the number of points associated to the reconstructed track. In the real algorithm the parameter space is divided into the pixels which accumulate the signal when crossed by transformation line. After the transformation the pixel space is searched for local maxima above some threshold which correspond to the track parameters.

This method is easy to generalized to higher dimensional space and (non-linear) track model. In the parameter space we can accumulate directly amplitude information from the detector. In this case we can skip the reconstruction of the space points from the clusters of adjacent channels as is often done, and we can directly transform the raw data. The disadvantage of this method when used for general tracking is the high dimensionality of the parameter space—five—therefore it is usually used with vertex constraint which lower the dimensionality to three. It can be reformulated using path integrals along the set of parameterized trajectories which is used in tracking as template method (however, the principle is the same). This method is often used in combination with Kalman filter, for the track seed finding.

LHC EXPERIMENTS

The LHC (Large Hadron Collider) [2] is an accelerator which brings protons and ions into head–on collisions at higher energies than ever achieved before (http://lhc-new-homepage.web.cern.ch/). It will be located at CERN (European Organisation for Nuclear Research), see http://www.cern.ch. Right now its construction is going on in the former, 27 km circumference, LEP tunnel (Large Electron Positron Collider) http://public.web.cern.ch/Public/ACCELERATORS/LepAcc.html. The magnets operate in superfluid Helium, they will have a two-in-one structure, that means both beams travel in the same gap, about 20 cm separated. Such large distance eases the design of the dipole and quadrupole magnets. In the 7 TeV/charge operation the dipole magnets will have to provide a field of about 8.3 T.

As was the LEP, the LHC will be feeded by the SPS accelerator (Super Proton Synchrotron) [3]. Protons and heavy ions (up to lead) will be injected at the energy 450 GeV/charge and accelerated to their colliding energy of 7 TeV/charge. The beams will provide a luminosity of 10^{34} cm^{-2}s^{-1} (protons) and 10^{27} cm^{-2}s^{-1} (Pb). Each bunch will consist of 1.1×10^{11} protons or 7×10^7 PB ions, the separation between two bunches will be about 25 ns for protons and 125 ns for Pb ions which corresponds to about 2800 proton bunches and almost 500 Pb bunches per ring. There will be four crossing points where the beam passes from one ring to the other and there the experiments will be situated.

Tracking in ALICE

ALICE (A Large Ion Collider Experiment) [4] is one of the experiments at the LHC optimized for the study of heavy-ion collisions, at a center-of-mass energy of 5.5 TeV per nucleon. The prime aim of this experiment is to probe in detail non-pertubative aspects of QCD, such as deconfinement and chiral symmetry restoration. Furthermore, of interest to the experiment are the newly emerging physics topics related to the study of QCD at very high-field strengths [5].

The strategy of the ALICE experiment is to combine almost exclusive measurement of particle production in the central region with spectroscopy of quarkonia states at central and intermediate rapidities and characterization of the event centrality. It therefore combines three major components

- the central barrel detectors
- the forward muon spectrometer
- the forward multiplicity and centrality detectors.

Central barrel detectors

Figure 1 shows the ALICE setup, the central part consists of the L3 (see `http://l3www.cern.ch/`) solenoidal magnet which hosts the ITS (Inner Tracking System); surrounded by the cylindrical TPC (Time Projection Chamber); the TRD (Transition Radiation Detector); the TOF array (Time-Of-Flight). Outermost are the single-arm electromagnetic calorimeter (PHOS) and ring imaging Čerenkov detector HMPID (High Momentum PID). The central barrel covers polar angles from 45° to 135° (pseudorapidity from −0.9 to 0.9) over the full azimuth.

Inner Tracking System

The basic functions of the inner tracker—secondary vertex reconstruction of hyperon and charm decays, particle identification and tracking of low-momentum particles and improvement of the momentum resolution—will be achieved with six barrels of high-resolution silicon detectors. Because of the high-particle density the innermost four layers need to be truly two-dimensional devices, i.e. silicon pixel detectors at 3.8 cm and 7.4 cm radius with a pixel size of 50×425 μm^2 and silicon drift detectors at 14 cm

FIGURE 1. Setup of the ALICE detector

and 24 cm radius with 300 μm anode pitch. The two outer layers at 39 cm and 45 cm will be equipped with double-sided silicon micro-strip detectors with strip pitches of 95 μm operating with amplitude readout. The outer four layers will use the analog readout for independent particle identification via dE/dx in the non-relativistic region. The inner tracking system will have a stand-alone tracking capability as a low-momentum particle spectrometer.

Time Projection Chamber

The TPC is the main tracking detector. It allows continuous tracking between 88 cm and 250 cm using about 160 pad rows. The pad sizes increase from 4×7.5 mm^2 over 6×10 mm^2 to 6×15 mm^2 with radius. This enables to to reconstruct up to 12 000 charged tracks within the acceptance. The inner radius of the TPC is given by the maximum acceptable hit density and the outer radius by the length required for a dE/dx resolution of better than 7%. With this resolution the TPC can serve, in addition to tracking, as a detector for electron identification up to momenta of 3 GeV/c.

Outer detectors

The TRD, placed in between the TPC and TOF, will be used for tracking as well. It is dedicated mainly to measure high-momentum electron tracks in order to detect Υ

decays.

Particle identification over a large part of the phase space and for different particles as pions, kaons and protons is realized by a large TOF array (about 100 m^2) at a radius of about 3.7 m with an intrinsic time resolution of about 100 ps.

The second PID system, the HMPID of smaller acceptance and at larger radii (4.7 m), will extend the accessible momentum range for inclusive particle spectra into the semi-hard region. It consists of a proximity-focussing RICH detector with liquid freon radiator, and a novel solid photocathode with pad readout.

The small-acceptance electromagnetic calorimeter PHOS will be located below the interaction region at about 5 m from the vertex, and will cover about 9 m^2 with 18k channels of scintillating PbWO$_4$ crystals. These very dense crystals are needed to cope with the large particle density, and to have sufficient light-output to allow readout with silicon photodiodes.

Forward detectors

The forward muon arm is designed to measure the complete spectrum of heavy quark resonances, i.e. J/ψ and Υ families. It will detect the decays of these resonances both in proton-proton and in heavy-ion collisions. The angular acceptance of the muon spectrometer is from 2° to 9° (pseudorapidity between 2.5 and 4). Its mass resolution will be better than 100 MeV at 10 GeV, sufficient to separate the Υ states. It consists of a composite absorber made of layers of both high- and low-Z materials, starting at 90 cm from the vertex, a large dipole magnet with a 3 Tm field integral placed outside the L3 magnet, and 10 planes of thin and high-granularity tracking stations. The second absorber at the end of the spectrometer and four more detector planes are used for muon identification and triggering. The spectrometer has to be shielded throughout its length by a dense absorber tube of about 60 cm outer diameter which surrounds the beam pipe.

The determination of the photon multiplicity and the measurement of the charged particle multiplicity will be performed with the FMD (Forward Multiplicity Detector) which will cover pseudorapidity up to 4 – 5. The ZDC (Zero Degree Calorimeter) will provide the fast trigger on the collision centrality.

Requirements

The LHC design luminosity for lead ions corresponds to an interaction rate of 8 kHz. The detectors shall work at rates up to 200 Hz. The produced particle multiplicities are difficult to predict, for central Pb–Pb collisions 2 000 to 8 000 charged particles per unit of rapidity are expected at mid-rapidity. This leads up to 20 000 primary and secondary charged particles in the TPC. This would result in 40 % occupancy at the innermost TPC radius and 15 % at the outermost radius of the TPC.

In oreder to resolve the different states in the Υ family, the mass resolution should be better than 100 MeV. Therefore the momentum resolution of 5 GeV/c particles should be better than 1.5 % . For low-momentum the precision better then 1 % should be realized.

Hadronic correlation observables place the highest demand on relative momentum and two-track resolution. To measure a source in the HBT analysis of the size of 25 fm the-two track separation has to reach a few MeV/c in momentum. For hadron particle identification a dE/dx resolution of 7 % is desirable, and for electrons at least 10 % should be reached. Good secondary vertexing capabilities are necessary for V0 and open charm particle detection.

The acceptance in has to be sufficient for event-by-event fluctuation analysis. For kaon to pion ratio fluctuation measurement about 250 identified kaons will be needed. The tracking efficiency for tracks with $p_t > 0.2$ GeV/c should be above 90 % but there should be some efficiency below as well.

Track finding

A classical approach for the tracking in the ALICE TPC was chosen. This means, before the tracking itself, two-dimensional clusters in pad-row–time planes have to be found. Then the positions of corresponding space points are reconstructed, which are interpreted as crossing points between the tracks and the center of the pad rows. First pre-clusters, groups of adjacent cells with a signal in a pad-row–time plane, are searched. If a pre-cluster has only one maximum, it is assumed that it was initiated by a single track, the center of gravity is stored as the space point. If there are several local maxima, the group of cells is split into more groups with respect to the saddle points. Individual center of gravity calculation leads to corresponding space points. The obtained average space point resolution is better than 1 mm.

Later the clusters should be fitted which will allows better unfolding of overlapping clusters. But for this, assumptions of the track parameters need to be done which could be made when reconstructing tracks in parallel with cluster finding.

The track finding is based on the Kalman filter. There are two disadvantages which affect the tracking in the ALICE TPC. First, clusters are reconstructed before applying the Kalman filter procedure. This is critical because due to the high occupancy a certain number of clusters get lost and others get displaced. Second, the tracking relies essentially on the determination of the good seeds to start a stable filtering procedure. Since the ALICE TPC is the main tracking detector, the seed has to be constructed using the TPC data themselves.

The tracking efficiency is defined as the ratio of the number of found good tracks to the number of findable tracks, while the fake probability is the ratio of found bad track to the number of findable tracks. In order to evaluate the tracking algorithm, nne have to choose certain reasonable definition what good track and findable track means. The track finding efficiency for TPC and TRD tracks amounts 98 % for charged particle densities of 1 300 per unit of rapidity and it drops down to 88 % for 8 300 charged particle per unit of rapidity, while the fake probability rises from 0 % to 2 % . No significant dependence of the tracking efficiency on pseudorapidity is observed. While for the lower assumed track density about 88 % of the clusters are found, at the higher density this number decreases to 65 % . In both cases tracks are mainly lost close to the inner TPC boundary.

The charged particle tracks found in the TPC should be connected to others detectors to improve the momentum resolution and particle identification. The dominant contribution of extrapolation errors comes from multiple scattering in the TPC field cage, high voltage degrader and vessel. The multiple scattering decreases with increasing momentum. Therefore first tracks with low curvature (high momentum) in the TPC are extrapolated to the outer layer of the ITS. The distance between the TPC and ITS is rather large and the track density inside the ITS is so high that simple continuation of the tracking procedure used for the TPC would be ineffective and if only the criterion of minimal χ^2 would be used for hit assignment there would be a high probability of incorrect assignments.

The first improvement with respect to the Kalman filter procedure is that all hits in a certain window with a reasonable χ^2 are accepted (not only the one with the best fit) and the tracking is done for all candidates through the whole ITS. After all information is known, the track with the most assigned hits and lowest overall χ^2 is chosen. The other change is that the vertex constraint is explicitly used. From the already known vertex position the track with its current parameters is propagated to the concerned ITS layer. This provides additional information to the Kalman procedure.

For the track finding and fitting procedure five of the maximal possible six hits are required. With an assumed not perfect detection efficiency the maximal reachable track reconstruction efficiency is about 94.5%. For tracks with transverse momenta above 400 MeV/c an efficiency only a few percent below the maximal achievable is obtained. For low-p_t tracks the efficiency is slightly worse because of the influence of multiple scattering.

Good momentum resolution is especially important for high-p_t electrons as described above. Simulations show that the required resolution cannot been achieved in the current setting at the highest expected track density, it is about 8% to 9% for $p_t = 5$ GeV/c particles. This is mainly due to the cluster overlapping which are more probable for higher particle densities. The resolution might be improved by refitting the clusters.

The required momentum resolution can be obtained when the TPC tracking will be combined with the ITS and TRD and the magnetic field will be increased from 0.2 T to 0.5 T. In this case resolution of 1.4% for $p_t = 5$ GeV/c particles. The ITS enhances the resolution of high-p_t tracks especially because of the increased track length.

Summary of ALICE tracking

The behavior of the tracking detectors is in general understood, the simulated results for the TPC are well within the specification, and the momentum resolution is as required. The track matching efficiency to the ITS is at low-momenta below expectation and the matching to the TRD is still under study. The concerns are that there is no precise tracking detector behind the TPC for its calibration, and the requirements for the drift detectors for alignment and calibration.

Other LHC experiments

Contrary to the ALICE experiment which is dedicated to heavy-ion physics the other experiments, namely ATLAS, CMS and LHCb, are designed to investigate proton-

proton interactions. Due to this they do not need to cope with as high track densities and they do not focus on low-momenta particles. On the other hand, they will run at much higher luminosities and will have to cover a larger rapidity range.

ATLAS experiment

ATLAS (A Toroidal LHC ApparatuS) is a general purpose experiment for recording proton-proton collisions at LHC. The detector design has been optimized to cover the largest possible range of LHC physics: searches for Higgs bosons and alternative schemes for the spontaneous symmetry-breaking mechanism; searches for supersymmetric particles, new gauge bosons, lepto-quarks, and quark and lepton compositeness indicating extensions to the Standard Model and new physics beyond it; studies of the origin of CP violation via high precision measurements of CP-violating B-decays; high-precision measurements of the third quark family, such as the top-quark mass and decay properties, rare decays of B-hadrons and spectroscopy of rarely produced B-hadrons.

The momentum resolution has to be excellent including the high momentum tracks and the pseudorapidity range up to 2.5 has to be covered. The whole detector has work at very rates corresponding to design luminosity, particle identification for electrons is required, secondary vertices shall be reconstructed, including those close to the interaction point, i.e. B-decays. Finally, everything has to be radiation tolerant.

The ATLAS detector includes an inner tracking detector inside a 2 T solenoid providing an axial field, electromagnetic and hadronic calorimeters outside the solenoid and in the forward regions, and barrel and end-cap air-core-toroid muon spectrometers.

ATLAS has two tracking systems. The inner detector is composed of semiconductor trackers (SCT) and transition radiation tracker (TRT) which both are placed in the barrel and in the two end-caps. The SCT consists of silicon pixel and silicon strip detectors. Three layers of pixel detectors will be situated in the barrel at radii of 4 cm, 10 cm and 13 cm and five disks will be placed on each end-cup. Their pixel size is 50×400 μm. Four layers of silicon strip detectors will surround the pixel detectors at radii of 30 cm, 38 cm, 45 cm and 52 cm. Nine wheels are placed on each end-cup side. Their strip pitch is 80 μm, the stereo angle amounts to 40 mrad. The TRT is made of straw tubes with 4 mm diameter, the position resolution is expected to be about 170 μm. It covers as a continuous tracker radii between 56 cm and 107 cm, and in each end-cup 18 wheels are placed.

The simulated tracking performance is very good thanks to robust continuous tracking capability even at high densities. The momentum resolution is excellent, it is of the order of 20 % over a wide range of pseudorapidity for high-p_t particles. The vertexing is satisfactory, the impact parameter resolution is better than 20 μm for high-p_t tracks.

CMS experiment

Since the CMS (Compact Muon Solenoid) experiment was designed for similar physics intentions as ATLAS, it also has similar requirements. The detector consists of two tracking systems, the inner tracker and the muon tracker. For the inner tracker the silicon solution was adopted. The inner layers of silicon detectors provide a resolution between 15 μm and 30 μm, the outer layers 40 μm to 60 μm. The muon system is integrated in the return yoke of the high-field solenoid magnet. The barrel muon system has the resolution of about 250 μm and the end-cap system between 75 μm and 150 μm.

The inner tracker of CMS consists of a large variety of detectors, the higher magnetic field (compared to ATLAS) leads to a better resolution. However there is, contrary to ATLAS, no continuous tracker. Probably this is the reason for significantly longer CPU time needed for the event reconstruction. The momentum resolution is excellent. The pixel size of 150×150 μm^2 used in the innermost silicon-pixel layers tries to catch the good impact parameter resolution in both direction which by fot not obvious. For the muon system a simple and robust solution was chosen.

LHCb experiment

The LHCb (Large Hadron Collider beauty experiment) detector is designed to study CP violation and other rare phenomena in decays of hadrons with heavy flavors, in particular B mesons. This is achieved by designing a detector which has very good trigger efficiencies for B-meson final states with both hadrons and leptons, capability of identifying kaons and pions in the required kinematic range and excellent proper decay time and mass resolutions. LHCb will be build as a wide angle single arm forward spectrometer.

Because of the high particle density close to the beam pipe, the LHCb tracking detector is split into inner and outer systems. The inner tracking chambers have dimensions of 60×40 cm^2. Drift chambers based on straw-tube technology are considered as baseline for the outer tracking detector. For the inner tracking system where high granularity is required, two options, namely silicon micro-strip detectors and Gaseous Electron Multiplier (GEM) chambers are being considered. The tracking system is placed in a dipole magnet of 1 T. It provides a spatial resolution better than 200 μm and a momentum resolution of about 0.3 %. This results in a mass resolution for B-meson better than 20 MeV. The very good resolution is achieved not only due to the tracking but but also due to the well optimized vertex detector.

REFERENCES

1. Billior, P., *Nucl.Instr.Meth.* **225**, 277 (1984)
2. Lefèvre, P., Pettersson, T., *The Large Hadron Collider, Conceptual Design* **CERN/AC/95-05(LHC)** (1995)
3. Collier, P. et. al. *The SPS as Injector for LHC : Conceptual Design* **CERN-SL-97-007-DI** (1997)
4. ALICE Collaboration, *Technical Proposal,* **CERN/LHCC/95-71** (1995)
5. Mueller, A.H., *Nucl.Phys.* **A 654**, 37c–54c (1999)

RELATIVISTIC HEAVY-ION COLLISIONS

Physics of Event Generators

K. Werner

SUBATECH, Université de Nantes - IN2P3/CNRS - EMN, Nantes, France

Abstract. An event generator for nuclear collisions is a microscopic model, obtained from extrapolating elementary interactions – as electron-positron annihilation, deep inelastic scattering, and proton-proton interactions – towards proton-nucleus and nucleus-nucleus scattering, by using Monte Carlo techniques.

In this paper, we will discuss the physical concepts behind such event generators. We first present some qualitative discussion of nuclear scattering, before discussing particle production and strings. We then discuss the parton model, and finally multiple scattering theory.

QUALITATIVE DISCUSSION OF NUCLEAR SCATTERING

Overview

Relativistic nuclei are Lorentz contracted, which means that the longitudinal dimension $2R$ is reduced to $2R/\gamma$, see Fig. 1, where R is the nuclear radius, and $\gamma = 1/\sqrt{1-(v/c)^2}$ the so-called gamma factor. At the heavy ion collider RHIC, we have $\gamma = 100$ and at LHC about $\gamma = 3000$, so relativistic contraction plays a very essential role.

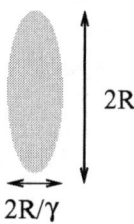

FIGURE 1. Nuclei are Lorentz contracted.

Considering the collision of two nuclei, there are first of all the primary interactions, when the two nuclei pass trough each other in a very short time, see fig 2. Since at very high energies the longitudinal size is due to the gamma factor almost zero (of the order of 0.1 fm at RHIC and 0.01 fm at LHC), all the nucleons of the projectile interact with all the nucleons of the target instantaneously. Many elementary interactions between nucleons in the two nuclei happen in parallel, resulting in many partons (quarks and gluons), moving mainly in longitudinal direction (pre-equilibrium). These partons interact and finally reach equilibrium, referred to as quark-gluon plasma (QGP). The system then expands, passing via a phase transition (or sudden crossover) into the hadronic phase. The density decreases further till the collision rate is no longer large

FIGURE 2. Nuclei pass through each other in a very short time.

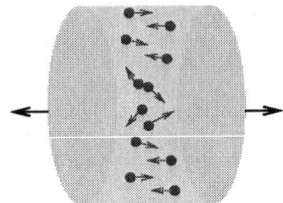

FIGURE 3. The partons interact and finally reach equilibrium (QGP).

enough to maintain chemical equilibrium, but there are still hadronic interactions till finally the particles "freeze out", i.e. they continue their way without further interactions.

Unfortunately there does not exist a single formalism being able to account for a complete nucleus-nucleus collision. Rather we have - at least for the moment - to divide the reaction into different stages, and try to understand the different stages as good as possible. These different stages are

- Initial stage,
- Pre-equilibrium stage,
- Quark-gluon plasma,
- Phase transition,
- Hadron gas,
- Non-equilibrium hadronic matter,
- Free hadrons.

Before discussing these stages one after the other, we introduce some useful variables, as there are the proper time τ and the space time rapidity η

$$\tau = \sqrt{t^2 - z^2}, \qquad \eta = \frac{1}{2} \ln \frac{t+z}{t-z},$$

and the transverse mass m_t and the rapidity y

$$m_t = \sqrt{E^2 - p_z^2}, \qquad y = \frac{1}{2} \ln \frac{E + p_z}{E - p_z}.$$

The proper time and the transverse mass have the property to be invariant under Lorentz transformations. The (space time) rapidity is additive under Lorentz boosts.

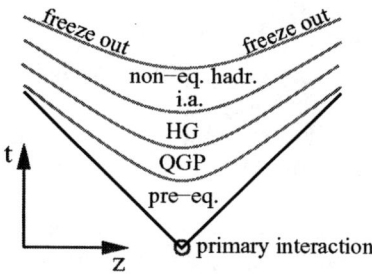

FIGURE 4. The different stages of heavy ion collisions.

Knowing that constant proper time represents hyperbolas in space-time $(t-z)$, one may present a space-time picture of the different stages of heavy ion collisions, as shown in Fig. 4.

Initial Stage

The understanding of the initial interactions is crucial for any theoretical treatment of a possible parton-hadron phase transition, the detection of which being the ultimate aim of all the efforts of colliding heavy ions at very high energies. Theoretical approaches have to consider the fact that the nuclear collision happens on a very short time scale, such that all nucleons of the target interact with all nucleons of the projectile practically instantaneous, see Fig. 5.

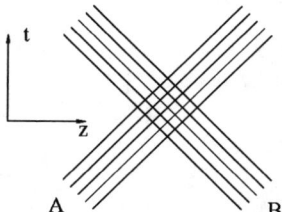

FIGURE 5. The initial stage of heavy ion collisions.

It is quite clear that coherence is crucial for the very early stage of nuclear collisions, so a real quantum treatment is necessary and any attempt to use a transport theoretical parton approach with incoherent quasi-classical partons should not be considered at this point. Also semi-classical hadronic cascades cannot be stretched to account for the very first interactions, even when this is considered to amount to a string excitation, since it is well known [1] that such a longitudinal excitation is simple kinematically impossible.

So what are the currently used fully quantum mechanical approaches? There are presently considerable efforts to describe nuclear collisions via solving classical Yang-Mills equations, which allows to calculate inclusive parton distributions [2]. This approach is to some extent orthogonal to ours: here, screening is due to perturbative processes, whereas we claim to have good reasons to consider soft processes to be at the origin of screening corrections.

Provided factorization works for nuclear collisions, on may employ the parton model, which allows to calculate inclusive cross sections as a convolution of an elementary cross section with parton distribution functions, with these distribution functions taken from deep inelastic scattering. Parton model based are for example Pythia [3] and HIJING [4].

Another approach is the so-called Gribov-Regge theory. This is an effective field theory, which allows multiple interactions to happen "in parallel", with the phenomenological object called "Pomeron" representing an elementary interaction. Using the general rules of field theory, on may express cross sections in terms of a couple of parameters characterizing the Pomeron. A disadvantage is the fact that cross sections and particle production are not calculated consistently: the fact that energy needs to be shared between many Pomerons in case of multiple scattering is well taken into account when calculating particle production (Monte Carlo applications), but energy conservation is not taken care of for cross section calculations. Models based on this approach are QGS [5], DPM [6, 7], and VENUS [8].

A new approach, called "Parton-based Gribov-Regge Theory" [1], solves some of the above-mentioned problems: one has a consistent treatment for calculating cross sections and particle production, considering energy conservation in both cases; one introduces hard processes in a natural way, and, compared to the parton model, one can deal with total cross sections without arbitrary assumptions. This model is incorporated in the NEXUS [1] event generator.

Pre-equilibrium Stage

The partons created in the primary interactions are certainly far from equilibrium, and is desirable to understand microscopically the equilibrium of the system, in other words the formation of a quark gluon plasma. This is a difficult task, since for example at RHIC energies there is still a large soft component. Nevertheless it is useful to study the evolution of partonic systems based on pQCD, ignoring soft physics.

The theoretical tool for this stage is the "parton cascade", which amounts to considering partons as classical particles which move on straight line trajectories, where binary interactions are defined via parton-parton cross sections calculated in the framework of perturbative QCD [9], see Fig. 6.

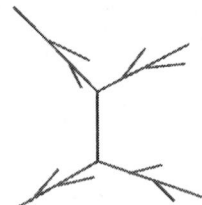

FIGURE 6. Partons, which have been produced initially, interact.

One has to carefully regard the range of validity of this approach: it is not meant to treat the primary interactions, where quantum mechanical interference should play a crucial role, so one may start the calculation once a system of incoherent classical partons has been established. On the other end, one should not stretch the perturbative treatment too far: perturbative calculations require large momentum transfer, which is not any more guaranteed if the interaction energy is getting too low.

Equilibrium Stage

We are now discussing the final stage of the collision, consisting of the QGP phase, the phase transition, and the hadron gas phase. We do not treat these three stages individually, because the known models treat usually more than just one stage.

The final aim of all the efforts in the field of ultra-relativistic heavy ion collisions is the creation of a thermalized system of quarks and gluons. Provided such an equilibrium has been established, one may use hydrodynamics, which is a macroscopic approach based on energy-momentum conservation and local thermal equilibrium. Hydrodynamical calculations have been used since a long time, either assuming particular symmetries and using analytical methods [10], or full 3-dimensional calculations numerical calculations [11]. Recently a new technique has been proposed, the so-called smoothed particle hydrodynamics [12], where fields $\rho(x)$ are represented by particles as $\rho_P(x) = \Sigma_b v_b \delta(x - x_b)$, and then smoothed:

$$\rho(x) \to \rho_{SP}(x) = \int \rho_P(x) W(x - x') dx' = \Sigma_b v_b W(x - x_b)$$

with some smoothing kernel W. The advantage is that the hydrodynamical equations are transfered into a system of ordinary differential equations, which can be solved by applying standard methods. In this way one may perform 3-dimensional calculations much faster than with traditional methods.

There are several attempts to treat at least the region around the phase transition in a microscopic way. A possibility is to apply transport theory based on the NJL model [13], which is an effective theory with a point-like interaction between two quarks (gluons are not considered explicitly). The model allows also for hadron production like quark plus anti-quark goes into meson plus meson. The dynamics is crucially affected by the density and temperature dependence of quark and hadron masses, one observes for example the formation of droplets of quark matter rather than homogeneous matter of lower density, since the latter one would imply higher quark masses.

A completely different hadronization scenario has been proposed based on the confinement mechanism [14], again ignoring gluons. Quarks are considered to be classical particles, their dynamics being determined by a classical Hamiltonian. The latter one contains a string potential and color factors, which force the quarks to form resonances, which subsequently decay into hadrons.

Another alternative approach is the hadronization via coalescence [15]. Again, starting from a quark-anti-quark plasma, hadronic resonances are formed based on coalescence, with a subsequent decay into hadrons.

Post-equilibrium Hadronic Stage

Once a purely hadronic system has been established, a microscopic treatment based on binary hadronic interactions is feasible. Here, hadrons propagate on classical trajectories and interact according to hadron-hadron scattering cross sections. If possible, parameterizations of measured cross sections are used. A couple of models have been constructed along these lines, like UrQMD [16, 17], ART [18], JAM [19]. Unfortunately, not all the necessary cross sections have been measured to a sufficient precision, and correspondingly, the above-mentioned approaches differ by using different model assumptions for the cross sections. We emphasize again that hadronic transport codes are a useful tool to treat the final stage of a heavy ion collision, but not for the primary interaction.

PARTICLE PRODUCTION, STRINGS

Particle production is relevant for pp, pA and the initial stage of AA collisions. However it is necessary to first study simpler systems as electron-positron annihilation.

The String Picture

For e^+e^- collisions, we have data available in a wide energy range (up to 200 GeV in the cms). Studying particle production, one observes (idealized) a rapidity plateau, see Fig. 7. When we move the reference system (for example from lab to rest frame),

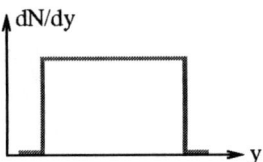

FIGURE 7. One observes (idealized) a rapidity plateau.

we observe a manifestation of boost invariance: the same rapidity distribution before and after the boost, see fig 8. What does boost invariance mean? Suppose an expanding

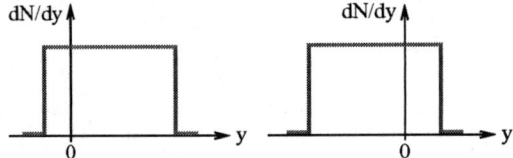

FIGURE 8. Manifestation of boost invariance

dynamical system such that some central part is at rest and the outer parts move away from the center, with increasing speed for larger distances, see Fig. 9(left). Now we perform a boost such that a different piece of the system is at rest. In the neighborhood of this region the system looks identical to the neighborhood of the point at rest before

FIGURE 9. Manifestation of boost invariance

the boost, see Fig. 9(right). In other words: the system is identical at all points in the corresponding local comoving frame.

What happens really? Electron and positron annihilate and form a virtual photon, then the virtual photon decays into a quark-antiquark pair, see Fig. 10(a). The quark and antiquark move apart from each other, see Fig. 10(b). But quarks and antiquarks cannot be observed individually! There is a gluon field acting between the two, whose energy is proportional to the separation distance. This object is called string, see Fig. 11. To

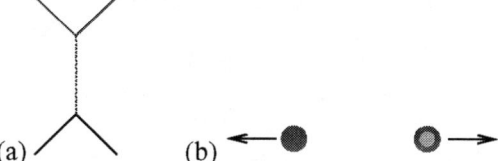

FIGURE 10. (a) Electron and positron annihilate and form a virtual photon, then the virtual photon decays into a quark-antiquark pair. (b) Quark and antiquark move apart from each other.

FIGURE 11. The quark-antiquark pair forms a string.

separate the quark from the antiquark, one need an infinite energy, which is impossible. The string breaks via quark-antiquark production, and these new string pieces are finally hadrons or resonances, see Fig. 12. String fragmentation is a boost invariant procedure

FIGURE 12. The string breaks via "quark-antiquark" production.

and provides exactly the situation discussed above: seen from a given point on the string, all the string pieces move away from this point with increasing speed towards the edges, as indicated in Fig. 13. In Fig. 14, we present the space-time picture of the string

before: ← ← ← → → →   after: ← ← ← ← → →

FIGURE 13. String fragmentation is a boost invariant procedure.

dynamics: at given proper time (hyperbola), the velocities of the string pieces (arrows) are such that they point all back to the origin and are longer towards the edges. This string decay provides a flat rapidity distribution.

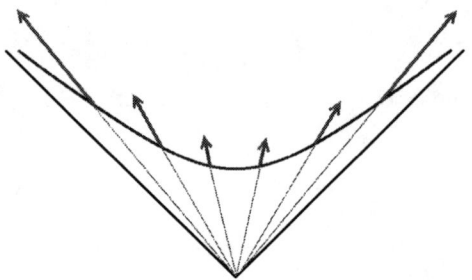

FIGURE 14. Space-time picture of string decay.

What is Really Done

After this qualitative discussion, let us discuss what is really done. A string can be considered as a two-dimensional surface in Minkowski space

$$X = X(r,t),$$

with r being a space-like and t a time-like parameter, see Fig. 15.

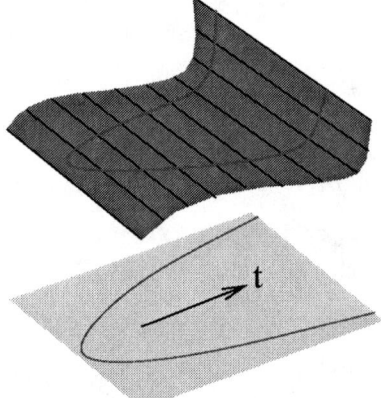

FIGURE 15. The string surface.

In order to obtain the equations of motion, we need a Lagrangian. It is obtained by demanding the invariance of the action with respect to gauge transformations. This way one finds [1] the Lagrangian of Nambu-Goto:

$$L = -\kappa\sqrt{(X'\dot{X})^2 - X'^2\dot{X}^2},$$

with "dot" and "prime" referring to the partial derivatives with respect to r and t, and with κ being the string tension. With this Lagrangian we get the Euler-Lagrange

equations of motion:
$$\frac{\partial}{\partial t}\frac{\partial L}{\partial \dot{X}_\mu} + \frac{\partial}{\partial r}\frac{\partial L}{\partial X'_\mu} = 0.$$

We use the gauge fixing
$$X'^2 + \dot{X}^2 = 0 \text{ and } X'\dot{X} = 0,$$

which provides a very simple equation of motion, namely a wave equation,
$$\frac{\partial^2 X_\mu}{\partial t^2} - \frac{\partial^2 X_\mu}{\partial r^2} = 0,$$

with the boundary conditions: $\partial X_\mu/\partial \sigma = 0$, $\sigma = 0, \pi$. The solution of the equation of motion (with initial extension zero) is
$$X^\mu(r,t) = X_0 + \frac{1}{2}\left(\int_{r-t}^{r+t} g^\mu(\xi)d\xi\right),$$

where g is the initial velocity, $g(r) = \dot{X}(r,t)_{t=0}$. Strings are classified according to the function g. Strings with piecewise constant g are called kinky strings, each segment being called kink, finally identified with perturbative partons. In Fig. 16, we show the evolution of a string generated in electron-positron annihilation (3 internal kinks).

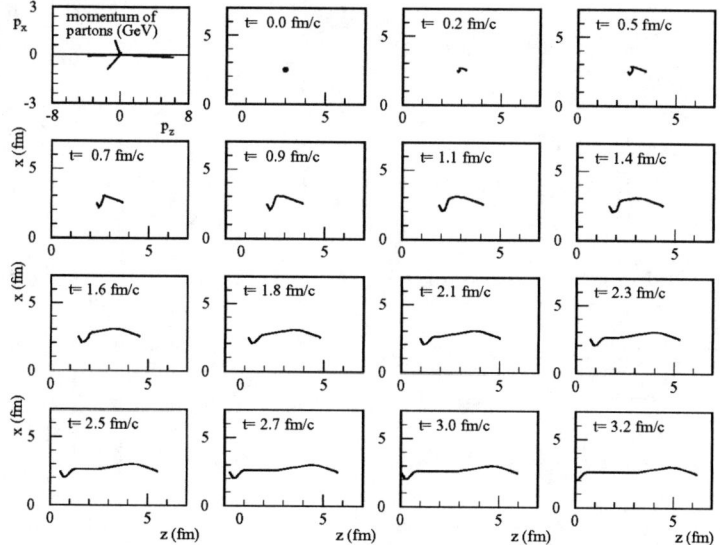

FIGURE 16. String evolution.

Results

We show some results for rapidity distributions from flat strings (no internal kinks) in Fig. 17. We observe a nice rapidity plateau, which gets broader with increasing energy. But the plateau height stays constant. Increasing the energy does not change the local properties of the string, the number of particles per unit of rapidity stays constant.

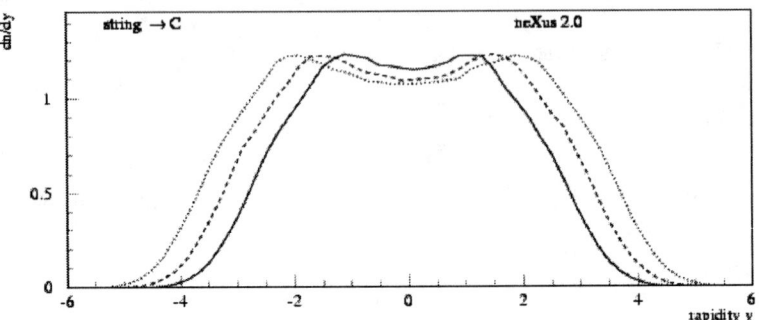

FIGURE 17. Rapidity distributions of flat strings at 14-22-34 GeV.

In real e^+e^- collisions, one has with increasing energy an increasing probability to have kinks, which makes the plateau rising with energy, as shown in Fig. 18, where we show the prediction of the string model together with experimental data from the TASSO [20], ALEPH [21], and OPAL [22, 23] collaborations. We show also longitudinal momentum fraction distributions for different energies.

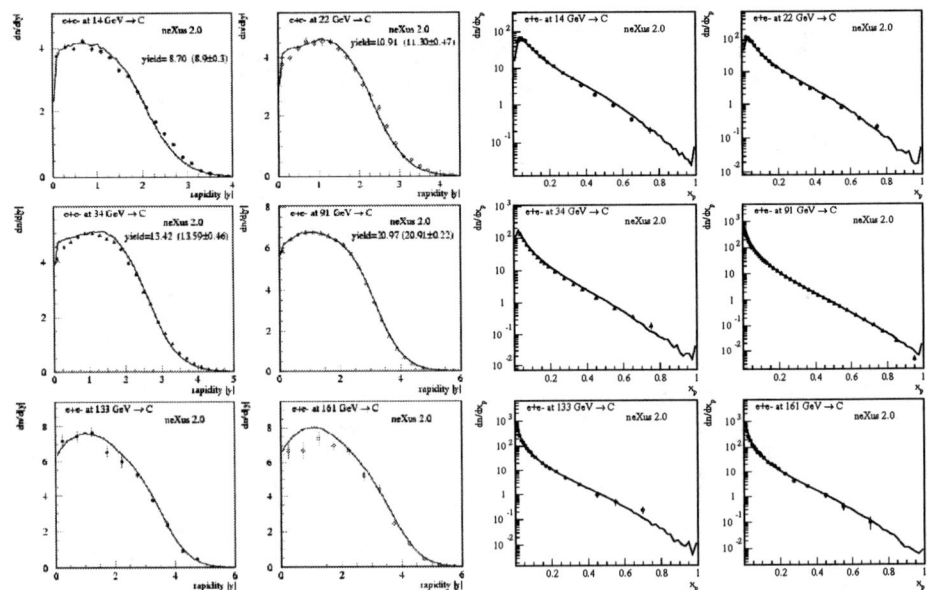

FIGURE 18. Rapidity and longitudinal momentum distributions: data (points) and NEXUS (line).

Hadron Flavors

There are some remarkable regularities among the hadrons, which became apparent in the early 1960s. The first is that the baryons fall into groups of multiplicity 1, 8, 10 (singlet, octet, decuplet). The mesons come in singlets and octets. See figs. 19, 20.

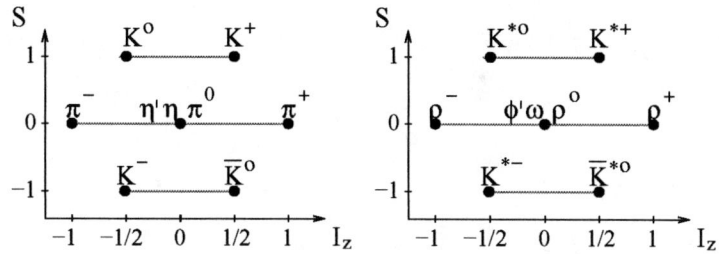

FIGURE 19. Meson octet plus singlet.

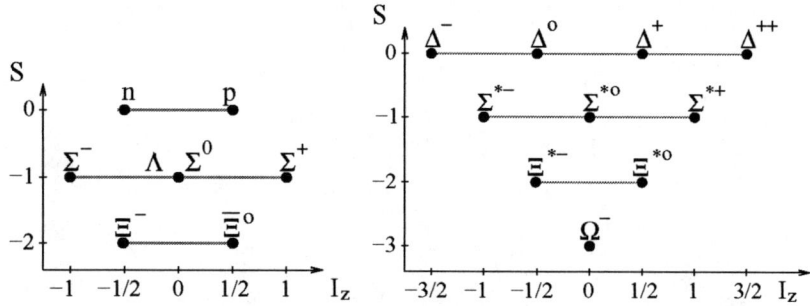

FIGURE 20. Baryon octet and decuplet.

These structures can be understood in the quark model for hadrons: the baryons are composed of three quarks, the antibaryons of three antiquarks and the mesons of a quark plus an antiquark. There are six flavors of quarks, the three most abundant ones being the u, d, and s flavor, the properties being given in table 1, shown in Fig. 21. In figs. 22 and 23, we indicate the quark content of the most frequent mesons and baryons. So the quark model can easily explain the striking regularities of the hadrons.

It will be the basis of all models of string fragmentation, as shown in Fig. 24. A string break is realized via quark-antiquark production, such that the quark-antiquark pair screens the color field. The string fragments consist then of quark-antiquark pairs, they

TABLE 1. Quark properties.

name	u	d	s
charge	2/3	-1/3	-1/3
strangeness	0	0	-1
isospin	1/2	-1/2	0

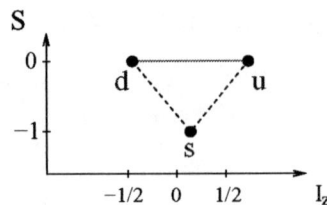

FIGURE 21. Quark properties.

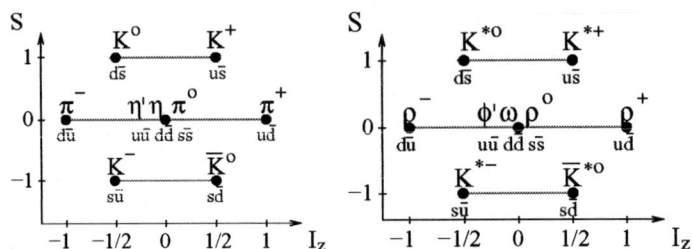

FIGURE 22. The quark content of mesons.

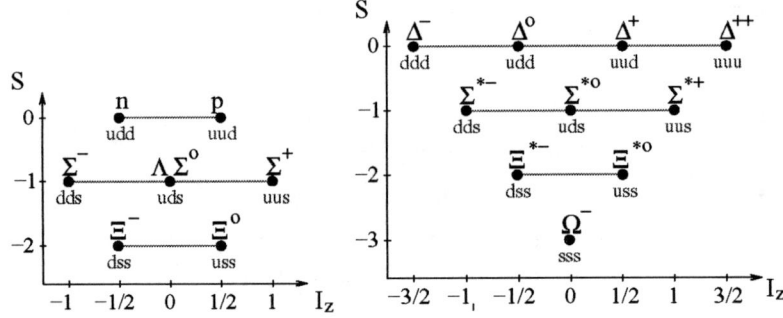

FIGURE 23. The quark content of baryons.

FIGURE 24. Hadron production from a quark-antiquark string.

are therefore mesons. It is also possible that the string breaks via diquark-antidiquark production, which amounts to baryon and antibaryon creation. In Fig. 25, we show some results concerning the production of identified hadrons in electron-positron annihilation. One observes that the string model works to a high precision.

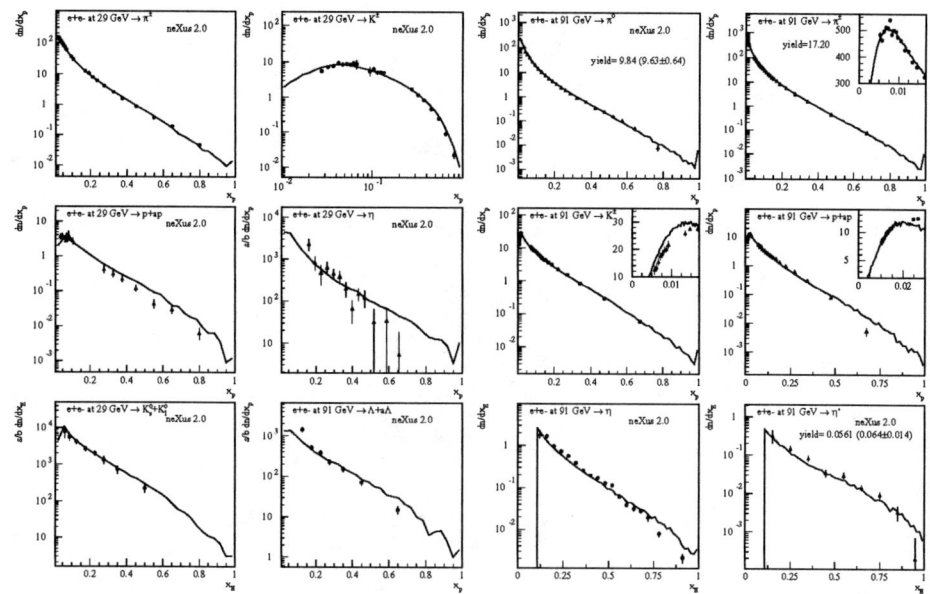

FIGURE 25. Hadron production in e^+e^- annihilation at 29 and 91 GeV.

PARTON MODEL

Whereas leptons are point-like in their behavior, it is not inconceivable that the quarks too enjoy this property. If we think of the hadrons as complicated "atoms" or "molecules" of quarks, then at high energies and momentum transfers, where we are probing the inner structure, we may discover a simple situation, with the behavior controlled by almost free, point-like constituents. The idea that hadrons possess a "granular" structure and that the "granules" behave as hard point-like, almost free (but nevertheless confined) objects, is the basis of Feynmans (1969) parton model.

The essence of the parton model is the assumption that, when a sufficiently high momentum transfer reaction takes place, the projectile, be it a lepton or a parton inside a hadron, sees the target as made up of almost free constituents, and is scattered by a single, free, effectively massless constituent.

Deep Inelastic Scattering

The historical way to study the hadronic structure was using point-like leptons as projectiles hitting a proton target. There is a basic difference compared to e^+e^- scattering: the proton is a composite particle, e^+, e^- are elementary particles. so one can probe the internal structure of the proton, see Fig. 26(left). In a lepton-proton scattering, one can measure the momentum distributions of constituents (partons) see Fig. 26(right). In lowest order of perturbation theory, the reaction is described by one photon exchange

FIGURE 26. lepton-proton scattering.

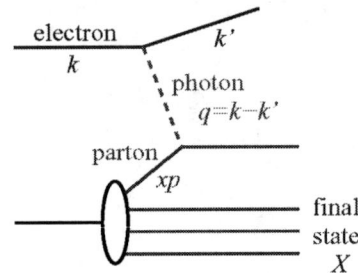

FIGURE 27. One photon exchange in a *ep* reaction.

diagram, see Fig. 27. Here k is known and k' is measured, so the momentum transfer q is known. One studies the cross section as function of two variables: the photon virtuality $Q^2 = -q^2$ and the Bjorken variable $x = Q^2/2pq$. Why Q^2? Because Q^2 sets the resolution scale: $\Delta x = 1/Q^2$. The bigger Q^2 the deeper one looks into the proton. Why x? Consider a parton with a momentum fraction z (momentum zp), see Fig. 28. In the

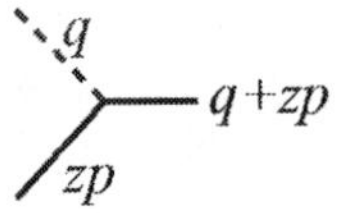

FIGURE 28. The photon-parton vertex.

reaction it received the transferred momentum q, so its new momentum is $q + zp$. But the parton is massless, so $q^2 + 2qzp + z^2p^2 = 0$, and therefore $z = x$. A reaction with a certain value of x probes a parton with momentum fraction $z = x$, which means that parton momentum distribution are measurable.

Let us do some kinematics: the virtual photon transfer being q and the initial proton momentum being p, the final state mass W is given as $W^2 = (p+q)^2$, see Fig. 29. We have,

$$W^2 = (q+p)^2 = q^2 + 2pq + p^2 \approx -Q^2 + Q^2/x,$$

which gives

$$W^2 = Q^2(\frac{1}{x} - 1)$$

So we arrive at an interesting result: small x corresponds to a big final state mass W.

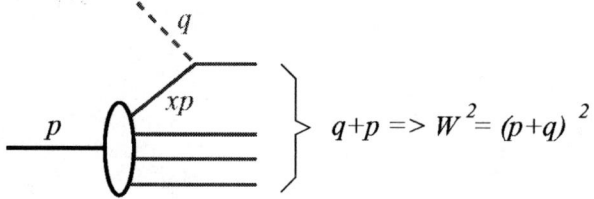

FIGURE 29. The total mass W of the hadronic final state.

One can write the ep cross section as

$$\frac{d\sigma^{ep}}{dQ^2 dx} = \frac{\alpha}{\pi Q^2 x}\left[\frac{1+(1-y)^2}{2}\sigma_T^{\gamma p} + (1-y)\sigma_L^{\gamma p}\right],$$

with $y = pq/pk$, and

$$\sigma_T^{\gamma p} = \frac{4\pi^2\alpha}{Q^2}(F_2 - F_L),\ \sigma_L^{\gamma p} = \frac{4\pi^2\alpha}{Q^2}F_L, F_L \ll F_2.$$

F_2 and F_L describe the proton structure as seen by the virtual photon, see Fig. 30.

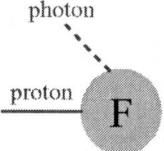

FIGURE 30. Proton structure functions.

A first look reveals $F_2(x, Q^2)$ to be only a function of x (scaling). This can be explained within the naive Parton Model, where the proton is considered to be a incoherent sum of partons (quarks) of flavor i, which are distributed according to parton distribution functions f_i, so

$$F_2(x, Q^2) = \sum_i e_i x f_i(x).$$

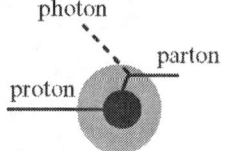

FIGURE 31. Parton model.

Looking more carefully, on finds that $F_2(x, Q^2)$ depends slightly on Q^2 (scale breaking). The partons are still distributed according to parton distribution functions $f_i(x, Q^2)$, which are now depending on a scale (defined by the probe). And we still have

$$F_2(x, Q^2) = \Sigma_i e_i x f_i(x, Q^2).$$

This formula takes in account the successive emissions of virtual partons, carrying less and less momentum. The photon virtualities, Q^2, are ordered down to some minimum value (this part is calculable in the framework of perturbative QCD). Below this minimum value, we have soft physics (non-perturbative regime), see Fig. 32.

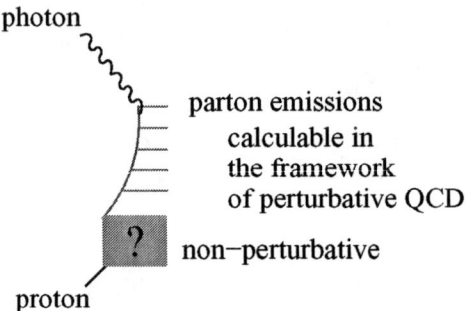

FIGURE 32. Successive parton emissions.

Let us consider the emission of the first (softest) parton being emitted from the non-perturbative area. If the parton carries a fraction x of the momentum of the proton, the mass of the soft object "proton minus parton" has a mass given by $m^2 = Q_0^2/x$, where Q_0 is a typical soft virtuality (of the order 1 GeV). This has interesting consequences: in case of sea quarks with distributions of the form $1/x$, one has typically small x and therefore a large mass m. For valence quarks, on the other hand, a $1/\sqrt{x}$ distribution provides in general large x values and therefore a small mass m. This means that in case of sea quarks, there is a large mass and small virtuality object between the proton and the parton, and we therefore have to consider two contributions to the structure functions, as indicated in Fig. 33. The large mass object is considered to be a Pomeron, to be discussed in detail later. The valence contribution provides a peak at large x and drops

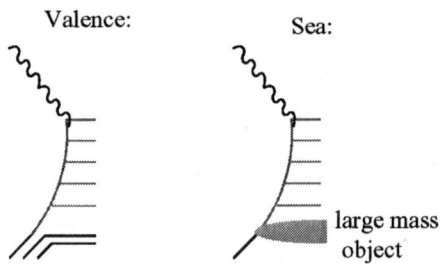

FIGURE 33. Two contributions for the proton structure function.

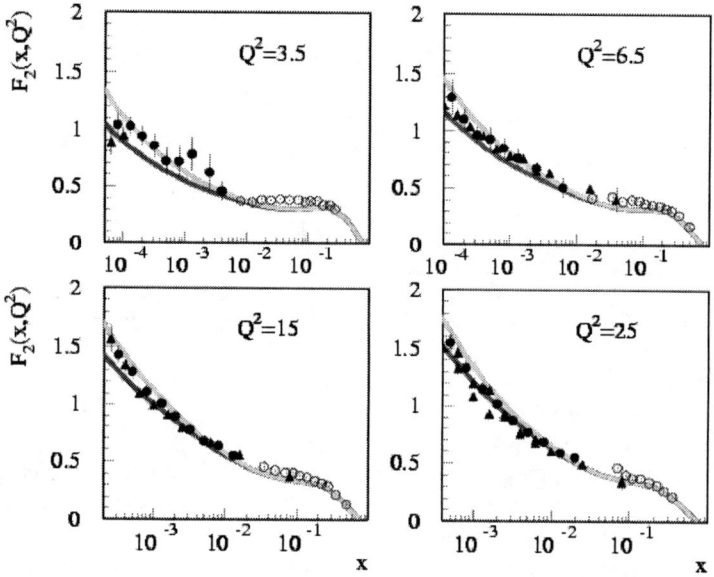

FIGURE 34. The structure function F_2.

fast for small values of x, whereas the sea contribution is important at small x and drops very fast towards large x. The sum of the two is shown in Fig. 34.

The Parton Model for pp

For *pp* interactions one uses the same concept as for lepton-proton scattering. Each proton contains partons distributed as $f_i(x, Q^2)$. Two of the partons (one from the first proton and the other from the second proton) interact via elementary interactions. The inclusive cross section for producing a parton k is

$$\frac{d\sigma^{pp \to k}}{dyd^2p_t} = \sum_{i,j} \int dx_1 dx_2 f_i(x_1, Q^2) f_j(x_2, Q^2) \frac{d\hat{\sigma}^{ij \to k}}{dyd^2p_t},$$

(see Fig. 35), where $f_i(x, Q^2)$ are the perturbative parton densities, measured by performing global fits of data taken from large sets of experiments as lepton-nucleon deep inelastic scattering and others. $d\hat{\sigma}/dyd^2p_t$ are partonic cross-section for the hard processes, calculable in perturbation theory. Here, one assumes universality of parton densities (independent of the hard processes, for large Q^2) and factorization (the possibility to separate the parton density functions from the partonic cross section). The assumption is based in the fact that hard processes ($Q \gg m_p$) occur in very short time ($\tau \sim 1/Q$), much lesser than the typical interaction times for the binding of the proton (hadron)

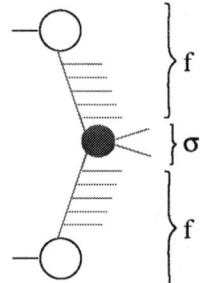

FIGURE 35. The inclusive cross section for producing partons.

($\tau \sim 1/m_p$). As a result, to study inclusive processes at large Q^2 it is sufficient to consider the interactions between the external probe and a single parton.

The model works well for most of the high energy physics experiments, but diverges for small transverse momentum. Why? Because soft (non-perturbative) physics enters. A solution is to introduce some lower limit (cutoff p_0) for the transverse momentum. Integrating over rapidity and the transverse momentum above the cutoff, we get the jet cross section $\sigma(p_0)$. Now another problem appears: the jet cross section grows very fast with energy, becoming finally bigger than the total one, see Fig. 36. The real solution

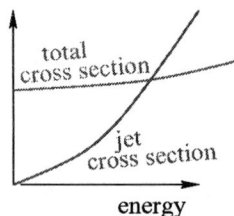

FIGURE 36. The jet cross section grows very fast with energy.

amounts to studying multiple scattering. The jet cross section is an inclusive one: several jets may contribute. So the parton model is very useful but is not the whole history. One needs a multiple scattering theory.

There is currently much discussion about saturation. What does it mean? Consider partons with transverse momentum p_0. Each one occupies a transverse area π/p_0^2, whereas the transverse area of the nucleon is πR_A^2. If the number $N_A(s, p_0^2)$ of partons is sufficiently high, they fill completely the transverse area of the nucleus. The relation

$$N_A(s, p_0^2)\pi/p_0^2 = \pi R_A^2 \text{ or } p_0^2 = N_A(s, p_0^2)/R_A^2$$

defines therefore the so-called saturation scale p_0^2. Since $N_A(s, p_0^2)$ increases with s and decreases with p_0^2, then the condition $p_0^2 = N_A(s, p_0^2)/R_A^2$ is solved by a function $p_0^2(s)$ which increases with s, such that at sufficiently high energy the scale $p_0^2(s)$ in in the perturbative domain.

MULTIPLE SCATTERING THEORY

Reminder: some Elementary Quantum Mechanics

Let us introduce some conventions. We denote elastic two body scattering amplitudes as $T_{2\to 2}$ and inelastic amplitudes corresponding to the production of some final state X as $T_{2\to X}$ (see Fig.37).

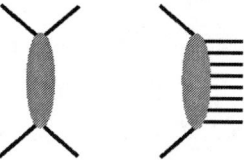

FIGURE 37. An elastic scattering amplitude $T_{2\to 2}$ (left) and an inelastic amplitude $T_{2\to X}$ (right).

As a direct consequence of unitarity on has $2\,\mathrm{Im}T_{2\to 2} = \Sigma_X (T_{2\to X})(T_{2\to X})^*$. The right hand side of this equation may be literally presented as a "cut diagram", where the diagram on one side of the cut is $(T_{2\to X})$ and on the other side $(T_{2\to X})^*$, as shown in Fig. 38. So the term "cut diagram" means nothing but the square of an inelastic

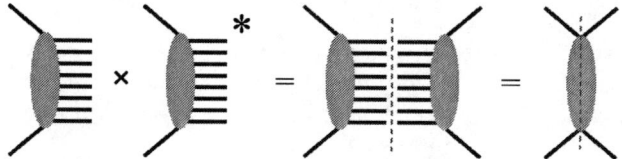

FIGURE 38. The expression $\Sigma_X (T_{2\to X}).(T_{2\to X})^*$ which may be represented as a "cut diagram".

amplitude, summed over all final states, which is equal to twice the imaginary part of the elastic amplitude. Based on these considerations, we introduce simple graphical symbols, which will be very convenient when discussing multiple scattering, shown in Fig. 39: a vertical solid line represents an elastic amplitude (multiplied by i, for convenience), and a vertical dashed line represents the mathematical expression related to the above-mentioned cut diagram (divided by $2s$, for convenience).

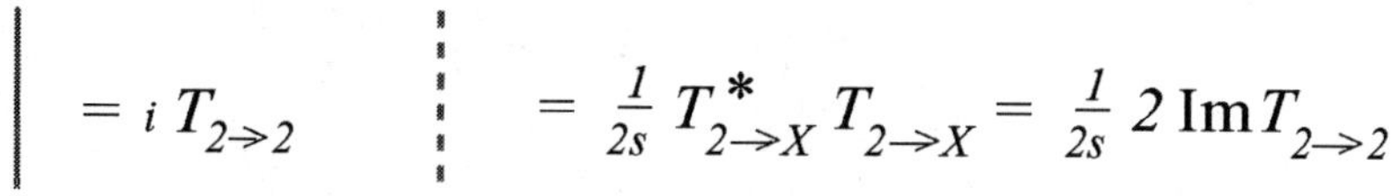

$$\big| = i\,T_{2\to 2} \qquad \vdots = \frac{1}{2s} T^*_{2\to X}\, T_{2\to X} = \frac{1}{2s}\,2\,\mathrm{Im}\,T_{2\to 2}$$

FIGURE 39. Conventions.

Elementary Interactions

Elementary nucleon-nucleon scattering can be considered as a straightforward generalization of photon-nucleon scattering: one has a hard parton-parton scattering in the

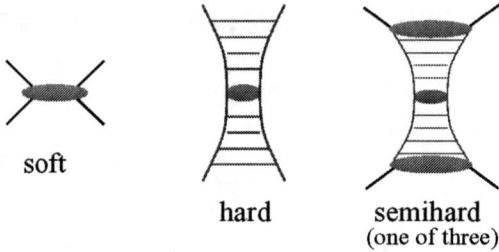

FIGURE 40. The elastic amplitude $T_{2 \to 2}$.

middle, and parton evolutions in both directions towards the nucleons. We have a hard contribution T_{hard} when the the first partons on both sides are valence quarks, a semi-hard contribution T_{semi} when at least on one side there is a sea quark (being emitted from a soft Pomeron), and finally we have a soft contribution, when there is no hard scattering at all (see Fig. 40). The total elementary elastic amplitude $T_{2 \to 2}$ is the sum of all these terms. We have a smooth transition from soft to hard physics: at low energies the soft contribution dominates, at high energies the hard and semi-hard ones, at intermediate energies (that is where experiments are performed presently) all contributions are important.

The multiple scattering theory will be based on these elementary interactions, the corresponding elastic amplitude $T_{2 \to 2}$ and the corresponding cut diagram, both being represented graphically by a solid and a dashed vertical line. We also refer to the solid line as Pomeron, to the dashed line as cut Pomeron.

Multiple Scattering

We first consider inelastic proton-proton scattering, see Fig. 41. We imagine an arbitrary number of elementary interactions to happen in parallel, where an interaction may be elastic or inelastic. The inelastic amplitude is the sum of all such contributions with at least one inelastic elementary interaction involved.

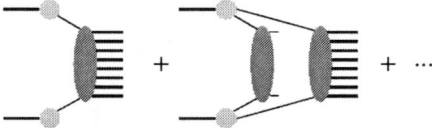

FIGURE 41. Inelastic scattering in pp.

To calculate cross sections, we need to square the amplitude, which leads to many interference terms, as the one shown in Fig. 42(a), which represents interference between the first and the second diagram of Fig. 41. Using the above notations, we may represent the left part of the diagram as a cut diagram, conveniently plotted as a dashed line, see

Fig. 42(b). The amplitude squared is now the sum over many such terms represented by solid and dashed lines.

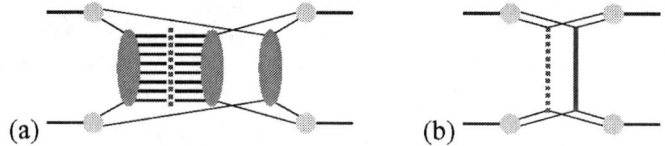

FIGURE 42. Inelastic scattering in pp. a) Amplitude, b) Squared amplitude makes

When squaring an amplitude being a sum of many terms, not all of the terms interfere – only those which correspond to the same final state. For example, a single inelastic interaction does not interfere with a double inelastic interaction, whereas all the contributions with exactly on inelastic interaction interfere. So considering a squared amplitude, one may group terms together representing the same final state. In our pictorial language, this means that all diagrams with one dashed line, representing the same final state, may be considered to form a class, characterized by $m = 1$ – one dashed line (one cut Pomeron) – and the light cone momenta x^+ and x^- attached to the dashed line (defining energy and momentum of the Pomeron). In Fig. 43, we show several diagrams belonging to this class, in Fig. 44, we show the diagrams belonging to the class of two inelastic interactions, characterized by $m = 2$ and four light-cone momenta $x_1^+, x_1^-, x_2^+, x_2^-$.

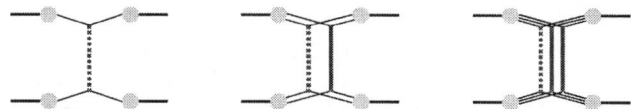

FIGURE 43. Class of terms corresponding to one inelastic interaction.

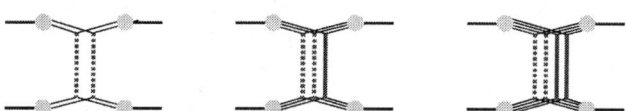

FIGURE 44. Class of terms corresponding to two inelastic interactions.

Generalizing these considerations, we may group all contributions with m inelastic interactions (m dashed lines = m cut Pomerons) into a class characterized by the variable

$$K = \{m, x_1^+, x_1^-, \cdots, x_m^+, x_m^-\}.$$

We then sum all the terms in a class K,

$$\Omega(K) = \sum \{\text{all terms in class } K\}.$$

The cross section is then simply a sum over classes,

$$\sigma_{\text{inel}}(s) = \sum_{K \neq 0} \int d^2 b \, \Omega(K).$$

Ω depends implicitly on the energy squared s and the impact parameter b. The individual terms $\int d^2 b \Omega(K)$, represent partial cross sections, since they represent distinct final states. They are referred to as topological cross sections.

The above concepts are easily generalized to nucleus-nucleus scattering, an example for a diagram representing a contribution to the squared amplitude is shown in Fig. 45.

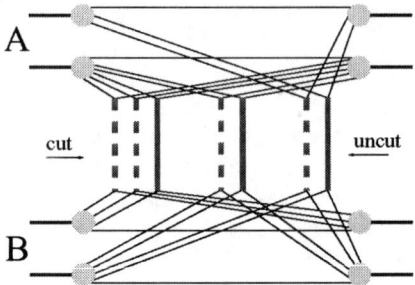

FIGURE 45. An interference term of total cross section of

We may also define classes, which correspond to well defined final states, in our notation a given number of dashed lines between nucleon pairs. We may number the pairs as 1, 2, 3, ... k ... , AB. We define m_k to be the number of inelastic interactions (cut Pomerons) of the pair number k. The μth of these m_k cut Pomerons is characterized by light cone momenta $x^+_{k\mu}, x^-_{k\mu}$. So a class may be characterized by

$$K = \{m_k, x^+_{k\mu}, x^-_{k\mu}\}.$$

We sum all terms in a class to obtain again a quantity called $\Omega(K)$, such that the cross section can be written as a sum over classes

$$\sigma_{\text{inel}}(s) = \sum_{K \neq 0} \int d^2 b \Omega(K),$$

as in the case of proton-proton scattering. Here, however, b is a multidimensional variable representing the impact parameter b_0 and the transverse distances b_k of all the nucleon-nucleon pairs. One can prove

$$\sum_K \Omega(K) = 1,$$

which is a very important result justifying our interpretation of $\Omega(K)$ to be a probability distribution for the configurations K. This provides also the basis for applying Monte Carlo techniques.

The function Ω is the basis of all applications of this formalism. It provides the basis for calculating (topological) cross sections, but also for particle production, thus providing a consistent formalism for all aspects of a nuclear collision.

Pomeron-Pomeron Interactions

So far, we consider the case where particle production from the individual elementary interactions is completely independent. At high energies with high particle densities this is not very realistic: particles emitted in one interaction could be absorbed in another one. In our language: we have to allow interactions of Pomerons, like the diagrams shown in Fig. 46.

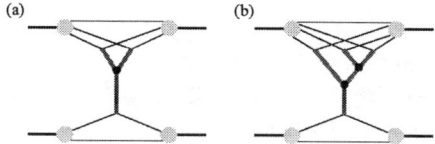

FIGURE 46. Pomeron-Pomeron interactions.

Such interactions are very important, being in particular responsible for screening (shadowing, saturation). If we assume for a moment that a Pomeron is roughly a parton ladder, then we we have the situation as shown in Fig. 47: independent Pomerons correspond to non-interacting parton ladders (left figure), whereas Pomeron interaction amount to interactions of partons from one ladder with the ones from the other one (right figure). It is clear: the more partons are produced, the more likely are such processes.

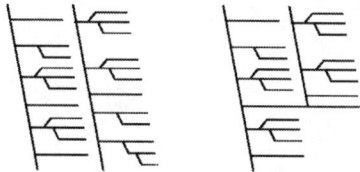

FIGURE 47. Pomeron-Pomeron interactions in parton language.

Also in case of Pomeron-Pomeron interactions, we are interested in particle production, and so we have to worry about cutting diagrams. Again, cut diagrams are the consequence of squaring amplitudes, i.e. multiplying an amplitude corresponding to some process with the complex conjugate amplitude corresponding to the same or some other process.

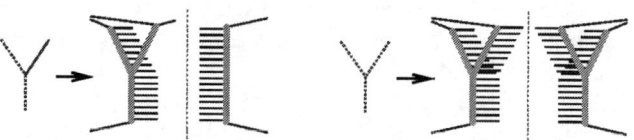

FIGURE 48. Cut diagrams as a result of squared amplitudes..

In Fig. 48, we show two examples: a ladder with an additional leg is multiplied with a simple ladder (left figure), and two ladders fused into one are multiplied with itself (right figure), We use again dashed and solid lines for cut and uncut diagrams. There are three cut diagrams of a Y diagram, as shown in Fig. 49: the lower leg is always cut; in addition, there may be none (a) or one (b) or two (c) of the upper leg being cut.

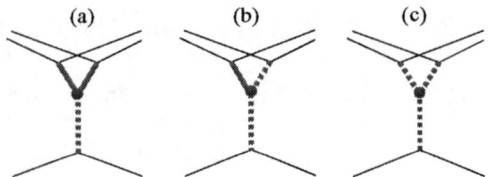

FIGURE 49. Cut Y diagrams.

An important property of this formalism is the so-called factorization. A dashed line corresponds to a cut Pomeron with given light cone momentum fractions x^+ and x^-. So the rapidity of the Pomeron is $1/2 \ln x^+/x^-$, the squared energy is sx^+x^-. Suppose this cut Pomeron represents a chain of particles with a typical transverse mass m. The range of rapidity is roughly given as $y^- < y < y^+$, with $y^\pm = \pm \ln(\sqrt{s}x^\pm/m)$. So we may assign a vertical rapidity scale and draw the dashed vertical lines exactly between y^+ and y^-.

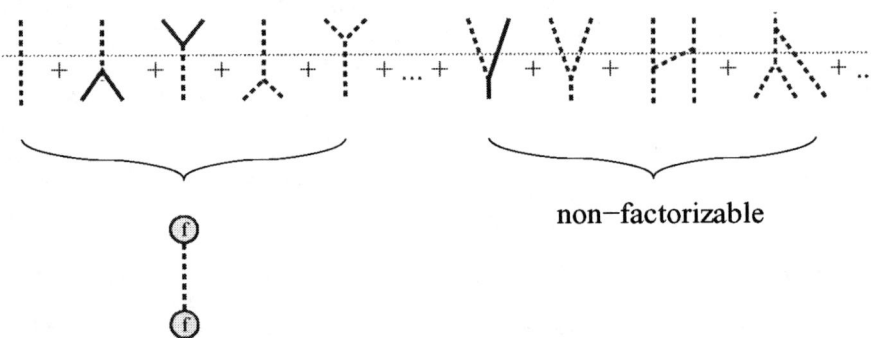

FIGURE 50. Factorization in pp scattering.

In Fig. 50, the diagrams have been plotted this way. The horizontal dashed line represents some given rapidity y. Due to some general rules, only those diagrams contribute to inclusive particle production at rapidity y, where exactly one line crosses the horizontal one. These are also the ones which factorize: they may be considered as a single line between two "blobs (f)", each blob being an infinite sum, providing thus a simple effective diagram. All the non-factorizable diagrams do not contribute to the inclusive cross sections.

In the same way, the structure function in deep inelastic scattering exhibits factorization, as shown in Fig. 51, with the same blob (f) as in pp scattering. This allows to write

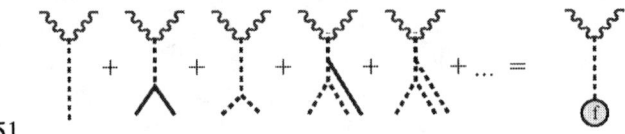

FIGURE 51. Factorization in deep inelastic scattering.

the inclusive cross section in pp as $f \times \hat{\sigma} \times f$, where $\hat{\sigma}$ represents the dashed line, and f is obtained from deep inelastic scattering. Essentially we recover here the parton model.

Does this mean that one can hide all these complicated multiple scattering features in one simple measurable function f? The answer is yes if one is only interested in calculating inclusive spectra. However, the situation is completely different when it comes to the total cross section, where we have to consider all diagrams. The above-mentioned cancellations concern only inclusive cross sections. In addition, for Monte Carlo applications, we need to evaluate topological cross sections, related to the probabilities to certain configurations (defined by the numbers of cut Pomerons). Here again, no cancellations apply, we need to consider all diagrams.

In this sense, the so-called eikonal approach is very questionable, where total and topological cross sections are calculated based on inclusive ones, neglecting all the non-factorizable contributions.

ACKNOWLEDGMENTS

I would like to thank Fuming Liu and C. Javier Solano for helping to prepare this manuscript.

REFERENCES

1. H. J. Drescher, M. Hladik, S. Ostapchenko, T. Pierog, and K. Werner, Phys. Rept. **350**, 93 (2001), hep-ph/0007198.
2. L. McLerran and R. Venugopalan, Phys. Rev. **D49**, 2233; 3352 (1994), hep-ph/9311205.
3. T. Sjostrand and M. van Zijl, Phys. Rev. **D36**, 2019 (1987).
4. X.-N. Wang, Phys. Rept. **280**, 287 (1997), hep-ph/9605214.
5. A. B. Kaidalov and K. A. Ter-Martirosyan, Phys. Lett. **117B**, 247 (1982).
6. A. Capella, U. Sukhatme, C.-I. Tan, and J. Tran Thanh Van, Phys. Rept. **236**, 225 (1994).
7. P. Aurenche et al., Phys. Commun. **83**, 107 (1994).
8. K. Werner, Phys. Rep. **232**, 87 (1993).
9. K. Geiger and B. Muller, Nucl. Phys. **B369**, 600 (1992).
10. G. Baym, B. L. Friman, J. P. Blaizot, M. Soyeur, and W. Czyz, Nucl. Phys. **A407**, 541 (1983).
11. D. H. Rischke, (1998), nucl-th/9809044.
12. C. E. Aguiar, T. Kodama, T. Osada, and Y. Hama, (2000), hep-ph/0006239.
13. P. Rehberg, L. Bot, and J. Aichelin, Nucl. Phys. **A653**, 415 (1999), hep-ph/9809565.
14. M. Hofmann et al., Phys. Lett. **B478**, 161 (2000), nucl-th/9908030.
15. P. Csizmadia and P. Levai, In *Trento 1999, Understanding deconfinement in QCD* 324- 330.
16. S. A. Bass et al., Prog. Part. Nucl. Phys. **41**, 225 (1998), nucl-th/9803035.
17. M. Bleicher et al., J. Phys. **G25**, 1859 (1999), hep-ph/9909407.
18. B.-A. Li and C. M. Ko, Phys. Rev. **C52**, 2037 (1995), nucl-th/9505016.
19. Y. Nara, Nucl. Phys. **A638**, 555C (1998), nucl-th/9802016.
20. TASSO, M. Althoff et al., Z. Phys. **C22**, 307 (1984).
21. ALEPH, R. Barate et al., Phys. Rep. **294**, 1 (1998).
22. OPAL, G. Alexander et al., Z. Phys. **C72**, 191 (1996).
23. OPAL, K. Ackerstaff et al., Z. Phys. **C75**, 193 (1997).

Baryon Production in Proton-Proton Collisions

F. M. Liu* and K. Werner*

SUBATECH, Université de Nantes - IN2P3/CNRS - EMN, Nantes, France

Abstract. Motivated by the recent rapidity spectra of baryons and antibaryons in pp collisions at 158 GeV and the $\bar{\Omega}/\Omega$ ratio discussion, we reviewed string formation mechanism and some string models. This investigation told us how color strings are formed in ultrarelativistic proton-proton collisions.

Recently, the NA49 collaboration has published [1] the rapidity spectra of p, Λ, Ξ as well as the corresponding antibaryons in pp interactions at 158 GeV. These measurements provide new insight into the string formation process. In the string picture, high energy proton-proton collisions create "excitations" in form of strings, being one dimensional objects which decay into hadrons according to longitudinal phase space. This framework is well confirmed in low energy electron-positron annihilation [2] where the virtual photon decays into a quark-antiquark string which breaks up into mesons(M), baryons(B) and antibaryons($\bar{B}$). An example of a $q-\bar{q}$ string fragmenting into hadrons is shown in Fig.1. Proton-proton collisions are more complicated due to the fact that even at 158 GeV proton-proton collisions are governed by soft physics, thus pQCD calculations can not be applied. And the mechanism of string formation is not clear, as will be discussed in the following.

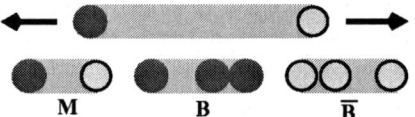

FIGURE 1. $e^+e^- \to \gamma^* \to q\bar{q}$. The $q\bar{q}$ string fragments into hadrons.

One may distinguish two classes of string models:

- Longitudinal excitation (LE) models: UrQMD [3], HIJING [4], PYTHIA [5], FRITIOF [6];
- Color exchange (CE) models: DPM [7], VENUS [2], QGS [8].

In the LE case the two colliding protons excite each other via a large transfer of momentum between projectile and target, Fig.2a. In contrast, the CE picture considers a color exchange between the incoming protons, leaving behind two octet states. Thus, a diquark from the projectile and a quark from the target, and vice versa, form color singlets. These are identified with strings, c.f. Fig.2b. The color exchange is a soft process. The transfer of momentum is negligible. The final result, two quark-diquark strings with valence quarks being their ends, however, is quite similar.

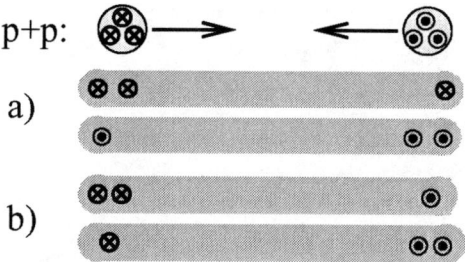

FIGURE 2. Two classes of string models for pp collisions are presented: a) longitudinal excition (LE). b) color exchange (CE).

How are baryons and antibaryons produced? The easiest way to obtain baryons is to break the strings via quark-antiquark pair production close to the valence diquark. Since the ingoing proton was composed of light quarks(qqq), the resulting baryon is of qqq or qqs type. Thus nucleons, Λs or Σs are formed. Since these baryons are produced at the string ends, they occur mainly close to the projectile rapidity or target rapidity (leading baryons).

Multi-strange baryons which consist of two or three strange quarks are produced near the quark end or the middle of the strings, via $s\bar{s}$-$\bar{s}s$ production. Therefore the distributions of multi-strange baryons are peaked around central rapidity. Thus, the corresponding yields of multi-strange baryons and their antiparticles should be comparable. A closer look reveals an interesting phenomenon: Theoretically one finds the ratio of yields [9]:

$$\overline{\Xi}^+/\Xi^- = 0.8 \sim 1.2.$$

Experimentally, however, $\overline{\Xi}^+$s are less frequent than expected. The ratio at midrapidity is [1]

$$\overline{\Xi}^+/\Xi^- = 0.44 \pm 0.08.$$

The situation for Ωs is even more extreme: from string models one gets [9]

$$\overline{\Omega}^+/\Omega^- = 1.6 \sim 1.9$$

at midrapidity. From extrapolating Λ and Ξ results (and from preliminary NA49 data) we expect [1]

$$\overline{\Omega}^+/\Omega^- = 0.5 \sim 0.8.$$

It seems impossible to get the $\overline{\Omega}^+/\Omega^-$ ratio smaller than unity from string models. As addressed in [9], this is due to the fact that the strings have a light quark (but not a strange quark) at the end, which disfavours multi-strange baryon production.

So is there something fundamentally wrong with string models? To answer this question we consider a new approach called Parton-Based Gribov-Regge theory. It is realised in the Monte Carlo program NEXUS 3 [10]. This model provides a self-consistent quantum mechanical treatment of high energy collisions. It is an effective theory based on effective elementary interactions. Multiple interactions happen in parallel in proton-proton collisions. The object exchanged in each elementary interaction is called Pomeron. A Pomeron can be seen as a (soft) parton ladder, which is attached to projectile and target via legs. Strings appear from cutting Pomerons. The spectators of each proton form a remnant.

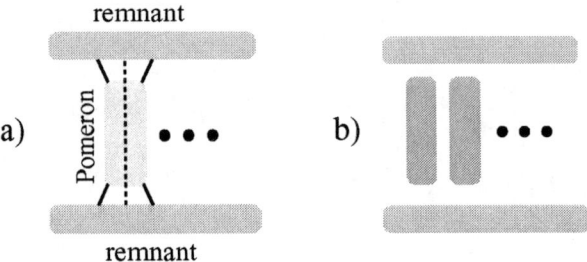

FIGURE 3. a) A typical collision configuration for pp collisions has two remnants and some Pomerons which stand for elementary interactions. Particles are produced from remnant decay and cut Pomerons. b) Each cut pomeron is regarded as two strings.

First the collision configuration is determined: i.e. the number of Pomerons exchanged between the projectile and target is fixed and the initial energy is shared between the Pomerons and the two remnants, c.f. Fig.3a. Two kinds of sources to produce particles exist, one is from remnant decay, the other is from each cut Pomeron. Hard, semihard and soft contributions are included in each Pomeron. The cut-off of virtuality between hard and soft processes, Q_0^2, is independent of the beam energy. The same formalism is used to calculate the cross section of each cut Pomeron and of each collision configuration. The former serves for the exclusive particle production, while the latter gives the weight of this collision configuration. Thus, energy sharing and particle production are treated consistently.

Each cut Pomeron is attached to projectile and target remnant through its two legs. The legs of cut Pomerons form color singlets, such as $q\bar{q}$, qqq or $\overline{qqq}$. The probability of qqq and $\overline{qqq}$ is controled by the parameter P_{qq}. As described above, strings are an intuitive way to understand how particles are produced from each cut Pomeron [10], each cut Pomeron is regarded as two strings, c.f. Fig.3b.

All the valence quarks stay in the remnants, wheras the string ends are represented by sea quarks. It is a natural idea to take quarks and antiquarks from the sea as string ends in NEXUS. Because in this way, an arbitary number of Pomeroms may be involved, all of them being identical. This point is different from the above-mentioned string models, where the string ends are valence quarks. Letting all the valence quarks stay in the remnants, complete symmetry concerning quarks and antiquarks at the string ends is given. This is different compared to the old version of NEXUS applied in [9]. Thus, the strings from cut Pomerons produce baryons and antibaryons in equal amounts. Whenever a quark or an antiquark is taken from the sea as a string end, a corresponding antiparticle is put in the remnant through the Pomeron leg attached to it to compensate the flavour. Thus, flavour conservation holds in every collision configuration.

The remnants have mass distribution $P(m^2) \propto (m^2)^{-\alpha}$, $m^2 \in (m_{min}^2, x^+s)$, where s is the squared CMS energy, m_{min} is the minimum mass of hadrons to be made from the remnant's quarks and antiquarks, and x^+ is the light-cone momentum fraction of the remnant which is determined in the collision configuration. The parameter α we use is 1.5. According to its number of quarks and antiquarks, a remnant emits mesons, antibaryons and baryons until the last one. The last one is considered as a string if its mass is bigger than a resonance plus a pion. The most simple and frequent collision configuration has two remnants and only one cut Pomeron represented by two $q-\bar{q}$ strings as in Fig.4a. Besides the three valence quarks, each remnant has additionally a quarks and an antiquark to compensate the flavours. For a remnant of such type, first a meson consisting of the antiquark and any of the four quarks is emitted, then the rest part of the remnant is treated as a baryon or a quark-diquark string, if the mass is bigger than the baryon and an additional meson. Thus, proton and Λ show strong leading particle effects (two wings in the rapidity spectra), but antibaryons and multi-strange baryons can not.

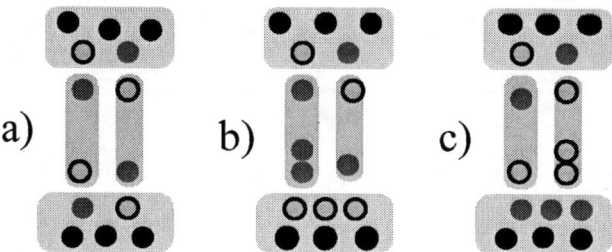

FIGURE 4. The most simple and frequent collision configuration has two remnants and only one cut Pomeron represented by two $q-\bar{q}$ strings. Besides the three valence quarks, each remnant has additionally a quarks and an antiquark to compensate the flavour. b) One of the $\bar{q}$ string-ends can be replaced by a qq string-end. c) With the same probability, one of the q string-ends can be replaced by a $\bar{qq}$ string-end. To compensate the flavour, one of the remnants now has six quarks. It decays into two baryons.

The leg of a cut Pomeron may be of qqq type with small probability P_{qq} (the mean multiplicity increases with P_{qq}), which means the corresponding string ends are a diquark and a quark. In this way we get quark-diquark (q-qq) strings from cut Pomerons. The qqq Pomeron leg has to be compensated by the three corresponding antiquarks in the remnant, as in Fig.4b. The ($3q3\bar{q}$) remnant may decay into three mesons (3M) or a baryon and an antibaryon (B+$\bar{B}$). Since the 3M mode is favored by phase space, we neglect B+$\bar{B}$ production here.

For symmetry reasons, the leg of a cut Pomeron is of $\bar{qqq}$ type with the same probability P_{qq}. This yields a $\bar{q}-\bar{qq}$ string and a (6q) remnant, as shown in Fig.4c. The (6q) remnant decays into two baryons. Since q-qq strings and $\bar{q}-\bar{qq}$ strings have the same probability to appear from cut Pomerons, baryons and antibaryons are produced equally. However, from remnant decay, baryon production is favored due to the initial valence quarks.

Fig.5 depicts the rapidity spectra of baryons and antibaryons from NEXUS 3.0 with $P_{qq} = 0.02$ (black solid lines). As a comparison, we also show the results from Pythia 6.2 simulations [5] (grey solid lines), and the preliminary data from the NA49 experiment [1] (points). In NEXUS 3.0, particles are produced by two sources: remnants (dotted lines) and cut Pomerons (dashed lines). Fig.5 demonstrates that NEXUS 3.0 describes

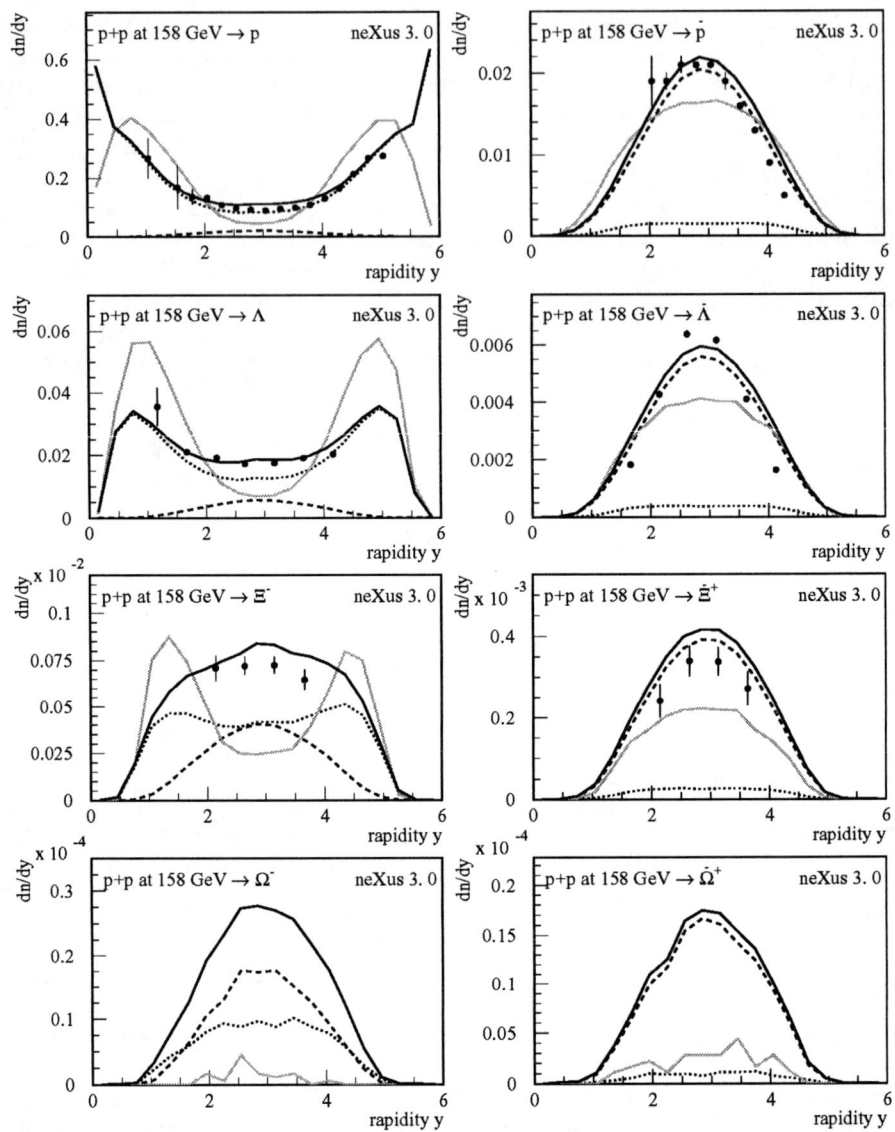

FIGURE 5. Rapidity spectra of baryons and antibaryons calculated from NEXUS 3.0 (remnant contribution: black dotted lines; Pomeron contribution: black dashed lines; sum: black solid lines), Pythia 6.2 (grey solid lines) and NA49 experiment [1] (points).

reasonably the rapidity spectra of baryons and antibaryons in pp collision at 158 GeV.

We also provide the particle yields at midrapidity, $y \in (y_{cm} - 0.5, y_{cm} + 0.5)$, from NEXUS 3.0, Pythia 6.2 and compare them to data in table 1.

TABLE 1. Particle yields at midrapidity in pp collisions at 158 GeV.

yield	NEXUS 3.0	Pythia 6.2	NA49 data
p	1.01×10^{-1}	4.85×10^{-2}	9.28×10^{-2}
$\bar{p}$	2.14×10^{-2}	1.64×10^{-2}	2.05×10^{-2}
Λ	1.65×10^{-2}	7.53×10^{-3}	1.79×10^{-2}
$\bar{\Lambda}$	5.86×10^{-3}	4.02×10^{-3}	5.57×10^{-3}
Ξ^-	7.45×10^{-4}	2.53×10^{-4}	7.08×10^{-4}
$\bar{\Xi}^+$	4.17×10^{-4}	2.20×10^{-4}	3.12×10^{-4}
Ω^-	2.68×10^{-5}	2.33×10^{-6}	–
$\bar{\Omega}^+$	1.69×10^{-5}	2.94×10^{-6}	–

From NEXUS 3.0, we get the ratios at midrapidity

$$\bar{\Xi}^+/\Xi^- = 0.56, \qquad \bar{\Omega}^+/\Omega^- = 0.71.$$

In conclusion, it seems that old string models fail to reproduce the experimental $\bar{\Xi}^+/\Xi^-$ and anticipated $\bar{\Omega}^+/\Omega^-$ ratio so far. The string formation mechanism as employed in NEXUS 3.0 is able to reproduce the experimental data nicely. The rapidity distributions of multi-strange baryons as well as Λs and protons can be understood. The main point is the fact that the final result of a proton-proton scattering is a system of projectile and target remnant and in addition (at least) two strings. All the valence flavour remains in the remnants, the string ends are composed of quarks and antiquarks from the sea only, with complete quark-antiquark symmetry. The present data does not seem to support the picture of having two strings only emerging from a proton-proton collision.

REFERENCES

1. K. Kadija, talk given at *Strange Quark Matter 2001*, Frankfurt, Germany; http://na49-pp.web.cern.ch/NA49-pp/
2. K. Werner, Phys. Rept. **232** (1993) 87.
3. M. Bleicher *et al.*, J. Phys. G **25** (1999) 1859. [arXiv:hep-ph/9909407].
4. X. N. Wang, Phys. Rept. **280** (1997) 287. [arXiv:hep-ph/9605214].
5. T. Sjostrand *et al.*, Comput. Phys. Commun. **135** (2001) 238. [arXiv:hep-ph/0010017].
6. H. Pi, Comput. Phys. Commun. **71** (1992) 173.
7. J. Ranft, Z. Phys. C **43** (1989) 439; A. Capella *et al.*, Phys. Rept. **236** (1994) 225.
8. A. B. Kaidalov *et al.*, Phys. Lett. **B 117** (1982) 247.
9. M. Bleicher *et al.*, arXiv:hep-ph/0111187.
10. H. J. Drescher *et al.*, Phys. Rept. **350** (2001) 93. [arXiv:hep-ph/0007198].

Hydrodynamics for Modeling Ultra-Relativistic Heavy-Ion Reactions

László P. Csernai[1,2]

[1] Section for Theoretical and Computational Physics, Department of Physics
University of Bergen, Allegaten 55, 5007 Bergen, Norway

[2] KFKI Research Institute for Particle and Nuclear Physics
P.O.Box 49, 1525 Budapest, Hungary

Abstract. In these lectures the recent status of hydrodynamic modeling of high energy heavy ion reactions will be reviewed. The relativistic hydrodynamic approach in this filed is used for almost 30 years by now and it became one of the best established theoretical and experimental methods. From our point of view the applicability of this method is vital, and we will discuss the initial and final stages of heavy ion reactions in great detail, in addition to the middle stages of the reaction where the relativistic hydrodynamic description provides an adequate and accurate description of the reaction.

INTRODUCTION

In these lectures many new results from recent years will be mentioned, which were achieved in collaboration with researchers at the University of Bergen: C. Anderlik, Ø. Heggø-Hansen, Z. Lázár, V. Magas, D. Molnár, A. Nyiri, K. Tamousiunas; at the University of Oulu: A. Keranen, J. Manninen; at the University of Sao Paulo: F. Grassi, Y. Hama; at the University of Rio de Janeiro: T. Kodama; at the University of Frankfurt: H. Stöcker, W. Greiner; at the Los Alamos National Laboratory: D.D. Strottman. These lecture notes were prepared with the help of young researchers from the University of Rio de Janeiro: G. Grise, L. Lima Silvia Portugal, B. Mattos Tavares, and D. de Paula, who contributed valuably to the manuscript.

We are looking for a way to describe ultra-relativistic heavy ion reactions using relativistic hydrodynamics. Not everything that is used for modeling relativistic heavy ion collisions can be extended for the ultra-relativistic case. One example of this is fluid dynamics: it's not obvious that fluid dynamics is adequate to describe systems with more than 1 A.GeV energy, because the basic collision mechanism is different. Thus we will talk, discuss an energy region that is not well known yet.

The interest in fluid dynamics came from the many applications and important results achieved in lower energy heavy ion reactions. To apply fluid dynamics we need local equilibrium and Equation of State (EoS), in other words we need a "stopping" of matter in contrast to transparency. Consequently from early on there was a large interest in studying the stopping power in ultra-relativistic collisions. At these energies, stopping power is a question, much more complex than in relativistic collisions. There exist a great number of works where this problem is investigated, but different mechanisms of the leading particle deceleration result in a substantial spread in the predicted magnitudes of the nuclear stopping power. Nuclear effects, such as density increase, rescattering, eventual phase transition or precritical behavior can modify the stopping power, mostly by increasing it. Fortunately, simple fluid dynamical models that assume complete stopping with well-chosen initial conditions can describe the final equilibrated stages of a collision, and so they are also good models for the ultra-relativistic cases. For the other stages of the reaction, we will present a Multi Module Model, where each stage of the reaction is described with a suitable theoretical approach.

The second immediate difficulty that we should study is phase transitions in finite systems. It's unclear if nuclear systems are so small that phase transitions can not be observed in them (or not in a simple way) and fluctuations may wash out (partly or fully) the signals of the phase transition. A conclusive opinion about this is quite premature, but is important to say that the connection between the models and the nuclear EoS should be firmly established before this final conclusion.

PHENOMENOLOGY OF HEAVY ION REACTIONS

There is a mounting amount of experience in modeling heavy ion collisions with different phenomenological approaches. For an introduction to the field and basic problems of modeling heavy ion reactions see ref. [1]. This lecture will have some parts which are discussed similarly but in greater detail in [1].

At SPS and RHIC more "stopping" is observed than was anticipated earlier, [2, 3] and the abundance of strange hadrons, particularly of anti-strange baryons, are enhanced. These observations led to modeling changes, new "energetic" degrees of freedom: string ropes, fused strings, quark clusters, were introduced in Monte-Carlo sting models to cope with these observations. These somehow indicate the occurrence of quark-gluon plasma effects or can be considered as its precursors.

Any model that we want to apply must fit what is already confirmed. First of all, it is well known by now that ultra-relativistic nuclear collisions show strong stopping, and this supports and justifies the use of one-fluid, fluid dynamical models, which assume a complete stopping. Of course fluid dynamics should be used only after stopping or local equilibration is achieved. At ultra-relativistic energies this does not happen instantly.

We should pay attention to basic principles of physics when modeling complicated processes as heavy ion collisions. A feature sometimes not given proper attention to is the entropy. During and after collision, the entropy must not decrease. This should be true to each "stage" or "Module" of our collision model, as well as, to the transition between stages or modules.

Equation of State

It is very important to know the behavior of the matter under high densities and temperatures, which can be reached in heavy ion reactions. The intermediate stages of a collision (with energies around ≈ 100 A· GeV) may produce a system involving several thousand particles, since the particle-antiparticle creation becomes easily possible. This system is, although, quite small it can already be sufficiently large for statistical and kinetic physics to be applicable. Since the heavy ion reaction is a highly dynamical process both the equilibrium and non-equilibrium properties of the matter can be studied. Of course equilibrium and non-equilibrium effects are not always easy to separate.

The thermodynamical properties of the matter in statistical equilibrium are described by an Equation of State. There are three interesting features of the EOS under investigation: i) the phase transition from continuous nuclear liquid into a nuclear vapor of fragments and nucleons, the so called nuclear liquid-gas phase transition, or the multifragmentation transition, ii) the compressibility of nuclear matter at and higher densities than the density of matter in ground state nuclei (n_0), and finally, iii) the phase transition to Quark-Gluon Plasma (QGP). While in the last two-three decades numerous results were reached in the first two topics, the main objective of the present research at ultra-relativistic energies (at SPS, RHIC and LHC) is the study of the properties of QGP

The non-equilibrium features of matter can also be studied in heavy ion reactions. Transport coefficients can be deduced from experimental results. On the microscopic level these are associated with in-medium nuclear cross sections. The reconciliation of several sorts of experiments and observations leads even to conclusions on the nucleon-nucleon interaction.

Here we will discuss the features of the phase transition to QGP only from the EoS point of view, and the sensitivity of the phase transition to the nuclear compressibility. The EoS provides only limited information about the nuclear matter: the static thermal equilibrium properties.

In the field of the equation of state of the quark gluon plasma most of the theoretical work is invested in the study of pure SU(N) Yang-Mills theory on the lattice. Most of these calculations, however, are so far restricted to zero net baryon density or zero chemical potential. To form a low or "zero" baryon density matter in the deconfined phase one expectedly needs extremely high energy. Relatively simple "phenomenological" theories on hand, are able to provide us with an equation of state (EoS) in the phase transition region for cold matter, for zero baryon charge at finite temperature, or in the complete phase space for finite density and temperature. These phenomenological equation of state studies can yield a good qualitative insight into the phase transition problem until a priori QCD calculations will be available in the complete density-temperature domain. There are even some advantages in the phenomenological approach: the results of these nuclear EoS studies can easily be incorporated into the phenomenological phase transition models.

Transport Theory and the Equation of State

If we do not have, even local equilibrium, the most frequent theoretical approach is kinetic or transport theory to describe the dynamics of the particles or the matter. We can use classical or quantum transport theories, in relativistic or non-relativistic description. In these theories we can determine the energy-momentum tensor, the four-currents of all conserved charges, we can introduce the definition(s) of flow velocity. If we do not have a perfect fluid different definitions are possible, like Eckart's or Landau's definitions.

Let us introduce the quantities, we need to describe the matter properties at relativistic or ultra-relativistic energies: the invariant scalars, n, e, p, baryon-density, energy-density and pressure, respectively. These are based on the baryon 4-current, N^μ, energy-momentum tensor, $T^{\mu\nu}$, 4-velocity, u^μ:

- Invariant scalar density n:
$$n \equiv N^\mu u_\mu, \tag{1}$$

- Invariant scalar energy density e:
$$e \equiv u_\mu T^{\mu\nu} u_\nu, \tag{2}$$

- Pressure tensor $P^{\mu\nu}$:
$$P^{\mu\nu} \equiv \Delta^\mu_\sigma T^{\sigma\tau} \Delta^\nu_\tau \tag{3}$$

$$P^{\mu\nu} = \underset{\substack{\text{hydrostatic}\\\text{pressure}\\\text{tensor}}}{-P\Delta^{\mu\nu}} + \underset{\substack{\text{viscous}\\\text{stress}\\\text{tensor}}}{\Pi^{\mu\nu}}, \tag{4}$$

$$T^{\mu\nu} = \underset{\substack{\text{reversible,}\\\text{part}}}{T^{\mu\nu\,(0)}} + \underset{\substack{\text{irreversible}\\(\text{dissipative})\text{ part}}}{T^{\mu\nu\,(1)}} \tag{5}$$

where

$$T^{\mu\nu\,(0)} = eu^\mu u^\nu - P\Delta^{\mu\nu} = (e+P)u^\mu u^\nu - Pg^{\mu\nu}, \tag{6}$$

which in the Local Rest frame yields

$$T^{\mu\nu\,(0)}_{(LR)} = \begin{pmatrix} e & 0 & 0 & 0 \\ 0 & P & 0 & 0 \\ 0 & 0 & P & 0 \\ 0 & 0 & 0 & P \end{pmatrix}. \tag{7}$$

The EoS describes the connection among the invariant scalars, $p = p(e,n)$, while the transport parameters determine the non-equilibrium parts of the pressure tensor and/or energy-momentum tensor.

Multi Module Model

A realistic and detailed description of an energetic heavy ion reaction requires a Multi Module Model, where the different stages of the reaction are each described with a suitable theoretical approach.[4] It is important that these modules are coupled to each other correctly: on the interface, all conservation laws should be satisfied and the entropy should not decrease.

Fluid dynamical models are widely used to describe ultra-relativistic heavy ion collisions. Their advantage is that one can vary flexibly the Equation of State of the matter, and test its consequences on the reaction dynamics and outcome. In energetic collisions of large heavy ions, especially if Quark-Gluon Plasma is formed in the collision, one fluid dynamics is a valid and good description for the intermediated stages of the reaction.

The initial and final, Freeze-Out (FO), stages of reaction are outside the domain of the applicability of the fluid dynamical model. After hadronization and FO, the matter is already dilute and can be described well with kinetic models. The initial stages are the most problematic. None of the theoretical models can unambiguously describe the initial stages. Since we cannot describe unambiguously the initial stages, we have to use some phenomenological models. There are two phenomenological models that are frequently used: the Landau model (which describes complete stopping) and the Bjorken model (which describes complete transparency).

Two Module Model

We want to build a Multi Module Model (in this case we will use two modules) for ultra-relativistic heavy ion collisions. We describe the initial state with the Effective String Rope Model (ESRM);[5] then the hydrodynamic evolution with a QGP EoS is used for the intermediate state. The hydro evolution stops at the FO hyper-surface, which is the boundary of the space-time domain where local equilibrium still persists before FO. Frequently this hyper-surface is approximated as a constant time hyper-surface in the CM frame of the collision.

On the other hand if we chose the FO hyper-surface as a constant time hyper-surface in the local rest frame of every fluid element, then the flow velocity does not change during the FO process, and the FO calculations can be done in a more simple way. We will find more details about the Freeze-out at the third part of these lectures.

ELEMENTS OF RELATIVISTIC KINETIC THEORY

The time development of the single-particle phase space distribution can be obtained by solving the Boltzmann Transport Equation, or some similar or equivalent transport equation. One can derive conservation laws for the four currents, and the energy-momentum tensor, and one can even evaluate the increase of the entropy density. All these are possible for any non-equilibrium configuration. One can even prove the "Boltzmann H-Theorem", which states that in an arbitrary configuration the entropy of the system increases, and a stationary, so called "equilibrium" state is reached when the entropy is at its maximum. See ref. [1], for a detailed introduction, and the most important definitions below.

The kinetic theory establishes a relationship between macroscopic and microscopic properties, by using a one particle distribution function $f(x,p)$. This function describes the phase space density of the particles and a kinetic equation or a transport equation governs the time development of $f(x,p)$. In most cases the kinetic description is adequate for dilute systems.

Kinetic Definition of Density, Energy, Momentum

The local density $n = n(x)$ as a function of space time coordinate x; ($N = n\Delta^3 x$) and the local particle flow $\vec{j} = \vec{j}(x)$, i.e. particle current across unit area in unit time, are elementary macroscopic quantities. From the above two quantities we can formally create a four vector:

$$N^\mu(x) = (n(x), \vec{j}(x))$$

We know, however, that $\vec{v} = \frac{\vec{p}}{p^0}$, so in a given reference frame the particle four flow can be expressed in terms of the distribution function:

$$N^\mu(x) = \int \frac{d^3p}{p^0} p^\mu \, f(x,p)$$

The energy-momentum tensor is another quantity characterizing the matter. $T^{00}(x)$ is the energy density. Since the energy of a particle in kinetic theory is p^0 in the kinetic theory:

$$T^{00} = \int d^3p \, p^0 \, f(x,p)$$

$$T^{0i} = \int d^3p \, p^0 v^i \, f(x,p) \quad \text{:energy flow}$$

$$T^{i0} = \int d^3p \, p^i \, f(x,p) \quad \text{:momentum density}$$

$$T^{ik} = \int d^3p \, p^i p^k \, f(x,p)$$

All these can be combined into the form:

$$\underset{\text{tensor}}{T^{\mu\nu}(x)} = \underset{\text{scalar}}{\int \frac{d^3p}{p^0}} \; \underset{\text{tensor}}{p^\mu p^\nu} \; \underset{\text{scalar}}{f(x,p)}$$

Local Rest Frame

In this reference frame we have $u^\mu = u^\mu_{(LR)} = (1,0,0,0)$. Since u^μ is time like there is always a Lorentz transformation which leads to the local rest frame (LR), in this frame, the operator of transverse projections is:

$$\Delta^{\mu\nu}_{(LR)} = \Delta_{\mu\nu\,(LR)} = diag(0, -1, -1, -1),$$

where:

$$\Delta^{\mu\nu} = g^{\mu\nu} - \frac{u^\mu u^\nu}{u^\rho u_\rho}$$

is the orthogonal projector to any time like four vector. When this projector acts on the flow velocity thus yields $\Delta^{\mu\nu} u_\mu = 0$

Eckart's Definition

According to this definition the local rest frame is tied to conserved particles or conserved charges like baryon charge (if there are any). If we know the particle four-flow, then the flow vector, an unit vector in the same direction is:

$$u^\mu = \frac{N^\mu}{(N^\nu N_\nu)^{\frac{1}{2}}} \quad \text{(Eckart)}$$

In general the energy or heat flow may have a different direction.

Landau's Definition

According to this definition the LR is tied to the energy flow:

$$u^\mu = \text{constant} \times T^{\mu\nu} u_\nu \quad \text{(Landau)}$$

This definition is applicable even if we do not have conserved charges, e.g. in the case of baryon free plasma. The LR system moves together with the energy (or heat), so there is no energy flow in this case with respect to the local rest frame, but in general there may exist baryon charge flow or other conserved charge flow.

Jüttner Distribution

The basic quantity of kinetic theory is the single particle phase space density. As we leaned in classical statistical physics the most frequently used distribution is the Maxwell-Boltzmann distribution, corresponding to classical ideal gases. Its relativistic generalization is the Jüttner distribution.

This $f^{\text{Juttner}}(p)$ is the distribution of particle momenta, p^μ, in a thermal system, which is moving with flow velocity u^μ, has a chemical potential μ and a temperature parameter T:

$$f^{\text{Juttner}}(p) = \frac{1}{(2\pi\hbar)^3} \exp\left(\frac{\mu - p^\nu u_\nu}{T}\right)$$

Since $(p^\nu u_\nu)_{LR} = p^0 = E$, and since $p^\nu u_\nu$ is an invariant scalar, consequently $f(x,p)$ is also an invariant scalar, i.e., it is the same in all reference frames.

Normalization

From the normalization requirement:

$$n = N^\mu u_\mu = u_\mu \int \frac{d^3 p}{p^0} p^\mu f(x,p) .$$

Since n is an invariant scalar, it can be evaluated in the local rest frame (n=constant), so

$$n = \frac{4\pi e^{\frac{\mu}{T}}}{(2\pi\hbar)^3} T^3 \int_{m/T}^\infty d\tau\, \tau \sqrt{\tau^2 - \left(\frac{m}{T}\right)^2} e^{-\tau}$$

$$n = \frac{e^{\frac{\mu}{T}}}{(2\pi\hbar)^3} 4\pi m^2 T K_2\left(\frac{m}{T}\right) ,$$

where K_2 is the modified Bessel function. This relation gives a connection between the chemical potential μ, and the density n. As we will see this is the stationary solution of the Boltzmann Transport Equation, maximising the entropy of the system.

RELATIVISTIC FLUID DYNAMICS

The equations of Fluid Dynamics (FD) can be most simply derived from the Boltzmann Transport Equation (BTE).

Boltzmann Transport Equation and Conservations Law

The relativistic BTE below, describes the time evolution of the single particle distribution function based on the assumptions that just two particle collisions are considered (binary collisions), the number of binary collisions at x is proportional to $f(x,p_1) \times f(x,p_2)$ and $f(x,p)$ is a smoothly varying function compared to mean free path. So, one can show that BTE has the form [see Chapter 3 of ref. [1]] :

$$p_k^\mu f_{k,\mu} = \sum_{l=1}^N C_{kl}(x,p_x)$$

where C is the collision integral, and k and l stand for different particle species or particle components. Then using microscopic conservation laws in the collision integral, from the BTE we can derive the differential form of conservation laws [see section 3.6 of ref. [1]]:

$$N^\mu_{,\mu} = \sum_{k=1}^{N} N^\mu_{k,\mu} = 0 \quad \text{:conservation of particle number}$$

$$Q^\mu_{,\mu} = \sum_{k=1}^{N} Q^\mu_{k,\mu} = 0 \quad \text{:conservation of charge}$$

$$T^{\mu\nu}_{,\mu} = \sum_{k=1}^{N} T^{\mu\nu}_{k,\mu} = 0 \quad \text{:conservation of energy and momentum}$$

where $_{,\mu}$ denotes $\partial/\partial\mu$ and there is a summation for indexes occurring twice. These equations are valid for any, non-equilibrium distribution! On the other hand they cannot be solved as they do not carry the information needed for the solution. They can be used only if we know the solution, by solving the BTE, and then we can evaluate the conserved currents and tensors.

Fluid Dynamics

These equations are also the equations of FD. But they are not considered as the consequence of BTE, rather these equations are **postulated**, as the differential forms of conservation laws. This can be done because such conservation laws can be derived in several other ways from other theoretical approaches also!

As we mentioned, these equations are not a closed set, because the energy momentum tensor and the particle four current should also be defined. Thus, we have two elementary levels of the fluid dynamical model:

In the Eulerian or perfect fluid dynamics we postulate the form of energy-momentum tensor as given by eq. (6) or (7), and an EoS given by the relation $P = P(e, n)$. This provides a closed set of partial differential equations that is solvable. It is a theoretical approach in its own right as any EoS, which is consistent with thermodynamics, can be used. One does not have to assume dilute systems with binary collisions, or the assumption of molecular chaos. The only requirement is the existence of local equilibrium, otherwise we would not have an EoS.

In the case of Navier-Stokes fluid dynamics, we assume that the energy momentum tensor also contains the dissipative part $T^{\mu\nu(1)}$, (see eq. (5)), which allows for small, first order deviations from local equilibrium. Thus, to have a solution we need not only the EoS, but also the transport coefficients that occur in the dissipative part of the energy-momentum tensor.

Boltzmann H-Theorem, Entropy and Dissipation

Boltzmann, in connection with the H-theorem introduced a kinetic definition of entropy for non-equilibrium distributions. The H-theorem shows that the evolution of the distribution function according to the BTE, and if the three conditions, - dilute system, with binary collisions, and the assumption of molecular chaos, - are applicable, then the change is such that the entropy always increases. When the distribution does not change any more, we have reached equilibrium, the entropy reached its maximum, and the distribution arrived to the Jüttner distribution. In this limiting case Boltzmann's entropy definition coincides with the standard thermodynamical definition of entropy first given by Clausius. What is even more interesting, from the resulting Jüttner distribution we can evaluate all thermodynamical quantities based on their kinetic definitions, and we obtain both the EoS of the ideal gas, $p = nT$, as well as the second law of thermodynamics, $Ts = e + P - \mu n$. See Assignment 3.10.c,d,e of ref. [1], and the lectures of Takeshi Kodama in this proceedings [6].

This means that if irreversible processes are present the entropy increases. In equilibrium the distribution is constant, the entropy is also constant, but it reached its maximum value.

It is important to know that this can be seen from within fluid dynamics also. In perfect fluid dynamics the flow is adiabatic, no entropy is produced. This can be proven using the equations of perfect fluid dynamics and standard thermodynamical relations. See Assignment 5.6.a in ref. [1]. Entropy production is strictly, and quantitatively connected to dissipative processes, see section 9.1.1 *ibid*. Perfect flow is adiabatic even in the presence of phase transitions, if the two phases are in thermal, mechanical (P), chemical and phase equilibrium all the time! On the other hand if the phase transition deviates from equilibrium this leads to entropy production. See Assignment 9.4.a in ref. [1].

As shock waves, detonations, deflagrations, are idealizations of very sharp or very rapid processes as discontinuities, the above conclusions are valid for discontinuities also. Nevertheless, frequently, dissipative processes can be neglected for most of the flow except the sharp or rapid changes. Thus, as an end effect the dissipative processes are frequently localized in sharp fronts or in hypersurfaces of discontinuities. One should not, however, forget that dissipation is due to transport properties and characteristic constants of phase transition dynamics, even if it happens in "discontinuities".

Perfect Fluid Dynamics

If the local momentum distribution deviates strongly from local kinetic equilibrium, the fluid dynamical approach (at least the one fluid one) is not applicable.

Classical FD models incorporate phase transitions in a trivial way, as the EoS is given for a phase transition. Even more so, the FD approach can describe systems out of the phase equilibrium, supplemented with dynamical equation describing the dynamics of phase transition, as local kinetic equilibrium in each phase is sufficient to apply FD [2, 3].

We assume that our system is not homogeneous, but the gradients are small, so local distribution can be an equilibrium distribution, e.g. a Jüttner distribution.

Now the local $n(x), e(x), s(x)$ and $P = P(n,e)$ are known, and we assume that in LR, $T^{\mu\nu}$ is diagonal. Now using the conservations law with introducing the apparent density as:

$$N \equiv n\gamma$$

the continuity equation takes the familiar form

$$(\partial_t + \vec{v}\, grad)N = -M\, \text{div}\, \vec{v}$$

similarly introducing

$$\vec{M} \equiv T^{0i} = w\gamma^2 \vec{v}$$
$$\varepsilon \equiv T^{00} = (e + P\vec{v}^2)\gamma^2$$

the energy and momentum conservation will take the form

$$(\partial_t + \vec{v}\, grad)\vec{M} = -\vec{M}(\text{div}\, \vec{v}) - gradP$$
$$(\partial_t + \vec{v}\, grad)\varepsilon = -\varepsilon\, \text{div}\, \vec{v} - \text{div}(P\vec{v})$$

Thus to solve the equations of relativistic fluid dynamics we have to solve the same partial differential equations as in the non-relativistic case. We cannot immediately apply the EoS to obtain the solution, as the EoS is given in terms of invariant scalars. So, we have to solve in addition the set of algebraic equations above, which connect N, $\vec{M}$, and ε to n, e, P and $\vec{v}$, for every fluid cell at every time step.

Important to note that the relativistic treatment is necessary not only at high velocities (i.e. if $v \sim c$) but also at low velocities when the pressure is not negligible compared to the energy density (i.e., $P \sim e$) as $w = e + P$ appears in the relativistic Euler equation! This is frequently the situation for ultra-relativistic gases, like the Stefan-Boltzmann or photon gas (or radiation pressure) where $P = e/3$. This is actually the indication of the fact that the constituents of the matter move with velocity c or close to it.

In principle the perfect fluid dynamics is absolutely unstable. Any spontaneous deviation from an exact solution, will grow exponentially, as their energy cannot be dissipated away [7]. In other words the Reynolds number tends to infinity and the system tends to turbulence. Viscosity is needed to stabilize the flow.

There are several numerical methods to obtain a solution. In all cases the solution is discretized in some way and a coarse graining is introduced. This coarse graining automatically leads to dissipative and transport properties, and the parameters of these transport processes can be determined by numerical experiments. The result is that all Computational Fluid Dynamical (CFD) models are dissipative or viscous, which is an advantage as the these are usually stable. By decreasing the cell size, the dissipative effects can be decreased in CFD, to a limit when instabilities, or turbulence occurs. As a side effect, CFD models lead to entropy production, as part of the kinetic energy is dissipated away due to the coarse graining or smoothing.

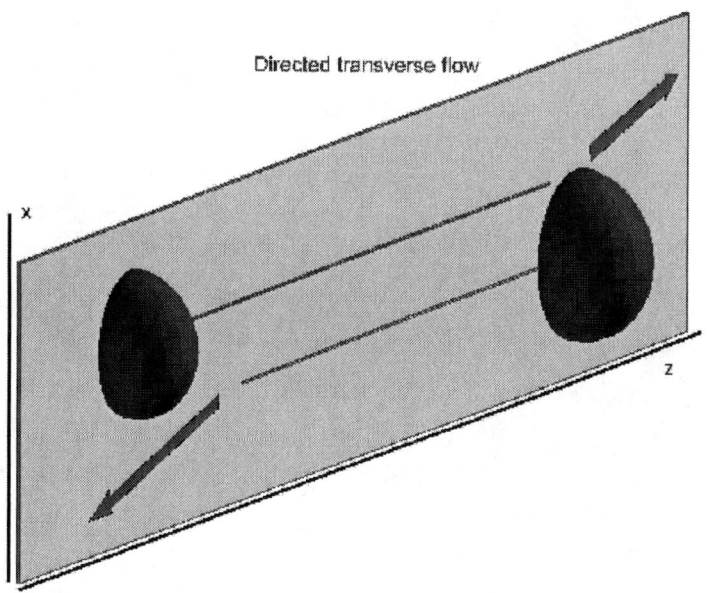

FIGURE 1. Directed transverse flow in the [x,z] reaction plane. At finite impact parameter b the Projectile and Target bounce off each other and their particles mode in the directions indicated by the arrows.

Observed Flow Patterns in Heavy Ion Reactions

Historically in the early 70s two basic flow patterns were considered, the collective "supersonic Mach cone" when a small projectile hits a large target centrally and the spherical expansion and explosion after a symmetric and central or nearly central collision, the so-called "Blast-Wave model". (See sections 6.3.2 and 7.8.1.a of ref. [1].) The Blast-Wave model is still used today due to its simplicity and to the fact that the flow is characterized by a single average number the "flow velocity".

The Mach cone was difficult to detect because exactly central collisions are not very probable, and very small projectiles have an overdamping effect. The probability to have a collision with impact parameter b is proportional with b^2! Thus, shocks can be more easily and more frequently observed in semi-central collisions. This was recognized in the first 3-dimensional simulations and the flow pattern was called "bounce-off" or "side-splash" or more recently "directed transverse flow", see Fig. 1.

Although, the directed transverse flow was predicted in the mid 70s already there was a decade of controversy about its existence, until it was accepted by the whole community in 1984 by the ground breaking first results of the Plastic Ball in Berkeley. See more details in sections 7.6 and 7.7 in ref. [1].

It is important to mention that the directed transverse flow was also a signal of the nuclear liquid-gas phase transition at lower energies. The directed transverse flow is caused by the collective pressure of the hot and dense matter, which deflects the particles.

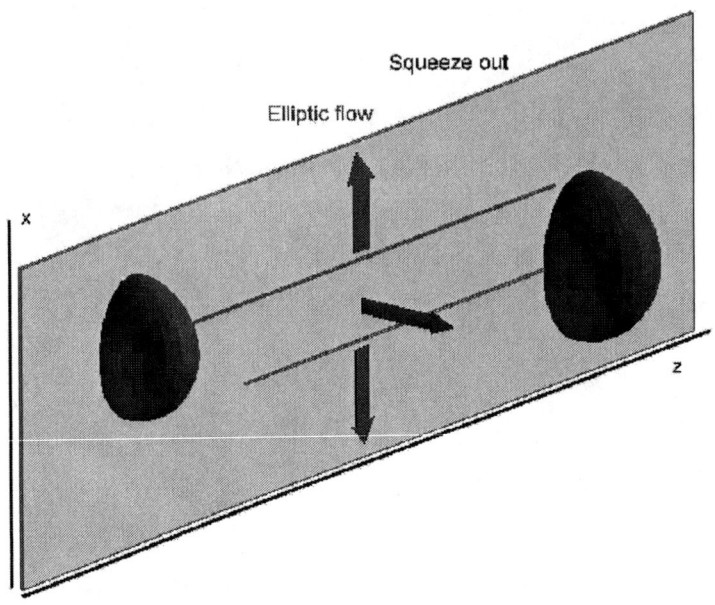

FIGURE 2. Squeeze-out flow component, orthogonal to the [x,z] reaction plane, indicated by one arrow. This was detected at BEVALAC energies. At ultra-relativistic energies (SPS and RHIC), due to the flat initial state lying in the [y,z]-plane, the flow is predominantly pointing into the ± [x]-direction, in the direction of the largest pressure gradient. This is indicated by the pair of arrows up and down.

At the phase transition this pressure is decreasing, and nuclear attraction between the projectile and target nuclei may overcome the repulsion of the pressure. This effect was observed at the MSU NSCL around 30-70 MeV/nucleon beam energy, where the directed transverse flow dropped down to zero, and then became negative. Similar drop of pressure, the "soft-point" is expected to be observed at the phase transition to QGP also.

Unfortunately, the transverse momentum arising from the collective pressure is not increasing with energy, because the pressure increase is modest, but the time and surface it can exercise the effect, are strongly reduced due to the Lorentz contraction. This general drop of transverse flow vs. the increasing longitudinal energy makes the measurement of directed flow difficult at SPS and RHIC energies. Thus to observe the soft point of the pressure in this anyway decreasing trend is not very promising.

After the success of the plastic ball the flow measurements became widespread and standard. Different quantitative measures were introduced to describe the flow quantitatively: Flow Tensor, Flow Angle, P_x vs. rapidity, v_1 vs. rapidity, v_2 vs. rapidity. Soon after the directed transverse flow was discovered, one could detect the matter squeezed out from between the target and projectile, orthogonal to the reaction plane, as indicated by the brown arrow in Figure 2.

As the collision energy increases the dominant flow patterns change. Due to the higher and higher particle densities and multiplicities the applicability of the fluid dynamical

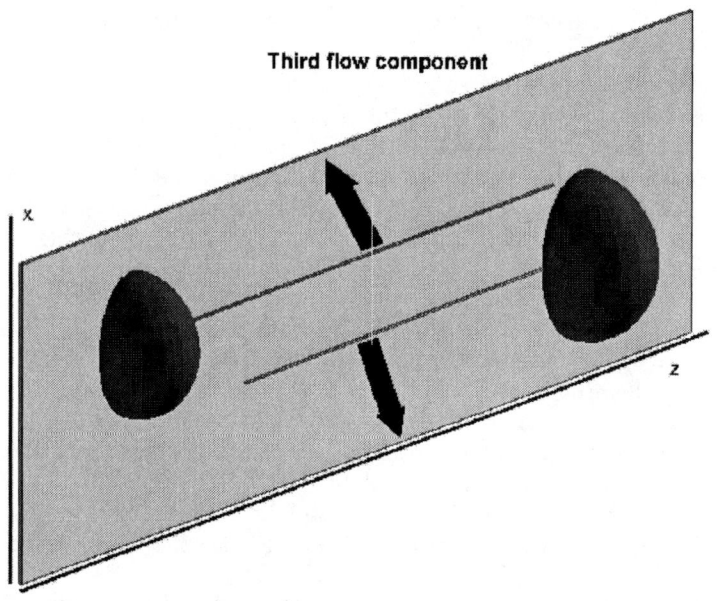

FIGURE 3. The third flow component in the reaction plane points in the direction indicated by the pair of arrows in the reaction plane. This is a consequence of the flat but tilted initial condition, of highly compressed matter, what we expect in QGP. In peripheral reactions shadowing may lead to similar effects, but impact parameter and energy dependence may separate the effects from each other.

model is becoming better.

If we anticipate an over-simplified collision dynamics, we may assume a Bjorken model type of transparency with a large extent of matter in the [z]-direction, and a lens shape participant region in the transverse, [x,y]-plane, which is flat in the [y]-direction. Consequently, the maximum of the pressure gradient points into the [x]-direction (up and down in Figure 2), and consequently the flow is expected to point in this direction also. This flow component is called the "Elliptic Flow" and indicated by a pink arrow in Figure 2.

At RHIC the first measurements of each experimental group showed unexpectedly strong elliptic flow which was an unexpected sensation of the first RHIC results. At colliders, although the rapidity gap between the projectile and target is large, the transversely emitted particles, i.e., particles with rapidities close to the CM rapidities can be measured. This luckily coincides with the fact that the elliptic flow is expected to show up at CM rapidities also!

Three dimensional reaction models show that more structure can be expected, in flow patterns, even if the interesting effects show up near to CM rapidities.

Figure 3 shows the direction of the "Third Flow component". [8] This is also in the reaction plane like the directed flow and the elliptic flow, but it points to the opposite direction than the directed flow. Therefore it is also called antiflow.

Determining the reaction plane at RHIC

At RHIC the directed transverse flow was not measured up to now. Different experimental difficulties were mentioned as the reason. In particular that the reaction plane could not be identified, although the so called v_2 parameter was measured.

This problem is partly due to the method used to determine v_2, which did not explicitly required the parallel determination of the reaction plane. It is, however, clear that the existence of strong elliptic flow indicates a strong collective correlation among the particles in every event: particles are preferably emitted upwards (+x) and downwards (-x) in the reaction plane and much less to the sides. Thus, if this correlation is detected and strong, the plane must be identifiable also! It is relatively easy to design methods, which work well statistically for very low event multiplicities.

Of course, to tell which is the target and projectile side of the plane may be difficult. If in a symmetric, A+A reaction one assumes, the idealized Bjorken picture with infinite length in the [z]-direction, and the lens shape source in the transverse plane, the target and projectile side cannot be distinguished in principle.

On the other hand, if one takes a more realistic three dimensional model, one will get both azimuthal asymmetry and forward/backward asymmetry. This can be used then to determine which side of the plane is which.

As v_2 is measured, the reaction plane, [x,z], is known, just the target/projectile side should be selected. This is not done due to the prejudice that the distribution of emitted particles is mirror-symmetric in CM:

$$f_{CM}(p_x, p_y, p_z) = f_{CM}(p_x, p_y, -p_z)$$

However, at finite impact parameters (2-15%) there is a fwd / bwd asymmetry, which utilize the following way to determine the reaction plane.

Let us calculate, event by event, the Q-vector (a la Danielewicz, and Odyniecz (1985), see in section 7.7 of ref. [1]):

$$Q_k = \sum_{ik} y_{CM,i} p_{x,i}$$

for all particles, i, of type k. Only the sign is relevant, as the plane is known already. This Q-vector will select the same side (e.g. projectile) in each event.

INITIAL STAGE - COHERENT YANG-MILLS MODEL

At highly ultra-relativistic energies, when QGP is formed, the pre collision initial state is far out of any thermal or mechanical equilibrium, and equilibrium can only be established in the QGP phase where the number of degrees of freedom are sufficiently large and the interaction frequency is also large, so that equilibration is expected to be established within one or a few fm/c. It means that the initial stage of the reaction is outside the domain of applicability of the fluid dynamical model.

Our goal in the present section is to show a model, based on the recent experiences gained in string Monte Carlo models and in parton cascades. One important conclusion of heavy ion research in the last decade is that the standard 'hadronic' string models fail to describe heavy ion experiments.

All string models had to introduce new energetic objects: string ropes [9, 10], quark clusters [11], fused strings [12], in order to describe the abundant formation of massive particles like strange antibaryons. Based on this, we describe the initial moments of the reaction in the framework of classical (or coherent) Yang-Mills theory following [13] assuming larger field strength (string tension) than in ordinary hadron-hadron collisions. In addition we satisfy all conservations law exactly.

We do not solve simultaneously the kinetic problem leading to parton equilibration, but assume that the arising friction is such that the heavy ion system will be an overdamped oscillator, i.e. yo-yoing of the two heavy ions will not occur. This assumption is based on recent string and parton cascade results.

Formulation of the Model

Our basic idea is to generalize the model developed in [13] for collisions of two heavy ions, and improve it by strictly satisfying conservation laws. First of all, we would create a grid in the [x,y] plane (z is the beam axes, [z,x] is the reaction plane). We will describe the nucleus-nucleus collision in terms of streak-by-streak collisions, corresponding to the same transverse coordinates, $\{x_i, y_j\}$. We assume that baryon recoil for both target and projectile arise from the acceleration of partons in an effective field $F^{\mu\nu}$, produced in the interaction. Of course, the physical picture behind this model should based on chromo-electric flux tube or string models, but for our purpose we consider $F^{\mu\nu}$ as an effective field. Phenomenological parameters describing this field must be fixed from comparison with experimental data.

Let describe the streak-streak collision:

$$\partial_\mu \sum_i T_i^{\mu\nu} = \sum_i F_i^{\nu\mu} n_{i\mu} , \qquad ((1))$$
$$\partial_\mu \sum_i n_\mu^i = 0 , \; i = 1, 2,$$

where n_μ^i is the baryon current of ith nucleus. (We are working in the Center of Rapidity Frame (CRF), which is the same for all streaks. The concept of using target and projectile reference frames has no advantage any more). $T^{\mu\nu}$ is a flux energy tensor. It consist of five parts, corresponding to both nuclei and free field energy (also divided into two parts) and one defines the QGP perturbative vacuum.

$$T^{\mu\nu} = \sum_i T_i^{\mu\nu} + T_{pert}^{\mu\nu} = \sum_i [e_i((1+c_0^2)u_i^\mu u_i^\nu - c_0^2 g^{\mu\nu}) + T_{F,i}^{\mu\nu}] + Bg^{\mu\nu}$$

where B is the bag constant, the equation of state is $P_i = c_0^2 e_i$, where e_i and P_i are energy density and pressure of QGP, and $F^{\mu\nu}$ has the form:

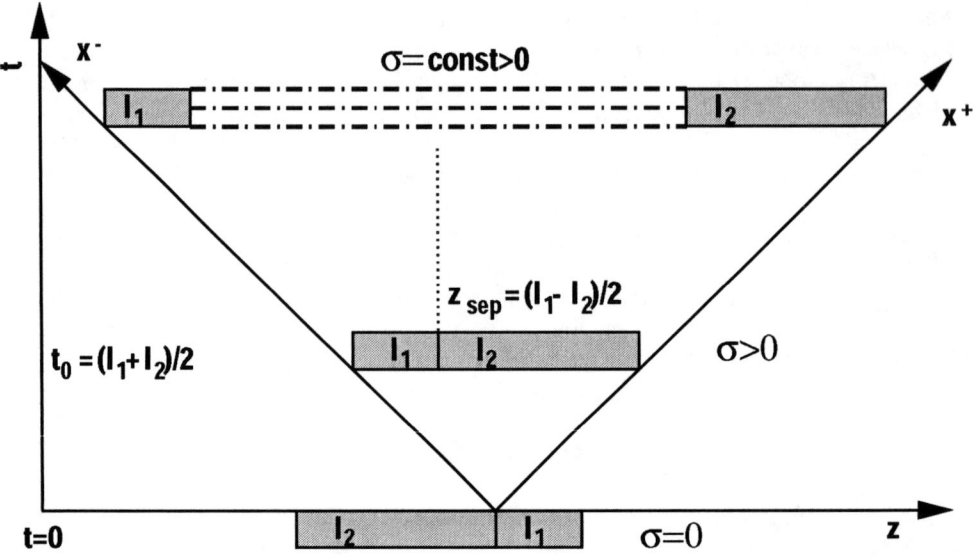

FIGURE 4. Streak-streak collision. The time of first touch of streaks is $t = 0$, and $t = t_0$ corresponds to complete penetration of streaks through each other. At this time strings are created, i.e., the string tension reaches an absolute value $\sigma = A \left(\frac{e_0}{m}\right)^2 n_0 \sqrt{l_1 l_2}$. The dash-dotted lines denote the chromo-electric field strings (it will be shown later that σ is so big that we can talk about several parallel strings or a "string rope"). From [5].

$$F^{\mu\nu} = \begin{pmatrix} 0 & -\sigma \\ \sigma & 0 \end{pmatrix}.$$

In our case the string tensions, σ_i, will have the same absolute value, σ, and opposite sign. σ_i will be constant in the space time region after string creation and before string decay. At the time of first touch of two streaks is no string tension. We assume that strings are created, i.e. the string tension achieves the value, σ, at time, $t = t_0$, corresponding to complete penetration of strings trough each other.

Conservation Laws - String Rope Creation

To know which quantities are really conserved, we can rewrite eq. (1) in the form:

$$\partial_\mu T^{\mu\nu} = \sum_i F_i^{\nu\mu} n_{i\mu} = \sum_i \partial^\mu (A_i^\nu n_{i,\mu}) - A_i^\nu \partial^\mu n_{i,\mu} - \partial^\nu (A_i^\mu n_{i,\mu}) + A_i^\mu \partial^\nu n_{i,\mu}$$

We can define a new energy-momentum tensor $\widetilde{T}^{\mu\nu}$, such that

$$\partial_\mu \widetilde{T}^{\mu\nu} = 0$$
$$\widetilde{T}^{\mu\nu} = \sum_i \widetilde{T}_i^{\mu\nu} + T_{pert}^{\mu\nu} = \sum_i (T_i^{\mu\nu} - A_i^\nu n_i^\mu + g^{\mu\nu} A_i^\alpha n_{i\alpha}) + B g^{\mu\nu}$$

Now the new conserved quantities are

$$Q_0 = \int \widetilde{T}^{00} dV = \sum_i \int_{\Omega_i} \widetilde{T}_i^{00} dV$$
$$Q_3 = \int \widetilde{T}^{03} dV = \sum_i \int_{\Omega_i} \widetilde{T}_i^{03} dV$$

Based on conservation of Q_0, Q_3 we can calculate rapidity, energy and baryon densities at the moment $t = t_0$, when the string with tension σ is created. These new quantities are used as initial condition for our different eqs. (1)

The initial dynamics using the effective string rope model led to an initial state with high energy density, reaching 20 GeV/fm^3 in the on the average and 60-80 Gev/fm^3 peak values in the middle. This coincides with other estimates based on the first flow results at RHIC. Due to the assumption of uniform energy and baryon charge distribution within each string the surface of the initial configuration did cut off to zero sharply at the surface. In the first 3 dimensional fluid dynamical calculations this (not very physical) assumption led to a huge pressure gradient at the surface, which created two sharp forward/backward moving, leading peaks, just as if we would have had a highly transparent collision. Otherwise, the initial state model led to a configuration which was similar to the initial state described in Landau's fluid dynamical model, with the difference that the relatively flat initial state was tilted. In a heavy ion reaction such a tilted initial state arises naturally as a consequence of the streak by streak momentum conservation, since at finite impact parameters streaks with different length and mass collide against each other, finally moving in the direction of the heavier streak.

Such a tilted flat initial state results in a "third flow component" which is observed at the highest SPS energies, and it was present in all FD model results which used QGP EoS.

Thus, we improved the initial state by also considering some expansion during the formation of the initial state. The expansion trajectories of the streak ends is shown in Fig. 5, while the resulting smoother initial energy distribution is shown in Fig. 6

Numerical 3-Dimensional CFD results for RHIC Energies

With the improved initial condition, the PIC-method used by the Bergen - Los Alamos collaboration was also upgraded to increase stability. As due to the extreme Lorentz contraction increasing the cell size was out of question, stability was increased by introducing randomly positioned "marker particles" (Lagrangian Fluid Cells) in the model. This reduced the sensitivity to grid related instabilities to a large extent. We also used a Bag Model EoS, which enabled supercooling, in contract to the earlier tabulated

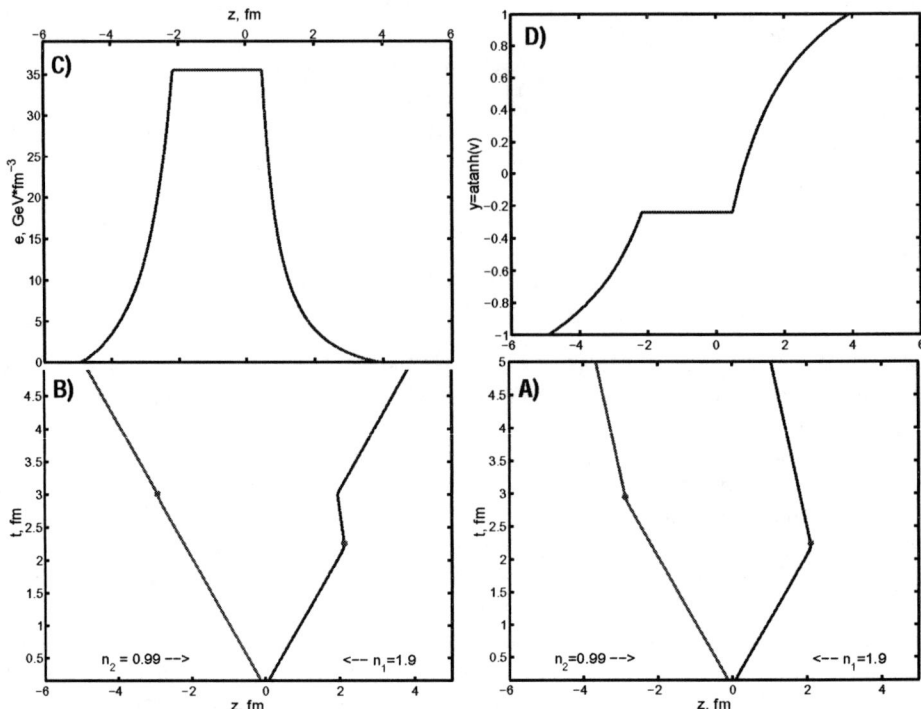

FIGURE 5. A) The typical trajectory of the ends of two initial streaks corresponding to the numbers of nucleons n_1 and n_2, $\varepsilon_0 = 65\ GeV/nucl$. Stars denote the points where $y_i = y_f$. From $t = t_0$ until these stars, the streak ends move according trajectories shown. Then the final streak starts to move like a single object with rapidity y_f, in CRF. B) The same situation as in subplot (A), but for expanding final streak. C) Shows $e(z)$ profiles of an expanding final streak ($t_h = 5\ fm$). We can clearly see three regions - two of forward and backward rarefaction waves and a middle where the initial energy density, e_f, is still preserved. D) Shows the rapidity profile of the expanding final streak ($t_h = 5\ fm$). We can clearly see three regions - two of forward and backward rarefaction waves and a middle where the initial rapidity, is still preserved. From [5].

EoS created by Maxwell Construction which assumed necessarily phase equilibrium. We had no hadronic matter EoS in the model at all, surface effects were neglected, and the Quark matter necessarily supercooled during the final expansion in the FD stage of the reaction.

The Third Flow Component

One of the dominant observables seen in all RHIC experiments as well as in all model calculations is the appearance of the collective "elliptic flow". The observed flow exceeded all previous estimates from string and parton cascade models. Simple 2+1 dimensional hydro models could achieve good agreement with the data by varying the

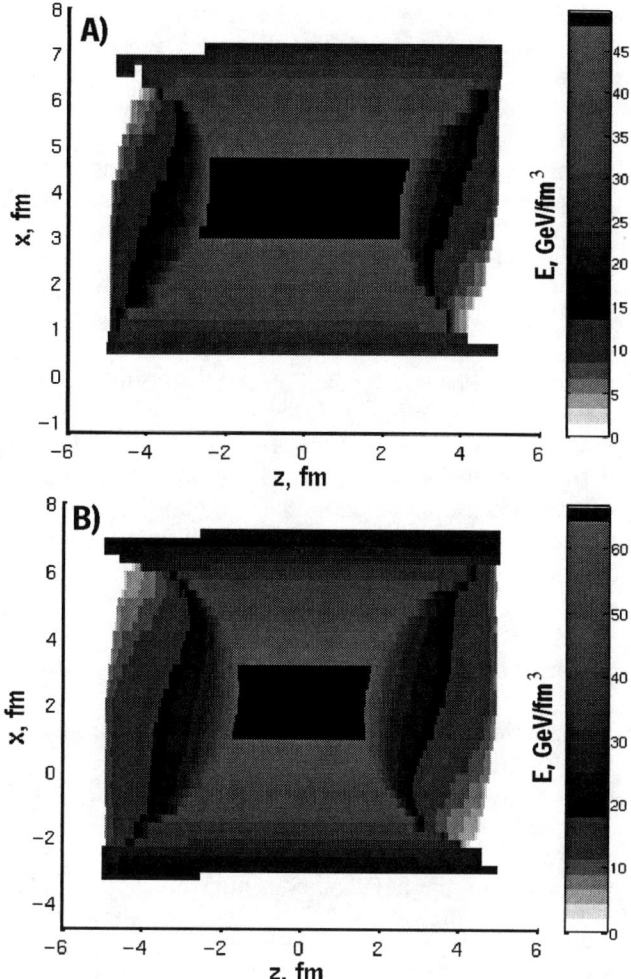

FIGURE 6. Au+Au collision at $\varepsilon_0 = 100\ GeV/nucl$, $E = T^{00}$ is presented in the reaction plane as a function of x and z for $t_h = 5\ fm/c$. Subplot A) $(b = 0.5 \cdot 2 R_{Au})$, subplot B) $(b = 0.25 \cdot 2 R_{Au})$. The QGP volume has a shape of a tilted disk and may produce a third flow component. From [5].

initial configuration and energy density.

Our 3+1 dimensional model was able to describe inherently 2 dimensional effects like the well known and discussed "Directed transverse flow", as well as the "Third flow component" which is also called as "Antiflow". There are actually two physical effects leading to similar but not identical flow patterns.

Shadowing in peripheral collisions leads to an enhanced emission of newly formed particles in the reaction plane, but in the direction opposite to the dominant "directed transverse flow" of the nucleons. This effect is called antiflow and well known for the kaons and in many cases it is also observed for pions. In very peripheral collisions nucleons also show a smaller peak in the same direction.

The reason for the third flow component is different, it does not arise from shadowing, but from a flat compressed and TILTED initial state. This is similar to the mechanism described by the Landau model, with the difference that the flat initial state now expands in the reaction plane orthogonal to the directed flow, while in the original Landau model the final expansion went in the beam direction. This effect is also a QGP signal, because to form a rather flat initial state we need a highly compressible matter like QGP with reduced pressure. When the Third Flow Component was first identified [8], it was also observed that ALL FD model calculations have predicted this effect with QGP EoS, independent of each other. At the same time the same models led to negative results with hadronic matter EoS!

To make accurate quantitative predictions on this and other flow effects we need a reliable third module describing hadronization and freeze out. Results from the mid 90s indicated that freeze out from QGP is a challenging process. Measurements, indicate short lifetime and small source size, while usual phase transition scenarios need long time and large volume increase to accommodate and balance the entropy drop which could be caused by the drop of the number of degrees of freedom when QGP hadronizes and the number of particles may decrease by a number of 2-3!

Simultaneous hadronization and Freeze Out

The problem of possible entropy decrease due to QGP hadronization was first recognized in ref. [14]. As the drop of the number of light degrees of freedom cannot be avoided, there are two ways to avoid this problem.

If one stays during hadronization close to phase equilibrium, letting the system cool gradually one has to wait for a large volume increase until the hadronization completes. According to the estimates on the homogeneous nucleation rates of QGP [15] this can be as much as 35-100 fm/c. SPS and RHIC correlation experiments contradict to this estimate when the experimental source size and life-time is less than 10 fm! Thus this mechanism is not applicable.

If we let the plasma supercool substantially, by 20-40% (!) then a rapid, non-thermal, hadronization is possible, however, in most of the cases the resulting hadron phase would be superheated above the critical temperature (!), so, hadronization would not work. There is, however, a tightly defined region of the phase space where the sudden hadronization is possible with entropy production and without superheating the hadronic matter.[14]

The mechanism cannot be a thermal, quasi-equilibrium process. Pure quantum field theoretical models, however, allow such rapid transitions as it was demonstrated first in ref. [16]. This, hadronization mechanism, gaining more and more popularity with new experimental results: Hadron abundances also show non-equilibrium features indicating rapid hadronization, e.g. large strange anti-baryon abundances.

In the followings we use the assumption of simultaneous hadronization and freeze out, from well supercooled QGP.

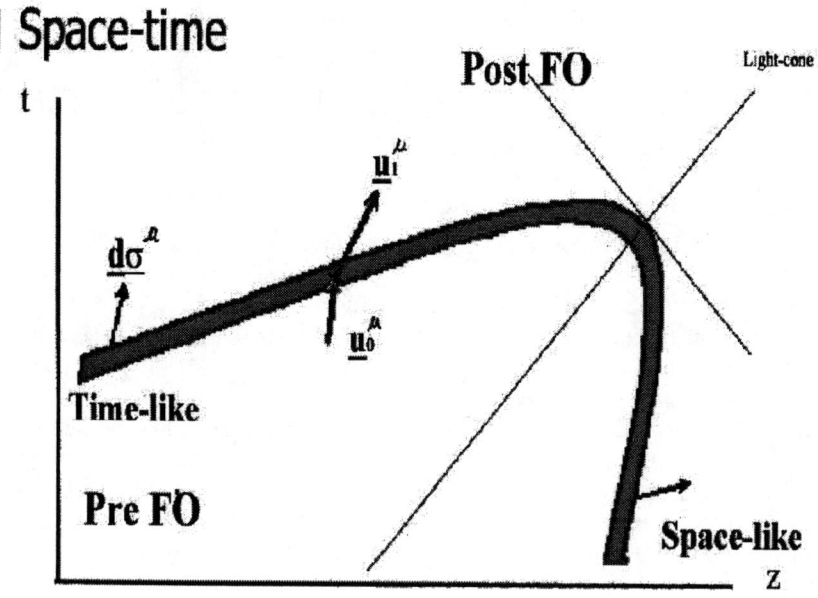

FIGURE 7. The freeze out surface

FREEZE-OUT AND HADRONIZATION

Freeze-out in String Models and the Idealized Freeze-out Layer

In continuum and fluid dynamical (FD) models, particles that leave the system and reach the detectors, can be taken into account via freeze-out (FO) or final break-up schemes, where the frozen-out particles are formed on a 3-dimensional hypersurface (Fig.7). A FO description is important ingredient in evaluation of two-particle correlation data, transverse-, longitudinal-, radial- and cylindrical-flow analyses, transverse momentum and transverse mass spectra, and many other observable.

In fact FO across a hypersurface is a discontinuity, where the EoS of the matter changes suddenly. This surface is an idealization of a layer of finite thickness (of the order of a mean free path or collision time) where the frozen out particles are formed and the interactions in the matter become negligible. The idealized, zero thickness limit of such a layer is called *idealized freeze out surface*.

Kinetic models for hadronic degrees of freedom indicate that such an idealization is meaningful only for collisions of massive heavy ions, such as Au+Au or Pb+Pb [17, 18]. If we include quark-gluon plasma in our reaction model with rapid final hadronization, coinciding with FO [14, 16], the applicability of the idealized surface becomes even better. In the case of FO directly from quark-gluon plasma (QGP) hadronization must always be completed by the end of the FO, as QGP, quarks and gluons never reach the detectors. So we must complete the transition to a single phase even if no thermal and/or mechanical equilibrium is established by the FO.

Nevertheless, the post FO non-equilibrium distribution should be evaluated in an adequate non-equilibrium dynamical model. A realistic and detailed FO model requires a consistent dynamical model which generates such a distribution. Present string and parton cascade models, may be used for this purpose with proper care, if they can handle both pre and post FO phases realistically.

The Cooper-Frye Freeze Out Model

The first attempt to solve the FO problem was made by Landau [19]. He just evaluated the flow velocity distribution at the freeze-out, and this distribution served as a basis of all observable. This approach was used in early FD simulations of heavy ion collisions, but it was not a very good one because the system loses energy in the local random thermal motion, so energy conservation is violated. Milekhin [20, 21] included thermal energy in evaluation of mesurables, adding random thermal velocities to the flow at FO. Although this procedure was still not exact. The FO model was improved later by Cooper and Frye [22] and this method is widely used ever since.

However, the method obtained by Cooper and Frye still did not provide a complete solution to the problem, and some typical mistakes appears in almost all FO calculations [23].

As we know, the FO across a hypersurface is a discontinuity and, the energy-momentum tensor changes discontinuously across this surface. If the flow is not orthogonal to this surface, the four-velocity of the flow will also change [14]. The invariant number of particles (world lines) crossing a surface element $d\sigma^\mu$ is

$$dN = N^\mu d\sigma^\mu \tag{8}$$

and the total number of all particles crossing this surface is

$$N = \int_S N^\mu \, d\sigma_\mu \,. \tag{9}$$

This total number, N, and the total energy and momentum are of course the same at both sides of the freeze out surface.

If we insert the kinetic definition of N^μ

$$N^\mu = \int \frac{d^3p}{p^0} \, p^\mu \, f_{FO}(x,p;T,n,u^\nu) \,,$$

into eq.(8) we obtain the Cooper-Frye formula:

$$E\frac{dN}{d^3p} = \int f_{FO}(x,p;T,n,u^\nu) \, p^\mu d\sigma_\mu \,, \tag{10}$$

where $f_{FO}(x,p;T,n,u^\nu)$ is the post FO phase space distribution of the frozen-out particles, which is not known from FD models. The problem is to choose its form correctly.

Matching Conditions

To evaluate measurable we have to use the parameters of the matter after the discontinuity. If we know the pre freeze out baryon current and energy-momentum tensor, N_0^μ and $T_0^{\mu\nu}$, we can calculate locally, across a surface element of normal vector $d\sigma^\mu$ the post freeze out quantities, N^μ and $T^{\mu\nu}$, from the matching conditions above [24]:

$$[N^\mu \, d\sigma_\mu] = 0 \, , \quad [T^{\mu\nu} \, d\sigma_\mu] = 0 \quad \text{and} \quad [S^\mu \, d\sigma_\mu] \geq 0. \tag{11}$$

Here $[A] \equiv A - A_0$. The first two equations are the conservation laws for N^μ and $T^{\mu\nu}$. The third one tell us that entropy always increase, where S^μ is the entropy 4-curent.

Post Freeze Out Distribution from Matching Conditions

Is usual assume that the pre FO momentum distribution is a local thermal one; boosted by the local collective flow velocity. Nevertheless, the post freeze out distribution need not be a thermal equilibrium distribution.

For time-like freeze out surfaces, let us assume a thermal distribution (Jüttner or the equivalent for relativistic quantum ideal gases) does not lead to any problem, because both p^μ and $d\sigma^\mu$ are time-like unit vectors, so $p^\mu d\sigma_\mu > 0$. As a consequence the integrand in eq.(10) is positive on both sides of the front, permitting its application.

Of course, the parameters T, n, u^μ, of the f_{FO} distribution, must be determined from the first two eq. of (11). The third eq. from (11) have to be used just to guarantee reliable solutions [24]. This important condition was missed in the original work [22] and in many applications since, leading to violation of conservation laws (unless the pre and post freeze out EOSs were identical).

In the space-like FO surface case the situation is more complicated. Here p^μ is time-like and $d\sigma^\mu$ is space-like, thus $p^\mu d\sigma_\mu$ can be positive or negative, and $d\sigma^\mu$ may point now both in post- and pre- FO directions. Thus, the integrand in eq.(10) may change sign, indicating that the distribution has contributions back and out of the front.

On the pre FO side particles may move in any direction and may rescatter on each other. However, on the post freeze out side, we do not allow re- and back- scattering any more. In other words a particle cannot scatter back if it has passed the FO front. So the post-FO distribution should have the form [24]:

$$f_{FO}^*(x,p;T,n,u^\nu,d\sigma^\mu) = f_{FO}(x,p;T,n,u^\nu)\Theta(p^\mu \, d\sigma_\mu) \tag{12}$$

where $\Theta(p^\mu \, d\sigma_\mu)$ is the step function. Thus, f_{FO}^* cannot be a Jüttner or other ideal gas distribution. Nevertheless, conservation laws have to be satisfied, even in this case.

Physically, space-like freeze out surfaces have a thickness and internal structure, which ensures that local thermal equilibrium is maintained on its pre FO side – involving back scattering from the front to the pre FO side – creating at the same time a momentum distribution on post FO side, with no particles moving in the front direction.

Taub Adiabat and Rayleigh Line for Arbitrary Hyper-surfaces

At this time we will present a simple FD solution for shock and deflagration waves, following Refs.[25, 1]. The shock can be approximated by a one-dimensional problem. Let us denote the two sides by "1" and "2", and the difference of a quantity, Q, by $[Q] = Q_2 - Q_1$

If the front is infinitely sharp (Eulerian FD) we have to consider a sharp surface where the flow variables are discontinuous. The surface of this discontinuity has a unit normal vector, Λ^μ, in the space-time. It can be space-like ($\Lambda^\mu \Lambda_\mu = -1$) or time-like ($\Lambda^\mu \Lambda_\mu = +1$).

Orthogonal to the front the energy and momentum flow should be identical on the two sides:

$$[T^{\mu\nu}\Lambda_\nu] = 0, \tag{13}$$

and the particle number should also be conserved:

$$[N^\mu \Lambda_\mu] = 0. \tag{14}$$

Eqs. (13) and (14) are the relativistic *Rankine-Hugoniot equations*, first obtained by A. Taub [26] (but only for space-like surfaces). The full derivation for both time-like and space-like surfaces was published nearly 40 years later [25], and in the meantime shocks and discontinuities across surfaces with time-like normals were considered unphysical.

The shock front (usually) propagates with $v_{shock} < 1$, i.e., the points of the front are in causal connection. However, this is not necessary, and $v_{shock} > 1$ is possible, because the discontinuity does not move with the matter, and does not carry information.

One can eliminate the velocity from eqs.(13) and (14) obtaining a scalar equation connecting the thermodynamical quantities on the two sides of the shock. The resulting equation is called the *shock or detonation adiabat*.

Let us use Eckart's definition. We can define

$$N^\mu = nu^\mu \tag{15}$$

where u^μ is the 4-velocity and n is the local density. Let us define the particle current across the surface:

$$j \equiv N^\mu \Lambda_\mu. \tag{16}$$

j is an invariant scalar, and according to eq.(14) it has the same value on both sides of the shock. It can also be written as $j \equiv nu^\mu \Lambda_\mu$ because outside the front there is no dissipation.

In the Local Rest frame (LR) of the front we have $\Lambda_{\mu\,LR} = (1, \vec{0})$ for time-like, and $\Lambda_{\mu\,LR} = (0, \vec{e})$ for space-like surfaces.

Eq.(13) is a 4-vector equation. To end up with the shock adiabat which is an equation connecting thermodynamical invariant quantities. We need two scalar equations projected out of eq.(13). So, we will need two projections: (i) Parallel projection to the surface (ii) Orthogonal projection to the surface

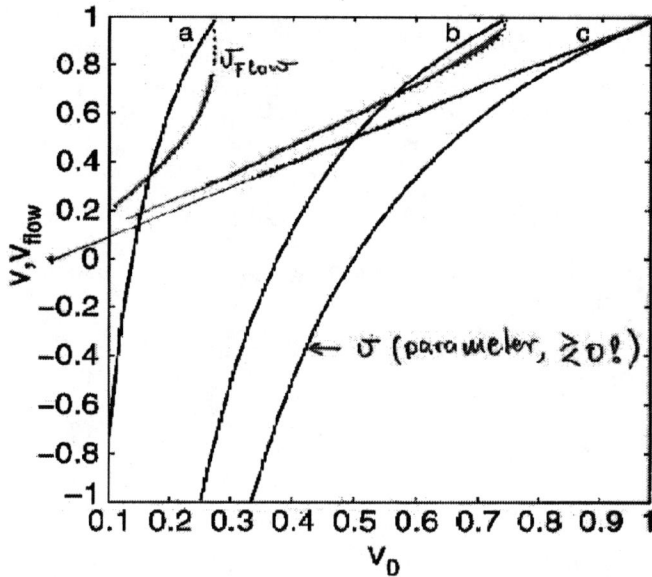

FIGURE 8. The change of flow velocity during freeze out. From [24].

1 Parallel projection.. Let us express the component of the conserved energy-momentum and baryon current, orthogonal to the surface, (parallel to its normal):

$$[T^{\mu\nu}\Lambda_\nu] = 0 \to [T^{\mu\nu}\Lambda_\nu\Lambda_\mu] = 0 \to [wu^\mu u^\nu \Lambda_\mu \Lambda_\nu - P\Lambda^\mu \Lambda_\mu] = 0 \quad (17)$$

where we used

$$T^{\mu\nu} = wu^\mu u^\nu - Pg^{\mu\nu} \quad (18)$$

and $w \equiv e + P$. Here e is the energy density, P is the pressure and $g^{\mu\nu}$ is the metric tensor. Using the alternative form of eq. (16) and inserting this into the eq. (17) we obtain:

$$\left[\frac{w}{n^2}j^2 - P\Lambda^\mu\Lambda_\mu\right] = 0 \to \left[\frac{w}{n^2}\right]j^2 - [P](\Lambda^\mu\Lambda_\mu) = 0.$$

This leads then to the equation of the *Rayleigh-line*:

$$j^2 = \frac{[P](\Lambda^\mu\Lambda_\mu)}{\vdots}w[X], \quad (19)$$

where $X \equiv \frac{w}{n^2}$, is the *generalized specific volume*. In the non-relativistic limit $X \longrightarrow m_0 V_{spec.}$. The $[P,X]$ plane corresponds to the $[P,V]$ plane in the non-relativistic limit. The equation of the Rayleigh line is a straight line on the $[P,X]$ plane. It gives the locus of the possible final state coordinates, P_2's and X_2's, if the initial state "1" is given. The slope of the Rayleigh line is given by the current across the front j. In heavy ion reactions j usually increases if the beam energy increases.

2 Orthogonal projection.. From the orthogonal projection of the 4-vector $T^{\mu\nu}\Lambda_\nu$ with the projector $\Delta_\Lambda^{\mu\nu} = g^{\mu\nu} - \frac{\Lambda^\mu \Lambda^\nu}{(\Lambda^\mu \Lambda_\mu)}$ we get a 4-vector tangent to the plane: $G^\tau \equiv T_{\mu\nu}\Lambda^\mu \Delta^{\nu\tau}$, and then a scalar equation by taking the norm $[G^\tau G_\tau] = 0$. Substituting equations (15),(16) and (18) on G^τ we obtain with a little algebra:

$$G^\mu = \frac{w}{n} j u^\mu - \frac{w}{n^2} j^2 \frac{\Lambda^\mu}{(\Lambda_\sigma \Lambda^\sigma)}.$$

Now the scalar equation $[G^\mu G_\mu] = 0$ can be solved. One of the cross terms and the last term cancel each other, so

$$\left[\frac{w^2}{n^2} j^2 u^\mu u_\mu - \frac{w^2}{n^4} j^4 \frac{\Lambda^\mu \Lambda_\mu}{(\Lambda_\sigma \Lambda^\sigma)^2}\right] = 0 \rightarrow \left[\frac{w^2}{n^2}\right] - [X^2] j^2 (\Lambda^\mu \Lambda_\mu) = 0.$$

This leads to the equation

$$j^2 = \frac{[wX]}{[X^2](\Lambda^\mu \Lambda_\mu)}. \tag{20}$$

Comparing eqs. (19) and (20) leads to the equation of *Taub adiabat*:

$$[P] = \frac{[wX]}{(X_2 + X_1)}. \tag{21}$$

Equation 21 is also called shock adiabat or relativistic Rankine-Hugoniot adiabat. The locus of the possible final states, "2", lies on the Taub adiabat.

If the initial state and the EoS of the final state are known the Taub adiabat with Rayleigh line determine the final state. If the final state is out of equilibrium, i.e., not a perfect fluid, this is not applicable.

As an example for the Taub adiabat, we can obtain it from the EoS of QCD plasma [1]. This is given by the equation:

$$(P + \frac{4}{3}B)(X - \frac{1}{3}X_0) = \frac{1}{3}X_0(w_0 - \frac{4}{3}B), \tag{22}$$

where B is the Bag Model's constant, and the index '0' represents the initial states. Another example can be shown obtained from the EOS of an ideal gas

$$(P + \frac{2}{3}P_0)(X - \frac{2}{3}X_0) = \frac{2}{3}X_0(e_0 - \frac{2}{3}P_0). \tag{23}$$

These two examples lead to adiabats represented hyperbolas.

Particles which Do Not Freeze Out: the Cut Jüttner Distribution

Returning to the problem proposed in section 3, the construction of the post freeze out distribution, f_{FO}^*, is an additional problem in the case of space-like freeze out fronts.

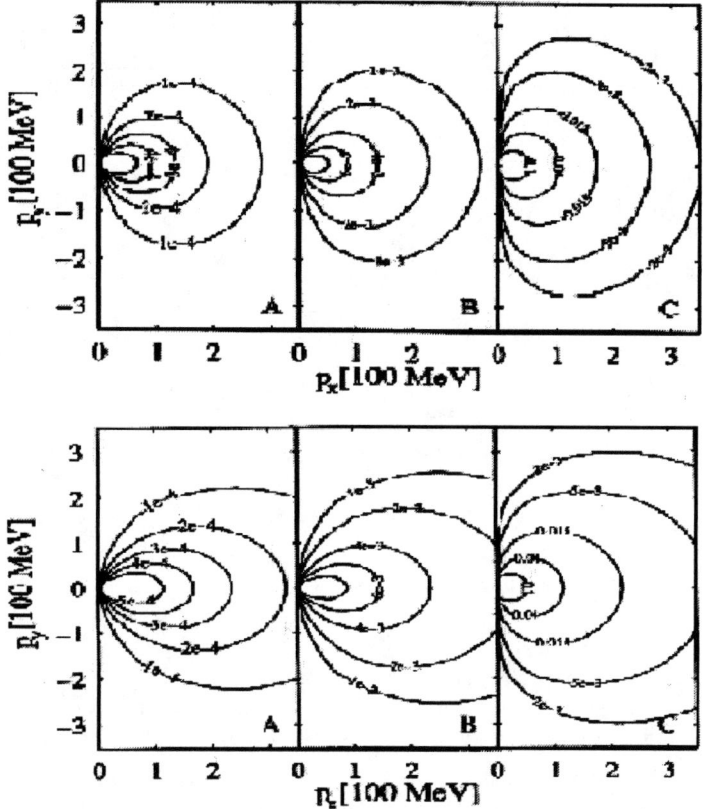

FIGURE 9. Post freeze out phase space distribution from kinetic theory. From [29].

We can calculate f_{FO}^* in appropriate transport theoretical approach [17, 18, 27, 28], and its parameters must be determined from conservation laws, eq. from (11). The obtained form of the distribution will, of course, be dependent on the features of the kinetic theory or model.

A simplified approach is to assume that f_{FO}^* from eq.(12) is a *cut Jüttner distribution*. Baryon current and energy momentum tensor can be evaluated in the rest frame of the non-cut, symmetrical Jüttner distribution (RFG) [30] moving with 4-velocity u_s^μ, which is not the same from the LR of f_{FO}^*. Of course, baryon 4-current and the energy-momentum tensor may then be Lorentz transformed into the LR of the post FO matter, which moves with $u^\mu = N^\mu/(N^\nu N_\nu)^{1/2}$ in the RFG, or into the LR frame of the freeze out front, where $d\sigma^\mu = (0,1,0,0)$ (see Ref.[23])

Now N^μ and $T^{\mu\nu}$, can be transformed to the frame where we want to evaluate the conservation laws, from eq.(11), and the parameters of the post FO, cut Jüttner distribution can be determined, satisfying the conservation laws.

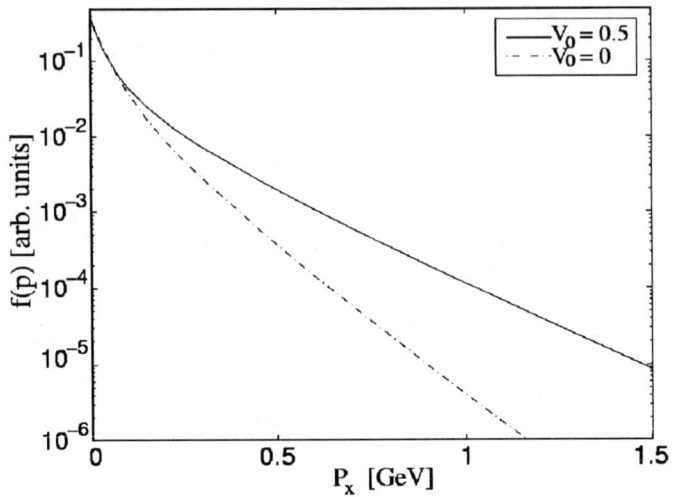

FIGURE 10. P_t distribution from kinetic freeze out model with rescattering. (Reprinted from Ref. [32], Copyright 1999, with permission from Elsevier Science.)

Finally (i) if we know the 5 parameters of the pre FO flow, and (ii) the local freeze out surface from kinetic considerations, then assuming that the post FO distribution $f^*_{FO}(\vec{p},x)$, is a cut Jüttner type, we can completely determine the parameters of the post FO matter from eq.(11).

Freeze Out from Kinetic Models

In this section we are going to demonstrate the kinetic model for a drastically oversimplified situation of a plane FO surface [31], that can reproduce a cut Jüttner distribution as a post-FO one.

Let us assume an infinitely long tube with its left half ($x < 0$) filed with nuclear matter; in the right vacuum is maintained. We can remove the dividing wall at $t = 0$, and then the mater will expand into the vacuum. By continuously removing particles at the right end of the tube and supplying particles on he left end, we can establish a stationary flow, where the particles will gradually freeze-out in an exponential rarefaction wave propagating to the left. We can describe it from the reference frame of the front (RFF). In this frame, we have a stationary supply of equilibrated matter from the left, and a stationary rarefaction front on the right, $x > 0$.

Now we assume that we have two components of our momentum distribution ($f_{free}(x,\vec{p})$ and $f_{int}(x,\vec{p})$), $f_{free}(x=0,\vec{p}) = 0$ and f_{int} is an ideal Jüttner distribution; Then f_{int} gradually disappears and f_{free} gradually increases as $x \to \infty$, according to the

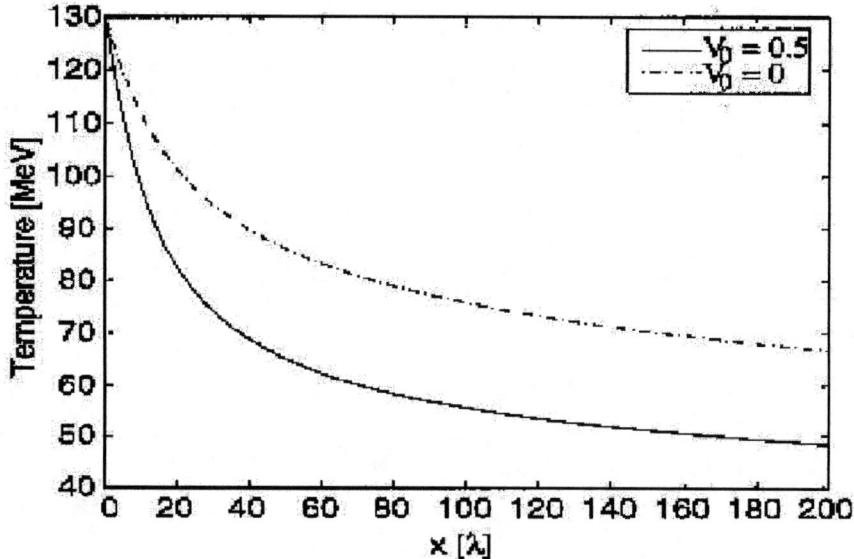

FIGURE 11. Change of temperature of the interacting matter component in gradual freeze out from a kinetic model with rescattering. From [29].

coupled differential equations:

$$\partial_x f_{int}(x,\vec{p})dx = -\Theta(p^\mu d_\sigma \mu) \frac{cos\theta_{\vec{p}}}{\lambda} f_{free}(x,\vec{p})dx$$

$$\partial_x f_{free}(x,\vec{p})dx = +\Theta(p^\mu d_\sigma \mu) \frac{cos\theta_{\vec{p}}}{\lambda} f_{free}(x,\vec{p})dx \quad (24)$$

where $cos\theta_{\vec{p}} = \frac{p^x}{p}$ in the RFF. It expresses the fact that particles with momenta orthogonal to the FO surface leave the system with bigger probability than particles emitted at an angle. This model allow cut Jüttner distribution as a post FO one but it does not allow the complete FO because f_{int} survives even if $x \to \infty$.

To improve our model we take into account re-scattering within the f_{int}, leading to re-thermalization and re-equilibration f_{int}. To simplify the dynamical description, we use relaxation time approximation. Thus the two component of f_{FO}^* now develop according to the differential equations

$$\partial_x f_{int}(x,\vec{p})dx = -\Theta(p^\mu d_\sigma \mu) \frac{cos\theta_{\vec{p}}}{\lambda} f_{free}(x,\vec{p})dx + [f_{eq}(x,\vec{p}) - f_{int}(x,\vec{p})] \frac{1}{\lambda'} dx \quad (25)$$

$$\partial_x f_{free}(x,\vec{p})dx = +\Theta(p^\mu d_\sigma \mu) \frac{cos\theta_{\vec{p}}}{\lambda} f_{free}(x,\vec{p})dx \quad (26)$$

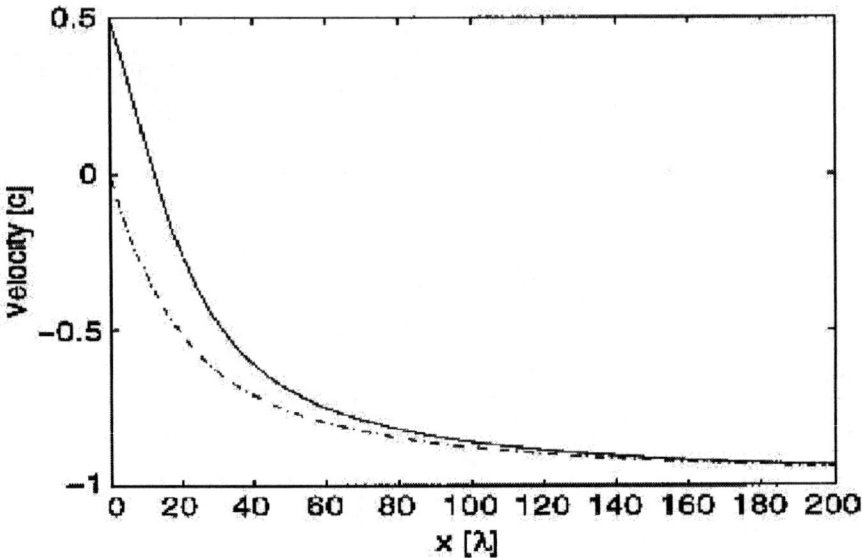

FIGURE 12. Change of velocity parameter of the interacting matter component during the gradual freeze out in a kinetic freeze out model with rescattering. From [33], reprinted with permission of Akadémiai Kiadó.

The interacting component of the momentum distribution, described by eq(25), shows the tendency to approach an equilibrated distribution with a relaxation length λ'. The f_{eq} distribution is not the same as the initial Jüttner distribution, due to energy, momentum and particle drain. However its parameters $n_{eq}(x), T_{eq}(x)$ and $u_{eq}^\mu(x)$, changes as required by conservation laws.

Let us assume that $\lambda' \ll \lambda$, i.e. re-thermalization is much faster than particles that FO or much faster than changes in $n_{eq}(x), T_{eq}(x)$ and $u_{eq}^\mu(x)$. So $f_{int} \approx f_{eq}$. For f_{eq} we assume *the spherical Jüttner* form at any x. In this case the changes of conserved quantities due to particle drain or transfer can be evaluated for an infinitesimal dx and new parameters of $f_{eq}(x+dx,\vec{p})$ can be found.

It is important to point out that, although for the spherical Jüttner distribution Landau and Eckart flow velocities are the same, the changes calculated from the loss of baryon and energy current, appear to be different $du_{i,E,RFG}^\mu(x) \neq du_{i,L,RFG}^\mu(x)$. This is clearly a consequence of the asymmetry caused by the FO process [29, 24]. Above we present some figures of FO nonthermal distribution, from our kinetic model.

SUMMARY

In conclusion we can state that hydrodynamic modeling of heavy ion reactions is alive and is better than ever. Clear hydrodynamic effects are seen everywhere, and from early on.

This indicates we are approaching a regime where collective matter type of behavior is dominant. We hope to gain more and more detailed information on QGP and its dynamical properties. Continued hard work is needed to exploit all possibilities, and the task of theoretical modeling and analysis is vital in future progress of the field.

ACKNOWLEDGMENTS

The author thanks the organizers for the kind hospitality and personal attention during the school. The large number of really motivated students was very positive and made this school really special.

REFERENCES

1. L.P. Csernai: *Introduction to Relativistic Heavy Ion Collisions*, Willey (1994).
2. L.P. Csernai, J.I. Kapusta, *Phys. Rev.* **D29** 2664. (1984).
3. L.P. Csernai, J.I. Kapusta, *Phys. Rev.* **D31** 2795. (1985).
4. V.K. Magas, L.P. Csernai, D.D. Strottman, *Phys. Rev.* **C64** 014901, (hep-ph/0010307) (2001).
5. V.K. Magas, L.P. Csernai, D. Strottman, Submitted to *Physical Review* C , (hep-ph/0202085) (2002).
6. T. Kodama: PASI Lectures in this proceedings, (2002).
7. L.D. Landau and E.M. Lifshitz: Hydrodynamics (Nauka, Moscow), Chapter 3. (1953).
8. L.P. Csernai and D. Röhrich, *Phys. Lett.* **B458** 454. (nucl-th/9908034) (1999).
9. T.S. Biró, H.B. Nielsen, J. Knoll, *Nucl. Phys.* B **245** 449. (1984).
10. H. Sorge, *Phys. Rev.* C **52** 3291. (1995).
11. K. Wegner, J. Aichelin, *Phys. Rev. Lett.* **76** 1027. (1996).
12. N.S. Amelin, M.A. Braun, C. Pajares, *Phys. Lett.* **B 306** 312 (1993), *Z. Phys.* **C 63** 507. (1993).
13. M. Gyulassy, L.P. Csernai: *Nucl. Phys.* **A 460** 723 (1986).
14. T. Csörgő and L.P. Csernai, *Phys. Lett.* **B333** 494. (hep-ph/9406365), (1994).
15. L.P. Csernai and J.I. Kapusta, *Phys. Rev.* **D46** 1379 (1992); and L.P. Csernai and J.I. Kapusta, *Phys. Rev. Lett.* **69** 737. (1992).
16. L.P. Csernai and I.N. Mishustin, *Phys. Rev. Lett.* **74** 5005. (1995).
17. L.V. Bravina, I.N. Mishustin, N.S. Amelin, J.P. Bondorf and L.P. Csernai, *Phys. Lett.* **B** 196. (1995).
18. L.V. Bravina, et al., *Heavy Ion Phys.* **5** 445 (1997).
19. L.D. Landau, Izv. Akad. Nauk SSSR **17** (1953).
20. G. Milekhin, Zh. Eksp. Teor. Fiz. **35** 1185 (1958); Sov. Phys.-JETP **35** 829 (1959).
21. G. Milekhin, Trudy FIAN **16** 51 (1961).
22. F. Cooper and G. Frye, Phys. Rev. **D 10** 186. (1974).
23. L.P. Csernai, Zs. Lázár and D. Molnár, *Heavy Ion Phys.* **5** 467. (1997).
24. Cs.Anderlik, L.P.Csernai, F.Grassi, W.Greiner, Y.Hama, T.Kodama, Zs.Lázár,V. Magas and H. Stöcker, *Phys. Rev.* **C59** 3309. (nucl-th/9806004) (1999).
25. L.P. Csernai, Sov. JETP **65** 216 (1987); Zh. Eksp. Theor. Fiz. **92** 379. (1987).
26. A.H. Taub, Phys. Rev. **74** 328. (1948).
27. F. Grassi, Y. Hama and T. Kodama, *Phys. Lett.* **B 355** 9. (1995).
28. F. Grassi, Y. Hama and T. Kodama, *Z. Phys.* **C 73** 153. (1996).
29. Cs. Anderlik, Z.I. Lázár, V.K. Magas, L.P. Csernai, H. Stöcker and W. Greiner, *Phys. Rev.*, **C59** 388, (nucl-th/9808024) (1999).
30. K.A. Bugaev, *Nucl. Phys.* **A606** 559. (1996).
31. V.K. Magas, Cs. Anderlik, L.P. Csernai, F. Grassi, W. Greiner, Y. Hama, T. Kodama, Zs. Lázár and H. Stöcker, *Nucl. Phys.* **A661** 596. (nucl-th/0001049) (1999).
32. V.K. Magas, et al. *Phys. Lett.* B **459** 33, (nucl-th/9905054); (1999).
33. V.K. Magas, Cs. Anderlik, L.P. Csernai, F. Grassi, W. Greiner, Y. Hama, T. Kodama, Zs. Lázár and H. Stöcker, *Heavy Ion Phys.* **9** 193. (nucl-th/9903045) (1999).

Non-equilibrium Hadrochemistry in QGP Hadronization

Johann Rafelski[*] and Jean Letessier[†]

[*]*Department of Physics, University of Arizona, Tucson, AZ, 85721*
[†]*Laboratoire de Physique Théorique et Hautes Energies
Université Paris 7, 2 place Jussieu, F–75251 Cedex 05*

Abstract. This survey offers an introductory tutorial for students of any age of the currently thriving field of hadrochemistry. We discuss the different chemical potentials, how the hadronic phase space is described and how one evaluates the abundance of hadrons at time of hadronization. We show that a rather accurate description of experimental data arises and we present results of fits to hadron yields at SPS and RHIC. We show that introduction of chemical non-equilibrium originating in a sudden hadronization of a QGP is favored strongly at SPS and is presently also emerging at RHIC. The low chemical freeze-out temperatures are consistent with the picture of single freeze-out scenario (chemical and thermal freeze-out coincide).

1. THERMAL EQUILIBRIUM

Hadronic interactions are strong, collisions are frequent, and along with Hagedorn [1, 2, 3], we expect formation of statistical equilibrium conditions. The formation of a space–time-localized fireball of dense matter is the key physical process occurring in high energy nuclear collisions. This said, the question is how such a fireball can possibly arise from a rather short sequence of individual reactions that occur when two, rather small, gas clouds of partons, clustered in nucleons, bound in the nucleus, collide? Indeed, at first sight, one would be led to believe that the small clouds comprising point-like objects would mutually disperse in the collision, and no localized, dense state of hadronic matter should be formed. It was suggested in some early work, that the two colliding 'eggs' should emerge from the high-energy interaction slightly 'warmed', but still largely 'unbroken'. In past few years some of our colleagues have made these eggs not only hot but also scrambled: they believe that somehow this hadronic fireball also develops a high degree of not only (kinetic) thermal, but also chemical equilibration. What does this mean precisely, and can this be true, is what we are going to study in depth here.

Two remarkable properties of hadronic interactions are responsible for deeply inelastic behavior, which could lead to localization of energy in space–time:

- the multiparticle production in hadron–hadron collisions; and
- the effective size of all hadrons expressed in term of their reaction cross sections.

What appears to be a thin system of point-like constituents is effectively already a volume-filling nucleon liquid, which will undergo, in a collision, a rapid self-multiplication with particle density rising and individual scattering times becoming progressively much shorter than the overall collision time. In historical context, the abundant particle production seen in high energy cosmic ray interactions, has lead Fermi to propose the statistical model of particle production [4]. Our present developments are a natural elaboration of this seminal work, toward a diversified field of many flavored hadron yields, allowing a precision level that a one parameter Fermi-model could of course not reach. Our objective is to account for rapid chemical non-equilibrium dynamics related to formation and sudden disintegration of a new phase of matter.

As the energy available in the collision is increased, the hadron particle/energy density will reach values at which the dissolution of the hadronic constituents into a common deconfined domain will become possible, and indeed must occur according to our knowledge about strong interactions. We do not really know whether deconfinement of hadrons is operating already at AGS energies [5], but there is today no experimental evidence that this low energy range suffices. In contradistinction, a significant number of results obtained at the top SPS energy range can be most naturally interpreted in terms of the formation of a deconfined space–time domain, and this becomes a much more convincing situation once RHIC results are considered. We note that, per participant, there are as many as 7–10 further hadrons produced at SPS energies. This implies that there are thousands of quarks and gluons in the space–time domain of interest, and hence consideration of a 'local' (in space–time) equilibrium makes good sense. However, we need to establish which equilibrium is reached, and if not fully, how close are we, and what the deviations from equilibrium tell us about the physics questions we are studying.

In order to establish locally in space and time thermal equilibrium, a rapid equipartition of energy among the different particles present has to occur. Thermal equilibrium can be achieved in principle solely by elastic scattering. Introduction of a (local) temperature T presupposes that thermal equilibrium has nearly been established. What is the mechanism for the establishment of kinetic momentum distribution equilibrium? The thermalization of the momentum distributions is driven by *all* scattering processes, elastic as well as inelastic, because all of them are associated with exchange of momentum and energy between particles. The *scattering time*, for particles of species i, is given in terms of the reaction cross section σ_{ij} with particle species j,

$$\tau_{i,\text{scatt}} = \frac{1}{\sum_j \langle \sigma_{ij} v_{ij} \rangle \rho_j}. \tag{1}$$

The sum in the denominator is over all available particle species with densities ρ_j, v_{ij} are the relative velocities, and the average is to be taken over the momentum distributions of the particle considered.

It is not hard to 'guesstimate' the time scale governing the kinetic equilibration in the QGP. The typical particle-collision time (the inverse of the collision frequency) is obtained from Eq. (1) above. Given the particle densities and soft reaction cross sections, with the relative velocity of these essentially massless components being the velocity of light c, we find for the quark–gluon plasma scattering time,

$$\tau_i^{\text{QGP}} = 0.2\text{–}2\,\text{fm}, \quad \text{with } \rho_i = 2\text{–}10\,\text{fm}^{-3},\ \sigma_i = 2\text{–}5\,\text{mb}, \tag{2}$$

as a range for different particles of type i, with the shorter time applying to the early high-density stage. This is about an order of magnitude shorter than the time scale for evolution of the fireball, which is derived from the spatial size of the colliding system: for the largest nuclei, in particular the Pb–Pb or Au–Au collisions, over a wide range of energy, we expect

$$\tau^{\exp} \simeq \frac{R_A}{c} \simeq 5\text{–}8\,\text{fm}/c. \tag{3}$$

The achievement of kinetic equilibrium must be visible in the thermal energy spectra, and this behavior, as we argued above, can be understood in qualitative terms using kinetic scattering theory for the case of nuclear collisions. However, it remains to date a mystery why in some important aspects thermal models succeed for the case of p–p reactions. In particular, the exponential fall off of particle spectra, suggesting thermal equilibrium, has been noted with trepidation for a considerable time.

Hagedorn evaluated this behavior in the experimental data some 35 years ago [1, 2, 3] and he developed the statistical bootstrap model which assumes a statistical phase-space distribution. Hagedorn called it *preestablished or preformed equilibrium*: particles are produced in an elementary interaction with a probability characterized by a universal temperature. We can today only speculate about the physical mechanisms.

In strong interaction physics it is possible that vacuum fluctuations interfere with particle production processes and could generate a preestablished thermal equilibrium distribution. In this case very little additional rescattering is needed for the development of thermal equilibrium. A possible mechanism is that the production of quark pairs by snapping strings is subject to a stochastic vacuum fluctuation force [6], which results in natural Boltzmann distribution of produced quark pairs at just the Hagedorn temperature, $T_H = 160$ MeV. Another informally discussed possibility is the presence of intrinsic chaotic dynamics capable of rapidly establishing kinetic equilibrium.

Sometimes, the fact that we do not fully understand thermalization in the p–p case is raised as an argument against the possibility of conventional equilibration in nuclear collisions. We do not think so. In fact, if the p–p case leads to thermal hadrons, we should have a yet better thermalization in the A–A case. Thus, a microscopic model that is adopted to extrapolate from p–p to A–A collisions should incorporate the concept of the hadronic preestablished thermal equilibrium, else it is not going to be fully successful.

The following stages are today believed to occur in heavy-ion collision:

1. The initial quantum stage.
   The formation of a thermalized state within τ_{th} is most difficult to understand, and is subject to intense current theoretical investigation. During the pre-thermal time, $0 \le t < \tau_{th}$, the properties of the collision system require the study both of quantum transport and of decoherence phenomena, a subject reaching far beyond the scope of this article. In this discussion, we assume that the thermal shape of a (quark, gluon) particle-momentum distribution is reached instantaneously compared with the time scales for chemical equilibration, see section 2. This allows us to sidestep questions regarding the dynamics occurring in the first moments of the heavy-ion interactions, and we explore primarily what happens after a time $\tau_0 \equiv \tau_{th} \simeq 0.25$–

1 fm/c. The value of τ_0 decreases as the density of the pre-thermal initial state increases, e.g., as the collision energy increases.

2. The subsequent chemical equilibration time.
   During the inter-penetration of the projectile and the target lasting no less than ~ 1.5 fm/c, diverse particle-production reactions occur, allowing the approach to chemical equilibrium by light non-strange quarks $q = u, d$. As the energy is redistributed among an increasing number of accessed degrees of freedom, the temperature drops rapidly.

3. The strangeness chemical equilibration.
   A third time period, lasting up to $\simeq 5$ fm/c, during which the production and chemical equilibration of strange quarks takes place. There is a reduction of temperature now mainly due to the expansion flow, though the excitation of the strange quark degree of freedom also introduces a non-negligible cooling effect.

4. The hadronization/freeze-out.
   The fireball of dense matter expands and decomposes into the final state hadrons, possibly in an (explosive) process that does not allow re-equilibration of the final-state particles. The dynamics is strongly dependent on the size of the initial state and on the nature of the equations of state.

Throughout these stages, a local thermal equilibrium is rapidly established and, as noted, the local temperature evolves in time to accommodate change in the internal structure and size, as is appropriate for an isolated physical system. We have a temperature that passes through the following nearly separable series of stages:

T_{th} the temperature associated with the initial thermal equilibrium,
↓ *evolution dominated mainly by production of q and $\bar{q}$*;
T_{ch} chemical equilibrium of non-strange quarks and gluons,
↓ *evolution dominated by expansion and production of s and $\bar{s}$*;
T_{s} condition of chemical equilibrium of u, d and s quark flavors,
↓ *expansion, dissociation by particle radiation*;
T_{f} temperature at hadron-abundance freeze-out,
↓ *hadron rescattering, reequilibration*; and
T_{tf} temperature at thermal freeze-out, $T_{\text{tf}} = T(\tau^{\text{exp}})$.

We encounter a considerable decrease in temperature. The entropy content of an evolving isolated system must increase, and this is initially related to the increase in the number of particles within the fireball and later also due to the increase in volume. However, in the later stages dominated by flow, the practical absence of viscosities in the quark–gluon fluid implies that there is little additional production of entropy. The final entropy content is close to the entropy content established in the earliest thermal stage of the collision at $t < \tau_0$, despite a drop in temperature by as much as a factor of two under current experimental RHIC conditions during the sequel evolution of the fireball.

Except for the unlikely scenario of a fireball not expanding, but suddenly disintegrating in the early stage, none of the temperatures discussed above corresponds to the temperature one would read off the (inverse) slopes of particle spectra. In principle, the freeze-out temperature determines the shape of emission and multiplicity of emitted particles. However, the freeze-out occurs within a local flow field of expanding matter and

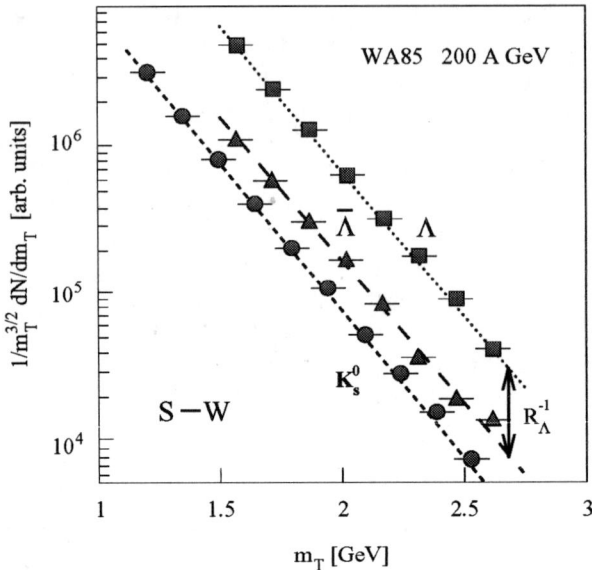

FIGURE 1. Strange particle spectra for Λ, $\overline{\Lambda}$, and K_S [7]. The line connecting the Λ and $\overline{\Lambda}$ spectra, denoted R_Λ^{-1}, shows how at fixed $m_\perp$ the ratio R_Λ of abundances of these particles can be extracted. Experimental WA85 results at $200A$ GeV [8, 9, 10].

the thermal spectrum is to be folded with the flow which imposes a Doppler-like shift of T_{tf}: we observe a higher temperature than is actually locally present when particles decouple from flowing matter (kinetic or thermal freeze-out). The observable temperature $T_\perp$ is related to the intrinsic temperature of the source:

$$T_\perp \simeq \frac{1+\vec{n}\cdot\vec{v}_{tf}}{\sqrt{1-\vec{v}_{tf}^2}} T_{tf} \rightarrow \sqrt{\frac{1+v_{tf}}{1-v_{tf}}} T_{tf}. \quad (4)$$

This relation must be used with caution, since it does not apply in the same fashion to all particles and has a precision rarely better than ±10%.

To check if thermalization (momentum equilibration) is established, we consider momentum distributions in the direction transverse to the collision axis. Under a Lorentz transformation along the collision axis, $p_\perp$ remains unchanged and thus

$$m_\perp = \sqrt{m^2 + \vec{p}_\perp^2},$$

is invariant. Transverse mass $m_\perp$-particle spectra are not directly distorted by flow motion of the fireball matter along the collision axis, and also no further consideration of the CM frame of reference is necessary, which in fixed target experiments is rapidly moving with respect to a laboratory observer.

In order to study the thermal properties in the fireball as 'reported' by the emitted particles, we analyze $m_\perp$ spectra of many different hadrons. The range of $m_\perp$, on the

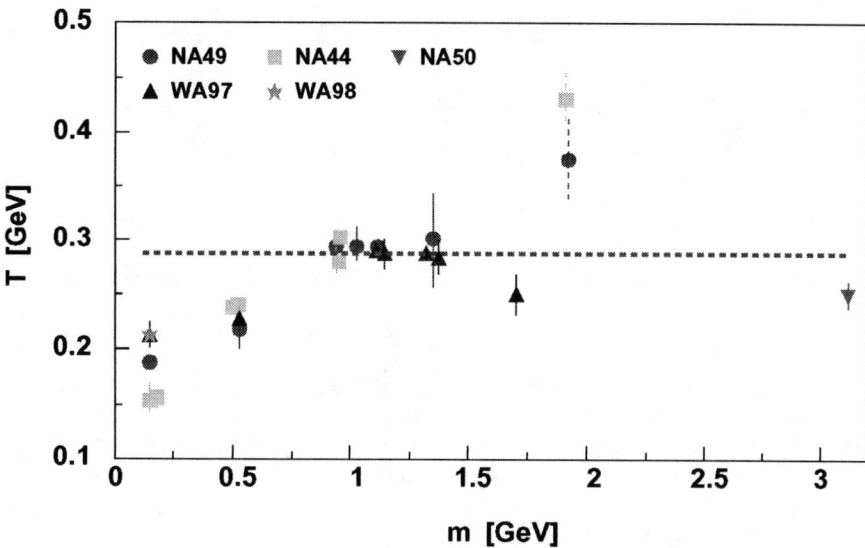

FIGURE 2. Inverse slopes $T_\perp$ observed in Pb–Pb interactions at $158A$ GeV as function of particle mass; symbols indicate the experiment from which data is drawn, as coded in the figure.

one hand, should not reach very large values, at which hadrons originating in hard parton scattering are relevant. On the other hand, we do study relatively small $m_\perp$, in order to avoid the non-exponential structure associated with transverse matter flow and unstable resonance decays.

The central-rapidity high-transverse-mass spectra of strange particles, K_s^0, $\overline{\Lambda}$, and Λ, given by the CERN–SPS WA85 collaboration [8, 9, 10], $m_\perp^{-3/2} dN_i/dm_\perp$, are shown in figure 1 on a semi-logarithmic display. The spectra can be fitted with a straight line. Similar results were also reported from the related work of the WA94 collaboration for S–S interactions [11]. We see, in figure 1, in the region of transverse masses presented, $1.5 \text{ GeV} < m_\perp < 2.6 \text{ GeV}$, that the behaviors of all three different particles feature the same inverse exponential slope, $T_\perp = 232 \pm 5$ MeV. This is not the actual temperature of the fireball, as noted in Eq. (4).

Presence of matter flow which generates the relatively high value of spectral slope is also responsible for differences in the observed values of $T_\perp$ for particles of widely different mass such as pions, kaons and nucleons considered in a similar $p_\perp$ interval. This is illustrated in figure 2. A remarkable feature is that many strange baryons are seen to have similar inverse slope, as is also seen in table 1.

2. CHEMICAL EQUILIBRIA

The average energy of each particle defines the local temperature T, good for all particles. A local chemical potentials σ_i need to be introduced for each kind of particle 'i' in

TABLE 1. Inverse slopes $T_\perp$ for various strange hadrons.

	Λ	$\overline{\Lambda}$	Ξ^-	$\overline{\Xi}^+$	$\Omega^- + \overline{\Omega}^+$	ϕ
Pb–Pb	289±2	287±4	286±9	284±17	251±19	305±15
S–W	233±3	232±7	244±12	238±16		

order to establish the particle density. We will see that it is often more convenient to use particle fugacity,

$$\Upsilon_i \equiv e^{\sigma_i/T}. \tag{5}$$

The statistical parameters T and σ_i (or Υ_i) express different types of equilibration processes in the hadron matter fireball, and in general there is a considerable difference in the time needed for the attainment of thermal and, respectively, chemical equilibrium.

The fugacity controls, independently of temperature, the yield of the particle species. For Boltzmann statistics, the fugacity Υ_i is multiplying the particle distribution, which in local rest frame assumes the usual form,

$$\frac{d^6 N_i}{d^3 p d^3 x} = g_i \frac{\Upsilon_i}{(2\pi)^3} e^{-E_i/T}. \tag{6}$$

For the Fermi/Bose quantum distributions the fugacity factor remains in front of the exponential,

$$\frac{d^6 N_i^{F/B}}{d^3 p d^3 x} = \frac{g_i}{(2\pi)^3} \frac{1}{\Upsilon_i^{-1} e^{E_i/T} \pm 1}, \tag{7}$$

The coefficient g_i is the degeneracy of the particle considered, it comprises the intrinsic properties such as spin and color of the particle.

The number of particles 'i' present is the result of momentum integral of the phase-space distribution. Thus the yield of particles depends on both temperature and the fugacity. However, fugacity practically does not control the shape of the momentum distribution, and thus it is generally said that it can be chosen to produce the required particle density, with temperature being the parameters determining the momentum distribution.

The allowed range of the (always positive) fugacities is not restrained for Boltzmann and Fermi particles. However for bosons, the condensation singularity of the distribution cannot be crossed, which requires that

$$\Upsilon_i \leq e^{m_i/T}. \tag{8}$$

As usual the $E_i(p) > m_i$ for all p and m_i is the particle rest mass. Almost never the exceptional value $\Upsilon_i = 1$ applies, with the exception of an chemically equilibrated gas of particles (such as photons) which do not carry any charge, i.e., a conserved, discrete quantum number.

As the number of particles evolves in a reaction, the fugacities usually change. For example, we begin with small initial yield of strange quarks and antiquarks in a deconfined quark-gluon plasma. Subsequent reactions of the type,

$$GG \to s\bar{s}, \qquad q\bar{q} \to s\bar{s}, \tag{9}$$

increase the yield of pairs of strange quarks and if we could cook the hadronic matter at a constant temperature T (thermal bath), than after some time known as the chemical relaxation time, the yield of strange quarks and antiquarks would approach the chemical equilibrium yields corresponding to the fugacity approaching unity,

$$\Upsilon_{i=s}(t) = \Upsilon_{i=\bar{s}}(t) \to 1. \tag{10}$$

The important message here is that the particle fugacities are time dependent and can evolve rapidly during the heavy-ion collision process. Description of the dynamical evolution of the fugacity is one of the challenges we are facing in order to understand the physics of hadronization.

Another important challenge which is arising, is due to the presence of two quite different types of chemical equilibria.

- In relativistic reactions, particles can be made as energy is converted to matter. Therefore, we can expect to approach slowly the *absolute* chemical equilibrium. Absolute chemical equilibrium is hard to grasp intuitively, since our instincts are distorted by the fact that a black body photon radiator is correctly assumed to be in absolute chemical equilibrium on the time scale defined by a blink of an eye. We characterize the approach to absolute chemical equilibrium by a fugacity factor γ_i for particle 'i'. We will evaluate the evolution of γ_i in the heavy-ion collision reaction as a function of time. Given its physical meaning it is often referred to as the phase-space occupancy.

- *Relative* chemical equilibration, the case commonly known in chemistry. It involves reactions that distribute or maintain a certain already existent element among different accessible compounds. Use of chemical potentials associated with conserved global properties, such as μ_b for baryon number, presupposes that the particular relative chemical equilibrium is present. For example production of quarks in pairs assures that there is always the same net flavor (thus baryon number) balance, and hence the relative flavor equilibrium is maintained. The chemical potential of antiparticles is in consequence the negative of that for particles, *provided* that we have introduced appropriate γ_i to control the absolute yield of particle-antiparticle pairs.

In general the fugacity of each individual particle will comprise the two chemical factors associated with the two different chemical equilibria. For example, let us look at the nucleon, and the antinucleon:

$$\Upsilon_N = \gamma_N e^{\mu_b/T}, \qquad \Upsilon_{\bar{N}} = \gamma_N e^{-\mu_b/T}. \tag{11}$$

Equivalently, we can keep separate chemical potentials for particles and antiparticles [12, 13],

$$\sigma_N \equiv \mu_b + T \ln \gamma_N, \qquad \sigma_{\bar{N}} \equiv -\mu_b + T \ln \gamma_N. \tag{12}$$

There is an obvious difference between the two chemical factors in Eq. (11): the number of nucleon-antinucleon pairs is associated with the value of γ_N but not with μ_b.

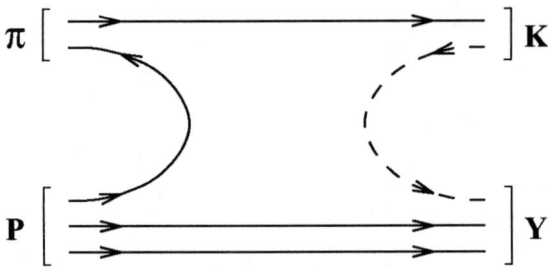

FIGURE 3. The production of strangeness in reactions of the type $\pi + N \to K + Y$ in the HG phase. Solid lines indicate the flow of light quarks and the disappearance of one $\bar{q}q$ pair, the dashed line is for the added $\bar{s}s$ pair.

This can be seen looking at the first law of thermodynamics, in this context written as:

$$\begin{aligned} dE &= -PdV + TdS + \sigma_N dN + \sigma_{\overline{N}} d\overline{N} \\ &= -PdV + TdS + \mu_b(dN - d\overline{N}) + T\ln\gamma_N(dN + d\overline{N}). \end{aligned} \quad (13)$$

To obtain the second form we have employed Eq. (12). We see that μ_b is the energy required to change the baryon number,

$$b \equiv N - \overline{N},$$

by one unit, while the number of nucleon-antinucleon pairs,

$$2N_{\text{pair}} \equiv N + \overline{N},$$

is related to γ_N. For $\gamma_N = 1$ the last term vanishes, at this point small fluctuation in number of nucleon pairs does not influence the energy of the system, we are have reached the absolute baryochemical equilibrium.

Presence of γ_N allows us to count and control within a theoretical description how many pairs of nucleons are added. When $\gamma_N \to 1$, we have as many as we would expect in absolute chemical equilibrium, of course establishing chemical equilibrium for antinucleons can take a long time. However, it is not uncommon to see in contemporary literature assumption of instantaneous absolute chemical equilibrium, $\gamma_i \to 1$, with particles 'instantaneously' reaching their absolute chemical equilibrium abundances. The argument presented is that such a simple model 'works', as it is capable to describe widely different particle yields qualitatively. This in fact is the observation which motivated Hagedorn's work. 40 years after, our foremost interest is to understand the deviations from this, in order to unravel the mechanism by which such a surprising result can arise.

In order to better understand the difference between absolute and relative chemical equilibrium, let us consider the abundance of strangeness in the baryon-rich hadronic gas (HG) phase. There is little 'strangeness' initially, strangeness is made in collision processes of the interacting hadronic matter. We begin far from absolute chemical-strangeness equilibrium. To make $s\bar{s}$ pairs in a HG phase, there are many possible reactions, classified usually as the direct- and associate-production processes. In the

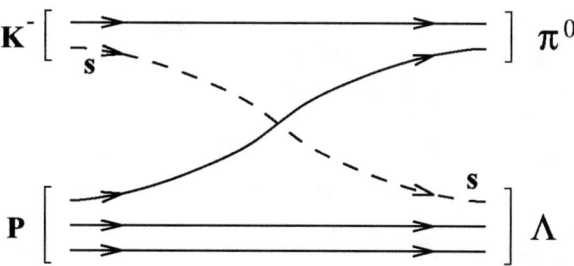

FIGURE 4. An example of a strangeness-exchange reaction in the HG phase: $K^- + p \to \Lambda + \pi^0$. Solid lines, flow of u and d quarks; dashed line, exchange of an s quark between two hadrons.

associate production process, a pair of strange quarks is shared between two existent hadrons, of which one is a baryon, typically a nucleon N, which becomes a hyperon Y:

$$\pi + N \leftrightarrow K + Y.$$

This situation is illustrated in figure 3.

In a direct-production process, a pair of strangeness-carrying particles is formed directly via annihilation of two mesons, adding a pair to the system:

$$\pi + \pi \leftrightarrow K + \overline{K}.$$

Here, a pair of strange particles is made in the form of a pair of kaons, K^+K^-.

With these two reaction types alone it could be that populations of strange mesons and baryons evolve differently. However, the meson carrier of the s quark, K^-, can exchange this quark, see figure 4, via fast exothermic reaction with a nucleon, forming a hyperon:

$$K^- + N \leftrightarrow \pi + Y.$$

This reaction establishes relative chemical equilibrium between mesons and baryons by being able to move the strange quark between these two different strangeness carriers, $s\bar{q}$ mesons and sqq baryons.

Reactions establishing the redistribution of existent flavor, or the abundance of some other conserved quantity, play a different role from the reactions that actually contribute to the formation of this flavor, or other quantum number, and facilitate the approach to absolute chemical equilibrium. Accordingly, the time constants for relaxation are different, since different types of reaction are involved.

Apart from the different relaxation times associated with the different types of thermal and chemical equilibria, there are different time scales associated with the different fundamental interactions involved. For example, the electro-magnetic interactions are considerably slower at reaching equilibrium than are the strong interactions governing the evolution of dense hadronic fireballs created in ultra-relativistic heavy-ion collisions. All the important time constants for relaxation in heavy-ion collisions arise from differences in mechanisms operating within the realm of strong interactions. Therefore, the separation of time scales is not as sharp as that between the different interactions, though a clear hierarchy arises and we have presented it above.

There are just a few chemical potentials which suffice in the study of hadronic matter. It is convenient to use the quark flavor for this purpose, since there is a continuity in the notation across the phase boundary of quark matter and hadronic matter. We thus use:

$$\lambda_u = e^{\mu_u/T}, \quad \lambda_d = e^{\mu_d/T}, \quad \lambda_s = e^{\mu_s/T}. \tag{14}$$

Since more often than not the light flavors u, d remain indistinguishable one often uses:

$$\mu_q = \frac{1}{2}(\mu_u + \mu_d), \quad \lambda_q^2 = \lambda_u \lambda_u. \tag{15}$$

Considering that three quarks make a baryon we also have

$$\mu_b = 3\mu_q, \quad \lambda_b = \lambda_q^3. \tag{16}$$

It is important to remember that the three quark flavors u, d, s carry baryon number. Thus when baryon density is evaluated this has to be appropriately allowed for, counting all three quark densities. This can be easily forgotten for strangeness, considering the usual definition of baryochemical potential, Eq. (16).

At this point, we recall the relationship of strange quark chemical potential and strangeness chemical μ_S potential (as opposed to strange quark chemical potential μ_s):

$$\mu_s = \frac{1}{3}\mu_b - \mu_S, \quad \lambda_s = \frac{\lambda_q}{\lambda_S}, \tag{17}$$

which expresses the fact that strange quarks carry negative strangeness and one third of baryon number. Consequently, hyperons have negative strangeness, antihyperons positive strangeness, $K^-(\bar{u}s)$ negative strangeness and thus it is called antikaon $\bar{K}$, and $K^+(u\bar{s})$ has positive strangeness, and it is the kaon K. This nomenclature has an obvious historical origin, when strangeness was first discovered, the $\bar{s}$ containing K^+ has been the particle produced.

Using Eq. (17) in statistical formulation allows to evaluate baryon number and strangeness by simply differentiating the grand canonical partition function $\mathscr{Z}$:

$$N_b = \lambda_b \frac{\partial \mathscr{Z}(\beta, \lambda_b, \lambda_S)}{\partial \lambda_b}, \quad N_S = \lambda_S \frac{\partial \mathscr{Z}(\beta, \lambda_b, \lambda_S)}{\partial \lambda_S}, \tag{18}$$

where as usual $\beta = 1/T$. Especially the computation of N_b is more cumbersome when $\lambda_i, i = u, d, s$, is used.

On the other hand, we believe that the use of fugacities which follow the yield of quarks, $\lambda_i, i = u, d, s$, is intuitive and easy in programing, mistakes are rather difficult to make. Particle yields can be easily followed and checked, and thus errors and omissions minimized. As an example, let us consider the ratio of a baryon and antibaryon with strangeness and for reasons which will become obvious in a moment we choose to look at the ratio $\overline{\Xi^-}(\bar{d}\bar{s}\bar{s})/\Xi^-(dss)$. Given the quark content and ignoring isospin asymmetry we find:

$$\frac{\overline{\Xi^-}}{\Xi^-} = \frac{\lambda_s^{-2}\lambda_q^{-1}}{\lambda_s^2 \lambda_q} = \lambda_s^{-4}\lambda_q^{-2} = \frac{e^{2\mu_S/T}e^{-\mu_b/T}}{e^{-2\mu_S/T}e^{\mu_b/T}} = e^{4\mu_S/T}e^{-2\mu_b/T}. \tag{19}$$

Since all Ξ resonances which contribute to this ratio are symmetric for particles and antiparticles, and possible weak interaction feed from $\overline{\Omega}(\bar{s}\bar{s}\bar{s})$ and, respectively $\Omega(sss)$ are small, these expressions are actually rather exactly giving the expected experimental ratio.

When we check the results presented by [14], we find for the freeze-out parameters stated there $T = 174$ MeV, $\mu_b = 46$ MeV and $\mu_S = 13.6$ MeV, a ratio $\overline{\Xi^-}/\Xi^- = 0.806$, which differs from the result $\overline{\Xi^-}/\Xi^- = 0.894$ given. While the difference appears small, when looked at in terms of the particle chemical potential this is a big effect: Considering that the Cascades have baryon number $b = 1$ and strangeness $S = -2$, we expect that the Ξ-chemical potential is $\mu_\Xi|_{\text{expected}} = \mu_b - 2\mu_S = 18.8$ MeV. On the other hand given the result seen in [14],

$$0.894 = \frac{\overline{\Xi^-}}{\Xi^-} = e^{-2\mu_\Xi/T}.$$

The particle chemical potential is $\mu_\Xi|_{\text{used}} = 9.75$ MeV. In order to approach using the stated potentials the answer presented in [14], we can try to use another quark fugacity definition,

$$\left.\frac{\overline{\Xi^-}}{\Xi^-}\right|_{\text{PBM}} = \frac{\lambda_s^2 \lambda_q^{-1}}{\lambda_s^{-2} \lambda_q} = e^{\frac{2}{3}\mu_b/T - 4\mu_S/T} = 0.87. \tag{20}$$

However, this redefinition is not consistent with the ratios of particles such as K^-/K^+ also seen in [14]. This example of a possible partial misunderstanding of the quantum numbers associated with different hadrons illustrates the danger of using anything but the (valance) quark fugacities in study of chemical properties of hadrons.

To complete the discussion of chemical potentials, we need to address the u–d asymmetry which is only noticeable for Pb–Pb interactions at SPS. We complement Eq. (15) by introducing

$$\mu_I = \mu_d - \mu_u, \quad \lambda_I = \frac{\lambda_d}{\lambda_u} \geq 1. \tag{21}$$

In the last inequality, we show the constraint arising given that nuclei have in general a greater neutron than proton number. The subscript I reminds us of isospin, we have also used $\delta\mu$ for μ_I in the past. A sensitive probe of λ_I is the ratio

$$\frac{\pi^-(\bar{u}d)}{\pi^+(u\bar{d})} = \lambda_I^2. \tag{22}$$

At RHIC, for $\sqrt{s_{NN}} = 130$ GeV, the outflow of projectile and target matter from the central rapidity region leaves a very small 'input' isospin asymmetry: with a baryon density per unit rapidity at $dN_b/dy \simeq 20$, but hundreds of mesons made, the valance quark asymmetry is negligible. However, at the top energy at SPS there is an effect at the level of 2–4% percent, since baryon number is retained in the fireball and the meson to baryon ratio is 8 times smaller, as measured in terms of the ratio of all negative hadrons,

$$h^- = \pi^- + K^- + \bar{p}, \tag{23}$$

to baryon number. At SPS, we have $h^-/b \simeq 2$, while at RHIC at $\sqrt{s_{NN}} = 130$ GeV, $h^-/b \simeq 16.5$, in both cases at central rapidity. We can estimate for SPS where $\lambda_q = 1.6$:

$$1.6 < \lambda_d < 1.66, \quad 1.54 < \lambda_u < 1.6.$$

The exact value of λ_d, λ_u depends on what form of matter (confined, deconfined) is the source of particles produced, and the influence of gluon fragmentation into quark pairs. The limiting values are found ignoring gluon fragmentation and evaluating the ratio of of all down and up quarks in hot quark matter, where quark density is $\rho_i \propto \mu_i T^2, i = u, d$. Thus we find

$$\frac{\rho_d}{\rho_u} = 1 + \delta = \frac{\mu_d}{\mu_u}, \quad \delta = \frac{n-p}{n+2p}, \qquad (24)$$

where $n(udd)$ and $p(uud)$ are the neutron, and respectively, proton input into the quark matter source. For heavy nuclei $\delta \simeq 0.156$, and we find:

$$\lambda_d = \lambda_q^{\frac{1}{1-\delta/2}}, \quad \lambda_u = \lambda_q^{\frac{1}{1+\delta/2}}. \qquad (25)$$

We now return to discuss further the characterization of chemical non-equilibrium. When we study the approach to chemical equilibrium for different hadrons, we can use three non-equilibrium parameters, $\gamma_u, \gamma_d, \gamma_s$, and equivalently, $\gamma_q^2 = \gamma_u \gamma_d$. In quark matter, these three factors express the approach to the expected chemical equilibrium yield by the quark abundances. Upon hadronization quarks are redistributes among all individual hadrons and the non-equilibrium abundances can be characterized by the same three factors only, since in all hadron formation reactions only quark-antiquark pairs of the same flavor can be formed. It is important to realize that even if there were no change of the quark pair number in hadronization, the values of $\gamma_u, \gamma_d, \gamma_s$ in hadron gas and quark matter must differ since the phase spaces have different size as we shall discuss further below. In general, we have to distinguish $\gamma_u^{QGP}, \gamma_d^{QGP}, \gamma_s^{QGP}$ from $\gamma_u^{HG}, \gamma_d^{HG}, \gamma_s^{HG}$. Moreover, as noted these sets of parameters differ due to hadronization of gluons into quark pairs.

We thus characterize the fugacities of all hadrons by six parameters. For baryons the typical examples are (considering protons $p(uud)$, antiprotons $\bar{p}(\bar{u}\bar{u}\bar{d})$, $\Lambda(uds)$, $\overline{\Omega}(\bar{s}\bar{s}\bar{s})$ as examples):

$$\Upsilon_p = \gamma_u^2 \gamma_d e^{2\mu_u+\mu_d}, \quad \Upsilon_{\bar{p}} = \gamma_u^2 \gamma_d e^{-2\mu_u-\mu_d}, \quad \Upsilon_\Lambda = \gamma_u \gamma_d \gamma_s e^{\mu_u+\mu_d+\mu_s}, \quad \Upsilon_{\overline{\Omega}} = \gamma_s^3 e^{-3\mu_s}. \quad (26)$$

The yield of mesons follows the same pattern (considering $\pi^+(u\bar{d})$, $\pi^-(\bar{u}d)$, $K^-(\bar{u}s)$, $\phi(\bar{s}s)$ as examples):

$$\Upsilon_{\pi^+} = \gamma_u \gamma_d e^{\mu_u-\mu_d}, \quad \Upsilon_{\pi^-} = \gamma_u \gamma_d e^{-\mu_u+\mu_d}, \quad \Upsilon_{K^-} = \gamma_u \gamma_s e^{-\mu_u+\mu_s}, \quad \Upsilon_\phi = \gamma_s^2. \quad (27)$$

It is important to note that this approach assumes that the relative population of heavier resonances is in chemical equilibrium with the lighter states, since we have the same value of Υ for all hadrons with the same valance quark content. For example $\Upsilon_p = \Upsilon_{\Delta^+}$. We thus realize that this approach solely focuses on the quark distribution and does not allow for the possibility that heavier resonances may simply not be populated. This method thus is most suitable for a hadronizing quark matter fireball, and may miss important features of a hadron fireball which never entered the deconfined phase. When conditions are present which let us believe that the population of important heavier resonance states such as $\Delta(qqq)$ could be suppressed beyond the thermal factor $\propto e^{m_\Delta/T}$ we have to restore a factor $\gamma_\Delta < \gamma_q^3$ which allows for such suppression. As long as this is not done, we in fact assume that the relative abundances of hadronic states of same quark content are in chemical equilibrium. The following study of deviations from chemical equilibrium is addressing a much more subtle question, namely how the overall number of (valence) quarks compares to the hadron phase-space chemical equilibrium expectations.

3. HOW NEAR TO CHEMICAL EQUILIBRIUM?

Is such a characterization of hadron abundances at all functioning, and is so, how close are we actually to absolute chemical equilibrium in heavy-ion reactions? To answer this question a systematic test was performed for the S–W/Pb $200A$ GeV experimental SPS results [15]. Particle yields were fitted allowing progressively greater and greater degree of chemical non-equilibrium. The results in table 2 show that the statistical significance is increasing with progressively greater chemical 'freedom'. Since the statistical significance which can be reached is satisfactory, it appears that the quark counting suffices to describe the yields of heavier hadron resonances.

We note, in the last raw in table 2, that while the yield of strangeness is still suppressed as compared to expectations based on absolute chemical equilibrium, the number of light quarks seen exceeds the count expected. Another way to say the same thing is that there is pion excess, or alternatively, entropy excess [16, 17, 18]. In fact there is a maximum value that γ_q can assume: consider again Eq. (7) for pions, assuming symmetry for u, d quarks:

$$f_\pi \equiv \frac{d^6 N_i^\pi}{d^3 p d^3 x} = \frac{3}{(2\pi)^3} \frac{1}{\gamma_q^{-2} e^{E_\pi/T} - 1}. \tag{28}$$

We see that for

$$\gamma_q \to e^{m_\pi/2T} = \gamma_q^c, \tag{29}$$

the Bose condensation becomes possible.

TABLE 2. Statistical parameters obtained from fits of data for S–Au/W/Pb collisions at $200A$ GeV, without enforcing conservation of strangeness [15].

λ_q	λ_s	γ_s	γ_q	χ^2/dof
1.52 ± 0.02	1^*	1^*	1^*	17
1.52 ± 0.02	0.97 ± 0.02	1^*	1^*	18
1.48 ± 0.02	1.01 ± 0.02	0.62 ± 0.02	1^*	2.4
1.49 ± 0.02	1.00 ± 0.02	0.73 ± 0.02	1.22 ± 0.06	0.90

* denotes fixed (input) values

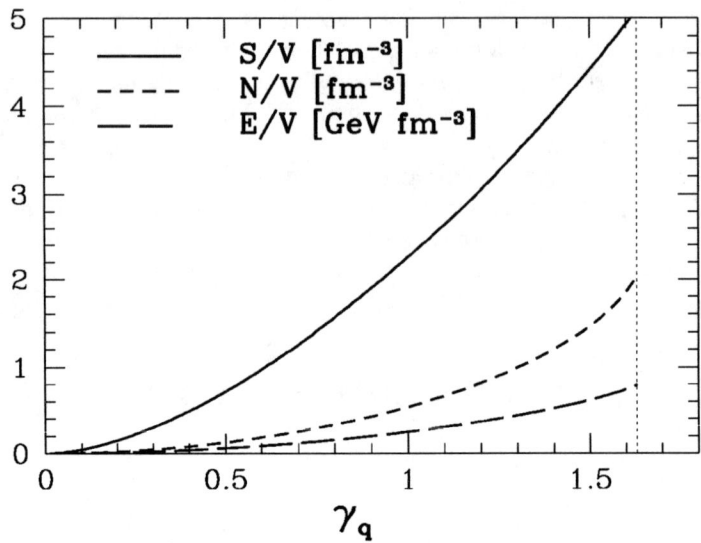

FIGURE 5. Pion-gas properties N/V for particles, E/V for energy, and S/V for entropy density, as functions of γ_q at $T = 142$ MeV.

The properties of the pion gas as function of γ_q are shown in figure 5, where the entropy, energy and particle number is expressed in terms of the momentum distribution function by:

$$\frac{S}{V} = \int d^3p[(1+f_\pi)\ln(1+f_\pi) - f_\pi \ln f_\pi], \qquad (30)$$

$$\frac{E}{V} = \int d^3p\sqrt{m_\pi^2 + p^2}f_\pi, \qquad (31)$$

$$\frac{N}{V} = \int d^3p f_\pi. \qquad (32)$$

There is significant rise in entropy content as γ_q grows toward the singular value. Excess entropy of a possibly color deconfined source can thus be squeezed in hadronization into a hadron phase comprising an over saturated pion gas. Indeed, when we perform an analysis of particles produced in the Pb–Pb 158A GeV reactions, as well as at RHIC, the value γ_q^c is always favored by a fit having a statistical significance [19].

When an analysis of the Pb–Pb collision system is performed allowing for quark pair abundance chemical non-equilibrium, we find that the light quark phase-space occupancy parameter γ_s prefers the maximum allowed γ_q^c, and this is also the numerical value preferred by the strange quark occupancy parameter γ_s. The freeze-out of particles is found at $T = 150$ MeV. The velocity of (transverse) expansion v_c (not further discussed here) is at half velocity of light. These results were obtained fitting all particles shown in table 3.

TABLE 3. WA97 (top) and NA49 (bottom) Pb–Pb 158A GeV-collision hadron ratios compared with phase-space fits.

Ratios	Reference	Experimental data	Pb$\vert^{s,\gamma_q}$	Pb$\vert^{\gamma_q}$
Ξ/Λ	[20]	0.099 ± 0.008	0.096	0.095
$\overline{\Xi}/\overline{\Lambda}$	[20]	0.203 ± 0.024	0.197	0.199
$\overline{\Lambda}/\Lambda$	[20]	0.124 ± 0.013	0.123	0.122
$\overline{\Xi}/\Xi$	[20]	0.255 ± 0.025	0.251	0.255
K^+/K^-	[21]	1.800 ± 0.100	1.746	1.771
K^-/π^-	[22]	0.082 ± 0.012	0.082	0.080
K_s^0/b	[23]	0.183 ± 0.027	0.192	0.195
h^-/b	[24]	1.970 ± 0.100	1.786	1.818
ϕ/K^-	[25]	0.145 ± 0.024	0.164	0.163
$\overline{\Lambda}/\bar{p}$	$y=0$		0.565	0.568
χ^2			1.6	1.15
$N; p; r$			9; 4; 1	9; 5; 1

Due to their exceptional inverse slope, see figure 2, the yields of both Ω and $\overline{\Omega}$ are known, to follow different systematic behavior and have not been considered in this fit. With the parameters here determined the yields of $\overline{\Omega}$ are under predicted. This excess yield originates at the lowest $m_\perp$, as we shall discuss below, see figure 10. The 'failure' of a statistical-hadronization model to describe yields of soft Ω and $\overline{\Omega}$ has several possible explanations. One is the possibility that an enhancement in production of Ω and $\overline{\Omega}$ is caused by pre-clustering of strangeness in the deconfined phase [26]. This would enhance the production of all multistrange hadrons, but most prominently the highly phase-space-suppressed yields of Ω and $\overline{\Omega}$. This mechanism would work only if pairing of strange quarks near to the phase transition were significant. Another possible mechanism of Ω and $\overline{\Omega}$ enhancement is the distillation of strangeness [27, 28], followed by breakup of strangelets (strangeness enriched quark drops) which could contribute to production of Ω and $\overline{\Omega}$. The decay of disoriented chiral condensates has recently been proposed as another source of soft Ω and $\overline{\Omega}$ [29].

In view of these pre- and post-dictions of the anomalous yield of Ω and $\overline{\Omega}$, and the difference in shape of particle spectra, we believe that one should abstain from introducing these particles into statistical-hadronization-model fits. We note that the early statistical descriptions of yields of Ω and $\overline{\Omega}$ have not been sensitive to the problems we described [30, 31]. In fact, as long as the parameter γ_q is not considered, it is not possible to describe the experimental data at the level of precision that would allow recognition of the excess yield of Ω and $\overline{\Omega}$ within statistical hadronization model. For example, a chemical-equilibrium fit, which includes the yield of Ω and $\overline{\Omega}$, has for 18 fitted data points with two parameters a $\chi^2/\mathrm{dof} = 37.8/16$ [32]. Such a fit is quite unlikely to contain all the physics, even if its appearance to the naked eye suggests that a very good description of experimental data as been achieved.

We also see, in table 4, the yield of strangeness obtained in the fit of the final state hadron phase-space, which with 0.7 strange pairs per baryon appears large. The question we next consider is what we should have expected in hadronization of a deconfined quark

TABLE 4. Upper section: the statistical model parameters which best describe the experimental results for Pb–Pb data seen in table 3. Bottom section: energy per entropy, anti-strangeness, and net strangeness of the full hadron phase-space characterized by these statistical parameters. In column two, we fix λ_s by the requirement of conservation of strangeness.

	$Pb|_v^{s,\gamma_q}$	$Pb|_v^{\gamma_q}$
T [MeV]	151 ± 3	147.7 ± 5.6
v_c	0.55 ± 0.05	0.52 ± 0.29
λ_q	1.617 ± 0.028	1.624 ± 0.029
λ_s	1.10^*	1.094 ± 0.02
γ_q	$\gamma_q^{c*} = e^{m_\pi/(2T_f)} = 1.6$	$\gamma_q^{c*} = e^{m_\pi/(2T_f)} = 1.6$
γ_s/γ_q	1.00 ± 0.06	1.00 ± 0.06
E/b [GeV]	4.0	4.1
s/b	0.70 ± 0.05	0.71 ± 0.05
E/S [MeV]	163 ± 1	160 ± 1
$(\bar{s}-s)/b$	0^*	0.04 ± 0.05

* indicates values resulting from constraints.

matter phase. We consider the ratio of the equilibrium density of strangeness, arising in the Boltzmann-gas limit, to the baryon density in a fireball of quark–gluon plasma:

$$\frac{\rho_s}{\rho_b} = \frac{s}{b} = \frac{s}{q/3} = \frac{\gamma_s^{QGP}}{\gamma_q^{QGP}} \frac{(3/\pi^2) T^3 W(m_s/T)}{\frac{2}{3}(\mu_q T^2 + \mu_q^3/\pi^2)}. \tag{33}$$

$W(x) = x^2 K_2(x)$ defines in Boltzmann limit the equilibrium strange-quark density, with $g_s = 6$. We assume that to a first approximation, perturbative thermal QCD corrections, cancel out in the ratio. For $m_s = 200$ MeV and $T = 150$ MeV, we have

$$\frac{s}{b} \simeq \frac{\gamma_s^{QGP}}{\gamma_q^{QGP}} \frac{0.7}{\ln \lambda_q + (\ln \lambda_q)^3/\pi^2}. \tag{34}$$

The relative yield s/b is mainly dependent on the value of λ_q. In the approximation considered, it is nearly temperature-independent, the result is shown in figure 6, as a function of $\lambda_q - 1$ (the variable chosen to enlarge the interesting region $\lambda_q \to 1$) for $\gamma_s^{QGP} = \gamma_q^{QGP} = 1$. At the top SPS energy where $\lambda_q \to 1.5$–1.6, we see that the equilibrium yield is at 1.5 pairs of strange quarks per participating baryon. Considering the experimental yield in table 4, we thus conclude $\gamma_s^{QGP} \sim 0.5$. Experimentally the directly accessible observable is the occupancy of the hadron-strangeness phase-space and we show in table 4, $\gamma_s^{HG} \simeq 1.6$. This thus implies that in the hadronization of a presumed QGP the smaller phase-space of hadrons is overpopulated by a QGP abundance, even when it is originally in the QGP phase well below the equilibrium value.

The above analysis demonstrated the importance of the study of quark-pair abundance chemical non-equilibrium features and how these are capable to refine our understanding of the QGP formation and hadronization.

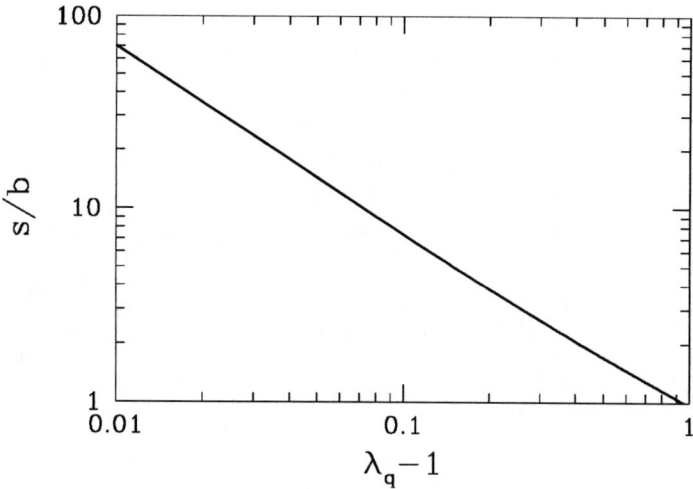

FIGURE 6. The yield of strangeness per baryon as a function of λ_q in equilibrated quark–gluon plasma.

4. QGP AND CHEMICAL ANALYSIS

An interesting feature seen in table 2 is that the value of λ_s converges to unity. This is not at all what could be expected for a equilibrated hadron gas. Namely, the value of λ_s arises from the requirement of conservation of strangeness. Specifically the net strangeness is given by,

$$0 = \langle n_s \rangle - \langle n_{\bar{s}} \rangle = \frac{T^3}{2\pi^2}\bigg[(\lambda_s\lambda_q^{-1} - \lambda_s^{-1}\lambda_q)\gamma_s\gamma_q F_K + (\lambda_s\lambda_q^2 - \lambda_s^{-1}\lambda_q^{-2})\gamma_s\gamma_q^2 F_Y $$
$$+ 2(\lambda_s^2\lambda_q - \lambda_s^{-2}\lambda_q^{-1})\gamma_s^2\gamma_q F_\Xi + 3(\lambda_s^3 - \lambda_s^{-3})\gamma_s^3 F_\Omega\bigg], \quad (35)$$

where we have employed the phase-space integrals for known hadrons,

$$F_K = \sum_j g_{K_j} W(m_{K_j}/T); \quad K_j = K, K^*, K_2^*, \ldots, \quad m \leq 1780 \text{ MeV},$$

$$F_Y = \sum_j g_{Y_j} W(m_{Y_j}/T); \quad Y_j = \Lambda, \Sigma, \Sigma(1385), \ldots, \quad m \leq 1940 \text{ MeV},$$

$$F_\Xi = \sum_j g_{\Xi_j} W(m_{\Xi_j}/T); \quad \Xi_j = \Xi, \Xi(1530), \ldots, \quad m \leq 1950 \text{ MeV},$$

$$F_\Omega = \sum_j g_{\Omega_j} W(m_{\Omega_j}/T); \quad \Omega_j = \Omega, \Omega(2250). \quad (36)$$

The g_i are the spin–isospin degeneracy factors, $W(x) = x^2 K_2(x)$, where K_2 is the modified Bessel function.

In general, Eq. (35) must be equal to zero since strangeness is a conserved quantum number with respect to the strong interactions, and no strangeness is brought into the reaction. The possible exception is dynamic evolution with asymmetric emission of strange and antistrange hadrons [28]. Eq. (35) can be solved analytically when the contribution of multistrange particles is small:

$$\lambda_s|_0 = \lambda_q \sqrt{\frac{F_K + \gamma_q \lambda_q^{-3} F_Y}{F_K + \gamma_q \lambda_q^3 F_Y}}. \tag{37}$$

We thus see that except for a very exceptional point where the kaon and hyperon strangeness phase-spaces for a given value of baryochemical properties are of same magnitude, the value of λ_s will not be unity. One can of course force the hadron multiplicity description to this value, but the prize one pays is a greatly reduced statistical significance [32, 33], where chemical equilibrium has been assumed. We note that for the value $\lambda_s = 1$, we can analytically solve Eq. (35), including the effect of multistrange hadrons and obtain:

$$\mu_b = 3T \ln(x + \sqrt{x^2 - 1}), \qquad 1 \leq x = \frac{F_K - 2\gamma_s F_\Xi}{2\gamma_q F_Y}. \tag{38}$$

In what situation is the value $\lambda_s \to 1$ natural? This value will arise when for all baryochemical conditions the strange and antistrange quark numbers can balance independently. This will in most cases be in the phase in which strange quarks can roam freely. In this case we have instead of condition Eq. (35),

$$0 = \langle n_s \rangle - \langle n_{\bar{s}} \rangle$$
$$= g \int \frac{d^3 p}{(2\pi)^3} \left(\frac{1}{1 + \gamma_s^{-1} \lambda_s^{-1} \exp\left(\frac{\sqrt{p^2 + m_s^2}}{T}\right)} - \frac{1}{1 + \gamma_s^{-1} \lambda_s \exp\left(\frac{\sqrt{p^2 + m_s^2}}{T}\right)} \right). \tag{39}$$

We note the change in the power of λ_s between these two terms, and recognize that this integral can vanish only for $\lambda_s \to 1$.

There is potentially small but significant asymmetry in λ_s due to the Coulomb charge present in baryon-rich quark matter: long-range electromagnetic potential $V_C \neq 0$ influence strange and antistrange particles differently, and a slight deviation $\lambda_s > 1$ is needed in order to compensate for this effect in the QGP phase. We have as generalization of Eq. (39) [34],

$$0 = \langle N_s \rangle - \langle N_{\bar{s}} \rangle$$
$$= g \int_{R_f} \frac{d^3 r d^3 p}{V(2\pi)^3} \left(\frac{1}{1 + \gamma_s^{-1} \lambda_s^{-1} e^{(E(p) - \frac{1}{3} V_C(r))/T}} - \frac{1}{1 + \gamma_s^{-1} \lambda_s e^{(E(p) + \frac{1}{3} V_C(r))/T}} \right), \tag{40}$$

which clearly cannot vanish for $V_C \neq 0$, in the limit $\lambda_s \to 1$. The volume integral is here over the fireball of size R_f. In the Boltzmann approximation one easily finds that

$$\tilde{\lambda}_s \equiv \lambda_s \ell_C^{1/3} = 1, \qquad \ell_C \equiv \frac{\int_{R_f} d^3 r \, e^{V/T}}{\int_{R_f} d^3 r}. \tag{41}$$

$\ell_C < 1$ expresses the Coulomb deformation of strange quark phase-space. ℓ_C is not a fugacity that can be adjusted to satisfy a chemical condition, since consideration of λ_i, $i = u, d, s$, exhausts all available chemical balance conditions for the abundances of hadronic particles, and allows introduction of the fugacity associated with the Coulomb charge of quarks and hadrons. Instead, ℓ_C characterizes the distortion of the phase-space by the long-range Coulomb interaction. This Coulomb distortion of the quark phase-space is naturally also present for u, d quarks, but appears less significant given that λ_u and λ_d are empirically determined. On the other hand, this effect eliminates much if not all the difference between μ_u and μ_d we have described above, since the quark abundance asymmetry arises naturally due to Coulomb effect — said differently, the Coulomb effect deforms the phase-space such that it is natural to have more d than u quarks and thus the asymmetry between λ_d and λ_u is reduced.

Choosing $T = 140$ MeV and $m_s = 200$ MeV, and noting that the value of γ_s is practically irrelevant since this factor cancels out in the Boltzmann approximation we find for $Z_f = 150$ that the value $\lambda_s = 1.10$ is needed for $R_f = 7.9$ fm, whereas for S–Au/W/Pb reactions, similar analysis leads to a value $\lambda_s = 1.01$. Chemical freeze-out at higher temperature, e.g., $T = 170$ MeV, leads for $\lambda_s = 1.10$ to somewhat smaller radii, which is consistent with the higher temperature used.

The fit of the Pb–Pb system seen in table 4 converged just to the value expected if QGP were formed. However, accidentally for $\gamma_q = \gamma_q^c$ this happens also to be where in hadronic gas strangeness balance is found. Thus in case of Pb–Pb the indication from chemistry that a QGP has been formed arises mainly from the fact that a significant entropy excess is available, (light quark-pair excess) and the strange quark-pair excess, seen the relatively high value of $\gamma_s \simeq \gamma_q$ in table 4. When compared to the yield expected from QGP, this implies that $\gamma_s^{HG}/\gamma_s^{QGP} \simeq 3$. To better understand this, we compare the phase-space of strangeness in quark–gluon plasma with that of the resulting hadronic gas. The absolute strangeness yields must be the same in both phases. We perform the comparison assuming that in a fast hadronization of QGP neither temperature T, nor the baryochemical potential μ_b, nor the reaction volume have time to change, the difference in the properties of the phases is absorbed in the change in the occupancy of phase-space, γ_i in the two phases.

We relate the two phase-space occupancies in HG and QGP, by equating the strangeness content in the two phases. One has to keep in mind that there is some additional production of strangeness due to gluon hadronization, however, this is not altering the argument presented significantly. On canceling out the common normalization factor $T^3/(2\pi^2)$, we obtain

$$\gamma_s^{QGP} V^{QGP} g_s W\left(\frac{m_s}{T^{QGP}}\right) \simeq \gamma_s^{HG} V^{HG} \left(\frac{\gamma_q \lambda_q}{\lambda_s} F_K + \frac{\gamma_q^2}{\lambda_q^2 \lambda_s} F_Y\right). \tag{42}$$

Here we have, without loss of generality, followed the $\bar{s}$-carrying hadrons in the hadronic gas phase-space, and we have omitted the contribution of multistrange antibaryons for simplicity. We now use the condition that strangeness is conserved, Eq. (37), to eliminate λ_s from Eq. (42), and obtain (making explicit which statistical

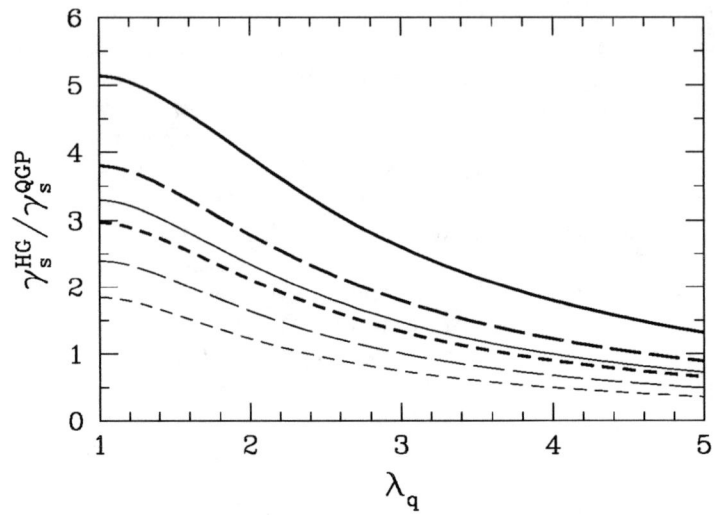

FIGURE 7. The HG/QGP strangeness-occupancy γ_s ratio in sudden hadronization as a function of λ_q. Solid lines, $\gamma_q^{HG} = 1$; long-dashed lines, $\gamma_q^{HG} = 1.3$; and short-dashed lines, $\gamma_q^{HG} = 1.6$. Thin lines are for $T = 170$ and thick lines for $T = 150$ MeV, for both phases.

parameter occurs in which phase)

$$\frac{\gamma_s^{HG}}{\gamma_s^{QGP}} \frac{V^{HG}}{V^{QGP}} = \frac{g_s W(m_s/T^{QGP})}{\sqrt{(\gamma_q F_K + \gamma_q^2 \lambda_q^{-3} F_Y)(\gamma_q F_K + \gamma_q^2 \lambda_q^3 F_Y)}}. \quad (43)$$

This ratio is shown in figure 7, and the two lines of particular interest are the thin solid line (chemical equilibrium $\gamma_q^{HG} = 1$ with $T = 170$ MeV in hadronic gas phase), and the short dashed thick line ($\gamma_q^{HG} = 1.6$ at $T = 150$ MeV) which correspond to the two hadronization scenarios which can be used to fit the SPS experimental results. We see that, near to the established value $\lambda_q \simeq 1.6$, both yield a squeeze by factor 3 for the ratio of the strangeness phase-space occupancies.

This result implies that hadronization of QGP is at SPS and RHIC accompanied by an increase by factor three in the value of γ_s and thus it is most important to allow in the study of particle yields for this of-equilibrium property, otherwise the tacit assumption has been made that QGP has not been formed. If at RHIC conditions are reached in which strangeness in QGP is closer to absolute chemical equilibrium, $\gamma_s^{QGP} \to 1$, even a more significant overpopulation by strangeness of the hadronic gas phase would result, $\gamma_s^{HG} \to 3$.

5. THERMAL FREEZE-OUT AND SINGLE FREEZE-OUT

The non-equilibrium features introduced into the chemical analysis reduce the chemical freeze-out temperature to values close to those expected for the thermal freeze-out, that is when the spectral shape of particles freezes out after at a low particle density also elastic hadron interactions cannot occur. One is thus tempted to ask the question if both freeze-out conditions (chemical and thermal) are not actually one and the same. The confirmation came when we found that the spectra and yields of hyperons, antihyperons and kaons were very well described within this scenario. However, we could not publish the results of a single freeze-out scenario at SPS [35], our scientific findings disagreed with personal opinion of than associate editor of The Physical Review Letters, Herr Ulrich Heinz. Our submission to PRL (dated March 8, 1999) occurred on the same day as the submission to Physics Letters B of the two-freeze-out chemical equilibrium, however this work was published [32], which as history has often shown makes a lot of difference.

Despite this setback we proceeded to enlarge on our single-freeze-out sudden hadronization picture for SPS, which in our opinion was strongly supported by many experimental facts, and we could present these results at meetings [36]. Specifically, the spectra of strange hadrons and anti-hadrons show universal slope, see figure 2, which supports such a single freeze-out scenario decisively. The understanding of pion spectra is very difficult since many (yet undiscovered) resonances contribute in a relevant fashion, and the soft part of the spectrum is particularly prone to 'deformation' by hadronization mechanisms and unknown matter flows [37]. Thus we report here on our precise study of spectra of kaons and hyperons [38].

The final particle distribution is composed of directly produced particles and decay products of heavier hadronic resonances:

$$\frac{dN_X}{dm_\perp} = \frac{dN_X}{dm_\perp}\bigg|_{\text{direct}} + \sum_{\forall R \to X+2+\cdots} \frac{dN_X}{dm_\perp}\bigg|_{R \to X+2+\cdots}. \quad (44)$$

$R(M, M_T, Y) \to X(m, m_T, y) + 2(m_2) + \cdots$, where we indicate by the arguments that only for the decay particle X we keep the information about the shape of the momentum spectrum.

In detail, the decay contribution to yield of X is:

$$\frac{dN_X}{dm_\perp^2 dy} = \frac{g_r b}{4\pi p^*} \int_{Y_-}^{Y_+} dY \int_{M_{T_-}}^{M_{T_+}} dM_T^2 \frac{M}{\sqrt{P_T^2 p_T^2 - \{ME^* - M_T m_T \cosh \Delta Y\}^2}} \frac{d^2 N_R}{dM_T^2 dY},$$
$$(45)$$

We have used $\Delta Y = Y - y$, and $\sqrt{s}$ is the combined invariant mass of the decay products other than particle X and $E^* = (M^2 - m^2 - m_2^2)/2M$, $p^* = \sqrt{E^{*2} - m^2}$ are the energy, and momentum, of the decay particle X in the rest frame of its parent. The limits on the integration are the maximum values accessible to the decay product X:

$$Y_\pm = y \pm \sinh^{-1}\left(\frac{p^*}{m_T}\right), \qquad M_{T_\pm} = M \frac{E^* m_T \cosh \Delta Y \pm p_T \sqrt{p^{*2} - m_T^2 \sinh^2 \Delta Y}}{m_T^2 \sinh^2 \Delta Y + m^2}.$$

The theoretical primary particle spectra (both those directly produced and parents of decay products) are derived from the Boltzmann distribution by Lorenz-transforming from a flowing intrinsic fluid element to the CM-frame, and integrating over allowed angles between particle direction and local flow.

We introduce in the current analysis two velocities: a local flow velocity v of the fireball matter where from particles emerge, and hadronization surface (breakup) velocity which we refer to as $v_f^{-1} \equiv dt_f/dx_f$. Particle production is controlled by the effective volume element, which comprises this quantity. In detail:

$$dS_\mu p^\mu = d\omega \left(1 - \frac{\vec{v}_f^{-1} \cdot \vec{p}}{E}\right), \quad d\omega \equiv \frac{d^3x d^3 p}{(2\pi)^3}. \tag{46}$$

The Boltzmann distribution we adapt has thus the form

$$\frac{d^2 N}{dm_T dy} \propto \left(1 - \frac{\vec{v}_f^{-1} \cdot \vec{p}}{E}\right) \gamma m_T \cosh y \, e^{-\gamma \frac{E}{T}\left(1 - \frac{\vec{v} \cdot \vec{p}}{E}\right)}, \tag{47}$$

where $\gamma = 1/\sqrt{1 - v^2}$. The normalization for each hadron type $h = X, R$ is $N^h = V_{\mathrm{QGP}} \Pi_{i \in h}^n \lambda_i \gamma_i$. We use the chemical parameters λ_i and γ_i, $i = q, s$, as obtained in the chemical analysis.

The experimental data we consider are published $m_\perp$ distribution [39], with additional information obtained about absolute normalization. This allowed us to perform the spectral shape analysis together with yield analysis, which indicates that the reaction volume is increasing as expected with the centrality of the reaction. This said, we will only focus here on the question if the shape of the spectra is consistent with the chemical freeze-out condition [38]. The best fit to the spectra in fact produces the temperature and transverse velocity in excellent agreement with those inferred from chemical analysis we have discussed above.

We show, in figure 8, the parameters determining the shape of the $m_\perp$ distributions, that is T, v, v_f, as function of the centrality for scattering bin 1, 2, 3, 4 with the most central bin being 4. The horizontal lines delineate range of result of the chemical analysis.

There is no indication of a significant, or systematic, change of T with centrality. This is consistent with the believe that the physics we are considering arises in all centrality bins explored by the experiment WA97 in Pb–Pb reactions at 158A GeV, i.e., for the number of participants greater than 60. Only most peripheral interactions produce a change in the pattern of strange hadron production [40, 41], and we are anticipating with deep interest the more peripheral hyperon spectra which should become available from experiment NA57. The magnitudes of the collective expansion velocity v and the break-up (hadronization) speed parameter v_f also do not show dependence on centrality, though within the experimental error, one could argue inspecting figure 8 (right) that there is systematic increase in transverse flow velocity v with centrality and thus size of the system. Such an increase is expected, since the more central events comprise greater volume of matter, which allows more time for development of the flow. Interestingly, it is in v, and not in T, that the slight change of spectral slopes noted in the presentation of the experimental data [39] is found.

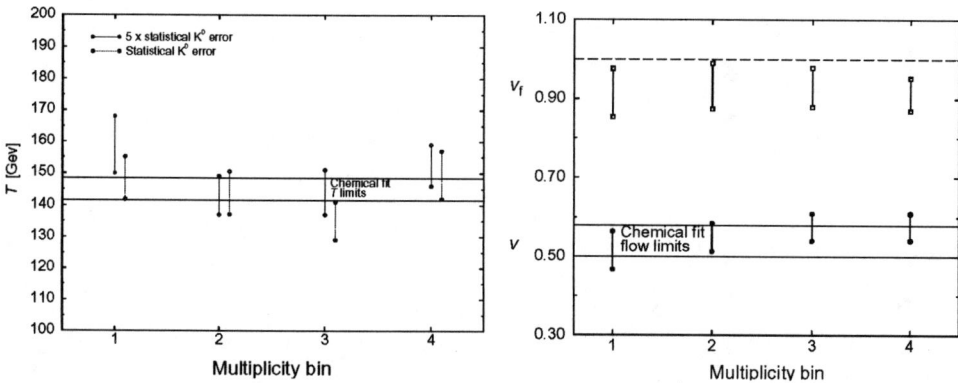

FIGURE 8. The thermal freeze-out temperature T (left), flow velocity v (bottom right), and breakup (hadronization hyper-surface-propagation) velocity v_f (top right) for various collision-centrality bins. The upper limit $v_f = 1$ (dashed line) and chemical-freeze-out-analysis limits for v (solid lines) are also shown. For the temperature, results obtained with increased error for kaon spectra are also shown.

The value of the beak-up (hadronization) speed parameter v_f shown in the top portion (right) of figure 8 is near to velocity of light which is consistent with the picture of a sudden breakup of the fireball. This hadronization surface velocity v_f was in the chemical fit fixed to be equal to v, as there was not enough sensitivity in purely chemical fit to determine the value of v_f.

It is important to explicitly check how well the particle $m_\perp$-spectra are reproduced. We group all centrality bin spectra and show, in figure 9, Λ, $\overline{\Lambda}$, Ξ and $\overline{\Xi}$. Overall, the description of the shape of the hyperon spectra is very satisfactory. In figure 10 on left we see for the four centrality bins the spectra for the sum $\Omega + \overline{\Omega}$. The two lowest $m_\perp$ data points are systematically under predicted. Some deviation at high $m_\perp$ may be attributable to acceptance uncertainties, also seen in the the Ξ results. This low $m_\perp$ enhancement of $\Omega + \overline{\Omega}$ is at the origin of the low value of the inverse slope and the associated excess of $\Omega, \overline{\Omega}$ compared to the chemical freeze-out analysis. Unlike with chemical analysis where the $\Omega, \overline{\Omega}$ have been omitted, the relatively large statistical errors allowed us to include the Ω-spectra in the fit procedure, which is dominated by the other hyperons and kaons. This allows to see how the deviation from the systematics established by other hyperons and kaons is arising. In fact the low $m_\perp$-bins of the $\Omega + \overline{\Omega}$ experimental spectrum with experimental yield excess at 1–2 s.d. translate into 3 s.d. deviations from the yields generated in the chemical analysis.

For kaons K^0 (figure 10 on right) the statistical errors are very small, and we find in a more in depth statistical analysis that they must be smaller than the systematic errors not considered. For this reason we have presented earlier in figure 8 results for temperatures obtained with both statistical, and 5 times greater than statistical error for kaons. This increased value was used in the fit with objective to asses the influence of the unknown systematic error. The stability of the result implies that statistically precise kaons do not overwhelm the fit procedure of hyperons, and/or that the hyperon and kaon results are consistent with the model employed.

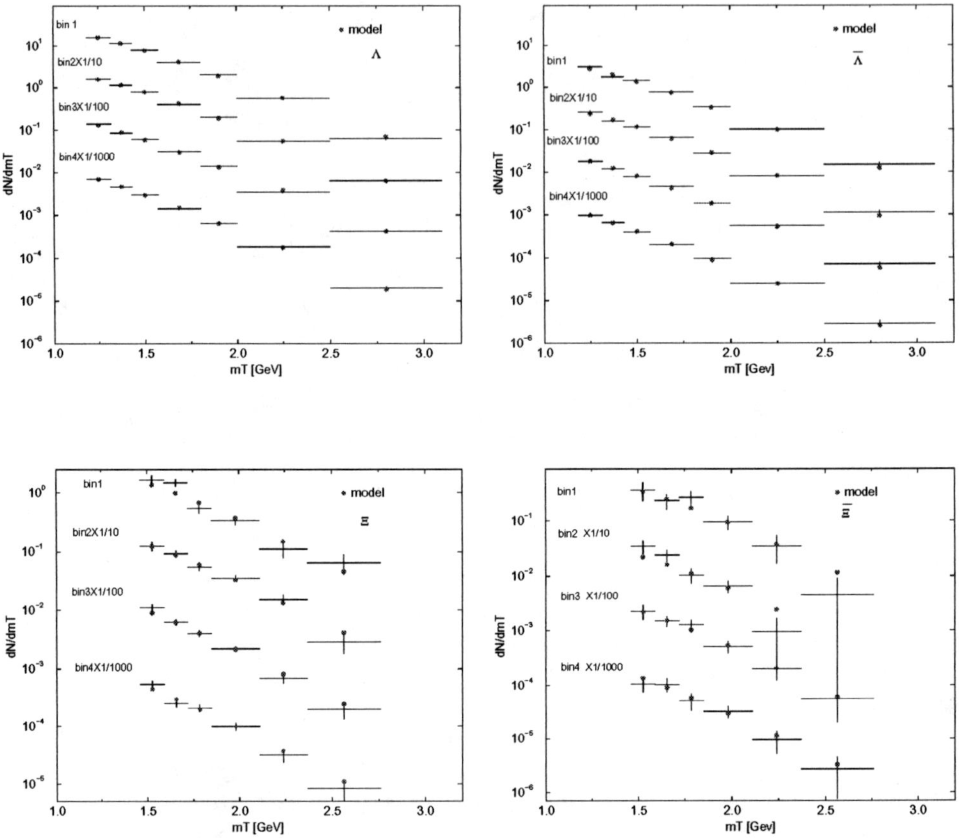

FIGURE 9. Thermal analysis m_T spectra: Λ (top left), $\overline{\Lambda}$ (top right) Ξ (bottom left), $\overline{\Xi}$ (bottom right).

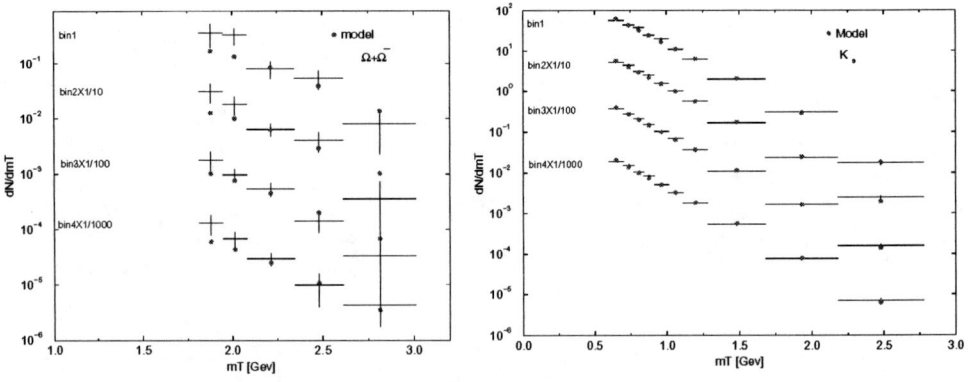

FIGURE 10. Thermal analysis m_T spectra: $\Omega + \overline{\Omega}$ (left) and K_s (right).

Overall, we see that the description of kaon and hyperon spectra with the parameters obtained in chemical freeze-out analysis if possible, indeed we conclude from the above study that the thermal analysis is proving the conclusion that both thermal and chemical freeze-out occur at the same condition of temperature and transverse velocity, as one would expect in sudden break up of a deconfined supercooled quark–gluon plasma. Strange hadrons at SPS originate in single freeze-out reaction.

6. RHIC-130 HADRON ABUNDANCE ANALYSIS, ROUND IIB

The strangeness yield observed in 158–200A GeV reactions corresponds to $\mathcal{O}(0.7)$-$s\bar{s}$-pairs of quarks per participant baryon. We estimate that this is corresponding to 50% QGP phase-space occupancy, thus considerably more extreme results on strangeness can arise at RHIC and with it greater anomalies in strange baryon and antibaryon enhancement. The experimental STAR and PHENIX collaboration results considered here were obtained at $\sqrt{s_{NN}} = 130$ GeV in the central-rapidity region, for the most central (5–7%) collision reactions. The particle production results available as of QM2002 meeting (Nantes, July 2002) allow a rather good understanding of the physical conditions established at the chemical hadron freeze-out at RHIC, and we use the opportunity to update our analysis.

Due to the approximate longitudinal scaling for the central rapidity results the effects of longitudinal flow at central rapidity cancels out and we can evaluate the full phase-space yields in order to obtain particle ratios. In our following analysis, we do not include natural results such as $\pi^+/\pi^- = 1$, which can be expected since the large hadron yield combined with the flow of baryon isospin asymmetry toward the fragmentation rapidity region assures this result to within a great precision, as we have discussed in section 2. We also do not fit the results for K^* and $\overline{K^*}$ since the reconstructed yields depend on the degree of rescattering of resonance decay products [42]. We consider here 19 particle ratios seen in table 5. We favor consideration of particle ratios, since this allows us to combine ratios from different experiments with slightly different trigger conditions, and reduces systematic errors. We have used published experimental ratios and also formed ratios of rapidity densities reported by one and the same experiment.

We present in table 5 in three last columns the results for both chemical equilibrium (last column) and non-equilibrium fits, *i.e.* in the letter case we fix the chemical parameters to their equilibrium value. We observe a considerable improvement in the statistical significance of the results of chemical non-equilibrium fits (see bottom line), as we have seen in our related earlier RHIC work [43, 44], in consistency with the situation at SPS. Next to the fitted results, we show in parenthesis the contribution to the error (χ^2) for each entry. We consider only statistical errors for the experimental results, since much of the systematic error should cancel in the particle ratios. However, we do not allow, when pion multiplicity is considered, that errors are smaller than $\simeq 8\%$, which is our estimated error in the theoretical evaluation of the pion yield due to incomplete understanding of the high mass hadron resonances. Some of the experimental results are thus shown with a theoretical error. When such an enlargement of the experimental error is introduced, a dagger as superscript appears in second column in table 5 below.

TABLE 5. Fits of central-rapidity hadron ratios at $\sqrt{s_{NN}} = 130$ GeV. Top section: experimental results, followed in middle by chemical parameters (results of fits), and the physical properties of the phase-space obtained from evaluation of final state hadron phase-space, and the fitting error. Columns: ratio considered, data value with reference, the non-equilibrium fit with 100% $\Xi \to Y$ cascading ($f_\Xi = 1$) and 40% $Y \to N$ ($f_\Lambda = 0.4$), the non-equilibrium fit with 40% $\Xi \to Y$ and 40% $Y \to N$, and in the last column, the chemical equilibrium fit with 40% cascading. The superscript * indicates quantities fixed by constrains and related considerations. The superscript † indicates the error is dominated by theoretical considerations. Subscripts Ξ, Λ mean that these values include weak cascading. In parenthesis we show the contribution of the particular result to the total χ^2.

	Data	Ref.	100% $\Xi \to Y$ 40% $Y \to N$	40% $\Xi \to Y$ 40% $Y \to N$	40% $\Xi \to Y$ 40% $Y \to N$
$\bar{p}/p$	0.71 ± 0.06	[45]	0.672(0.4)	0.678(0.3)	0.689(0.1)
$\bar{\Lambda}_\Xi/\Lambda_\Xi$	0.71 ± 0.04	[46]	0.759(1.0)	0.748(0.9)	0.757(1.4)
$\bar{\Xi}/\Xi$	0.83 ± 0.08	[47]	0.794(0.2)	0.804(0.1)	0.816(0.0)
K^-/K^+	0.87 ± 0.07	[48]	0.925(0.6)	0.924(0.6)	0.934(0.8)
$K^-/\pi^\pm$	$0.15 \pm 0.02^\dagger$	[48]	0.159(0.2)	0.161(0.3)	0.150(0.0)
$K^+/\pi^\pm$	$0.17 \pm 0.02^\dagger$	[48]	0.172(0.0)	0.174(0.1)	0.161(0.2)
Λ_Ξ/h^-	$0.059 \pm 0.004^\dagger$	[46]	0.057(0.3)	0.050(5.1)	0.045(11.9)
$\bar{\Lambda}_\Xi/h^-$	$0.042 \pm 0.004^\dagger$	[46]	0.043(0.0)	0.037(1.3)	0.034(3.8)
Λ_Ξ/p	0.90 ± 0.12	[45]	0.832(0.3)	0.691(3.0)	0.491(11.6)
$\bar{\Lambda}_\Xi/\bar{p}$	0.93 ± 0.19	[45]	0.929(0.0)	0.763(0.8)	0.539(4.2)
$\pi^\pm/p_\Lambda$	9.5 ± 2	[48]	9.4(0.0)	9.2(0.5)	7.6(22.8)
$\pi^\pm/\bar{p}_\Lambda$	13.4 ± 2.5	[48]	13.7(0.1)	13.4(0.0)	10.9(7.9)
Ξ^-/π	$0.0088 \pm 0.0008^\dagger$	[47, 49]	0.0096(1.0)	0.0103(3.6)	0.0067(7.1)
Ξ^-/h^-	0.0085 ± 0.0015	[47, 49]	0.0079(0.1)	0.0084(0.0)	0.0054(4.3)
$\bar{\Xi}^-/h^-$	0.0070 ± 0.001	[47, 49]	0.0063(0.5)	0.0068(0.1)	0.0044(6.7)
Ξ^-/Λ	0.193 ± 0.009	[49]	0.195(0.1)	0.196(0.1)	0.132(45.2)
$\bar{\Xi}^-/\bar{\Lambda}$	0.221 ± 0.011	[49]	0.213(0.6)	0.214(0.4)	0.144(48.7)
Ω/Ξ^-			0.205	0.21	0.18
$\bar{\Omega}/\bar{\Xi}^-$			0.22	0.23	0.20
$\bar{\Omega}/\Omega$	0.95 ± 0.1	[49]	0.87(0.7)	0.88(0.5)	0.89(0.4)
$\bar{p}/h^-$			0.046	0.049	0.063
ϕ/K^-	0.15 ± 0.03	[49]	0.178(0.9)	0.185(1.3)	0.146(0.0)
T			140.1 ± 1.1	142.3 ± 1.2	164.3 ± 2.2
γ_q^{HG}			1.64^*	1.63^*	1^*
λ_q			1.070 ± 0.008	1.0685 ± 0.008	1.065 ± 0.008
μ_b [MeV]			28.4	28.3	31.0
$\gamma_s^{HG}/\gamma_q^{HG}$			1.54 ± 0.04	$1.54 \pm 0.0.04$	1^*
λ_s			1.0136^*	1.0216^*	1.0196^*
μ_S [MeV]			6.1	6.4	7.1
E/b [GeV]			35.0	34.6	34.8
s/b			9.75	9.7	7.2
S/b			234.8	230.5	245.7
E/S [MeV]			148.9	150.9	141.5
χ^2/dof			7.1/(19−3)	19/(19−3)	177.2/(19−2)

As is today well understood, the high yields of hyperons require significant corrections for unresolved weak decays. Some experimental results are already corrected in this fashion: the weak cascading corrections were applied to the most recent p and $\bar{p}$ results by the PHENIX collaboration [45], and in the Ξ/Λ and $\overline{\Xi}/\overline{\Lambda}$ ratio of the STAR collaboration we use here [47, 49]. However, some of the results we consider are not yet corrected [48, 46], and are indicated in the first column in table 5 by a subscript Λ or Ξ.

The subscript Λ means that the cascading of hyperons into nucleons has to be included while fitting the particle ratio considered, and we have found that a 35-40% cascading ($f_\Lambda = 0.35$-0.4) is as expected favored by the fits of pion to nucleon ratio. Similarly, the subscript Ξ indicates that the Ξ cascading into single strange hyperons $Y = \Sigma, \Lambda$ was also not corrected for in the considered result. Here we find empirically that full acceptance of the Cascades into singly strange hyperons is favored, again by both equilibrium and non-equilibrium fits ($f_\Xi \to 1$). We also include in the Ξ and Λ yield with the appropriate branching the weak decay of Ω, with the experimental acceptance fraction being the same as in the decay of Ξ ($f_\Omega = f_\Xi$).

In other words, in the results shown in table 5, the hyperon yields are:

$$\Lambda_\Xi = \Lambda_{th} + \Sigma^0_{th} + f_\Xi(2\Xi^-_{th} + \Omega_{th}), \quad \overline{\Lambda}_\Xi = \overline{\Lambda}_{th} + \overline{\Sigma}^0_{th} + f_\Xi(2\overline{\Xi^-}_{th} + \overline{\Omega^-}_{th});$$

$$p_\Lambda = p_{th} + f_\Lambda 0.77 \Lambda_\Xi, \quad \bar{p}_\Lambda = \bar{p}_{th} + f_\Lambda 0.77 \overline{\Lambda_\Xi};$$

$$\Xi^- = \Xi^-_{th} + 0.08 \Omega^-_{th}, \quad \overline{\Xi^-} = \overline{\Xi^-}_{th} + 0.086 \overline{\Omega^-}_{th}.$$

Here subscript 'th' indicates statistical model yields. The factor '2' preceding $\Xi^-_{th}, \overline{\Xi^-}_{th}$ allows for the decay of $\Xi^0, \overline{\Xi^0}$ which are assumed to be equally abundant. However, there is considerable isospin asymmetry in the decay pattern, and the branching ratio weight 0.77 arises as follows: 64% of Λ and 51.6% of Σ^+ decay to protons. Statistical model evaluation shows that Σ^+ (the only isospin channel decaying into $p, \bar{p}$) are produced at the level of 25% of $\Lambda + \Sigma^0$ (which one usually calls Λ), considering within the statistical model that a significant fraction of all Λ originates in the $\Sigma^*(1385)$ resonance.

Below the fit results we show the statistical parameters which are related to each fit. The results shown in the table 5 are obtained minimizing in the space of 3 parameters: the chemical freeze-out temperature T, and 2 chemical parameters λ_q, γ_s, the value of γ_q is set at its maximal value $\gamma_q^2 = \gamma_q^{c2} = e^{m_\pi/T}$ and the value of λ_s is derived from the strangeness conservation constraint. We note that even without this requirement the fit converges to local strangeness neutrality within a few percent.

In table 5 the last column presents the results of chemical equilibrium fit with 40% cascading. We note that several particle yileds are not properly described and hence a large χ^2 results. However, on a logarithmic scale only results involving $\Lambda, \overline{\Lambda}$ would be clearly visible as a discrepancy. The second and third last column show result of chemical non-equilibrium fit, the second last with 40% cascading which yields a good fit, and the third last with an increased 100% acceptance for the Ξ-cascading gives the smallest χ^2, and has a very high confidence level.

Our current results, when compared to our earlier effort [43, 44] show a 7% reduction in the freeze-out temperature both for equilibrium and non-equilibrium case. Such lower T reduces selectively the relative yield of baryons. The required level of proton and hyperon yields are now reached at lower temperature due to the introduction of strange baryon weak cascading, or respectively, of experimental results which are corrected for cascading. The range of temperature now seen agrees better with the expectations we had considering the effect of the fast expansion of QGP [50]. The chemical equilibrium freeze-out temperature, $T = 164$ MeV, is in agreement with results of chemical equilibrium single freeze-out model of Broniowski and Florkowski [51, 52]. We believe that the case for single freeze-out particle production at RHIC will be statistically stronger with introduction of chemical non-equilibrium. We have seen this happen in section 5 for the SPS results.

Returning to discuss results seen in the bottom of table 5 we note that the specific strangeness content $s/b \simeq 10$ in chemical non-equilibrium fits is 35% greater than the chemical equilibrium result. This originates in $\gamma_s/\gamma_q \simeq 1.5$ (i.e. $\gamma_s \simeq 2.5$). Figure 6 ahows that the chemically equilibrated QGP phase-space has 10-14 pairs of strange quarks per baryon at $1.062 < \lambda_q < 1.078$. This implies that at RHIC-130 we have $\gamma_s^{QGP} \simeq 0.85 \pm 0.15$. This in turn *theoretically* implies that $\gamma_s^{HG} \simeq 3(0.85 \pm 0.15) = 2.5 \pm 5$ given the enhancement of the HG phase space occupancy by a factor three seen in figure 7. Thus the specific yield of strangeness and strangeness occupancy we measure in the HG after hadronization is consistent with QGP properties with nearly saturated strangeness phase space. Even so, a further enhancement of γ_s can be expected in the 200 GeV RHIC run.

Once the comparison N-N experimental results will become available, the enhancement of the production of multistrange baryons by this extraordinary strangeness abundance must exceed the high values seen at SPS by a factor $(1.5\text{-}2.5)^{n_s}$, where n_s is strangeness content of the hadron. An enhancement by a factor 50–300 of the $\Xi, \overline{\Xi}, \Omega, \overline{\Omega}$ yields is difficult to explain without new physics, and is natural for the fast hadronizing baryonpoor QGP phase.

In closing we would like to emphasize that the chemical non-equilibrium description of hadronization process is statistically highly favored at SPS and RHIC. Both the particle spectra and the HBT measurement of space-time size are favoring the sudden hadronization of a relatively small, short lived and rapidly expanding matter fireball, which reaction picture necessitates presence of chemical non-equilibrium. We hope that the reader who studies this report gains further the impression that a non-equilibrium chemical analysis of heavy-ion particle yields offers profound insight into the physical properties of the dense hadronic matter formed in the relativistic heavy-ion collisions.

ACKNOWLEDGMENTS

Work supported in part by a grant from the U.S. Department of Energy, DE-FG03-95ER40937, and by NSF grant INT-0003184. Laboratoire de Physique Théorique et Hautes Energies, LPTHE, at University Paris 6 and 7 is supported by CNRS as Unité Mixte de Recherche, UMR7589.

REFERENCES

1. R. Hagedorn. *Suppl. Nuovo Cimento*, **3**, 147, 1965.
2. R. Hagedorn and J. Ranft. *Suppl. Nuovo Cimento*, **6**, 169, 1968.

3. R. Hagedorn. Statistical bootstrap model. In *Cargese Lectures in Physics*, volume 5. Pergamon Press, Oxford, 1973.
4. E. Fermi. *Prog. Theor. Phys.*, **5**, 570, 1950.
5. J. Letessier, J. Rafelski, and A. Tounsi. *Phys. Lett.* B, **328**, 499, 1994.
6. A. Bialas. *Phys. Lett.* B, **466**, 301, 1999.
7. J. Rafelski, J. Letessier, and A. Tounsi. *Acta. Phys. Pol.* B, **27**, 1037, 1996.
8. D. Evans et al., WA85 collaboration. *Nucl. Phys.* A, **566**, 225c, 1994.
9. D. Evans et al., WA85 collaboration. In *Strangeness in Hadronic Matter: S'95*, p. 234, American Institute of Physics Proceedings Series, vol. **340**, New York, 1995.
10. D. Di Bari et al., WA85 collaboration. *Nucl. Phys.* A, **590**, 307c, 1995.
11. S. Abatzis et al., WA94 collaboration. *Phys. Lett.* B, **354**, 178, 1995.
12. T. Matsui and H. Satz. *Phys. Lett.* B, **178**, 416, 1986.
13. T. Matsui, B. Svetitsky, and L. D. McLerran. *Phys. Rev.* D, **34**, 783, 1986.
14. P. Braun-Munzinger, D. Magestro, K. Redlich, and J. Stachel. *Phys. Lett.* B, **518**, 41, 2001.
15. J. Letessier and J. Rafelski. *Phys. Rev.* C, **59**, 947, 1999.
16. J. Letessier, A. Tounsi, U. Heinz, J. Sollfrank, and J. Rafelski. *Phys. Rev. Lett.*, **70**, 3530, 1993.
17. J. Letessier, A. Tounsi, U. Heinz, J. Sollfrank, and J. Rafelski. *Phys. Rev.* D, **51**, 3408, 1995.
18. M. Gaździcki. *Z. Phys.* C, **66**, 659, 1995.
19. J. Rafelski, J. Letessier, and G. Torrieri. *Phys. Rev.* C, **64**, 54 907, 2001.
20. I. Králik, et al., WA97 collaboration. *Nucl. Phys.* A, **638**, 115, 1998.
21. C. Bormann, et al., NA49 collaboration. *J. Phys.* G, **23**, 1817, 1997.
22. F. Siklér et al., NA49 collaboration. *Nucl. Phys.* A, **661**, 45c, 1999.
23. P. G. Jones et al., NA49 collaboration. *Nucl. Phys.* A, **610**, 188c, 1996.
24. H. Appelshäuser et al., NA49 collaboration. *Phys. Rev. Lett.*, **82**, 2471, 1999.
25. S. V. Afanasev et al., NA49 collaboration. *Phys. Lett.* B, **491**, 59, 2000.
26. J. Rafelski. *Phys. Rep.*, **88**, 331, 1982.
27. C. Greiner, P. Koch, and H. Stöcker. *Phys. Rev. Lett.*, **58**, 1825, 1987.
28. J. Rafelski. *Phys. Lett.* B, **190**, 167, 1987.
29. J. I. Kapusta and S. M. H. Wong. *Phys. Rev. Lett.*, **86**, 4251, 2001.
30. F. Becattini, M. Gaździcki, and J. Sollfrank. *Eur. Phys. J.* C, **5**, 143, 1998.
31. J. Letessier, J. Rafelski, and A. Tounsi. *Phys. Lett.* B, **410**, 315, 1997.
32. P. Braun-Munzinger, I. Heppe, and J. Stachel. *Phys. Lett.* B, **465**, 15, 1999.
33. P. Braun-Munzinger, J. Stachel, J. P. Wessels, and N. Xu. *Phys. Lett.* B, **365**, 1, 1996.
34. J. Letessier and J. Rafelski. *J. Phys.* G, **25**, 295, 1999.
35. J. Rafelski and J. Letessier. *Phys. Lett.* B, **469**, 12, 1999.
36. J. Rafelski and J. Letessier. *Acta Phys. Pol.* B, **30**, 3559, 1999.
37. J. Letessier, G. Torrieri, S. Hamieh, and J. Rafelski. *J. Phys.* G, **27**, 427, 2001.
38. G. Torrieri and J. Rafelski. *New J. Phys.*, **3**, 12, 2001.
39. F. Antinori et al., WA97 collaboration. *Eur. Phys. J.* C, **14**, 633, 2000.
40. S. Kabana et al., NA52 collaboration. *Nucl. Phys.* A, **661**, 370c, 1999.
41. S. Kabana et al., NA52 collaboration. *J. Phys.* G, **25**, 217, 1999.
42. C. Markert, G. Torrieri, and J. Rafelski. Strange hadron resonances: Freeze-out probes in heavy-ion collisions. In this volume, 2002.
43. J. Rafelski. *J. Phys.* G, **28**, 1833, 2002.
44. J. Rafelski and J. Letessier. *Nucl. Phys* A, **702**, 304, 2002.
45. C. Adcox et al., PHENIX collaboration. *E-print nucl-ex/0204007*, 2002.
46. C. Adler et al., STAR collaboration. *E-print nucl-ex/0203016*, 2002.
47. J. Castillo et al., STAR collaboration. Mid-rapidity multi-strange baryon production at $\sqrt{s_{NN}} = 130$ gev. In this volume, 2002.
48. C. Adcox et al., PHENIX collaboration. *Phys. Rev. Lett*, **88**, 242301, 2002.
49. July 2002 Presentations at Nantes. Quark matter 2002,. Nuclear Physics A, 2003.
50. J. Rafelski and J. Letessier. *Phys. Rev. Lett.*, **85**, 4695, 2000.
51. W. Broniowski and W. Florkowski. *Phys. Rev. Lett.*, **87**, 272302, 2001.
52. W. Broniowski and W. Florkowski. *Phys. Rev.* C, **65**, 064905, 2002.

Nonlinear Behavior of Quarkonium Formation and Deconfinement Signals

R. L. Thews

Department of Physics, University of Arizona, Tucson, AZ 85718 USA

Abstract. We anticipate new features of quarkonium production in heavy ion collisions at RHIC and LHC energies which differ from a straightforward extrapolation of results at CERN SPS energy. General arguments indicate that one may expect quarkonium formation rates to increase more rapidly with energy and centrality than the production rate of the heavy quarks which they contain. This is due to new formation mechanisms in which independently-produced quarks and antiquarks form a bound quarkonium state. This mechanism will depend quadratically on the total number of initially-produced heavy quark pairs, and becomes numerically significant only at RHIC and LHC energy. When viewed as a signal of color deconfinement, a transition from suppression to enhancement may be observed. Explicit model calculations are presented, in which one can follow striking variations of final quarkonium production within a range of parameter space.

INTRODUCTION

The production of heavy quarkonium states in high energy hadronic interactions proceeds through creation of the corresponding flavor heavy quark-antiquark pair. Given the heavy quark mass to provide a perturbative scale, one can employ a perturbative QCD calculation for this initial process [1]. The spectrum of heavy quarkonium states can be described by essentially non-relativistic heavy quarks interacting via a static potential. Phenomenological extraction of the potential leads to a linear rise at large separation, which provides the observed quark confinement. This potential can also be directly calculated via lattice methods [2], and confirms these general properties. These lattice methods can also be utilized for QCD at finite temperature, which reveals that at sufficiently high T the QCD spectrum will change from confined hadrons to colored degrees of freedom. The corresponding heavy quark static potential in this case shows a decrease in the long range part, and disappears above the deconfinement temperature. The goal of high energy heavy ion collision experiments is to create a region of space-time within which these finite temperature predictions can be tested.

A signature of color deconfinement which utilizes heavy quarkonium production rates was proposed more than 15 years ago [3]. One invokes the argument that in the deconfined region where the color-confining force has disappeared, a heavy quark and antiquark cannot form a quarkonium bound state, and they may diffuse away from each other to separations larger than typical hadronic dimensions. As the system cools and the confining potential reappears, these heavy quarks will not be able to "find" each other and form heavy quarkonium. They will then bind with quarks which are close by to them at hadronization. Since these quarks are predominantly the lighter u, d, and s flavors, they will most likely form a final hadronic state with "open" heavy flavor. The result will be a decreased population of heavy quarkonium relative to those which would have formed if a region of deconfinement had not been present. This scenario as applied to the charm sector is known as J/ψ suppression. There are now extensive data on J/ψ production using nuclear targets and beams. Results from the NA50 experiment at the CERN SPS reveal an "anomalous" suppression, prompting claims that this effect could be the expected signature of deconfinement [4]. Measurements at higher energies of RHIC, and eventually at the heavy ion runs at LHC, should be able to provide enough information to either support or refute these claims. Straightforward extension of the deconfinement scenarios to these higher energies anticipate that J/ψ suppression would be virtually complete at all centralities [5].

The purpose of this work is to point out that there will be another consequence of the increased beam energy for the suppression scenario. This is because one expects that multiple pairs of charm-anticharm quarks will be produced in the initial partonic stage of the collision. Perturbative QCD estimates predict about 10 charm pairs at RHIC energy, and several hundred pairs at LHC [6]. This situation provides a "loophole" in the Matsui-Satz argument, since there will be many more heavy quarks in the interaction region with which to combine. In order for this to happen, however, one must invoke a physical situation in which quarkonium states can be formed from *all combinations* of the heavy quark pairs. If this is possible, then the rate of quarkonium formation from N initially-produced quark-antiquark pairs would initially be proportional to N^2. If the total number of quarkonium states remains a small fraction of the total number of pairs, then the final population will retain this quadratic dependence. Although still small compared with N, this number can be much larger than the "ordinary" expectations which are linear in N.

In the next section, we go through the pQCD methods and results which lead to the large N-values at high energy. The variation of these quantities with centrality is also explored. The following section presents generic arguments for the properties of quarkonium formation in any physical situation for which the quadratic process is allowed. The last two sections present results specific to two physical models which share some of these properties.

HEAVY QUARK PRODUCTION IN A-A COLLISIONS

The calculation of heavy quark production by hadrons is based on perturbative QCD processes at the parton level. The perturbative approach requires a large scale to justify the expansion in powers of α_s, which is provided by the heavy quark mass. One then uses hadronic structure functions measured in other reactions (e.g. Deep Inelastic Scattering and Drell-Yan) plus factorization to calculate the hadronic cross section. The general expression is of the form

$$\sigma(s, m_Q, \mu_F, \mu_R) = \sum_{i,j} \int dx_1 \int dx_2 F_i(x_1, \mu_F) F_j(x_2, \mu_F) \hat{\sigma}_{ij}(\hat{s}, m_Q, \mu_R) \qquad (1)$$

where i and j label the initial state partons, $F(x, \mu_F)$ are the structure functions, evaluated at a factorization scale μ_F, and $\hat{\sigma}_{ij}$ are the partonic cross sections for producing

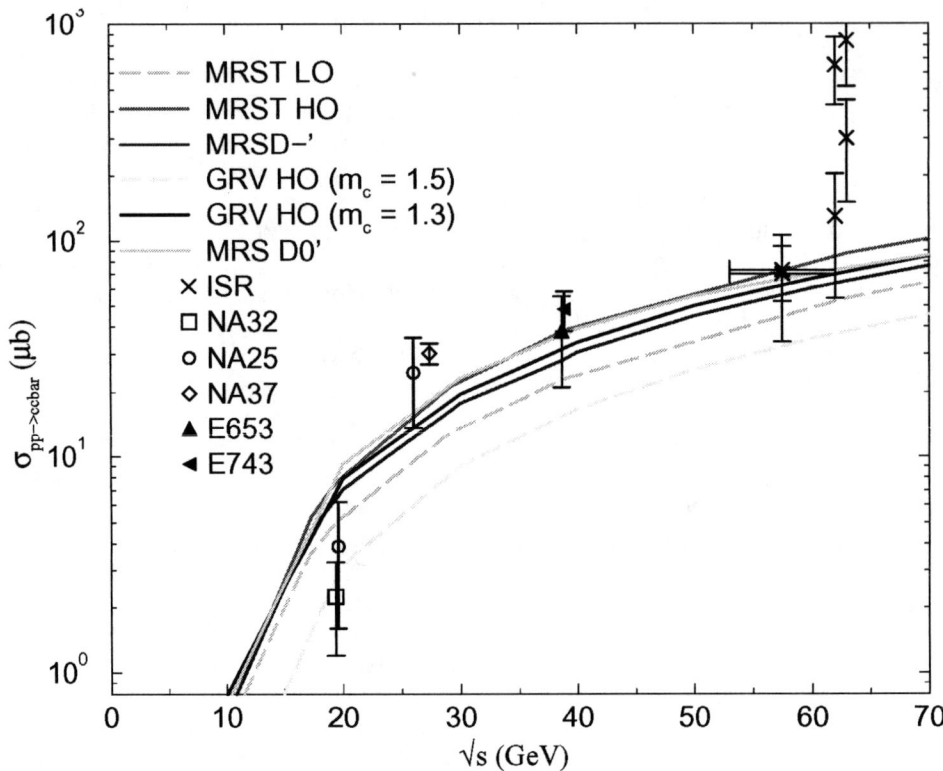

FIGURE 1. pQCD calculated cross section for $pp \to c\bar{c}$.

a heavy quark-antiquark pair, which depend on the partonic process subenergy $\hat{s} = x_1 x_2 s$, the heavy quark mass m_Q and the renormalization scale μ_R.

The partonic cross sections have been calculated [7],[8],[9] to lowest order (LO), in which quark-antiquark annihilation and gluon-gluon fusion and annihilation into $Q\bar{Q}$ contribute to order $\alpha_s^2(\mu_R)$, and also in next to leading order (NLO), where real gluons emitted in the final states of the LO processes, plus processes involving quark or antiquark plus gluon in the initial state, plus virtual loop corrections all contribute to order $\alpha_s^3(\mu_R)$. There is additional dependence on μ_R in NLO, in the form of terms proportional to $ln(\mu_R^2/m_Q^2)$. In practice one usually takes $\mu_R = \mu_F \equiv \mu$, but in general they are independent parameters of the calculation, along with m_Q. In addition, there are several sets of parton distribution functions which can be utilized. In order to constrain these parameters, a comprehensive comparison with existing charm production data in N-N and π-N reactions was undertaken [6]. For a recent update of this procedure, see Reference [10].

The charm mass m_c was allowed to vary between 1.2 and 1.5 GeV, and μ was varied between m_c and $2m_c$. The results for the cross section $pp \to c\bar{c}$ in the energy interval 10 GeV $< \sqrt{s} <$ 70 GeV are shown in Figure 1.

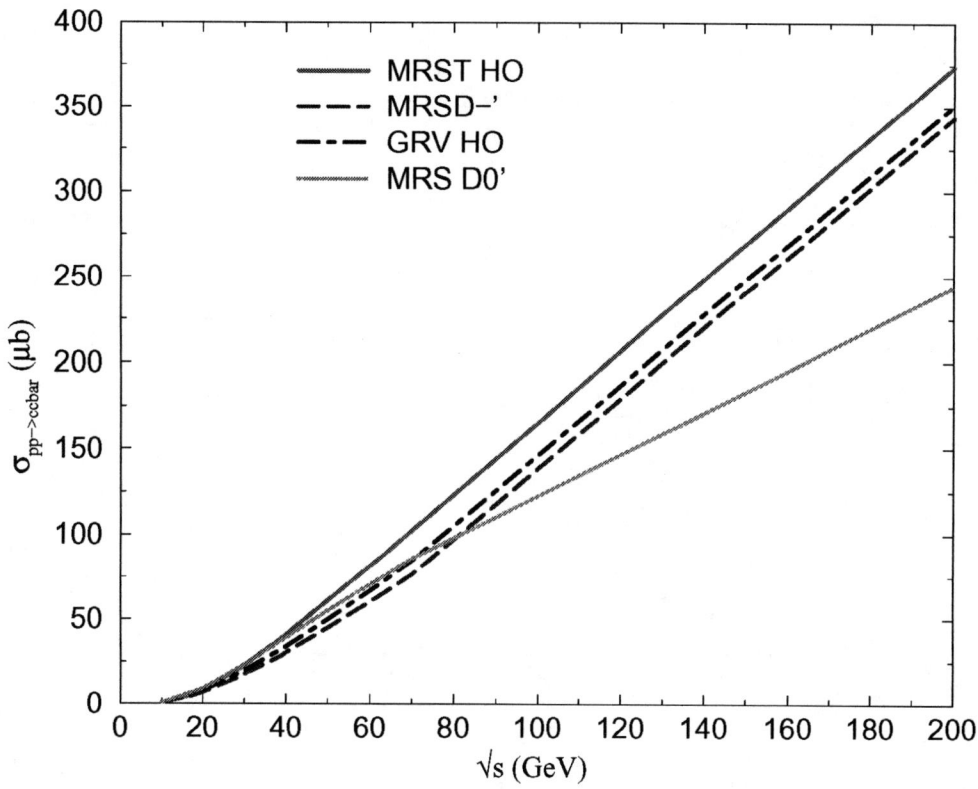

FIGURE 2. pQCD calculated cross section for $pp \to c\bar{c}$ extrapolated to RHIC energy.

Results from the original structure function sets MRSD0 and MRSD-' are shown, and supplemented by the more recent sets of structure functions MRST HO and GRV HO. The difference in predictions between these sets is in general smaller than the experimental uncertainties, except for the GRV with large m_c and the low order set MRST LO, which is only shown for comparison (for consistency one must use the NLO version of the structure functions together with NLO partonic cross sections). However, we must remember that the x-values probed in these calculations are always greater than $m_c/\sqrt{s}$, which is large enough such that all structure functions are very well constrained by the DIS and DY data. What one may worry about is the large change in cross section between the LO alone and the total LO + NLO. The ratio of these values (the K-factor) typically varies between 1.5 and 2.5 in this energy region. Thus one might expect that higher order terms in the perturbative expansion (NNLO) might also be as significant as the NLO. In this case, a satisfactory fit to the data would probably require different values of the mass and scale parameters, thus altering the predictions at higher energy.

The extrapolation of these calculations up to RHIC energies is shown in Figure 2. We show only the four structure function sets which agree with the low energy data. One

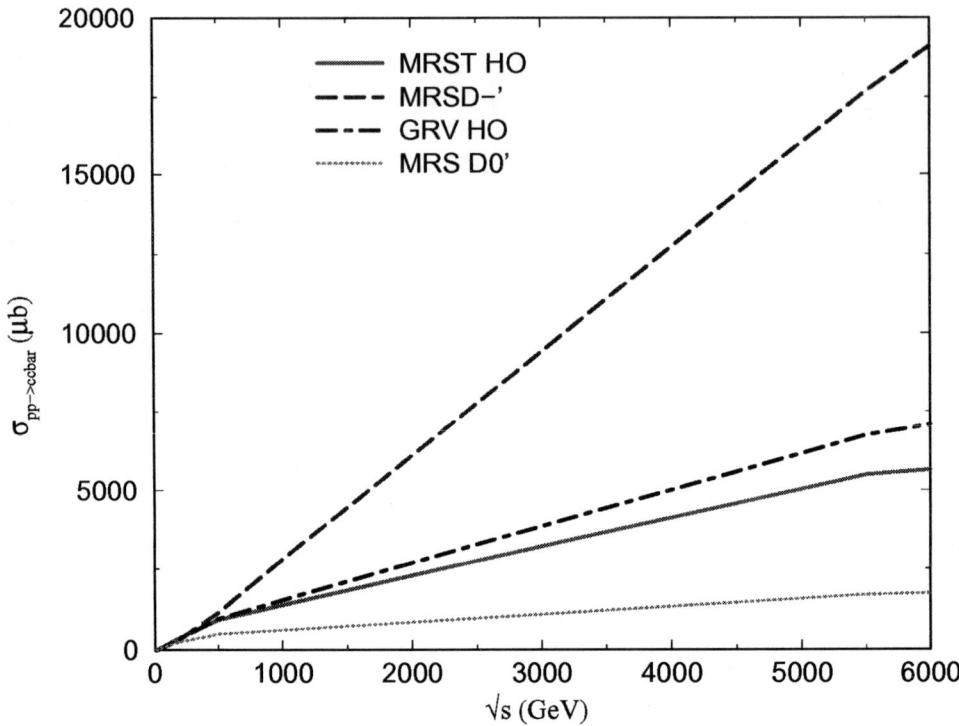

FIGURE 3. pQCD calculated cross section for $pp \to c\bar{c}$ extrapolated to LHC energy.

sees that there is some divergence at the highest $\sqrt{s} = 200$ GeV. However, this plot is on a linear scale, which maximizes the appearance of the differences.

Finally, we show in Figure 3 the extrapolation of these calculations to the LHC heavy ion energy region. The scale is again linear, and there is almost a factor of 10 difference between the old structure function sets. This can be attributed to the low x-values probed at $\sqrt{s} = 5500$ GeV, down to about $x \approx 3 \times 10^{-3}$. The more recent structure functions include DIS data from HERA which is sensitive to these low x-values, and we see that both the MRST and GRV HO sets follow each other much more closely. Note however, that we have not used the complete set of modern structure functions and parameter space values which fit the low energy data. The calculations in Reference [10] which do include all of these possibilities predict a range which differs by a factor of 2.3 between highest and lowest values.

We now use these cross sections to predict the number of heavy quark pairs which will be produced in heavy ion collisions. In the simplest case when we assume that the heavy ion collision is just an incoherent superposition of N-N collisions (no shadowing corrections), we only need calculate the integrated N-N luminosity in a single A-A collision. For a central collision between identical nuclei, geometry tells us this should be of the form $A^2/\pi R_A^2 \propto A^{\frac{4}{3}}$.

For a more general calculation, one needs a parameterization of the nuclear density $\rho_A(\vec{s},z)$, where z is the coordinate along the beam direction and $\vec{s}$ is the 2-dimensional position vector in the transverse plane. Then one can calculate a nuclear thickness function

$$T_A(\vec{s}) = \int dz \, \rho_A(\vec{s},z) \qquad (2)$$

which is normalized to the total nuclear number

$$\int d^2\vec{s} \, T_A(\vec{s}) = A. \qquad (3)$$

Then consider a collision of two nuclei which are incident along paths parallel to the z-axis separated in the transverse plane by a vector $\vec{b}$ (the magnitude of $\vec{b}$ is the impact parameter b). The integrated N-N luminosity is then just the product of the two nuclear thickness functions integrated over each overlap point in the transverse plane. This is called the nuclear overlap function:

$$\int d^2\vec{s} \, T_A(\vec{s}) \, T_B(\vec{b}-\vec{s}) \equiv T_{AB}(b) \qquad (4)$$

which by axial symmetry can only depend on the impact parameter b.

Calculations using standard nuclear density profiles produce a typical value for heavy ion (e.g. Pb-Pb or Au-Au) of $T_{AA}(b=0) \approx 30 mb^{-1}$. This leads to the estimates of $N_{c\bar{c}}$ for central collisions at the various experimental facilities. Taking average values for the open charm cross section calculations, we obtain $N_{c\bar{c}}$(b=0) = 0.2 (SPS), 10 (RHIC200), and 200 (LHC). The energy dependence and variation with structure function set are shown in Figures 4 and 5.

The first RHIC measurement of open charm has been reported at this institute [11],[12]. From observation of high-p_T electrons by the PHENIX Collaboration from the 130 GeV run, an equivalent value of the p-p open charm cross section can be extracted. For central collisions, the reported value is $\sigma(pp \to c\bar{c}) = 380 \pm 200 \mu b$. Although the uncertainties are still quite large, we can already draw some conclusions and make some interesting speculations.

The corresponding $N_{c\bar{c}}$ values for b = 0 collisions are overlayed on Figure 4 and shown in Figure 6. One sees that the magnitude of the calculated values are consistent with the measurement within errors. However, the central value is well above the calculated values, even exceeding the nominal prediction at the higher 200 GeV energy. A simple extrapolation of this central point to $\sqrt{s}$ = 200 GeV would imply $N_{c\bar{c}}$ between 15 and 20. This is substantially above the nominal estimate of 10, and could enhance the expected nonlinear effects for J/ψ formation. This situation is somewhat surprising, since these values are extracted just from nuclear geometry and calculations using structure functions of nucleons. One might expect that there will be a depletion of gluons in a heavy nucleus relative to free nucleons, similar to the shadowing of quark structure functions observed for DIS with nuclear targets [13]. These speculations, of course, are only relevant assuming the eventual uncertainties of the measured value will converge toward the current central value.

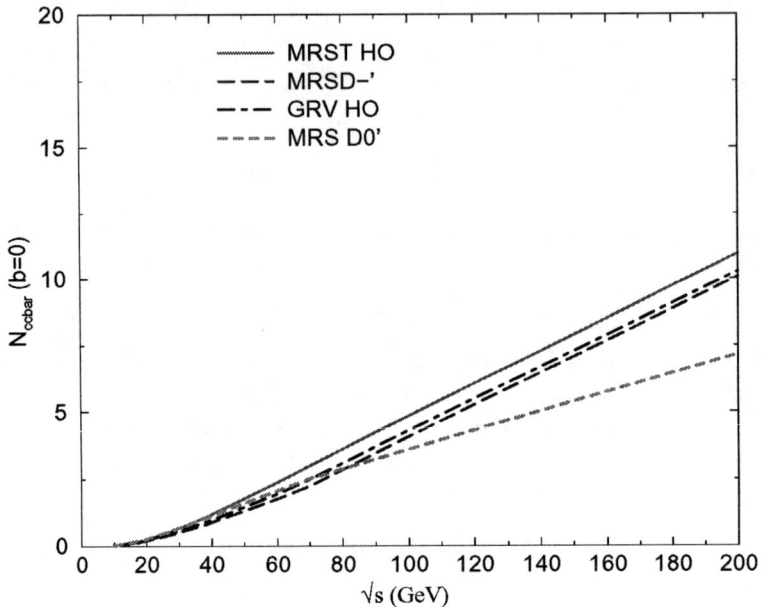

FIGURE 4. Energy dependence of central $N_{c\bar{c}}$ in RHIC region.

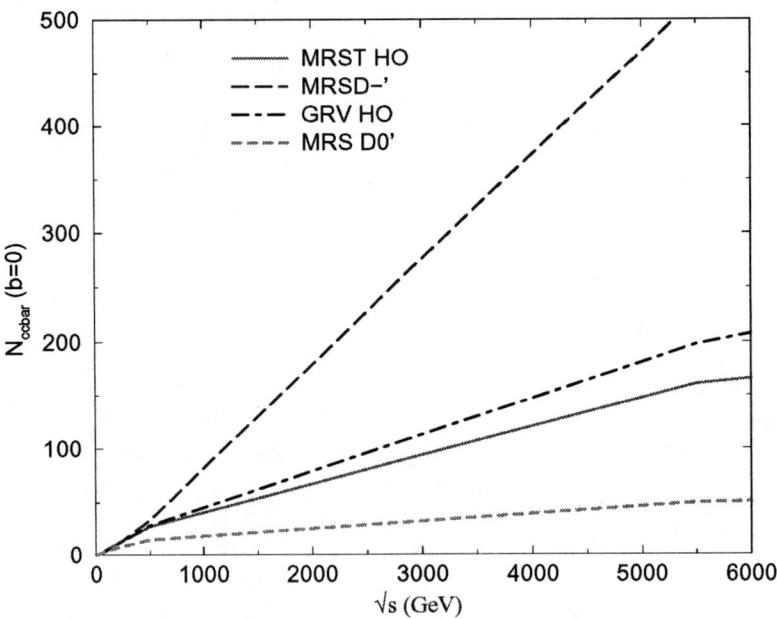

FIGURE 5. Energy dependence of central $N_{c\bar{c}}$ in LHC region.

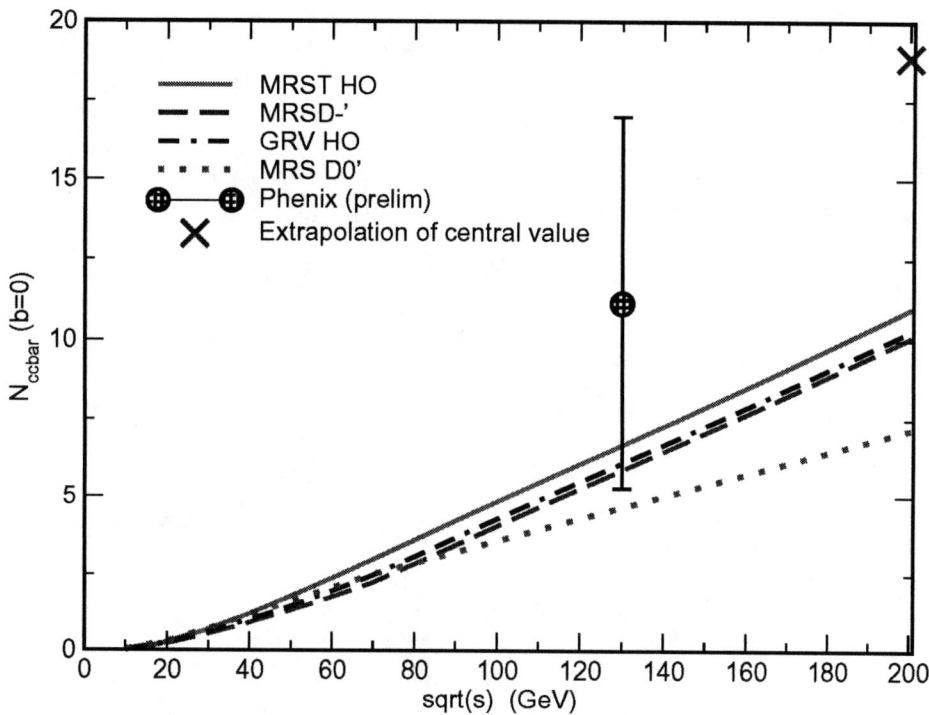

FIGURE 6. Comparison of PHENIX measurement with pQCD calculations of central $N_{c\bar{c}}$.

We can also use the overlap function $T_{AA}(b)$ to predict the centrality dependence. However, the impact parameter b is not directly measurable. Instead, the number of participant nucleons N_p is generally used as a connection between nuclear geometry and experimental measurables. Most experiments measure the transverse energy E_T of each event and relate this to centrality. For this, one needs a model for N_p. One popular choice is the wounded nucleon model [14], in which every nucleon which undergoes at least one inelastic collision is called "wounded" and is counted as a participant nucleon. The utility is reinforced by the experimental observation that N_p and E_T are linearly-related over quite a wide range in centrality [15]. We show in Figure 7 the calculated dependence of N_w on impact parameter, using standard nuclear density profiles and an inelastic N-N cross section of 50 mb. A similar behavior to that of the nuclear overlap function T_{AA} is evident. Also shown for future reference (dotted line) is the wounded nucleon density in impact parameter space (evaluated at the center of the overlap area). We can then recast the centrality dependence of $T_{AA}(b)$ in terms of the participant (or in this case wounded) nucleon number. This is shown in Figure 8.

One sees that there is a power law dependence, $T_{AA} \propto N_w^{4/3}$. This relationship could have been anticipated from the general arguments concerning equivalent N-N integrated luminosity, but it is pleasing to verify in an explicit model.

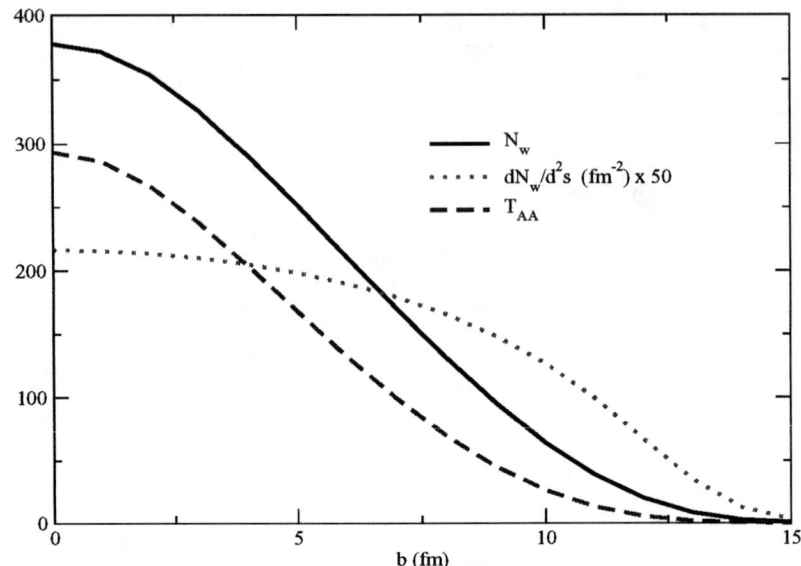

FIGURE 7. Wounded nucleon and nuclear overlap dependence on impact parameter.

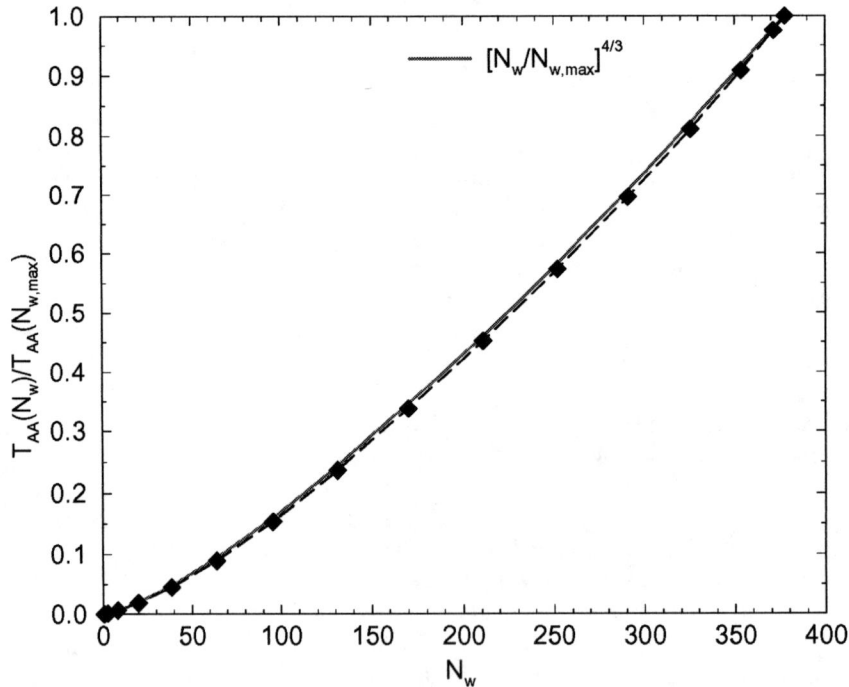

FIGURE 8. Centrality dependence of T_{AA} in terms of N_w.

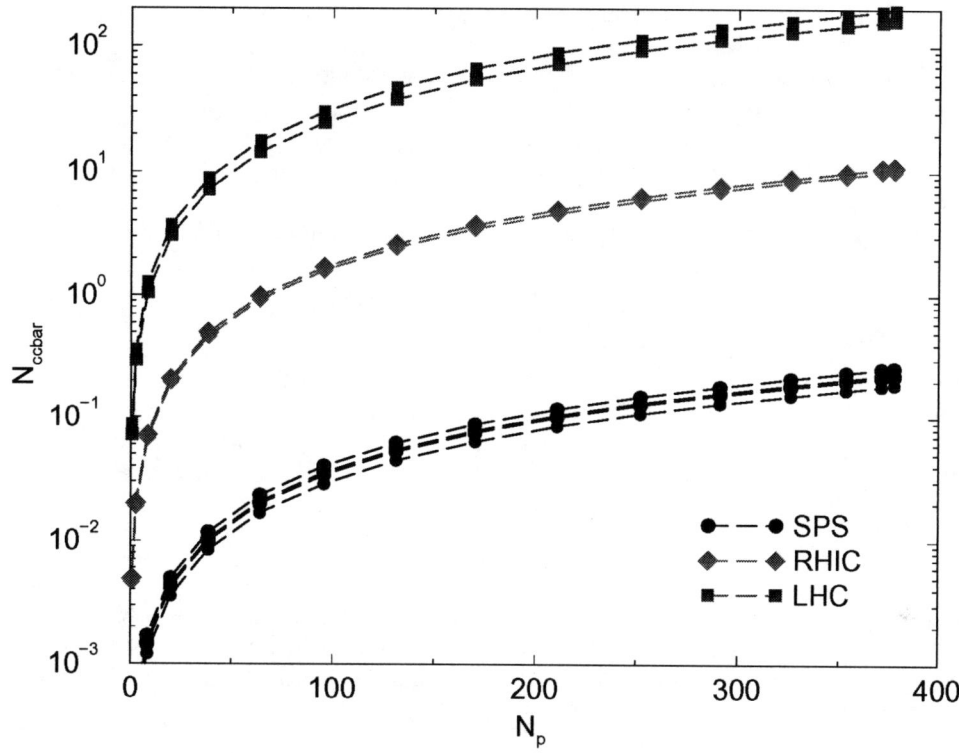

FIGURE 9. Centrality dependence of $N_{c\bar{c}}$.

Finally, one can plot the predictions for $N_{c\bar{c}}$ as a function of centrality, using participant nucleon number N_p as a label. This is shown in Figure 9 for SPS, RHIC200, and LHC energies. The multiple points in each curve come from using several structure function sets. The decrease with centrality just follows the power law behavior of T_{AA}, and one sees that at sufficiently peripheral collisions the average number of charm pairs produced will decrease below unity for all energies considered. This behavior will be a very useful constraint on the models to be considered.

QUARKONIUM FORMATION FROM UNCORRELATED PAIRS

As shown in the previous section for heavy ion collision energies at RHIC and LHC, the initial number of heavy quark pairs produced in each collision will be qualitatively different than the number produced at SPS energies. To be specific, let us consider charm quarks and the subsequent production/formation of J/ψ. Typically, the number of charm quark pairs is expected to be of the order of ten at RHIC and several hundred at LHC in the most central collisions [6]. Let us attempt to extract features of J/ψ formation from the initially-produced charm quarks which are independent of detailed dynamics [16].

We consider scenarios in which the formation of J/ψ is allowed to proceed through any combination of one of the N_c charm quarks with one of the $N_{\bar{c}}$ anticharm quarks which result from the initial production of $N_{c\bar{c}}$ pairs in a central heavy ion collision. This of course would be expected to be valid in the case that a space-time region of color deconfinement is present, but is not necessarily limited to this possibility. For a given charm quark, one expects then that the probability $\mathscr{P}$ to form a J/ψ is proportional to the number of available anticharm quarks relative to the number of light antiquarks,

$$\mathscr{P} \propto N_{\bar{c}}/N_{\bar{u},\bar{d},\bar{s}} \approx N_{c\bar{c}}/N_{ch}. \tag{5}$$

In the second step we have replaced the number of available anticharm quarks by the total number of pairs initially produced, which assumes that the total number of bound states formed remains a small fraction of the total. Also, we normalize the number of light antiquarks by the number of produced charged hadrons. Since this probability is generally very small, one can simply multiply by the total number of charm quarks N_c to obtain the number of J/ψ expected in a given event.

$$N_{J/\psi} \propto N_{c\bar{c}}^2/N_{ch}, \tag{6}$$

where the use of the initial values $N_{c\bar{c}} = N_c = N_{\bar{c}}$ is again justified by the relatively small number of bound states formed. For an ensemble of events, the average number of J/ψ per event is calculated from the average value of initial charm $< N_{c\bar{c}} >$, and we neglect fluctuations in N_{ch}.

$$< J/\psi > = \lambda (< N_{c\bar{c}} > +1) < N_{c\bar{c}} > /N_{ch}, \tag{7}$$

where we place all dynamical dependence in the parameter λ.

One can extend this formula to the case where J/ψ formation is effective not over the entire rapidity range Y_{total}, but only if the quark and antiquark are within the same rapidity interval Δy. There is a significant simplification if the rapidity dependence of the charm quark pairs and the charged hadrons (or equivalently the light antiquarks) are the same. In this case the entire effect is just the replacement $< N_{c\bar{c}} > +1 \to < N_{c\bar{c}} > +Y_{total}/\Delta y$. However, one must remember that the prefactor λ will in general contain some dependence on the size of the rapidity interval.

The essential property of this result is that the growth with energy of the term quadratic in total charm [6] is expected to be much stronger than the growth of total particle production in heavy ion collisions [17]. J/ψ production without this quadratic mechanism is typically some small energy-independent fraction of total initial charm production [18], so that we can expect the quadratic formation to become dominant at high energy.

We show numerical results in Table 1 for these quantities with a prefactor λ of unity. Estimates for the charm and particle numbers are very approximate, but serve to show the anticipated trend with energy. At SPS, this formation mechanism is most probably insignificant. At RHIC it is comparable with "normal" formation, while at LHC one might expect it to be dominant. Of course, the exact result will depend on the details of the physics which controls the formation.

TABLE 1. Comparison of J/ψ formation variation with energy

	SPS	RHIC	LHC
$\sqrt{s}$ (GeV)	18	200	5500
$<N_{c\bar{c}}>$	0.2	10	200
N_{ch}	1350	3250	16500
$<N_{J/\psi}>$	0.00018	0.034	2.4
$N_{J/\psi}^{initial}$	0.0012	0.06	1.2

We can also estimate the centrality dependence of J/ψ production from Equation 7. The number of charm quark pairs should obey

$$N_{c\bar{c}} \propto N_p^{4/3} \tag{8}$$

from the properties of the nuclear overlap function. The number of hadrons produced generally scales with the number of participants,

$$N_{ch} \propto N_p^{\alpha}, \tag{9}$$

where one has measured values $\alpha = 1.07 \pm 0.04$ at SPS [19] and $\alpha = 1.13 \pm 0.05$ at RHIC-130 [20]. At LHC, one might anticipate that hadron production would become dominated by QCD minijets [17], so that $\alpha \approx 4/3$. However, a comparison of RHIC results at 130 and 200 GeV does not indicate a dramatic effect in this energy range. [21],[22]

Given these values, one can predict

$$N_{J/\psi} \propto N_p^{\beta=\frac{8}{3}-\alpha} \tag{10}$$

for collisions in which $<N_{c\bar{c}}> \gg 1$, and

$$N_{J/\psi} \propto N_p^{\beta=\frac{4}{3}-\alpha} \tag{11}$$

for collisions in which $<N_{c\bar{c}}> \ll 1$.

One can make an indirect check of the prediction at SPS energy, utilizing the NA50 data [4] on (J/ψ)/DY as a function of transverse energy E_T. Since there is a linear relationship between E_T and N_p over almost the entire measured range, we can just multiply the measured ratios at each E_T by the expected centrality dependence of the Drell-Yan process, $E_T^{4/3}$. This is shown in Figure 10. A power-law fit to the resulting points yields an exponent $\beta \approx 0.77$. This value is significantly higher than one would predict from the generic arguments above, since at SPS with $<N_{c\bar{c}}> \ll 1$, $\beta \approx \frac{4}{3} - \alpha \approx 0.26$. Figure 11 illustrates this feature.

As a reference, the expected centrality dependence for initial J/ψ production is shown, which includes the hard production process followed by normal nuclear absorption. The parameterization of the absorption cross section is taken from [23], which leads to an effective power law exponent $\beta \approx 1.2$ The solid line shows the centrality

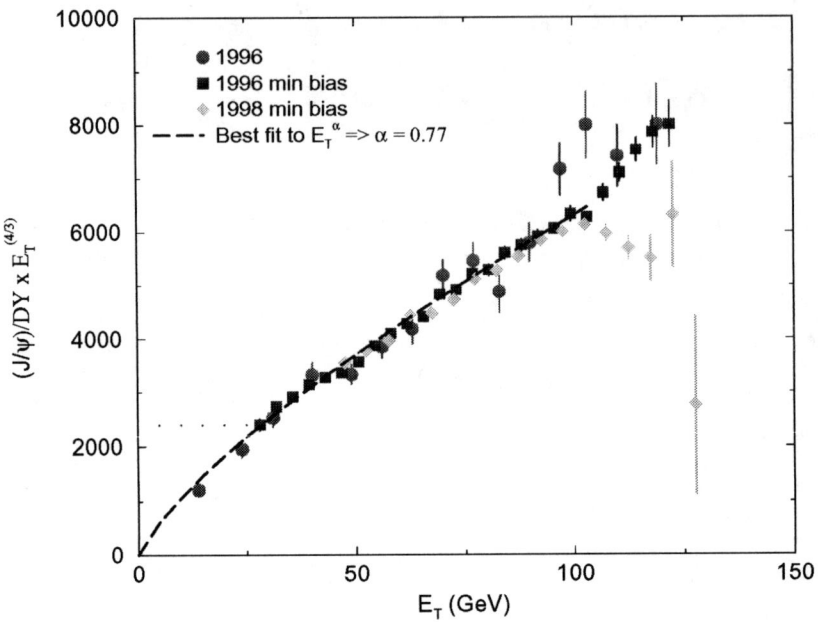

FIGURE 10. Centrality dependence of J/ψ inferred from measured ratio J/ψ/Drell-Yan by NA50 at SPS.

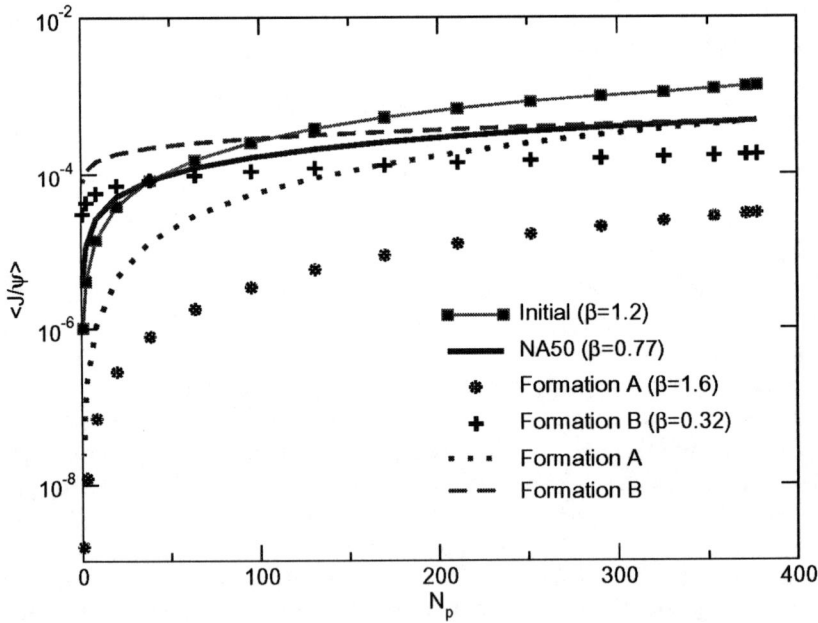

FIGURE 11. Generic J/ψ production at SPS

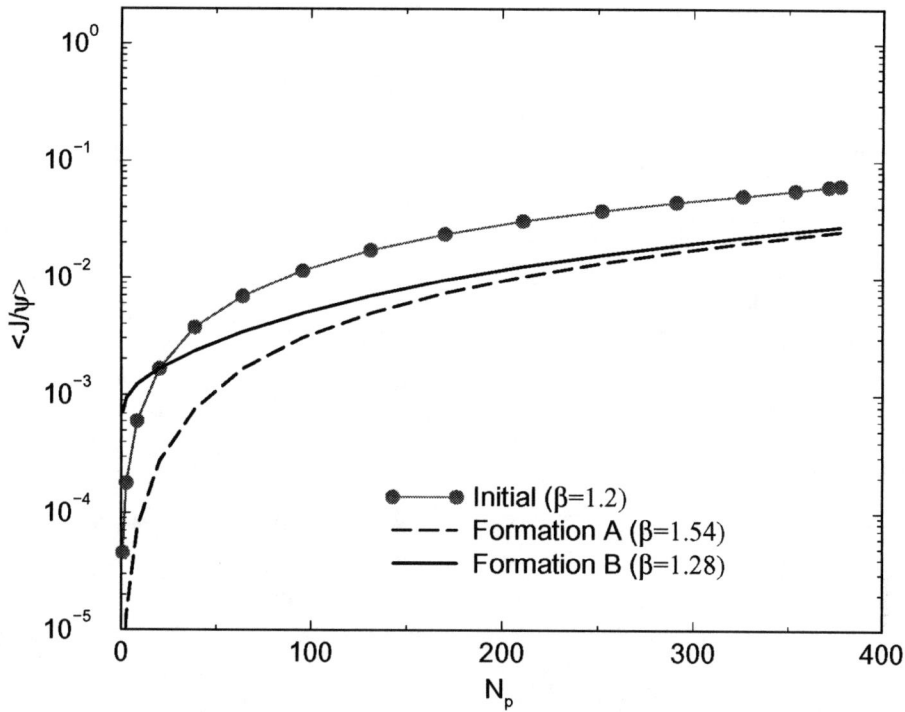

FIGURE 12. Generic J/ψ production at RHIC

dependence implied by the NA50 data described above, and is normalized such that it coincides with the initial production value for sufficiently peripheral collisions. The predictions of the generic quadratic formation formula are shown by the stars (labeled A). Of course, the initial charm production at SPS energies is certainly small enough such that the linear dependence is dominant. This effect is shown by the plus symbols (labeled B), which include both the quadratic and linear terms. A single power law fit to this composite curve yields $\beta = 0.32$, again indicative that the linear term with its own $\beta = 0.26$ is dominant. Also shown by the dotted and dashed lines are these same curves, but normalized for central collisions to coincide with the "NA50" curve for ease of comparison. It is clear that the centrality dependence implied by the NA50 data is not reproduced by the generic expectations. In this respect it is fortunate that the expected magnitude of J/ψ formation implied by the generic arguments is likely to be insignificant at SPS energies.

The corresponding results at RHIC energy are shown in Figure 12. Again the generic formation curves use the prefactor $\lambda = 1$. Here we have used the expected charm yield of 10 pairs per central collision at 200 GeV, but used the measured α value at 130 GeV. At RHIC, one expects to see the quadratic dependence for central collisions (Curve A) gradually convert to linear dependence as appropriate for the small number of $N_{c\bar{c}}$ for peripheral collisions. Curve B shows the combination of linear and quadratic

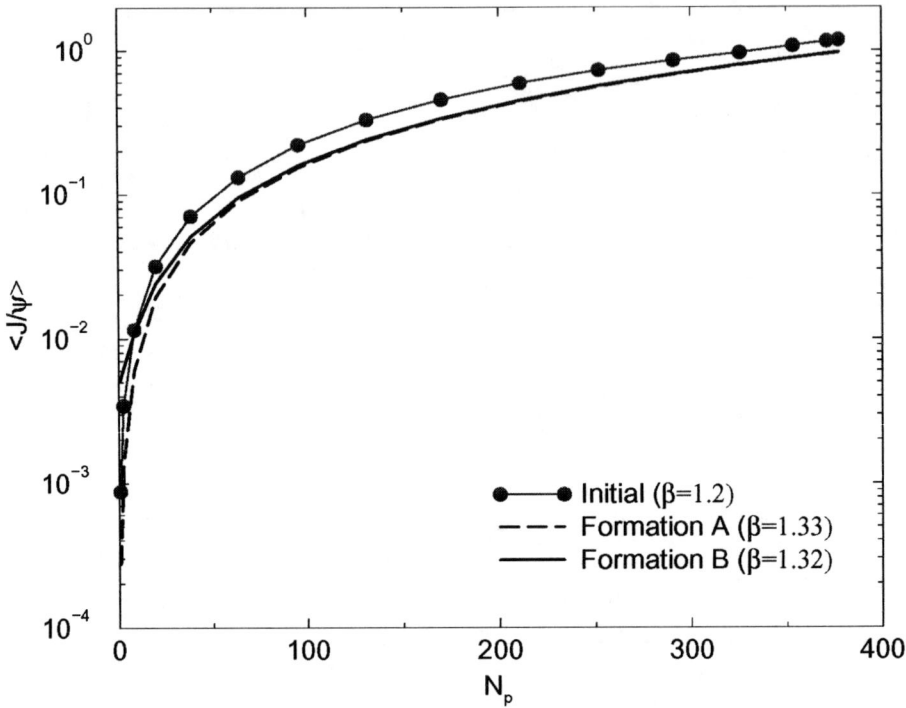

FIGURE 13. Generic J/ψ production at LHC

components, which is fit by a single power $\beta = 1.28$ (indicates the quadratic component dominates over most of the centrality range). Figure 13 contains the same calculations at LHC energy. Here the purely quadratic formula (B) is dominant over the entire centrality region, since the number of charm pairs for central collisions is so large (≈ 200). We have assumed at LHC energy that the particle production centrality dependence has increased as appropriate for total domination by minijets. If this is not correct, the β values will be even higher but the dominance of the quadratic term will remain.

We compile in Figure 14 the expected centrality dependence for all energies considered. The absolute magnitudes correspond to the prefactor $\lambda = 1$. Also shown is the dependence implied by the NA50 data, which is normalized to coincide with the generic SPS curve for the most central point. Certainly the centrality dependence at RHIC and LHC will be crucial for the interpretation of any enhanced J/ψ production.

STATISTICAL HADRONIZATION MODEL

This model is motivated by the success of attempts to explain the relative abundances of light hadrons produced in high energy interactions in terms of the predictions of a hadron gas in chemical and thermal equilibrium [24]. Such fits, however, are not able to describe

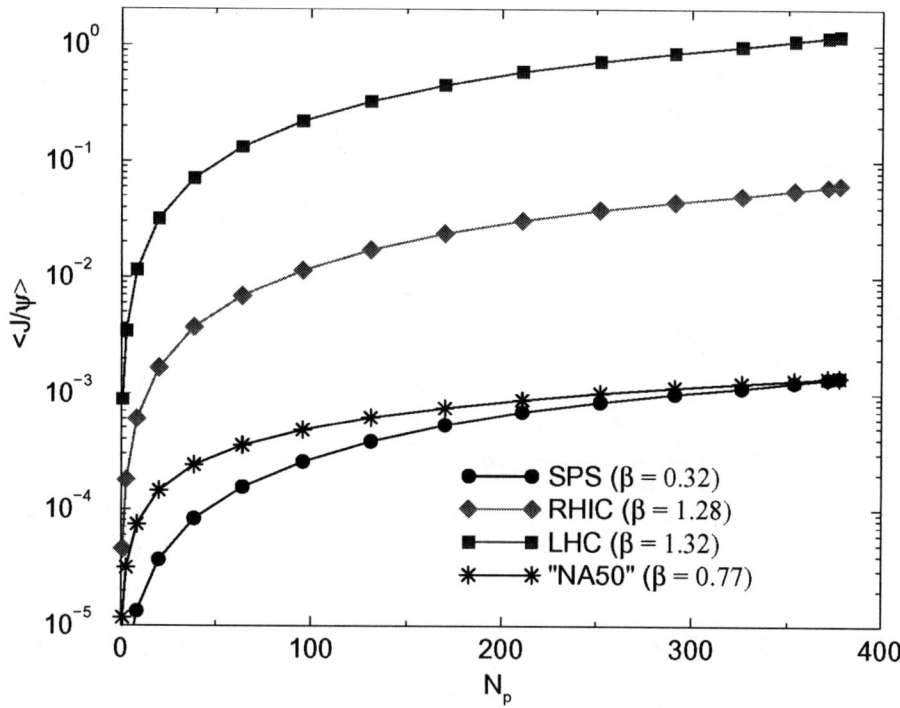

FIGURE 14. Centrality dependence of generic J/ψ production with dynamical factor $\lambda = 1$.

the abundances of hadrons containing charm quarks. This can be understood in terms of the long time scales required to approach chemical equilibrium for heavy quarks. However, it is expected that for high energy heavy ion collisions the initial production of charm quark pairs exceeds the number expected at chemical equilibrium as determined by the light hadron abundances. As an illustration, we show in Figure 15 the density of charm quarks in equilibrium over a range of temperatures. This is compared with the charm quark density which results from distributing the initially-produced charm quark pairs over the volume of a deconfined region. Each line in the figure corresponds to a different initial temperature. The decrease in the density as temperature decreases is due to the expansion of the deconfined region. (We use nominal RHIC conditions for central collisions, $N_{c\bar{c}} = 10$, initial volume $V_o = \pi R^2 \tau_o$ with $\tau_o = 1$ fm, and isentropic longitudinal expansion, VT^3 = constant.) One sees that at temperatures below typical deconfinement transitions, the initial charm densities all exceed that expected for thermal equilibrium. (This will always happen at some finite T, since the decrease of equilibrium density is exponential while the initial densities only decrease due to power law volume expansion.)

The statistical hadronization model assumes that at hadronization these charm quarks are distributed into hadrons according to chemical and thermal equilibrium, but adjusted by a factor γ_c which accounts for oversaturation of charm density. One power of this

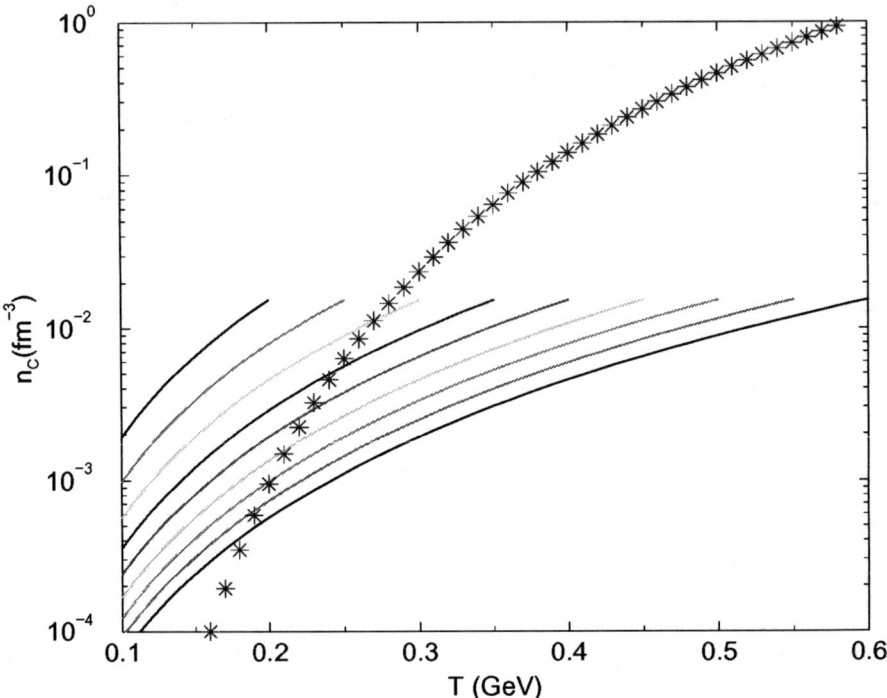

FIGURE 15. Charm quark density.

factor multiplies a given thermal hadron population for each charm or anticharm quark contained in the hadron. Thus the relative abundance of J/ψ to that of D mesons, for example, will be enhanced in this model. The enhancement factor is determined by conservation of charm, again using the time scale argument to justify neglecting pair production or annihilation before hadronization.

$$N_{c\bar{c}} = \frac{1}{2}\gamma_c N_{open} + \gamma_c^2 N_{hidden}, \qquad (12)$$

where N_{open} is the number of hadrons containing one charm or anticharm quark and N_{hidden} is the number of hadrons containing a charm-anticharm pair. (The contribution of multiply-charmed baryons or antibaryons are generally neglected since their large mass leads to very small thermal densities.) Note that the actual particle numbers, not just the densities, are required in this approach. Thus the volume of the thermal system is an additional parameter which must be included.

For most applications, N_{hidden} in Equation 12 can be neglected compared with N_{open} due to the hadronic mass differences. Thus the charm enhancement factor is simply

$$\gamma_c = \frac{2N_{c\bar{c}}}{N_{open}}. \qquad (13)$$

This is easily seen to predict a quadratic dependence of the population of hidden charm hadrons. Using the thermal densities allows one to calculate the prefactor λ from the previous section.

$$N_{J/\psi} = \gamma_c^2 N_{J/\psi}^{thermal} \tag{14}$$

It is important to note at this time that $N_{J/\psi}^{thermal}$ includes the thermal population of all hidden charm states which decay into the observed J/ψ. Since all of the individual terms are multiplied by the same charm factor γ_c^2, the statistical hadronization model predicts that all ratios of various hidden charm state populations are identical to those predicted by the thermal densities alone. It was first noted in Reference [25] that the measured ratio Ψ'/Ψ for heavy ion interactions at SPS was quite close to that expected in thermal equilibrium at temperature close to the deconfinement transition for sufficiently central collisions.

We replace the one remaining factor of system volume by the ratio of charged particle number to density to compare with the generic expectations in Equation 6.

$$N_{J/\psi} = 4 \frac{n_{ch} n_{J/\psi}}{n_{open}^2} \frac{N_{c\bar{c}}^2}{N_{ch}} \tag{15}$$

This is the result contained in the initial formulation of the statistical hadronization model [26], where the goal was to compare with results of the NA50 experiment on J/ψ production. It was soon realized [27], [28] that for such an application, an important correction must be applied to Equation 12. We have tacitly assumed up to now that the thermal particle numbers N_{open} and N_{hidden} have been calculated in the grand canonical formalism, and that they are large enough such that charm conservation is satisfied by the average values with fluctuations suppressed by these large particle numbers. However, one knows that at SPS energies the average thermal charm numbers per collision are much less than unity for all collision centrality. Even at RHIC and LHC, sufficiently non-central events will always involve small particle numbers. In these cases, one cannot satisfy exact charm conservation on the average, and one must utilize the canonical formalism to calculate the thermal particle numbers. This is obvious in the limiting case where each collision produces either one or zero charm quark pairs, and they can go only into either one hidden charm hadron or one charm and one antJcharmH hadron [29].

There is a simple correction factor which can be applied to calculate the canonical particle number N_{can} from the grand canonical particle number N_{gc} for a thermal system in which the conserved quantity (charm in this case) population in baryons can be neglected (justified by large masses).

$$N_{can} = N_{gc} \frac{I_1(N_{gc})}{I_0(N_{gc})}. \tag{16}$$

which just involves the modified Bessel functions I_n [30]. One uses this expression with $N_{gc} = \gamma_c N_{open}$ to revise Equation 12. Note that there is no canonical correction for N_{hidden}, which has zero total charm.

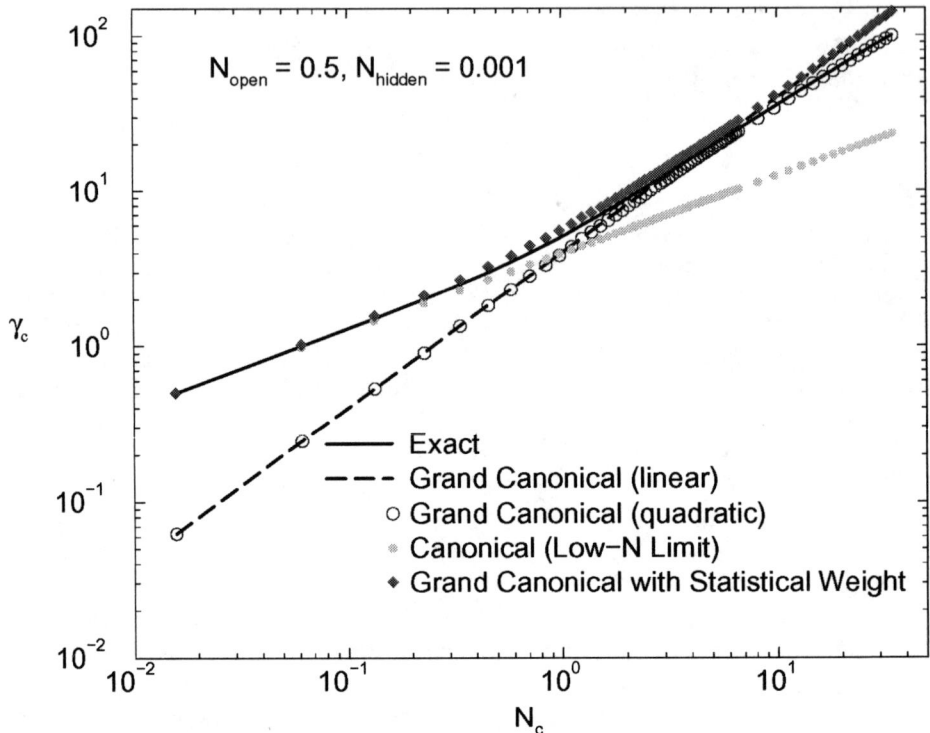

FIGURE 16. Relation between charm enhancement factor and total number of charm quarks for statistical hadronization model.

$$N_{c\bar{c}} = \frac{1}{2}\gamma_c N_{open}\frac{I_1(\gamma_c N_{open})}{I_0(\gamma_c N_{open})} + \gamma_c^2 N_{hidden}, \quad (17)$$

In the limit of large N_{gc}, the ratio of Bessel functions approaches unity, and one recovers the grand canonical result. In the opposite limit when N_{gc} approaches zero, the ratio of Bessel functions goes to $\frac{1}{2}N_{gc}$, and the solution for the charm enhancement factor is

$$\gamma_c \to \frac{2\sqrt{N_{c\bar{c}}}}{N_{open}}. \quad (18)$$

The net effect in this limit is then just to change the dependence on $N_{c\bar{c}}$ in Equation 15 from quadratic to linear.

In general, a solution for γ_c as a function of $N_{c\bar{c}}$ must be obtained numerically. Such a solution is shown in Figure 16, using some specific values of N_{open} and N_{hidden} for charm. One sees the quadratic behavior in the large $N_{c\bar{c}}$ limit and also the linear behavior in the small $N_{c\bar{c}}$ limit. Also shown on this plot is a curve which follows the exact solution for Equation 12, but with $N_{c\bar{c}}$ replaced by $(N_{c\bar{c}}(N_{c\bar{c}}+1))^{\frac{1}{2}}$.

It is interesting to note that this replacement allows the grand canonical solutions to incorporate the behavior of the canonical corrections to an impressive degree of accuracy. It appears that the ad hoc substitution above could be motivated by an averaging procedure for $N_{c\bar{c}}^2$, which has been noted previously for the two limiting cases [28]. However, apparent general validity of this procedure requires further study, and may involve the properties of the kinetic equations which describe the approach to equilibrium.

This formalism was originally applied to the NA50 measurement of J/ψ production in fixed-target heavy ion collisions at the CERN SPS. There remain uncertainties in the absolute magnitude of J/ψ yields, which has lead to different approaches in the literature. One method is to assume knowledge of $N_{c\bar{c}}$ from measurements in N-N interactions scaled up to heavy ion interactions as appropriate for a point-like process, Then the J/ψ can be predicted, including the centrality dependence. The other approach takes $N_{c\bar{c}}$ as a parameter to be fixed by the measured J/ψ yields. In both cases [27], [31], a common conclusion appears to emerge. One must require that the magnitude of charm production must be a factor of 3-5 greater than that inferred from N-N interactions, and the centrality dependence must increase much more rapidly than the $N_p^{\frac{4}{3}}$ expected for the pQCD production process. This conclusion may be related to the observation of an excess of dileptons in the intermediate mass region by NA50 [32] for which one source could be enhanced charm production. This situation underscores the need for separate measurements of both total charm and charmonium production in order to test the production mechanisms. Experiment NA60 is expected to provide this information at SPS.

Next we look at applications of the statistical hadronization model at RHIC and LHC. Since the general properties of this model obey the generic expectations of the previous section, we can utilize the generic expectations for centrality dependence. The overall magnitudes are determined by the thermal parameters (including volume), and are generally taken from existing thermal fits to light hadron species. We choose to show the ratio $< J/\psi >$/charm, which is less sensitive to the normalization of total charm production the absolute number of J/ψ.

Figure 17 shows several applications for RHIC conditions.

The centrality dependence is modeled by the number of nucleon participants, and one sees the change in shape due to the transition between canonical and grand canonical formalism. The absolute magnitudes of $< J/\psi >$/charm are comparable with the initial production estimates of a fraction of a percent, indicating that this process may overwhelm suppression for central collisions at RHIC. The lowest curve is the calculation of Reference [31], which includes the centrality dependence. The next higher is from Reference [33] which only included the most central collision point. The same is true for the highest curve from Reference [27], which uses rapidity intervals of width $\Delta y = 1$ as a requirement for the quarks to form the J/ψ. I have completed the calculation of implied centrality dependence for these two calculations. Note that the highest curve exhibits a stronger rise for lower centrality events than the others, since with total charm quark numbers limited by one unit of rapidity, the region which receives a substantial canonical ensemble correction extends further toward very central events.

Corresponding information is shown in Figure 18 for LHC energy. Here we have completed the centrality dependence for one case considered in the literature [27], and

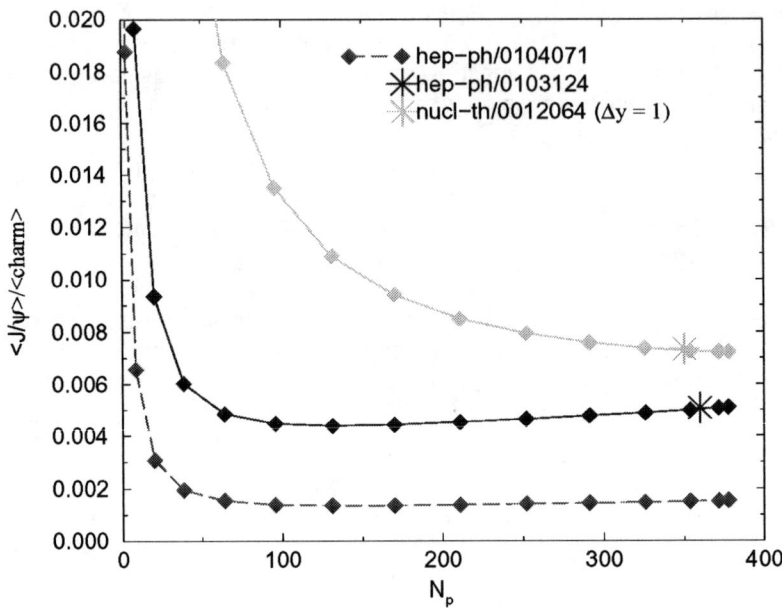

FIGURE 17. Ratio $<J/\psi>$ over initial charm at RHIC for several applications of the statistical hadronization model.

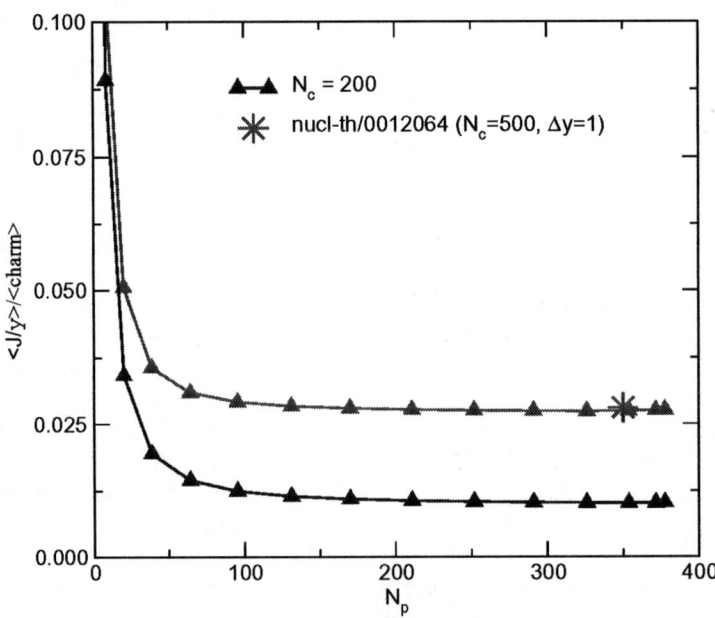

FIGURE 18. Ratio $<J/\psi>$ over initial charm at LHC for two applications of the statistical hadronization model.

contrast it with the generic calculation with no constraint on rapidity interval. Typical magnitudes are factors of 3-5 above the corresponding predictions for RHIC, indicating a strong enhancement of J/ψ formation in the statistical hadronization model.

KINETIC FORMATION MODEL

In this model [34],[35], we investigate the possibility to form J/ψ directly in a deconfined medium. The formation will take advantage of the mobility of initially-produced charm quarks in a spatial region of deconfinement. Then one expects that interactions can occur between a charm quark which was produced along with its anticharm partner in one of the initial nucleon-nucleon collisions, and an anticharm quark which was produced with its own charm partner in an entirely different initial nucleon-nucleon collision. Thus all combinations of a charm plus antichiarm quarks in the initial $N_{c\bar{c}}$ are allowed to participate in the formation of charmonium states. Of course, there is an upper limit of $N_{c\bar{c}}$ itself on the total number of charmonium states which can be formed, but in practice this limit will never be approached. Since the rates of formation are quadratic in the number of *unbound* charm pairs, one anticipates that the final charmonium population will be approximately quadratic in the *initial* value $N_{c\bar{c}}$. In this respect, it then fits in with the generic expectations previously derived based on probabilities of quark number combinations. However, the additional dependence of that generic expectation on hadron production will come about in an entirely different manner.

For the purposes of this study, we consider a physical picture of deconfinement in which the "standard" quarkonium suppression mechanism is via collisions with free gluons in the deconfined medium [36]. The dominant formation process, in which a quark and an antiquark in a relative color octet state are captured into a color singlet bound quarkonium state and emit a color octet gluon, is simply the inverse of the the breakup reaction which is responsible for the suppression. It is then an inevitable consequence of this picture of suppression that the corresponding formation process must also take place.

At this point one might ask about the effect of color screening in this picture. One view might be that the most deeply bound quarkonium states can still exist above the deconfinement temperature, and that this formation mechanism (and of course the competing dissociation mechanism) will only exist for temperatures above the deconfinement temperature (since we require mobile heavy quarks) but still below some critical temperature T_{screen} which defines the point at which the quarkonium state can no longer exist. In this case the new formation mechanism will just modify the dissociation effectiveness, although in the case of very large numbers of quark pairs this modification will be sufficient to actually change the sign of the effect. In this study we advocate an alternate viewpoint. In this viewpoint, both the color screening mechanism and the gluon dissociation reaction are the same physical phenomenon, but manifest themselves in two different limiting cases. In the limit of very large time scales, the screening view is appropriate, since it assumes that the heavy quarks are subject to a static potential which determines the bound state spectrum. In the limit of very small time scales, the quarkonium states cannot be expected to follow the fluctuations of the color fields due to the light quarks and gluons, and it is assumed that a collisional description will be appropriate. Work is underway [37] on quantifying these arguments. Initial results indicate that the time necessary to completely screen away a deeply bound quarkonium state is comparable to estimated lifetimes for a deconfined state in heavy ion collisions.

Given the above assumptions, we consider the dynamics of charm quark pairs and J/ψ's in a region of color deconfinement populated by a thermal density of gluons. The time evolution if the J/ψ population is given by the rate equation

$$\frac{dN_{J/\psi}}{d\tau} = \lambda_F N_c N_{\bar{c}} [V(\tau)]^{-1} - \lambda_D N_{J/\psi} \rho_g, \qquad (19)$$

with ρ_g the number density of gluons, τ the proper time and $V(\tau)$ the volume of the deconfined spatial region. The reactivities $\lambda_{F,D}$ are the reaction rates $\langle \sigma v_{\text{rel}} \rangle$ averaged over the momentum distribution of the initial participants, i.e. c and $\bar{c}$ for λ_F and J/ψ and g for λ_D. The gluon density is determined by the equilibrium value in the QGP at each temperature. For simplicity, it is assumed to be spatially homogeneous, as are the charm quark and J/ψ distributions.

We enforce exact charm conservation in solving this equation, but in practice this means that the number of charm quarks N_c and anticharm quarks $N_{\bar{c}}$ are always approximately equal to the number of initial pairs $N_{c\bar{c}}$ for the following reasons: (a) Reactions in which charm quark-antiquark pairs annihilate into light quarks or gluons are small since the charm density due to initial charm is less than the thermal equilibrium value over during most of the time; (b) Production of additional charm quark pairs from interactions of light quarks and gluons is negligible during the time of deconfinement [38]; (c) Formation of other states in the charmonium spectrum is a small fraction of initial charm (as is the fraction of J/ψ itself); and (d) Disappearance of single charm quarks or antiquarks via formation of open charm mesons is effectively reversed immediately because the time scale for dissociation of these large states with small binding energy is typically less than a fraction of a fermi.

To illustrate this last point and set the relevant time scales, we show in Figure 19 the dissociation rates for thermal gluons on various states of quarkonium, as a function of gluon temperature. The reaction cross section used will be discussed below.

The behavior with binding energy and spatial bound state size is evident. Certainly the open charm mesons will have typical dissociation times much less than 1 fm. It is also clear that the heavy quarkonium bound states have dissociation times between 1 and 100 fm for typical QGP temperatures, making this process relevant during the time of deconfinement. Note that the temperature dependence arises from two effects. The initial rapid increase in the low temperature region starts when average gluon energies are able to overcome the dissociation threshold, and the continued rise for large temperature is due to the continuing increase in gluon density.

We allow the system to undergo a longitudinal isentropic expansion, which fixes the time-dependence of the volume $V(\tau) = V_o \tau / \tau_o$. The expansion is taken to be isentropic, $VT^3 = $ constant, which then provides a generic temperature-time profile.

It is evident that the solution of Equation 19 grows quadratically with initial charm $N_{c\bar{c}}$, as long as the total $J/\psi \ll N_{c\bar{c}}$. In this case we can write an analytic expression

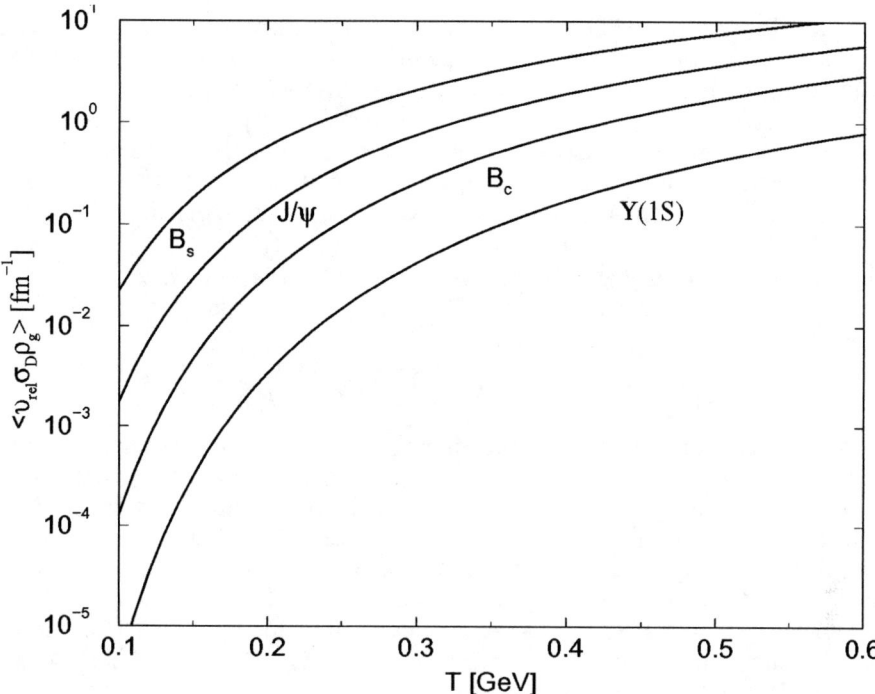

FIGURE 19. Dissociation rates of various bound states of charm due to interactions with thermal gluons.

$$N_{J/\psi}(\tau_f) = \varepsilon(\tau_f) \times [N_{J/\psi}(\tau_0) + N_{c\bar{c}}^2 \int_{\tau_0}^{\tau_f} \lambda_F [V(\tau)\varepsilon(\tau)]^{-1} d\tau], \quad (20)$$

where τ_f is the hadronization time determined by the initial temperature (T_0 is a variable parameter) and final temperature (T_f ends the deconfining phase). The function $\varepsilon(\tau_f) = e^{-\int_{\tau_0}^{\tau_f} \lambda_D \rho_g d\tau}$ would be the suppression factor in this scenario if the formation mechanism were neglected. During the remainder of these calculations, we concentrate on solutions in which the initial number of produced J/ψ is zero. Then the additional term due to the new formation mechanism represents the final number of J/ψ which result from a competition between the formation and dissociation reactions during the lifetime of the deconfined region. Note that this number is always positive, since one cannot dissociate more bound states than are formed.

We can then compare this additional term with what is anticipated from the generic considerations summarized in Equation 7. The quadratic factor $N_{c\bar{c}}^2$ is present as expected. The normalizing factor of N_{ch} does not appear automatically. (Remember that this factor was obtained in the statistical hadronization model by replacing a factor of system volume V by the ratio of N_{ch} over the thermal charged hadron density.) In the kinetic model result, the additional formation term in Equation 20 also has a volume term present. This volume is present to account for the decreasing charm quark density during expansion. It also has some other essential differences. First, the kinetic model volume is time-dependent, and is integrated over the duration of deconfinement. Second, the transverse area of the volume is determined not just by nuclear geometry, but by the dynamics which determine over which region there will be deconfinement. Finally, the factor $\varepsilon(\tau_f)/\varepsilon(\tau)$ due to dissociation processes during deconfinement will play a role. These differences will be seen explicitly when the centrality dependence is considered.

For our quantitative estimates, we utilize a cross section for the dissociation of J/ψ due to collisions with gluons which is based on the operator product expansion [39],[40]:

$$\sigma_D(k) = \frac{2\pi}{3} \left(\frac{32}{3}\right)^2 \left(\frac{2\mu}{\varepsilon_o}\right)^{1/2} \frac{1}{4\mu^2} \frac{(k/\varepsilon_o - 1)^{3/2}}{(k/\varepsilon_o)^5}, \qquad (21)$$

where k is the gluon momentum, ε_o the binding energy, and μ the reduced mass of the quarkonium system. This form assumes the quarkonium system has a spatial size small compared with the inverse of Λ_{QCD}, and its bound state spectrum is close to that in a nonrelativistic Coulomb potential. These assumptions are somewhat marginal for the charmonium spectrum, but should be better satisfied for the upsilon states.

The magnitude of the cross section is controlled just by the geometric factor $4\mu^2$, and its rate of increase in the region just above threshold is due to phase space and the p-wave color dipole interaction. This same cross section is utilized with detailed balance factors to calculate the primary formation rate for the capture of a charm and anticharm quark into the J/ψ.

We use parameter values for thermalization time $\tau_0 = 0.5$ fm, initial volume $V_0 = \pi R^2 \tau_0$ with R = 6 fm, deconfinement temperature $T_f = 150$ MeV, and a wide range of initial temperatures 200 MeV $< T_0 <$ 600 MeV. We begin by showing some results which assume also thermal charm quark momentum distributions, but this will be generalized later.

Shown in Figure 20 is one typical time evolution of J/ψ taken from our numerical solutions. Also shown are the time dependence of the formation and dissociation reactions. falls due to the decrease in charm quark density with the expanding volume. The dissociation rate starts out at zero since there are no initial J/ψ's, but then jumps up to follow the J/ψ population. This rate also eventually decreases due to the drop in average gluon energy related to the falling temperature. The net number of J/ψ formed continues to increase with time as the formation and dissociation rates both slowly decrease. One might wonder at what point one would reach equilibrium, where the two rates would cancel exactly. The answer is that in this case we have controlled the temperature by external means, and equilibrium is never attained during the lifetime of the deconfinement. If we had assumed a constant temperature and volume, there would of course be an equilibrium point beyond which the J/ψ population would become constant.

Figure 21 shows the final J/ψ population as a function of $N_{c\bar{c}}$ for a range of values which include that expected for central collisions at RHIC.

For these calculations, we also allowed the initial J/ψ number to be nonzero. The solid, short dashed, and dashed lines correspond to $N_{J/\psi}(\tau_o) = 0, 1$, and 2, respectively. Variation within these line types is from variation of the initial temperature parameter T_o,

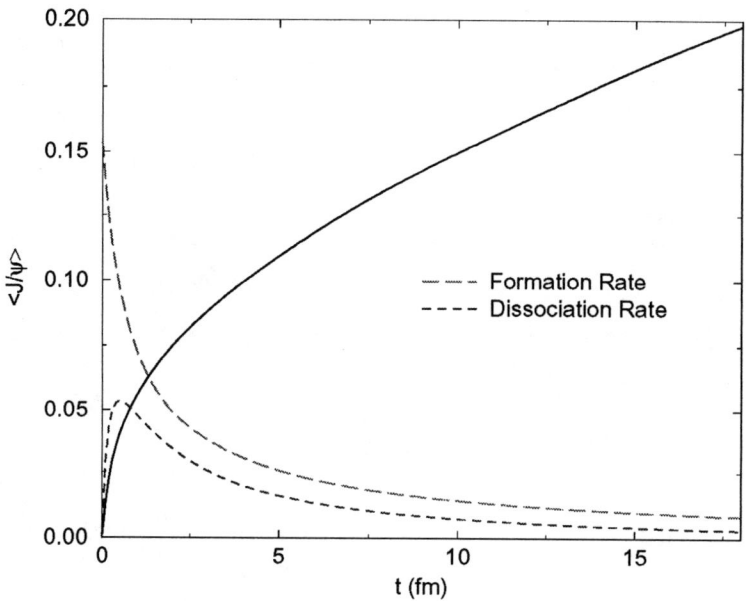

FIGURE 20. Time dependence of formation and dissociation rates and total $< N_{J/\psi} >$ at RHIC energy.

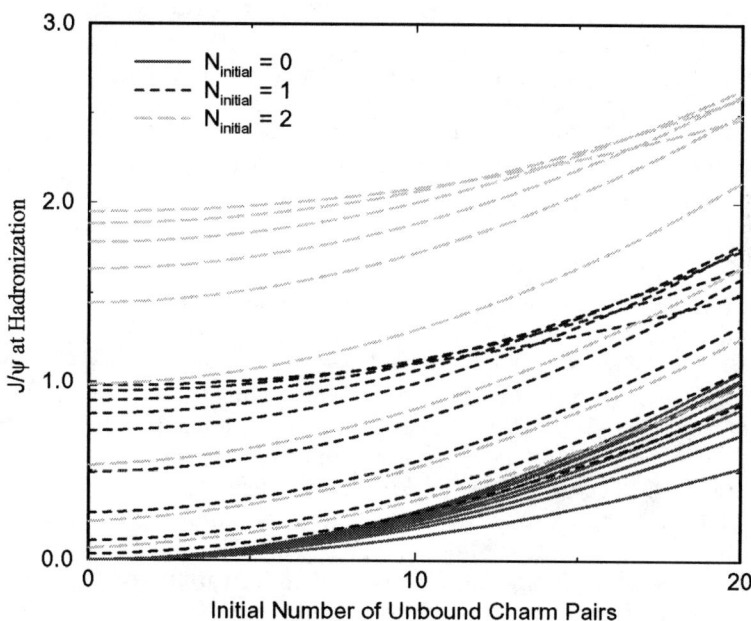

FIGURE 21. Quadratic dependence of $< J/\psi >$ as a function of total initial charm.

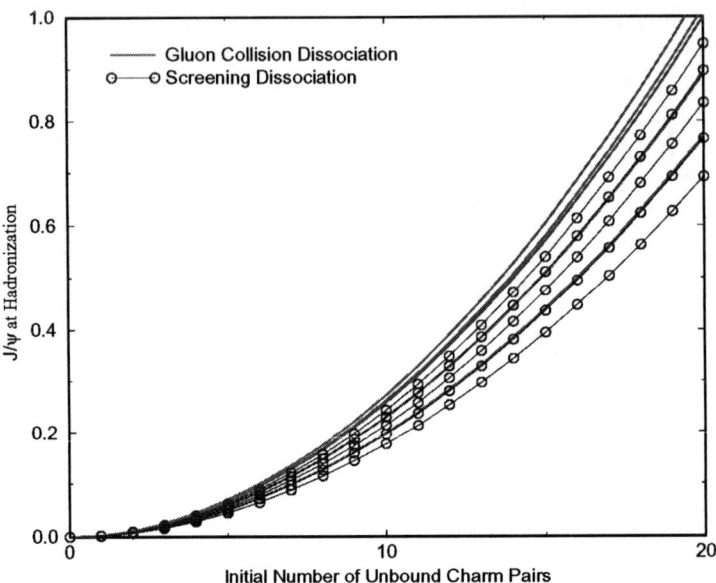

FIGURE 22. Comparison of screening and collisional dissociation scenarios.

which also controls the volume expansion and lifetime. One sees the expected quadratic dependence on $N_{c\bar{c}}$ for all parameter choices. The solid curves are the quantity of main interest, where the initial J/ψ is zero. At the expected $N_{c\bar{c}} = 10$, one sees final J/ψ in the range of 0.1 to 0.3 per central collision. This is at or above the number one would expect from 10 initial ccbar pairs in N-N interactions, and suggests that the formation process will be competitive at RHIC energy. The two sets of dashed curves show how the dissociation of the initial population adds to the final number, and the expected temperature dependence of this dissociation.

At this point we digress somewhat to examine the effects of varying some of our parameters or assumptions.

- In our model of a deconfined region, we have used the vacuum values for masses and binding energy of J/ψ, and assumed that the effects of deconfinement are completely included by the dissociation via gluon collisions. For a complementary viewpoint, we have also employed a deconfinement model in which the J/ψ is completely dissociated when temperatures exceed some critical screening value T_s. Below that temperature, the new formation mechanism will still be able to operate, and we use the same cross sections and kinematics.

We find that for $T_s = 280$ MeV, the final J/ψ population is approximately unchanged. This behavior is shown in Figure 22. The solid lines are our previous yield curves using gluon collision dissociation for various T_o, and the circles show the corresponding results with a screening cutoff temperature $T_s = 280$ MeV. One finds that for lower values of T_s the formation mechanism produces fewer J/ψ, since the most effective formation period is during the initial times when the total

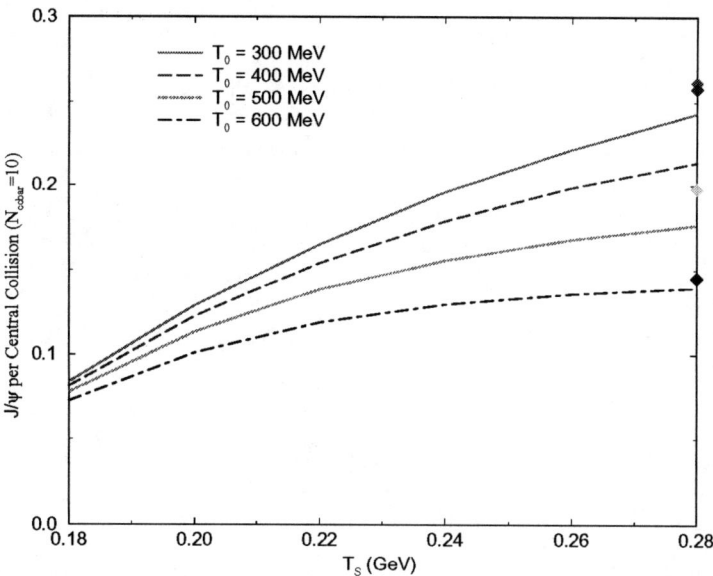

FIGURE 23. Predictions of kinetic model for $<J/\psi>$ with maximum temperature bounded by screening.

volume is small and the corresponding charm quark densities are large. Shown in Figure 23 is the variation of the final yields with T_s, for various temperature-time profiles. One sees that one could get reductions in formation up to factors of 2 or 3 in this scenario.

- The validity of the cross section used assumes strictly nonrelativistic bound states, which is somewhat marginal for the J/ψ. All of the existing alternative models predict larger values for this cross section. If we arbitrarily increase the cross section by a factor of two, or alternatively set the cross section to its maximum value (1.5 mb) at all energies, we find an increase in the final J/ψ population of about 15%. This occurs because the kinetics always favors formation over dissociation, and a larger cross section just allows the reactions to approach completion more easily within the lifetime of the QGP. This behavior is shown in Figure 24.
- A nonzero transverse expansion will be expected at some level, which will reduce the lifetime of the QGP and reduce the efficiency of the new formation mechanism. We have calculated results for central collisions with variable transverse velocity, and find a decrease in the equivalent λ factor of about 15% for each increase of 0.2 in the transverse velocity. This behavior is shown in Figure25.

The new formation mechanism exhibits a significant sensitivity to the charm quark momentum distribution, as might be anticipated. In the calculation of the formation reactivity λ_F, momenta of any given pair will determine the reaction energy which in turn determines the effective value of the cross section. In addition, the charm quark energies enter into the formation rate through the usual formulas for non-collinear

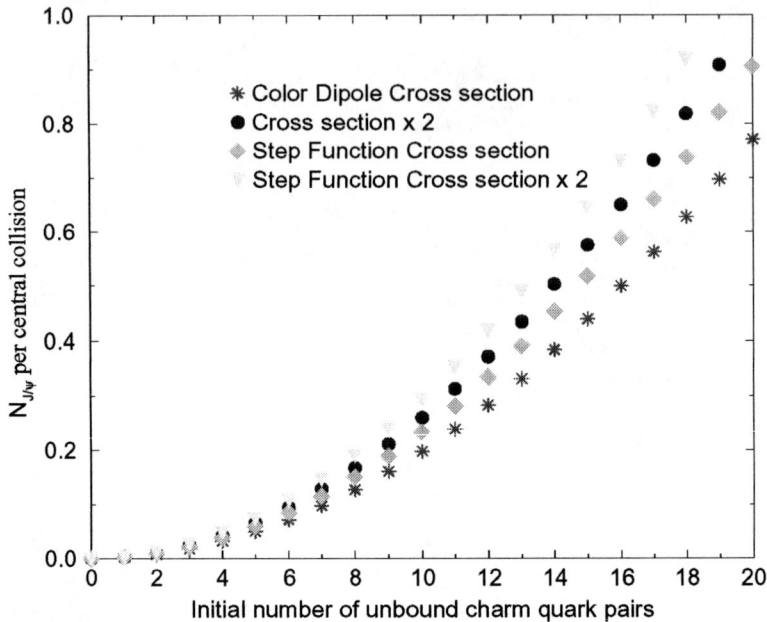

FIGURE 24. Variation of formation results with input cross section.

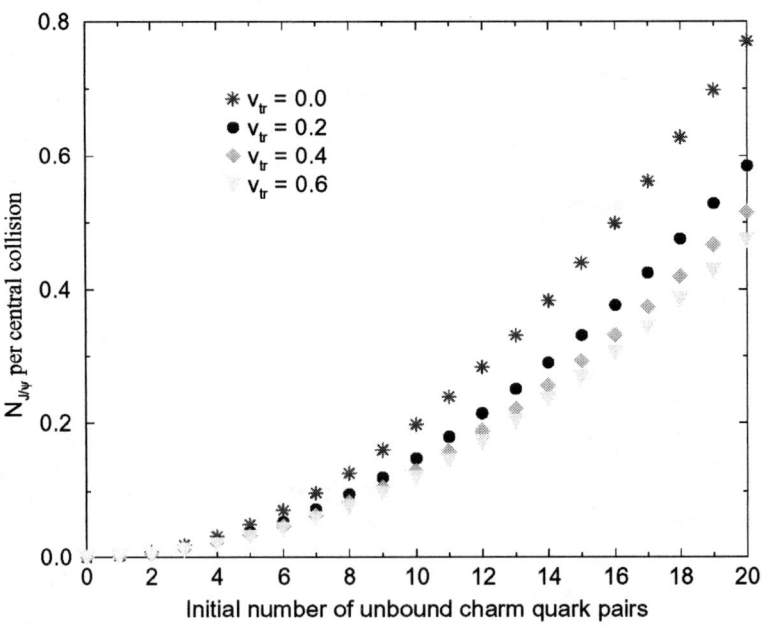

FIGURE 25. Variation of formation results with nonzero transverse expansion.

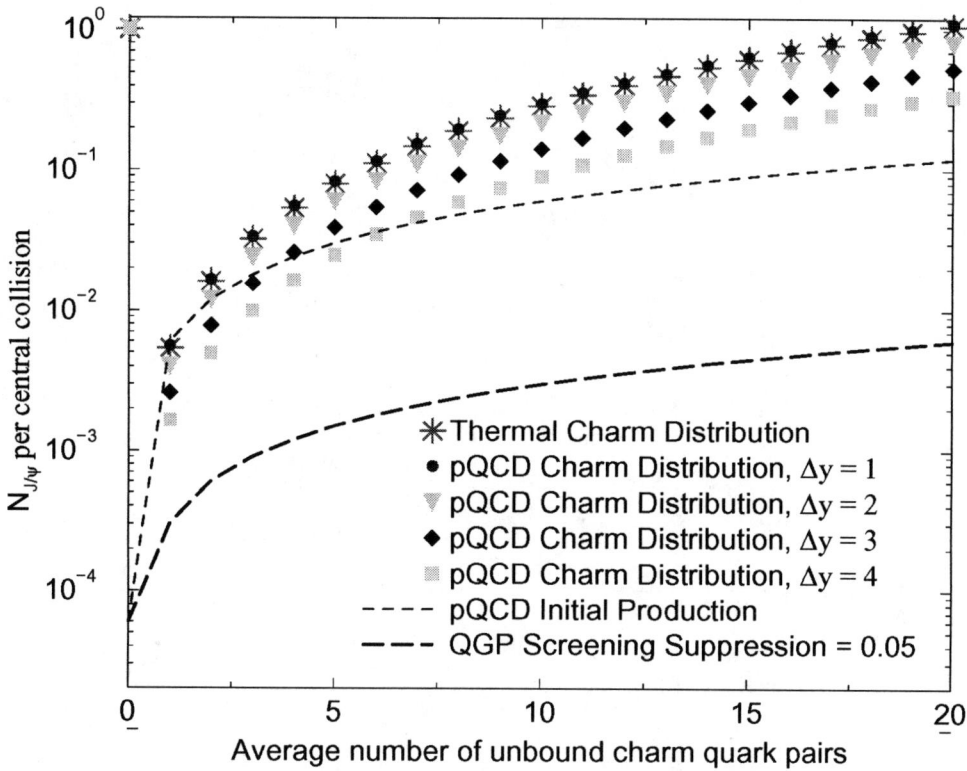

FIGURE 26. Predictions of kinetic model variation of $<J/\psi>$ due to charm momentum distributions.

reactions. Thus we consider a large range of possibilities for these distributions. At one extreme, we model the distribution to simulate the initial production distribution from the pQCD calculations [6]. The transverse momentum p_T distribution is taken to be Gaussian with a width of 1 GeV^2. The rapidity dependence is taken as flat, with the width of the plateau $\Delta y = 4$. We then allow for thermalization and energy loss processes in the deconfinement region by reducing the Δy, terminating at one unit. This corresponds approximately to the charm quark momentum distribution if complete thermalization were attained for all charm quarks in the central rapidity unit.

The results of the calculations are shown in Figure 26. Shown are curves for final $N_{J/\psi}$ for central RHIC collisions as a function of $N_{c\bar{c}}$. All of the curves for various charm quark momentum distributions use zero initial J/ψ and an initial temperature $T_o = 300$ MeV to specify the gluon density and the lifetime of the deconfined region. Also shown for reference is the number of *initial* J/ψ which would be produced without any nuclear or final state effects, which we approximate by $0.01 N_{c\bar{c}}$ [18]. The lowest curve just scales the initial production curve down by a factor 0.05, which is a typical suppression factor for RHIC conditions estimated from applications of suppression-only

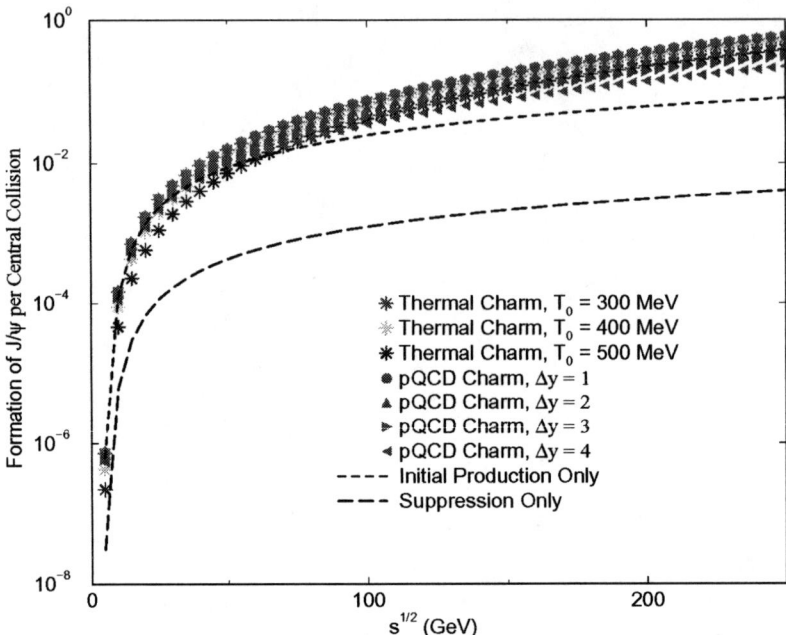

FIGURE 27. Predictions of kinetic model variation of $<J/\psi>$ as a function of $(\sqrt{s})_{NN}$ for RHIC conditions.

deconfinement models [5]. It is seen that over even this wide range of charm quark momentum distributions, the formation predictions are above the initial production estimate, thus indicating an *enhancement* in J/ψ. (This enhancement is even more extreme if the base is taken to be the suppressed initial number without any formation mechanism.)

We also show Figure 27 the equivalent energy dependence of these results, using RHIC conditions for the deconfinement properties but replacing $N_{c\bar{c}}$ by $\sqrt{s}$ from the pQCD expectations.

The centrality dependence will provide another prediction of this model. The most significant effect of varying centrality is to sweep through a range of $N_{c\bar{c}}$ from the centrality dependence of initial production as modeled by $T_{AA}(b)$, which we will change to the participant number dependence $T_{AA}(N_p)$. However, a number of other components of the kinetic model will change with centrality, through the effect of nuclear geometry on the initial conditions and spatial properties of the deconfined region.

The initial temperature T_o is expected to decrease with increasing impact parameter b, due to a decrease in the local energy density. We model the energy density in terms of the local density of participant nucleons in the transverse plane $n_w(b, s=0)$, which is shown in Figure 7. The dependence is then

$$T_o(b) = T_o(0)[n_w(b,s=0)/n_w(0,s=0)]^{\frac{1}{4}} \tag{22}$$

This dependence however is not very significant until the very peripheral region is reached.

One also needs the initial transverse size of the deconfined region. We model this area as the ratio of the participant number to the local density of participants, effectively using the fall-off of density to define the effective area within which the total number of participants would result if their density remained at its maximum value. Again this is not an absolute statement, since we normalize all of the centrality-dependent quantities to their values at b = 0.

$$A_T(b) = A_T(b=0)[N_w(b)n_w(0,s=0)/N_w(0)n_w(b,s=0)] \qquad (23)$$

The results are most conveniently displayed in terms of the ratio $N_{J/\psi}/N_{c\bar{c}}$, which eliminates the trivial dependence on one power of $N_{c\bar{c}}$ expected in any physical mechanism for production of J/ψ. The centrality dependence of this ratio should be proportional to $N_p^{4/3-\alpha}$, where now α contains the net effect of the kinetic model centrality dependence.

The results for RHIC conditions are shown in Figure 28. The canonical $N_{c\bar{c}} = 10$ value for central collisions is assumed, and required to vary with centrality as previously discussed. One sees the increase with large centrality due to the quadratic behavior of formation, which is a characteristic signature of this type of mechanism. It contrasts nicely with the initial production curve, which has the opposite behavior due to nuclear absorption of the initially-produced J/ψ. (The corresponding suppression-model curves would decrease even more rapidly with increasing centrality). The behavior for very peripheral collisions is probably an artifact of the procedure to determine the volume of the deconfinement region, and should not be taken at face value. One also sees the range of absolute values which result from variation of the charm quark momentum distribution. For sufficiently central collisions, all of these distributions predict final J/ψ which exceed the initial production, i.e. enhancement.

Also shown for comparison are the statistical hadronization model calculations previously presented in Figure 17. The magnitudes are somewhat lower for the cases considered, but probably can be made compatible with some variation in parameters. The primary difference would appear to be the sharp increase for peripheral collisions due to the canonical corrections for small particle numbers.

Figure 29 shows the corresponding predictions for LHC. Here we have used $N_{c\bar{c}} = 200$ for a central Pb-Pb collision at LHC. The range of charm quark momentum distributions has been increased to include up to $\Delta y = 7$ to account for the increased energy. It is clear that the absolute magnitudes of $N_{J/\psi}/N_{c\bar{c}}$ are larger than at RHIC. This is due to the increase in $N_{c\bar{c}}$ as contained in the quadratic dependence of the new formation mechanism.

SUMMARY

Expectations based on general grounds for enhanced formation of heavy quarkonium in relativistic heavy ion collisions have been verified in two different models. In particular, one expects at RHIC and LHC to see an enhancement in the heavy quarkonium for-

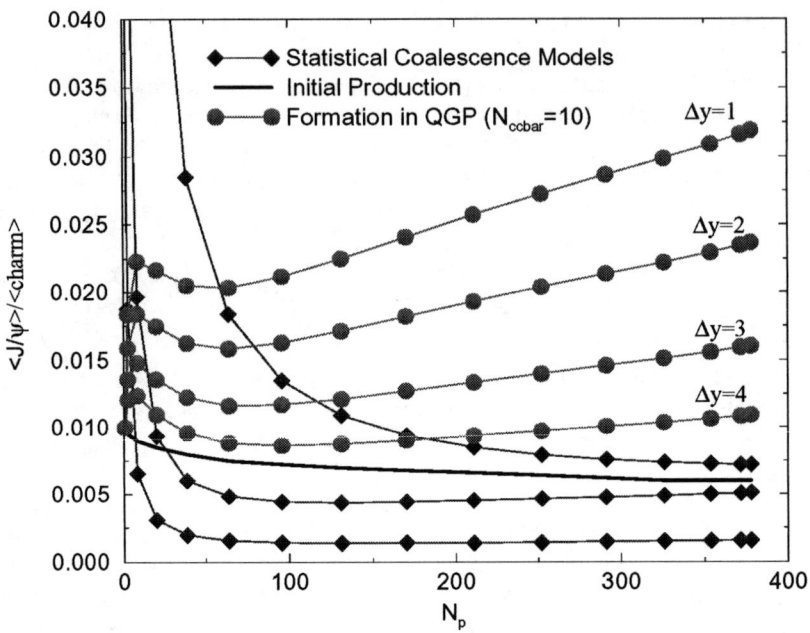

FIGURE 28. Predictions of kinetic model for $<J/\psi>$ over initial charm at RHIC energy.

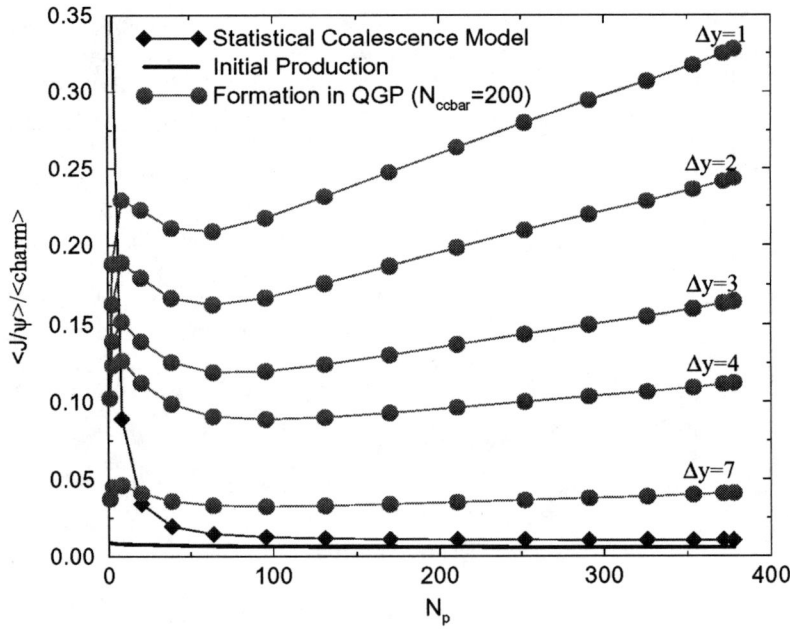

FIGURE 29. Predictions of kinetic model for $<J/\psi>$ over initial charm at LHC energy.

mation rate, even when compared to unsuppressed production via elementary nucleon-nucleon collisions in vacuum. The magnitude of this effect is expected to grow with the centrality of the heavy ion collision, just opposite to the predictions of various suppression scenarios. The physics bases for these models, however, are quite distinct. Their differences should manifest themselves in details of the magnitudes and centrality dependence. In this regard, it is essential to have a simultaneous measurement of open flavor production to serve as an unambiguous baseline.

ACKNOWLEDGMENTS

This work has been supported by U.S. DOE grant FG03-95SC306700 and by NSF grant INT-0003184.

REFERENCES

1. Altarelli, G., Lectures on Perturbative QCD, these proceedings.
2. Karsch, F., Lectures on Lattice QCD at Finite Temperature, these proceedings.
3. Matsui, T., and Satz, H., *Phys. Lett.* **B178**, 416 (1986).
4. M. C. Abreu *et al.*, NA50 Collaboration, *Phys. Lett.* **B477**, 28–36 (2000).
5. Vogt, R., *Nucl Phys.* **A661**, 250c (1999).
6. McGaughey, P. L., Quack, E., Ruuskanen, P. V., Vogt, R., and Wang, X.-N., *Int. J. Mod. Phys.* **A10**, 2999–3042 (1995).
7. Nason, P., Dawson, S., and Ellis, R. K., *Nucl. Phys.* **B303**, 607 (1988).
8. Mangano, M. L., Nason, P., and Ridolfi, G., *Nucl.Phys.* **B405**, 507–535 (1993).
9. Frixone, S., Mangano, M. L., Nason, P., and Ridolfi, G., *Nucl.Phys.* **B431**, 453–483 (1994).
10. Vogt, R., "Systematics of heavy quark production at RHIC", hep-ph/0203151 (2002).
11. Zajc, W. A., Lectures on Phenix at RHIC, these proceedings.
12. K. Adcox *et al.*, PHENIX Collaboration, *Phys. Rev. Lett.* **88**, 192303 (2002).
13. Eskola, K., Kolhinen, V., and Salgado, C., *Eur. Phys. J.* **C9**, 61–68 (1999).
14. Białas, A., Bleszyński, M., and Czyz, W., *Nucl. Phys.* **B111**, 461 (1976).
15. Margetis, S. *et al.*, NA49 Collaboration, *Nucl. Phys.* **A590**, 355c (1995).
16. Thews, R. L., *Nucl. Phys.* **A702**, 341–345 (2002).
17. Wang, X.-N., and Gyulassy, M., *Phys. Rev. Lett.* **86**, 3496–3499 (2001).
18. Gavai, R., Kharzeev, D., Satz, H., Schuler, G., Sridhar, K., and Vogt, R., *Int. J. Mod. Phys.* **A10**, 3043–3070 (1995).
19. Agarwal, M. M. *et al.*, WA98 Collaboration, *Eur. Phys. J.* **C18**, 651–663 (2001).
20. Adcox, K. *et al.*, PHENIX Collaboratiion, *Phys. Rev. Lett.* **87**, 052301 (2001).
21. Back, B. B. *et al.*, PHOBOS Collaboration, *Phys. Rev. Lett.* **88**, 022302 (2002).
22. Bearden, I. G. *et al.*, BRAHMS Collaboration, nucl-ex/0112001.
23. Kharzeev, D., Lourenco, C., Nardi, M., and Satz, H., *Z. Phys.* **C74**, 307–318 (1997).
24. Braun-Munzinger, P., Magestro, D., Redlich, K., and Stachel, J., *Phys. Lett.* **B518**, 41–46 (2001).
25. Sorge, H., Shuryak, E., and Zahed, I., *Phys. Rev. Lett.* **79**, 2775–2778 (1997).
26. Braun-Munzinger, P., and Stachel, J., *Phys. Lett.* **B490**, 196–202 (2000).
27. Braun-Munzinger, P., and Stachel, J., *Nucl. Phys.* **A690**, 119–126 (2001).
28. Gorenstein, M., Kostyuk, A., Stocker, H., and Greiner, W., *Phys. Lett.* **B509**, 277–282 (2001).
29. Redlich, K., Koch, V., and Tounsi, A., *Nucl. Phys.* **A702**, 326–335 (2002).
30. Cleymans, J., Redlich, and Suhonen, E., *Z. Phys.* **C51**, 137–141 (1991).
31. Gorenstein, M., Kostyuk, A., Stocker, H., and Greiner, W., hep-ph/0110269.
32. Abreu, M. C. *et al.*, NA38/50 Collaboration, *Nucl. Phys.* **A698**, 539–542 (2002).
33. Grandchamp, L., and Rapp, R., *Phys. Lett.* **B523**, 60–66 (2001).

34. Thews, R. L., Schroedter, M., and Rafelski, J., *Phys. Rev.* **C63**, 054905 (2001).
35. Thews, R. L., and Rafelski, J., *Nucl. Phys.* **A698**, 575–578 (2002).
36. Kharzeev, D., and Satz, H., *Phys. Lett.* **B334**, 155–162 (1994).
37. Sucipto, E., and Thews, R. L., work in progress.
38. Rafelski, J., Letessier, J., and Tounsi, A., *Heavy Ion Phys.* **4**, 181–192 (1996).
39. Peskin, M. E., *Nucl Phys.* **B156**, 365 (1979).
40. Bhanot, G., and Peskin, M. E., *Nucl Phys.* **B156**, 391 (1979).

Nuclear Effects in the Drell-Yan Process

J. Raufeisen

Los Alamos National Laboratory, MS H846, Los Alamos, NM 87545, USA

Abstract. In the target rest frame and at high energies, Drell-Yan (DY) dilepton production looks like bremsstrahlung of massive photons, rather than parton annihilation. The projectile quark is decomposed into a series of Fock states. Configurations with fixed transverse separations in impact parameter space are interaction eigenstates for pp scattering. The DY cross section can then be expressed in terms of the same color dipole cross section as DIS. We compare calculations in this dipole approach with E772 data and with next-to-leading order parton model calculations. This approach is especially suitable to describe nuclear effects, since it allows one to apply Glauber multiple scattering theory. We go beyond the Glauber eikonal approximation by taking into account transitions between states, which would be eigenstates for a proton target. We calculate nuclear shadowing at large Feynman-x_F for DY in proton-nucleus collisions and compare to E772 data. Nuclear effects on the transverse momentum distribution are also investigated.

1. DY DILEPTON PRODUCTION IN PP SCATTERING

Although cross sections are Lorentz invariant, the partonic interpretation of the microscopic process depends on the reference frame. As pointed out in [1], in the target rest frame, Drell-Yan (DY) dilepton production should be treated as bremsstrahlung, rather than parton annihilation (see also [2]). The space-time picture of the DY process in the target rest frame is illustrated in Fig. 1. A quark (or an antiquark) from the projectile hadron radiates a virtual photon on impact on the target.

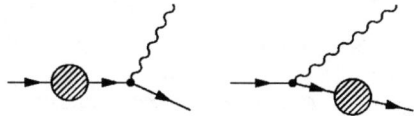

FIGURE 1. A quark (or an antiquark) inside the projectile hadron scatters off the target color field and radiates a massive photon. The subsequent decay of the γ^* into the lepton pair is not shown.

A salient feature of the rest frame picture of DY dilepton production is that at high energies and in impact parameter space the DY cross section can be formulated in terms of the same dipole cross section as low-x_{Bj} DIS. The cross section for radiation of a virtual photon from a quark after scattering on a proton, can be written in factorized light-cone form [1, 2, 3],

$$\frac{d\sigma(qp \to \gamma^* X)}{d\ln\alpha} = \sum_{T,L} \int d^2\rho \, |\Psi^{T,L}_{\gamma^* q}(\alpha,\rho)|^2 \sigma_{q\bar{q}}(\alpha\rho), \tag{1}$$

similar to the case of DIS. Here, $\sigma_{q\bar{q}}$ is the cross section [4] for scattering a $q\bar{q}$-dipole off a proton which depends on the $q\bar{q}$ separation $\alpha\rho$, where ρ is the photon-quark transverse separation and α is the fraction of the light-cone momentum of the initial quark taken away by the photon. For shortness, we do not explicitly write out the energy dependence of $\sigma_{q\bar{q}}$. We use the standard notation for the kinematical variables, $x_1 - x_2 = x_F$, $\tau = M^2/s = x_1 x_2$, where x_F is the Feynman variable, s is the center of mass energy squared of the colliding protons and M is the dilepton mass. In (1) T stands for transverse and L for longitudinal photons.

The physical interpretation of (1) is similar to the DIS case. The projectile quark is expanded in the interaction eigenstates. We keep here only the first eigenstate,

$$|q\rangle = \sqrt{Z_2}|q_{bare}\rangle + \Psi^{T,L}_{\gamma^* q}|q\gamma^*\rangle + \ldots, \tag{2}$$

where Z_2 is the wavefunction renormalization constant for fermions. In order to produce a new state the interaction must distinguish between the two Fock states, *i.e.* they have to interact differently. Since only the quarks interact in both Fock components the difference arises from their relative displacement in the transverse plane. If ρ is the transverse separation between the quark and the photon, the $\gamma^* q$ fluctuation has a center of gravity in the transverse plane which coincides with the impact parameter of the parent quark. The transverse separation between the photon and the center of gravity is $(1-\alpha)\rho$ and the distance between the quark and the center of gravity is correspondingly $\alpha\rho$. Therefore, the argument of $\sigma_{q\bar{q}}$ is $\alpha\rho$. More discussion can be found in [5].

The transverse momentum distribution of DY pairs can also be expressed in terms of the dipole cross section [3]. The differential cross section is given by the Fourier integral

$$\frac{d\sigma(qp \to \gamma^* X)}{d\ln\alpha d^2 q_T} = \frac{1}{(2\pi)^2} \int d^2\rho_1 d^2\rho_2 \exp[i\vec{q}_T \cdot (\vec{\rho}_1 - \vec{\rho}_2)]\Psi^*_{\gamma^* q}(\alpha,\vec{\rho}_1)\Psi_{\gamma^* q}(\alpha,\vec{\rho}_2)$$
$$\times \frac{1}{2}\left\{\sigma_{q\bar{q}}(\alpha\rho_1) + \sigma_{q\bar{q}}(\alpha\rho_2) - \sigma_{q\bar{q}}(\alpha(\vec{\rho}_1-\vec{\rho}_2))\right\}. \tag{3}$$

after integrating this expression over the transverse momentum q_T of the photon, one obviously recovers (1).

The LC wavefunctions can be calculated in perturbation theory and are well known [2, 5]. The dipole cross section on the other hand is largely unknown. Only at small distances ρ it can be expressed in terms of the gluon density. However, several successful parameterizations exist in the literature, describing the entire function $\sigma_{q\bar{q}}(x,\rho)$, without explicitly taking into account the QCD evolution of the gluon density. We use the parameterization by Golec-Biernat and Wüsthoff [6] for our calculations, Fig. 2. This parameterization vanishes $\propto \rho^2$ at small distances, as implied by color transparency [4] and levels off exponentially at large separations.

In Fig. 2, we compare to E772 data [7] on low x_2 DY dilepton production [8]. Most of the data are quite well described without any K-factor, which does not appear in this approach since higher order corrections are supposed to be parameterized in $\sigma_{q\bar{q}}(\rho)$. Moreover, the calculation in the dipole approach agrees with the next-to-leading order (NLO) parton model calculation at low x_2. Note that the dipole approach is valid only at low x_2 [5]. At large x_2, this approach is not applicable and differs strongly from the parton model calculation. The disagreement between the data and both of the calculations in some points is probably due to systematic errors in the measured cross section. Preliminary E866 data [9] agree well with the NLO parton model calculation.

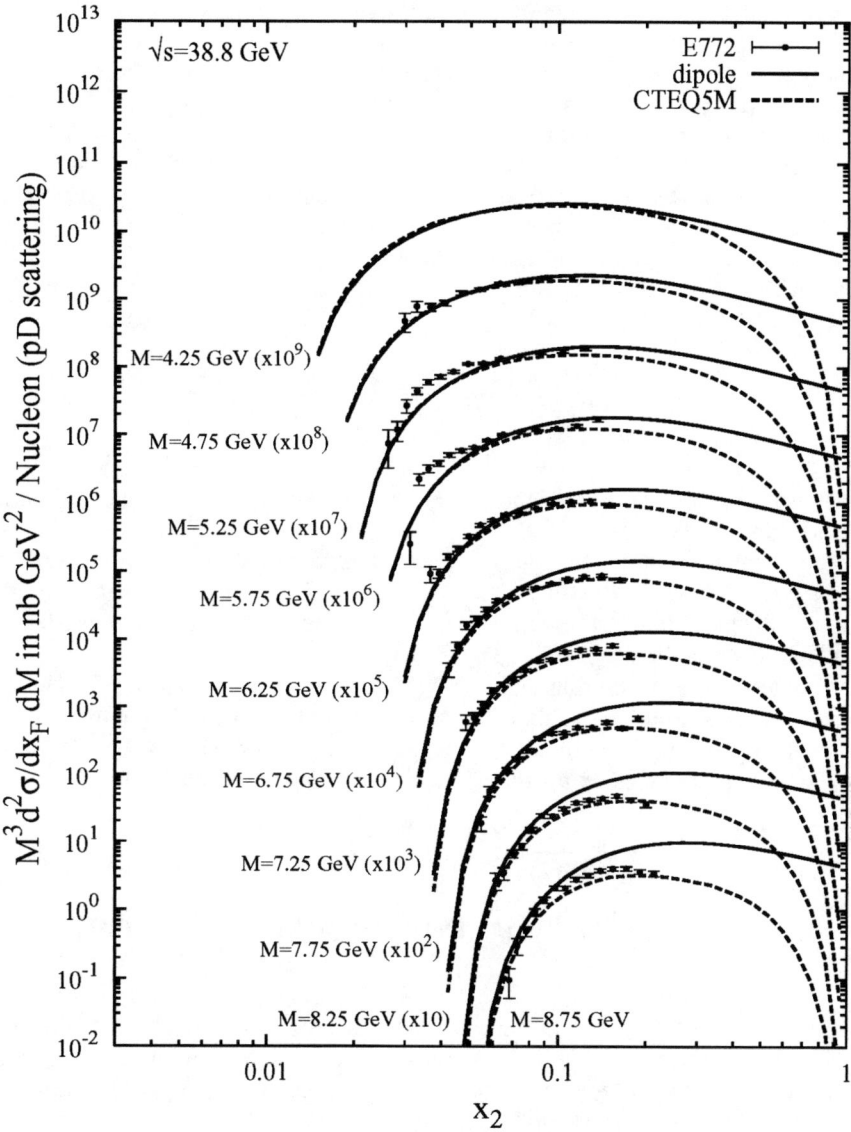

FIGURE 2. The points represent the measured DY cross section in pD scattering from E772 [7]. The solid curve is calculated in the dipole approach, while the NLO parton model calculation (using CTEQ5M parton distributions [10]) is shown as dashed curve. The dipole approach is valid only at small x_2.

2. PROTON-NUCLEUS (PA) SCATTERING

The main advantage of the dipole approach is its easy generalization to nuclear targets. Furthermore, it also includes some higher twist effects that are important in multiple scattering, and it provides insight into the physical mechanisms underlying nuclear effects, which are not easily accessible in the parton model [11].

Shadowing in DY is an interference phenomenon due to multiple scattering of the projectile quark inside the nucleus. In the target rest frame, where DY dilepton production is bremsstrahlung of massive photons, shadowing is the Landau-Pomeranchuk-Migdal (LPM) effect. These interferences occur (Fig. 1), because photons radiated at different longitudinal coordinates z_1 and z_2 are not independent of each other. Thus, the amplitudes have to be added coherently. Destructive interferences can occur only if the longitudinal distance $z_2 - z_1$ is smaller than the so called coherence length l_c, which is the time needed to distinguish between a quark and a quark with a γ^* nearby. It is given by the uncertainty relation,

$$l_c = \frac{1}{\Delta P^-} = \frac{1}{m_N x_2} \frac{(1-\alpha)M^2}{q_T^2 + (1-\alpha)M^2 + \alpha^2 m_q^2}. \tag{4}$$

Here, ΔP^- is the light-cone energy denominator for the transition $q \to q\gamma^*$ and q_T is the relative transverse momentum of the $\gamma^* q$ Fock state. For $z_1 - z_2 > l_c$, the radiations are independent of each other.

An immediate consequence of this is that l_c has to be larger than the mean distance between two scattering centers in the nucleus (~ 2 fm in the nuclear rest frame). Otherwise, the projectile quark could not scatter twice within the coherence length and no shadowing would be observed.

We develop a Green function technique [3], which allows one to resum all multiple scattering terms, similar to Glauber theory, and in addition treats the coherence length exactly. The formalism is equivalent to the one proposed in [12] for the LPM effect in QED. Our general expression for the nuclear DY cross section reads

$$\begin{aligned}\frac{d\sigma(qA \to \gamma^* X)}{d\ln\alpha} &= A\frac{d\sigma(qp \to \gamma^* X)}{d\ln\alpha} - \frac{1}{2}\text{Re} \int d^2b \int_{-\infty}^{\infty} dz_1 \int_{z_1}^{\infty} dz_2 \int d^2\rho_1 \int d^2\rho_2 \\ &\times \left[\Psi_{\gamma^* q}(\alpha,\rho_2)\right]^* \rho_A(b,z_2)\, \sigma_{q\bar{q}}(\alpha\rho_2)\, G(\vec{\rho}_2,z_2|\vec{\rho}_1,z_1) \\ &\times \rho_A(b,z_1)\, \sigma_{q\bar{q}}(\alpha\rho_1)\, \Psi_{\gamma^* q}(\alpha,\rho_1).\end{aligned} \tag{5}$$

The first term is just A times the single scattering cross section, where A is the nuclear mass number. The second term is the shadowing correction. The impact parameter is b and the nuclear density is ρ_A. The Green function G describes, how the bremsstrahlung-amplitude at z_1 interferes with the amplitude at z_2.

To make the meaning of Eq. (5) more clear, let us first consider a limiting case for G. In the simplest case, the coherence length, Eq. 4, is infinitely long and only the double scattering term is taken into account. Then $G(\vec{\rho}_2,z_2|\vec{\rho}_1,z_1) = \delta^{(2)}(\vec{\rho}_1 - \vec{\rho}_2)$ and one of the ρ integrations can be performed. The δ-function means that at very high energy (infinite coherence length) the transverse size of the $\gamma^* q$ Fock-state does not vary during propagation through the nucleus, it is frozen due to Lorentz time dilatation. Furthermore, partonic configurations with fixed transverse separations in impact parameter space were identified a long time ago [4] in QCD as interaction eigenstates. This is the reason, why we work in coordinate space. Namely, in coordinate space, all multiple scattering terms can be resummed and in the limit of infinite l_c one obtains

$$G^{frozen}(\vec{\rho}_2, z_2 | \vec{\rho}_1, z_1) = \delta^{(2)}(\vec{\rho}_1 - \vec{\rho}_2) \exp\left(-\frac{\sigma_{q\bar{q}}(\alpha \rho_1)}{2} \int_{z_1}^{z_2} dz \rho_A(b, z)\right). \quad (6)$$

The frozen approximation is identical to eikonalization of the dipole cross section in Eq. (1). Thus, the impact parameter representation allows a very simple generalization from a proton to a nuclear target, provided the coherence length is infinitely long.

At Fermilab fixed-target energies ($\sqrt{s} = 38.8$ GeV for E772), this last condition is not fulfilled and one has to take a finite l_c into account. The problem is however, that l_c, Eq. (4), depends on the relative transverse momentum q_T of the $\gamma^* q$-fluctuation which is the conjugate variable to the size ρ of this Fock-state and therefore completely undefined in ρ-representation. The quantum mechanically correct way to treat the q_T^2 in Eq. (4) is to represent it by a two-dimensional Laplacian Δ_T in ρ-space. The Green function which contains the correct, finite coherence length and resums all multiple scattering terms fulfills a two-dimensional Schrödinger equation with an imaginary potential,

$$\left[i\frac{\partial}{\partial z_2} + \frac{\Delta_T(\rho_2) - \eta^2}{2E_q \alpha(1-\alpha)} + \frac{i}{2}\rho_A(b, z_2) \sigma_{q\bar{q}}(\alpha \rho_2)\right] G(\vec{\rho}_2, z_2 | \vec{\rho}_1, z_1)$$
$$= i\delta(z_2 - z_1) \delta^{(2)}(\vec{\rho}_2 - \vec{\rho}_1), \quad (7)$$

where $\eta^2 = (1-\alpha)M^2 + \alpha^2 m_q^2$. For details of the derivation, we refer to [3].

The imaginary potential accounts for all higher order scattering terms. The Laplacian implies that the Green function is no longer proportional to a δ-function. This means the size of the $\gamma^* q$ fluctuation is no longer constant during propagation through the nucleus. One can say that an eigenstate of size ρ_1 evolves to an eigenstate of size $\rho_2 \neq \rho_1$, so transitions between eigenstates occur.

Calculations with Eqs. (5) and (7) are compared to E772 data [13] in Fig. 3. Note that the coherence length l_c at E772 energy becomes smaller than the nuclear radius. Shadowing vanishes as x_2 approaches 0.1, because the coherence length becomes smaller than the mean internucleon separation. It is therefore important to have a correct description of a finite l_c in this energy range. The eikonal (frozen) approximation, Eq. (6), does not reproduce the vanishing shadowing toward $x_2 \to 0.1$. The curves in Fig. 3 are somewhat different from the ones in [14], because we used a different parameterization of the dipole cross section. Note that for heavy nuclei, energy loss [15] leads to an additional suppression of the DY cross section.

Nuclear effects on the q_T-differential cross section calculated at RHIC energy are shown in Fig. 4. See [3] for details of the calculation. The differential cross section is suppressed at small transverse momentum q_T of the dilepton, where large values of ρ dominate. This suppression vanishes at intermediate $q_T \sim 2$ GeV. The Cronin enhancement that one could expect in this intermediate q_T region [14] is suppressed due to gluon shadowing [16].

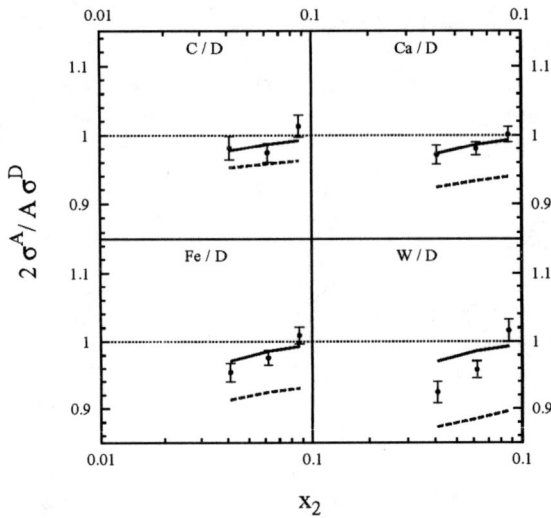

FIGURE 3. Comparison between calculations in the Green function technique (solid curve) and E772 data at center of mass energy $\sqrt{s} = 38.8$ GeV. for shadowing in DY. The dashed curve shows the eikonal (frozen) approximation, which is not valid at this energy, any more.

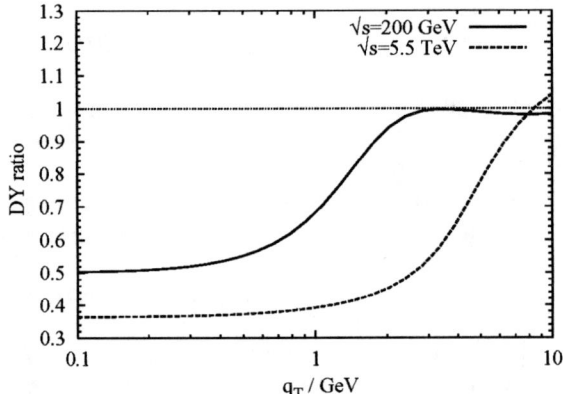

FIGURE 4. Nuclear effects on the DY transverse momentum distribution at RHIC and LHC for dilepton mass $M = 4.5$ GeV and Feynman $x_F = 0.5$.

A nuclear target provides a larger momentum transfer than a proton target and harder fluctuations are freed, which leads to nuclear broadening. Note, that not the entire suppression at low q_T is due to shadowing. Some of the dileptons missing at low q_T reappear at intermediate transverse momentum. At very large transverse momentum nuclear effects vanish.

3. SUMMARY

We express the DY cross section in terms of the cross section for scattering a $q\bar{q}$ dipole off a proton. This is the same dipole cross section that appears in DIS. At low x_2 and for proton-proton scattering, calculations in the dipole approach agree with calculations in the NLO parton model. Some E772 data points are not well described by either of the approaches, which is probably due to a systematic error in the measured cross section.

At very high energy, the dipole approach is easily extended to nuclear targets by eikonalization. At lower fixed target energies (E772) the eikonal approximation is no longer valid, because the size of a Fock state varies during propagation through the nucleus. Therefore, transitions between interaction eigenstates (i.e. partonic configurations with fixed transverse separations) occur.

We develop a Green function technique, which takes variations of the transverse size into account and resums all multiple scattering terms as well. For light nuclei, calculations with the Green function technique are in good agreement with DY shadowing data from E772. For heavier nuclei, also energy loss becomes important.

We have also calculated nuclear effects in the transverse momentum distribution of DY pairs at RHIC energy. The DY cross section is suppressed at low transverse momentum. The expected Cronin enhancement at intermediate $q_T \sim 2$ GeV is reduced because of gluon shadowing. Nuclear effects vanish at very large q_T.

ACKNOWLEDGMENTS

The author wishes to thank the Gesellschaft für Schwerionenforschung (GSI), Darmstadt, Germany for financial support during this summer school. This work was supported in part by the U.S. Department of Energy at Los Alamos National Laboratory under Contract No. W-7405-ENG-38.

I am indebted to Jörg Hüfner, Mikkel Johnson, Boris Kopeliovich, Jen-Chieh Peng and Alexander Tarasov for valuable discussion.

REFERENCES

1. Kopeliovich, B.Z., proc. of the workshop Hirschegg '95: *Dynamical Properties of Hadrons in Nuclear Matter,* Hirschegg January 16-21, 1995, ed. by Feldmeyer, H., and Nörenberg, W., Darmstadt, 1995, p. 102 (hep-ph/9609385).
2. Brodsky, S.J, Hebecker, A., and Quack, E., *Phys. Rev. D* **55**, 2584 (1997).
3. Kopeliovich, B.Z, Schäfer, A., and Tarasov, A.V., *Phys. Rev. C* **59**, 1609 (1999), extended version in hep-ph/9808378.
4. Zamolodchikov, A.B., Kopeliovich, B.Z., and Lapidus, L.I., *Sov. Phys. JETP Lett.* **33**, 612 (1981).
5. Kopeliovich, B.Z., Raufeisen, J., and Tarasov, A.V., *Phys. Lett. B* **503**, 91 (2001).
6. Golec-Biernat, K., and Wüsthoff, M., *Phys. Rev. D* **59**, 014017 (1999); *Phys. Rev. D* **60**, 114023 (1999).
7. E772 collab., McGaughey, P.L., et al., *Phys. Rev. D* **50**, 3038 (1994); erratum *Phys. Rev. D* **60**, 119903 (1999).
8. Raufeisen, J., Peng, J.-C., and Nayak, G.C., paper in preparation.

9. Webb, J.C., *et al.* [E866/NuSea Collaboration], proceedings of the *7th Conference, Intersections of Particle and Nuclear Physics*, Parsa, Z., and Marciano, W.J., eds., AIP Conference Proc., vol. 549, Melville, NY, USA, 2000.
10. Lai, H.L., *et al.* [CTEQ Collaboration], *Eur. Phys. J. C* **12**, 375 (2000) [hep-ph/9903282].
11. Eskola, K.J., Kolhinen, V.J., Ruuskanen, P.V., and Thews, R.L., arXiv:hep-ph/0108093.
12. Zakharov, B.G., *Phys. Atom. Nucl.* **61**, 838 (1998).
13. E772 collab., Alde, D.M., et al., *Phys. Rev. Lett.* **64**, 2479 (1990);
14. Kopeliovich, B.Z., Raufeisen, J., and Tarasov, A.V., arXiv:hep-ph/0104155.
15. Johnson, M.B., *et al.* [FNAL E772 Collab.], *Phys. Rev. Lett.* **86**, 4483 (2001) [hep-ex/0010051]; Johnson, M.B., *et al.*, *Phys. Rev. C* **65**, 025203 (2002) [arXiv:hep-ph/0105195]; Moss, J.M., *et al.*, arXiv:hep-ex/0109014.
16. Kopeliovich, B.Z., Raufeisen, J., Tarasov, A.V., and Johnson, M.B., arXiv:hep-ph/0110221.

Strange Hadron Resonances: Freeze-Out Probes in Heavy-Ion Collisions[1]

C. Markert*, G. Torrieri† and J. Rafelski†

*Yale University, New Haven, Connecticut 06520
†Department of Physics, University of Arizona, Tucson, AZ 85721

Abstract. Hyperon resonances are becoming an extremely useful tool allowing the study of the properties of hadronic fireballs made in heavy ion collisions. Their yield, compared to stable particles with the same quark composition, depends on hadronization conditions. The resonance's short lifetime makes them ideal probes of the fireball chemical freeze-out mechanisms. An analysis of resonance abundance in heavy ion collisions should be capable of distinguishing between possible hadronization scenarios, in particular between sudden and gradual hadronization. In this paper, we review the existing SPS and RHIC experimental data on resonance production in heavy ion collisions, and discuss in terms of both thermal and microscopic models the yields of the two observed resonances, K^* and $\Lambda(1520)$. We show how freeze-out properties, namely chemical freeze-out temperature and the lifetime of the interacting hadron phase which follows, can be related to resonance yields. Finally, we apply these methods to SPS and RHIC measurements, discuss the significance and interpretations of our findings, and suggest further measurements which may help in clarifying existing ambiguities.

INTRODUCTION AND MOTIVATION

The experimental measurement of short-lived hadron resonances can potentially be very useful in clarifying some of the least understood aspects of heavy ion collisions. In general, evolution of a hot hadronic system proceeds according to Fig. 1. When mesons and baryons emerge from a pre-hadronic state, presumably quark gluon plasma, their abundances are expected to be fixed by hadronization temperature and chemical fugacities. This stage of fireball evolution is commonly known as chemical freeze-out. After initial hadronization, the system may evolve as an interacting hadron gas. At a certain point (which can vary according to particle species), thermal freeze-out, where hadrons stop interacting, is reached.

A quantitative understanding of the above picture is crucial for any meaningful analysis of the final state particles. Many probes of deconfinement are most sensitive when the dense hadron matter fireball breakup is sudden and re-interaction time short or non-existent. Final state particles could, however also emerge remembering relatively little about their primordial source, having been subject to re-scattering in purely hadronic gas phase. In fact, theoretical arguments have been advanced in support of both a sudden reaction picture [1, 2] and a long re-interaction timescale [3]. Both pictures have

---

[1] This survey combines separate oral contributions made at the meeting

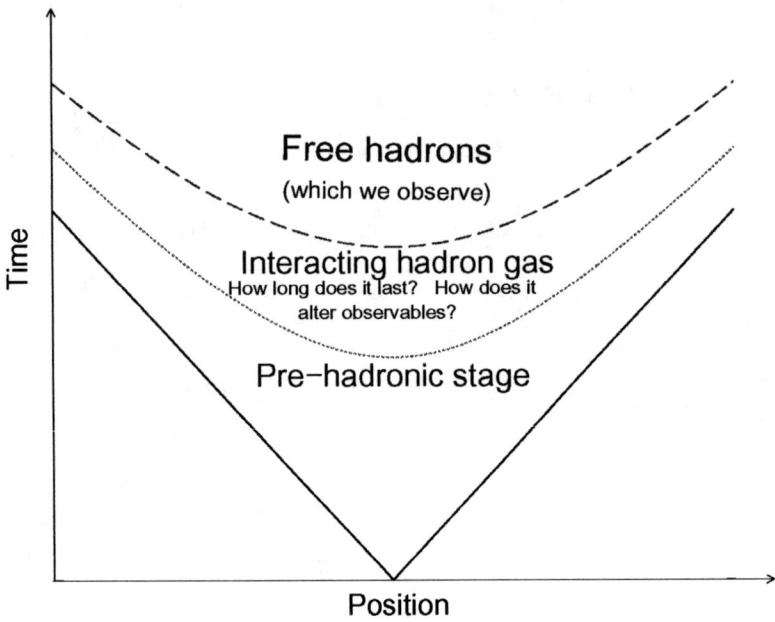

FIGURE 1. Stages of the space-time evolution of a heavy ion collision. At a certain moment in proper time (known as chemical freeze-out), hadrons emerge. The system then evolves as an interacting hadron gas, until thermal freeze-out, the point at which all elastic interactions cease as well.

been applied to phenomenological fits of hyperon abundances and distributions [4, 5].

It is apparent that hyperon resonances can be crucial in resolving this ambiguity: their initial abundance, compared to stable particles with the same quark composition, depends primarily to the temperature at hadron formation. However, the observed abundance will be potentially quite different, and will strongly depend on the lifetime of the interacting hadron gas phase. Resonances can only be observed via invariant mass reconstruction, and their short lifetime means that they can decay within the interacting hadron gas (Fig. 2). In this case, the decay products can undergo re-scattering, and emerge from the fireball with no memory about the parent resonance. Thus, the observable resonance abundance is sensitive to precisely those parameters needed to distinguish between the sudden and gradual freeze-out models.

This paper offers a pedagogical description of what we know about resonance production in heavy ion collisions. We start with a review of presently available experimental data, and the open questions which arise. We then proceed to describe how to calculate the initial resonance abundance and the effect of re-scattering in terms of the hadronization temperature and the lifetime of the interacting hadron gas. We show how these two parameters can be extracted from the experimental observations. Finally, we discuss possible answers to experimental challenges raised in the first section, and suggest ways by which these questions could be resolved by further measurements. This survey incorporates recently published experimental measurements [6, 7, 8, 9, 10] and theoretical results [13, 14, 15].

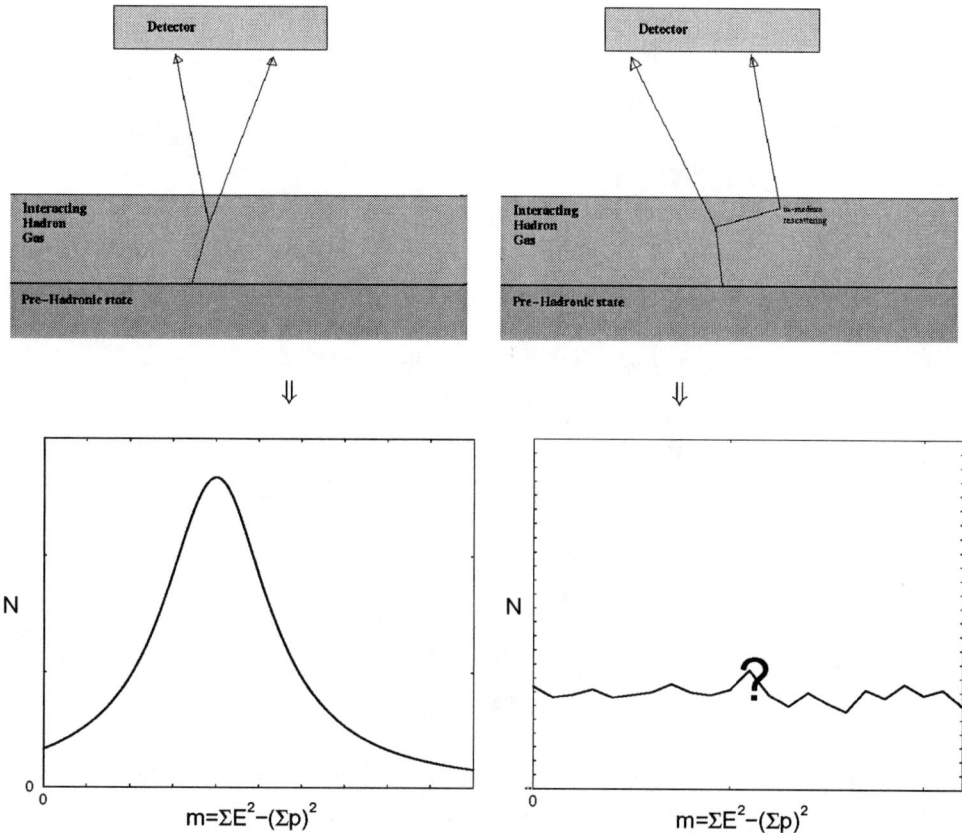

FIGURE 2. Re-scattering can inhibit resonance reconstruction. Left: in case the resonance decay products reach the detector without further interactions, their invariant mass distribution should yield a clear peak at the resonance mass, with the yield corresponding to resonance abundance. Right: should these decay products undergo re-scattering before reaching the detector, the signal may be indistinguishable from the background caused by unrelated particle pairs. Generally a weakening by the medium of the resonance invariant mass signal must be allowed for.

EXPERIMENTAL MEASUREMENTS

Resonance Reconstruction

Resonances are detected through invariant mass reconstruction of their decay products. The channels relevant for the here considered $K^{*0}, \Lambda(1520)$ resonances are:

$$\overline{K}^{*0}(892) \to \pi^+ + K^-,$$
$$K^{*0}(892) \to \pi^- + K^+,$$

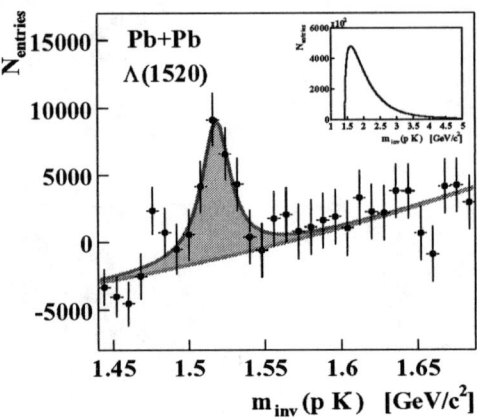

FIGURE 3. Λ(1520) mass plot after mixed event background subtraction with a Breit-Wigner fit from NA49 at $\sqrt{s} = 17$ GeV, insert plot: Mass plot before mixed event background subtraction [8].

$$\Lambda(1520) \rightarrow p + K^-,$$

and the proposed Σ^* measurement would be performed via the observation of the decay:

$$\Sigma^*(1385) \rightarrow \Lambda + \pi.$$

At present, NA49 reported a $\overline{K^*} \rightarrow K^- \pi^+$ signal [6], while STAR published an average of K^* and $\overline{K^*}$ yields [7].

Both the STAR [16] and the NA49 [17] experiments perform this reconstruction based on data collected in their large volume *Time Projection Chambers* (TPC) detectors. Charged particle momenta are obtained by a measurement of their trajectories in a uniform magnetic field, while particle identification is done evaluating the rate of energy loss dE/dx in the TPC [18]. A centrality trigger selects the 14% (STAR) and 5% (NA49) most central inelastic interactions. The decay daughter candidates are selected via their momenta and dE/dx. The resonance signal is obtained by the invariant mass reconstruction of each pair combination and the subtraction of a mixed event background estimated by combining candidates from different events (Figure 3). The multiplicity is obtained after applying correction factors for acceptance, particle identification cut, reconstruction efficiency and branching ratio.

Determination of Width and Mass

The measured width and mass of Λ(1520) and K* in heavy ion collisions is in agreement within the statistical errors with the values from the Particle Data Group (table 1).

TABLE 1. Resonance mass and width

collision	Energy GeV	particle	mass MeV/c²	σ_{mass} MeV/c²	width MeV/c²	σ_{width} MeV/c²	reference
Pb+Pb	17.2	$\Lambda(1520)$	1518.1	2.0	22.7	6.5	[8]
Au+Au	130	K^{*0}	893.0	3.0	58.0	15.0	[10]
Au+Au	130	$\overline{K^{*0}}$	896.0	4.0	63.0	11.0	[10]
PDG		$\Lambda(1520)$	1519.5	1.0	15.6	0.6	[11]
PDG		K^{*0} and $\overline{K^{*0}}$	896.1	0.27	50.7	15.0	[11]

TABLE 2. Resonance yields

collision	Energy GeV	particle	yield	reference
p+p	17.2	$\Lambda(1520)$	0.0121 ± 0.003	[6, 8]
p+Pb	17.2	$\Lambda(1520)$	0.0072 ± 0.003	[6]
Pb+Pb	17.2	$\Lambda(1520)$	1.45 ± 0.40	[6, 8]
Pb+Pb	17.2	K* 5-10% most central	12.88 ± 4.53	[6]
Au+Au	130	K* $(dN/dy)_{y<|0.5|}$	10.0 ± 0.8	[10, 7]
Au+Au	130	$\Lambda(1520)$	< 4.2 *	[9]

* at 95% confidence level

Within the statistical error there are no width broadening or mass shifts observed. While there is no calculation available that estimates the width and mass profile of a resonance by taking the medium density evolution during the expansion into account, shifts in width and mass would be expected in a lengthy hadron gas phase [19]. A prediction based on relativistic chiral SU(3) dynamic calculations [20, 21] gives a 100 MeV broadening of the $\Lambda(1520)$ resonance and a mass shift of about 100 MeV to lower masses in medium at a density of $\rho = 0.17$ fm^{-3}.

The fact that the $\Lambda(1520)$ NA49 signal from Pb+Pb reactions shows no width broadening within the errors, indicates that only the decay products coming from the vacuum $\Lambda(1520)$ decay are observed. Any medium modified $\Lambda(1520)$ resonances must thus decay rapidly within the medium, with decay products rescattering, or, alternatively no such $\Lambda(1520)$-medium interaction is present. To distinguish these alternatives one needs to consider the yields of $\Lambda(1520)$.

Particle Multiplicities

Experimental results for K* and $\Lambda(1520)$ resonance production in heavy ion collisions are available from NA49 experiment in central Pb+Pb and p+p at $\sqrt{s} = 17$ GeV and from STAR experiment in central Au+Au $\sqrt{s_{NN}} = 130$ GeV collision energy and are shown in table 2.

Comparing $\Lambda(1520) \to pK$ yields at different experiments, we see a marked decrease

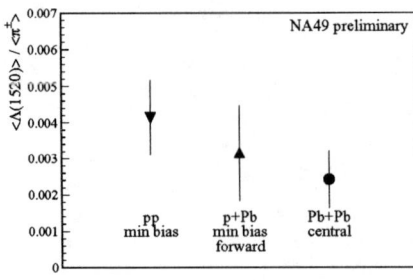 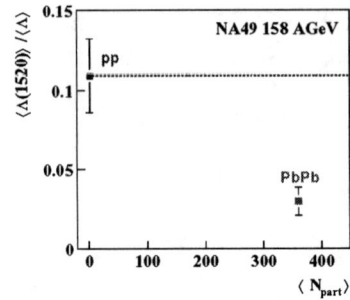

FIGURE 4. Left: $\Lambda(1520)/\pi^+$ as a function of participants for p+p, p+Pb and Pb+Pb collisions at 158 AGeV (Reprinted from Ref. [6], Copyright 2002, with permission from Elsevier Science). Right: $\langle\Lambda(1520)\rangle/\Lambda$ ratio as a function of number of participants. Data points are p+p and central Pb+Pb collisions from NA49.

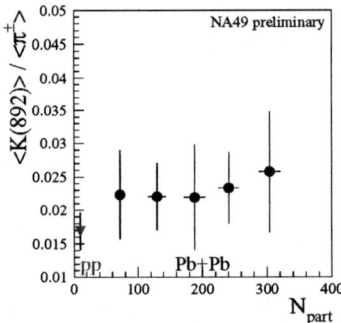

FIGURE 5. $K^*(892)/\pi$ as a function of participants for p+p and Pb+Pb collisions at 158 AGeV.

from elementary p+p to Pb+Pb collisions at 17.2 GeV, which is even more pronounced in the $\Lambda(1520)/\Lambda$ ratio (figure 4 left and right). This suggests that heavy ion collisions are not merely superpositions of elementary p+p collisions. Here we can ask the first question. "Why is the $\Lambda(1520)$ signal from the invariant mass reconstruction in the K-p decay channel actually suppressed in heavy ion collisions, whereas all other hyperons are enhanced?" Moreover, the K* over π production at the same energy indicates no such signal decrease from p+p to Pb+Pb collisions, shown in figure 5. The second question is now "Why is no comparable suppression observed in the K^*?" A hint for the signal loss could be due to secondary interactions is the measurement of the $\Lambda(1520)$ production in p+Pb collisions which seems to be decreasing. However the errors are to large to say this conclusively (figure 4 left).

For the collision energy of $\sqrt{s_{NN}} = 130$ GeV there is only an upper limit for the $\Lambda(1520)$, namely 4.2 at 95% confidence level (2σ). The expected multiplicity from extrapolated elementary p+p reactions including an addition factor of 2 for strangeness enhancement (taken from Λ at SPS [22]) is ~ 7.7. The upper limit estimate indicates

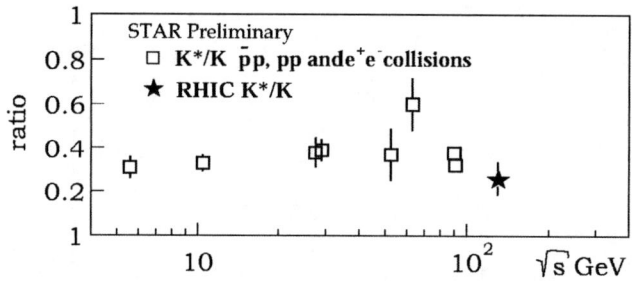

FIGURE 6. K*⁰/K measured in different colliding systems at different energies. For elementary e^+e^- and pp collisions in comparison [23, 24, 25, 26, 27, 28, 30, 31, 32] with the heavy ion collision Au+Au at RHIC energies $\sqrt{s_{NN}} = 130$ GeV [7].

TABLE 3. Particle yields from thermal model predictions

collision	Energy GeV	particle	predicted yield	reference
Pb+Pb	17.2	Λ(1520)	3.48	[35]
Pb+Pb	17.2	Λ(1520)	5.20	[3]
Au+Au	130	K^*/h^-	0.037	[12]

that we see at RHIC energies the same trend of signal loss as at SPS energies. The K*/K ratio at RHIC energies fits in within errors to the ratios from elementary p+p and e^++e^- collisions figure 6, that leads us to the conclusion that there is no indication of K* signal loss related to K yield. Our conclusions from the results from SPS and RHIC energies are in agreement.

A recent UrQMD calculation [33] predicts for the K* a signal loss of 66 % and for the Λ(1520) 50 % at SPS energies due to re-scattering of the decay products (figure 7). At RHIC energies the signal loss is on the order of 55% for K* and 30 % for Λ(1520) [34]. The K* yield at A+A collisions for SPS and RHIC energies shows no indication of signal loss and therefore it is not in agreement with the predicted signal loss from the UrQMD model. This means that the mechanism of re-scattering alone can not explain the signal loss of the Λ(1520) with respect to the K* production.

As table 3 shows, the agreement of the data with thermal model predictions presents a considerable problem as well. At SPS energies, thermal model calculations over-predict the Λ(1520) yield by a factor of two [36], while the upper limit measurement of Λ(1520) from STAR is 30% lower than the predicted value. However, published thermal model-calculations significantly under-predict the K^* (although, as will be seen in the next section, the observed K^* is consistant with thermal production at a lower freezeout temperature).

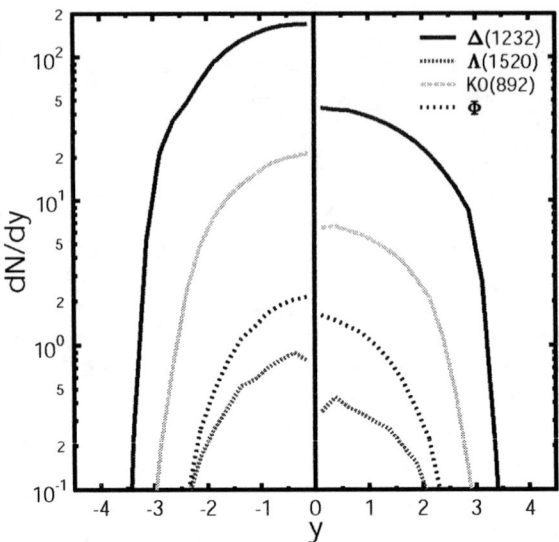

FIGURE 7. Urqmd calculation. Rapidity distributions of produced resonances (left half), and of the resonances which remain observable after thermal freezeout (Reprinted from Ref. [33] Copyright 2002, with permission from Elsevier Science).

MODELING RESONANCES IN HEAVY ION COLLISIONS

Direct production at hadronization

We assume that at hadronization, a volume element will be at thermal and chemical equilibrium in its local rest frame. This means hadrons produced directly from a medium at temperature T (much lower than the particle mass, as in all cases considered here) fill the available phase space according to the relativistic Boltzmann distribution:

$$dN \propto g d^3 p f(\lambda, E, T), \qquad (1)$$

$$f(\lambda, E, T) = \lambda e^{-E/T}. \qquad (2)$$

Here g is the statistical degeneracy and $\lambda = e^{\mu/T}$ is the hadron's fugacity. Since a hadron is a composite state of flavored quarks, it is natural to parametrize its fugacity as a product of the flavor fugacities of each constituent. As a further step, each quark's fugacity can be decomposed into a term λ_i controlling the net flavor ($q - \bar{q}$) and an occupancy number γ_i which determines the number of quark-antiquark pairs. Therefore,

$$\lambda = \prod_{q=1}^{n} \lambda_i \gamma_i, \qquad (3)$$

$$\lambda_q = \lambda_{\bar{q}}^{-1}, \qquad (4)$$

$$\gamma_q = \gamma_{\bar{q}}, \qquad (5)$$

and the precise values of λ_i, γ_i can be determined from global particle abundance fits [2].

We can now change to rapidity (y) and transverse mass (m_T) coordinate system, where transverse and longitudinal directions are defined with respect to the beam. (The advantage of this system is that m_T is invariant between the lab and the center of mass frames, while the rapidity transforms additively).

$$dN = gm_T^2 \cosh(y) dm_T dy f(\lambda, m_T \cosh(y), T) \tag{6}$$

$$m_T = \sqrt{m^2 + p_T^2} \tag{7}$$

$$y = \frac{1}{2} \ln\left(\frac{E + p_L}{E - p_L}\right) \tag{8}$$

To get particle ratios in a certain region of phase space collective expansion (flow) as well as freeze-out geometry has to be included. The most general way to do this is to introduce a space-time hypersurface (a 3-D surface in 4-D space-time) $d\Sigma_\mu$, which will transform as a 4-vector. This vector can either be time-like, corresponding to a freeze-out surface evolving in time, or space-like, corresponding to a freeze-out occurring simultaneously across the fireball volume. The total number of emitted particles will then be

$$N_i = \int d\Sigma_\mu(x) j^\mu(x). \tag{9}$$

We can describe the current 4-vector j^μ in terms of the hadron density in the volume element at rest, (see Eq. 1)

$$j^\mu = n_{\text{rest}} u^\mu = f(\lambda, E_{\text{local}}, T) \frac{p^\mu}{E}, \tag{10}$$

where E_{local} is the energy in the local rest frame with respect to flow, and u^μ is the volume element's 4-velocity.

In the limit of a homogeneous fireball (Temperature and chemical potential constant in space) and full phase space, the total number of produced particles is, as expected, independent of flow and hadronization geometry, since in this case Eq. (9) becomes

$$N = \int d\Sigma_\mu(x) j^\mu(x) = \int d\Sigma_\mu(x) u^\mu(x) \int d^3p f(\lambda, E_{\text{local}}, T) = V \int d^3p f(\lambda, E_{\text{local}}, T). \tag{11}$$

We can now find N analytically

$$N = V\lambda \int d^3p \, e^{-(p^2+m^2)/T} = \frac{1}{2\pi^2} \lambda m^2 T K_2(m/T) \tag{12}$$

remembering the properties of Bessel functions

$$K_n(x) = \frac{2^n n!}{(2n)!} x^{-n} \int_0^\infty dz (z^2 + x^2)^{-\frac{1}{2}} z^{2n} e^{-\sqrt{z^2+x^2}}, \tag{13}$$

If the fireball is not homogeneous, or we want to only calculate the yield within the experimental acceptance limits, we need to use a more general result. Putting Eqs. (9)

and (10) together one gets the relativistically invariant Cooper-Frye formula [40]

$$E\frac{d^3N}{d^3p} = \frac{1}{2\pi m_T}\frac{dN}{dy dm_T} = \int d\Sigma_\mu p^\mu f(\lambda, p_\mu u^\mu, T)\theta(\Sigma_\mu p^\mu), \qquad (14)$$

where the step function $\theta(\Sigma_\mu p^\mu)$ has been introduced to eliminate unphysical emission in the direction opposite to expansion [41, 42]. Eq. (14) can be integrated to yield the total particle number over an arbitrary phase space region

$$N = \int \frac{d^3p}{E} \int d\Sigma_\mu p^\mu f(\lambda, p_\mu u^\mu, T)\theta(\Sigma_\mu p^\mu) \qquad (15)$$

Many of the parameters used in these formulae can be eliminated by judicious choice of observables. For instance, the ratio of a hyperon resonance to its ground state is independent of the chemical potential, since all chemical potential terms cancel out. Suitable ratios include Σ^*/Λ, $\Lambda(1520)/\Lambda$, K^{*0}/K^+, $\overline{K^{*0}}/K^-$. A particular case is the $K^0/\overline{K^0}$ system: While the K^{*0} decays long before it has a chance of oscillate, the experimentally observed K_S is a superposition of K^0 and $\overline{K^0}$. Suitable average are $\overline{K^{*0}}/K^-$ (measured by NA49 [6]) K^{*0}/K^+ and $(K^* + \overline{K^*})/(K^+ + K^-)$ (measured by STAR [7]).

Moreover, we found that for these ratios (more generally for ratios of particles with comparable masses), effects due to flow and surface geometry cancel out to a very good approximation and Eq. (12) gives the produced particle ratios to a good accuracy.

Feed down from resonance decays

Many of the hyperons considered here can be produced either directly at hadronization or in the decay of a resonance produced earlier. For instance,

$$\frac{\Lambda(1520)}{\Lambda_{\text{total}}} = \frac{\Lambda(1520)}{\Lambda + (\Sigma^0 \to \Lambda\gamma) + (\Sigma^* \to \Lambda\pi) + (\Sigma^* \to \Sigma^0\pi \to \Lambda\pi\gamma)}. \qquad (16)$$

Unless a resonance will be directly reconstructed by the experiment, its decay products will be a component in the hyperon abundances (albeit with a subtly different momentum distribution). This contribution, therefore, will need to be taken into account through calculation. As in the case of direct production, all kinematic effects are integrated out in a calculation of the total yield, and one simply sums each term in the resonance contribution as a term of the form given in Eq. (12). If a finite acceptance in phase space needs to be taken into account, however, a distribution of decay products will need to be computed from the statistical distribution of resonances.

We assume that, in a decay of the form,

$$R \to 1 + 2 + ..., \qquad (17)$$

any dynamical effects in the decay (the S-matrix) average out over a statistical sample of many resonances. In other words, in the rest frame comoving with the "average" resonance, the distribution of the decay products will be isotropic. The rate of particles of type 1 (as in Eq. (17)) produced with momentum $\vec{p}_1^*$ in the frame at rest w.r.t. the resonance will then simply be given by the Lorenz-Invariant phase space factor of a particle of mass M_1 and momentum $\vec{p}_1^*$ within a system with center of mass energy equal to the resonance mass M_R.

$$\frac{d^3N_1}{d^3p_1^*} = b \int \prod_{i=2}^{n} \frac{d^3p_i^*}{2E_i^*} \delta(\sum_{i=2}^{N} p_i^* - p_1^*) \delta(\sum_{i=2}^{N} E_i^* - E_1^* - M_R), \quad (18)$$

where b is the branching ratio of the considered decay channel. All that is left is to change coordinates from the resonance's rest frame (p^*, E^*) to the lab frame (p, E).

If more than two bodies are considered, this calculation becomes more involved [45]. For the general N-body case, it is better left to Monte Carlo methods [46]. In the case of the 2-body decay, the situation is greatly simplified by the fact that p^*, E^* are fixed by energy-momentum conservation to,

$$E_1^* = \frac{1}{2M_R}(M_R^2 - m_1^2 - m_2^2), \quad (19)$$

$$p_1^* = -p_2^* = \sqrt{E_1^{*2} - m_1^2}. \quad (20)$$

Putting the constraints in Eq. (19) into Eq. (18) one gets, after some algebra [44]

$$\frac{dN}{dm_{T1}^2 dy_1} = \frac{b}{4\pi p_1^*} \int_{Y-}^{Y+} dY_1 \int_{M_T-}^{M_T+} dM_{T1}^2 J \frac{d^2N_R}{dM_{TR}^2 dY_R}, \quad (21)$$

$$J = \frac{M_R}{\sqrt{P_{TR}^2 P_{T1}^2 - (M_R E_R^* - M_{TR} m_{T1} \cosh \Delta Y)^2}}, \quad (22)$$

$$\Delta Y = Y_R - y_1. \quad (23)$$

J is simply the Jacobian of the transformation from the resonance rest frame to the lab frame, and the limits of the kinematically allowed integration region are:

$$Y_\pm = y_1 \pm \sinh^{-1}\left(\frac{p_1^*}{m_{T1}}\right),$$

$$M_T^\pm = M_R \frac{E_R^* m_{T1} \cosh(\Delta Y) \pm p_{T1}\sqrt{p_1^{*2} - m_{T1}^2 \sinh^2(\Delta Y)}}{m_{T1}^2 \sinh^2(\Delta Y) + m_1^2}.$$

We now apply the methods outlined here to calculate ratios at hadronization for a range of experimentally observable resonances, both at full and central rapidity ($|y| \leq 0.5$). The total ratio was calculated using the analytical formula in Eq. (12), while for mid-rapidity equations in Eq. (15) and (21) were used. In the midrapidity calculation we used a uniform spherically symmetric expanding fireball, with a hadronization surface used earlier by fits in [5] and [44],

$$d\Sigma_\mu = (1, -\frac{\partial t_f}{\partial r}\vec{e}_r)d^3\vec{r}, \quad (24)$$

TABLE 4. Resonances contributing to Λ and K production, with their degeneracies, rest-frame momentum (p^*) and possibility for experimental reconstruction

g	Reaction	p^* MeV	branching	visible?
≈ 4	$\Sigma^{*0}(1385) \to \Sigma^0 \pi^0$	127	$\approx 4\%$	No
8	$\Sigma^{*\pm}(1385) \to \Lambda \pi^{\pm}$	208	88%	Yes
4	$\Sigma^{*0}(1385) \to \Lambda \pi^0$	208	88%	No
2	$\Sigma^0 \to \Lambda \gamma$	74	100%	No
4	$\Lambda(1520) \to N\overline{K}$	244	45%	Yes
3	$K^{*0}(892) \to K^+ \pi^-$	291	67%	Yes

where $\vec{e}_r$ is the unit vector in the radial direction and $\partial t_f / \partial r$ is a constant. We mentioned earlier that flow and hadronization surface details cancel out to a very good approximation when hadron masses are comparable. In fact, when we varied $\partial t_f / \partial r$, the results did not change by more than a percentage point. Table 4 summarizes the decay processes considered in our analysis and their parameters (Clebsh-Gordon coefficients have been used to estimate decays such as $(N^{*0} \to N^+ \pi^-)/(N^{*0} \to N^0 \pi^0)$).

In Fig. 8 we show the relative thermal production ratios at chemical freeze-out over the entire spectrum of rapidity and m_T (solid lines) and central rapidity range (dashed lines). The sensitivity of resonance yields to hadronization temperature is apparent for all resonances under consideration. In particular, the Σ^* emerges as a very promising candidate for further study. For example, at the lowest current estimates ($T \simeq 100$ MeV) of the final break up temperature in 158A GeV SPS collisions 33% of Λ are arising from primary Σ^*, the percentage rises to slightly more than 50% if chemical freeze-out occurs at $T = 190$ MeV. Both SPS and STAR detectors are capable to measure hyperon and resonance yields well within the precision required to distinguish between the two limiting cases.

as we discussed earlier, the experimentally observed K^* yield is compatible with the thermally produced ratio for a range of hadronization temperatures, but the $\Lambda(1520)$ seems very suppressed. To discuss this further, an estimation of the effect of rescattering on resonance abundances is necessary.

RESCATTERING

As explained in the introduction, direct observation of resonances relies on invariant mass reconstruction. Therefore, to calculate the observed resonance abundances, rescattering after hadronization will have to be taken into account. This can be looked at within a microscopic model of hadronic matter such as uRQMD. As we have shown earlier, uRQMD failed to describe experimental data if a long re-interaction period and initial thermal model yields were assumed. However, as yet there was no detailed study

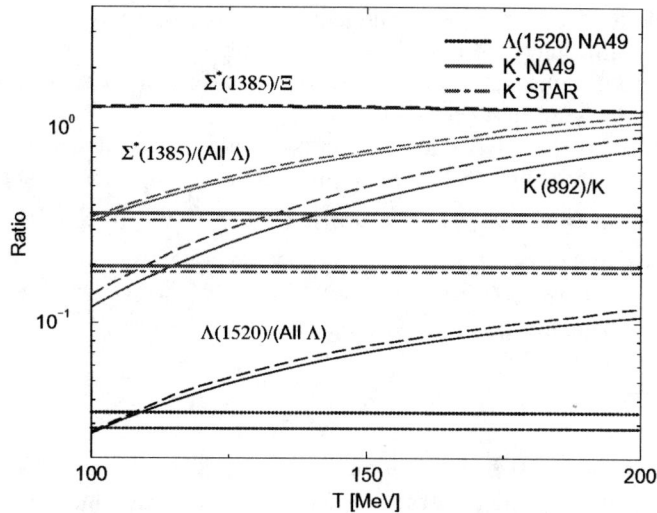

FIGURE 8. Temperature dependence of ratios of Σ^*, K^{*0} and $\Lambda(1520)$ to the total number of observed K Λ s and Ξ s. Branching ratios are included. Dashed lines show the result for a measurement at central rapidity $\Delta y = \pm 0.5$. The experimental measurements (horizontal lines, not Σ^*/Ξ) included in the diagram were presented in the first part of this paper. For the Ξ yields in the diagram, chemical potentials were taken from [2]. It should be noted that the K^*/K ratio is actually $\overline{K^*}/K^-$ for the NA49 measurement, and an average $(K^* + \overline{K^*})/(K^+ + K^-)$ for STAR.

of the dependence of the resonance yields on the lifetime of the interacting hadron gas phase (the interval between chemical and thermal freeze-out) as well as the chemical freeze-out temperature.

Here, we present such a study using a "back of the envelope" model which nevertheless seems to provide an acceptable qualitative description of the propagation of resonances and their decay products through opaque matter.

We first note in Fig. 8, that the relative Σ^*/Ξ signal is remarkably independent (within 5%) of Temperature. This is because the $\Xi^*(1530)$ contribution cancels nearly exactly the thermal suppression of the Ξ originating in the $\Xi - \Sigma^*$ mass difference. This effect could be used for a direct estimate of the Σ^* lost through rescattering, even without knowledge of the freeze-out temperature, should the chemical parameters (λ_i and γ_i) be independently known. A simple test of sudden hadronization model consists in measurement of the ratio Σ^*/Ξ. If it is significantly smaller than unity, we should expect a re-equilibration mechanism to be present. Otherwise sudden hadronization probably applies, since Σ^* emerge from chemical freeze-out without undergoing interactions.

We can, however, go further and use the suppression of the considered resonances as a tool capable of yielding a quantitative estimate of the lifetime of the interacting hadron gas phase. We consider the decay of a generic resonance Y^*

$$Y^* \to Y\pi, \quad (25)$$

in a gas of pions and nucleons. We shall assume that one interaction of either Y or π

is sufficient for that resonance to be undetectable, and that the decay products travel through the medium with speed v_i (where i can mean either Y or π). The interaction probability is proportional to v_i, the interaction cross-section of the decay product with each particle in the hadronic medium ($\sigma_{ij}(v_i)$, where j can refer to either pions, Kaons, nucleons or antinucleons. Note that the cross-section itself depends, in a generally complicated way, on the incident momentum, and hence on v_i), and the particle density in the fireball ρ_j. ρ_j is increased by a factor $\gamma_i = 1/\sqrt{1-v_i^2}$ due to Lorentz-contraction, and decreases as time passes because of the fireball's collective expansion (parametrized by the flow velocity v_{flow}, assumed to be of the order of the relativistic sound speed $c/\sqrt{3}$.) The time dependence of the densities will therefore be,

$$\rho_j(t) = \gamma_i \rho_{0j} \left(\frac{R}{R + v_{flow}t} \right)^3, \qquad (26)$$

and ρ_{0j}, the density of j at hadronization, can be calculated from the chemical freeze-out temperatures and chemical potentials. Putting everything together, the rescattering reaction rate is

$$P_i = \sum_{v_i} \left[\sigma_{i\pi}(v_i)\rho_{0\pi} + \sigma_{iK}(v_i)\rho_{0K} + \sigma_{iN}(v_i)\rho_{0N} + \sigma_{i\bar{N}}(v_i)\rho_{0\bar{N}} \right] (\gamma v)_i \left(\frac{R}{R + v_{flow}t} \right)^3, \qquad (27)$$

If we use the average,

$$\sum_{v_i} \sigma(v_i)v_i\gamma_i \simeq <\sigma><\gamma_i v_i> = <\sigma>\frac{p_i}{m_i}, \qquad (28)$$

(where p and m are the resonance's momentum and mass) Eq. (27) becomes,

$$P_i = \left[\langle \sigma_{i\pi} \rangle \rho_{0\pi} + \langle \sigma_{iK} \rangle \rho_{0K} + \langle \sigma_{iN} \rangle \rho_{0N} + \langle \sigma_{i\bar{N}} \rangle \rho_{0\bar{N}} \right] \frac{p_i}{m_i} \left(\frac{R}{R + v_{flow}t} \right)^3, \qquad (29)$$

Neglecting in-medium resonance regeneration and particle escape from the fireball, the population equation describing the scattering loss abundance (N_i) is:

$$\frac{dN_i}{dt} = \frac{1}{\tau}N_{N^*} - N_i P_i, \quad i = 1, 2 \qquad (30)$$

$$\frac{dN_{N^*}}{dt} = -\frac{1}{\tau}N_{N^*} \qquad (31)$$

The required nucleon and antinucleon density at hadronization, ρ_{0N} is obtained through Eq. (12)

$$\rho_{0N} = \frac{g}{(2\pi\hbar c)^3} 4\pi m^2 (\lambda_q \gamma_q)^3 T K_2(\frac{m}{T}), \qquad (32)$$

We consider the nucleons to have a mass of $\simeq 1$ GeV, and a degeneracy of six, to take the p,n and the thermally suppressed but higher degeneracy Δ contributions into account.

TABLE 5. Scattering model parameters

$\sigma_{\pi N}$ (mb)	σ_{KN}	$\sigma_{\pi\pi}$	$\sigma_{\pi K}$	σ_{NN}	$\sigma_{\bar{N}N}$
24	20	40	20	24	50

Γ_{Σ^*}	$\Gamma_{\Lambda(1520)}$	$\Gamma_{K^{*0}(892)}$
35 MeV	15.6 MeV	50 MeV

escape rate (fm^{-3})	negligible
v	0.5
R(fm)	8·145/T [MeV]
μ_b	220 MeV

the pion density at hadronization is computed in the massless particle limit, leading to

$$\rho_{0\pi} = \frac{\pi^2}{90}T^3, \tag{33}$$

The model presented here is remarkably insensitive to the individual cross-sectional areas. The values we used in the calculation are given in table 5, but order-of-magnitude variations of the more uncertain cross-sections did not produce variations of more than 30%. Similarly, the value of the initial fireball radius R_0 (which is constrained by the entropy per baryon) does not significantly affect the final ratios. This reassures us that had we used a more exact approach than the approximations in Eq. (28), the qualitative features of our model would not have changed. The results, however, exhibit a very strong dependence on both the Temperature (which fixes the initial resonance yield as well as the hadron density of the fireball) and fireball lifetime (in a short-lived fireball not many resonances decay, so their products do not get a chance to rescatter). Figure 9 shows the dependence of the $\Lambda(1520)/\Lambda$, Σ^*/Λ and $K^{*0}(892)/K$ on the temperature and lifetime of the interacting phase. It is clear that, given a determination of the respective signals to a reasonable precision, a qualitative distinction between the high temperature chemical freeze-out scenario followed by a rescattering phase and the low temperature sudden hadronization scenario can be made. We also note that despite the shorter lifetime of the Σ^* and higher pion interaction cross section, more Σ^* decay products should be reconstructible than in the $\Lambda(1520)$ case, at all but the highest temperatures under consideration. This reinforces our proposal that the Σ^* is a very good candidate for further measurement.

Diagrams such as those in Fig. 9 still contain an ambiguity between temperature and lifetime of the interacting hadron gas phase. A low observed ratio can either mean a low freeze-out temperature or a lot of rescattering in a long re-interaction phase. However, this ambiguity can be resolved by looking at a selection of resonances, with different masses and lifetimes. Fig. 10 shows how the initial temperature and the lifetime of the re-interaction phase decouple when two resonance ratios are measured simultaneously. A data point on diagrams such as those in Fig. 10 is enough to measure both the hadronization temperature and to distinguish between the sudden freeze-out scenario and a long re-interaction phase. The plots in Fig. 10 can also be used as consistency checks for

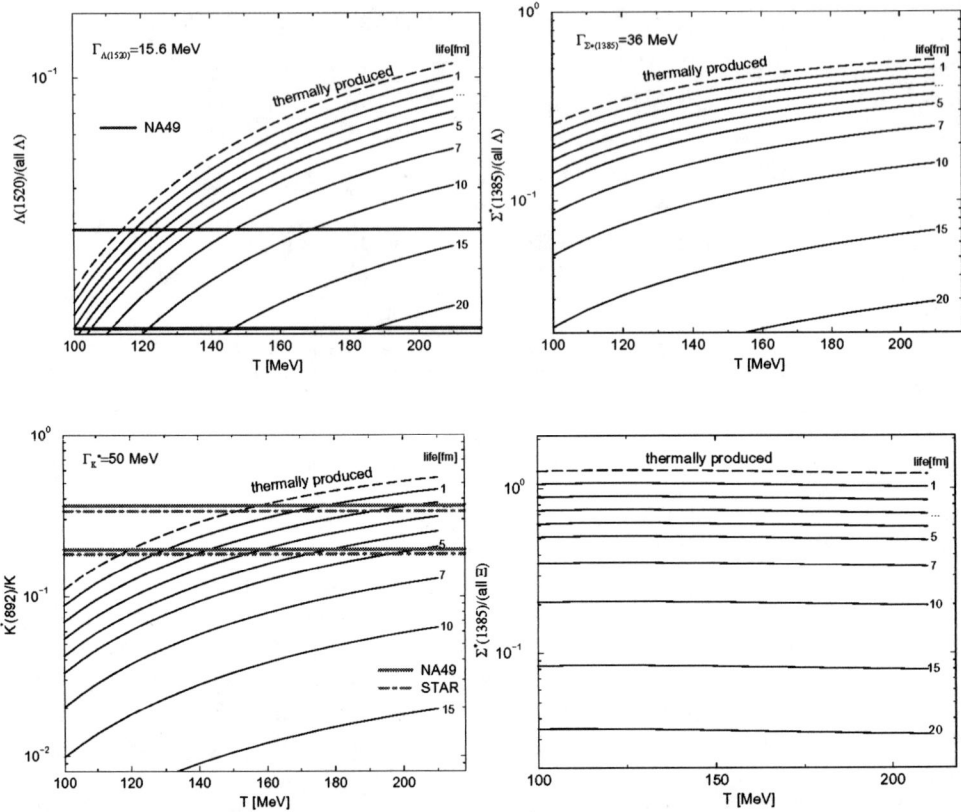

FIGURE 9. Produced (dashed line) and observable (solid lines) ratios $\Lambda(1520)/$(total Λ), $\Sigma^*/$(total Λ), K^{*0}/K and $\Xi/$(total Λ). The solid lines correspond to evolution after chemical freeze-out of 1,2,3,4,5,7,10,15,20 fm/c, respectively. The values at time zero (chemical freezeout) were taken from Fig. 8. See Fig. 8 caption for the meaning of K, K^*.

the model: For example, the near independence of Σ^*/Ξ on temperature means that the equal temperature lines in the Σ^*/Ξ vs Σ^*/Λ diagram are nearly insensitive to the details of the rescattering model. Moreover the mass differences and lifetimes of the Σ^*, K^* combine in such a way as to make the $\Sigma^*/$(all Λ) vs $K^{*0}(892)/$(all K^-) diagram fold into a very narrow band. Any serious shortcoming within our rescattering model would be revealed if the observed particle ratios stray from this band.

DISCUSSION

The first experimental results on short-lived resonances have raised more questions than answers. The low $\Lambda(1520)$ multiplicity measured by NA49, together with its non-detection at STAR, can not be understood with a thermal model exclusively. A $\Lambda(1520)$

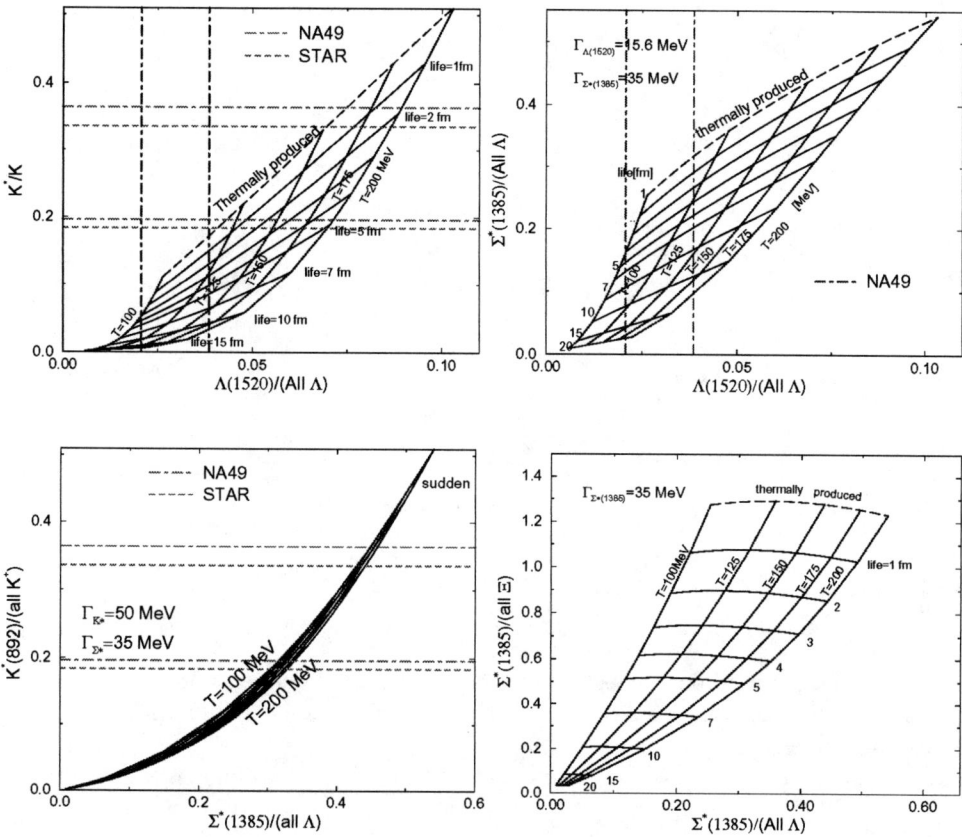

FIGURE 10. How temperature and fireball lifetime decouple when two resonances of different masses and widths are compared. A point on any of the above diagrams is potentially sufficient to fix both of these quantities. The experimental measurements are discussed in the first section. See Fig. 8 caption for the meaning of K,K^*.

suppression factor of at least 2 is needed to reproduce the measured multiplicity. However, this suppression does not appear to result from simple scattering with in-medium particles, described by models such as uRQMD. Such suppression should manifest itself much more strongly in case of the K^*, which has a much shorter lifetime. On the contrary, no comparable depletion has been found in the case of the K^* either at SPS or RHIC energies. Moreover, no broadening of the resonance width, which should accompany a strongly interacting medium, has been observed. If these results are incorporated in the diagrams of Fig. 10, the apparent result is sudden freeze-out with an unphysically low (~ 100 MeV) temperature. In analysing these results it should be kept in mind that the $\Lambda(1520)$ is a very unusual particle: Unlike most other hadronic resonances ($\Sigma^*, K^*, \Delta, \rho$ ecc.), its high spin is due not to valence quark spin configuration but to the fact that the $\Lambda(1520)$ s valence quarks are believed to be in an L=1 state. The greater space separation of L=1 wave-functions means that the $\Lambda(1520)$ is especially suscepti-

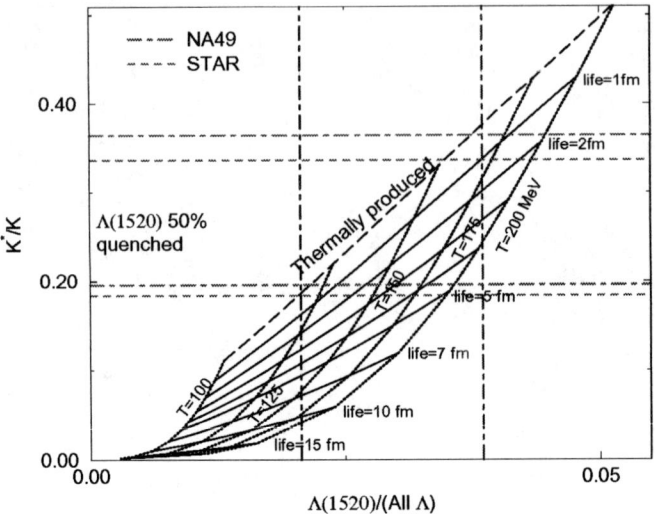

FIGURE 11. K^*/K vs $\Lambda(1520)/\Lambda$ assuming half of the $\Lambda(1520)$ s are suppressed in-medium. See Fig. 8 caption for the meaning of K, K^*.

ble to in-medium modifications. As Fig. 11 shows, a 50% suppression of the $\Lambda(1520)$ signal at hadronization would mean the data is perfectly compatible with the sudden freeze-out model described in [1]

Some proposed mechanisms which could lead to a $\Lambda(1520)$ suppression (Some suggested by Berndt Muller at the meeting [49]) are:

- If hadrons form in a quark coalescence process, the $\Lambda(1520)$ would be suppressed for the same reason that L=1 (p-shell) electrons are never captured by nuclei [50]: The coalescence probability is dependent on the wave-function overlap at the formed particle's location $\sim |\Psi(x=0)|^2$. The overlap of two particles goes down for higher L states, and is 0 for odd L=1,3,... states. However, this effect should also manifest itself in proton-proton collisions, which serve as basis of the AA-reaction yield. We can thus exclude this effect.
- Similarly, the larger spatial extent of the $\Lambda(1520)$ means it is more susceptible to in-medium screening of the strong potential. This mechanism invoked for Charmonium suppression can also suppress states such as the $\Lambda(1520)$.
- In-medium effects can mix the $\Lambda(1520)$ $|S=3/2, I=1/2\rangle$ with the Σ^* $|S=3/2, I=3/2\rangle$ state. Reactions such as

$$\pi + \Lambda(1520) \to \Sigma^* \qquad (34)$$

can deplete the $\Lambda(1520)$ states without a significant effect on the (much more abundant) Σ^* population [14]. This is analogous to the phenomenon of quenching, well known in atomic resonances [51]: Resonance radiation from many gases is

known to be suppressed by in-medium collisions, which prevent photon emission by converting the excitation energy of the resonance into kinetic energy.
- Spin-orbit interactions due to in-medium chromomagnetic fields can also mix the $\Lambda(1520)$ with the Σ^*.
- Isospin conservation forbids the decay $\Lambda(1520) \to \Lambda\pi$ but not $\Lambda(1520) \to \Lambda\sigma$, which is forbidden in vacuum since $m_\Lambda + m_\sigma > m_{\Lambda(1520)}$. At a temperature high enough for (partial) chiral symmetry restoration to take place, this may no longer be true, and the $\Lambda(1520) \to \Lambda\sigma$ decay would become possible. In the same way, partial restoration of chiral symmetry may be suppressing the $K^* \to K\pi$ decay channel.
- Finally, in a long re-interaction phase there is a possibility that the K^* will be regenerated. Processes such as

$$K\pi \to K^* \to K\pi$$

will lead to an enhancement of observed K^*. Processes which regenerate the $\Lambda(1520)$, such as

$$pK \to \Lambda(1520) \to pK$$

are considerably less likely, due to the suppression of higher ($L > 0$) partial waves.

The last effect can be properly taken into account by a more elaborate rescattering model, which takes all of the microscopic reaction dynamics into account. The other mechanisms are rather complex and uncertain, though for the RHIC system, potential screening, spin-orbit mixing and σ mass change can be explored in a $\mu_B = 0$ lattice calculation. However, these effects also depend on the unusual characteristics of the $\Lambda(1520)$ resonance. It seems to us that all these consideration suggest that other resonances should be studied to constrain the fireball freeze-out properties.

The non-suppression of the K^* makes it likely that other hadron resonances can also be detectable. The diagrams shown in the previous section imply that the Σ^* should be abundantly produced, and it has many characteristics which would make it a logical next step in the study of resonances produced in heavy ion collisions. Other proposed targets of observation are the ρ and even the $\Sigma^0 \to \Lambda\gamma$.

The sudden freeze-out scenario can be thoroughly tested by the detection of non-strange baryon resonances ($N^*(1440)$ and $\Delta(1230)$). The larger widths of these particles, as well as the larger scattering cross-section in hadronic matter of their decay products, mean that the very detectability of these particles in terms of an invariant mass analysis would be strong evidence of a very fast hadronization process.

ACKNOWLEDGMENTS

Work supported in part by a grant from the U.S. Department of Energy, DE-FG03-95ER40937, and by NSF grant INT-0003184. Ch. Markert is supported by the Humboldt foundation.

REFERENCES

1. J. Rafelski, J. Letessier *Phys. Rev. Lett.*, **85** (2000) 4695, and references therein.
2. J. Letessier and J. Rafelski, *Int. J. Mod. Phys.* E **9** (2000) 107, and references therein.
3. P. Braun-Munzinger, I. Heppe and J. Stachel, *Phys. Lett.* B **465** (1999) 15.
4. S.V. Akkelin, P. Braun-Muntzinger, Yu. M. Sinyukov arXiv:nucl-th/0111050.
5. G. Torrieri and J. Rafelski, *New Jour. Phys.* **3** (2001) 12.
6. V. Friese [NA49 Collaboration], *Nucl. Phys.* A **698** (2002) 487.
7. P. Fachini for the STAR Collaboration, SQM2001, *J. Phys.* G (2002) at press.
8. C. Markert for the NA49 Collaboration, poster presentation at QM2001.
9. C. Markert for the STAR Collaboration, SQM2001, *J. Phys.* G (2002) at press.
10. Z. Xu for the STAR Collaboration, *Nucl. Phys.* A **698** (2002) 607.
11. Particle Data Group, *Eur. Phys. J.* C **3** (1998).
12. P. Braun-Munzinger, D. Magestro, K. Redlich and J. Stachel, *Phys. Lett.* B **518** (2001) 41.
13. G. Torrieri and J. Rafelski, hep-ph/0112195, SQM2001, *J. Phys.* G (2002) at press.
14. J. Rafelski, J. Letessier and G. Torrieri, *Phys. Rev.* C **64** (2001) 054907.
15. G. Torrieri and J. Rafelski, *Phys. Lett.* B **509**, 239 (2001).
16. K.H. Ackermann, *et al.*, *Nucl. Phys.* A **661** (1999) 681.
17. S.V. Afanasiev, *et al.*, *Nucl. Instr. Meth.* A **430** (1999) 210.
18. For a more detailed discussion of experimental methods see contributions of J. Harris, and of K. Safarik in this volume.
19. J. Aichelin, G. Torrieri, and F. Gastineau, this volume.
20. M.F.M. Lutz and C.L. Korpa, nucl-th/0105067.
21. M.F.M. Lutz, E.E. Kolomeitsev and C.L. Korpa, SQM2001, *J. Phys.* G (2002) at press.
22. M. Gaździcki, M.I. Gorenstein, *Acta Phys. Polon.* B **30** (1999) 2705.
23. J. Canter *et al.*, *Phys. Rev.* D **20** (1979) 1029.
24. H. Albrecht *et al.*, *Z. Phys.* C **61** (1994) 1.
25. M. Aguilar-Benitz *et al.*, *Z. Phys.* C **50** (1991) 405.
26. M. Derrick *et al.*, *Phys. Lett.* B **158** (1985) 519.
27. D. Drijard *et al.*, *Z. Phys.* C **9** (1981) 293.
28. T. Akesson *et al.*, *Nucl. Phys.* B **203** (1982) 27.
29. A. Mischke for the NA49 Collaboration, SQM2001 *J. Phys.* G (2002) at press.
30. K. Abe *et al.*, *Phys. Rev.* D **59** (1990) 052001.
31. Y.J. Pei, *Z. Phys.* C **72** (1996) 39.
32. P.V. Chliapnikov, *Phys. Lett.* B **470** (1999) 263.
33. J. Aichelin, *Phys. Lett.* B **530** (2002) 81.
34. M. Bleicher private communication.
35. F. Becattini, M. Gaździcki, J. Sollfrank, *Eur. Phys. J.* C **5** (1998) 143.
36. F. Becattini, SQM2001, *J. Phys.* G (2002) at press.
37. S. Soff *et al.*, *J. Phys.* G **27** (2001) 449.
38. G. Torrieri and J. Rafelski, *Phys. Lett.* B **509** (2001) 239.
39. G. Torrieri and J. Rafelski, SQM2001, *J. Phys.* G (2002) at press.
40. F. Cooper and G. Frye, *Phys. Rev.* D **10** (1974) 186.
41. K. A. Bugaev, *Nucl. Phys.* A **606** (1996) 559.
42. C. Anderlik *et al.*, *Phys. Rev.* C **59** (1999) 3309.
43. M.A.C. Lamont, SQM2001, *J. Phys.* G (2002) at press.
44. E. Schnedermann, J. Sollfrank and U. Heinz NATO ASI series B, Physics vol. **303** (1995) 175.
45. E. Byckling and K. Kajantie, *Particle Kinematics*, Wiley (1973).
46. R. Kleiss, W.J.Stirling, *Nucl. Phys.* B, **385** (1992) 413.
47. A. Baldini, *et al* (eds), *Total cross-sections for reactions of high energy particles* Landolt-Börnstein numerical data and functional relationships in science and technology Springer (1990).
48. B.R. Martin, D. Morgan and G. Shaw, *Pion-Pion Interactions in Particle Physics* Academic Press, (New York 1976).
49. B. Muller, remarks in discussion session at this meeting.
50. R. Eisberg and R. Resnick, *Quantum physics of atoms, molecules, nuclei, and particles*, Wiley (1985).
51. A. Ruark and H. Urey, *Atoms, molecules, and quanta* McGraw-Hill (1964).

Critical Review of Quark Gluon Plasma Signatures

H. Stöcker[*], J. Berger[†], U. Eichmann[*], S. Salur[**], S. Scherer[*] and D. Zschiesche[*]

[*]*Institut für Theoretische Physik, Johann Wolfgang Goethe-Universität, D-60054 Frankfurt am Main, Germany*
[†]*Institut für Kernphysik, Johann Wolfgang Goethe-Universität, D-60486 Frankfurt am Main, Germany*
[**]*A.W. Wright Nuclear Structure Laboratory, Yale University, New Haven, CT*

Abstract. We discuss the uniqueness of often proposed experimental signatures for quark matter formation in relativistic heavy ion collisions, using insight gained from non-equilibrium models (three-fluid hydrodynamics and the hadronic transport model UrQMD).

It is demonstrated that these two models – although they do treat the most interesting early phase of the collisions quite differently (thermalizing QGP vs. coherent color fields with virtual particles) – both yield a reasonable agreement with a large variety of the available heavy ion data.

Hadron/hyperon yields, including J/Ψ meson production/suppression, strange matter formation, dileptons, and directed flow (bounce-off and squeeze-out) are investigated. Observations of interesting phenomena in dense matter are reported.

However, we emphasize the need for systematic future measurements to search for simultaneous irregularities in the excitation functions of several observables in order to come close to pinning the properties of hot, dense QCD matter from data.

INTRODUCTION

In the last few years researchers at Brookhaven and CERN have succeeded to measure a wide spectrum of observables with heavy ion beams, $Au+Au$ and $Pb+Pb$. While these programs continue to measure with greater precision the beam energy-, nuclear size-, and centrality dependence of those observables, it is important to recognize the major milestones passed thus far in that work. The experiments have conclusively demonstrated

- stopping and directed collective transverse and longitudinal flow of baryons and mesons – in and out of the impact plane, both at AGS and SPS energies –,
- hadronic resonance production,
- strangeness enhancement,
- the existence of strong nuclear A dependence of, among others, J/ψ and ψ' meson production and suppression.
- dilepton-enhancement below and above the ρ meson mass, and

These observations support that a novel form of "resonance matter" at high energy- and baryon density has been created in nuclear collisions. The global multiplicity and transverse energy measurements prove that substantially more entropy is produced in $A+A$ collisions at the SPS than simple superposition of $A \times pp$ would imply. Multiple initial and final state interactions play a critical role in all observables. The high mid-rapidity baryon density (stopping) and the observed collective transverse and directed flow patterns constitute one of the strongest evidence for the existence of an extended period ($\Delta \tau \approx 10$ fm/c) of high pressure and strong final state interactions. The enhanced ψ' suppression in $S+U$ relative to $p+A$ also attests to this fact. The anomalous low mass dilepton enhancement shows that substantial in-medium modifications of multiple collision dynamics exists, probably related to in-medium collisional broadening of vector mesons. The non-saturation of the strangeness (and anti-strangeness) production shows that novel non-equilibrium production processes arise in these reactions. Finally, the centrality dependence of J/ψ absorption in $Pb+Pb$ collisions presents further hints towards the non-equilibrium nature of such reactions.

Is there evidence for the long sought-after quark-gluon plasma that thus far has only existed as a binary array of predictions inside teraflop computers?

As we will discuss, it is too early to tell definitely – notwithstanding the combined results of the CERN-SPS program as announced in February 2000.

Theoretically there are still too many "scenarios" and idealizations to provide a satisfactory answer. Recent results from microscopic transport models as well as macroscopic hydrodynamical calculations differ significantly from predictions of simple thermal models, e. g. in the flow pattern. Still, these non-equilibrium models provide reasonable predictions for the experimental data. We may therefore be forced to rethink our concept of what constitutes the deconfined phase in ultra-relativistic heavy-ion collisions. Most probably it is not a blob of thermalized quarks and gluons. Hence, a quark-gluon plasma can only be the source of *differences* to the predictions of these models for hadron ratios, the J/Ψ meson production, dilepton yields, or the excitation function of transverse flow. And there are experimental gaps such as the lack of intermediate mass $A \approx 100$ data and the still limited number of beam energies studied thus far, in particular between the AGS and SPS. In the meantime, the field is at the doorstep of the next milestones: $A+A$ at $\sqrt{s} = 30-200$ AGeV has begun at RHIC/BNL in 2000, while the SPS (and the planned GSI-SIS/200) program continues to investigate the lower energy range $\sqrt{s} = 6-12$ AGeV (30-80 AGeV in the laboratory frame).

Organization of this review

This review is organized as follows: We first introduce and discuss two Non-equilibrium models used in the investigation of ultra-relativistic heavy ion collisions, the 3-fluid hydrodynamical model and the UrQMD hadronic transport model. Then we present nuclear stopping power as a fingerprint of the creation of compressed hadronic matter. We discuss flow signatures as probes of the pressure reached in the fireball, and particle spectra and ratios as probes of temperature and chemical properties of the fireball. As strong hints at the transient existence of deconfined quark matter, we review strangeness enhancement and the ideas of strangelets and hypermatter. As a further possible signal of the quark-gluon plasma, we then discuss the suppression of the J/ψ. Dileptons are presented as the (next to direct photons) unique probe to look directly at the heart of the fireball, and the experimental results and their interpretation are reviewed. We conclude our presentation with an outlook at the possible end of short distance physics and the predictive power of QCD: the conceivable creation of black holes due to the effects of large extra dimension in heavy ion collisions at the planned LHC machine.

NON-EQUILIBRIUM MODELS FOR THE STUDY OF ULTRA-RELATIVISTIC HEAVY ION COLLISIONS

In the present survey we employ two sharply distinct non-equilibrium models for relativistic heavy ion collisions, namely the macroscopic 3-fluid hydrodynamical model [1] and the Ultra-relativistic Quantum Molecular Dynamical model, UrQMD [2].

The first model assumes that a projectile- and a target fluid interpenetrate upon impact of the two nuclei, creating a third fluid (in the present version baryon free, see, however, [3]) via new source terms in the continuity equations for energy- and momentum flux. Those source terms are taken from energy- and rapidity loss measurements in high energy pp-collisions. The equation of state (EoS) of this model assumes equilibrium only in each fluid separately and allows for a first order phase transition to a quark gluon plasma in fluid 1, 2 or 3, if the energy density in the fluid under consideration exceeds the critical value for two phase coexistence. Pure QGP can also be formed in every fluid separately, if the energy density in that fluid exceeds the maximum energy density for the mixed phase.

The UrQMD model, on the other hand, assumes an independent evolution of hadrons, strings, and constituent quarks and diquarks in a non-equilibrium multi-particle system. The collision terms in this system of coupled Boltzmann equations (partial differential/integral equations) are taken from experimental data, where available, and otherwise from additive quark model and string phenomenology.

What is the role of partonic degrees of freedom in heavy ion reactions at the SPS?

Fig. 1 shows the time evolution of the energy density ε in central $Pb+Pb$ reactions at 160 AGeV as obtained within *a)* the parton cascade approach VNI [4], *b)* the UrQMD model [5]. It can be seen that in both models and at early times of the collision, a large fraction of the energy density is contained in partonic degrees of freedom (VNI) or to nearly equal parts in constituent diquarks and quarks from the strings and in virtual hadrons. This (virtual) "partonic" phase in $Pb+Pb$ reactions at 160 AGeV is, however, not to be identified with an equilibrated QGP. Note that the absolute values differ by a factor 2 in the two models and depend heavily on the rapidity cuts imposed to discriminate between virtual free streaming and interacting matter.

A sharp transition from a partonic state to the hadronic phase can be described within a non-relativistic, classical microscopic framework [6], as shown in Fig. 2. The critical temperature here is determined by the strength of the quark color potential.

This model, the quark molecular dynamics (qMD) [7], can be used to investigate the detailed hadronization dynamics from a hot and dense system of quarks to a gas of baryonic resonances, baryons and mesons.

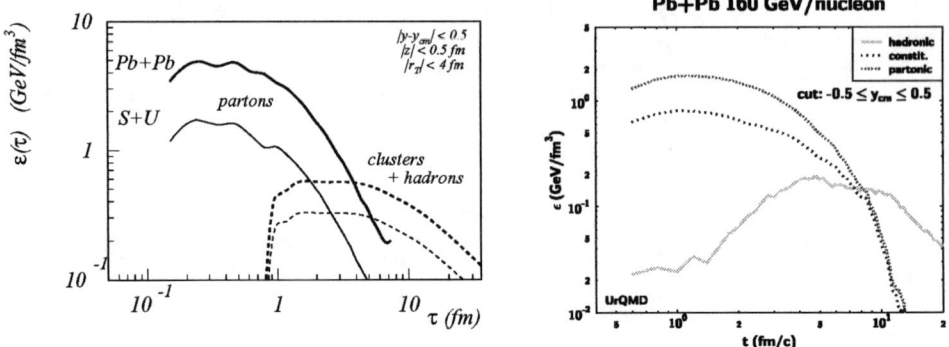

FIGURE 1. (Left:) Parton cascade (VNI, from [4]) and (Right:) UrQMD results for the time evolution of energy density in central $Pb+Pb$ reactions at 160 AGeV. At an early stage, most of the energy is contained in the partonic degrees of freedom (VNI) or in constituents (UrQMD).

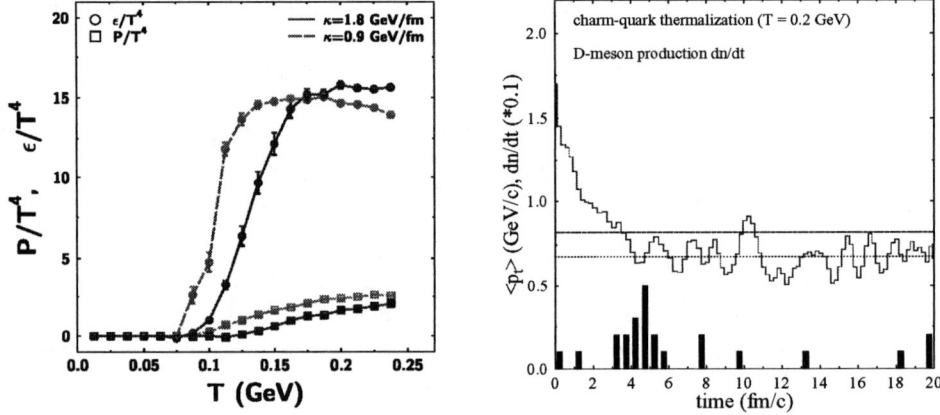

FIGURE 2. (Left) Energy density and pressure in a multi quark system interacting via a linear potential. (Right) D meson formation in the same model, showing charm quark thermalization.

CREATION OF COMPRESSED HADRONIC MATTER: NUCLEAR STOPPING POWER.

To study the properties of highly excited hadronic or quark gluon matter, normal nuclear matter has to be strongly compressed or heated. How is this possible? The answer was proposed 30 years ago [8, 9]: With high-energy nucleus-nucleus collisions. The idea was that the occurrence of multiple collisions converts longitudinal momentum into transverse momentum and secondary particles, leading to the creation of a zone of high energy density. Nuclear shock waves have been suggested as a primary mechanism of creating high energy densities in collisions with $\sqrt{s} \leq 20$ GeV [8, 9, 10]. The resulting

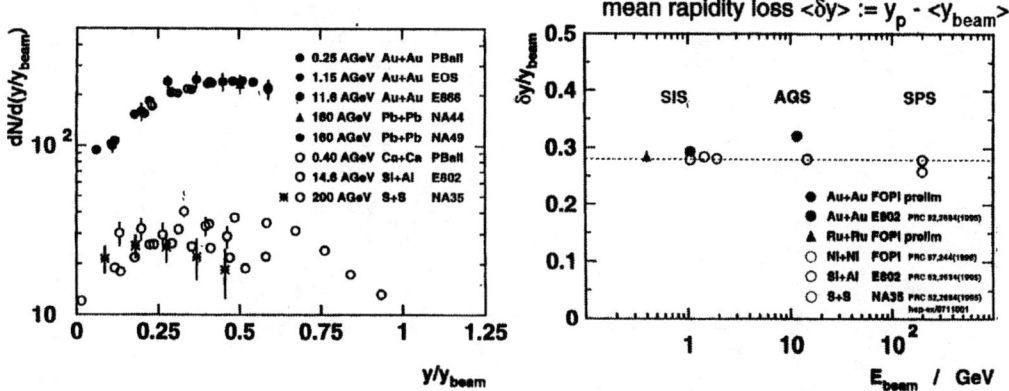

FIGURE 3. Left: Scaled rapidity $dN/d(y/y_{beam})$ distribution as a function of scaled rapidity (y/y_{beam}). Right: Compilation of mean rapidity loss versus beam energy at SIS, AGS and SPS energies.

energy density should be high enough to produce a state of high density resonance matter [9, 11, 12] or even a quark-gluon plasma [13]. For CERN SPS energies estimates for the achieved energy density at proper time $\tau_0 \approx 1\,\mathrm{fm}/c$ are of the order $\varepsilon \approx 1 - 5\,\mathrm{GeV}/\mathrm{fm}^3$. Extrapolations to BNL RHIC energies suggest that energy densities up to 20 GeV/fm^3 at $\rho \approx 2\rho_0$ may be reached.

The degree of stopping the incident nucleus suffers, when colliding with another nucleus (*nuclear stopping power*) manifests itself in a shift of the rapidity-distributions of the incident nucleons towards mid-rapidity. That means: The shape of the baryon rapidity distribution can give clear indications on the degree of stopping and the onset of critical phenomena. Due to the strong dependence of the baryon rapidity distribution on the baryon–baryon cross section [14, 15, 16], a rapid change in the shape of the scaled $dN/d(y/y_p)$ distribution with varying incident beam energy is a clear signal for new degrees of freedom which show up during the reaction (i.e. deconfinement), e.g. due to phenomena such as critical scattering [17]. The width of the $dN/d(y/y_p)$ distribution for baryons is inversely proportional to their cross section.

As an example on the left of figure 3 the measured scaled rapidity $dN/d(y/y_{beam})$ distribution as a function of scaled rapidity (y/y_{beam}) is shown for several collision systems at various energies. One observes a clear dependence of the shift in the scaled rapidity distribution on the collision systems. For the large Systems (Au+Au, Pb+Pb) a significant shift occurs, i.e. strong stopping power is observed, while for the smaller systems (e.g. Ca+Ca) only little stopping occurs. That means: Large systems are best suited for the creation of a zone of very high energy density. This is because during the collision of large systems more collisions occur and therefore the incident nucleons lose more energy that is deposited in the central reaction zone. This finding is independent of the collision energy. Figure 3 (right) shows the mean rapidity loss versus beam energy. It stays constant in the shown energy region.

The experimental situation can be summarized as follows: At AGS and SPS an extensive investigation of the nuclear stopping power is near completion. Proton-proton

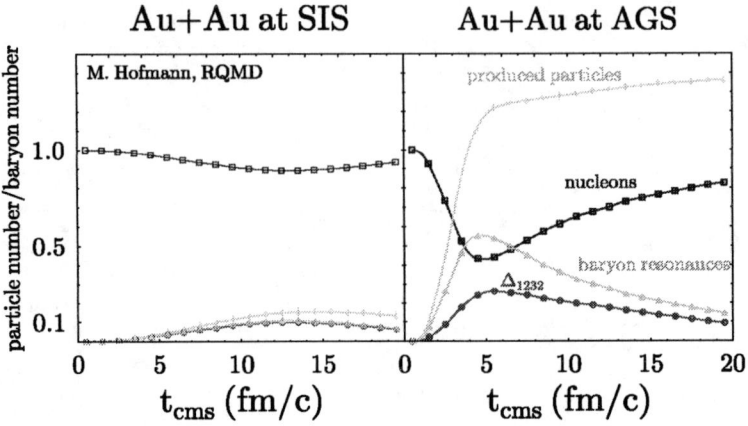

FIGURE 4. Time evolution of particle multiplicities (scaled with the number of incident nucleons) for central Au+Au collisions at 1 GeV/nucleon (SIS) and at 10.6 GeV/nucleon (AGS). At SIS energies, only about 10% of the nucleons are excited to resonances whereas at AGS energies the degree of excitation exceeds 50%. For a time-span of up to 10 fm/c the baryons are in a state of Δ−matter. The figure has been taken from [25].

[19] and peripheral nucleus-nucleus interactions at AGS [20, 21] and SPS [22] energies yield a forward–backward peaked dN/dy distribution in the C.M. frame, and a low degree of baryon stopping. A higher degree of stopping is observed for central collisions of intermediate mass nuclei (Si+Si at AGS, S+S at SPS): The rapidity distribution is flat at C.M. rapidities, two broad bumps are observed between projectile/target and C.M. rapidities respectively [20, 21, 22]. The heaviest collision systems (gold and lead respectively) exhibit the largest stopping power and thus correspond to the creation of the highest baryon densities: The form of the measured baryon rapidity distributions shows experimentally that the central rapidity region up to $E_{lab} \sim 200$ GeV/nucleon is not net-baryon free, in contrast to what had been expected in most early papers. Rather strong stopping as assumed first in hydrodynamic model studies [23, 24] is observed. Therefore, results of theoretical analyses, which rely heavily on a net-baryon free mid-rapidity region with zero baryo-chemical potential have to be taken with care. The quantitative measurements of the A-dependent stopping of baryons is one of the most important results of the AGS and SPS measurements. In summary, the experimental results demonstrate that highly excited dense matter is formed at mid-rapidity. They prove that a new state of elementary matter has been created.

What are the properties of this new state of matter? Hadronic transport model calculations indicate that the excited state of baryonic matter is dominated by the Δ_{1232} and other baryon resonances, i.e. the deposited energy in the central rapidity region is mainly transformed into these new degrees of freedom. E.g. for central Au+Au collisions at AGS a long apparent lifetime (> 10 fm/c) and a rather large volume (several hundred fm^3) for this resonance-matter state [25] (see figure 4) is predicted.

The energy densities of $\varepsilon \approx 3$-20 GeV/fm^3 estimated for SPS and RHIC indicate that part of the system may have entered the predicted state of deconfinement [18]. However,

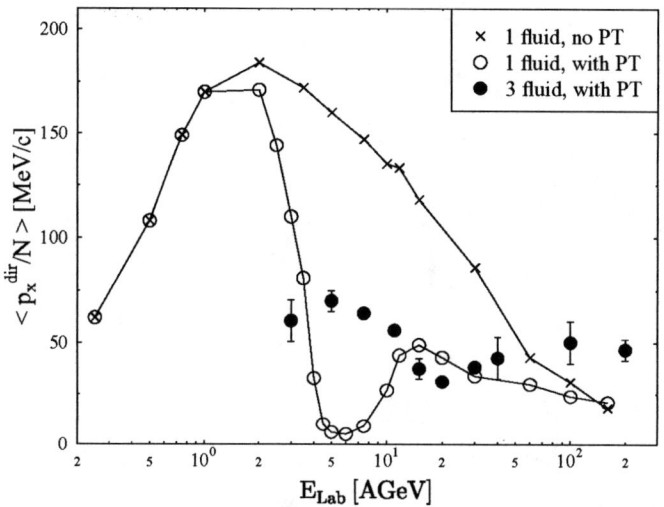

FIGURE 5. The excitation function of transverse flow as obtained from one fluid hydrodynamics with (open circles) and without (crosses) a first order phase transition[31], and the results of the three-fluid hydrodynamical model (filled circles). The drop of p_x due to the softening of the EoS is shifted to $E_{\text{lab}} \approx 20$ AGeV.

hadronic transport models predict or reproduce the measured rapidity distributions, if baryon and meson rescattering and particle production via string decay [26, 27, 28, 29, 2] are included. Also hadronic models which include multi–quark droplets [30] above $\varepsilon_{\text{crit}}$ seem to give similar results. Therefore up to now, the inclusive central distributions do not give a clear and decisive answer to the question of whether this matter is predominantly of hadronic or quark nature. The inclusion of string excitations, collisions and decays are a first step towards modeling the parton/quark substructure of hadrons. These models go beyond what one would term purely hadronic model and should give more insight in the structure and properties of the matter in the highly compressed and heated central interaction zone of high energy nucleus-nucleus collisions.

FLOW SIGNATURES: PROBING THE PRESSURE OF THE FIREBALL

Collective Flow and the softening of the EoS

The excitation function of collective transverse flow is the earliest predicted signature for probing compressed nuclear matter. Transverse collective flow depends directly on the pressure $p(\rho, S)$, i.e. the EoS. The flow excitation function is sensitive to phase transitions [23] by a collapse of the directed transverse flow [32, 33]. This is commonly referred to as *softening* of the EoS [31].

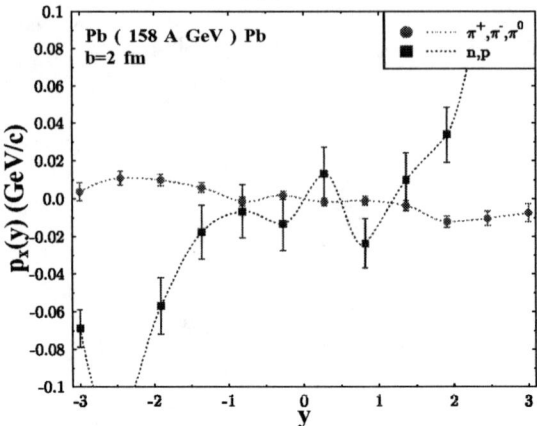

FIGURE 6. Theoretical predictions for transverse collective flow in $Pb+Pb$ collisions as obtained with the microscopic UrQMD transport model.

An observation of a local minimum in the excitation function of the transverse directed flow would thus be an unambiguous signal for a first order phase transition in dense matter. It's experimental measurement would serve as strong evidence for a QGP, if that phase transition is of first order.

Recent calculations within three-fluid hydrodynamics [34] show a shift in the drop of transverse flow to higher energies, $E_{\text{lab}} \approx 20$ AGeV, see Fig. 5. Experimentally, the recent discovery of proton flow and pion anti-flow at the SPS is in line with UrQMD and RQMD predictions (Fig. 6, and [35]).

Let us now discuss the results obtained from hadronic probes, such as observed particle ratios transverse momentum spectra and ratios, enhancement of strange baryons, light mesons, and production of J/Ψ.

Observed hadrons include feeding by the decay of resonances.

PARTICLE SPECTRA AND RATIOS: PROBING THE THERMAL AND CHEMICAL PROPERTIES OF THE FIREBALL

Thermal or statistical interpretations of the particle production in high energy collisions of elementary particles and heavy ions have been carried out for a long time [36, 37, 38, 39, 40, 41, 42, 43]. In the case of heavy-ion collisions a range of particle ratios measured at SIS, AGS, SPS and now also RHIC for different energies and different colliding systems have been fitted with a noninteracting gas calculation. The measured ratios reflect the relative particle numbers at the point of chemical freeze-out, i.e. the point when inelastic collision cease and therefore the chemical composition of the system is not changed anymore. E.g. noninteracting gas fits yield a chemical freeze-out curve in the $T - \mu$ plane (see fig 7,[44]).

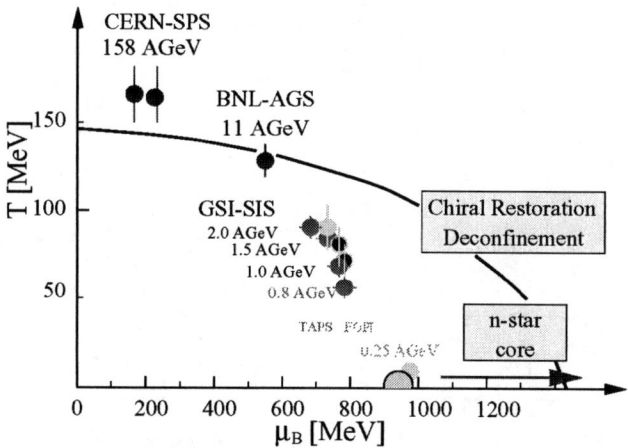

FIGURE 7. Compilation of ideal gas freeze-out conditions based on results of Ref. [44]. Solid line: graphic rendition of the expected phase boundary.

The measured particle spectra can be connected to the thermal or kinetic freeze-out, i.e. the point when no more elastic collisions occur.

The underlying model is as follows: One assumes a thermalized system with a constant density $\rho(r)$ (box profile), a constant temperature $T(r)$ and a linear radial and longitudinal flow velocity profile $\beta_\perp(r), \beta_\parallel(r)$. These parameters are assumed to be the same for all hadrons/fragments. At some time $t^{\text{break-up}}$ and density $\rho^{\text{break-up}}$, the system decouples as a whole (a horizontal freeze-out in the $T(z)$-plane) and the particles are emitted instantaneously from the whole volume of the thermal source. A complete loss of memory results, due to thermalization – the emitted particles carry no information about the evolution of the source. If one wants to use the inverse slope parameter T as thermometer [23], the feeding from Δ's etc., as well as the radial flow need to be incorporated into the analysis. The same holds for the use of d/p, π/p etc. as an entropymeter[45]. In addition, the proper Hagedorn volume correction can be applied [46]. A two parameter fit (μ_q, T, μ_s is fixed by strangeness conservation) to the hadronic freeze-out data describes the experimental results well, if feeding is included (see e.g. [44]). The particle multiplicities at the chemical freeze-out are calculated using an ideal gas approach, i.e. all particles are on mass-shell. Does this compatibility with a thermal source proof volume emission from a globally equilibrated source? Are the obtained temperatures and chemical potentials unique or strongly model dependent? How do interactions change the deduced freeze-out values?

The values for T and μ_B that were obtained using the ideal gas analyses of particle ratios can be used as input for a $SU(3)$ chiral mean-field model [47] extended to finite temperatures [48]. This model self-consistently contains scalar and vector interactions that change the properties of the hadrons in the medium. Furthermore the model shows a transition to a chirally restored phase. The nature of this transition in the model depends

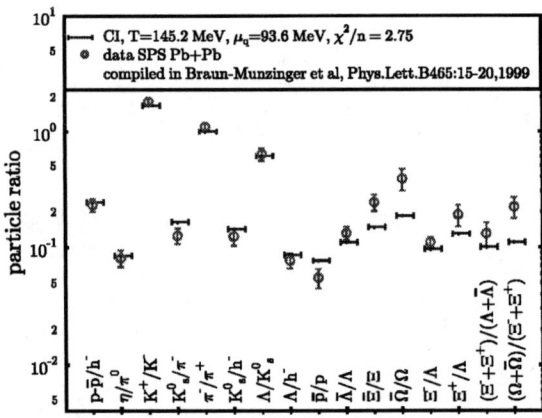

FIGURE 8. Fit of hadron ratios from the chiral model with a first order phase transition (CI) to data from $Pb+Pb$ collisions at SPS. The obtained values of T and μ allow the prediction of further ratios. T is much lower than results from the ideal hadron gas approach.

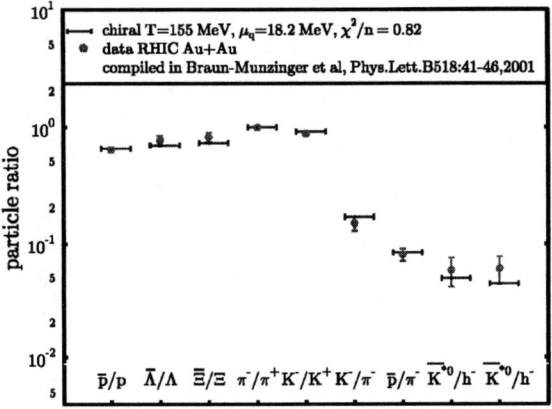

FIGURE 9. Fit of hadron ratios from the chiral model to data from $Au+Au$ collisions at RHIC.

on the included degrees of freedom and the chosen parameters [49]. Feeding from the decay of higher resonances is included. One finds that in such a model with a first order phase transition the ideal gas model values $T = 168$ MeV and $\mu_B = 266 MeV$ [44] lead to strong deviations from the experimental data and one has to re-fit the T and μ values. The obtained values for SPS Pb+Pb at SPS are: $T \approx 145$MeV and $\mu_B \approx 281$MeV. While the chemical potential is nearly unchanged due to the interactions, a decrease of $\Delta T \approx 20$MeV in the temperature results. Furthermore the analyses shows that no fit is possible above the phase transition temperature, what is realistic, since above T_C the particle densities are very high and the system is not dilute at all. Using the re-fitted T, μ values, the hadronic chiral model satisfactorily reproduces the measured particle ratios for Pb+Pb at SPS (fig. 8). The larger χ^2/n value is caused by the neglect of final weak

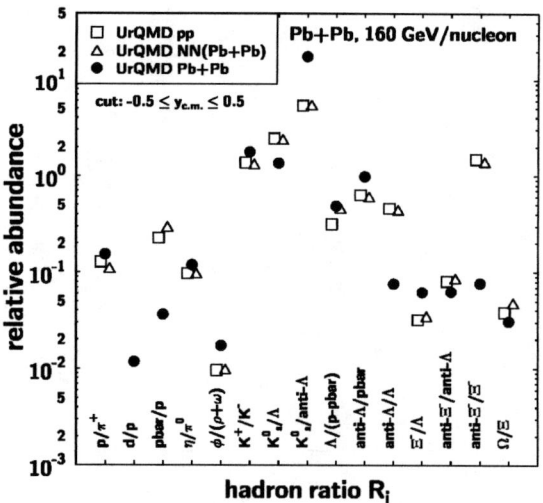

FIGURE 10. UrQMD prediction for hadron ratios in $Pb+Pb$ collisions at mid-rapidity (full circles), compared to a superposition of pp, pn and nn reactions with the isospin weight of the $Pb+Pb$ system (open triangles), i.e. a first collision approach.

interactions, as has already been shown by [50].

For RHIC the situation is similar: Ideal gas analyses yield freeze-out temperatures in the range of $T = 160 - 180$MeV, the chiral model with a first order phase transition yields $T \approx 150$MeV. The baryon chemical potential μ_B in both cases is around 40MeV. The agreement between data and calculated ratios is even improved, as can be seen from fig. 9.

Other parameterizations of the chiral model yield different T, μ_B values, but comparable quality of the fits [51]. That means: Not only ideal gas models, but also equilibrium models which account for the interactions in the hot and dense central region of relativistic heavy ion collisions can reproduce the measured particle ratios. The obtained chemical potentials seem to be robust but the temperatures strongly depend on the used model. This has also been shown in [52]. Therefore one must conclude that the extracted freeze-out values from different thermal and chemical equilibrium approaches are not unique but strongly model dependent.

The microscopic UrQMD transport model is in good agreement with the measured hadron ratios of the system $S+Au$ at CERN/SPS [53]. A thermal model fit to the calculated ratios yields a temperature of $T = 145$ MeV and a chemical potential of $\mu_B = 165$ MeV. However, these ratios exhibit a strong rapidity dependence. Thus, thermal model fits to data may be distorted due to different acceptances for the individual ratios.

Hadron ratios for the system $Pb+Pb$ are predicted by UrQMD and can be fitted by a thermal model with $T = 140$ MeV and $\mu_B = 210$ MeV (Fig. 10). Analyzing the results of non-equilibrium transport model calculations by an equilibrium model may, however, be not meaningful.

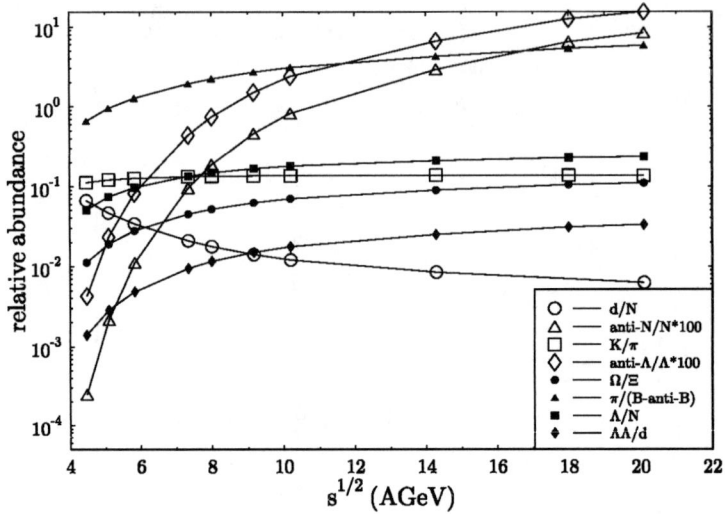

FIGURE 11. The excitation function of various particle ratios as calculated from the S/A values obtained from the three-fluid model. Feeding due to decays of resonances is taken into account.

There is a problem in the definition of equilibrium in itself: Do heavy ion collisions ever reach a thermalized system? Or are there transient steady states off equilibrium [54]? Due to the rapid dynamics of the system, the assumption of detailed balance is not fulfilled in the initial stage. This drives the system into a steady state far from equilibrium, but stationary in time. This steady state is easily visible in an enhanced production of light mesons, as compared to thermal models.

During the initial off-equilibrium stage of energetic nuclear collisions, a large amount of entropy can be produced [10]. The subsequent expansion is, on the other hand, often assumed to be nearly isentropic. The entropy produced during the compression stage is closely linked to the finally observable relative particle yields.

This entropy production can be calculated [55] within three-fluid hydrodynamics. The entropy per net participating baryon, S/A, saturates rapidly as a function of CM-time and is essentially time independent for later times when the freeze-out is reached. The chemical composition of the fireball is given by the net baryon density, the net (zero) strangeness of the system, and the specific entropy S/A, as described for the thermal model above.

The hadron ratios thus obtained are shown in Fig. 11. At AGS and SPS energies, they are quite close to the data [56, 54]. For such a simple estimate of hadron production in nuclear collisions, deviations from the experimental ratios by up to factors of two have to be expected. Nevertheless, it is clear from Fig. 11 that the simultaneous measurement of various hadron ratios, like $\pi/(B-\overline{B})$, d/N and, in particular, $\overline{B}/B$ (provided antibaryons also reach chemical equilibrium) allows to determine the produced entropy in the energy range between the AGS and the SPS. In contrast, the K/π-ratio is practically constant. The total specific entropy S/A produced within the three-fluid model is consistent with the S/A values extracted from data using relative particle yields from the thermal model.

One finds $S/A = 11$ for AGS and $S/A = 38$ for SPS energies.

The excitation function of the specific entropy $S/A(\sqrt{s})$ does not exhibit any threshold signatures of the phase transition to the QGP incorporated in the EoS. This is due to the gradual transition through the wide coexistence region in the energy density between $E_{\text{lab}} \approx 10\text{--}100$ AGeV.

STRANGENESS ENHANCEMENT: HINTS AT DECONFINED QUARK MATTER.

Let us now turn to multi-strange signals. In nucleon nucleon collisions, the production of particles containing strange quarks is strongly suppressed as compared to the production of particles with u and d quarks due to the higher mass of the $s\bar{s}$ quark pair [57]. It has been speculated that the yield of strange and multi-strange mesons, (anti-) baryons and anti-hyperons ($\bar{\Lambda}, \bar{\Sigma}, \bar{\Xi}$ and $\bar{\Omega}$) should be enhanced in the presence of a QGP. This can be understood as follows. The strangeness production rate is described by the Schwinger-factor [57] as long as no chemical equilibrium has been reached. This factor shows a mass dependence like $A\exp(-km_q)$, where A and k are some constants and m_q is the quark mass. So the production of heavier quarks like the strange quark is strongly suppressed compared to that of light quarks. On the other hand if the system can reach chemical equilibrium, the yield of different quarks and anti quarks are almost the same. It is now believed that the QGP-phase exists long enough for the chemical equilibrium to establish, so up, down and strange quarks reach a similar ratio. However if no QGP is produced, the chemical equilibrium of the hadronic phase has a much lower strangeness content. So the idea is that one can distinguish the production of a hadronic phase from the QGP phase by looking at the strangeness of the produced particles.

However only the end of the whole process is observable. To be able to interpret high strangeness ratio as a QGP signal it has to be supposed that matter passes through the hadronic phase fast enough to make equilibration impossible. Only than it is possible to discern the QGP production from the production of a hadron phase.

The study of (multi)strange hyperons by the WA97 [58] and the NA49 collaborations show an enhancement of strangeness production for central collisions when studying the centrality dependence of various strange particle yields (Λ, Ξ, Ω) in $Pb + Pb$ collisions at 158 AGeV as compared to $p + Pb$ collisions at 158 AGeV. The centrality is given as the extrapolated number of participant nucleons N_{part}. We propose as centrality variable the number of produced pions N_{π^-}. N_{part} shows a nonlinear behavior with the volume of the participant zone, while N_{π^-} shows perfect participant scaling. Scaling has been observed for central collisions ($N_{\text{part}} \geq 100$).

The UrQMD calculations (Fig. 12, right) show scaling. The hyperon to π^- ratio is depicted in Fig. 12 (right) as a function of impact parameter b. For central collisions, all ratios change only moderately, thus an approximate linear scaling of the hyperon yield with pion number N_{π^-} is observed. For peripheral collisions, the ratios decrease. The ratios vary with a factor of 2 to 5 for different impact parameters depending on the hyperon and its strangeness content. The three-fluid hydrodynamical model with an EoS with a first order phase transition to a QGP yields constant ratios (Fig. 12 left). Note the

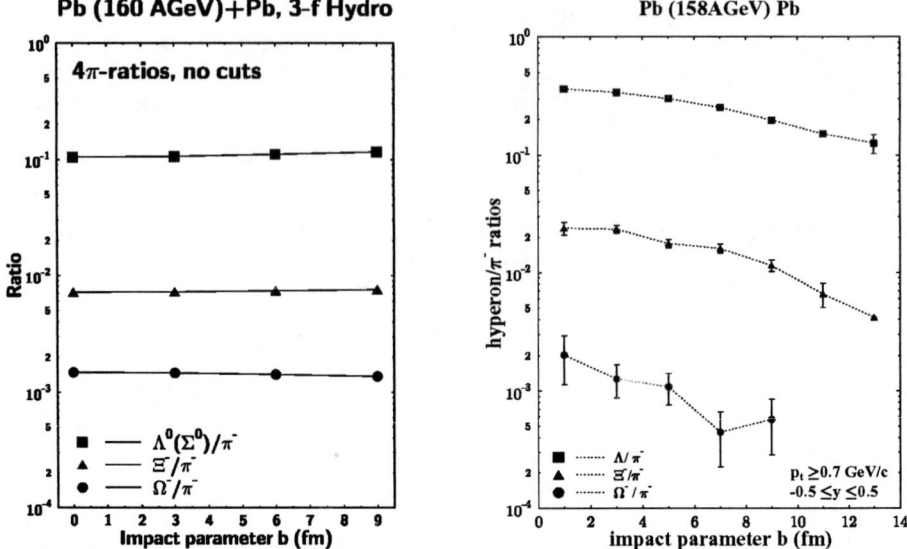

FIGURE 12. Hyperon to π^- ratio as a function of impact parameter b, as obtained from the three-fluid hydrodynamical model (left) and the UrQMD model (right). In the UrQMD model, the observed strangeness enhancement is already a natural consequence of ordinary hadronic rescattering.

substantial differences in the Ξ/π-ratios between the two predictions. The QGP which will be created at AGS, RHIC or SPS energies will most likely be characterized by non-zero chemical potential μ_d and mu_u. This will cause an increase in the densities of u and d quarks with respect to the densities of the s and $\bar{s}$ quarks. Due to the increase of the density of the u and d quarks, or the decrease of the s and $\bar{s}$ quarks, the $\bar{s}$ quarks are more likely to combine with u or d quarks. And similarly s quarks combine with $\bar{u}$ or $\bar{d}$. There will be a change in the ratio of $K^{\mp}/\pi$ during the QGP case.

The strangeness fugacity γ_s is introduced in thermo-chemical approach to account for incomplete chemical equilibration. It has been also compared to the measured ratios and the connected thermo-dynamical variables with a hadron gas scenario or a QGP scenario with some hadronization. However there are some certain drawbacks. One is the strange particle abundances after freeze out, are very close to those of the fully equilibrated hadron gas at the same entropy content. The reason can be explained by the volume of the hadron gas of the same total energy which must be larger due to the smaller number of available degrees of freedom. As a result one expects that the abundance of strange quarks is diluted during the hadronization process. This effect can be seen in many hadronization models. The computation of particle abundances in the QGP and the hadron gas scenario are mostly based on the assumption of chemical and thermal equilibrium. However for the hadronic case these assumptions cannot be justified since it has been shown that the strangeness equilibration time exceeds the reaction time of a heavy ion collision by at least one order of magnitude.

Strangeness production in the hadronic scenario is a non-equilibrium process. During the pre-equilibrium stages, typical longitudinal momenta are much higher than in the case of a thermal momentum distribution which leads to enhanced strangeness production. It's final "equilibrium" temperature is therefore only partly connected to the measured strange particle yields and spectra.

STRANGELETS AND HYPERMATTER

The observed abundant production of strange baryons at AGS and SPS energies led people to speculate about implications for hypermatter (multi-hyperon clusters or strange quark droplets) formation [59, 60, 61, 62, 63]. Speculations about the existence of such objects, with baryon numbers $B > 100$, have been around for decades, in particular within astrophysics. Such states are allowed for by the standard model, although so far their existence has not been proven in nature, e.g. in the form of strange neutron stars. *Quark matter* systems with $A > 1$ are unstable, if they only consist of u and d quarks, due to the large Fermi energy of these non-strange quarks. The system's energy may be lowered by converting some of the u and d quarks into s quarks (i.e. introducing a new degree of freedom). The energy gain may over-compensate the high mass of the s quarks and thus strange quark matter (SQM) may be absolutely stable [62]. Hadrons with $B > 1$ and $S < 0$ have been considered even before the advent of QCD [59, 60]. However, first the development of the MIT Bag Model [64] allowed to model such states. Long hypermatter lifetimes (for hundreds of quarks and a strangeness per baryon ratio in the order of one) have been predicted, up to 10^{-4} seconds [61]. Further detailed investigations of small pieces of strange quark matter, so called *strangelets*, reveal possible (meta)stability for $B > 6$ [62, 63]. Figure 13 shows possible creation scenarios for strangelets and strange nuclei. The right hand side shows the clustering of strange baryons to strange nuclei, while the left hand side shows the distillation of strange quark matter from a quark gluon plasma.

The simplest *strangelet* is the H−dibaryon with zero charge, $B = 2$ and $S = -2$, which consists of $2u, 2d$ and $2s$ quarks, followed by the *strange quark*−α with $6u, 6d$ and $6s$ quarks [62, 65]. For a QGP − hadron fluid first order phase transition with nonzero baryo-chemical potential, a mechanism analogous to associated kaon-production yields an enriched population of s quarks in the quark-gluon phase, while the $\bar{s}$ quarks drift into the hadron phase [66, 67]. This strangeness separation results in the distillation of metastable *strangelets* only if the Bag constants are very small, $B < 180$ MeV/fm^{-3} [66]. Experimentally *strangelets* are distinguishable from normal nuclei due to their very small or even negative charge to mass ratio. The most interesting candidates for long-lived *strangelets* are lying in a valley of stability which starts at the *quark*−α and continues by adding one unit of negative charge, i.e. (A,Z)=(8,-2),(9,-3)...[68]. Recent calculations indicate that positively charged *strangelets* seem only to exist for $A > 12$ and very low bag parameters [68]. There exist, however, other forms of hypermatter with similar properties as *strangelets*: hyperclusters or MEMO's (metastable exotic multi-hyperon objects) consist of multiple Λ, Σ and Ξ hyperons [69], and possibly also nucleons. The double-Λ hypernucleus $^{6}_{\Lambda\Lambda}$He has been observed long ago [70]. Properties

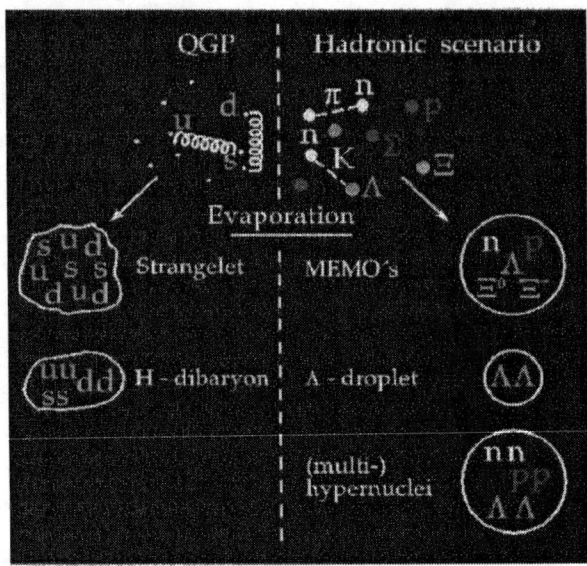

FIGURE 13. Possible formation scenarios for strange matter. The left hand side shows the formation of strange quark matter from the quark gluon plasma. The right hand side shows the clustering of strange baryons to strange nuclei

of MEMOs have been estimated using the Relativistic Mean Field model. MEMOs can contain multiple negatively charged hyperons, therefore they may also have zero or negative charge-to-mass-ratios.

MEMOs or hyperclusters could form a doorway state to *strangelet* production, or vice versa: MEMOs may coalescence in the high multiplicity region of the reaction. If strangelets are stronger bound than "conventional" confined MEMOs, the latter may transform into *strangelets*. The cross sections for production of MEMOs in relativistic heavy ion collisions rely heavily on model parameters (e.g. in the in the coalescence model p_0 and r_0). The predicted yields are typically $< 10^{-8}$ per event [69, 71].

Strangelet searches are underway at the AGS [72, 73, 74] and SPS [75, 76, 77, 78, 79]. So far no long lived ($\tau > 10^{-7}$ s) *strangelets* have been unambiguously identified – the upper limits for the production cross sections established by the experiment are still consistent with theoretical predictions for short lived MEMOs since they cannot be tested in the present long flight path experiments. There has been a report of one candidate with $Z = -1$, $N/Z = 7.4$ GeV and $\tau > 85$ μs [79, 77, 80]. Therefore this exciting topic awaits more experimental effort.

Current experiments are designed to detect *strangelets* with a small charge-to-mass ratio and rather long lifetime ($\tau \geq 12$ μs in the case of [77, 78]). The present experimental setups are hardly sensitive to the most promising long-lived and negatively charged *strangelet* candidates beyond the *strange quark–α*. Unfortunately, plans for extending experiment E864 at the AGS to look for highly charged strangelets with $B > 10$ could not be followed because the AGS fixed target heavy ion program was put to rest.

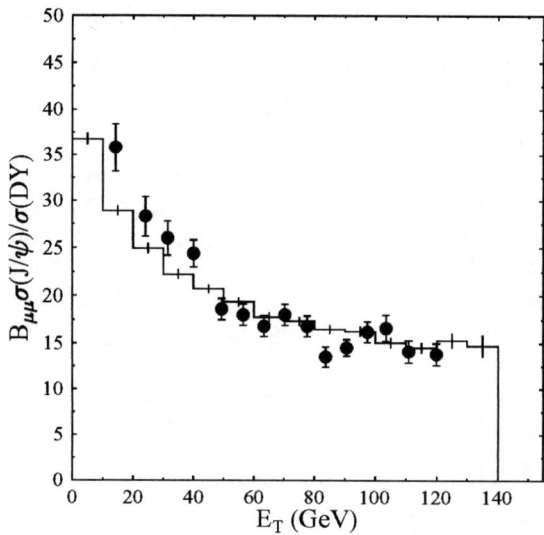

FIGURE 14. The ratio of J/ψ to Drell-Yan production as a function of E_T for $Pb+Pb$ at 160 GeV. The experimental data are from Ref. [86], the histogram is a UrQMD calculation [87]. No scaling factor has been applied to the x-axis for either the calculations or the data.

Future experiments at collider energies (STAR at RHIC and ALICE at LHC) will be sensitive for short-lived metastable hypermatter, too [81, 82, 83].

J/ψ SUPPRESSION: A CLEAR SIGNAL OF THE QGP?

Debye screening of heavy charmonium mesons in an equilibrated quark-gluon plasma may reduce the range of the attractive force between heavy quarks and antiquarks [84]. Mott transitions then dissolve particular bound states, one by one. NA38 found evidence of charmonium suppression in light ion reactions. Then also in $p+A$ such suppression was observed. New preliminary $Pb+Pb$ data of NA50 show "anomalous" suppression.

One of the main problems in the interpretation of the observed suppression as a signal for deconfinement is that non-equilibrium dynamical sources of charmonium suppression have also been clearly discovered in $p+A$ reactions, where the formation of an equilibrated quark-gluon plasma is not expected. A recent development is the calculation of the hard contributions to the charmonium- and bottomium-nucleon cross sections based on the QCD factorization theorem and the non-relativistic quarkonium model [85]. Including non-perturbative contributions, the calculated $p+A$ cross section agrees well with the data. Whereas these descriptions of nuclear absorption can account for the $p+A$ observation, the corrections needed for an extrapolation to $A+A$ reactions are, however, not yet under theoretical control.

Purely hadronic dissociation scenarios have been suggested [88, 89, 90] which could account for J/ψ and ψ' suppression without invoking the concept of deconfinement ("comover models"). Suppression in excess to that due to preformation and nuclear absorption is ascribed in such models to interactions of the charmonium mesons with "comoving", but probably off-equilibrium, mesons and baryons, which are produced copiously in nuclear collisions. Fig. 14 shows an UrQMD calculation which employs a microscopic free streaming simulation for J/ψ production and a microscopic transport calculation for nuclear and comover dynamics as well as for rescattering [87]. The dissociation cross sections are calculated using the QCD factorization theorem [85], feeding from ψ' and χ states is taken into account, and the $c\bar{c}$ dissociation cross sections increase linearly with time during the formation of the charmonium state. Taking into account the non-equilibrium "comovers" ($\sigma_{meson} \approx 2/3 \sigma_{nucleon}$), the agreement between theory and data is reasonable (Fig. 14). New, unpublished data agree better with the model predictions, but the high and low E_T regions remain to be studied carefully in the experiment. At present, no ab initio calculation does predict sudden changes in the suppression. In fact, from three-fluid calculations, even with QGP phase included, only a moderate change of the average and local energy density with bombarding energy is predicted. This seems to strongly speak against drastic threshold effects in the charmonium production.

The strong dependence of these results on details, such as the treatment of the formation time or the time dependent dissociation cross section, remain to be studied further. Furthermore, quantum effects such as energy dependent formation and coherence lengths must be taken into account [91] before definite statements can be made with regard to the nature of the J/ψ suppression. Interpretations of the data based on plasma scenarios are also increasingly evolving away from the original Mott transition analog [92, 93].

Hence, the theoretical debate on the interpretation of the pattern of charmonium suppression discovered by NA38/NA50 at the SPS is far from settled. It is not clear whether the suppression is the smoking gun of non-equilibrium dynamics or deconfinement. It is not likely to be due to simple Debye screening.

The major goal of further theoretical work is not to continue to try to rule out more "conventional" explanations, but to give positive proof of additional suppression by QCD-calculations which actually *predict* the E_T-dependence of the conjectured signature. Consistency tests and a detailed simultaneous analysis of all other measured observables are needed, if at least the same standards as for the present calculations are to be hold up.

DILEPTON PRODUCTION

Beside results from hadronic probes, electromagnetic radiation – and in particular dileptons – offer an unique probe from the hot and dense reaction zone: here, hadronic matter is almost transparent.

Dileptons can carry information on the thermodynamic state of the medium at the moment of production. Since the dileptons interact only electromagnetically they can leave the hot and dense reaction zone basically undistorted.

The main background contributions stem from pion annihilation, resonance decays

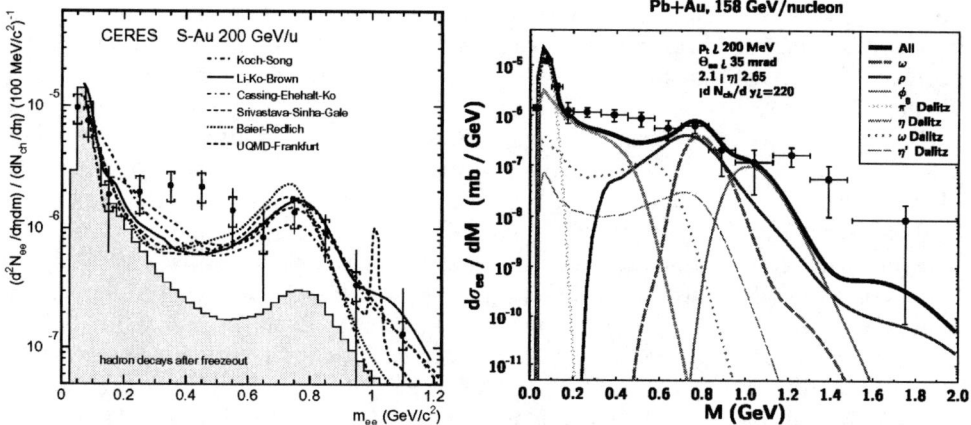

FIGURE 15. Left: inclusive e^+e^- mass spectra in 200 GeV/nucleon S+Au collisions as measured by the CERES collaboration [94], data originally presented in similar form in [95]. The shaded area depicts hadronic contributions from resonance decays. The data are compared with calculations based on a purely hadronic scenario [96, 97, 98, 99, 100]. Right: UrQMD prediction copmpared with data from [101] for Pb+Au at 160 GeV/nucleon.

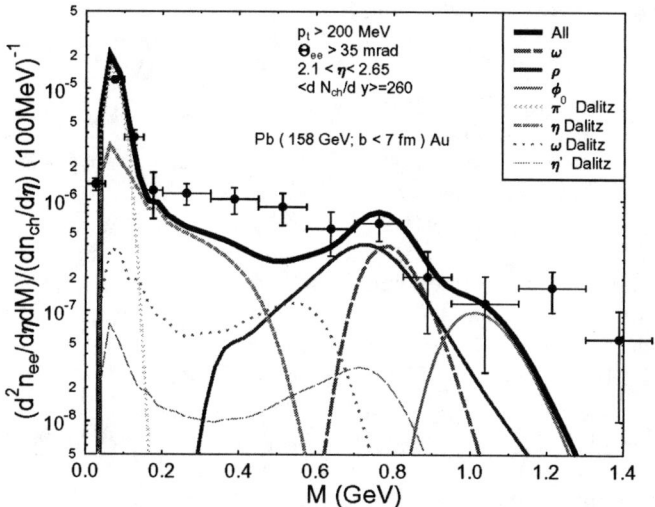

FIGURE 16. Microscopic calculation of the dilepton production in the kinematic acceptance region of the CERES detector for $Pb+Au$ collisions at 158 GeV. No in-medium effects are taken into account. Plotted data points are taken at CERES in '95.

[102, 103, 104, 105, 106] (two pions can annihilate, forming either a virtual photon or a rho meson – both may then decay into a dilepton) and $\pi - \rho$ interactions [107, 108] at low dilepton masses and Drell–Yan processes [109, 110] at high masses. Furthermore meson resonances such as the rho-, omega- or phi- meson may be produced directly or in the decay of strings and heavier resonances. As all of those vector mesons carry the same quantum numbers as the photon, they may decay directly into a dilepton. Resonances can also emit dileptons via Dalitz decays. The Drell–Yan process describes the annihilation of a quark of one hadron with an anti-quark (in proton proton collisions from the sea of $\bar{q}$) of the other hadron, again resulting in a virtual photon which decays into a dilepton. The open charm contribution to the dilepton mass spectrum has been estimated to be negligible for low dilepton masses [111] at the CERN/SPS. At RHIC and LHC energies, however, charm contributions dominate the dilepton mass spectrum above 2 GeV [112].

Most original calculations on dileptons as signals of a QGP at CERN/SPS energies focused on masses below the rho meson mass [113, 114, 115, 116, 117, 103, 118, 119, 120, 121]. The current understanding of hadronic background contributions [102, 104, 105, 106] shows that most probably dileptons originating from a QGP are over-shined by hadrons, with the possible exception of masses around 1 to 1.5 GeV [122, 123] where the rates from a plasma (at very high temperatures around 500 MeV) may suffice to be visible. At higher masses, the yield of Drell–Yan processes from first nucleon nucleon collisions most probably exceeds that of thermal dileptons from a QGP. Finite baryochemical potential will, at a given energy density, reduce the number of dileptons emitted from a QGP [124, 125, 126], due to the dropping temperature in that system.

The dependence of the yield of high mass dileptons on the thermalization time is still a point of open debate [127, 128]. The parton cascade [129] and other models of the early equilibration phase [128, 130] predict an excess of dileptons originating from an equilibrating QGP over the Drell–Yan background in the mass range between 5 and 10 GeV. Then the early thermal evolution of the deconfined phase could be traced in an almost model independent fashion [131].

The secondary dilepton production via quark-antiquark annihilation has also been studied on the basis of a hadronic transport code (UrQMD [2]). Here, one obtains a realistic collision spectrum of secondary hadrons for SPS energies. Using parton distribution functions and evaluating the contributions of all individual hadronic collisions one finds that meson-baryon interactions enhance the mass spectrum at mid-rapidity below masses of 3 GeV considerably [110]. Preresonance interactions are estimated to enhance this secondary yield by up to a factor of 5.

Dileptons can be measured at CERN in form of dimuons by the HELIOS3, NA38 and NA50 [132, 133, 134, 135] collaborations and in form of electron pairs by the CERES collaboration [94]. Dimuons exhibit an excess in AA collisions in the mass range $0.2 < M < 2.5$ GeV/c^2 up to the J/Ψ, as compared to pp and pA collisions. For dielectrons an excess is observed in the low–mass region $0.2 < M < 1.5$ GeV/c^2, again relative to pp and pA collisions (c.f. figure 15).

Both, the dielectron as well as the dimuon data seem to be compatible with a hydrodynamic approach assuming the creation of a thermalized QGP [97]. Hadronic transport calculations are not able to fully reproduce the observed excess [96, 136, 137]. However, at least part of the observed enhancement of lepton pairs at intermediate and low masses might be either caused by the previously neglected source of secondary Drell-Yan processes [110] or by contributions of heavy mesons, such as the a_1 [138].

The observed enhancement of the dilepton yield at intermediate invariant masses ($M_{e^+e^-} > 0.3$ GeV) received great interest: it was prematurely thought that the lowering

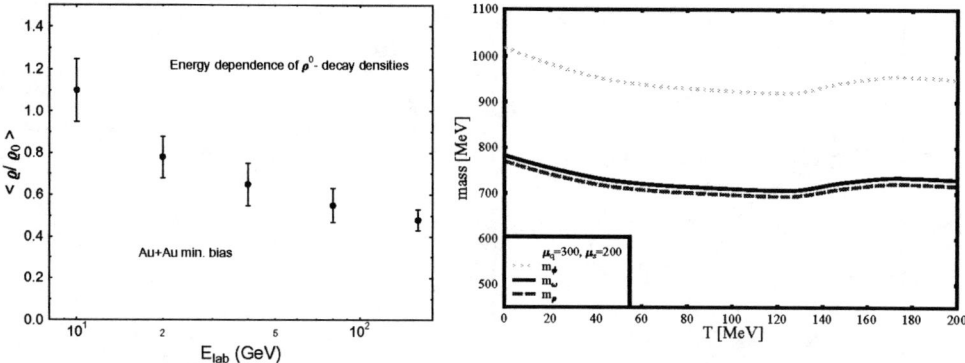

FIGURE 17. (Left) The mean "freeze-out" density at the location of ρ meson decays in $Au+Au$ collisions. (Right) The mass of the ρ-meson as obtained from the chiral model at different temperatures and finite density.

of vector meson masses is required by chiral symmetry restoration (see e.g. [139] for a review). However, there seems no theoretical support for this speculation. Calculations within a chiral $SU(3)$ mean-field approach [47] show only a modest dependence of temperature of the mass of the ρ meson (Fig. 17, right). AA-data are compatible with broadening spectral functions found in pure hadronic calculations of the scattering on the constituents of the excited matter (see e.g. [140]). The present data do not allow to draw definite conclusions.

Fig. 16 shows a microscopic UrQMD calculation of the dilepton production in the kinematic acceptance region of the CERES detector for $Pb+Au$ collisions at 158 GeV. This is compared with the '95 CERES data [101]. Aside from the difference at $M \approx 0.4$ GeV there is a strong enhancement at higher invariant masses. It is expected that this discrepancy at $m > 1$ GeV could be filled up by direct dilepton production in meson-meson collisions [141] as well as by the mechanism of secondary Drell-Yan pair production proposed in [110].

The mean "freeze-out" density at the location of ρ meson decays in $Au+Au$ collisions is shown in Fig. 17 (left) for different incident energies [142]. From AGS to CERN energies, there is a decrease of the baryonic density, indicating that baryonic modifications to the ρ meson are better studied at energies of $20-40$ AGeV. The low baryon densities at high energies will make it hard to explain the CERES data by ρ meson modifications of nucleonic origin alone.

CREATION OF BLACK HOLES IN LARGE HADRON COLLIDER

An exciting new idea, marking the end of short distance physics as we know it from perturbative QCD, may be the production of black holes at energies reached in collisions at the LHC.

This thrilling possibility is a consequence of reasoning about the existence of extra dimensions beyond the usual, well-known 4 space-time dimensions. These ideas have

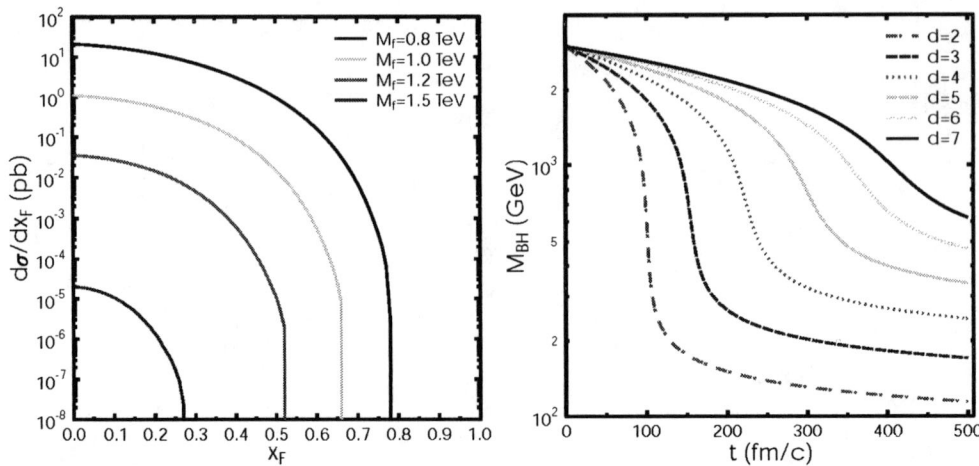

FIGURE 18. (Left:) The Feynman-x distribution of black holes for $M \geq M_f$ TeV in proton-proton interactions. (Right:) Time evolution of the mass of evaporating black holes produced at LHC. Different lines correspond to different numbers of extra dimensions d, see [143] (unpublished).

been motivated by string theories, and they offer a unique opportunity to tackle the so called hierarchy problem, the large gap between the electroweak energy scale ($m_W = 10^3$ GeV) and the Planck scale as we know it ($m_{Planck} = 10^{19}$ GeV).

In scenarios with d compact, so called large extra-dimension (LXD, "large" means here scales up to millimeter – in contrast to the tiny compactification radii in usual string or Kaluza-Klein theories in the order of the Planck length), the fundamental energy scale m_f might be as big as m_W, reaching the TeV region [144, 145, 146]. But since only gravity would propagate in the extra dimensions, while the fields and particles of the standard model would be constrained to the well known 4 space-time dimensions, this would allow the large value of our 4-dimensional Planck mass m_{Planck}.

The possibility that the fundamental energy scale, m_f, might be as low as the electroweak scale m_W makes the future high energy colliders CLIC, LHC and TESLA capable of producing black holes. A black hole would be produced whenever the energy of the collision were concentrated in a region of space small enough to be trapped inside the Schwarzschild horizon corresponding to the energy involved. However, due to Hawking radiation, the black hole decays again, spitting out all kinds of standard model particles. The problem is to find signatures of the creation and subsequent evaporation of the black hole.

For non-spinning black holes, figure 18 depicts the momentum distribution of black holes produced in pp collisions[147]. Many of the black holes are formed in scattering processes of valence quarks. The dependence of the cross section on the number of extra dimensions is less than 10%.

The time evolution of the black hole can be seen in Figure 18 for different numbers of compactified space-like extra dimensions. Extra dimensions lead to an increase in the lifetime of the black hole. It can be seen from the calculations depicted in figure 18 that for $d = 2$, a black hole with mass $\sim$ TeV exists for at least 100 fm/c [148]. Afterwards the mass drops below the fundamental scale.

The effects of LXDs allow for the creation of 10^7 black holes with lifetime τ=(10-600)fm/c per year at LHC. Hawking radiation is also suppressed with large extra dimensions. The prediction of the suppression of jets with high p_T due to the formation of black holes in the final state will be a clear signature for extra dimensions. Maybe with LHC particle physicists will be able to compete with astronomers in the field of observing black holes.

OUTLOOK

At the CERN/SPS new data on flow, electro-magnetic probes, strange particle yields (most importantly multi-strange (anti-)hyperons) and heavy quarkonia are interesting to follow closely. Simple energy densities estimated from rapidity distributions and temperatures extracted from particle spectra indicate that initial conditions could be near or just above the domain of deconfinement and chiral symmetry restoration. Still the quest for an *unambiguous* signature remains open.

Directed flow has been discovered – now a flow excitation function, filling the gap between 10 AGeV (AGS) and 160 AGeV (SPS), would be extremely interesting: look for the softening of the QCD equation of state in the coexistence region. The investigation of the physics of high baryon density (e.g. partial restoration of chiral symmetry via properties of vector mesons) is presently not accessible due to the lack of dedicated accelerators in the $10 - 200$ AGeV regime. The achieved and planned 40 AGeV and 80 AGeV runs at CERN are an absolute necessity into this new direction, but it can only be a first step.

However, dedicated accelerators would be mandatory to explore these intriguing effects in the excitation function. It is questionable whether this key program will actually get support at CERN. In this respect, the planned new machine at GSI, SIS/200, will be of extreme importance. Also the excitation function of particle yield ratios ($\pi/p, d/p, K/\pi...$) and, in particular, multi-strange (anti-)hyperon yields, can be a sensitive probe of physics changes in the EoS. The search for novel, unexpected forms of $SU(3)$ matter, e.g. *hypermatter*, *strangelets* or even *charmlets* is intriguing. Such exotic QCD multi-meson and multi-baryon configurations would extend the present periodic table of elements into hitherto unexplored dimensions. A strong experimental effort should continue in that direction.

Experiments and data on ultra-relativistic collisions are essential in order to motivate, guide, and constrain theoretical developments. They provide the only terrestrial probes of non-perturbative aspects of QCD and its dynamical vacuum. The understanding of confinement and chiral symmetry – and the possibility of even gaining glimpses of the realm of quantum gravity – remains one of the key questions at the beginning of this millennium.

ACKNOWLEDGMENTS

This work was supported by DFG, GSI, BMBF, Graduiertenkolleg Theoretische und Experimentelle Schwerionenphysik, the A. v. Humboldt Foundation, and the J. Buchmann Foundation.

REFERENCES

1. Brachmann, J., Dumitru, A., Maruhn, J. A., Stocker, H., Greiner, W., and Rischke, D. H., *Nucl. Phys.*, **A619**, 391–412 (1997).
2. Bass, S. A., Belkacem, M., Bleicher, M., Brandstetter, M., Bravina, L., Ernst, C., Gerland, L., Hofmann, M., Hofmann, S., Konopka, J., Mao, G., Neise, L., Soff, S., Spieles, C., Weber, H., Winckelmann, L. A., Stocker, H., Greiner, W., Hartnack, C., Aichelin, J., and Amelin, N., *Prog. Part. Nucl. Phys.*, **41**, 225–370 (1998).
3. Rosenhauer, A., Csernai, L. P., Maruhn, J. A., and Greiner, W., *Phys. Scripta*, **30**, 45–51 (1984).
4. Geiger, K., *Nucl. Phys.*, **A638**, 551c–554c (1998).
5. Weber, H., et al., *to be publ.* (2002).
6. Hofmann, M., Bleicher, M., Scherer, S., Neise, L., Stocker, H., and Greiner, W., *Phys. Lett.*, **B478**, 161–171 (2000).
7. Scherer, S., Hofmann, M., Bleicher, M., Neise, L., Stocker, H., and Greiner, W., *New Jour. Phys.*, **3**, 8 (2001).
8. Scheid, W., Ligensa, R., and Greiner, W., *Phys. Rev. Lett.*, **21**, 1479–1482 (1968).
9. Chapline, G. F., Johnson, M. H., Teller, E., and Weiss, M. S., *Phys. Rev.*, **D8**, 4302–4308 (1973).
10. Scheid, W., Muller, H., and Greiner, W., *Phys. Rev. Lett.*, **32**, 741–745 (1974).
11. Boguta, J., *Phys. Lett.*, **B109**, 251 (1981).
12. Li, Z.-X., Mao, G.-J., Zhuo, Y.-Z., and Greiner, W., *Phys. Rev.*, **C56**, 1570–1575 (1997).
13. Bjorken, J. D., *Phys. Rev.*, **D27**, 140–151 (1983).
14. Hartnack, C., Zhuxia, L., Neise, L., Peilert, G., Rosenhauer, A., Sorge, H., Aichelin, J., Stocker, H., and Greiner, W., *Nucl. Phys.*, **A495**, 303c–320c (1989).
15. Berenguer, M., et al., *J. Phys.*, **G18**, 655–679 (1992).
16. Schmidt, W., Katscher, U., Waldhauser, B., A. Maruhn, J., Stocker, H., and Greiner, W., *Phys. Rev.*, **C47**, 2782–2793 (1993).
17. Gyulassy, M., and Greiner, W., *Annals Phys.*, **109**, 485–527 (1977).
18. Harris, J. W., and Muller, B., *Ann. Rev. Nucl. Part. Sci.*, **46**, 71–107 (1996).
19. Blobel, V., et al., *Nucl. Phys.*, **B69**, 454–492 (1974).
20. Abbott, T., et al., *Phys. Rev.*, **C50**, 1024–1047 (1994).
21. Barrette, J., et al., *Phys. Rev.*, **C50**, 3047–3059 (1994).
22. Baechler, J., et al., *Phys. Rev. Lett.*, **72**, 1419–1422 (1994).
23. Stocker, H., and Greiner, W., *Phys. Rept.*, **137**, 277–392 (1986).
24. Clare, R. B., and Strottman, D., *Phys. Rept.*, **141**, 177–280 (1986).
25. Hofmann, M., Mattiello, R., Sorge, H., Stocker, H., and Greiner, W., *Phys. Rev.*, **C51**, 2095–2098 (1995).
26. Sorge, H., Keitz, A. v., Mattiello, R., Stocker, H., and Greiner, W., *Phys. Lett.*, **B243**, 7 (1990).
27. von Keitz, A., Winckelmann, L., Jahns, A., Sorge, H., Stocker, H., and Greiner, W., *Phys. Lett.*, **B263**, 353–358 (1991).
28. Sorge, H., Berenguer, M., Stocker, H., and Greiner, W., *Phys. Lett.*, **B289**, 6–11 (1992).
29. Pang, Y., Schlagel, T. J., and Kahana, S. H., *Phys. Rev. Lett.*, **68**, 2743–2746 (1992).
30. Aichelin, J., and Werner, K., *Phys. Lett.*, **B300**, 158–162 (1993).
31. Rischke, D. H., Pursun, Y., Maruhn, J. A., Stocker, H., and Greiner, W., *Heavy Ion Phys.*, **1**, 309 (1995).
32. Hofmann, J., Stocker, H., Heinz, U. W., Scheid, W., and Greiner, W., *Phys. Rev. Lett.*, **36**, 88–91 (1976).

33. Hartnack, C., Stocker, H., and Greiner, W., ",", in *Proceedings of the Nato Advanced Study Institute on the Nuclear Equation of State*, Plenum Press, 1990.
34. Brachmann, J., et al., *Phys. Rev.*, **C61**, 024909 (2000).
35. Sorge, H., von Keitz, A., Mattiello, R., Stoecker, H., and Greiner, W., *Phys. Lett.*, **B243**, 7–12 (1990).
36. Fermi, E., *Prog. Theor. Phys.*, **5**, 570–583 (1950).
37. Landau, L. D., *Izv. Akad. Nauk SSSR Ser. Fiz.*, **17**, 51–64 (1953).
38. Hahn, D., and Stocker, H., *Nucl. Phys.*, **A452**, 723–737 (1986).
39. Braun-Munzinger, P., Heppe, I., and Stachel, J., *Phys. Lett.*, **B465**, 15–20 (1999).
40. Rafelski, J., and Letessier, J., *nucl-th/9903018* (1999).
41. Becattini, F., Cleymans, J., Keranen, A., Suhonen, E., and Redlich, K., *Phys. Rev.*, **C64**, 024901 (2001).
42. Yen, G. D., and Gorenstein, M. I., *Phys. Rev.*, **C59**, 2788–2791 (1999).
43. Braun-Munzinger, P., Magestro, D., Redlich, K., and Stachel, J., *Phys. Lett.*, **B518**, 41–46 (2001).
44. Braun-Munzinger, P., and Stachel, J., *Nucl. Phys.*, **A638**, 3–18 (1998).
45. Siemens, P. J., and Kapusta, J. I., *Phys. Rev. Lett.*, **43**, 1486–1489 (1979).
46. Gorenstein, M. I., Rischke, D. H., Stocker, H., Greiner, W., and Bugaev, K. A., *J. Phys.*, **G19**, L69–L75 (1993).
47. Papazoglou, P., Zschiesche, D., Schramm, S., Schaffner-Bielich, J., Stocker, H., and Greiner, W., *Phys. Rev.*, **C59**, 411–427 (1999).
48. Zschiesche, D., P. Papazoglou, P., Schramm, S., Beckmann, C., Schaffner-Bielich, J., Stocker, H., and Greiner, W., *Springer Tracts Mod. Phys.*, **163**, 129 (2000).
49. Zschiesche, D., Schramm, S., Stocker, H., and Greiner, W., *Phys. Rev.*, **C65**, 064902 (2002).
50. Michalec, M., *nucl-th/0112044* (2001).
51. Zschiesche, D., et al., *to be publ.* (2002).
52. Michalec, M., Florkowski, W., and Broniowski, W., *Phys. Lett.*, **B520**, 213–216 (2001).
53. Bass, S. A., Belkacem, M., Brandstetter, M., Bleicher, M., Gerland, L., Konopka, J., Neise, L., Spieles, C., Soff, S., Weber, H., Stocker, H., and Greiner, W., *Phys. Rev. Lett.*, **81**, 4092–4095 (1998).
54. Bravina, L. V., I. Gorenstein, M. I., Belkacem, M., Bass, S. A., Bleicher, M., Brandstetter, M., Hofmann, M., Soff, S., Spieles, C., Weber, H., Stocker, H., and Greiner, W., *Phys. Lett.*, **B434**, 379–387 (1998).
55. Reiter, M., Dumitru, A., Brachmann, J., Maruhn, J. A., Stocker, H., and W., G., *Nucl. Phys.*, **A643**, 99–112 (1998).
56. Braun-Munzinger, P., Stachel, J., Wessels, J. P., and Xu, N., *Phys. Lett.*, **B344**, 43–48 (1995).
57. Wong, C.-Y., *Introduction to High-Energy Heavy-Ion Collisions*, World Scientific, 1994.
58. Andersen, E., et al., *Phys. Lett.*, **B433**, 209–216 (1998).
59. Ivanenko, D. D., and Kurdgelaidze, D. F., *Astrophys.*, **1**, 251 (1965).
60. Bodmer, A. R., *Phys. Rev.*, **D4**, 1601–1606 (1971).
61. Chin, S. A., and Kerman, A. K., *Phys. Rev. Lett.*, **43**, 1292 (1979).
62. Farhi, E., and Jaffe, R. L., *Phys. Rev.*, **D30**, 2379 (1984).
63. Witten, E., *Phys. Rev.*, **D30**, 272–285 (1984).
64. Chodos, A., Jaffe, R. L., Johnson, K., Thorn, C. B., and Weisskopf, V. F., *Phys. Rev.*, **D9**, 3471–3495 (1974).
65. Michel, F. C., *Phys. Rev. Lett.*, **60**, 677–679 (1988).
66. Greiner, C., Koch, P., and Stocker, H., *Phys. Rev. Lett.*, **58**, 1825–1828 (1987).
67. Lukacs, B., Zimanyi, J., and Balazs, N. L., *Phys. Lett.*, **B183**, 27 (1987).
68. Schaffner-Bielich, J., Greiner, C., Diener, A., and Stocker, H., *Phys. Rev.*, **C55**, 3038–3046 (1997).
69. Schaffner, J., Stocker, H., and Greiner, C., *Phys. Rev.*, **C46**, 322–329 (1992).
70. Prowse, D. J., *Phys. Rev. Lett.*, **17**, 782–785 (1966).
71. Baltz, A. J., et al., *Phys. Lett.*, **B325**, 7–12 (1994).
72. Shiva Kumar, B., *Nucl. Phys.*, **A590**, 29c–45c (1995).
73. Rotondo, F. S., et al., *Nucl. Phys.*, **A610**, 297c–305c (1996).
74. Barish, K. N., *Heavy Ion Phys.*, **4**, 423–428 (1996).
75. Borer, K., et al., *Phys. Rev. Lett.*, **72**, 1415–1418 (1994).
76. Dittus, F., et al., *Nucl. Phys.*, **A590**, 347c–354c (1995).

77. Klingenberg, R., et al., *Nucl. Phys.*, **A610**, 306c–316c (1996).
78. Appelquist, G., et al., *Phys. Rev. Lett.*, **76**, 3907–3910 (1996).
79. Beringer, J., et al., "Search for strangelets and production of antinuclei in Pb+Pb collisions at CERN," in *Structure of Vacuum and Elementary Matter*, edited by H. Stocker, A. Gallmann, and J. Hamilton, 1996, international Conference on Nuclear Physics at the Turn of Millennium, Wilderness / George, South Africa.
80. Ambrosini, G., et al., *Nucl. Phys.*, **A638**, 411–414 (1998).
81. Harris, J. W., *Nucl. Phys.*, **A566**, 277c–285c (1994).
82. Schukraft, J., "Little bang at big colliders: Heavy ion physics at LHC and RHIC," in *Structure of Vacuum and Elementary Matter*, edited by H. Stocker, A. Gallmann, and J. Hamilton, 1996, international Conference on Nuclear Physics at the Turn of Millennium, Wilderness / George, South Africa.
83. Spieles, C., et al., *Phys. Rev. Lett.*, **76**, 1776–1779 (1996).
84. Matsui, T., and Satz, H., *Phys. Lett.*, **B178**, 416 (1986).
85. Gerland, L., Frankfurt, L., Strikman, M., Stocker, H., and Greiner, W., *Phys. Rev. Lett.*, **81**, 762–765 (1998).
86. Romana, A., "Anomalous J/ψ suppression in Pb+Pb collisions," in *Proceedings of XXXIIIrd Rencontres de Moriond*, 1998.
87. Spieles, C., Vogt, R., Gerland, L., Bass, S. A., Bleicher, M., Frankfurt, L., Strikman, M., Stocker, H., and Greiner, W., *Phys. Rev.*, **C60**, 054901 (1999).
88. Neubauer, D., Sailer, K., Muller, B., Stocker, H., and Greiner, W., *Mod. Phys. Lett.*, **A4**, 1627 (1989).
89. Gavin, S., and Vogt, R., *Nucl. Phys.*, **B345**, 104–124 (1990).
90. Gavin, S., and Vogt, R., *Phys. Rev. Lett.*, **78**, 1006–1009 (1997).
91. Hufner, J., and Kopeliovich, B., *nucl-th/9606010* (1996).
92. Kharzeev, D., Lourenco, C., Nardi, M., and Satz, H., *Z. Phys.*, **C74**, 307–318 (1997).
93. Satz, H., *Nucl. Phys.*, **A642**, 130–142 (1998).
94. Agakishiev, G., et al., *Phys. Rev. Lett.*, **75**, 1272–1275 (1995).
95. Drees, A., *Nucl. Phys.*, **A610**, 536c–551c (1996).
96. Winckelmann, L. A., et al., *Nucl. Phys.*, **A610**, 116c–123c (1996).
97. Srivastava, D. K., Sinha, B., Pal, D., Gale, C., and Haglin, K., *Nucl. Phys.*, **A610**, 350c–357c (1996).
98. Cassing, W., Ehehalt, W., and Ko, C. M., *Phys. Lett.*, **B363**, 35–40 (1995).
99. Li, G.-Q., Ko, C. M., and Brown, G. E., *Phys. Rev. Lett.*, **75**, 4007–4010 (1995).
100. Koch, V., and Song, C., *Phys. Rev.*, **C54**, 1903–1917 (1996).
101. Agakishiev, G., et al., *Phys. Lett.*, **B422**, 405–412 (1998).
102. Cleymans, J., Redlich, K., and Satz, H., *Z. Phys.*, **C52**, 517–526 (1991).
103. Koch, P., *Z. Phys.*, **C57**, 283–304 (1993).
104. Gale, C., and Lichard, P., *Phys. Rev.*, **D49**, 3338–3344 (1994).
105. Song, C., and Ko, C., C. M.and Gale, *Phys. Rev.*, **D50**, R1827–R1831 (1994).
106. Winckelmann, L. A., Sorge, H., Stocker, H., and Greiner, W., *Phys. Rev.*, **C51**, 9–11 (1995).
107. Baier, R., Dirks, M., and Redlich, K., *Phys. Rev.*, **D55**, 4344–4354 (1997).
108. Murray, J., Bauer, W., and Haglin, K., *Phys. Rev.*, **C57**, 882–888 (1998).
109. Drell, S. D., and Yan, T.-M., *Phys. Rev. Lett.*, **25**, 316–320 (1970).
110. Spieles, C., Gerland, L., Hammon, N., Bleicher, M., Bass, S. A., Stoecker, H., Greiner, W., Lourenco, C., and Vogt, R., *Eur. Phys. J.*, **C5**, 349–355 (1998).
111. Braun-Munzinger, P., Miskowiec, D., Drees, A., and Lourenco, C., *Eur. Phys. J.*, **C1**, 123–130 (1998).
112. Gavin, S., McGaughey, P. L., Ruuskanen, P. V., and Vogt, R., *Phys. Rev.*, **C54**, 2606–2623 (1996).
113. Shuryak, E. V., *Phys. Lett.*, **B78**, 150 (1978).
114. Domokos, G., and Goldman, J. I., *Phys. Rev.*, **D23**, 203 (1981).
115. Kajantie, K., and Miettinen, H. I., *Zeit. Phys.*, **C9**, 341 (1981).
116. Kajantie, K., and Miettinen, H. I., *Z. Phys.*, **C14**, 357–362 (1982).
117. Chin, S. A., *Phys. Lett.*, **B119**, 51–56 (1982).
118. Cleymans, J., and Fingberg, J., *Phys. Lett.*, **B168**, 405 (1986).
119. Cleymans, J., Fingberg, J., and Redlich, K., *Phys. Rev.*, **D35**, 2153 (1987).
120. Siemens, P. J., and Chin, S. A., *Phys. Rev. Lett.*, **55**, 1266–1268 (1985).

121. Seibert, D., *Phys. Rev. Lett.*, **68**, 1476–1479 (1992).
122. Ruuskanen, P. V., *Nucl. Phys.*, **A525**, 255–268 (1991).
123. Ruuskanen, P. V., *Nucl. Phys.*, **A544**, 169–182 (1992).
124. Ko, C. M., and Xia, L. H., *Phys. Rev. Lett.*, **62**, 1595–1598 (1989).
125. Dumitru, A., et al., *Phys. Rev. Lett.*, **70**, 2860–2863 (1993).
126. He, Z.-J., and Zhang, J.-J., *J. Phys.*, **G21**, L49–L55 (1995).
127. Kapusta, J., McLerran, L. D., and Kumar Srivastava, D., *Phys. Lett.*, **B283**, 145–150 (1992).
128. Kampfer, B., and Pavlenko, O. P., *Phys. Lett.*, **B289**, 127–131 (1992).
129. Geiger, K., and Kapusta, J. I., *Phys. Rev. Lett.*, **70**, 1920–1923 (1993).
130. Shuryak, E. V., and Xiong, L., *Phys. Rev. Lett.*, **70**, 2241–2244 (1993).
131. Strickland, M., *Phys. Lett.*, **B331**, 245–250 (1994).
132. Mazzoni, M. A., *Nucl. Phys.*, **A566**, 95c–102c (1994).
133. Masera, M., *Nucl. Phys.*, **A590**, 93c–102c (1995).
134. Lourenco, C., et al., *Nucl. Phys.*, **A566**, 77c–85c (1994).
135. Gonin, M., et al., *Nucl. Phys.*, **A610**, 404c–417c (1996).
136. Ko, C. M., Li, G.-Q., Brown, G. E., and Sorge, H., *Nucl. Phys.*, **A610**, 342c–349c (1996).
137. Bratkovskaya, E. L., and Cassing, W., *Nucl. Phys.*, **A619**, 413–446 (1997).
138. Li, G.-Q., Brown, G. E., Gale, C., and Ko, C. M., *nucl-th/9712048* (1997).
139. Koch, V., *Int. J. Mod. Phys.*, **E6**, 203–250 (1997).
140. Cassing, W., Bratkovskaya, E. L., Rapp, R., and Wambach, J., *Phys. Rev.*, **C57**, 916–921 (1998).
141. Li, G.-Q., and Gale, C., *Phys. Rev.*, **C58**, 2914–2927 (1998).
142. Winckelmann, L. A., *Ultrarelativistische Quanten-Molekular-Dynamik: Ein Modell zur Simulation von Schwerionen-Reaktionen*, Ph.D. thesis, Johann Wolfgang Goethe Universität (1996).
143. Hossenfelder, S., Hofmann, S., Bleicher, M., and Stocker, H., *hep-ph/0109085* (2001).
144. Arkani-Hamed, N., Dimopoulos, S., and Dvali, G. R., *Phys. Lett.*, **B429**, 263–272 (1998).
145. Arkani-Hamed, N., Dimopoulos, S., and Dvali, G. R., *Phys. Rev.*, **D59**, 086004 (1999).
146. Antoniadis, I., Arkani-Hamed, N., Dimopoulos, S., and Dvali, G. R., *Phys. Lett.*, **B436**, 257–263 (1998).
147. Bleicher, M., Hofmann, S., Hossenfelder, S., and Stocker, H., *hep-ph/0112186* (2001).
148. Hofmann, S., et al., *hep-ph/0111052* (2001).

Hot Hadronic Matter with the NJL Model

J. Aichelin*, G. Torrieri† and F. Gastineau*

*SUBATECH, Ecole des Mines, 4 Rue Alfred Kastler, F-44070, Nantes, Cedex 03, France
†Department of Physics, University of Arizona, Tucson, AZ 85721

Abstract. We give a review of the Nambu-Jona-Lasinio (NJL) model applied to systems at finite temperature and density. We show how to calculate certain quantities relevant to the study of hot hadronic matter, such as the quark condensate and hadron masses. We then present the results of some of these calculations, and describe how the NJL model can be used in a dynamical description of the hadronic fireball.

INTRODUCTION AND MOTIVATION

In the study of heavy ion collisions, one crucial yet not well understood problem is the behavior of hadrons in a hot and dense medium. At present, there are strong experimental indications [1] as well as theoretical reasons to believe that in-medium hadron properties change considerably from their measured vacuum counterparts. A partial chiral symmetry restoration, expected at moderately high temperatures, can lead to a significant change in both baryon and meson masses, as well as their interaction cross sections within a hadron gas. The importance of understanding these phenomena, and their dependence on temperature and baryonic density, is apparent: Many quark-gluon plasma signatures rely on particles produced during the phase transition to "normal" hadronic matter. These particles go through a (possibly long) reinteraction before reaching the observer, and hence any information they possessed about the deconfined phase might be modified by in-medium behavior. Furthermore, any meaningful study of properties of the QGP phase will require comparison with systems in which a quark gluon plasma was not produced, and all evolution happens within hot but confined (or semi-confined) hadronic matter. Analyzing these effects quantitatively, however, has proven to be extremely difficult. In medium hadronic matter inhabits the low energy (large α_s) QCD regime, so we can not develop a perturbative series from the basic QCD Lagrangian. Lattice calculations have produced some useful results, but the computing power and numerical difficulties associated with the calculation of many useful quantities (such as the variation of hyperon masses with temperature) are still a big problem. Moreover, lattice methods can not be applied to systems with finite nuclear density (see appendix), or to study dynamical systems (such as an expanding fireball with interacting hadrons).

Effective field theory provides a possible way out: Under this approach, one uses a Lagrangian which respects the fundamental symmetries of QCD, but where the actual gluon degrees of freedom have been integrated out. The remaining degrees of freedom will then be chosen to reproduce observed QCD features (such as Chiral symmetry breaking) and generate observed hadron interactions up to the energy at which the color degree of freedom becomes relevant.

Most effective QCD theories integrate out both the quark and gluon degrees of freedom completely, and operate with mesons and baryons as fundamental objects. Such an approach will not help to describe the temperature and density dependence of the hadron masses, since in such theories these masses will be constants to be phenomenologically adjusted, rather than dynamically generated quantities to be calculated. We therefore need an effective theory which preserves quarks as fundamental degrees of freedom, and describes hadrons as quark bound states.

THE NAMBU-JONA-LASINIO MODEL

Perhaps the most successful attempts to write an effective description of QCD without sacrificing the quark degrees of freedom has been the Nambu-Jona-Lasinio (NJL) model. [2] [3] [4].

It reduces all interactions of colored objects to point-like interactions. One then writes down all the Lagrangian terms necessary to describe the observed quark bound states ($q\bar{q}$ for scalar, vector and isoscalar mesons, and qq for baryons), keeping track of the global symmetries of QCD (SU(2) for two u and d quarks and SU(3) if the s quark is included). The resulting (SU(3)) Lagrangian has the form

$$L_{\text{NJL}} = L_{\text{Kinetic}} + L_{\text{Singlet}} + L_{\text{Multiplet}}, \quad (1)$$

$$L_{\text{Kinetic}} = \bar{q}i(\partial_\mu \gamma^\mu - m_f)q, \quad (2)$$

$$L_{\text{Singlet}} = g_s[(\bar{q}q)^2 + (\bar{q}i\gamma^5 \lambda_F^A q)^2], \quad (3)$$

$$L_{\text{Multiplet}} = g_D[(\bar{q}_c i\gamma_5 \lambda_F^A \lambda_C^A q)(\bar{q}i\gamma_5 \lambda_F^A \lambda_C^A q_c]. \quad (4)$$

where the antisymmetric $\lambda_{F,C}^A$ s are SU(3) representations(Gell-Man matrices) corresponding, respectively, to color and flavor) and g_S, g_D are constants. The term L_{Singlet} generates meson (quark-antiquark, SU(3) singlet states), while the colored $L_{\text{Multiplet}}$, corresponding to a quark-quark flavor-antisymmetric interaction, is needed to describe baryons.

In principle, the relation between g_S and g_D can be fixed from the underlying QCD symmetries, by a procedure known as Fierz transformation (described in detail in [3]). However, for the subsequent calculations we prefer to keep them as free parameters, adjusting their values to better describe experimental data.

The NJL model is non-renormalizable, and has to be regularized by a cutoff Λ in all momentum integrals (corresponding to the energy at which gluons can not be ignored). Moreover, confinement has to be put in by hand, by just calculating terms corresponding to the observed hadronic states (The next sections will show, however, that certain features of confinement do emerge dynamically from the model). Nevertheless, as a perturbative effective theory, it has proven capable to describe the properties of both mesons [6] and baryon [7] states. In the next sections, we will show that it manages to reproduce the main features and bound states of QCD. We will then describe how to calculate NJL observables at finite temperature and chemical potential, and proceed to apply the model to study the thermal evolution of meson and baryon masses. Finally, we will show how to use the NJL model in a quantum molecular dynamics approach to describe the time evolution of a hadronic fireball.

CHIRAL SYMMETRY

Gap equation

One of the more attractive features of the NJL model is that it can fully describe Chiral symmetry as a spontaneously broken symmetry in a quantitative way. It can be easily verified that L_{NJL} is invariant under

$$q \to e^{i\lambda^\alpha \theta_\alpha \gamma_5/2} q \qquad (5)$$

provided that the bare quark masses are 0. However, this symmetry is broken by the NJL ground state. This violation can proven using either a BCS-like variational approach or through loop corrections to the quark's self-energy. We will use first approach (both proofs are presented in detail by [3]).
We will work with the singlet part of the $m_f = 0$ single flavour NJL Hamiltonian density

$$\mathcal{H}_{NJL} = -i\bar{q}\gamma \cdot \nabla q - g_s[(\bar{q}q)^2 + (\bar{q}_f i\gamma_5 q)^2] \qquad (6)$$

and the usual Fourier expansion of the field into particle and antiparticle creation and annihilation operators.

$$q = \int \frac{d^3p}{(2\pi)^3} \sum_s (a_p^s u^s(p) e^{-ip\cdot x} + b_p^{+s} v^s(p) e^{ip\cdot x}), \qquad (7)$$

$$\bar{q} = \int \frac{d^3p}{(2\pi)^3} \sum_s (b_p^s \bar{v}^s(p) e^{-ip\cdot x} + a_p^{+s} \bar{u}^s(p) e^{ip\cdot x}), \qquad (8)$$

where u and v are the particle and antiparticle spinors and a and b^+ are respectively the particle destruction and antiparticle creation operators. Analogously to the BCS ground state, but keeping in mind that the interaction term in the Hamiltonian is particle-antiparticle (rather than the two alike particles which make up a Cooper pair), one introduces the following test ground State:

$$|vac\rangle = \prod_{p,s}[\cos(\theta(p)) + s\sin(\theta(p))a^+(p,s)b^+(-p,s)]|0\rangle \qquad (9)$$

where s is the spin.
This vacuum corresponds to a condensate of 0 total momentum particle antiparticle pairs, annihilated by the Bogoliubov-transformed operators

$$B(p,s) = \cos(\theta(p))a(p,s) + s\sin(\theta(p))b^+(-p,s), \qquad (10)$$
$$D(p,s) = \cos(\theta(p))b(p,s) - s\sin(\theta(p))a^+(-p,s). \qquad (11)$$

After some algebra, it is found (counting color and flavor degeneracies) that

$$\langle \text{vac}|\mathscr{H}_{\text{NJL}}|\text{vac}\rangle = -2N_c N_f \int \frac{d^3 p}{2\pi^3} p \cos(\phi) - 4g_s (N_c N_f)^2 [\int \frac{d^3 p}{2\pi^3} \sin(\phi)]^2 \quad (12)$$

where $\phi = 2\theta$.

If one compares this to $\langle 0|\mathscr{H}_{\text{NJL}}|0\rangle$ (the computation of which is left as an exercise), we see that the chosen state is indeed the "true" vacuum, in the sense that

$$\langle \text{vac}|\mathscr{H}_{\text{NJL}}|\text{vac}\rangle \leq \langle 0|\mathscr{H}_{\text{NJL}}|0\rangle \quad (13)$$

All that is left is to find the minimum energy, by minimizing $\langle \text{vac}|\mathscr{H}_{\text{NJL}}|\text{vac}\rangle$ with respect to the variational parameter ϕ. One than finds the minimization condition

$$p \tan(\phi(p)) = 4g_s N_c N_f \int \frac{d^3 q}{(2\pi)^3} \sin\phi(q) = m^*. \quad (14)$$

where m^* is a constant independent of momentum. Rearranging and using trigonometry identities, we can transform Eq. (14) into the equivalent of a gap equation.

$$m^* = 4g_s N_c N_f \int \frac{d^3 p}{(2\pi)^3} \frac{m^*}{\sqrt{p^2 + m^{*2}}} \quad (15)$$

which can be solved in terms of g_s and $N_{f,c}$. It is apparent that the $|\text{vac}\rangle$, in Eq. (9), is not invariant under the transformation given in Eq. (5). It is also easy to show (left to the reader as an exercise!) that $g_s \langle \text{vac}|q\bar{q}|\text{vac}\rangle = m^*(\neq 0!)$.

Evaluation of condensates

We can generalize the result of the previous section to an NJL model with light bare quark masses ($m_q \ll \langle \text{vac}|q\bar{q}|\text{vac}\rangle$) and different flavors. The extra mass will simply show up as a first order perturbative shift in m^*. However, due the flavor mass difference, the values of the different flavour condensates (m^* s) are not necessarily the same. We therefore define

$$<u\bar{u}> = \alpha, \quad (16)$$
$$<d\bar{d}> = \beta, \quad (17)$$
$$<s\bar{s}> = \gamma. \quad (18)$$

We proceed to use the mean field approximation, replacing quark-quark interactions by the interaction of each quark with a "field" given by the other quarks in the medium.

$$q_i \bar{q}_i \rightarrow <q_i \bar{q}_i>, \quad (19)$$
$$q_i \bar{q}_i q_j \bar{q}_j \rightarrow <q_i \bar{q}_i> q_j \bar{q}_j + <q_j \bar{q}_j> q_i \bar{q}_i - 2 <q_i \bar{q}_i><q_j \bar{q}_j>. \quad (20)$$

The effective Hamiltonian becomes

$$H^{MF} = \int d^3x \sum_{f=u,d,s} [\bar{q}_f i\gamma^0 \partial_0 q_f + 2g_s(\alpha^2 + \beta^2 + \gamma^2)]. \tag{21}$$

In other words, quarks will dynamically acquire an effective mass term corresponding to the value of the condensate $<q\bar{q}>$ of that flavor.

$$M_u = m_u - 2g_s\alpha^2, \tag{22}$$
$$M_d = m_d - 2g_s\beta^2, \tag{23}$$
$$M_s = m_s - 2g_s\gamma^2. \tag{24}$$

This quantity, and it's evolution in temperature and chemical potential, can be calculated from the g_S constant in the Lagrangian by minimizing the free energy of the system

$$\hat{\Omega} = -Ttr[e^{\frac{\hat{H}-\mu\hat{N}}{T}}] \tag{25}$$

with respect to $<q\bar{q}>$.

Using the NJL Hamiltonian together with the field q defined as in Eq. (7), and

$$\hat{N} = \hat{n}_q - \hat{n}_{\bar{q}} = a^+(p,s,f)a(p,s,f) - b^+(p,s,f)b(p,s,f) \tag{26}$$

one gets [5]

$$<q_f\bar{q}_f> = -M_f \frac{N_C}{\pi^2} \int_0^{\Lambda_{QCD}} dp p^2 [1 - f(E+\mu,T) - f(E-\mu,T)], \tag{27}$$

$$f(x,T) = \frac{1}{1+e^{x/T}}. \tag{28}$$

Fig. 1 shows how the quark masses evolve with temperature and chemical potential. It is clear that low temperature breaking of Chiral symmetry, and it's restoration at high T (the model gives $\sim$ 230 MeV) and/or μ is very well reproduced by the NJL model.

MESONS

A meson as a $q\bar{q}$ bound state

Scalar (σ,η) and pseudoscalar (π,K) mesons are described by the second term of $L_{singlet}$ (Eq. (3)). As bound states, they appear as poles in the scattering matrix between quarks and anti-quarks, corresponding, to first order, to the sum of the diagrams in Fig. 2.

The following results are computed for pseudoscalar mesons, but can be generalized easily to describe scalars [3] (an additional term in $\mathscr{L}_{NJL}$ is needed for vector modes such as ρ). Summing the infinite series diagrams, and putting the result to be equal to the meson propagator, one obtains Eq. (29)

$$iU^{ij} = i\gamma_5\tau_i \frac{g_{\pi qq}^2}{k^2 - m_\pi^2} i\gamma_5\tau_j = i\gamma_5\tau_i g_s^2(1 + g_s^2\Pi + (g_s)^4\Pi^2 + ...)i\gamma_5\tau_j = i\gamma_5\tau_i \frac{g_s^2}{1-g_s^2\Pi} i\gamma_5\tau_j \tag{29}$$

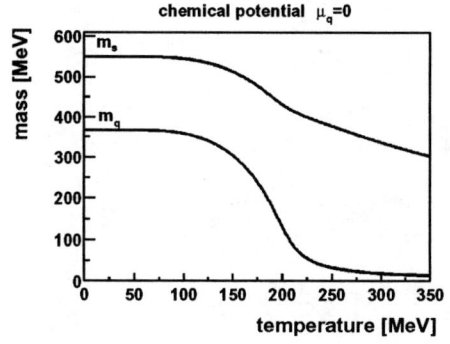 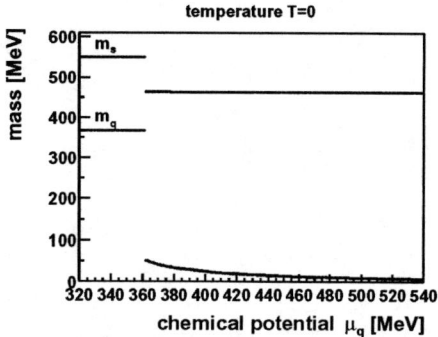

FIGURE 1. The mass of strange and light quarks as a function of the temperature (lhs) and as a function of the light quark chemical potential (rhs)

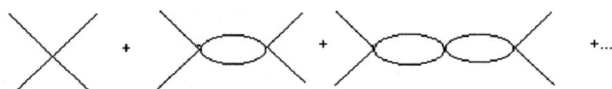

FIGURE 2. NJL leading order diagrams for $q\bar{q}$ pairs

where τ_i are SU(3) elements. The Π term corresponds to the one loop irreducible diagram (Fig. 3).

$$\Pi(k) = \int \frac{d^3p}{(2\pi)^3 E} Tr[i\gamma_5 \tau^i S(p,T,\mu) i\gamma_5 \tau^j S(k-p,T,\mu)], \qquad (30)$$

where $S(p,T,\mu)$ is a quark propagator corresponding to momentum p, temperature T and chemical potential μ (see appendix and [14]).

At the pole, we get

$$1 - g_s^2 \Pi(k^2 = m_\pi^2) = 0 \qquad (31)$$

Taylor expanding Eq. (29) around $k^2 = m_\pi^2$ and using Eq. (31)

$$RHS \sim \frac{g_s^2}{1 - g_s^2 \Pi^2} \simeq \frac{g_s^2}{1 - g_s^2 \Pi^2(k^2 = m_\pi^2) + (\frac{\partial \Pi}{\partial k^2})_{k^2 = m_\pi}(k^2 - m_\pi^2)}, \qquad (32)$$

$$= (\frac{\partial \Pi}{\partial k^2})^{-1}_{k^2 = m_\pi} \frac{-i}{k^2 - m_\pi^2}. \qquad (33)$$

Equating Eq. (33) with the LHS of Eq. (29) fixes $g_{\pi qq}$ in terms of the NJL model's fundamental constants.

$$g_{\pi qq}^2 = (\frac{\partial \Pi}{\partial k^2})^{-1}_{k^2 = m_\pi^2}. \qquad (34)$$

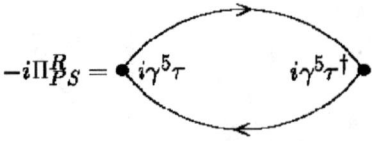

FIGURE 3. 1 loop irreducible diagram from $L_{singlet}$ term

Meson masses in terms of quark masses

At T=0, feynman rules in $\mathscr{L}_{singlet}$ result in the following expression for $\Pi(k^2)$ in terms of $m^* = <q\bar{q}>$.

$$\Pi(k^2) = -4iN_CN_f \int \frac{d^4p}{(2\pi)^4} \frac{m^{*2} - p^2 + \frac{1}{4}k^2}{[(p+\frac{1}{2}k^2)^2 - m^{*2}][(p-\frac{1}{2}k^2)^2 - m^{*2}]}. \qquad (35)$$

After some algebra and variable substitutions [3],

$$\frac{1}{i}\Pi(k^2) = 4N_CN_f \int \frac{d^4p}{(2\pi)^4} \frac{1}{p^2 - m^{*2}} - 2N_CN_f k^2 I(k^2), \qquad (36)$$

where

$$I(k^2) = \int \frac{d^4p}{(2\pi)^4} \frac{1}{[(p+\frac{1}{2}k^2)^2 - m^{*2}][(p-\frac{1}{2}k^2)^2 - m^{*2}]}. \qquad (37)$$

Now, we know from Eq. (15) that the first term in Eq. (36) is equal to $m^* - m_0 = <q\bar{q}>$. Putting this into 36 and rearranging we get

$$1 - 2g_s^2 \Pi(k^2) = \frac{m_0}{m^*} + 4ig_s^2 N_C N_f k^2 I(k^2) \qquad (38)$$

Since by Eq. (31) the left hand side vanishes for $k = m_{\pi^2}$ we finally get

$$m_\pi = -\frac{m_0}{m^*} \frac{1}{2ig_S N_c N_f I(m_\pi^2)} \qquad (39)$$

The equation above can be solved numerically to get the meson mass in terms of the quark condensate. $I(K^2)$ can also be generalized according to the rules given in the appendix, so that the temperature and chemical potential dependence of the π and K mesons can be calculated. Adjusting the dynamical quark masses and the coupling constants, one obtains the fits in table 1.

Fig. 4 shows how the mass of the π and K mesons (as well as the scalar σ and axial η) and the masses of their constituents vary with temperature. We note that, while the theory is not, strictly speaking, confining, something like the deconfinement phase transition does emerge dynamically.

TABLE 1. Meson properties from the NJL model

Particle	q	s	π	K
Theory (MeV)	422	627	137	550
Measured (MeV)	N/A	N/A	137	495
$\frac{\text{Theory} - \text{experiment}}{\text{experiment}}$ (%)	N/A	N/A	0	11

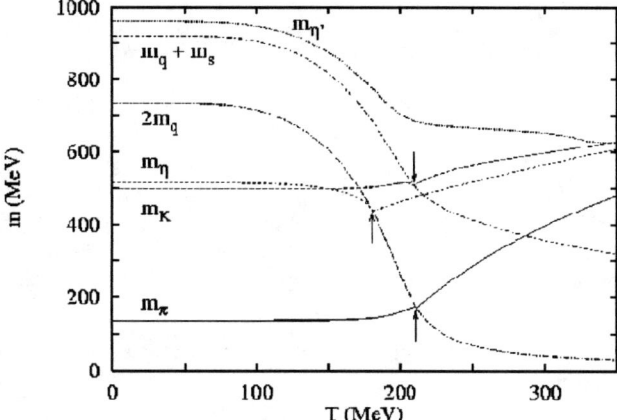

FIGURE 4. Evolution of meson masses with T

As can be seen, the π and K masses stay roughly constant until a T of roughly 200 MeV, at a value considerably below than the masses of their constituents. At low temperature, therefore, there is a strong tendency for quarks and antiquarks to bind into mesons. As T rises, however, π and K begin to get heavier, until their masses become bigger than the masses of their constituent quarks ($T \sim 212 MeV$). At this point, which can be thought of as the onset of deconfinement, pions and Kaons become unstable.

The σ and π masses, and their temperature dependencies, were also calculated on the lattice [9]. The same qualitative behavior, and very similar quantitative values to Fig. 4 were observed. We can therefore be confident in the capability of the NJL model to describe mesons in low energy.

BARYONS

Diquarks

Baryons are described within the NJL model as the bound state of a quark and a diquark. Diquarks are states arising from the $L_{\text{Multiplet}}$ term in the NJL model Lagrangian

(Eq. (4)). These states do not correspond to physical degrees of freedom, since only quark-diquark bound states (ie the baryons) will be included when calculating observables (this is what is meant by putting confinement in the NJL model by hand). The constants in the Lagrangian are principally given by the Fierz transformations, but slightly modified to reproduce observable baryon masses.

$L_{\text{Multiplet}}$ couples the quark field q to $\overline{q}_c = C\overline{q}$. The charged conjugate operator C simply replaces the particle operators (a and a^+ in Eq. (7)) by their corresponding antiparticle operators (b and b^+). Therefore,

$$q_c = \int \frac{d^3p}{(2\pi)^3} \sum_s (b_p^s v^s(p)e^{-ip.x} + a_p^{+s} u^s(p)e^{ip.x}), \tag{40}$$

$$\overline{q}_c = \int \frac{d^3p}{(2\pi)^3} \sum_s (a_p^s \overline{v}^s(p)e^{-ip.x} + b_p^{+s} \overline{u}^s(p)e^{ip.x}). \tag{41}$$

The form of the coupling means only scalar diquarks, antisymmetric in flavor and spin are allowed. The diquark bound state appears as a pole in the scattering matrix obtained by summing a series of diagrams which has exactly the same form as the one giving rise to meson states. However, the fact that $L_{\text{Multiplet}}$ couples q to q_c implies that $\gamma_5 \tau^i \to \Omega^i$ where Ω^i is a representation of $SU(3)_{\text{color}} \otimes SU(3)_{\text{flavor}} \otimes i\gamma^5 C$. Putting all these terms together in a perturbation series and summing in the same way as in the meson section, one obtains the polarization tensor

$$\Pi^{ij}(k) = \int \frac{d^3p}{(2\pi)^3 E} \overline{\Omega}^i_{\alpha,\beta} S^{\alpha\delta}(p,T,\mu) S^{\beta\gamma}(k-p,T,\mu) \overline{\Omega}^j_{\delta\gamma}, \tag{42}$$

and the scattering matrix

$$T(k)_{\alpha\beta\delta\gamma} = \Omega^a_{\alpha\beta} \frac{2iK_{ab}}{1+\Pi K_{ab}} \overline{\Omega}^b_{\delta\gamma}, \tag{43}$$

$$K_{ab} = g_D. \tag{44}$$

Again, the bound state is a pole in the scattering matrix so

$$|1 + K\Pi(q^2 = m_D^2)| = 0. \tag{45}$$

Taylor-expanding $T(k)$ around $k = m_D$ yields the equation linking quark-quark-Diquark coupling constant with the NJL parameters (g_D and the quark masses),

$$g_{Dqq}^{-2} = \left(\frac{\partial \Pi(k^2)}{\partial k^2}\right)_{k^2 = m_D^2} \tag{46}$$

in the same way as for mesons.

Since Diquarks are not physical states, and g_D itself is a "fictitious" quantity, we can not calculate anything physically meaningful from their properties. Diquark terms, however, will appear in the construction of baryonic states.

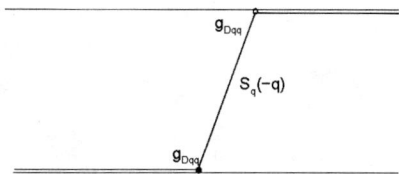

FIGURE 5. Simplest quark-Diquark interaction

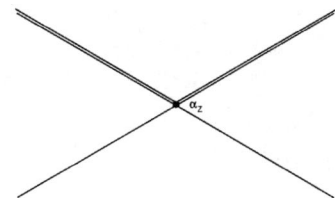

FIGURE 6. Pointlike quark-Diquark interaction

Baryons

We are now ready to construct the baryon as the bound state of a quark with a Diquark. The first order diagram which corresponds to such a bound state is the exchange of a quark between a Diquark state and a single quark, corresponding to the diagram in Fig. 5.

If the exchanged quark momentum is of the order of the NJL cut-off (Λ), this interaction reduces to a point interaction (Fig. 6) with amplitude

$$Z^{\beta\gamma}_{bc} = g_{Dqq}\Omega^{\beta\delta}_a \alpha_Z \overline{\Omega}^{\gamma\alpha}_b g_{Dqq}, \tag{47}$$

and a coupling constant $\alpha_Z = -\frac{i}{M_q}$. This approximation is obviously good for very low T, since chiral symmetry breaking enhances the mass of all quarks by an amount $\sim \Lambda$. However, the results presented below become questionable at higher temperature, unless only cases where the strange quark is exchanged are considered. Once the point-like quark-diquark interaction has been defined, the procedure to arrive to baryon masses is the familiar one:

One defines a scattering matrix to first order as the sum of all diagrams as in Fig. 7, and defines the bound state as a singularity in the obtained expression.

$$|1 - \Pi(m_B^2)Z| = 0, \tag{48}$$

$$\Pi(P) = i \int \frac{d^3p}{(2\pi)^3 E} \text{Tr}\left[\tau_i \overline{\Omega}^{\alpha\delta}_c g_{Dq_1q_2} iS^{\delta\gamma}_q(P-p,T,\mu) iS^{cd}_D(p,T,\mu) \Omega^{\gamma\beta}_d g_D \tau_j\right]. \tag{49}$$

Using the quark masses obtained before, and adjusting g_D to reproduce the observed baryon masses, one gets the best-fit results in table 2. It is apparent that the NJL model can provide a good description of observed (T=0 $\mu = 0$) baryons.

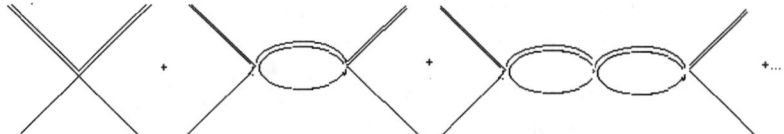

FIGURE 7. First-order quark-Diquark series

TABLE 2. Baryon properties from the NJL model

Particle	N	Λ	Σ	Ξ
Theory (MeV)	938	1083	1181	1280
Measured (MeV)	938	1116	1190	1315
$\frac{\text{Theory} - \text{experiment}}{\text{experiment}}$ (%)	0	3	0.7	2.3

We can thus be confident that it will provide at least a qualitative description when it is extended to finite temperatures and chemical potentials. Figures 8 and 9 show how masses change, respectively, at finite T and μ.

All baryon masses decrease with increasing temperature, a result expected from chiral symmetry restoration, as M_q, M_s descend to their bare values. Comparing the baryon masses to the masses of quarks and Diquarks, we see that, just like in the meson case, something like low temperature confinement and high temperature deconfinement emerge: At T=0 baryons are much lighter than their constituents, but this changes as T increases. However, we get a particle-specific crossover rather than a phase transition: nucleons and Λ hyperons get unstable already at $T \approx 220$ MeV, whereas the Ξ is stable longer due to its large strangeness content(since m_s is not so small, it does not change too much with T). Σ decays significantly before all other particles. Because of it's small binding energy, it becomes unstable as soon as the quark masses start to decrease.

Similarly, all masses decrease and baryons become unstable for a given density: $\rho/\rho_0 \approx 2.6$ for the nucleon ($M_{p,n} \approx 425$ MeV) and Λ ($M_\Lambda \approx 850$ MeV); $\rho/\rho_0 \approx 1.7$ for Σ ($M_\Sigma \approx 950$ MeV). The value of the nucleon mass at $\rho/\rho_0 = 1$ is about 700 MeV. Encouragingly, these results are in good agreement with infinite nuclear matter (eg., Walecka or HQT model [10]) calculations.

Finally, it should be noted that for "reasonable" temperatures (much less than the phase transition) and nuclear densities, mass modifications and baryon-antibaryon splittings are not larger than 30 %.

MOLECULAR DYNAMICS

The previous sections show that the NJL model is capable of calculating static properties of hadrons, and the evolution of these properties with temperature. However, unlike lattice calculations, the NJL model can also be used to calculate scattering cross-sections. Because of this, it can be used to model dynamical processes such as the evolution of a

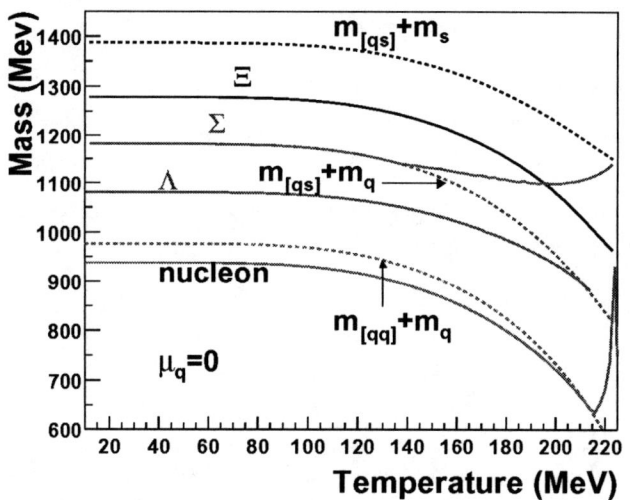

FIGURE 8. Temperature dependence of the baryon masses (solid line). The dashed line represents the sum of the two baryon's constituent masses (Diquark + quark)

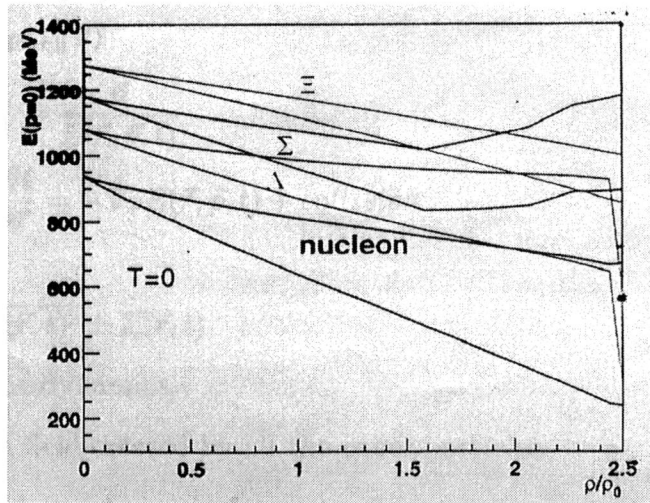

FIGURE 9. Density dependence of the nucleon, Σ, Λ, Ξ as well as their antiparticles. ρ_0 is the normal nuclear density

"colored" plasma (hot hadron gas where the color degrees of freedom have not quite unfrozen), and it's hadronization into "cold" mesons and baryons. What follows is a rough introduction to what is, by itself, a vast and rapidly evolving field. The interested reader should go to [11] [12] for further details.

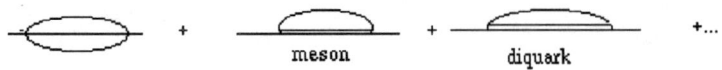

FIGURE 10. Perturbation series for a quark's self-energy

We take the molecular dynamics approach and treat the system semiclassically. We can then write down Hamilton's equations

$$\frac{d\vec{r}}{dt} = \frac{\partial \mathcal{H}}{\partial \vec{p}}, \tag{50}$$

$$\frac{d\vec{p}}{dt} = -\frac{\partial \mathcal{H}}{\partial \vec{r}} + I_{\text{collision}}, \tag{51}$$

where $I_{\text{collision}}$ is necessary to model discontinuous momentum changes due to particle-particle scattering. We then assume that quarks and hadrons are weakly interacting perturbations of the medium described in section 2, and hence their wave function is, to a good approximation, a wave packet.

$$\Psi = \prod_i \Psi_i, \tag{52}$$

$$\Psi_i(\vec{r},\vec{p},t) = \frac{1}{(2\pi L)^{3/2}} e^{-(\vec{r}-\vec{r}_i(t))^2 L^2} e^{-(\vec{p}-\vec{p}_i(t))^2/L^2}. \tag{53}$$

If we then use the Hartree-Fock approximation $\mathcal{H} =< \Psi|\mathcal{H}|\Psi >$, and put in the Hamiltonian from Eq. (6), we obtain

$$\frac{\partial \mathcal{H}}{\partial \vec{p}} = \frac{\vec{p}}{E}, \tag{54}$$

$$\frac{\partial \mathcal{H}}{\partial \vec{x}} = -\nabla E + I_{\text{collision}}, \tag{55}$$

$$E = \sqrt{\vec{p}(t)^2 + m^{*2}(\vec{r},t) + \Sigma(r,t)}, \tag{56}$$

where Σ is the sum of all the diagrams contributing to a particle's self-energy besides m^* (As in Fig 10).

All that we need now, is to calculate $I_{\text{collision}}$. This quantity can be inferred from the scattering cross-sections σ_k through a semiclassical Monte Carlo process [12]: For each pair of particles, the coordinates and momenta are boosted into the center of mass frame. In this frame, the particle's separation $\vec{x}_i - \vec{x}_j$ is decomposed into components parallel and perpendicular to the relative momentum $\vec{p}_i - \vec{p}_j$ The allowed scattering processes are selected, and the total cross-section,

$$\sigma_{\text{tot}} = \sum_k \sigma_k, \tag{57}$$

is then computed.

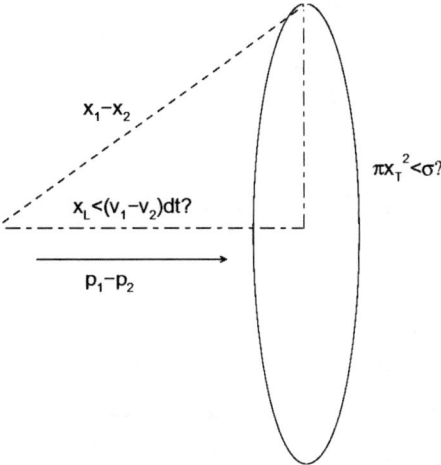

FIGURE 11. Diagrammatic representation for the collision criterion

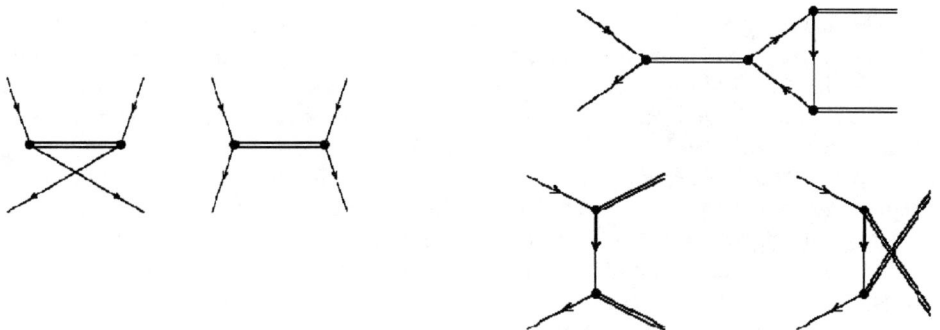

FIGURE 12. Scattering and meson exchange diagrams

The conditions for scattering to occur are then (see Fig. 11):

- the longitudinal distance should be less than the distance traveled by the particle in one time step,
- the transverse distance should be smaller than $\sqrt{\sigma_{tot}/\pi}$.

If a pair of particles satisfy this criterion, they will interact. Their interaction channel and the momentum distribution of the resulting products will be selected at random, with the probabilities given by $\sigma(k)$ and $\partial\sigma(k)/\partial\Omega$. All that is left now, is the tedious but simple task of implementing the allowed scattering for quarks, mesons, baryons, Diquarks ecc. through the NJL Feynman diagrams. At present, only quark-quark elastic interactions and meson processes, such as those in Fig. 12 have been included, although the inclusion of Diquarks and baryons will follow in the future.

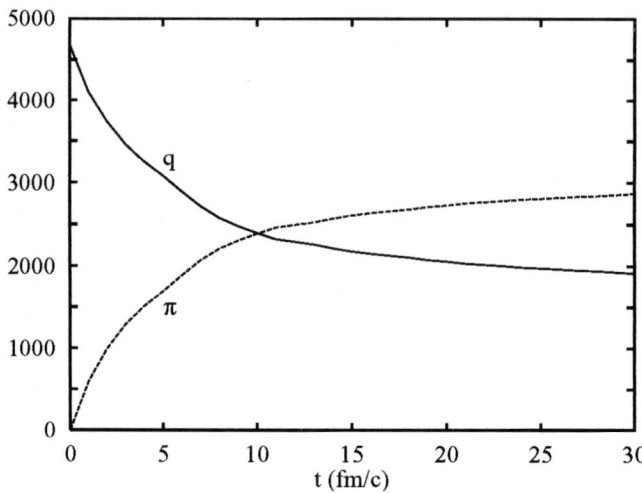

FIGURE 13. Quarks and mesons with time

Following these rules, the system may be evolved from the initial conditions (number of particles, size, momentum distribution ecc.). As expected, provided the initial temperature is high (~ 200 MeV) quarks and antiquarks are dominant degrees of freedom, and mesons only make their appearance as short-lived unstable resonances. As the system expands and cools, though, pions become more and more dominant eventually absorbing all of the free quarks (Fig. 13). It should be noted that the plasma itself tends not to expand, as pions tend to be created at the surface of the plasma and end up radiating away (Fig. 14). Rapidity distributions remain relatively constant in time. (Fig. 15).

MPEG files with the evolution of a model such as the one described here can be obtained from: http://star.physics.yale.edu/IntroHI.html

It should be remembered that the logo used in the CERN official announcement of the discovery of a new state of matter [13] (Fig. 16) was just an image of an NJL-type simulation.

CONCLUSIONS

The Nambu-Jona-Lasinio model has been proven to be a very useful effective field theory of QCD. While the transition from full QCD has to be put in by hand, the NJL model can reproduce satisfactorily the main low-energy features of QCD, such as hadron masses and chiral symmetry breaking, and their evolution on the nuclear phase diagram. Unlike lattice calculations, however, observables in this model can be calculated via a perturbation expansion. This means its results can be calculated analytically, and used in dynamical simulations. For these reasons, it is expected that NJL based calculations will continue to constitute a useful tool in the study of hot and dense strongly interacting matter.

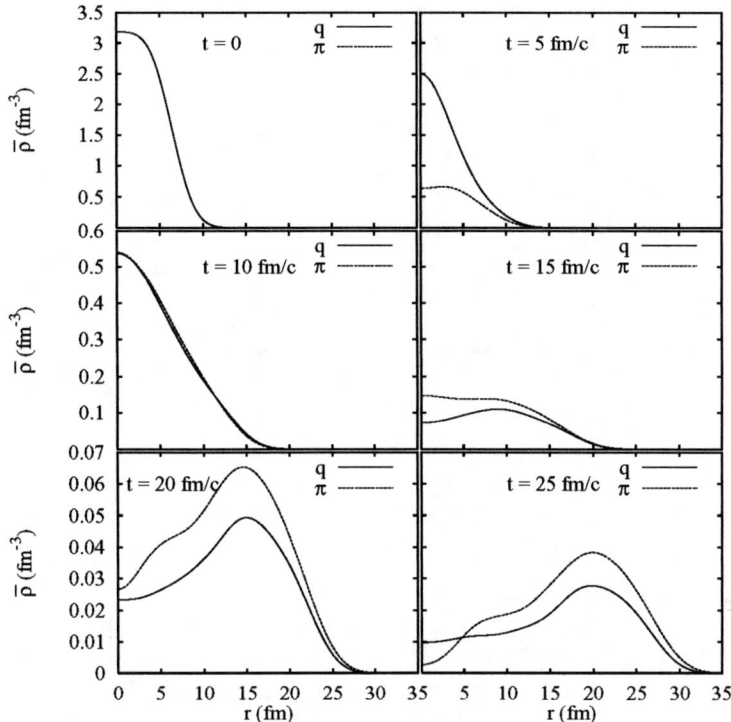

FIGURE 14. space distribution of quarks and mesons

APPENDIX

A few words on finite temperature field theory

The expectation value for an operator $\hat{O}$ in quantum field theory is

$$<\hat{O}>_{QFT} = Tr(\hat{O}e^{i\int dxdt \mathscr{L}})/Z, \quad (58)$$
$$Z = Tr(e^{i\int dxdt \mathscr{L}}), \quad (59)$$

where the trace becomes a functional integral in the continuum limit. it can be seen, by inspection, that this quantity can be related to the $\hat{O}$'s thermal expectation value

$$<\hat{O}>_{Thermal} = Tr(\hat{O}e^{\frac{\mathscr{H}-\mu\hat{N}}{T}}/Z), \quad (60)$$
$$Z = Tr(e^{\frac{\mathscr{H}-\mu\hat{N}}{T}}). \quad (61)$$

We are interested in finding an expression for the thermal average of a particle's Green's function $S(x-x',t-t') =< T[\psi(x,t)\psi^+(x',t')] >_{Thermal}$ in terms of it's quantum field theory equivalent. Once we do this, we will be able to generalize Feynman rules

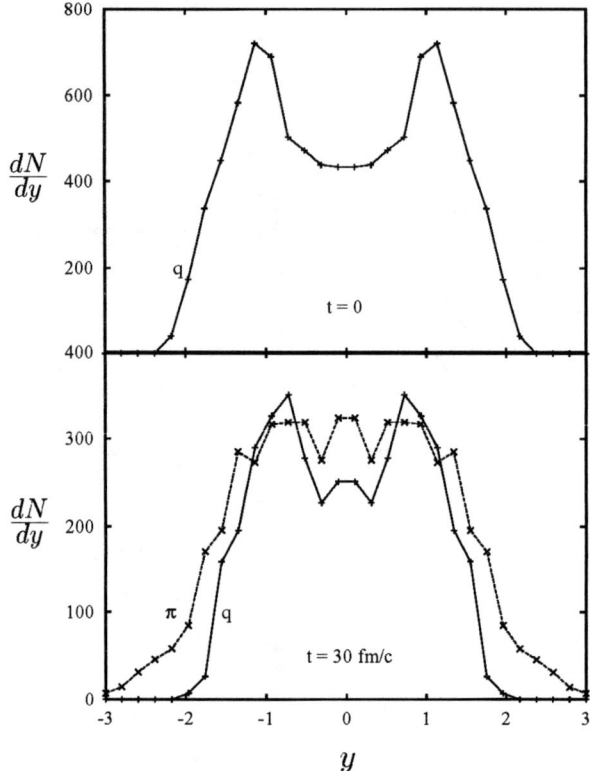

FIGURE 15. rapidity distribution with time

to any temperature and chemical potential. Doing this in practice can be very difficult, both analytically and numerically (especially at both finite temperature and chemical potential). For a detailed technical description of how to perform this calculation for the 1-loop terms in the Green's function $S(k^2, \mu, T)$ used to calculate bound state masses (Fig. 3 ecc.) see [14]. In this section we will limit ourselves to introducing the basic concepts and explaining where the difficulties lie.

Case $\mu = 0$

At $\mu = 0$, it can be seen that Eq. (58) reduces to Eq. (60) if

- $\mathscr{L} \to \mathscr{H}$,
- $t \to i/T$.

The second condition will allow us to evaluate the desired quantities by simply computing the QFT Green's functions in imaginary time. It can be shown in a few lines

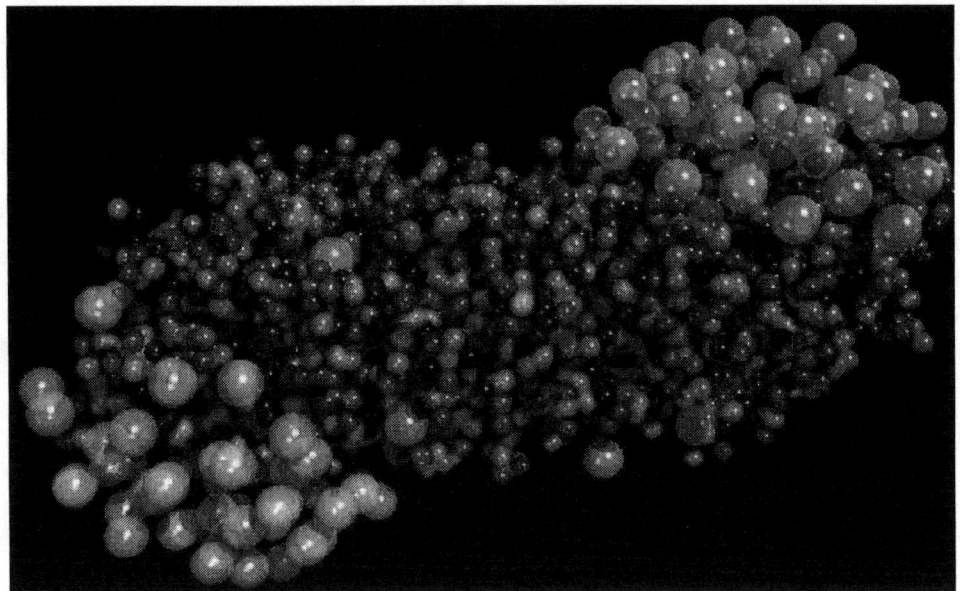

FIGURE 16. Mpeg frame showing fireball evolution

[8] that this leads to a periodicity condition in time

$$S(0,x_0,\tau,x) = \pm S(\frac{1}{T},x_0,\tau,x) \qquad (62)$$

(respectively for bosons and fermions), The momentum space propagator will therefore be given by:

$$S(p,T) = \int_0^\infty dt \int d^3x e^{i(Et-px)} S(t-t_0, x-x_0) \to \sum_{n=-\infty}^{\infty} \int d^3x \int d\tau e^{iv_n\tau} S(x,\tau)_{T=0}, \qquad (63)$$

$$S(x-x',t-t') = T \sum_{n=-\infty}^{\infty} e^{-iv_n(t-t')} \int \frac{d^3p}{(2\pi)^3} e^{ip(x-x')} S(p,w_n)_{T=0}, \qquad (64)$$

$$v_n^{\text{fermions}} = 2\pi(n+1)T, \qquad (65)$$

$$v_n^{\text{Bosons}} = 2\pi nT. \qquad (66)$$

For instance, a quark propagator such as those considered in sections dealing with meson and baryon masses would be extended to non-zero temperature as:

$$\frac{1}{\gamma_0.E - \gamma.p - m} \to \sum_{n=-\infty}^{n=\infty} \frac{1}{i2\pi(n+1)\gamma_0/T - \gamma.p - m}, \qquad (67)$$

which can not be handled quite as easily as it's T=0 counterpart.

finite μ

When chemical potential is not 0,

$$\mathcal{H} \to \mathcal{H} - \mu\hat{N} = \mathcal{H} - \mu\psi\psi^+ = \mathcal{H} - \psi\mu\gamma^0\overline{\psi}. \tag{68}$$

This can be implemented by modifying the fermion's kinetic Lagrangian (and, hence, it's propagator) to

$$L_{\text{Kinetic}} \to \overline{\psi}i(\delta_\mu\gamma^\mu - m_0 + \gamma^0\mu)\psi, \tag{69}$$

$$S(p,\mu) \to \frac{1}{i\gamma_\mu\partial^\mu - m + \gamma_0\mu}. \tag{70}$$

Unfortunately, the extra μ term means that Wick-rotating Eq. (58) will lose it's usefulness, since it will merely yield a different complex object. (this is why lattice calculations are generally not possible at $\mu \neq 0$). In addition, in theories with vacuum condensation (this includes the NJL model), $\psi\overline{\psi} \neq 0$ will shift μ. Therefore, to do any perturbation theory, we must first calculate this dynamically generated and generally non-perturbative quantity. For these reasons, unless one works under certain limits (such as $\mu \sim \Lambda_{\text{QCD}}$) in general quantitative calculations at $\mu \neq 0$ are very hard to do. The fact that the NJL model is one of the few theories where such calculations are possible is another strong point in it's favor.

REFERENCES

1. Quark Matter 2001 *Nucl. Phys.* **A698** (2002)
2. Nambu Y and Jona-Lasinio G (1961) *Phys. Rev.* **122** (1961)
3. S. P. Klevansky, *Rev. Mod. Phys.* **64** (1992) 649.
4. T. Hatsuda and T. Kunihiro *Phys. Rep.* **247** (1994) 221-367
5. R. Nebauer and J. Aichelin, arXiv:hep-ph/0101289.
6. S. Klimt, M. Lutz, U. Vogl and W. Weise, *Nucl. Phys.* A **516**, 429 (1990).
7. A. Buck, R. Alkofer and H. Reinhardt, *Phys. Lett.* B **286**, 29 (1992).
8. Fetter and Walecka, Quantum theory of many particle systems (McGraw-Hill)
9. Laermann, QM 1986 presentation
10. B. D. Serot and J. D. Walecka, *Int. J. Mod. Phys.* E **6**, 515 (1997) [arXiv:nucl-th/9701058].
11. S. P. Klevansky, A. Ogura and J. Hufner, *Annals Phys.* **261**, 37 (1997) [arXiv:hep-ph/9708263].
12. P. Rehberg, L. Bot and J. Aichelin, *Prog. Part. Nucl. Phys.* **42** (1999) 323.
13. See the web page: http://www.cern.ch/CERN/Announcements/2000/NewStateMatter/ Text of the scientific consensus view of the spokesmen of CERN experiments is also available as E-print: nucl-th/0002042, *"Evidence for a New State of Matter: An Assessment of the Results from the CERN Lead Beam Programme"*, compilation by U. Heinz, and M. Jacob.
14. P. Rehberg and S. P. Klevansky, *Annals Phys.* **252**, 422 (1996) [arXiv:hep-ph/9510221].

Purpose and Physics of Event-by-Event Analysis

Thomas A. Trainor

*CENPA 354290
University of Washington
Seattle, WA 98195
trainor@hausdorf.npl.washington.edu*

Abstract. Event-by-event analysis is the study of event-wise and ensemble correlation structure in the hadronic final state of heavy ion collisions. Statistical and topological techniques are applied to large event populations to reveal fluctuations and correlations deviating from a null hypothesis. Correlations can arise from initial-state scattering, incomplete equilibration and from critical correlations developed during traversal of the QCD phase boundary. The goal of this analysis is a phenomenological description of full QCD at and above the phase boundary.

INTRODUCTION

The QCD phase boundary is defined on the thermodynamic state space $(T, \mu_B, \mu_S, ...)$ of equilibrated bulk nuclear matter under global constraints. On one side of this boundary at some asymptotic distance lies the idealization of perturbative QCD, color deconfinement and an uncorrelated Stefan-Boltzmann gas of colored quarks and gluons (partons) wherein QCD symmetries are restored. On the other side lies the highly-correlated hadronic regime: a resonance gas of colorless hadrons and broken symmetries described approximately by strong-interaction phenomenology. During a relativistic heavy ion collision of sufficient energy a small system, far from equilibrium, approaches this boundary from the p-QCD side, expanding, cooling and traversing the phase boundary in a complex way. Partially restored symmetries are rebroken and hadrons reform. The boundary is of finite width, and its structure may not conform to classical bulk-matter thermodynamic concepts.

Only the hadronic final state is accessible to observation. In event-by-event (EbyE) analysis we search for remnants of the deconfined system and phase-boundary traversal embedded in the correlation structure of this final state [1, 2]. After a period of initial searching we have found that these remnants are present, but at small amplitudes whose detection and study require precision analysis techniques. The correlations we observe are those that survive rapid boundary traversal and equilibration processes in partonic and hadronic cascades. Recent progress in EbyE analysis has included not only the development of new correlation analysis techniques but also a reformulation of the basic questions of phase-boundary traversal, equilibration and critical phenomena appropriate for small systems. In some cases standard many-body and statistical concepts must be modified to accommodate this interesting problem.

In this paper I begin with the relationship between fluctuations and correlations and proceed to an analysis of p_t fluctuations with a differential variance measure based on the central limit theorem (CLT). I then relate variance comparison measures to the correlation integral, which creates a connection between variance differences and correlation measures, specifically the autocorrelation density. I contrast the correlation structure in two-point spaces with evidence for incomplete equilibration in single-particle distributions by comparison with the Lévy distribution. I consider centrality dependence of correlation in the context of an initial-state two-nucleon space which supplements the final-state multihadron concept. I discuss some difficulties with experiment comparisons and recommend certain simplifying techniques. I then consider two-point distributions in axial (η, ϕ) phase space and the formation of joint autocorrelation densities for certain charge combinations which provides the most dispersive access to axial multiparticle correlations. Finally, I summarize some physics topics that are central to event-by-event analysis.

FLUCTUATIONS AND CORRELATIONS

Fluctuations and correlations are closely-related aspects of a structured measure distribution defined on a primary space (*e.g.*, particle density on a many-body phase space, of which momentum space is directly observable and aspects of configuration space may be inferred). Distributions of several hadron species in the primary momentum space for each collision event carry information on collision dynamics in the form of multiparticle correlations. We can extract this information directly from multiparticle densities in higher-dimensional q-particle product spaces (correlation analysis), or bin the primary space itself and study frequency distributions of the bin contents and their moments (statistical or 'fluctuation' analysis). The basic tools of correlation analysis are the scale-dependent correlation integral, autocorrelation density and mixed-q-tuple reference. The basic tools of fluctuation analysis are frequency-distributions, their moments and the central limit reference.

The effects we seek are typically small compared to uninteresting statistical structures resulting from finite event multiplicities. Thus, we require a differential approach and precision references. Correlation references are constructed with *mixed* q-tuples formed from different events, as opposed to *sibling* q-tuples formed from single events. Fluctuation references are based on the central limit theorem, representing the assumptions (independent samples from a fixed parent distribution) implicit in the formation of a mixed q-tuple correlation reference. Correlation and fluctuation measures are related algebraically by successive integrations. As in time-dependent signal analysis such integrations represent a tradeoff between bandwidth reduction (and possible signal loss) on the one hand and statistical 'noise' rejection on the other.

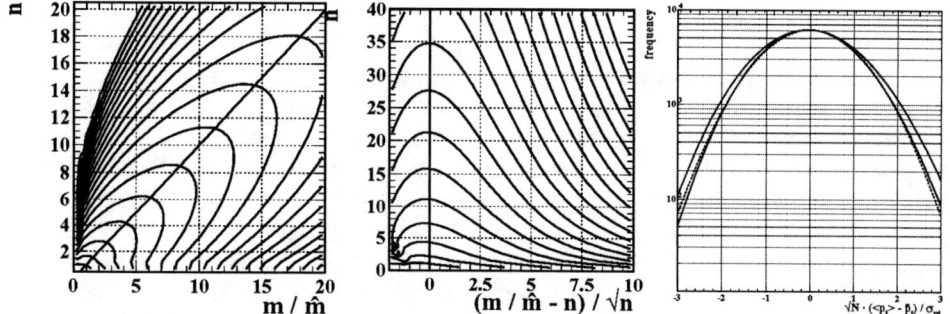

FIGURE 1. The left panel shows a CLT reference distribution corresponding to a random binning of a uniform measure and random point set. The middle panel shows a transformation of this distribution to a minimally-biased variable. The right panel shows a comparison among real data, a gamma reference distribution and a gaussian reference using this variable.

The degree of integration which optimizes the signal-to-noise ratio varies from case to case. Direct study of correlation structure in q-tuple spaces (*e.g.*, two-particle correlations) is the most differential technique, having maximum bandwidth. Relative correlation amplitudes in a heavy-ion central collision are typically 1 *permil*. A first integration is the projection of a q-particle density onto a difference ('scale') variable yielding an autocorrelation density. The autocorrelation density may or may not retain all the correlation information, but substantial reduction of statistical noise yields an improved S/N ratio. The next integration, on the scale or difference variable, is within a factor the *total variance*. The relative signal amplitude for this differential fluctuation measure may be as high as 10-20 *percent*.

In summary, the total variance (a fluctuation measure) is the definite integral of an autocorrelation density (a correlation measure) which is in turn the projective integral of a two-point density. This integration sequence provides the algebraic relationship between fluctuation measures and correlation measures. Greater precision may be gained by integration at the possible cost of loss of correlation information. A comprehensive analysis considers all aspects of the correlation/fluctuation system over all accessible scales. The basic strategy is projection from a multidimensional multiparticle momentum space to visually and algebraically accessible subspaces which best reveal significant correlation structures. We now proceed with a fluctuation analysis based on variance differences and proceed by several steps to the correlation structure of two-point distributions.

VARIANCE COMPARISONS

The left panel in Fig. 1 represents the random partition of a space containing a uniform measure density and a random point set ('particles'). The contours represent the probability distribution of partition-element contents on $(m/\hat{m}, n)$, where $\hat{m}$ is a single-particle ensemble mean. This joint distribution is the basic reference for the central limit theorem. If partition elements of a specific 'volume' are selected as a condition the resulting 2D peaked conditional distribution on $(m/\hat{m}, n)$ represents a 'grand canonical' ensemble. If partition elements with a specific point number are selected the result is a 'canonical' ensemble. Its 1D conditional distribution on measure m is a gamma distribution. Alternatively, if partition elements with a fixed amount of measure m are specified the resulting 1D conditional distribution on point number n is a Poisson distribution.

If the continuous measure is $m \to p_t$, the total transverse momentum in a heavy-ion collision event or subset thereof, the mean $<p_t> \equiv p_t/n$ is a temperature estimator. It is then tempting to transform the CLT reference distribution in the left panel to the space $(<p_t>, n)$ to be used as a reference for event-by-event 'temperature' fluctuations. However, *ratios* of random variables exhibit measure bias, wherein mean values of statistical estimators are systematically displaced from the parent value they estimate (in this case a parent 'temperature'). The underlying bias mechanism is variation of the distribution width with event multiplicity and consequent dependence of the $<p_t>$ variance on multiplicity fluctuations. To minimize measure bias one must design statistical measures which minimize sensitivity to multiplicity fluctuations. The *difference* variable $(m/\hat{m} - n)/\sqrt{n}$ with corresponding reference distribution shown in the middle panel of Fig. 1 is minimally biased. The additional factor $\sqrt{n}$ greatly reduces sensitivity to multiplicity fluctuations: the distribution width in the center panel is by construction independent of n.

The distribution in the right panel of Fig. 1 is on the equivalent p_t variable $\sqrt{n}(<p_t> - \hat{p}_t) = (p_t - n\hat{p}_t)/\sqrt{n}$ and represents real-data $<p_t>$ fluctuations with 10% excess *rms* relative width (upper solid curve) compared to a CLT reference gamma distribution (lower solid curve). Also shown is a gaussian distribution with the same reference *rms* width (dashed curve) for comparison: the gamma distribution and data have nonzero higher cumulants, especially skewness, whereas the gaussian by definition does not. The variance comparison measure corresponding to this case is

$$\Delta\sigma^2_{p_t,n} \equiv \overline{n(<p_t> - \hat{p}_t)^2} - \sigma^2_{\hat{p}_t} \qquad (1)$$

which is minimally biased for heavy-ion collisions. Other variance comparisons with the same structure are: $\Delta\sigma^2_n \equiv \overline{(n-\bar{n})^2}/n - 1$, $\Delta\sigma^2_{n_+,n_-} \equiv \overline{(n_+ - n_-)^2/(n_+ + n_-)} - 1$ and $\Delta\sigma^2_{p_t} \equiv \overline{(p_t - \bar{n}\hat{p}_t)^2/n} - \sigma^2_{\hat{p}_t}$. The overlines in these expressions indicate averages over all bins in an event ensemble, where a bin may be the entire detector acceptance, or fractions of the acceptance at smaller scale as discussed in the next section. This approach is bin-based rather than event-based to provide a more general analysis framework compatible with scaling studies.

The alternative p_t fluctuation measure $\overline{(<p_t> - \hat{p}_t)^2} - \sigma^2_{\hat{p}_t}/\bar{n}$ [3] based on the $<p_t>$ *ratio* seems to be a more straightforward differential variance measure, but exhibits a multiplicative bias of up to *two orders of magnitude* for the low multiplicities encountered with small-scale binnings or peripheral AA collisions. No measure is bias free, and the amount of bias depends on correlation details.

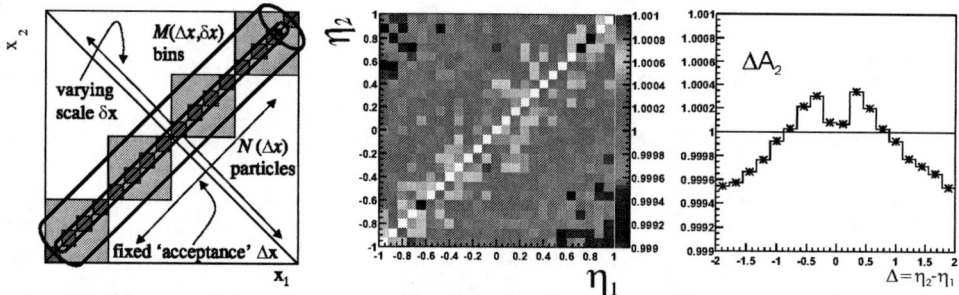

FIGURE 2. The left panel shows a two-point space partitioned by bins at two scales. The integration volume for a correlation integral is defined by two coaxial cylinders approximated by the scaled binnings. The center panel shows a two-point density on $\eta \otimes \eta$ space, illustrating lack of structure on the sum variable (η_Σ) and substantial structure on the difference variable (η_Δ). The right panel shows the projection of the two-point density onto the difference variable ($\Delta = \eta_1 - \eta_2$): the autocorrelation density ratio.

THE CORRELATION INTEGRAL

Statistical analysis of collision data is based on binning a primary space (*e.g.*, a subset of the single-particle momentum space), typically bounded by a detector acceptance aperture or simply acceptance. The bin size is the 'scale' of the binning, ranging from some limiting resolution scale to the acceptance scale. In a 'vertical' analysis a single bin centered on the acceptance is varied in size, in a 'horizontal' analysis multiple bins cover the full acceptance at each scale. This terminology is derived from factorial-moment analysis. The previous section describes a variance comparison at the acceptance scale, which can then be extended to study the scale dependence of fluctuations and probe more detailed aspects of two-particle correlations. A direct relationship between fluctuations and correlations can be established through the correlation integral.

The horizontal-analysis binning strategy approximates a correlation integral [4] in a q-tuple space, as shown for pairs in the left panel of Fig. 2, where x represents a momentum component: m_t, η, ϕ. In this two-particle space the accessible interval Δx on the sum variable is determined by the detector acceptance. The difference or scale variable δx is the bin size. The integration domain of a correlation integral is bounded by cylinders (binnings) at two scales. The normalized correlation integral for an object distribution from the resolution scale to scale δx within acceptance Δx is approximated by bin distribution moments in the form [5]

$$C_{2,obj}(\Delta x, \delta x) \approx 1/M(\Delta x, \delta x) \cdot \overline{m^2}(\delta x)/\overline{m}^2(\delta x) \qquad (2)$$

If the distribution is uncorrelated then $\overline{m^2} = \overline{m}^2$, and we have the reference integral $C_{2,ref}(\Delta x, \delta x) = 1/M(\Delta x, \delta x)$. The mean bin multiplicity is $\bar{n}(\delta x) = N(\Delta x)/M(\Delta x, \delta x)$, and the single-particle inclusive mean is denoted $\hat{m}$. With these definitions we can express a CLT variance comparison in terms of correlation integrals and relate it algebraically to two-particle densities, particularly the autocorrelation density.

$$\Delta\sigma_m^2(\delta x_1, \delta x_2) = \hat{m}^2(N(\Delta x) - 1) \cdot \Delta C_2(\Delta x, \delta x_1, \delta x_2) \qquad (3)$$

where the last factor is the difference between object and reference correlation integrals over the scale interval $[\delta x_1, \delta x_2]$.

For a distribution on the two-point space $x_1 \otimes x_2$ transformation to linear combinations $x_\Sigma \equiv x_1 + x_2$ and $x_\Delta \equiv x_1 - x_2$ simplifies a correlation analysis, because correlation structure in nuclear collisions is typically more slowly varying on the sum variable and more rapidly varying on the difference variable, with reflection symmetry about the latter (cf Fig. 2, middle panel). The projection of a two-point density $\rho_{2,m}$ onto the difference variable x_Δ is the autocorrelation density A_2, which may retain most or all of the correlation information in the two-point density while substantially reducing statistical noise. We can write the correlation integral in terms of the autocorrelation density as

$$\begin{aligned} C_{2,obj}(\Delta x, \delta x) &\equiv \int_0^{\Delta x} dx_\Sigma \int_0^{\delta x} dx_\Delta\, \rho_{2,m}(x_1, x_2) \Big/ \int_0^{\Delta x} dx_\Sigma \int_0^{\Delta x} dx_\Delta\, \rho_{2,m}(x_1, x_2) \\ &= \int_0^{\delta x} dx_\Delta\, A_{2,obj}(\Delta x, x_\Delta) \Big/ \int_0^{\Delta x} dx_\Delta\, A_{2,ref}(\Delta x, x_\Delta) \end{aligned} \qquad (4)$$

The correlation integral provides data compression and dimension reduction (making visualization easier or even possible), often with little loss of information. The variance difference can then be expressed in terms of a scale integral of the autocorrelation density

$$\Delta\sigma_m^2(\delta x_1, \delta x_2) = 1/N(\Delta x) \cdot \int_{\delta x_1}^{\delta x_2} dx_\Delta\, \Delta A_2(\Delta x, x_\Delta) \qquad (5)$$

where $\Delta A_2 = A_{2,obj} - A_{2,ref}$ and $\int_0^{\Delta x} dx_\Delta\, A_{2,ref}(\Delta x, x_\Delta) \to N(N-1)\hat{m}^2$. This provides a direct connection between fluctuation analysis and correlation analysis through the autocorrelation density, and aids in the physical interpretation of fluctuation measurements.

LÉVY DISTRIBUTIONS AND 1D P_T SPECTRA

Variance excess revealed in a CLT variance comparison is a manifestation of net two-particle correlations in an object distribution relative to a reference, as we have seen. We can also ask about the structure of the single-particle distribution itself which may reveal further aspects of incomplete equilibration. Exponential Boltzmann factors describe a fully equilibrated system in which a uniform continuous measure is randomly partitioned, a result of equilibration.

Recently, G. Wilk et al. [6] have introduced the Lévy distribution to describe processes in which a *correlated* continuous measure is randomly partitioned, for example decaying systems with fluctuating lifetimes, particle ranges with fluctuating cross sections and Boltzmann factors for partially-equilibrated many-body systems

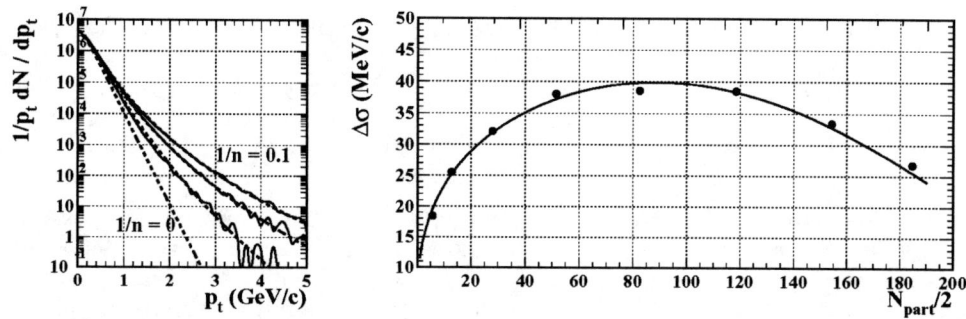

FIGURE 3. The left panel shows four Lévy distributions (dashed curves) and three Hijing p_t distributions (solid curves) for central Au-Au collisions under three conditions: jets on - no quench (top), default and jets off - Cronin on (bottom). The Lévy curve for $1/n = 0$ is a Maxwell-Boltzmann distribution for comparison. The right panel shows a preliminary STAR result for the centrality dependence of nonstatistical p_t fluctuations in Au-Au collisions [8].

– systems in which a dissipative process is substantial but *incomplete* and residual correlations remain. In each case a sample distribution deviates from its expected exponential Boltzmann form to that of a Lévy distribution. The exponent parameter n of the Lévy distribution represents fluctuations in the sampled measure (described by a gamma distribution) with relative variance given by $1/n \approx \sigma_m^2/\bar{m}^2$.

p_t distributions for high-energy collisions can be described approximately by

$$1/p_t \, dN/dp_t = C / \left(1 + \beta_0 \sqrt{m_0^2 + p_t^2}/n(p_t,\sqrt{s})\right)^{n(p_t,\sqrt{s})} \quad (6)$$

which is a Lévy distribution if the exponent parameter n is independent of p_t. The Lévy distribution represents a pathway state in the equilibration process resulting from substantial but incomplete equilibration. If $1/n \to 0$ the equilibrium exponential Boltzmann factor results. More generally, a p_t distribution may be more structured, with a soft-QCD ('thermal') component at lower p_t and separate perturbative-QCD-related 'power-law' component - $(p_t/p_0)^n$ - at higher p_t, requiring a p_t- and $\sqrt{s}$-dependent exponent as in Eq. (6). The 'power-law' p_t dependence encountered in very high energy unequilibrated elementary collisions should not be confused with the Lévy distribution. A careful differential analysis with a Lévy-distribution reference can reveal important details about the equilibration process in complex A-A collisions.

Fig. 3 shows in the left panel Hijing distributions [7] for central events and three different simulation conditions (see figure caption) and four Lévy distributions, three of which well describe the Hijing results. The fourth (lowest) curve is included to illustrate the Maxwell-Boltzmann limit of the Lévy distribution for $1/n \to 0$. For $1/n \approx 0.1$ we have $\sigma_T/\bar{T} \approx 0.3$: relative temperature variation within each event is roughly 30 *percent*. Thus, one could conclude from these Hijing results that central heavy ion collisions at RHIC energies are fluctuation dominated, but substantial dissipation is present. Part of EbyE analysis involves understanding the structure of these intra-event fluctuations and distinguishing them from true event-wise temperature fluctuations.

CENTRALITY DEPENDENCE

Correlation structure observed in the final state may have multiple sources, including initial-state scattering and critical correlations. The centrality dependence of correlation structure is an essential diagnostic tool for distinguishing among sources. If excess fluctuations are generated by stochastic initial-state binary collisions there are specific implications for centrality dependence. On the other hand, central events may contain critical correlations associated with a phase transition which do not occur in peripheral events. An expectation for initial-state scattering can be derived from the Glauber model. From a Glauber analysis one observes that the number of binary collisions varies approximately as the 1/3 power of the participant number (actually $\approx N_{part}^{0.355}$). Thus, a component of nonstatistical fluctuations with this centrality dependence would suggest stochastic initial-state binary collisions as one source of final-state correlations.

We generalize the two-point correlation-space concept to include initial-state nucleons in a two-nucleon correlation space, with indices n_a, n_b for nucleons in the two A-A collision partners. Centrality variation reveals the correlation structure in this space, suggested by terms like 'wounded nucleon' and 'binary collision' which characterize different 'correlation lengths.' This picture of centrality dependence extends the concept of correlation integral to the initial state. We can then generalize the CLT variance comparison to include a scale integral in the initial-state participant-index space

$$\Delta \sigma_m^2(N_{part}, \delta x_1, \delta x_2) \approx \hat{m}^2 dN/d\eta|_{pp} R_{AA/pp}(N_{part}) \cdot \int_0^{N_{part}} d(n_b - n_a)$$
$$\int_{\delta\eta_1}^{\delta\eta_2} d(x_2 - x_1) \Delta A_2(n_b - n_a, x_2 - x_1)/\bar{A}_2 \quad (7)$$

interpreting centrality variation as a form of scale variation.

EXPERIMENT COMPARISONS

An example of the centrality dependence of excess p_t fluctuations is shown in the right panel of Fig. 3. We observe with the STAR detector at RHIC a significant excess of charge-independent p_t fluctuations with nontrivial centrality dependence [8]. There is then the question how to compare this result with previous SPS observations and with other experiments at RHIC. The difficulty with such comparisons lies in the complex energy and acceptance/scale dependencies of correlations. There is no simple, model-independent comparison method. Each comparison is effectively a model test. We can however adopt certain practices to simplify the comparison problem.

First, we can normalize A-A centrality variations to p-p results inferred from extrapolation of peripheral A-A data. A-A results can thus be factored into A-A/p-p ratios for centrality dependence and p-p data obtained directly and from A-A extrapolations, just as in particle production studies. This method approximately factors $\sqrt{s}$ variations from the centrality variation of correlations at fixed $\sqrt{s}$. It is important to use an A-A extrapolation rather than p-p collision results to maximize common-mode reduction of detector systematics, which change with different beams and operating conditions. Detectors with different acceptances make measurements at different scales, and correlation physics is scale dependent. Thus, simply scaling correlation data by multiplicity or acceptance aperture sizes to make experiment comparisons is incorrect. Measurements should be made for each detector at its maximum scale aperture (acceptance) and possibly over a range of smaller scales. Simple consistency checks between experiments can be made on *overlapping* scale intervals. But, comparisons between different scales, whether for one detector or several, test physical models of correlation scale dependence.

THE BASIC η-ϕ TWO-POINT SPACES

Two-particle correlations in p-p axial phase space have been studied extensively with CERN ISR and Fermilab fixed-target experiments [9], and provide limiting cases for the centrality dependence of two-particle correlations in A-A collisions. These structures serve as an extrapolation limit for peripheral A-A. Previous p-p studies are incorporated phenomenologically in Lund-model string fragmentation rules. Whereas correlation effects are of order unity in p-p collisions, the effects we observe in central A-A collisions are at the *permil* level because of combinatoric dilution in high-multiplicity events. Great care is thus required in constructing correlation measures and references. Events must be ordered by global characteristics to construct multiparticle references. This technique maintains an essential distinction between *extramural* correlation structure among events and *intramural* correlation structure within events. To minimize artifacts two-particle density ratios are formed from sibling pairs and from mixed pairs formed from 'nearest neighbor' events in an event characteristic space.

Correlations are studied for like-sign and unlike-sign pairs. We construct the basic two-particle momentum spaces $m_t \otimes m_t$, $\eta \otimes \eta$ and $\phi \otimes \phi$ for each particle charge combination. Spaces are constructed by first forming inclusive sibling- and mixed-pair densities on a two-point space from the same event set using event ordering [10], so that mixed pairs are formed from 'nearest-neighbor' event pairs in a space spanned by several global event variables such as multiplicity and primary vertex position. The density ratio of sibling to mixed pairs is then formed. Deviations from unity in this ratio are typically 1 *permil* for central A-A events. Results for $\phi \otimes \phi$ and $\eta \otimes \eta$ are shown in Fig. 4 (first two panels) for central Au-Au collisions in the STAR detector. The ϕ result clearly shows the dominant $cos(2\phi_\Delta)$ contribution from elliptic flow [11].

As noted above, most of the structure in these distributions appears on the

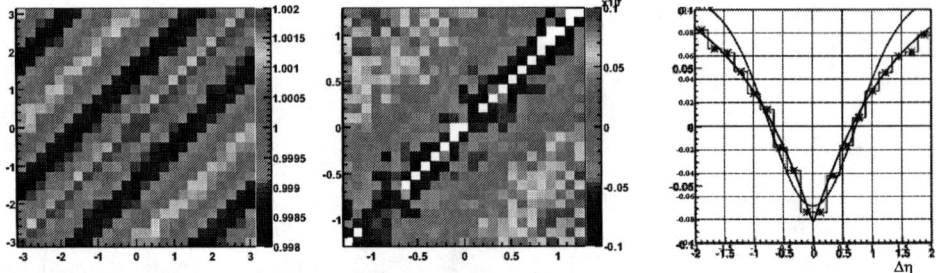

FIGURE 4. The left and center panels are sibling/mixed two-point density ratios for central Au-Au and for $\phi \otimes \phi$ and $\eta \otimes \eta$ spaces respectively. Refer to the text for discussion of charge combinations. The right panel is a projection of the center panel onto its difference variable - the autocorrelation density. The autocorrelation (histogram and points) is compared to gaussian and exponential distributions (curves).

difference variables ϕ_Δ and η_Δ. Thus, projection onto these variables retains most of the correlation information, improves statistical S/N and provides the dimension reduction required for practical visualization. An autocorrelation density difference for the $\eta \otimes \eta$ space is shown in the right-most panel.

CHARGE PAIRS AND JOINT AUTOCORRELATIONS

Two-point spaces can be formed for each of four combinations of + and − charges. However, these can be combined to give a more compact representation. Under the assumption of charge symmetry (++ = −− and +− = −+) we can combine distribution pairs to form like-sign (LS) and unlike-sign (US) distributions. This facilitates comparison with previous p-p analyses which also adopted this procedure. However, the linear combinations 'charge independent' (CI ≡ LS + US) and 'charge-dependent' (CD ≡ LS − US) are better able to distinguish different correlation mechanisms. The left panel in Fig. 4 represents the CI charge combination for $\phi \otimes \phi$ while the center panel represents the CD combination for $\eta \otimes \eta$. The right panel represents the autocorrelation density ratio for the CD combination in the $\eta \otimes \eta$ space: the projection of the center panel onto its difference variable η_Δ. The so-called balance function [12] is an approximation to this autocorrelation ratio based on the conditional-distribution concept.

Given that most of the correlation information is retained in 1D autocorrelation densities we can form 2D *joint* autocorrelations taking two-point spaces in pairs: $\eta \otimes \phi$, $\eta \otimes m_t$ and $m_t \otimes \phi$. The joint autocorrelations for CI and CD charge combinations maximize the dispersion of different correlation mechanisms within the practical limitations of visualization and numerical analysis. Examples are shown in the first two panels of Fig. 5 for CD and CI combinations of the $\eta \otimes \phi$ joint autocorrelation. ϕ_Δ is the vertical variable in each panel and η_Δ is the horizontal variable. New correlation structures appear in these 2D joint distributions which are not visible in the individual autocorrelations on a single difference variable.

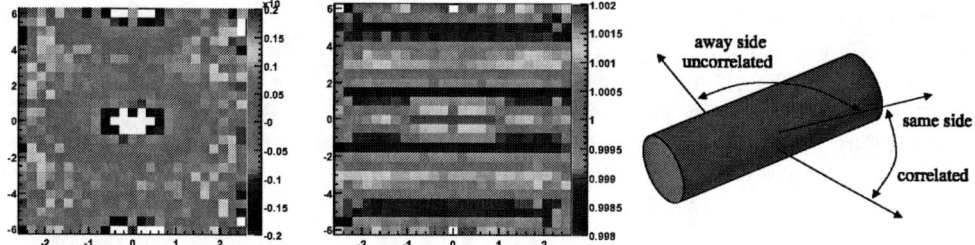

FIGURE 5. The left and center panels are joint autocorrelation density ratios on $\phi \otimes \eta$ for CD and CI charge combinations observed with the STAR detector [13]. The right panel is a sketch of a boost-invariant collision volume which defines away-side and same-side pairs.

Detailed examination of the correlation structure in Fig. 5 indicates that there is qualitatively different structure for away-side pairs and same-side pairs as defined by the relative azimuthal emission angle of the pair. As illustrated in the right panel of this figure, same-side pairs fall in the interval $|\phi_\Delta| \leq \pi/2$, while away-side pairs fall in the complementary interval $\pi/2 < |\phi_\Delta| \leq \pi$. We see in Fig. 5 that there is little CD or CI correlation structure for away-side pairs in central Au-Au collisions, whereas there are substantial structures evident for same-side pairs. This trend depends only weakly or not at all on the pseudorapidity difference η_Δ. It appears to be a global property of central collisions over a pseudorapidity difference interval of at least two units. This result suggests the presence of source opacity during a substantial part of the kinetic freezeout interval. The observed azimuthal anisotropy in two-particle angular correlations near mid rapidity (more or less correlation for same-side or away-side pairs) is apparently caused by scattering of one pair partner in the interior of the approximately boost-invariant collision volume after chemical freezeout.

PHYSICS OF E-BY-E ANALYSIS

The heavy ion collision is a violent and incomplete equilibration process in which correlation structure is both generated and attenuated by competing processes in the overall trend to greater entropy and less correlation. As with other aspects of nature, the large-scale entropic trend toward reduced correlation may be served locally by *increased* correlation (*e.g.*, turbulence vortices, biological organisms, hadron formation). A general collision model is that of a complex cascade where dynamical degrees of freedom vary over time and space, from nuclei to nucleons to partons to prehadrons to hadrons. During much of the collision a description in terms of specific particle types is not efficient. A more generic and portable basis is required: the abstract concept of correlation.

We consider several categories of correlation sources: initial-state binary collisions over a range of scales (Cronin N-N scattering, parton scattering, nucleus-nucleus scattering), quantum correlations, hadronic resonances, critical correlations, each of which has a different scale of influence, different centrality dependence and different $\sqrt{s}$ dependence. Correlations are also attenuated by rescattering in partonic and hadronic matter, which can be described as a dissipative medium, and by rapid traversal of the phase boundary, in which correlation is both attenuated and enhanced in a coalescence process - again a balance between global entropy increase and local entropy reduction.

CONCLUSIONS

Until recently, the description of heavy-ion collisions has been based on classical many-body concepts: a multiparticle phase space admitting different chemical species and described by cascade Monte Carlos, equilibrium statistical models or hydrodynamic continuum models. These asymptotic cases are not valid in general, although they provide essential guidance. As precision experimental results are forthcoming these models must fail in the details. In the last few years, event-by-event analysis has emerged as an alternative model-independent approach in which the collision final state is characterized phenomenologically by its correlation content, both event-wise and on an ensemble basis. For complex systems this may be the ultimate approach, a broader context in which particles and continua are limiting cases. First RHIC heavy-ion collisions have shown a wealth of correlation detail in the initial stages of analysis whose interpretation is just beginning.

ACKNOWLEDGMENTS

I thank the PASI Board of Directors for their kind invitation to participate in a very stimulating physics environment and the Brazilian local organizing committee for their warm hospitality in an excellent venue. This work was supported in part by USDOE grant DE-FG03-97ER41020

REFERENCES

1. "Probing Small Length Scales in Heavy Ion Collisions with Event-by-event Correlation Analysis," T.A. Trainor New Frontiers in Soft Physics and Correlations on the Threshold of the Third Millennium, Torino, IT, 12-17 June, 2000, Eds. A. Giovannini and R. Ugoccioni, Nucl. Phys. B (Proc. Suppl.) (2001) 92.
2. "Recent Results from STAR EbyE Analysis at RHIC," T.A. Trainor, Proceedings of the 17th Winter Workshop on Nuclear Dynamics, Park City, UT, 10-17 March, 2001, Eds. W. Bauer and G. Westfall, Kluwer Academic (New York).
3. S. A. Voloshin, V. Koch and H. G. Ritter, nucl-th/9903060.
4. P. Lipa, H. C. Eggers and B. Buschbeck, hep-ph/9304208, Phys. Rev. D **53**, 4711 (1996).
5. T.A. Trainor, hep-ph/0001148.
6. G. Wilk, Z. Wlodarczyk, hep-ph/0002145, Phys. Rev. Lett **84** (2000) 2770.
7. X. N. Wang and M. Gyulassy, nucl-th/9502021, collisions," Phys. Rev. D **44**, 3501 (1991).
8. J.G. Reid (STAR Collaboration), Nucl. Phys. **A698** (2002), 611c-614c and private communication.

9. T. Kafka *et al.*, Phys. Rev. **D16** (1977) 1261, J. Whitmore, Phys. Repts. **27** (1976) 187-272.
10. J.G. Reid, T.A. Trainor, Nucl. Inst. and Meth. **A457** (2001) 378-383.
11. K. H. Ackermann *et al.* [STAR Collaboration], nucl-ex/0009011, Phys. Rev. Lett. **86**, 402 (2001)
12. S. A. Bass, P. Danielewicz and S. Pratt, nucl-th/0005044, functions," Phys. Rev. Lett. **85**, 2689 (2000).
13. A. Ishihara (STAR collaboration), private communication.

LATIN-AMERICAN
RESEARCH

QCD/QGP in Latin America 2002

Erasmo Ferreira

Instituto de Fisica, U.F.R.J., Rio de Janeiro

Abstract. We report on some aspects of the activities in Hadron Physics and Relativistic Nuclear Physics in some Latin American countries.

Latin America is a continent with large distances, many nations — some big, some small — formed with people with many common cultural similarities, and nearly the same language, but is not at all homogeneous. Specifically, in Physics research activities there are huge differences.

The distances are large, and connections by travelling are very expensive, particularlay in view of the general scarcity of funds. Collaboration inside the geographical area is painful and inefficient. The Web now helps a lot, but it has not yet become available to all.

In our branch of physics, it is difficult to grow up to international level, starting from zero. Small groups, based on isolated individuals, can start and disappear after some time. The concentration of experimental means in a few laboratories require mobility and a minimum of local support, which is often outside reality.

Academic positions are not open to young physicists that work in this field, because small institutions are afraid of having to sustain groups with requirements of frequent international contacts. Thus these institutions decide rather to allow the development only of condensed matter, materials, classical subjects, viewing easier and immediate applications.

The formation of hadron physics communities occur in very concentrated forms. For example, in enormous Brazil, hadron physics groups practically exist in only 3–4 cities. The attraction of students to the area suffers from the rather dark perspectives they observe. In Brazil we may count 0.44 students for one physicist in the area, while the average of all areas is 0.64 students per physicist.

Of course these characteristic features of high energy physics are not exclusive of Latin America, but the crucial problem is to reach the critical point for sustained development. In a region with only small groups, even trivial discontinuities may cause a rupture with the past effort of construction. Programs in national, continental and international scale are then crucial for permanent self-sustained development.

As for continental organization, there is an institution, called CLAF — Centro Latino Americano de Fisica (Latin American Center for Physics) — which is a multinational body, with 13 member states, connected to UNESCO in its origin. CLAF becomes 40 year old in this year 2002, and has played along these years an important role in the coordination and implementation of activities in all these countries. Unfortunately CLAF has never had enough money to correspond to all expectations, but has done a lot in running systems of scholarships, sponsoring meetings (such as the present one), talking to governments (for instance, about the importance of science for development!), providing advices, and so on. The present Director of CLAF, Luis Masperi, is a man of our area, gave a seminar on Strangelets and a talk about CLAF programs during the meeting.

Unfortunately only a rather small number of Latin American physicists (except for Brazilians, of course) attended this meeting. The few ones who were present had a chance to describe to the other participants what is going on in their countries.

In general the situation is not brilliant. In the smaller countries, due to lack of a critical mass needed for self nourishing and continuity, and also due to economic and political effects, research groups that once existed, are barely remembered today. This is particularly true of nuclear/particle physics areas, and fortunately some progress has ocurred in condensed matter and other more accessible areas. In many countries, graduate studies at the PhD level are not offered. Young people that obtain doctor degrees in other countries are unable to return for a proper job, or do so changing the field of work. This is too frequent and dramatic for those few who have worked in experiments in the big laboratories.

In spite of these difficulties, we cannot accept that the situation is hopeless for physicists belonging to the small Latin American countries who wish to work in areas which are essentially based on accelerator experiments. The fatal difficulties to obtain scientific information are now dramatically reduced with the web communication (although this is still very poor in many places). Projects of national, regional or international scope, with regular funding lasting for some years should exist effectively, and reach a large enough number of physicists. ICTP is an example of a fruitful initiative in this direction.

EXPERIMENTAL PROGRAMS

We mention below some programs of participation of Latin-American physicists in experiments related to the main interest of this meeting. Several other programs are running well and actively, in high-energy physics, in Fermilab and CERN, and with the Auger project for cosmic rays. The information given here is certainly incomplete, and refers only to heavy ion collisions and related subjects. As far as we know, only Mexico and Brazil have groups participating officially in the international collaboration projects in RHIC/LBL and LHC/CERN.

STAR at RHIC

A group (five physicists, led by A.S. de Toledo) of the Nuclear Physics Department of University of São Paulo is an official member of the STAR (Solenoidal Tracker at RHIC) project, designed to study the plasma formed in heavy ion collisions. The group develops and implements the Silicon Drift Detectors (SVT), working also in the data analysis to identify strange mesons and baryons. Other activities of the group are in the reconstruction of particle trajectoties with the TPC (Time Projection Chamber) and in the study of particles with large transverse momentum using the STAR calorimeter. Previously, and until now, the group collaborates in the fixed target experiments E864 and E896 of AGS.

PHOENIX at RHIC

This project gives continuation to the on-going collaboration LIP–BNL between the Laboratório de Instrumentação e Partículas (LIP) of the University of São Paulo, the Brookhaven National Laboratory (BNL), and the European Laboratory for Particle Physics (CERN), in

i) the study of nucleus-nucleus collisions at ultra-relativistic energies (Project PHENIX–BNL), and

ii) the development of instrumentation for high energy physics (Project PHENIX–BNL and Project ATLAS–BNL/CERN).

The studies of nucleus–nucleus collisions at ultra-relativistic energies — an extension to considerably higher energies of the studies done at AGS energies by experiments such as E814 and E877, of which the LIP group was a member — will be conducted with the large detector of Project PHENIX installed at one of the crossing points of BNL's Relativistic Heavy Ion Collider (RHIC). The primary goal of the PHENIX collaboration is the detection and characterization of the quark-gluon plasma through the detection of photons, hadrons and lepton pairs.

Physics topics include searches for Quark-Gluon Plasma signals through light vector meson production, inclusive particle spectra, J/ψ suppression, Hanbury-Brown-Twiss source size measurements and particle multiplicity vs. energy density.

The instrumentation development includes

a) Time Expansion Chambers for Project PHENIX;

b) Liquid Argon ionization calorimeters for Project ATLAS;

c) Signal processing systems for electronic detectors in a high rate environment for Projects PHENIX (TEC Chambers) and Project ATLAS (EM Barrel Calorimeter).

ALICE at LHC

A Mexican group is an official participant of the LHC/CERN project ALICE, that will study heavy ion physics. They have the responsability to build the VOL detector, which is a set of fast scintillators to be used as a trigger in the experiment. The group will also work in the theory of ultrarelativistic heavy ions.

Members of the group, and their roles are as follows.

1) Instituto de Ciencias Nucleares, UNAM — Alejandro Ayala, Juan Carlos DOlivo, Eleazar Cuautle, Lukas Nellen. Theoretical description of experiments. Collaboration envisaged with other groups, both experimental (at RHIC) and theoretical.

2) Instituto de Fisica, UNAM — Arturo Menchaca, Arnulfo Martinez, Ernesto Belmont. Construction of hardware (the VOL detector).

3) CINVESTAV Merida — Guillermo Contreras. Simulation and Monte Carlo studies for the physics processes in ALICE. Investigation of the possible observation of jets in Pb Pb (and other ions) collisions.

4) CINVESTAV Mexico — Gerardo Herrera Corral, Luis Manuel Montano. Construction of silicon detectors and theoretical studies in ultrarelativistic heavy ion physics. At the moment, investigation of the Λ polarization as a possible signature of the quark-gluon plasma.

5) Institute of Physics, Universidad Michoacana de San Nicolas Hidalgo — Luis M. Villasenor. Working in the electronics of the detector and in the simulation of its responses.

There is an agreement between CERN and CONACyT (Mexican government research agency), which may be used to support the specific collaboration in ALICE. A Millenium project, supported by the World Bank is in operation, giving the main support for the activities of our Mexican colleagues in ALICE.

While LHC is under construction, Mexican groups wish to participate in RHIC physics.

There is in Mexico a strong community of nuclear physicists, working in traditional topics of nuclear structure and nuclear reactions, without direct connection with QCD and QGP. ICN-UNAM is now interested to start experimental programs, and has decided to participate in ALICE. The implementation of this idea will have strong effect in the research activities in Mexico.

ATLAS at LHC

UFRJ (Brazil) works in CERN since 1988 in hadronic calorimetry and in ATLAS detector since the start of the project in the 90's.

The activities of the group in ATLAS are:

1) Eletronics: Responsability by design, production and tests of the analogical adder circuits to be used in the first level trigger in ATLAS hadronic calorimeter. These circuits are constructed and tested in Brazil.

2) Simulations: Development of algorithms for the identification of particles and patterns using neural networks in the hadronic calorimeter.

3) Software: Studies related to the GRID for the ATLAS detector ; development of tools for the remote data analysis through the web.

4) Trigger: development of algorithms for the first level trigger.

5) Studies of the possibility to investigate heavy ion collisions, specifically proton-Pb, with a center-of-mass energy 8850 GeV/nucleon.

THEORETICAL WORK ON HADRON PHYSICS AND HIGH ENERGY NUCLEAR PHYSICS

Theoretical research in hadronic and nuclear physics has well established tradition in the larger countries of the continent.

Mexico has strong theoretical groups in traditional low energy nuclear physics, and changes of research lines towards fundamental QCD and high energy processes occur steadily. An important effort in this direction has been described in the ALICE experiment session of this report.

Argentina in Buenos Aires and La Plata, and Chile in Valparaiso and Santiago, have small active groups doing research in hadron physics.

In Brazil the communities are of good size, with about 100 practioneers, and well organized. Two series of international workshops are being held for 13 years, besides regular internal collaboration meetings. The situation is described in detail in the proceedings of the workshop Hadron Physics 2000 (E.F., World Scientific Publ., 2000, p.219-227). Several theoretical groups work directly on the physics of relativistic heavy ion physics in Brazil. In particular, the collaboration between Rio de Janeiro and Sao Paulo has been running for more than 10 years. The main subjects are relativistic hydrodynamics and its numerical implementation (SPH method), mechanism of particle production (freeze-out), HBT interferometry, quantum transport theories, equation of state, statistical models, QCD sum-rules, effective theories of QCD such as chiral models, etc. Applications to high energy astrophysics and cosmology are also studied by some groups. Recently a lattice QCD program started. These collaborations are supported mainly by state agencies that support research (FAPESP, FAPERJ) and by the federal agency CNPq. In Rio de Janeiro, the groups are mainly in Federal University of Rio de Janeiro (UFRJ), Federal University of Fluminense (UFF), State University of Rio de Janeiro (UERJ) and Centro Brasileiro de Pesquisas Fisicas, accounting for approximately 20 researchers and corresponding graduate students.In São Paulo, groups are at the University of São Paulo (USP) and State University of São Paulo (UNESP-IFT), also with around 20 researchers and their graduate students. UNICAMP, in the city of Campinas (São Paulo State) has a group on hadronic collisions. In the south, there is a group in the Universidade Federal do Rio Grande do Sul, with ramifications in nearby centers. Smaller and active groups live in the states of Santa Catarina (UFSC) and Paraiba (UFPb).

ACKNOWLEDGMENTS

We thank the experimental colleagues and T. Kodama for essential information here presented.

Λ^0 Polarization and QGP Production in High-Energy Heavy-Ion Collisions

A. Ayala*, E. Cuautle*, G. Herrera[†] and L. M. Montaño[†]

*Instituto de Ciencias Nucleares, Universidad Nacional Autónoma de México, Apartado Postal 70-543, México Distrito Federal 04510, México.
[†]Centro de Investigación y de Estudios Avanzados, Apartado Postal 14-740, México Distrito Federal 07000, México.

Abstract. We study the polarization of Λ^0's produced in ultra-relativistic heavy-ion collisions. We argue that when the density of participants in the reaction is below the critical value for the production of a Quark-Gluon Plasma (QGP), recombination processes dominate the Λ^0 production cross section and hence, Λ^0's are polarized. However, when the density of participants is above the critical value for QGP formation, coalescence processes dominate, giving rise to a reduction in the Λ^0 polarization. For densities below the critical density for QGP production, we describe Λ^0 polarization in terms of the DeGrand-Miettinen model correcting for the effects introduced by multiple scattering of the produced Λ^0's within the nuclear environment.

INTRODUCTION

High-energy nuclear collisions are expected to produce states of matter subject to extreme conditions, the most exciting of such is the Quark-Gluon Plasma (QGP) where the elementary QCD degrees of freedom are not confined within the volume of a hadron but rather, over a larger volume on the order of the nuclear size. To identify the phase transition from ordinary nuclear matter to the QGP in this kind of collisions, a number of experimental observables have been proposed [1]. At the same time, it has also been recognized that no single signal can, by itself, provide clear cut evidence for the existence of the QGP and that it is only through the combined analysis of all possible signals that the production of such state of matter can be firmly established. This realization has stimulated the search for new probes to aid in the understanding of the properties of the complex environment produced in heavy-ion collisions at high energy.

Among the first proposed signatures to reveal the production of a QGP in high-energy nuclear collisions was the qualitative study of the change in the polarization properties of Λ^0 hyperons as compared to that observed in proton-proton collisions [2, 3]. Surprisingly, ever since the idea was put forward, there has been a lack of a model to quantitatively realize it in this kind of environment.

For proton-induced reactions, it has long been established [4] that the spin of the Λ^0 is carried by the produced s-quark and that the u- and d-quarks can be thought of as being coupled into a diquark with zero total angular momentum and isospin. A quantitative description of hyperon polarization properties in proton- and meson-induced reactions that includes this observation, has been attained by the semi-classical model of DeGrand and Miettinen [5] (see also Ref. [6]). In this model, the hyperon polarization is due to a

Thomas precession effect during the quark recombination process of slow (sea) s-quarks and fast (valence) ud-diquarks.

In the case of relativistic nucleus-nucleus collisions, the expectation is that, Λ^0's coming from the zone where the critical density for QGP formation has been achieved, are produced through the coalescence of independent slow sea u-, d- and s-quarks and are emitted via an evaporation-like process. Consequently, these *plasma* created Λ^0's should show zero polarization [3].

In this paper, we propose a method to extract information from Λ^0 polarization measurements in ultra-relativistic heavy-ion collisions as a means to determine the production of a QGP. We study a model where the Λ^0's produced in the zone where the critical density for QGP formation is reached are unpolarized while the Λ^0's produced in the rest of the reaction zone are produced polarized, in the same way as they are in free nucleon-nucleon reactions. We use the DeGrand-Miettinen Model (DMM) to describe the polarization of these latter Λ^0's. The effects on the polarization arising from multiple scattering are introduced in terms of a sequential model of collisions [7, 8]. We also discuss possible depolarization effects introduced by spin-flip interactions of Λ^0 within the nuclear medium. For a more elaborated discussion, we refer the reader to Ref. [9].

Λ^0 POLARIZATION IN THE DEGRAND-MIETTINEN MODEL

Since the discovery of a substantial Λ^0 transverse polarization in high energy hadron-nucleon [10], nucleon-nucleon [11] and nucleon-nucleus [12] reactions, a large amount of theoretical and experimental activity has been devoted to understanding the origin of such polarization. The fact that Λ^0's were produced with a significant polarization while $\bar{\Lambda}^0$'s were unpolarized was, at first, surprising. Several models were proposed to explain that phenomena. One of the most successful is the DMM [5]. In this semi-classical model, the Λ^0 polarization results from a Thomas precession of the spin of the s-quark in the recombination process. The u- and d-quarks are assumed to form a diquark in a state with total angular momentum $J = 0$ (and isospin $I = 0$) and thus, the s-quark is entirely responsible of the spin of the Λ^0.

The recombining quarks in the projectile must also carry a transverse momentum given, approximately by half of the transverse momentum of the outgoing hadron, while the other half is carried by the s-quark. Since the x distribution of the (sea) s-quarks peaks at very low values and is very steep, a Λ^0 must get most of its longitudinal momentum from the valence ud-diquark momentum. In the process, the s-quark is, on average, accelerated but, since its transverse momentum is different from zero, the force $\vec{F}$ felt by the quark is not parallel to its velocity $\vec{\beta}$, giving rise to Thomas precession. Notice that in the model, the polarization asymmetry arises as a consequence of a strong momentum ordering where the s-quark is (on average) slow and the ud-diquark is (on average) fast. This ordering does not happen when the recombination involves only sea s- u- or d-quarks or antiquarks, as is the case of antihyperon production or, more important to our purposes, a Λ^0 originated from a QGP.

The polarization of a Λ^0 in proton-proton collisions is given in the DMM by

$$\mathscr{P}^{\text{REC}} = -\frac{12}{\Delta x_0 M^2} \left(\frac{1 - 3\xi(x)}{[1 + 3\xi(x)]^2} \right) p_T, \tag{1}$$

where

$$M^2 = \left[\frac{m_D^2 + p_{TD}^2}{1 - \xi(x)} + \frac{m_s^2 + p_{Ts}^2}{\xi(x)} - m_{\Lambda^0}^2 - p_T^2\right] \quad (2)$$

with m_D, p_{TD} (m_s, p_{Ts}) the mass and transverse momentum of the *ud*-diquark (*s*-quark), m_{Λ^0} and p_T the mass and transverse momentum of the Λ^0, Δx_0 a distance scale, on the order of the proton radius, characterizing the recombination length scale and $\xi(x) \equiv x_s/x = (1-x)/3 + 0.1x$ the ratio of the longitudinal momentum fraction of the *s*-quark to the longitudinal momentum fraction of the Λ^0 with respect to the beam proton. Using this parametrization, DeGrand and Miettinen obtain a good description of experimental data [5]. If, on the other hand, use is made of a recombination model [13] and $\xi(x)$ explicitly computed, the results for the polarization do not show a drastic change. We therefore use here, for the sake of simplicity, the linear parametrization for $\xi(x)$ given by DeGrand and Miettinen. To complete the overall parametrization of $\mathscr{P}^{\text{REC}}$, we also take $m_D = 2/3$ GeV, $m_s = 1/2$ GeV and $p_{Ts}^2 = p_{TD}^2 = (1/4)p_T^2 + \langle k_T^2 \rangle$ with $\langle k_T^2 \rangle = 0.25$ GeV^2 [5].

Λ^0 PRODUCTION AND POLARIZATION IN ULTRA-RELATIVISTIC HEAVY-ION COLLISIONS

In a QGP the $s\bar{s}$ pair production can be reasonably well described by the lowest order QCD processes $q + \bar{q} \to s + \bar{s}$ and $g + g \to s + \bar{s}$ [14]. To simplify the analysis, we assume that in this environment, the Λ^0's are the sole products of the subsequent *s*-quark recombination. The enhancement of strangeness production in a QGP plasma leads to an enhancement of hyperon production. However, Λ^0 formation in this environment should not show a strong ordering for the momenta of any of the *u*-, *d*- or *s*-quarks, and consequently, according to the DMM, these Λ^0's should not be polarized [15].

On the other hand, in the interaction region where nucleon-nucleon collisions take place but the density is not high enough to deconfine quarks and gluons, Λ^0's would be produced by recombination of (ud)-diquarks, coming from the interacting nucleons and, *s* quarks coming from the sea.

The total cross section σ_{Λ^0} for Λ^0 production is then given by these two components; $\sigma_{\Lambda^0} = \sigma_{\Lambda^0}^{\text{REC}} + \sigma_{\Lambda^0}^{\text{QGP}}$

Λ^0 production from recombination

To describe Λ^0 production by recombination we write the production cross section at impact parameter b in the collision of ions A and B as

$$\frac{1}{\sigma_{\Lambda^0}^{NN}} \frac{d^2 \sigma_{\Lambda^0}^{\text{REC}}}{d^2 b} = T_{AB}(b), \quad (3)$$

where $\sigma_{\Lambda^0}^{NN}$ is the Λ^0 production cross section in nucleon-nucleon collisions, which we take as $\sigma_{\Lambda^0}^{NN} = 3.2$ mb [16] and $T_{AB}(b)$ is given by

$$T_{AB}(z,b) = \int d^2s\, T_A(z,\mathbf{s})\, T_B(z,\mathbf{s}-\mathbf{b}), \tag{4}$$

with T_A and T_B given in terms of

$$T_A(z,\mathbf{s}) = \int_{-z/2}^{z/2} \rho_A(z',\mathbf{s})dz' \tag{5}$$

extending the limits of integration over z in Eq. (5) to $[-\infty,\infty]$. For $\rho_A(\mathbf{r})$ we use the standard Woods-Saxon density profile

$$\rho_A(\mathbf{r}) = \frac{\rho_0}{1+e^{(r-R_A)/a}}, \tag{6}$$

with $R_A = 1.1 A^{1/3}$ fm, $a = 0.53$ fm and ρ_0 fixed by normalization giving $\rho_0 = 0.17$ fm^{-3} for ^{208}Pb.

We assume that each sub-collision produces final state particles in the same way as in free nucleon reactions and thus we can estimate the number of Λ^0's produced by recombination from Eq. (4). However, in order to exclude the zone where the density of participants n_p is above the critical density n_c to produce a QGP, we rewrite Eqs. (3) and (4) as

$$\frac{d^2\sigma_{\Lambda^0}^{REC}}{d^2b} = \sigma_{\Lambda^0}^{NN} \int T_B(\mathbf{b}-\mathbf{s})T_A(\mathbf{s})\theta[n_c - n_p(\mathbf{s},\mathbf{b})]d^2s, \tag{7}$$

where $n_p(\mathbf{s})$ is the density of participants at the point $\mathbf{s}$ in the transverse plane and θ is the step function.

The density of participants per unit transverse area during the collision of nucleus A on nucleus B, at an impact parameter vector $\mathbf{b}$, has a profile given by [17]

$$n_p(\mathbf{s},\mathbf{b}) = T_A(\mathbf{s})[1 - e^{-\sigma_{NN}T_B(\mathbf{s}-\mathbf{b})}] + T_B(\mathbf{s}-\mathbf{b})[1 - e^{-\sigma_{NN}T_A(\mathbf{s})}], \tag{8}$$

where σ_{NN} is the nucleon-nucleon inelastic cross section which we take as $\sigma_{NN} = 32$ mb. The total number of participants N_p at impact parameter b is

$$N_p(b) = \int n_p(\mathbf{s},\mathbf{b})d^2s. \tag{9}$$

Following the reasoning in Ref. [17], we choose $n_c = 3.3$ fm^{-2}, which results from the observation of a substantial reduction of the J/ψ yield in Pb – Pb collisions at the SPS.

Λ^0 production from a QGP

The average number of strange quarks produced in a QGP scales with the number of participants N_p^{QGP} in the collision roughly as [18]

$$\frac{\langle s \rangle}{N_p^{QGP}} = c N_p^{QGP}. \tag{10}$$

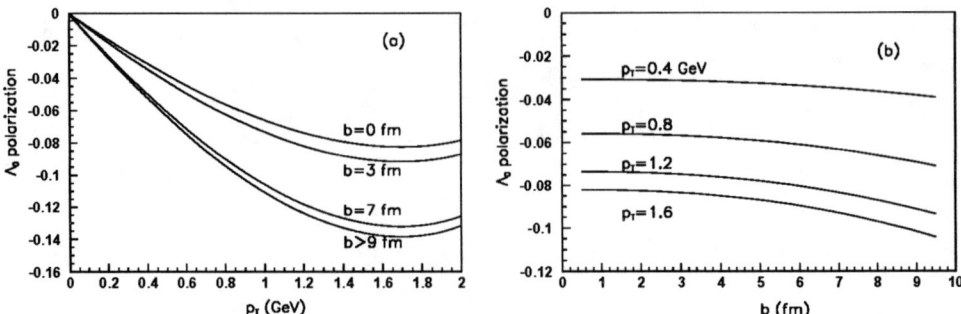

FIGURE 1. Λ^0 polarization by recombination (*a*) as a function of p_T for different values of impact parameter b and (b) as a function of b for different values of p_T.

Assuming for the sake of simplicity that, as a result of hadronization, only Λ^0's are obtained from these produced s-quarks, we can estimate the number of Λ^0's originating in the QGP. N_p^{QGP}, as a function of the impact parameter is given, using Eq. (9), as

$$N_p^{\text{QGP}}(b) = \int n_p(\mathbf{s},\mathbf{b})\theta[n_p(\mathbf{s},\mathbf{b}) - n_c]d^2s. \quad (11)$$

The proportionality constant c in Eq. (10) can be read off from Fig. 5a in Ref. [18]. Depending on the precise value of α_s and for ^{208}Pb – ^{208}Pb collisions, c is found in the range

$$0.001 \leq c \leq 0.005. \quad (12)$$

Therefore, we take Eq. (10) as representing the behavior of the differential cross section, $d^2\sigma_{\Lambda^0}^{\text{QGP}}/d^2b$, namely

$$\frac{d^2\sigma_{\Lambda^0}^{\text{QGP}}}{d^2b} = c\left[N_p^{\text{QGP}}(b)\right]^2. \quad (13)$$

Λ^0 polarization

Recall that the Λ^0 polarization asymmetry $\mathscr{P}$ is defined as the difference between the number of Λ^0's produced with spin pointing along and opposite to the normal of the production plane. In terms of the differential cross sections, $\mathscr{P}$ is given by [19]

$$\mathscr{P} = \left[\frac{d^2\sigma_{\Lambda^0\uparrow}}{d^2b} - \frac{d^2\sigma_{\Lambda^0\downarrow}}{d^2b}\right] \Big/ \left[\frac{d^2\sigma_{\Lambda^0\uparrow}}{d^2b} + \frac{d^2\sigma_{\Lambda^0\downarrow}}{d^2b}\right], \quad (14)$$

where $d^2\sigma_{\Lambda^0\uparrow}/d^2b$ and $d^2\sigma_{\Lambda^0\downarrow}/d^2b$ are the spin up and spin down (with respect to the normal of the production plane) differential cross sections, respectively. In the scenario where Λ^0's originate from two different processes, one must take into account the corresponding contribution to the polarization.

As we have argued, Λ^0's coming from the QGP regions are expected to be produced with their spins isotropically oriented and thus these do not contribute to the net Λ^0 polarization. We define

$$f(b) = \left[\frac{d^2\sigma_{\Lambda^0}^{QGP}}{d^2b}\right] \Big/ \left[\frac{d^2\sigma_{\Lambda^0}^{RBC}}{d^2b}\right] \qquad (15)$$

as the ratio of the differential cross sections for Λ^0 production from the QGP and from recombination processes.

Equation (14) can be written in terms of $f(b)$ as

$$\mathscr{P} = \frac{\mathscr{P}^{RBC}}{[1+f(b)]}, \qquad (16)$$

where

$$\mathscr{P}^{RBC} = \left[\frac{d^2\sigma_{\Lambda^0\uparrow}^{RBC}}{d^2b} - \frac{d^2\sigma_{\Lambda^0\downarrow}^{RBC}}{d^2b}\right] \Big/ \left[\frac{d^2\sigma_{\Lambda^0\uparrow}^{RBC}}{d^2b} + \frac{d^2\sigma_{\Lambda^0\downarrow}^{RBC}}{d^2b}\right], \qquad (17)$$

given in the DMM by Eq. (1), is the Λ^0 polarization that would be observed in the absence of Λ^0's produced by a QGP. Figure 1 shows the polarization obtained from Eqs. (1) and (16). Figure 1(a) shows the polarization as a function of p_T for different values of b. Figure 1(b) shows the polarization as a function of b for different values of p_T.

DEPOLARIZATION EFFECTS IN NUCLEUS-NUCLEUS COLLISIONS

In the case of nucleus-nucleus collisions, the effects that can possibly produce a diminishing of the Λ^0 polarization have been enumerated in Refs. [3]. In addition to Λ^0 production from a QGP these are: (i) secondary Λ^0's produced by pion-nucleon scattering and (ii) secondary scattering of leading Λ^0's with nucleons within the interaction zone.

Λ^0's originating from pion-nucleon scattering are mainly produced beyond the free nucleon-nucleon phase space kinematical limit and at low (laboratory) momenta. It is thus possible to set kinematical constraints in the reconstruction of these Λ^0's and therefore exclude them from the polarization analysis. Here, we concentrate on the effects introduced by secondary scattering of Λ^0's with nucleons in the zone where the density of participants is below the threshold for a QGP formation.

Secondary elastic scattering of Λ^0's with nucleons can influence the final polarization measurements by producing (a) a shift in the Λ^0 longitudinal momentum, (b) a shift in the Λ^0 transverse momentum and (c) a flip in the original Λ^0 spin direction. We proceed to quantitatively discuss each one of these effects.

Longitudinal momentum shift

Multiple elastic scattering can be cast in terms of a sequential model [7, 8], where it is assumed that the average energy of a particle after $n+1$ collisions is given in terms of the average energy after n collisions by

$$\langle E_{\Lambda^0}\rangle_{n+1} = (1-I)\langle E_{\Lambda^0}\rangle_n, \tag{18}$$

where I is the inelasticity coefficient [21]. In the high energy limit $(p \gg m_{\Lambda^0})$

$$\langle E_{\Lambda^0}\rangle_n \simeq \langle p_{\Lambda^0}\rangle_n = p\langle x\rangle_n \tag{19}$$

where p is the initial momentum value of the nucleon that produced (through recombination) the Λ^0 and x the fraction of the initial longitudinal Λ^0 momentum to the nucleon longitudinal momentum. Equations (18) and (19) can be combined to find the average value of x after n collisions as

$$\langle x\rangle_n = (1-I)^n x. \tag{20}$$

To find the average value $\langle x(z,b)\rangle$ after the produced Λ^0 has traveled a longitudinal distance z, in collisions with impact parameter b, recall that the distribution probability $P_n(z,b)$ for n-collisions in reactions with impact parameter b is Poissonian, that is

$$P_n(z,b) = \frac{1}{n!}\bar{N}^n(z,b)e^{-\bar{N}(z,b)}, \tag{21}$$

with

$$\bar{N}(z,b) = \sigma^{tot}_{\Lambda^0 N} T_A(z,b/2), \tag{22}$$

being the average number of collisions after the Λ^0 has traveled a longitudinal distance z, given in terms of the total Λ^0-nucleon cross section, which we take as $\sigma^{tot}_{\Lambda^0 N} \simeq 25$ mb and T_A given by Eq. (5). Notice that the argument of the function T_A in Eq. (22) referring to the location of the trajectory of the Λ^0 in the transverse planes has been approximated as the geometrical average in these planes.

Therefore, $\langle x(z,b)\rangle$ is given by

$$\langle x(z,b)\rangle = xe^{-I\bar{N}(z,b)}. \tag{23}$$

It can be shown [8] that $I = \lambda/2$, where $\lambda \simeq \sigma^{inel}_{\Lambda^0 N}/\sigma^{tot}_{\Lambda^0 N}$ with $\sigma^{tot}_{\Lambda^0 N}$ and $\sigma^{inel}_{\Lambda^0 N}$ the total and inelastic Λ^0-nucleon cross sections, respectively. Although these should be taken as the cross sections in nuclear matter, such information has not been experimentally obtained up until now. Nevertheless, from data at low and intermediate energies, it is known that the free cross sections follow the behavior of the corresponding nucleon-nucleon cross sections [22]. Assuming that this is also the case within nuclear matter, we extrapolate the existing data on free Λ^0-nucleon interactions to high energies [20] and obtain $\lambda \simeq 0.4$ which is consistent with the corresponding value obtained for nucleon-nucleon collisions within nuclear matter [8].

Transverse momentum shift

Let $Q(\mathbf{s_f}, \mathbf{s_i}, z)$ be the probability that the produced Λ^0 ends up traveling in the direction $\mathbf{s_f}$ after traveling a longitudinal distance z, having been produced moving in direction $\mathbf{s_i}$. Let $q_n(\mathbf{s_f}, \mathbf{s_i})$ be the probability that the Λ^0 ends up traveling in the direction $\mathbf{s_f}$ after n-collisions with nucleons, having been produced moving in direction $\mathbf{s_i}$. The vectors $\mathbf{s_f}$ and $\mathbf{s_i}$ can be thought of as two dimensional unit vectors, given the isotropy of the angular distribution in each collision. Since the distribution probability for n-collisions is given by Eq. (21), $Q(\mathbf{s_f}, \mathbf{s_i}, z)$ is given as [23]

$$Q(\mathbf{s_f}, \mathbf{s_i}, z) = \sum_{n=0}^{\infty} P_n(z, b) q_n(\mathbf{s_f}, \mathbf{s_i}). \tag{24}$$

The average change in the Λ^0 momentum direction is computed from the r.m.s value of the total dispersion angle, α, given by

$$\alpha = \Gamma \sqrt{\bar{N}(z, b)}, \tag{25}$$

where $\mathbf{s} = \mathbf{s_f} - \mathbf{s_i}$, since Q depends only on such difference. It is now a simple matter to express the average value of the Λ^0 transverse momentum after having traversed a longitudinal distance z within the nuclear medium. In the high-energy limit, where $\cos \alpha \sim 1$, we can write

$$\langle p_T(z, b) \rangle \simeq p_T e^{-I\bar{N}(z,b)} \cos\left\{ \Gamma \sqrt{\bar{N}(z, b)} \right\}, \tag{26}$$

where p_T is the transverse momentum that the Λ^0 carried originally and we have used Eq. (25).

Γ can be estimated using the data on angular distributions of free Λ^0-nucleon elastic scattering. For intermediate energies such data exists and Γ can be read off from the parametrization of the angular distribution for $p_{\Lambda^0} > 800$ MeV/c in terms of Legendre polynomials with $l = 1, 2, 3$ in Ref. [24]. By doing so it is easy to get that $\Gamma \simeq 1.2$ rad., which represents a large value. However, we expect that at higher energies, elastic dispersion takes place with smaller values of Γ. In fact, as the energy of the produced Λ^0 increases from about 1 GeV to a few tenths of GeV, its deBroglie wavelength $\bar{\lambda}$ decreases from about 0.1 fm to 0.01 fm. Since the transverse size d of the scatterers (nucleons) is about 1 fm, we can estimate that [23] $\Gamma \sim \bar{\lambda}/d = 0.01$. Hereafter, we use this value of Γ in our calculations.

To incorporate the effects of the shift in momentum experienced by Λ^0's traveling in the nuclear medium for densities below the critical density for QGP formation, we recall that, according to the above analysis, a Λ^0 produced within a given phase space cell labeled by the pair of values (x, p_T), is scattered into a different phase space cell, which, on average, is labeled by the pair of values $(\langle x \rangle, \langle p_T \rangle)$. Omitting from the analysis the spin-flip depolarization effects, the original Λ^0 polarization computed from Eq. (1) is preserved but corresponds (on average) to a detected Λ^0 with momenta labels $(\langle x \rangle, \langle p_t \rangle)$. This is shown in Fig. 2 where, for the purposes of comparison, we plot the polarization

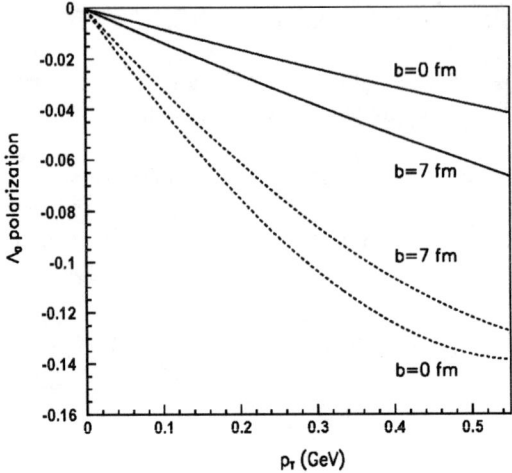

FIGURE 2. Comparison of Λ^0 polarization values with (lower curves) and without (upper curves) the transverse momentum shift due to multiple scattering in the nuclear medium below the critical density for QGP formation.

with and without considering the shift in transverse momenta due to multiple scattering, computed with the explicit values of $I = 0.2$ and $\Gamma = 0.01$ for two different values of the impact parameter b.

Spin flip

The existing data on Λ^0 production by meson-induced reactions on light nuclei [25] show that the effects on the Λ^0 polarization produced by final state interactions (Λ^0-nucleon scattering) are small. Since, within a light nucleus, the average number of collisions experimented by a produced Λ^0 is $\bar{N} < 1$, it is to be expected that the effects, if any, reflect the depolarization involving the spin interactions.

Let us first note [26] that the spin-orbit interactions cannot contribute to Λ^0 depolarization, given that this interaction is parity-conserving. Since depolarization requires spin-flip, the only interaction capable to produce it is the spin-spin interaction.

To quantify the degree of depolarization in a single Λ^0-nucleon collision, it is customary to introduce the polarization transfer coefficient [27] D, which, in the case of forward scattering, expresses the final polarization $\mathscr{P}'$ in terms of the original polarization $\mathscr{P}$ as

$$\mathscr{P}' = D\,\mathscr{P}. \qquad (27)$$

In a multiple scattering scenario, such as the one considered here, the average depolarization due to two-body Λ^0-nucleon interactions can be written as

$$\langle \mathscr{P}'(z,b) \rangle = \mathscr{P} e^{-\bar{N}(z,b)(1-D)}. \qquad (28)$$

Assuming that the spin-spin interactions are isotropic, D can be expressed as

$$D = \frac{|\bar{V}|^2 + |S_\Lambda|^2 + |S_N|^2 - |\Delta|^2}{|\bar{V}|^2 + |S_\Lambda|^2 + |S_N|^2 + 3|\Delta|^2}, \tag{29}$$

where $\bar{V}$, S_Λ, S_N and Δ represent the amplitudes for the spin-independent, Λ^0 spin-orbit, nucleon spin-orbit and spin-spin interactions, respectively, appearing in the expression for the two-body Λ^0-nucleon potential [28].

The above parameters can be reasonably constrained at low energies (a few MeV's) from the spectral analysis of hypernuclei levels [29], however, no further experimental data exists at higher energies. Nevertheless, it is clear that, when in the range of energies of interest, the parameters appearing in Eq. (29) are measured, D can be considered to be a constant smaller than unity. Therefore, for a given impact parameter, the spin-flip interactions have a simple effect on the depolarization of Λ^0's produced in nucleus-nucleus collisions. For the purposes of this work and until experimental information becomes available, we will omit these effects from the analysis.

CONCLUSIONS

In conclusion, we propose to study the polarization of Λ^0's as a means to identify the production of deconfined matter in ultra-relativistic nucleus-nucleus collisions. We have studied a model where Λ^0's are produced by two competing mechanisms, namely, recombination type of processes, below the critical density for QGP formation, where we expect that Λ^0's are produced polarized, and coalescence type of processes, above the critical density for QGP formation, where we expect that Λ^0's are unpolarized. The overall polarization detected would thus depend on the relative contribution of each process to the overall Λ^0 yield.

To describe the polarization of Λ^0's produced by recombination, we use the DMM, accounting for the effects introduced by multiple elastic scattering experienced by the produced Λ^0's in the nuclear environment. Multiple scattering is responsible for two distinct effects: a momentum shift where the polarization of the produced Λ^0's is preserved but their final momenta change and a depolarization due to spin flip interactions.

We have used the existing data on Λ^0-nucleon interactions to obtain the inelasticity parameter I and have estimated the average dispersion angle per collision Γ. We have also given an explicit expression for the depolarization coefficient, in terms of the parameters describing the two-body Λ^0-nucleon potential. These last parameters have yet to be measured at high energies.

Though our analysis is as quantitative as the existing data allow it [30], there is no doubt that in order for the model to have a stronger predictive power, more accurate data on Λ^0-nucleon interactions at high energies are needed, which, together with an increasing interest on finding new probes for the existence of a QGP, can make Λ^0 polarization measurements a powerful analyzing tool in high energy heavy-ion collisions.

ACKNOWLEDGMENTS

Support for this work has been received in part by CONACyT under the ICM grant number 35792-E and by DGAPA-UNAM under the PAPIIT grant number IN108001.

REFERENCES

1. See U.W. Heinz, *Hunting Down the Quark Gluon Plasma in Relativistic Heavy-Ion Collisions*, Proceedings of the Conference on Strong and Electroweak Matter (SEWM 98), Copenhagen Denmark, Eds. J. Ambjorn, P. Damgaard, K. Kainulainen and K. Rummukainen (World Scientific Publ. Co., Singapore, 1999), 81.
2. N. Angert *et al.*, Proceedings of the Conference on Quark Matter Formation and Heavy Ion Collisions, Eds. M. Jacob and H. Satz (World Scientific Publ. Co., Singapore, 1982), 557.
3. A.D. Panagiotou, Phys. Rev. C **33**, 1999 (1986), Int. J. Mod. Phys. A **5**, 1197 (1990).
4. B.E. Bonner *et al.*, Phys. Rev D **38**, 729 (1988).
5. T.A. DeGrand and H.I. Miettinen, Phys. Rev. D **23**, 1227 (1981), *ibid* **24**, 2419 (1981); T. Fujita and T. Matsuyama, *ibid* **38**, 401 (1988); T.A. DeGrand, *ibid* **38**, 403 (1988).
6. Y. Kitsukawa and K. Kubo, Prog. Theo. Phys. **103**, 1173 (2000).
7. R.C. Hwa, Phys. Rev. Lett. **52**, 492 (1984).
8. L.P. Csernai and J.I. Kapusta, Phys. Rev. D **29**, 2664, *ibid* **31**, 2795 (1985).
9. A. Ayala, E. Cuautle, G. Herrera and L.M. Montaño, Phys. Rev. C **65**, 024902 (2002).
10. M. Pepin *et al.*, Phys. Lett. **B26**, 35 (1967).
11. P. Aahlin, *et al.*, Lett. Nuov. Cimento **21**, 236 (1978).
12. G. Bunce *et al.*, Phys. Rev. Lett. **36**, 1113 (1976).
13. G. Herrera, J. Magnin, L.M. Montaño and F.R.A. Simão, Phys. Lett. **B382**, 201 (1996).
14. J. Rafelski and B. Müller, Phys. Rev. Lett. **48**, 1066 (1982), *ibid* **56**, 2334 (1986); T. Biró and J. Zimányi, Phys. Lett. **B113**, 6 (1982), Nucl. Phys. **A395**, 525 (1983).
15. G. Herrera, L.M. Montaño Phys. Lett. **B381**, 337 (1996); G. Herrera, Rev. Mex. Fis. **43-S1**, 29 (1997).
16. J. Ranft, Act. Phys. Pol. **B 10**, 911 (1979).
17. J.P. Blaizot and J.Y. Ollitrault, Phys. Rev. Lett. **77**, 1703 (1996).
18. J. Letessier, J. Rafelski and A. Tounsi, Phys. Lett. **B389**, 586 (1996).
19. J. C. Anjos, G. Herrera, J. Magnin, and F.R.A. Simão, Phys. Rev. D **56**, 394 (1997).
20. See *Total Cross Sections for Reactions of High Energy Particles* (Landolt-Börnstein), New Series Vol. **I/12 a** and **I/12 b**, Ed. H. Schopper (1988).
21. J. Hüfner and A. Klar, Phys. Lett. **B145**, 167 (1984).
22. J.M. Hauptman, J.A. Kadyk and G.H. Trilling, Nucl. Phys. **B125**, 29 (1977).
23. B.J. Uscinski, *The Elements of Wave Propagation in Random Media*, McGraw-Hill International Book Company (1977).
24. J.A. Kadyk, G. Alexander, J.H. Chan, P. Gaposchkin and G.H. Trilling, Nucl. Phys. **B27**, 13 (1971).
25. S. Ajimura *et al.*, Phys. Rev. Lett. **68**, 2137 (1992).
26. H.E. Conzett, Phys. Rev. C **48**, 924 (1993).
27. J. Bystricky, F. Lehar and P. Winternitz, J. Phys. (Paris) **39**, 1 (1978).
28. D.J. Millener, A. Gal, C.B. Dover, R.H. Dalitz, Phys. Rev. C **31**, 499 (1985).
29. M. May *et al.*, Phys. Rev. Lett. **47**, 1106 (1981); M. May *et al.*, *ibid* **78**, 4343 (1997); K. Tanida *et al.*, Nucl. Phys. **A684**, 560c (2001).
30. For a recent experimental study on the polarization properties of Λ^0's produced in 11.6 A GeV/c Au-Au collisions at the AGS, see R. Bellwied (E896 Collaboration), Nucl. Phys. A698, 499c (2002).

Soft and Hard QCD in Charmonium Production

C. Brenner Mariotto*, M. B. Gay Ducati* and G. Ingelman[†]

*Inst. of Physics, Univ. Fed. do Rio Grande do Sul, Box 15051, CEP 91501-960 Porto Alegre, Brazil
[†]High Energy Physics, Uppsala University, Uppsala, Sweden and DESY, Hamburg, Germany

Abstract. Hard and soft QCD dynamics are both important in charmonium hadroproduction, as presented here through a next-to-leading order QCD matrix element calculation combined with the colour evaporation model. Observed x_F and $p_\perp$ distributions of J/ψ in hadroproduction are reproduced. Quite similar results can also be obtained with a Monte Carlo event generator where $c\bar{c}$ pairs are instead produced through leading order matrix elements and the parton shower approximation of higher order processes. The soft dynamics may alternatively be described by the soft colour interaction model. We also discuss the relative rates of different charmonium states and introduce an improved model for mapping the continuous $c\bar{c}$ mass spectrum on the physical charmonium resonances.

The theoretical description of charmonium production separates the hard and soft parts of the process based on the factorisation theorem in QCD. Thus, we first consider the perturbative production of a $c\bar{c}$ pair at the parton level and then the non-perturbative formation of a bound charmonium state [1].

Perturbative QCD (pQCD) should be applicable for $c\bar{c}$ production, since the charm quark mass m_c is large enough to make $\alpha_s(m_c^2)$ a small expansion parameter. The leading order (LO) processes are $gg \to c\bar{c}$ and $q\bar{q} \to c\bar{c}$. The next-to-leading order (NLO) processes, i.e. $\mathcal{O}(\alpha_s^3)$, include the emission of a third parton and virtual corrections (where divergences are properly cancelled). The full NLO matrix elements, with explicit charm quark mass, are available in a computer program [2] giving total and differential cross sections.

An alternative description of the pQCD production of $c\bar{c}$ pairs is given by the PYTHIA [3] Monte Carlo, where all LO QCD $2 \to 2$ processes are included with their corresponding matrix elements and the incoming and outgoing partons may branch as described by the DGLAP equations. A $c\bar{c}$ pair can then be produced as described by the LO matrix elements for $q\bar{q} \to c\bar{c}$ and $gg \to c\bar{c}$ (with explicit m_c dependence) or in a gluon splitting $g \to c\bar{c}$ in the parton shower.

The main free parameter is the charm quark mass m_c, taken as $m_c = 1.5$ GeV in the NLO program and $m_c = 1.35$ GeV in PYTHIA. In both approaches, the factorization and renormalization scales are taken as the average transverse mass of the c and $\bar{c}$.

The formation of bound hadron states occurs through processes with small momentum transfers such that α_s is large and prevents the use of perturbation theory. The lack of an appropriate method to calculate non-perturbative processes, forces us to use phenomenological models to describe the formation of charmonium states from perturbatively produced $c\bar{c}$ pairs. The Color Evaporation Model (CEM) [4] and the Soft Colour Interaction (SCI) model [5, 6] are based on a similar phenomenological approach, where soft colour interactions can change the colour state of a $c\bar{c}$ pair from an octet to a singlet. They employ the same hard pQCD processes to produce a $c\bar{c}$ pair regardless of its spin state. A colour singlet $c\bar{c}$ pair with an invariant mass below the threshold for open charm ($m_{c\bar{c}} < 2m_D$) will then form a charmonium state.

In CEM [4, 7, 8, 9] the exchange of soft gluons is assumed to give a randomisation of the colour state. This implies a probability $1/9$ that a $c\bar{c}$ pair is in a colour singlet state and produces charmonium if its mass is below the threshold for open charm production, $m_{c\bar{c}} < 2m_D$. The fraction of a specific charmonium state i, relative to all charmonia, is given by a non-perturbative parameter ρ_i ($\rho_{J/\psi} = 0.4 - 0.5$) [4].

In SCI [5, 6, 10] it is assumed that colour-anticolour, corresponding to non-perturbative gluons, can be exchanged between partons emerging from a hard scattering and hadron remnants. The unknown probability to exchange a soft gluon between parton pairs is given by a phenomenological parameter R. These colour exchanges lead to different topologies of the confining colour string-fields and thereby to different hadronic final states after hadronisation. The mapping of $c\bar{c}$ pairs, with mass below the threshold for open charm production, is here made based on spin statistics resulting in a fraction of a specific quarkonium state i with total angular momentum J_i given by $f_i = \frac{\Gamma_i}{\sum_k \Gamma_k}$, where $\Gamma = (2J_i + 1)/n_i$ including a suppression of radially excited states through the main quantum number n_i. This model was found to give a correct description of the different heavy quarkonium states observed at the Tevatron [6].

The complete models are formed by adding the CEM or SCI models for the soft processes to any of the descriptions for the hard pQCD processes. The first model we label **CEM-NLO** and is the combination of the CEM model with the NLO program. The second model is **CEM-PYTHIA**, where CEM has been implemented in PYTHIA version 5.7 [3]. The third model, **SCI-PYTHIA**, is to use the SCI model as implemented in PYTHIA 5.7. Further ingredients are the intrinsic $k_\perp$, due to the Fermi motion of partons inside the initial state hadrons, and soft p_T in soft gluon exchange that neutralize color. Both effects are modelled by a gaussian distribution of width $0.6 - 0.8$ GeV used in PYTHIA and in the NLO program.

Comparing these three models we can separate different effects. With CEM implemented in the NLO program and in PYTHIA, we can compare the pQCD contributions, namely NLO versus LO plus the parton shower approximation of higher orders. Having SCI and CEM implemented in PYTHIA, we can explicitly compare these two non-perturbative models and see to what extent they can account for observed soft effects.

Detailed comparisons between the models have been done as well as extensive comparison with data, both from fixed target experiments and the Tevatron collider [1]. Here we limit ourselves to proton beams. The targets are different nuclei, but the experimental results are rescaled to the cross section per nucleon. Thus we compare directly with our models which do not include any nuclear effects but treat hadron-nucleon interactions.

Fig. 1 shows x_F and $p_\perp$ distributions of the produced J/ψ for proton beams of different energies. As can be seen, the data are approximately reproduced, both in shape and normalization, by all three models. Looking into the details of the x_F distributions, one can observe that the model curves fall less steeply than the data and therefore overshoot somewhat at large x_F. The observed $p_\perp$ distribution is better reproduced, with only small differences between the models.

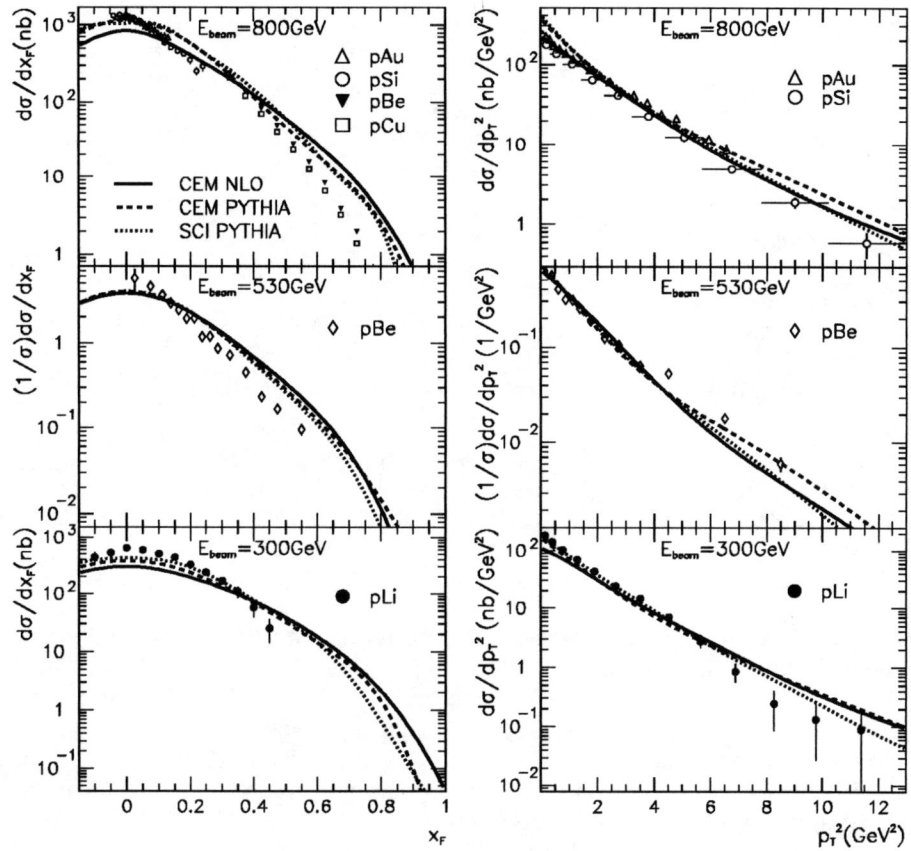

FIGURE 1. Distributions in x_F and $p_\perp^2$ of J/ψ produced with proton beams of energies 800, 530 and 300 GeV on fixed target. Data [11, 12, 13, 14] compared to CEM based on NLO pQCD matrix elements, and CEM and SCI based on LO matrix elements plus parton showers in the PYTHIA Monte Carlo.

Having CEM combined with different treatments of the pQCD production of $c\bar{c}$, we can now investigate pQCD effects in more detail. Fig. 2 illustrates this for the case of 800 GeV proton energy, similar conclusions can also be drawn for other energies and beam particles. For the x_F distribution in Fig. 2a, the full NLO result and that based on LO+PS agree reasonably well. The NLO corrections are very important, as we see by comparing the LO and the full NLO results. In the LO+PS result, however, the PS contribution is unimportant for the overall cross section which is dominated by the LO $c\bar{c}$ production. The agreement with the NLO result is here obtained by using a lower charm mass, $m_c = 1.35$ GeV. We have cross-checked this within the NLO program, where the full result is essentially reproduced by the LO part if this lower mass value is used. This demonstrates that the NLO correction is essentially an overall K-factor from soft and virtual corrections. For the $p_\perp$ distributions in Fig. 2b, the NLO program gives

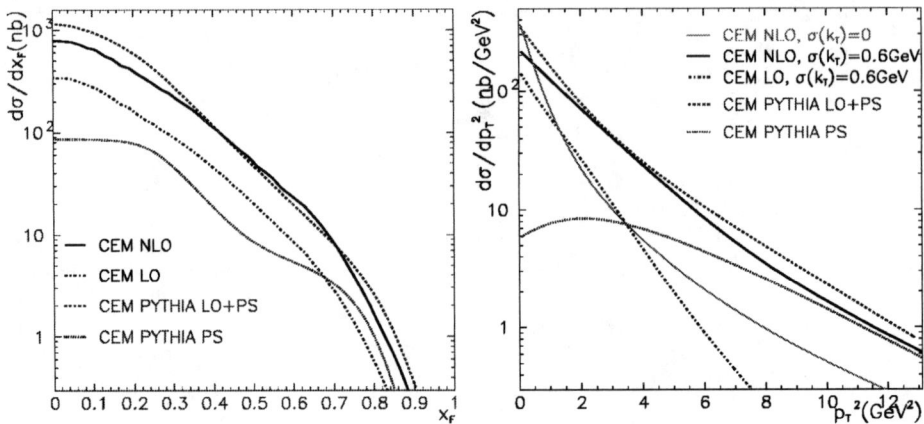

FIGURE 2. Distributions in x_F and $p_\perp^2$ of J/ψ (in 800 GeV proton on proton as in Fig. 1) from variations of the pQCD treatment. CEM based on the NLO program with $m_c = 1.5$ GeV: NLO and LO matrix elements, NLO with no intrinsic $k_\perp$. CEM based on PYTHIA with $m_c = 1.35$ GeV: LO matrix elements plus parton showers (PS) and PS contribution shown separately.

a $p_\perp$ distribution with a much larger tail at large $p_\perp$, but it is still substantially affected by the inclusion of the intrinsic $k_\perp$ at the limited values of $p_\perp$ accessible at fixed target energies. The $p_\perp$ distribution resulting from the LO+PS in the PYTHIA approach, is at high-$p_\perp$ dominated by $c\bar{c}$ from gluon splittings in the partons showers, whereas the bulk of the cross section comes from the low-$p_\perp$ region where the LO diagrams dominate. The total LO+PS result, which also includes a gaussian intrinsic $k_\perp$ with the same width 0.6 GeV, agrees quite well with the NLO result.

Data on ψ' production provide an additional testing ground for the models, which produce all charmonium states with the same dynamics. A comparison made in [1] shows that all models account quite well for the shape of the distributions. The proper normalization of CEM is obtained by chosing $\rho_{\psi'} = 0.066$. The spin statistics used in SCI predicts only a factor two suppression of ψ', and must be lowered by an additional factor four in order to reproduce the data. This has prompted us to develop a more elaborate model for turning $c\bar{c}$ pairs into different charmonium resonances [1], which is briefly described here.

The $c\bar{c}$ pair is produced in a pQCD process with a continuous distribution of its invariant mass $m_{c\bar{c}}$ and must be mapped onto the discrete spectrum of charmonium states. The soft interactions that turn the pair into a colour singlet and form the state, may very well change its mass by a few hundred MeV, which is the typical scale of the soft interactions. We model this by a gaussian smearing of a few hundred MeV. The probability to end up in a specific resonance, shown in Fig. 3, is then proportional to the superposition of this gaussian with the resonance peak, times the corresponding spin-statistics factor. The smearing of $m_{c\bar{c}}$ across the threshold $2m_D$ for open charm, implies non-zero contributions for charmonium also above the $D\bar{D}$ threshold as well as some open charm production for $m_{c\bar{c}}$ originally below this threshold.

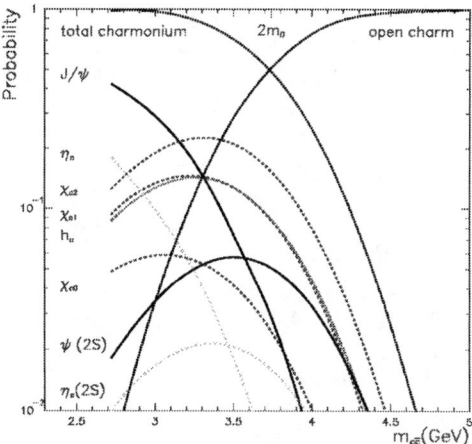

FIGURE 3. Probability distributions for the different charmonium states as obtained in the model with gaussian smearing ($\sigma_{sme} = 400\,MeV$). The resulting total probability for charmonium production and the remainder as open charm production are also shown.

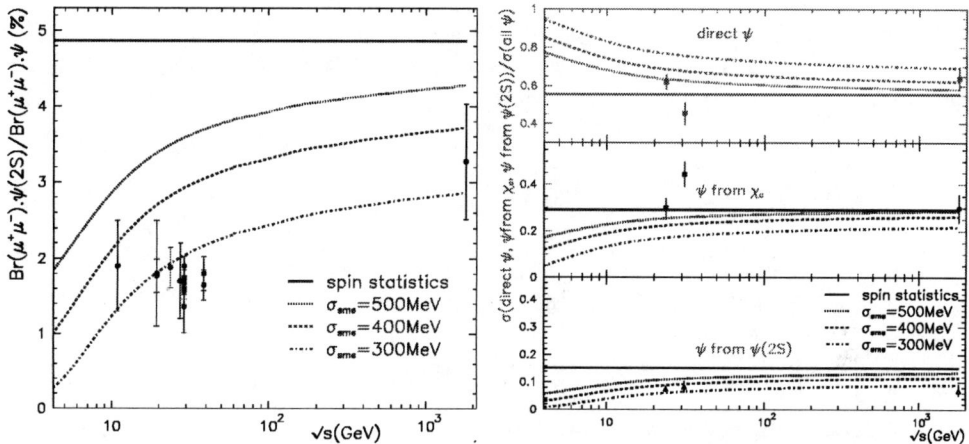

FIGURE 4. The ratio of ψ' to J/ψ production (times their branching ratios for decay into $\mu^+\mu^-$) (left) and fractions of J/ψ produced directly, and coming from the decay of χ_c and ψ' states (right) in hadron-hadron interactions of cms energy $\sqrt{s}$. Data [15, 16, 17, 18] compared to simple spin statistics and to our model with different gaussian smearing widths applied to CEM.

By folding these charmonium probability functions with the distribution in $m_{c\bar{c}}$ obtained from pQCD, one gets the cross section for a given charmonium state. Applying this mapping procedure to the CEM and SCI models we obtain the results in Fig. 4. As opposed to the simple spin statistics factor, this model gives a reasonable description of the observed ratio of ψ' to J/ψ production and fractions of J/ψ produced directly, coming from decays of χ_c states and from ψ'. In particular, the model gives a characteristic energy dependence of the kind indicated by the data.

In summary, both hard and soft QCD dynamics play important roles in the production of charmonium states in hadronic interactions. The $c\bar{c}$ pair production in pQCD have substantial higher order contributions, with a factor two increase of the total cross section from NLO corrections. These come mainly from soft and collinear gluon emissions combined with virtual corrections and can be effectively accounted for by an overall K-factor. This supports to the use of the PYTHIA Monte Carlo with LO matrix elements and a reduced charm quark mass to increase the cross section correspondingly. The high $p_\perp$ tail of the cross section is, however, dominated by higher order tree diagrams in the NLO matrix elements and in the parton showers of the Monte Carlo approach.

The non-perturbative formation of J/ψ, can be described by the Colour Evaporation Model and the Soft Colour Interaction model, where $c\bar{c}$ pairs in a colour octet state can turn into a colour singlet state by soft gluon exchange. A simple spin statistics factor is not sufficient for a proper description of other charmonium states, but our elaborated model to map $c\bar{c}$ pairs onto the physical charmonium states improves this situation.

To conclude, the main features of hadroproduction of charmonium can be described in these models combining pQCD and effects of soft colour exchanges. This shows, in particular, that these models for the soft QCD dynamics contain the essential effects and therefore improve our understanding of non-perturbative QCD.

This work was partially financed by Fundação Coordenação de Aperfeiçoamento de Pessoal de Nível Superior (CAPES), Brazil, and by the Swedish Natural Science Research Council.

REFERENCES

1. C. Brenner Mariotto, M.B. Gay Ducati, G. Ingelman, hep-ph/0111379, EPJC in press, 2001.
2. M.L. Mangano, P. Nason, G. Ridolfi, *Nucl. Phys.* B **373**, 295, 1992.
3. T. Sjöstrand, *Computer Phys. Commun.* **82**, 74, 1994.
4. J.F. Amundson et al., *Phys. Lett.* B **390**, (1997) 323; O.J.P. Éboli, E.M. Gregores, F. Halzen, *Phys. Rev.* D **60**, 117501, 1999.
5. A. Edin, G. Ingelman, J. Rathsman, *Phys. Lett.* B **366**, 371 (1996); *Z. Phys.* C **75**, 57, 1997.
6. A. Edin, G. Ingelman, J. Rathsman, *Phys. Rev.* D **56**, 7317, 1997.
7. R. Gavai et al., *Int. J. Mod. Phys.* A **10**, 3043, 1995.
8. G.A. Schuler, R. Vogt, *Phys. Lett.* B **387**, 181, 1996.
9. M.B. Gay Ducati, C.B. Mariotto, *Phys. Lett.* B **464**, 286, 1999.
   M.B. Gay Ducati, V.P. Goncalves, C.B. Mariotto, *Phys. Rev.* D **65**, 037503, 2002.
10. R. Enberg, G. Ingelman, N. Timneanu, *J. Phys. G: Nucl. Part. Phys.* **26**, 712, 2002.; *Phys. Rev.* D **64**, 114015, 2001.
11. M.H. Schub et al, *Phys. Rev.* D **52**, 1307, 1995; T. Alexopoulos et al., *Phys. Rev.* D **55**, 3927, 1997.
12. L. Antoniazzi et al., *Phys. Rev.* D **46**, 4828, 1992.
13. M.S. Kowitt et al., *Phys. Rev. Lett.* **72**, 1318, 1994.
14. A. Gribushin et al., *Phys. Rev.* D **62**, 012001, 2000.
15. M.C. Abreu et al., *Phys. Lett.* B **444**, 516, 1998, and references therein.
16. F. Abe et al, *Phys. Rev. Lett.* **75**, 4358, 1995; **79**, 572, 1997; **79**, 578, 1997.
17. L. Antoniazzi et al., *Phys. Rev. Lett.* **70**, 383, 1993.
18. V. Koreshev et al., *Phys. Rev. Lett.* **77**, 4294, 1996.

Unitarity Corrections and Structure Functions

M. B. Gay Ducati and M. V. T. Machado

Instituto de Fisica, Universidade Federal do Rio Grande do Sul
Av. Bento Gonçalves, 9500. CEP 91510-970, Porto Alegre, RS Brazil

Abstract. We have studied the color dipole picture for the description of deep inelastic process, mainly the structure functions which are driven directly by the gluon distribution. Estimates are obtained using the Glauber-Mueller dipole cross section in QCD encoding the corrections due to the unitarity effects which are associated with the saturation phenomenon. Frame invariance is verified in the calculations when analysing the experimental data.

INTRODUCTION

In the kinematical region of small proton momentum fraction x, the gluon is the main parton driving the behavior of the deep inelastic quantities. The standard QCD evolution [1] furnishes a powerlike growth for the gluon distribution and related observables. This result leads, at first glance, to the unitarity violation at asymptotic energies, requiring a sort of control. In the partonic language, at the infinite momentum frame, the small x region corresponds to the high parton density winsdom. The latter is connected with the black disk limit of the proton target and with the parton recombination phenomenon. These issues can be addressed through a non-linear dynamics beyond the usual DGLAP formalism. The complete knowledge about the non-linear dynamical regime plays an important role in the theoretical description of the reactions in the forthcoming experiments RHIC and the LHC, where these effects are enhanced by the high energies reached or by nuclear probes.

The description of DIS in the color dipole picture is somewhat intuitive, providing with a simple representation in contrast to the involved one from the Breit (infinite momentum) frame. Considering small values of the Bjorken variable x, the virtual photon fluctuates into a $q\bar{q}$ pair (dipole) with fixed transverse separation r at large distances upstream of the target and interacts in a short time with the proton. More complicated configurations should be considered for larger transverse size systems, for instance the photon Fock state $q\bar{q}$ + gluon. An immediate consequence from the lifetime of the pair ($l_c = 1/2m_p x$) to be bigger than the interaction one is the factorization between the photon wavefunction and the cross section dipole-proton in the $\gamma^* p$ total cross section. The wavefunctions are perturbatively calculable, namely through QED for the $q\bar{q}$ configuration [2] and from QCD for the $q\bar{q}G$ one [3]. The effective dipole cross section should be modeled and it includes perturbative and non-perturbative content. However, since the interaction strength relies only on the configuration of the interacting system the dipole cross section turns out to be universal and may be employed in a wide variety of small x processes.

We have taken into account a sound formalism providing the unitarity corrections to the DIS at small x, namely the Glauber-Mueller approach in QCD. It was introduced by A. Mueller [4], who developed the Glauber formalism to study saturation effects in the quark and gluon distributions in the nucleus considering the heavy onium scattering. Later developments obtained an evolution equation taking into account the unitarity corrections (perturbative shadowing), generating a non-linear dynamics which is connected with higher twist contributions. Its main characteristic is to provide a theoretical framework for the saturation effects, lying on the multiscattering of the pQCD Pomeron. In this contribution we report our studies considering the parton saturation formalism to describe the observables driven by the gluonic content of the proton at the color dipole picture [5]. The inclusive structure function F_2 is calculated, disregarding the fairly approximations commonly considered in previous calculations [6]. The structure functions F_L and $F_2^{c\bar{c}}$ are also presented using the Glauber-Mueller approach and rest frame in comparison with the experimental data.

THE DIS AT THE REST FRAME AND GLAUBER-MUELLER APPROACH

The rest frame physical picture is advantageous since the lifetime of the photon fluctuation and the interaction process are well defined [7]. The more simple case is the quark-antiquark state (color dipole), which is the leading configuration for small transverse size systems. The well known coherence lenght is expressed as $l_c = 1/(2xm_p)$, where x is the Bjorken variable and m_p the proton mass. An important consequence of this formulation is that the photoabsortion cross section can be derived from the expectation value of the interaction cross section for the multiparticle Fock states of the virtual photon weighted by the light-cone wave functions of these states [2]. That cross section can be cast in the quantum mechanical factorized form,

$$\sigma_{T,L}^{\gamma^* p}(x, Q^2) = \int d^2\mathbf{r} \int_0^1 dz \, |\Psi_{T,L}(z, \mathbf{r})|^2 \, \sigma^{\text{dipole}}(x, z, \mathbf{r}), \qquad (1)$$

The formulation above is valid even beyond perturbation theory, since it is determined from the space-time structure of the process. The $\Psi_{T,L}(z, \mathbf{r})$ are the photon wavefunctions (for transverse T, and longitudinal L, polarizations) describing the pair configuration; z and $1 - z$ are the fractions of the photon's light-cone momentum carried by the quark and antiquark of the pair, respectively. The transverse separation of the pair is $\mathbf{r}$. The explicit expressions for the wavefunctions are well known,

$$|\Psi_T(z, \mathbf{r})|^2 = \frac{6\alpha_{\text{em}}}{4\pi^2} \sum_i^{n_f} e_i^2 \left\{ [z^2 + (1-z)^2] \varepsilon^2 K_1^2(\varepsilon r) + m_q^2 K_0^2(\varepsilon r) \right\} \qquad (2)$$

$$|\Psi_L(z, \mathbf{r})|^2 = \frac{6\alpha_{\text{em}}}{4\pi^2} \sum_i^{n_f} e_i^2 \left\{ 4Q^2 z^2 (1-z)^2 K_0^2(\varepsilon r) \right\}. \qquad (3)$$

Lets define the auxiliary variable $\varepsilon^2 = z(1-z)Q^2 + m_q^2$, with m_q the light quark mass, and K_0 and K_1 are the Mc Donald functions of rank zero and one, respectively. The quantity σ^{dipole} is interpreted as the cross section of the scattering of the effective dipole with fixed tranverse separation **r** [2]. The most important feature of the dipole cross section is its universal character, namely it depends only on the transverse separation **r** of the color dipole. The dependence on the external probe particle, i.e., the photon virtuality, relies in the wavefunctions. In general, an ansätz for the effective dipole cross section is obtained and the the process is analized in the impact parameter space. The main feature of the current models in the literature is to interpolate the physical regions of small transverse separations (perturbative QCD picture) and the large ones (Regge-soft picture). Here we have used the Glauber-Mueller approach to determine the dipole cross section, with the advantage of providing the corrections required by unitarity in an eikonal expansion. For the large r region, we choose to follow a similar procedure from the saturation model (GBW) [8], namely saturating (r-independent constant value) the dipole cross section at this region.

Now, we shortly show the main results from the Glauber-Mueller approach. Considering the scattering amplitude dependent on the usual Mandelstan variables s and t, now written in the impact parameter representation **b**,

$$a(s,\mathbf{b}) \equiv \frac{1}{2\pi} \int d^2\mathbf{q}\, e^{-i\mathbf{q}\cdot\mathbf{b}} \mathscr{A}(s,t=-q^2). \tag{4}$$

the corresponding total and elastic cross sections (from Optical theorem) are rewritten in the impact parameter representation (**b**) as

$$\sigma_{tot} = 2\int d^2\mathbf{b}\, Im\, a(s,\mathbf{b}); \qquad \sigma_{el} = \int d^2\mathbf{b}\, |a(s,\mathbf{b})|^2, \tag{5}$$

The most important property when treating the scattering in the impact parameter space is the simple definition for the unitarity constraint [6]. If the real part of the scattering amplitude vanishes at the high energy limit, corresponding to small x values, the solution to the that constraint is

$$a(s,\mathbf{b}) = i\left[1-e^{-\frac{1}{2}\Omega(s,\mathbf{b})}\right]; \qquad \sigma_{tot} = 2\int d^2\mathbf{b}\left[1-e^{-\frac{1}{2}\Omega(s,\mathbf{b})}\right], \tag{6}$$

where the opacity Ω is an arbitrary real function and it should be determined by a detailed model for the interaction. The opacity function has a simple physical interpretation, namely $e^{-\Omega}$ corresponds to the probability that no inelastic scatterings with the target occur. To realize the connection with the Glauber formalism, the opacity function can be written in the factorized form $\Omega(s,\mathbf{b}) = \Omega(s)\,S(\mathbf{b})$, considering $S(\mathbf{b})$ normalized as $\int d^2\mathbf{b}\,S(\mathbf{b}) = 1$ (for a detailed discussion, see i.e. [9]).

We identify the opacity $\Omega(s \approx Q^2/x;\mathbf{r}) = \sigma^{\text{nucleon}}(x,\mathbf{r})$. The ($q\bar{q}$ pair) dipole-proton cross section is well known [6, 9] and in double logarithmic approximation (DLA) has the following form

$$\sigma^{q\bar{q}}_{\text{nucleon}}(x,r) = \frac{\pi^2 \alpha_s(\tilde{Q}^2)}{3} r^2 x G(x,\tilde{Q}^2) \tag{7}$$

with the r-dependent scale $\tilde{Q}^2 = r_0^2/r^2$. Considering Eq. (7) one can connect directly the dipole picture with the usual parton distributions (gluon), since they are solutions of the DGLAP equations. In our case, we follow the calculations in Ref. [6, 9] and consider the effective scale $\tilde{Q}^2 = 4/r^2$. From the above expression, one obtains a dipole cross section satisfying the unitarity constraint and a framework to study the unitarity effects (saturation) in the gluon DGLAP distribution function. Hence, hereafter we use the Glauber-Mueller dipole cross section given by

$$\sigma_{dipole}^{GM} = 2\int d^2\mathbf{b} \left(1 - e^{-\frac{1}{2}\sigma_{nucleon}^{q\bar{q}}(\mathbf{x},\mathbf{r})S(\mathbf{b})}\right). \tag{8}$$

In order to perform numerical estimates one needs to define the profile function $S(b)$. This function contains information about the angular distribution in the scattering. We have chosen a simple gaussian shape in the impact parameter space, $S(b) = \frac{A}{\pi R_A^2}e^{-b^2/R_A^2}$, where A is the atomic number and R_A is the target radius. We will keep this notation although we are only concerned with the nucleon case. The R_A^2 value should be determined from the data, ranging between $5-10$ GeV^{-2} for the proton case [6]. Here, we have used the value ($R_A^2 = 5$ GeV^{-2}) obtained from a good description of both inclusive structure function and its derivative [10]. Such a value corresponds to significative unitarity corrections to the standard DGLAP input even in the current HERA kinematics.

In the calculations we have used the GRV94 parametrization [11]: bearing in mind that $Q^2 = 4/r^2$, its evolution initial scale $Q_0^2 = 0.4$ GeV^2 allows to scan dipole sizes up to $r_{cut} = \frac{2}{Q_0}$ GeV^{-1} (= 0.62 fm). For recent parametrizations, where $Q_0^2 \sim 1$ GeV^2 ($r_{cut} \approx 0.4$ fm), the uncertain due to nonperturbative content in the calculations would increase. An additional advantage is that GRV94 does not include non-linear effects to the DGLAP evolution since it was obtained from rather large x values, i.e. this ensures that GRV94 does not include unitarity corrections in the initial scale. To proceed, for the large r region, we choose the following ansatz: the gluon distribution is frozen at scale r_{cut}, namely $xG(x, \tilde{Q}_{cut}^2)$. Then, for the large distance contribution $r \leq r_{cut}$ the gluon distribution reads as

$$xG(x, Q^2 \leq Q_0^2) = \frac{Q^2}{Q_0^2}xG(x, Q^2 = Q_0^2), \tag{9}$$

leading to the correct behavior $xG(x, Q^2) \sim Q^2$ as $Q^2 \to 0$.

OBTAINING THE STRUCTURE FUNCTIONS

The structure function F_2

First, we perform estimates for the structure function F_2 at the rest frame considering the Glauber-Mueller dipole cross section [5]. The expression, with the explicit integration limits on photon momentum fraction z and transverse separation r is,

$$F_2(x, Q^2) = \frac{Q^2}{4\pi^2 \alpha_{em}} \int_0^\infty d^2\mathbf{r} \int_0^1 dz \left(|\Psi_T(z,\mathbf{r})|^2 + |\Psi_L(z,\mathbf{r})|^2\right) \sigma_{dipole}^{GM}(x, \mathbf{r}^2). \tag{10}$$

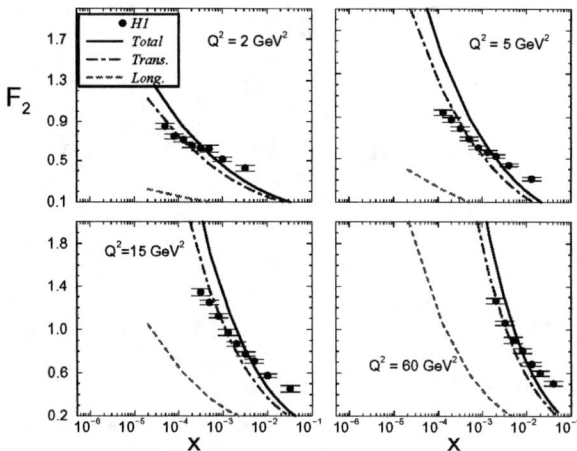

FIGURE 1. The Glauber-Mueller (GM) result for the $F_2(x,Q^2)$ structure function. It is shown the transverse contribution (dot-dashed), the longitudinal one (dashed) and total one (solid line).

In the Fig. (1) one shows F_2 for representative virtualities Q^2 from the latest H1 Collaboration measurements [12]. The longitudinal and transverse contributions are shown separately. An effective light quark mass (u,d,s quarks) was taken, with the value $m_q = 300$ MeV, and the target radius is considered $R_A^2 = 5$ GeV^{-2}. It should be stressed that this value leads to larger saturation corrections rather than using radius ranging over $R_A^2 \sim 8-15$ GeV^{-2}. The soft contribution comes from the freezing of the gluon distribution at large transverse separation as discussed at the previous section.

From the plots we verify a good agreement in the normalization, however the slope seems quite steep. This fact is due to the modeling for the soft contribution and it suggests that a more suitable nonperturbative input should be taken. To clarify the role played by the soft nonperturbative contribution to F_2, in the Fig. (2) we plot separately the perturbative contribution and parametrize the soft contribution introducing the non-perturbative structure function $F_2^{\text{soft}} = \mathscr{C}_{\text{soft}} x^{-0.08} (1-x)^{10}$ [13], which is added to the perturbative one. The soft piece normalization is $\mathscr{C}_{\text{soft}} = 0.22$. Accordingly, we have used just shadowing corrections for the quark sector, taking into account only the transverse photon wavefunction and zero quark mass. The integration on the transverse separation is taken over $1/Q^2 \leq r^2 \leq 1/Q_0^2$, with $Q_0^2 = 0.4$ GeV^2 for leading order GRV94 gluon distribution. This leads to a residual contribution to the soft piece which would come from the transverse separations $r^2 < 1/Q^2$. It is again verified that the soft contribution is important at small virtualities and decreasing as it gets larger. The data description is quite successful.

Concluding, we have a theoretical estimate, i.e. no fitting procedure, of the inclusive structure function $F_2(x,Q^2)$ through the Glauber-Mueller approach for the dipole cross section, detecting a non negligible importance of a suitable input for the large dipole size region.

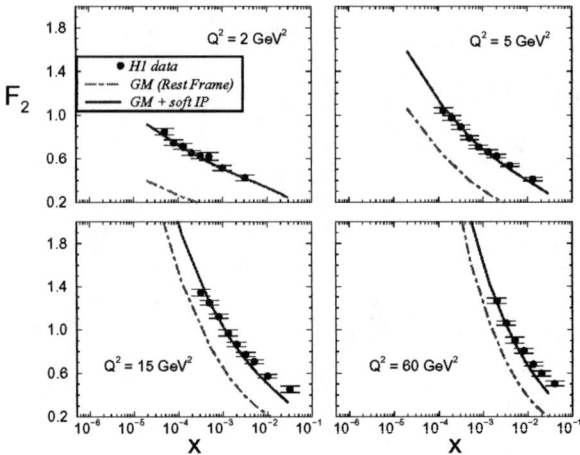

FIGURE 2. The Glauber-Mueller prediction for the F_2 structure function in the rest frame. For sake of comparison, one uses quark sector ($R_A^2 = 5$ GeV^{-2}, $m_q = 0$) and only transverse wavefunction. Radius integration $1/Q^2 < r^2 < 1/Q_0^2$ and soft Pomeron added ($F_2^{\text{soft}} = \mathscr{C}_{\text{soft}} x^{-0.08}(1-x)^{10}$).

The structure function F_L

From QCD theory, the structure function F_L has a non-zero value due to the gluon radiation, as is encoded in the Altarelli-Martinelli equation (see [14]), considering the Breit frame. Experimentally, the determination of the F_L is quite limited, providing few data points. Most recently, the H1 Collaboration has determined the longitudinal structure function through the reduced double differential cross section, where the data points were obtained consistently with the previous measurements, however being more precise and lying into a broader kinematical range [12].

In Fig. (3) we present the estimates for the F_L structure function, in representative virtualities as a function of x [5]. For the calculations, it was considered light quarks (u, d, s) with effective mass $m_q = 300$ MeV and the target radius $R_A^2 = 5$ GeV^{-2}. The large r region is considered by the freezening of the gluon distribution at this region. Our expression for the observable is then,

$$F_L(x,Q^2) = \frac{Q^2}{4\pi^2 \alpha_{\text{em}}} \int_0^\infty d^2\mathbf{r} \int_0^1 dz |\Psi_L(z,\mathbf{r})|^2 \sigma_{\text{dipole}}^{GM}(x,\mathbf{r}^2). \qquad (11)$$

The behavior is quite consistent with the experimental result, either in shape as in normalization. The quantity is less sensitive to the non-perturbative content than F_2. A better description can be obtained by fine tunning the target size or the considered gluon distribution function, however it should be stressed that the present prediction is parameter-free and determined using the dipole picture taking into account unitarity (saturation) effects in the effective dipole cross section [5]. We verify that the rest frame calculation, taking into account the dipole degrees of freedom and unitarity effects produces similar conclusions to those ones using the Breit system. For instance, in

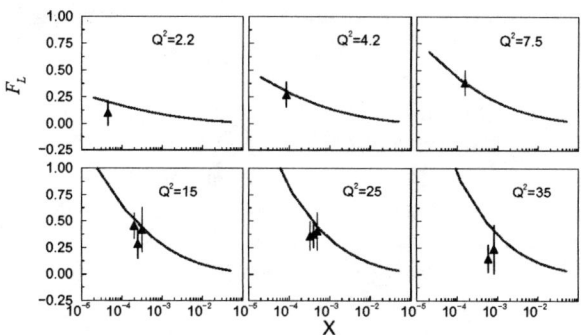

FIGURE 3. The Glauber-Mueller estimates for the F_L structure function. One uses light quarks ($m_q = 300$ MeV), target size $R_A^2 = 5$ GeV^{-2} and frozen gluon distibution at large r. Data from H1 Collaboration [12].

a previous work [14], the unitarity corrections to the longitudinal structure function were estimated in the laboratory frame considering the Altarelli-Martinelli equation, with unitarized expressions for F_2 and $xG(x, Q^2)$, obtaining that the expected corrections reach up to 70 % as $\ln(1/x) = 15$, namely on the kinematical corner of the upcoming THERA project.

The higher twist corrections to the longitudinal structure function have been pointed out. For instance, Bartels et al. [15] have calculated numerically the twist-four correction founding they are large for F_T and F_L, however having opposite signs. This fact leads to remaining small effects to the inclusive structure function by almost complete cancellation between those contributions. The higher twist content is analyzed considering the model [8] as initial condition. Concerning F_L, it was found that the twist-four correction is large and has negative signal, concluding that a leading twist analysis of F_L is unreliable for high Q^2 and not too small x. The results are in agreement with the simple parametrization for higher twist (HT) studied by the MRST group in Ref. [16], where $F_2^{HT}(x, Q^2) = F_2^{LT}(x, Q^2)(1 + \frac{D_2^{HT}(x)}{Q^2})$. The second term would parametrize the higher twist content. In our case, the unitarity corrections provide an important amount of higher twist content, namely it takes into account some of the several graphs determining the twist expansion.

The structure function $F_2^{c\bar{c}}$

In perturbative QCD, the heavy quark production in electron-proton interaction occurs basicaly through photon-gluon fusion, in which the emitted photon interacts with a gluon from the proton generating a quark-antiquark pair. Therefore, the heavy quark production allows to determine the gluon distribution and the amount of unitarity (saturation) effects for the observable. In particular, charmed mesons have been measured at deep-inelastic at HERA and the corresponding structure function $F_2^{c\bar{c}}(x, Q^2)$ is defined from the differential cross section for the $c\bar{c}$ pair production.

Experimentally, the measurements of the charm structure function are obtained by measuring mesons $D^{*\pm}$ production [17]. The function $F_2^{c\bar{c}}(x,Q^2)$ shows an increase with decreasing x at constant values of Q^2, whereas the rise becomes sharper at higher virtualities. The data are consistent with the NLO DGLAP calculations. Concerning the ratio $R^{c\bar{c}} = F_2^{c\bar{c}}/F_2$, the charm contribution to F_2 grows steeply as x diminishes. It contributes less than 10% at low Q^2 and reaches to about 30 % for $Q^2 > 120$ GeV^2 [17].

Once more the color dipole picture will provide a quite simple description for the charm structure function in a factorized way. Now, the Glauber-Mueller dipole cross section is weighted by the photon wavefunction constituted by a $c\bar{c}$ pair with mass m_c. Our expression for the charmed contribution in deep inelastic is thus written as

$$F_2^{c\bar{c}}(x,Q^2) = \frac{Q^2}{4\pi^2 \alpha_{em}} \int_0^\infty d^2\mathbf{r} \int_0^1 dz \left(|\Psi_T^{c\bar{c}}(z,\mathbf{r})|^2 + |\Psi_L^{c\bar{c}}(z,\mathbf{r})|^2 \right) \sigma_{dipole}^{GM}(x,\mathbf{r}^2) \quad (12)$$

where $|\Psi_{T,L}^{c\bar{c}}(z,\mathbf{r})|^2$ is the probability to find in the photon the $c\bar{c}$ color dipole with the charmed quark carrying fraction z of the photon's light-cone momentum with T, L polarizations. For the correspondent wavefunctions, the quark mass in Eqs. (2,3) should be substituted by the charm quark mass m_c. Here, we should take care of the connection between the Regge parameter $x = (W^2 + Q^2)/(Q^2 + 4m_q^2)$ and the Bjorken variable x_{Bj}. For calculations with the light quarks these variables are equivalent, however for heavier quarks the correct relation is [18]: $x_{Bj} = x(Q^2/Q^2 + 4m_c^2)$.

In Fig. (4) we show the estimates for the charm structure function as a function of x_{Bj} at representative virtualities [5]. In our calculations, it was used charm mass $m_c = 1.5$ GeV, target size $R_A^2 = 5$ GeV^{-2} and frozen gluon distibution at large r. We have verified small soft contribution, decreasing as the virtuality rises. There is a slight sensitivity to the value for the charm mass, increasing the overall normalization as m_c diminishes. Such a feature suggests that the charm mass is a hard scale suppressing the non-perturbative contribution to the corresponding cross section. This conclusion is in agreement with the recent BFKL color dipole calculations of Nikolaev-Zoller [18] and those from Donnachie-Dosch [19].

Regarding the Breit system description, in Ref. [14] it was found strong corrections to the charm structure function, which are larger than those of the F_2 ones. Considering the ratio $R_2^c = F_2^{c\,GM}(x,Q^2)/F_2^{c\,DGLAP}(x,Q^2)$, the corrections predicted by the Glauber-Mueller approach would reach up to 62 % at values of $\ln(1/x) \approx 15$ (THERA region). Then, an important result is a large deviation of the standard DGLAP expectations at small x for the ratio $R^{c\bar{c}} = F_2^{c\bar{c}}/F_2$ due to the saturation phenomena (unitarization). With our calculation [5] one verifies that it is obtained a good description of data in both reference systems, suggesting a consistent estimation of the unitarity effects for that quantity.

ACKNOWLEDGMENTS

The authors thanks the Organizers of this lively meeting.

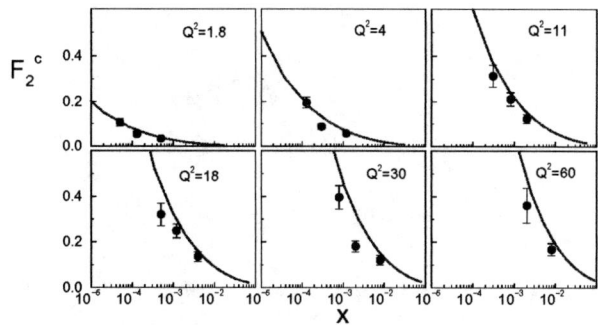

FIGURE 4. The Glauber-Mueller result for the $F_2^{c\bar{c}}$ structure function as a function of Bjorken variable x at fixed virtualities (in GeV^2). One uses charm mass $m_c = 1.5$ GeV, target size $R_A^2 = 5$ GeV^{-2} and frozen gluon distibution at large r. Data from ZEUS Collaboration [17] (statistical errors only).

REFERENCES

1. Yu.L. Dokshitzer. *Sov. Phys. JETP* **46**, 641 (1977);
   G. Altarelli and G. Parisi. *Nucl. Phys.* **B126**, 298 (1977);
   V.N. Gribov and L.N. Lipatov. *Sov. J. Nucl. Phys* **28**, 822 (1978).
2. N.N. Nikolaev, B.G. Zakharov, *Z. Phys.* **C49**, 607 (1991); *Phys. Lett.* **B260**, 414 (1991); *Z. Phys.* **C53**, 331 (1992).
3. M. Wüsthoff, *Phys. Rev.* **D56**, 4311 (1997).
4. A. H. Mueller, Nucl. Phys. **B335**, 115 (1990).
5. M.B. Gay Ducati, M.V.T. Machado, *Phys. Rev.* **D** (to appear) (2002) [hep-ph/0111093].
6. A.L. Ayala, M.B. Gay Ducati and E.M. Levin, *Nucl. Phys.* **B493**, 305 (1997); *Nucl. Phys.* **B511**, 355 (1998).
7. V. Del Duca, S.J. Brodsky, P. Hoyer, *Phys. Rev.* **D46**, 931 (1992).
8. K. Golec-Biernat, M. Wüsthoff. *Phys. Rev.* **D59**, 014017 (1999); *Phys. Rev.* **D60**, 114023 (1999).
9. E.Gotsman, E.Levin, U. Maor, *Nucl. Phys.* **B493**, 354 (1997).
10. M.B. Gay Ducati, V.P. Gonçalves, *Phys. Lett.* **B487**, 110 (2000);
11. M. Gluck, E. Reya, A. Vogt, *Z. Phys.* **C67**, 433 (1995).
12. H1 Collaboration, *Eur. Phys. J.* **C21**, 33 (2001).
13. A.L. Ayala, M.B. Gay Ducati, E.M. Levin, *Eur. Phys. J* **C8**, 115 (1999).
14. A.L. Ayala, M.B. Gay Ducati, V.P. Gonçalves, *Phys. Rev.* **D59**, 054010 (1999).
15. J. Bartels, K.Golec-Biernat, K. Peters. *Eur. Phys. J.* **C17**, 121 (2000).
16. A.D. Martin, R.G. Roberts, W.J. Stirling, R.S. Thorne, *Phys. Lett.* **B443**, 301 (1998);
17. ZEUS Collaboration, *Eur. Phys. J.* **C12**, 35 (2000).
18. N.N. Nikolaev, V.R. Zoller, *Phys. Lett.* **B509**, 283 (2001).
19. A. Donnachie, H.G. Dosch, *Phys. Lett.* **B502**, 74 (2001); *Phys. Rev.* **D65**, 014019 (2002).

Particle Emission in Hydrodynamics: A Problem Not Yet Solved

F. Grassi

Instituto de Física, Universidade de São Paulo, C. P. 66318, 05315-970 São Paulo-SP, Brazil

Abstract. A survey of various mecanisms for particle emission in hydrodynamics is presented.

INTRODUCTION

Hydrodynamical and thermal models have been used extensively to describe data from relativistic nuclear collisions. The usual mecanism for particle emission in hydrodynamics is freeze out so I will use it as a point of comparison. I will start in section 2, reminding what it is, some of its problems and ways outs.

There is another particle emission scenario which is a small extension of this idea of sudden freeze out: the separate chemical and thermal freeze outs scenario. It has become used a lot e.g. to analyse data. So I will discuss in section 3 what it is, its alternatives and how to incorporate it in hydrodynamics.

Continuous emission is a mecanism for particle emission that we proposed some years ago. As the very name suggests, it is not "sudden" like the usual freeze out mecanism. I will explain what it is precisely in section 4, how well it describes data compared to freeze out and its alternatives (still based on the notion of "continuity").

Finally I will conclude in section 5.

SUDDEN FREEZE OUT

To apply hydrodynamics to relativistic heavy ion collisions, we make a strong assumption: a fluid is created in local chemical and thermal equilibrium for a short while, in a small region. It is thought to be a reasonable assumption because many particles are created in such collisions: hundreds in a central S+S collision at the SPS [1], few to several thousands for heavy projectiles at SPS and RHIC [2, 3, 4, 5, 6]. In addition, one usually assumes that the fluid is perfect, so its energy-tensor is simpler and the hydrodynamical equations to be solved also are simpler.

The hydrodynamical equations are conservation equations. Given initial conditions and an equation of state, we can solve them and get temperature, chemical potentials and fluid velocity for various times and locations in the fluid. After obtaining this, one has to handle the problem of how particles are emitted by this fluid.

The usual picture called freeze out is as follows. When a certain criterium for "dilution" of matter is reached, for example the temperature has decreased to a certain low value, particles suddenly stop interacting and fly freely towards the detectors.

The way to implement this mathematicaly is to use the so-called Cooper-Frye formula [7]

$$\frac{Ed^3N}{dp^3} = \int_{T_{f.out}} d\sigma_\mu p^\mu f(x,p). \qquad (1)$$

To compute particle spectra, one defines a freeze out surface, for exemple the surface of constant temperature $T_{f.out}$ and count particles as they cross this surface. $d\sigma_\mu$ is the normal vector to this surface, p^μ the particle momentum and f its distribution function. Usually one assumes a Bose-Einstein or Fermi-Dirac distribution for this f.

An important point to note for the following is that since particles are assumed to stop interacting when they cross the freeze out surface, they will carry to the detectors information about the freeze out surface only. So in this approach one believes that observables such as momentum spectra, abundances, only give information on the freeze out conditions.

This sudden freeze out approach is often used however it is known to have some bad features. I will mention two. First when using the Cooper-Frye formula, we sometime meet negative terms ($d\sigma_\mu p^\mu \leq 0$) corresponding to particles re-entering the fluid. However since they presumably had stopped interacting (being in the frozen out region), they should not re-enter the fluid and start interacting again. So usually one removes these negative terms from the calculations as being unphysical. However by doing this, one removes baryon number, energy and momentum from the calculation and violates conservation laws. It is not a negligible problem, for example it can be a 20% overestimate of particle number using SPHERIO [8]. There are some ways to avoid these violations as described by L.Csernai but none is completely satisfying[9, 10].

The second problem is the following: do particles really *suddenly* stop interacting when they reach a certain hypersurface? Intuitively no, this must happen over a mean free path. This is corroborated by results from simulations of microscopical models[11, 12, 13]: *the shape of the region where particles last interacted is generally not a sharp surface as assumed for sudden freeze outs*. Some exceptions might be heavy particles in heavy systems or the phase transition hypersurface. We come back to this problem in section 4 but for the moment, we set it aside and turn to a description that has become very popular over the past few years.

SEPARATE CHEMICAL AND THERMAL FREEZE OUTS

The following picture has emerged (see e.g. [14, 15]) from a study of collisions at SIS, AGS, SPS energies, with a variety of targets and projectiles (for a compilation see e.g. [16]). It is also expected to hold at RHIC and LHC energies. The dense and hot fluid expands and cools until chemical freeze-out occurs for some species of particles. Namely these particles stop having inelastic collisions so that their abundances are frozen. Therefore by studying the abundances of these chemically frozen particle species, one learn the conditions at chemical freeze-out. For example, at CERN, $T_{ch.f.} \sim 160-200$ MeV [17]. The fluid goes on cooling until thermal freeze-out happens. Precisely, particles stop having elastic interactions and so the shape of their momentum distribution is fixed.

Therefore by studying these spectra, one gets information about the conditions at thermal freeze-out. For example at CERN, using various types of particles [18] or combining information about the spectrum of a single particle species and its Bose-Einstein correlations [19, 20], one extracts $T_{th.f.} \sim 100 - 140$ MeV. There are some deviations to this picture. For example, some particles like the Ω may undergo both freeze-outs almost together and early (due to their small cross section).

I think at this point it is useful to mention that within the sudden freeze out scenario, there exist alternative descriptions of the data. The first one is explosive emission presented by J.Rafelski[21] and being incorporated in an hydrodynamics code by L.Csernai[9]. Another possibility which is more speculative but if true, extremely interesting, is the following. Hadron masses can be affected by medium effects, so some species may be more or less easy to produce than if they had their vaccum masses. It is possible to reproduce CERN ratios with temperature for chemical freeze out about equal to temperature for thermal freeze out, so a single freeze out might be necessary[22]. However it is still necessary to study other types of data such as transverse mass distributions before drawing a more final conclusion.

Though data call for separate chemical and thermal freeze outs, no hydrodynamical code so far includes chemical and thermal freeze outs self-consistently. Namely, the effect of the early chemical freeze out on the fluid expansion is never taken into account. In [23], we discuss how to incorporate separate chemical and thermal freeze-outs in a hydrodynamical code.

Separate conservation equations must be written for all particles that make early chemical freeze out.

To simplify the discussion, we use a known and simple hydrodynamical model, Bjorken one-dimensional boost invariant model[24]. Before chemical freeze-out, the fluid evolution is governed by the hydrodynamical equations

$$\frac{\partial \varepsilon}{\partial t} + \frac{\varepsilon + p}{t} = 0 \tag{2}$$

$$\frac{\partial n_B}{\partial t} + \frac{n_B}{t} = 0. \tag{3}$$

The last equation can be solved easily

$$n_B(t) = \frac{n_B(t_0) t_0}{t}. \tag{4}$$

The first equation must be completed by the choice of an equation of state $p(n_B, \varepsilon)$, for the pressure as function of the net baryon density and energy density. We suppose that the fluid is a gas of non-interacting resonances.

When the fluid temperature has decreased to some temperature $T_{ch.f.}$, (which corresponds to a certain time $t_{ch.f.}$), some particle species get their abundances frozen. To fix ideas, we suppose that Λ and $\bar{\Lambda}$ are in this situation. Then in addition to the above hydrodynamical equations, we introduce separate conservation laws for these two types of particles for time $t > t_{ch.f.}$, namely

$$\frac{\partial n_\Lambda}{\partial t} + \frac{n_\Lambda}{t} = 0 \tag{5}$$

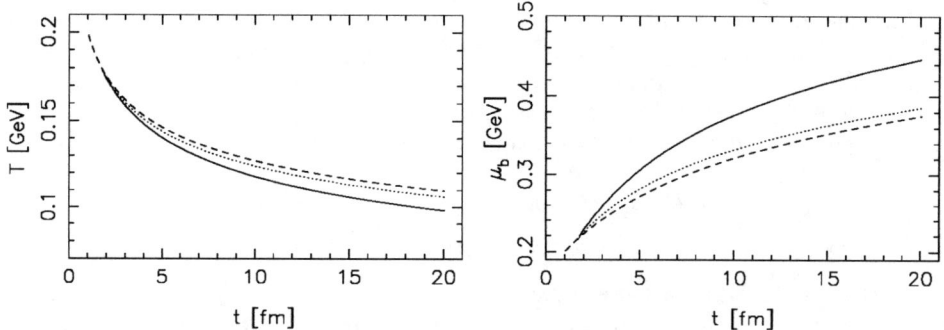

FIGURE 1. μ_B and T as function of time in the case where all particles have simultaneous freeze-outs (dashed line) and (I) all strange particles in basic multiplets make an early chemical freeze-out (continuous line), (II) all strange particles except K and K^*'s make an early chemical freeze-out (dotted line).

$$\frac{\partial n_{\bar{\Lambda}}}{\partial t} + \frac{n_{\bar{\Lambda}}}{t} = 0. \qquad (6)$$

These equations have solutions of the same form as (4) but with t_0 substituted by t_{ch}. Therefore what remains to be done is to solve the energy-momentum equation (2) with a *modified equation of state*, to account for the particles who make an early chemical freeze-out. The scheme presented above can easily be generalized to particles making chemical freeze-out at different times (using different $t_{ch.f.}$) and particles doing chemical and thermal freeze-outs together (whose contribution drop out of the equation of state).

For illustration, we present results using in the equation of state, the basic multiplets of resonances (pseudoscalar meson octet plus singlet, vector meson octet plus singlet, baryon octet and baryon decuplet) and supposing that the early chemical freeze-out occurs at 180 MeV. In figure 1, we compare the behavior of T and μ_B as function of t, obtained from the hydrodynamical equations using the modified equation of state and the unmodified one. We note that if more particles undergo early chemical freeze-out, stronger effects for $T(t)$ and $\mu_B(t)$ are seen. We concentrate on I of figure 1 thereon. We see that the deviations between I and the unmodified equation of state case increase with time. In particular from this figure, if the thermal freeze-out occurs at 110 MeV, the thermal freeze-out time is 13 fm for the modified equation of state and 20 fm for the unmodified one; the corresponding baryonic potentials are not very different, 405 and 375 MeV respectively. If the thermal freeze-out occurs earlier, say at 140 MeV, the difference in the thermal freeze-out times would be much less. An immediate consequence of this, is that the thermal freeze-out volume (in our case simply proportional to time) may be much smaller for the modified equation of state. This leads us to the main point of this section. *If the chemical and thermal freeze-out temperatures are very different* (in our simplified case, 180 and 110 MeV) *or if many particle species make an early chemical freeze out* (e.g. also pions), *it is important to take into account the effect of the early chemical freeze-out on the equation of state* to make predictions for observables which depend on thermal freeze-out volumes, for example particle abundances and eventually particle correlations.

If the chemical and thermal freeze-out temperatures are not very different (say 180 and 140 MeV), or if few particle species make the early freeze out, one can proceed as follows. One can use an unmodified equation of state in a hydrodynamical code and to account for early chemical freeze-out of species i, when the number of type i particles was fixed, use a modified Cooper-Frye formula

$$\frac{Ed^3N_i}{dp^3} = \frac{N_i(T_{ch.f.})}{N_i(T_{th.f.})} \times \int_{S_{th.f.}} d\sigma_\mu p^\mu f(x,p). \quad (7)$$

The second factor on the right hand side is the usual one and it gives the shape of the spectrum at thermal freeze-out, the first factor is a normalizing term introduced such that upon integration on momentum p, the number of particles of type i is $N_i(T_{ch.f.})$. Using HYLANDER-PLUS [25], we showed that that both the shapes of $m_\perp$ spectra and abundances can be reproduced for $T_{ch.f.} = 176$ MeV and $T_{th.f.} = 139$ MeV, while simultaneous freeze-outs at $T_{ch.f.} = T_{th.f.} = 139$ MeV would yield the correct shapes but too few particles. Therefore, results with HYLANDER-PLUS and a modified Cooper-Frye formula (7) support the separate freeze-outs picture. However in this code, $T_{th.f.}$ is fixed to 139 MeV while as already mentioned, some data seem to imply lower thermal freeze-out temperatures. In the next generation of hydrodynamical codes, it is desirable to consider a wider range of $T_{th.f.}$.

CONTINUOUS EMISSION

The notion that particle emission does not necessarily occur on a three dimensional surface but may be continuous was incorporated in a hydrodynamical description in [26]. In this model, the fluid is assumed to have two components, a free part plus an interacting part and its distribution function reads

$$f(x,p) = f_{free}(x,p) + f_{int}(x,p). \quad (8)$$

f_{free} counts all the particles that last scattered earlier at some point and are at time x^0 in $\vec{x}$. f_{int} describes all the particles that are still interacting (i.e. that will suffer collisions at time $> x^0$). The invariant momentum distribution is then

$$Ed^3N/dp^3 = \int d^4x D_\mu[p^\mu f_{free}(x,p)]. \quad (9)$$

$D_\mu[p^\mu f_{free}(x,p)]$ is a covariant divergence in general coordinates and d^4x is the invariant volume element. A priori formula (9) is sensitive to the whole fluid history and not just to freeze out conditions as in formula (1). We expect that heavy particles, due to thermal suppression, will bring information on early times, when they are more numerous. Slow particle will probe late times, when they finally are in diluted matter and make their last collision. Light particles such as the pion should probe the whole fluid history.

Our main task is to compute f_{free}. In previous works, 1) we wrote $f_{free} = \mathcal{P}f$ and $f_{int} = (1 - \mathcal{P})f$ or $f_{free} = \mathcal{P}/(1 - \mathcal{P})f$, where $\mathcal{P}$ is the proportion of free particles,

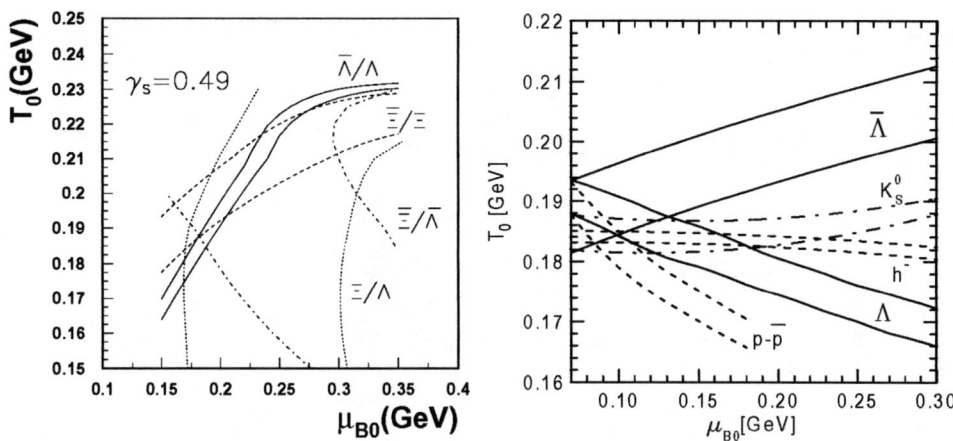

FIGURE 2. Window in the initial conditions that permits to reproduce WA85 ratios (left) with for a resonance gas with volume correction and NA35 abudances (right) for a resonance gas.

or alternatively the the probability to escape without collision, 2) we assumed that f_{int} can be approximated by a thermal distribution. This last assumption is rather strong so alternative formulations are under study[27]. Finally, we also made a third assuption: f_{int} is approximately thermal only until a fraction $\mathscr{P} \leq \mathscr{P}_F$ of particles are free. This leads to:

$$Ed^3N/dp^3 \sim \frac{1}{1-\mathscr{P}_F} \int_{\mathscr{P}=\mathscr{P}_F} f_{th}(x,p) p^\mu d\sigma_\mu \qquad (10)$$

Over the years we have made comparison of the predictions of continuous emission with a large number of data, using a simplified Bjorken hydrodynamics[26, 28]. We can use particle ratios from WA85 or NA35 abundances to fix the initial conditions of the fluid evolution. We find in figure 2 that $T_0 \sim 180-200$ MeV, $\mu_{B0} \sim 100-200$ MeV, $\gamma_0 \sim 0.5-0.7$. This paralels the use of abundances or ratios in the model of section 3, to fix chemical freeze out variables. However while in that model, the use of transverse mass spectra fixes thermal freeze out variables, we do not have such a freedom in the continuous emission scenario. Namely initial conditions have been fixed by the ratios or abundances and have to be used to predict transverse mass spectra. This is not a trivial test. Results are shown in figure 3. For light projectiles, we found that spectra of a variety of particles can indeed be reproduced. For heavy projectiles, Λ and $\bar{\Lambda}$ spectra, come out well, for Ξ and $\bar{\Xi}$ not too well, and for the Ω badly. This is not a surprise. The same problem happens for the freeze out model, where one assumes that the Ω must have decoupled both chemically and thermally early. In our case, we had supposed for simplicity that the cross sections needed to compute $\mathscr{P}$ were the same for all particles. Taking more realistic values, we can reproduce rather well the inclinations as seen in figure 4. Another piece of data which is important is the pion abundance. As a general rule, freeze out models, while they can reproduce strange particle abundances, underpredict the pion abundance. This was first noted by [29] in a study of NA35 data and emphasized

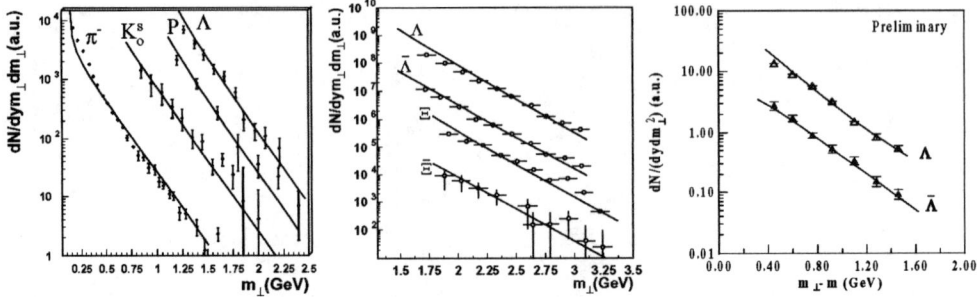

FIGURE 3. With initial conditions determined as in figure 2, transverse mass spectra are computed and compared with data: a) NA35 b) WA94 c) WA97.

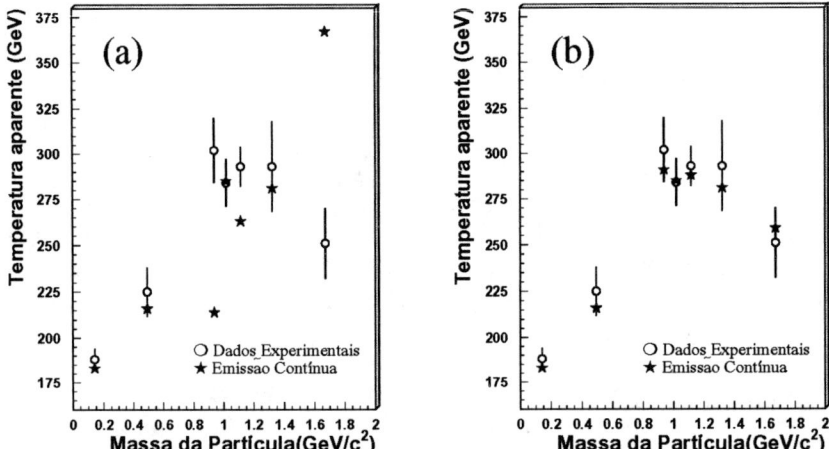

FIGURE 4. Comparison of experimental data and continuous emission predictions for effective temperatures: left all cross section equal, right more realistic cross sections.

by [30, 31] in an analysis of the WA85 strange particle ratios and EMU05 specific net charge $D_q \equiv (N^+ - N^-)/(N^+ + N^-)$ (with N^+ and N^-, the positive and negative charge multiplicity respectively). A similar problem arises with the Pb+Pb data from NA49 [32, 33]. Various possible improvements have been suggested so that a hadronic gas could yield both the correct strange particle and pion multiplicities: sequential freeze out[34], hadronic equation of state with excluded volume corrections[35, 36, 37, 33] non-zero pion chemical potential[29, 36, 32], equilibrated plasma undergoing sudden hadronization and immediate decoupling[30, 31, 38]. Given this dificulty of the more basic freeze out model, it is interesting to study pion production in the continuous case. Contrarely to the freeze out case, we expect that some entropy will be created, due to fact that e.g. we have continuous re-thermalization, an irreversible process. This excess entropy may appear as additional pions. In table 1, we show that with the initial conditions determined in figure 1, various abundances are reproduced, inclusive the pion one.

TABLE 1. Comparison between NA35 midrapidity abundances and prediction from continuous emssion for initial conditions determined in figure 1. For comparison results are also shown for freeze out at $T_{f.out} = T_0$ and $\mu_{bf.out} = \mu_{b0}$.

	experimental value	continuous emission	freeze out
Λ	1.26±0.22	0.96	0.92
$\bar{\Lambda}$	0.44±0.16	0.29	0.46
$p - \bar{p}$	3.2±1.0	3.12	1.32
h^-	26±1	27	15.7
K_S^0	1.3±0.22	1.23	1.06

It is however surprizing that the abundance of pion is so well reproduced, as it is not obvious that enough entropy would be created. This is still under investigation[28]. Finally, we mention that we also made comparison of Bose-Einstein corralations as predicted by a freeze out model and by continuous emission and have initiated study of continuous emission by a quark-gluon plasma. From these studies, we see that continuous emission is as compatible with data as freeze out, perhaps more if one thinks of the Ω temperature or pion abundance.

Before closing this section, we mention another type of approach which goes into the direction to incorporate the possibility that particle emission is a continuous process. Various groups have matched a hydrodynamical code describing the equilibrated fluid with a microscopical code for the emission the hadrons. This was pioneered by Bass and Dumitru[39], also studied by Teaney et al.[40], and a similar approch is planned by the Nantes group using Nexus for the initial conditions[41], SPHERIO for the hydro evolution[8, 42] and URQMD for the emision.

CONCLUSION

As must be obvious, given the number of approaches that exist to treat particle emisson and that I tried to review in the previous sections, particle emission in hydrodynamics is not a solved problem. Moreover *it is a problem that deserves careful attention because it is important to draw conclusions about dense matter.* To illustrate this simply, we remind that the pion abundance in the basic freeze out model with a hadron gas is too low and might mean that a quark-gluon plasma has been created. On the other side, in the continuous emission case, higher abundances of pions appear naturally.

ACKNOWLEDGMENTS

I take this opportunity to thank my collobarators over the past few years: from Bergen, Cs.Anderlik, L.P.Csernai, Zs.I.Lázar, V.K.Magas, from Frankfurt, H.Stöcker and W.greiner, from Rio, T.Kodama and from São Paulo, Y.Hama, T.Osada, S.Padula and O.Socolowski Jr. This work was partially supported by FAPESP (2000/04422-7,2000/05769-0) and CNPq (proc. 300054/92-0).

REFERENCES

1. K.Safarik, these proceedings.
2. F.Antinori, these proceedings.
3. W.Busza, these proceedings.
4. C.Roland, these proceedings.
5. J.Harris, these proceedings.
6. W.Zajc, these proceedings.
7. F. Cooper and G. Frye, Phys.Rev. D10 (1974) 186.
8. T.Osada, these proceedings.
9. L.Csernai, these proceedings.
10. Cs.Anderlik, L.P.Csernai, F.Grassi, W. Greiner, Y.Hama, T.Kodama, Zs.I.Lázár, V.K. Magas & H.Stöcker, Phys. Rev.C 59 (1999) 338. V.K. Magas, Cs.Anderlik, L.P.Csernai, F.Grassi, W. Greiner, Y.Hama, T.Kodama, Zs.I.Lázár & H.Stöcker, Heavy Ion Physics 9 (1999) 192; Phys. Lett. B459 (1999) 33; Nucl.Phys. A661 (1999) 596-599
11. L.Bravina et al., Phys.Lett. B354 (1995) 196; Phys. Rev.C60 (1999) 044905.
12. H.Sorge, Phys. Lett. B373 (1996) 16.
13. S.Bass et al., Phys. Rev.C69 (1999) 021902.
14. E.V.Shuryak, Nucl. Phys. A661 (1999) 119c.
15. U.Heinz, Nucl. Phys. A661 (1999) 141c.
16. J.Cleymans and K.Redlich, Phys. Rev. C60 (1999) 054908.
17. J.Sollfrank, J.Phys. G23 (1997) 1903.
18. N.Xu et al., NA44 collaboration, Nucl. Phys. A610 (1996) 175c.
19. U.E.Wiedermann, B.Tomásik and U.Heinz, Nucl. Phys. A638 (1997) 475c.
20. U.E.Wiedermann, Nucl. Phys. A661 (1999) 68c.
21. J.Rafelski, these proceedings.
22. D.Zschiesche et al., nucl-th/0101047.
23. N.Arbex, F.Grassi, Y.Hama & O.Socolowski Jr., Phys.Rev. C64 (2001) 064906.
24. J.D. Bjorken, Phys. Rev. D27 (1983) 140.
25. N. Arbex, U. Ornik, M. Plümer and R. Weiner, Phys.Rev. C55 (1997) 860.
26. F. Grassi, Y. Hama & T. Kodama, Phys. Lett. B355, 9 (1995); Z. Phys. C73, 153 (1996).
27. Yu.M.Sinyukov, S.V.Akelin & Y.Hama, nucl-th/0201015v4.
28. F.Grassi & O.Socolowski Jr., Phys. Rev. Lett. 80, 1170 (1998); J.Phys. G 25 (1999) 331; ibid. 339; F.Grassi, Y.Hama, T.Kodama, & O.Socolowski Jr. to be submitted; F.Grassi, Y.Hama, S.Padula & O.Socolowski Jr., Phys. Rev. C62 (2000) 044904.
29. N.J. Davidson et al., Z. Phys. C56, 319 (1992).
30. J. Letessier et al., Phys. Rev. Lett. 70, 3530 (1993).
31. J. Letessier et al., Phys. Rev. D51, 3408 (1995).
32. B.Tomášik,U.A.Wiedemann and U.Heinz, nucl-th/9907096.
33. G.D.Yen and M.I.Gorenstein, Phys.Rev.C59 (1999) 2788.
34. J.Cleymans et al., Z. Phys. C58 (1993) 347.
35. R.A.Ritchie et al. Z.Phys. C75 (1997) 535.
36. G.D.Yen et al. Phys. Rev.C 56 (1997) 2210.
37. G.D.Yen and M.Gorenstein, Phys. Rev. C59 (1999) 2788.
38. J.Letessier and J.Rafelski Phys. Rev. C59 (1999) 947.
39. Bass and Dumitru, Phys. Rev. C61 (2000) 064909.
40. Teaney et al. Phys. Rev. Lett.86 (2001) 4783.
41. K.Werner, these proceedings.
42. C.E. Aguiar, T. Kodama, T. Osada & Y. Hama, J.Phys. G27 (2001) 75; Nucl. Phys.A698 (2002) 639c.

The F_2^D Logarithmic Slope and the Saturation Phenomena

M. V. T. Machado

Instituto de Fisica, Universidade Federal do Rio Grande do Sul
Av. Bento Gonçalves, 9500. CEP 91510-970, Porto Alegre, RS. Brazil

Abstract. The logarithmic slope of the diffractive structure function, F_2^D, is a potential observable scanning the hard and soft contributions in diffraction, allowing to disentangle the QCD dynamics. We report our calculations concerning this quantity, in particular the estimates emerging from the saturation model applied to diffraction dissociation.

THE DIFFRACTIVE SLOPE

The measurements of the derivative quantity F_2-slope on Q^2 allowed a renewed interest of testing the matching of hard and soft approaches and provided constraints for the saturation formalisms. The reported turnover on the x dependence was readily associated with the transition region between the interplaying domains. The most recent determinations of the slope still deserve a better theoretical description, keeping a turnover pattern at fixed c.m.s. energy W. When we focus in diffractive DIS (DDIS), i.e. the structure function F_2^D, the situation is far from clear: initially considered as a predominantly soft process, the experimental results suggest that the diffractive cross section at HERA contains hard and soft components. Fortunately, if we consider the pQCD approach, diffraction stands a more profitable field to study saturation effects than the inclusive case. This comes from the fact that in DDIS the interaction probes larger dipole size configuration (soft content) than in the DIS reaction. We have proposed a derivative quantity, the F_2^D slope, which would help to disentangle the underlying dynamics in DDIS, settling its validity range, if such observable is measured. In Ref. [1] one perform studies of the proposed quantity for two sound models, representing the essential features in both Regge and pQCD formalisms. It was found that important deviations in the predictions between the models emerge, considering the available kinematic spectrum in diffractive DIS [2]. Here we report in particular the calculations considering the saturation pQCD model applied to diffraction dissociation and its comparison with the non-satured QCD approach, performed in details in Ref. [3].

In perturbative QCD, the $\gamma^* p$ process is described in terms of the photon splitting into a $q\bar{q}$ pair, far upstream of the nucleon, which then scatters in the proton. This reaction is mediated by the one gluon exchange which turns out into a multi-gluon one when the saturation region is approached. For transverse and longitudinally polarized photons the

$\gamma^* p$ cross section in the color dipole picture has the form

$$\sigma_{T,L}(x, Q^2) = \int d^2\mathbf{r} \int_0^1 d\alpha\, |\Psi_{T,L}(\alpha, \mathbf{r})|^2\, \hat{\sigma}(x, r^2), \qquad (1)$$

where $\Psi_{T,L}$ are the photon wavefunctions. The $q\bar{q}$ cross section, $\hat{\sigma}$, contains strong non-perturbative contributions which should be modeled. We choose the saturation model[4], which reproduces the experimental results at both inclusive and diffractive electroproduction. The dynamics of saturation is present in the effective dipole cross section in an eikonal way $\hat{\sigma}(x, r^2) = \sigma_0 \left[1 - \exp\left(-\frac{r^2}{4R_0^2(x)}\right)\right]$, where the x-dependent saturation scale is $R_0(x) = \frac{1}{Q_0}\left(\frac{x}{x_0}\right)^{\lambda/2}$. The parameters were determined from HERA data on F_2 and photoproduction cross section with $x < 10^{-2}$. Going to DDIS, the elastic scattering of the $q\bar{q}$ pair dominates the diffractive $\gamma^* p$ process for not too large values of the diffractive mass M_X. Instead, at large M_X the emission of a gluon becomes the leading contribution. The diffractive cross section is written in the color dipole picture, given the slope B_D, as

$$\sigma^D(x, Q^2) = \frac{1}{B_D}\frac{1}{16\pi} \int d^2\mathbf{r} \int_0^1 d\alpha\, |\Psi_{T,L}(\alpha, \mathbf{r})|^2\, \hat{\sigma}^2(x, r^2). \qquad (2)$$

The diffractive structure function is derived from Eq. (2) in the momentum space. We numerically calculated the F_2^D slope, using the same parameters values from Ref.[4]. In Fig. (1), one shows the $x_{I\!P}$ behavior for the approaches with (GW)[4] and without saturation (BW)[5], at typical β values. We analyze in particular the transition region between hard and soft dynamics ($Q^2 \sim 1.5 - 9$ GeV^2). The saturation model produces a transition between positive and negative slope values at low $\beta = 0.04$, while presents a positive slope for medium and large β. The pQCD approach without saturation[5], shows a positive slope for the whole Q^2 and $x_{I\!P}$ range. In the pQCD models without saturation, an ad hoc cutoff in the transverse momentum loop integration is inserted, as well as the energy dependence of the unintegrated gluon distribution. In the saturation model, instead, the saturation radius $R_0(x)$ gives the infrared cutoff (the saturation momentum scale) and determines the energy dependence. If this scale is large (1-2 GeV^2), then the resulting process is not soft and can be completely calculated using pQCD methods. Therefore, the saturation model extends the pQCD approach towards lower Q^2 values. We conclude that, the difference between the behaviors predicted by these two models for the $x_{I\!P}$ spectrum, mainly in the region of small β and medium Q^2, is large and which should allow to discriminate the dynamics in future experimental analyzes.

ACKNOWLEDGMENTS

The author thanks the Organizers of this lively meeting and the enjoyable discussions with J. Raufeisen, J. Osborne, E. Luna and M.A. Betemps. I am grateful to V.P. Gonçalves who collaborated in a joint way to obtain the results presented here.

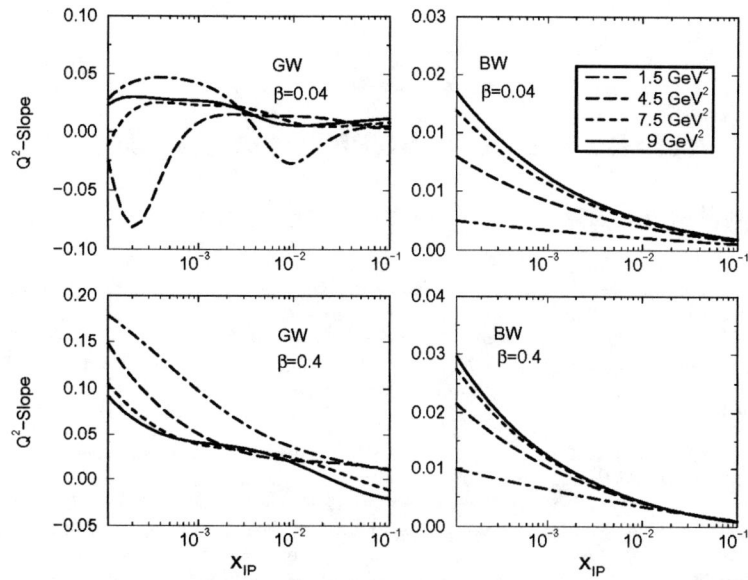

FIGURE 1. The $x_{I\!P}$ dependence on the logarithmic slope for the pQCD model (BW) and the the saturation model (GW) presented at typical β values.

REFERENCES

1. M.B. Gay Ducati, V.P. Gonçalves, M.V.T. Machado, *Phys. Lett.* **B506**, 52 (2001).
2. M.B. Gay Ducati, V.P. Gonçalves, M.V.T. Machado, in Proc. IXth Blois Workshop, Pruhonice near Prage, June 9-15 (2001) [hep-ph/0110025]; in Proc. XXXIth ISMD, Datong, China, Sep. 1-7 (2001) [hep-ph/0111444].
3. M.B. Gay Ducati, V.P. Gonçalves, M.V.T. Machado, *Nucl. Phys.* **A697**, 767 (2002).
4. K. Golec-Biernat, M. Wüsthoff, *Phys. Rev.* **D59**, 014017 (1999); *Phys. Rev.* **D60**, 114023 (1999).
5. J. Bartels, M. Wüsthoff, *J. Phys. G: Nucl. Part. Phys.* **22**, 929 (1996); J. Bartels, C. Royon, *Mod. Phys. Lett.* **A14**, 1583 (1999).

Small Quark Stars in the Chromodielectric Model

M. Malheiro*, E. O. Azevedo*, L. G. Nuss*, M. Fiolhais[†] and A. Taurines**

*Instituto de Física, Universidade Federal Fluminense, Av. Litorânea, 24210-310 Niterói, Brazil
[†]Departamento de Física and Centro de Física Computacional, Universidade de Coimbra,
P-3004-516 Coimbra, Portugal
**Instituto de Física, Universidade Federal do Rio Grande do Sul CP 15051, 91501-970
Porto Alegre, Brazil

Abstract. Equations of state for strange quark matter in beta equilibrium at high densities are used to investigate the structure (mass and radius) of compact objects. The chromodielectric model is used as a general framework for the quark interactions, which are mediated by chiral mesons, σ and $\vec{\pi}$, and by a confining chiral singlet dynamical field, χ. Using a quartic potential for χ, two equations of state for the same set of model parameters are obtained, one with a minimum at around the nuclear matter density ρ_0 and the other at $\rho \sim 5\rho_0$. Using the latter equation of state in the Tolman-Oppenheimer-Volkoff equations we found solutions corresponding to compact objects with $R \sim 5-8$ km and $M \sim M_\odot$. The phenomenology of recently discovered X-ray sources is compatible with the type of quark stars that we have obtained.

INTRODUCTION

Various effective models using quarks as fundamental dynamical fields, originally designed for the nucleon, have also been successfully used to describe infinite quark matter, and the resulting equations of state (EOS's) have been applied to investigate the structure of compact stars [1–6].

The chromodielectric model (CDM) [7–9], for example, provides a reasonable phenomenology for the nucleon [10, 11] and also allows us to obtain EOS's for dense quark matter. The model yields soliton solutions representing single baryons with three quarks dynamically confined by a scalar field, χ, whose quanta can be assigned to 0^{++} glueballs. When it is applied to quark matter in two or three flavors [12–14] the resulting EOS turns out to be relatively soft at large densities.

Using a quadratic potential for the χ field, Drago et al. [15, 16] applied the CDM to describe the inner part of neutron stars, obtaining masses in the range $1-2M_\odot$ and radii of the order 10 km or higher, with a small hadron crust of 2 km. In this work we consider an extension of the model used in Ref. [15], taking quartic instead of quadratic potentials. In addition to the structures found by Drago et al., the quartic model predicts another type of compact objects made out of quarks only, smaller and denser than neutron stars.

From the observational point of view, the recent discovery of X-ray sources, by the Hubble and Chandra telescopes, increased the plausibility that these sources might be strange quark stars [17–19]. In particular, the isolated compact object RX J1856.5-3574 with a small radius does not show evidence of spectral lines or edge features [20, 21], reinforcing the conjecture for the existence of stars made out of strange matter. The phenomenology of these objects seems to be compatible with the small and dense quark stars reported in this work.

THE MODEL

CDM Lagrangian

We write the CDM Lagrangian in the form [8–10]

$$\mathcal{L} = \mathcal{L}_q + \mathcal{L}_{\sigma,\pi} + \mathcal{L}_{q-\text{meson}} + \mathcal{L}_\chi , \qquad (1)$$

where

$$\mathcal{L}_q = i\bar\psi\gamma^\mu\partial_\mu\psi , \qquad \mathcal{L}_{\sigma,\pi} = \tfrac{1}{2}\partial_\mu\hat\sigma\partial^\mu\hat\sigma + \tfrac{1}{2}\partial_\mu\hat{\vec\pi}\cdot\partial^\mu\hat{\vec\pi} - W(\hat{\vec\pi},\hat\sigma) , \qquad (2)$$

and $W(\hat{\vec\pi},\hat\sigma)$ is the Mexican hat potential. In the u, d sector the quark-meson interaction is described by

$$\mathcal{L}_{q-\text{meson}} = \frac{g}{\chi}\bar\psi(\hat\sigma + i\vec\tau\cdot\hat{\vec\pi}\gamma_5)\psi . \qquad (3)$$

The last term in (1) contains the kinetic and the potential piece for the χ-field:

$$\mathcal{L}_\chi = \tfrac{1}{2}\partial_\mu\hat\chi\partial^\mu\hat\chi - U(\hat\chi) . \qquad (4)$$

The potential term is

$$U(\chi) = \frac{1}{2}M^2\chi^2\left[1 + \left(\frac{8\eta^4}{\gamma^2} - 2\right)\frac{\chi}{\gamma M} + \left(1 - \frac{6\eta^4}{\gamma^2}\right)\frac{\chi^2}{(\gamma M)^2}\right] , \qquad (5)$$

where M is the χ mass. The parameterization used in (5) allows for a physically meaningful interpretation of the parameters γ and η: $U(\chi)$ has a global minimum at $\chi = 0$ and a local one at $\chi = \gamma M$, and $U(\gamma M) = \eta^4 M^4$ (see Fig. 1). The height of the local minimum, $B = (\eta M)^4$, is interpreted as a "bag pressure" and this is used to fix the parameters in $U(\chi)$. Assuming the wide range $0.150 \leq B^{1/4} \leq 0.250$ GeV, one has $0.08 \leq \eta \leq 0.15$, using $M = 1.7$ GeV. We note that γ is not a free parameter since the quartic term of $U(\chi)$ must be positive and the cubic term negative, which implies $\gamma^2 \geq 6\eta^4$. In the soliton sector of the model, best nucleon properties are obtained for $G = \sqrt{gM} \sim 0.2$ GeV (only G matters for the nucleon properties) and we keep that G in our quark matter calculations.

In order to study strange quark matter, we add to the interaction Lagrangian (3) the term [22]

$$\mathcal{L}_{s-\text{meson}} = \frac{g_s}{\chi}\bar\psi_s\psi_s . \qquad (6)$$

accounting for the coupling between the strange quark and the χ field.

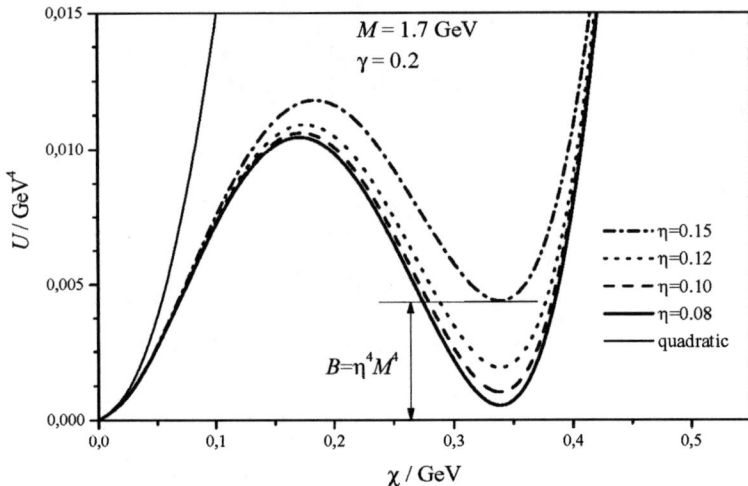

FIGURE 1. Quartic potential, Eq. (5), for fixed M and γ, in dependence of χ. For comparison, the quadratic potential ($\gamma \to \infty$) is also shown. The value of the potential at the second minimum can be interpreted as a bag pressure, B.

Strange quark matter

In order to study strange quark matter in beta equilibrium, an electron gas must also be considered. The mean-field energy per unit volume for strange quark matter in the CDM (plus electrons) is given by

$$\varepsilon = \alpha \sum_{f=u,d} \int_0^{k_f} \frac{d^3k}{(2\pi)^3} \sqrt{k^2 + m_f(\sigma, \chi)^2} + \alpha \int_0^{k_s} \frac{d^3k}{(2\pi)^3} \sqrt{k^2 + m_s(\chi)^2}$$
$$+ 2\int_0^{k_e} \frac{d^3k}{(2\pi)^3} \sqrt{k^2 + m_e^2} + U(\chi) + \frac{m_\sigma^2}{8 f_\pi^2}(\sigma^2 - f_\pi^2)^2, \qquad (7)$$

where the first two terms refer to quarks and the third one to the electrons, all described by plane waves. The degeneracy factor is $\alpha = 6$ (for spin and color). The last term is the Mexican hat potential (with $\vec{\pi} = 0$ and $f_\pi = 93$ MeV). The k_i in (7) are the Fermi momenta of quarks and electrons.

The quark masses in (7) are [22]: $m_{u,d} = g_{u,d}\sigma/(\chi f_\pi)$ and $m_s = g_s/\chi$ with the coupling constants given by $g_u = g(f_\pi + \xi_3)$, $g_d = g(f_\pi - \xi_3)$ and $g_s = g(2f_k - f_\pi)$ [$\xi_3 = -0.75$ MeV, $f_K = 113$ MeV].

A variational principle applied to the energy density, Eq. (7), leads to two gap equations for σ and χ. In the interior of a compact star the matter should satisfy both the electrical charge neutrality and chemical equilibrium. These conditions should supplement the gap equations, and altogether we have a system of six algebraic equations to solve at each baryon density $\rho = (\rho_u + \rho_d + \rho_s)$ [here, $\rho_i = \alpha k_i^3/(6\pi)^2$ stand for each flavor density]. The solution of the system of equations are the meson fields, σ and χ,

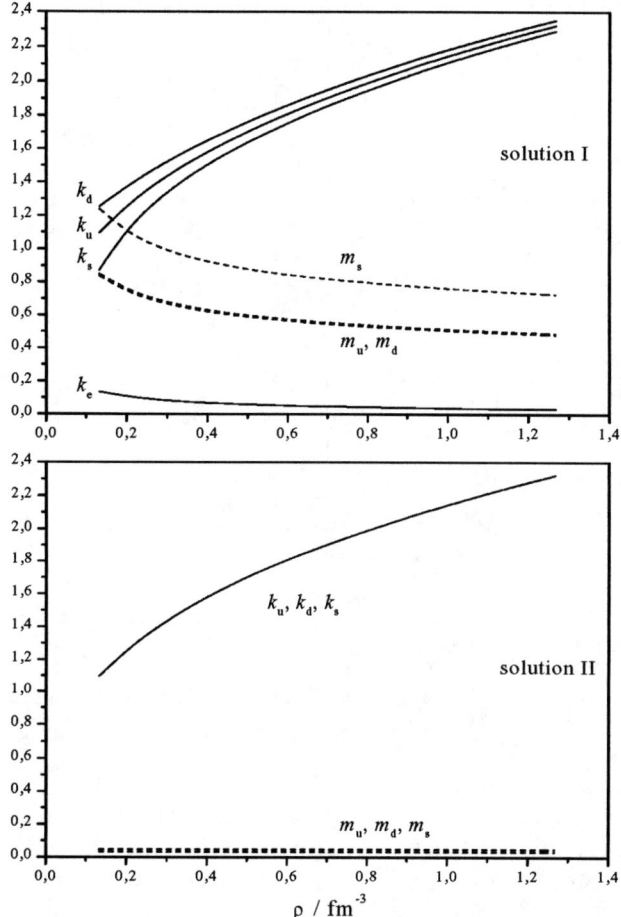

FIGURE 2. Fermi momenta (solid lines) and quark masses (dashed lines) for solution I and II in dependence of the baryon density. For solution II the quark masses and the electron Fermi momentum almost vanish. The model parameters are $g = 0.023$ GeV, $M = 1.7$ GeV, $\gamma = 0.2$ and $\eta = 0.12$. The vertical scales are in fm^{-1}.

and the Fermi momenta, k_u, k_d, k_s and k_e. For the same set of model parameters we found two stable solutions, hereafter denoted by I and II, shown in Fig. 2.

For both solutions σ is always close to f_π. In solution I, the χ field is a slowly increasing function of the density, remaining always smaller than ~ 0.05 GeV. For such a small χ, the quartic potential and the quadratic potential are indistinguishable (see Fig. 1), thus, in practice, solution I corresponds to the one obtained and used by Drago et al. [15] in the framework of the quadratic potential. Due to the smallness of the χ field, quark masses are large and the system is in a chiral broken phase. The solution II exhibits a large confining field, $\chi \sim \gamma M$ (local minimum of U), independent of the density. The

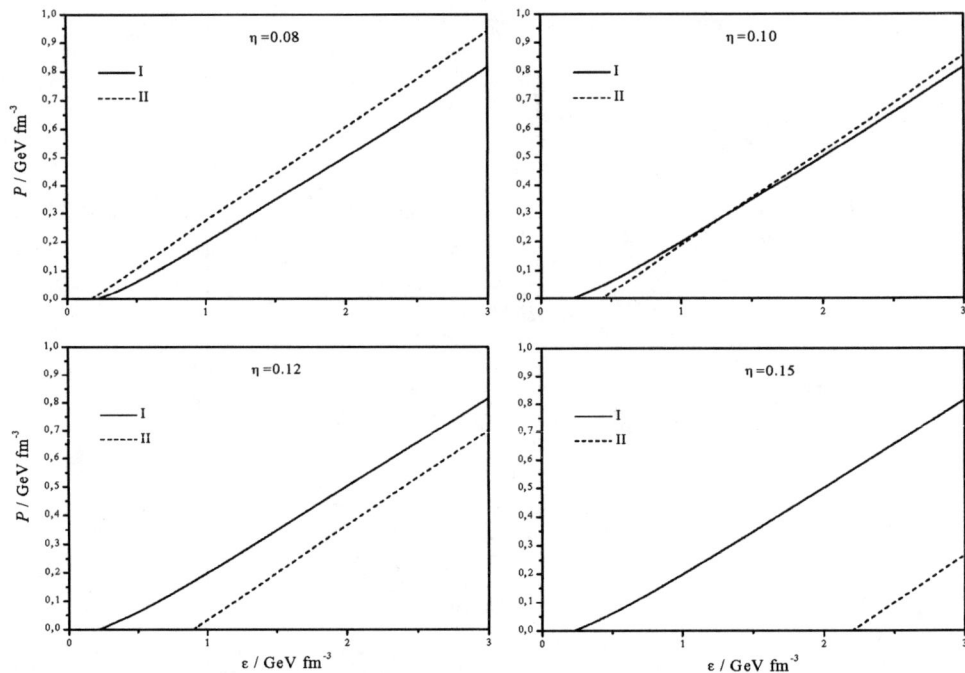

FIGURE 3. Pressure versus energy density for the two types of solution for various parameters η.] The other model parameters are $M = 1.7$ GeV, $g = 0.023$ GeV and $\gamma = 0.02$.

resulting quark masses are similar for the three flavors and very close to zero (chiral restored phase). Therefore, the chemical potentials in solution II are dominated by the Fermi momentum contribution, $\mu_u \simeq \mu_d \simeq \mu_s$ and $\mu_e \simeq 0$, i.e. in solution II there are almost no electrons. Besides solutions I and II, there is an additional *unstable* solution corresponding to $\chi \sim \gamma M/2$ [local maximum of $U(\chi)$].

Equations of state

The energy per baryon number as a function of the baryon density (EOS) is readily evaluated for each solution. We stress that EOS-I is not sensitive to γ and η (since χ is small), just depends on G. The saturation density occurs at a relatively low density ($\rho = 1.2 - 1.6\rho_0$) and the behavior of EOS-I at intermediate densities, is similar to the hadronic EOS's (see Ref. [13] for the two flavors sector). The EOS-II is also insensitive to γ, but does depend on η [in fact, the dependence is on $(\eta M)^4$, as we have already pointed out]: the energy per baryon number increases with η and so does the saturation density. For $\eta \sim 0.12$ the saturation density is $\rho \sim 5\rho_0$ and the energy per baryon number is some 20% higher than for solution I at its saturation density.

In Fig. 3 we present the EOS's in the form of pressure $P = \rho^2 \frac{\partial}{\partial \rho}\left(\frac{\varepsilon}{\rho}\right)$ versus energy density plots for solutions I and II and various η. It is worth noticing that our results with two distinct EOS predicted by the CDM, are consistent with the results from perturbative QCD: it is remarkable indeed that, for $\eta \sim 0.12$ the CDM reproduces accurately the EOS's recently obtained in a perturbative QCD calculation (compare our third panel of Fig. 3 with figure 1 of Ref. [23]).

Regarding energetics, both phases are almost degenerated at high densities and have similar shapes in the $P \times \varepsilon$ plane even at intermediate densities (or energy densities) in the narrow range $0.1 \leq \eta \leq 0.12$. In that region of η, one solution is not clearly lower in energy than the other. However, we should point out that they correspond to two different χ values and for the system to undergo a transition from the chiral restored to the chiral broken phase it has to go through a high potential energy barrier. In a 3D plot of the energy per baryon number versus (ρ, χ) the stable solutions correspond to two distinct "valleys", and the unstable solution mentioned at the end of the previous section corresponds to the top of the barrier between the two valleys.

QUARK STARS

In order to investigate the structure of stars we solved the Tolman-Oppenheimer-Volkoff (TOV) equation using the two EOS. Since EOS-I is identical to the one using a quadratic potential, it leads to stars that have the same phenomenology as the hybrid stars obtained by Drago et al. [15]: $R \sim 10 - 12$ km, a hadron crust and a mass $M \sim 1 - 2M_\odot$. At low densities, hadronization occurs and an hadronic equation of state should be used, replacing EOS-I.

Since EOS-II saturates at a high density and, in addition, the system is not likely to undergo a transition to solution I, one should not perform any connection to the hadronic sector: the EOS-II alone generates a new family of strange quark stars. In Fig. 4 it is shown the mass-radius relation for different values of η. These quark stars are smaller and denser in comparison with those resulting from EOS-I. For $\eta \sim 0.115$ (and $M = 1.7$ GeV, yielding $B^{1/4} \sim 0.195$ GeV) one obtains a maximum radius $R \sim 6$ km and a corresponding mass $M \sim 0.9 M_\odot$, which are the fitted radius and mass for the nearby compact object RX J 1856.5-3754 [20, 21]. According to our calculation, such star has a central density of $10\rho_0$ (ρ_0 is the nuclear matter density) and a central energy density $\varepsilon \sim 3 \times 10^{15}$ g/cm^3. At the edge, the density drops to $5\rho_0$ and $\varepsilon \sim 1.35 \times 10^{15}$ g/cm^3. The ratio ε/ρ remains approximately constant inside the star. The maximum period of the star, computed using the expression given in Ref. [24], is ~ 0.4 ms. From Fig. 4 one concludes that the mass-radius relation for these strange small stars mainly depends on the height of the local minimum of the χ potential.

CONCLUSIONS

Using a mean-field variational method we obtained two solutions for homogeneous strange matter in beta equilibrium using the CDM with a quartic potential, with its parameters fixed in the nucleon sector and to yield a reasonable bag constant.

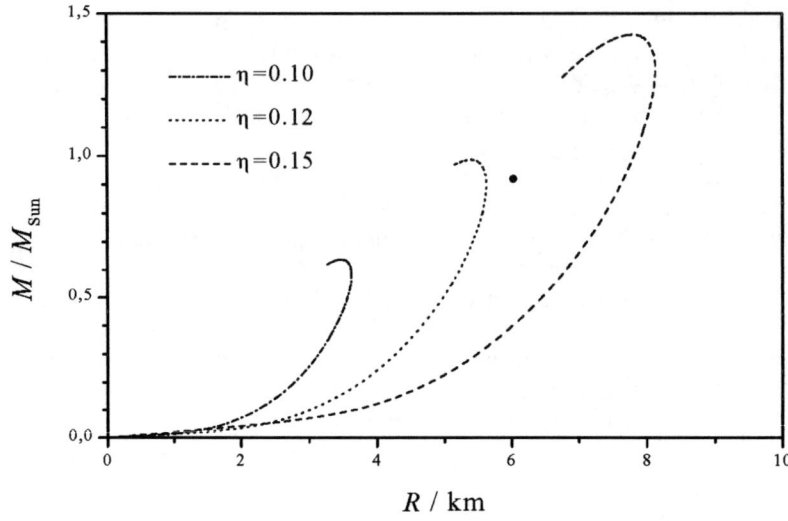

FIGURE 4. Mass versus radius for the pure quark stars (solution II) in the CDM model. The dot indicates the maximum radius star for $\eta = 0.115$. For the other model parameters see caption of Fig. 3.

One solution is similar to the already known solution for quadratic potentials in the CDM, with massive quarks. In the other solution, quarks are almost massless. The EOS for both solutions are similar to those found recently in the framework of perturbative QCD.

The pure quark stars emerging from the chiral symmetric solution are small and dense compact objects, showing a phenomenology compatible with the nearby X-ray source whose mass and radius was recently fitted [20, 21].

ACKNOWLEDGMENTS

This work was supported by FCT (POCTI/FEDER program), Portugal and by CNPq/ICCTI through the Brazilian-Portuguese scientific exchange program. We thank G. Marranghello and B. Garcia for some useful discussions. E.O.A. and L.G.N. acknowledge the support of the PIBIC/CNPq program for young researchers.

REFERENCES

1. N. K. Glendenning, *Compact Stars – Nuclear Physics, Particle Physics, and General Relativity* (Springer, New York, 1997)
2. F. Weber, *Pulsars as Astrophysical Laboratories for Nuclear and Particle Physics*, (IOP, UK, 1999)
3. H. Heiselberg, M. Hjorth-Jensen, *Phys. Rept.* **328**, 237 (2000)
4. M. Hanauske, L. M. Satarov, I. N. Mishustin, H. Stocker, W. Greiner, *Phys. Rev. D* **64**, 043005 (2001)
5. E. Witten, *Phys. Rev. D* **30**, 272 (1984); C. Alcock, E. Farhi, A. Olinto, *Astrophys. J.* **310**, 261 (1986); P. Haensel, J. L. Zdunik, R. Schaeffer, *Astron. Astrophys.* **160**, 121 (1986)

6. A. R. Taurines, C. A. Z. Vasconcellos, M. Malheiro, M. Chiapparini, *Phys. Rev. C* **63**, 065801 (2001)
7. H. B. Nielsen, A. Patkós, *Nucl. Phys. B* **195**, 137 (1982)
8. H. J. Pirner, *Prog. Part. Nucl. Phys.*, **29**, 33 (1992)
9. M. K. Banerjee, *Prog. Part. Nucl. Phys.*, **31**, 77 (1993)
10. T. Neuber, M. Fiolhais, K. Goeke, J. N. Urbano, *Nucl. Phys. A* **560**, 909 (1993)
11. A. Drago, M. Fiolhais, U. Tambini, *Nucl. Phys. A* **609**, 488 (1996)
12. S. K. Ghosh, S. C. Phatak, *Phys. Rev. C* **52**, 2195 (1995)
13. A. Drago, M. Fiolhais, U. Tambini, *Nucl. Phys. A* **588**, 801 (1995)
14. W. Broniowski, M. Čibej, M. Kutschera, M. Rosina, *Phys. Rev. D* **41**, 285 (1990)
15. A. Drago, U. Tambini, M. Hjorth-Jensen, *Phys. Lett. B* **380**, 13 (1996); *Prog. Part. Nucl. Phys.* **36**, 407 (1996)
16. A. Drago, A. Lavagno *Phys. Lett. B* **511**, 229 (2001)
17. I. Bombaci, A. V. Thampan, B. Datta, *Astrophys. J.* **541**, L71 (2000)
18. X. D. Li, I. Bombaci, M. Dey, J. Dey, E. P. J. van den Heuvel, *Phys. Rev. Lett.* **83**, 3776 (1999)
19. M. Dey, I. Bombaci, J. Dey, S. Ray, B. C. Samanta, *Phys. Lett. B* **438**, 123 (1998)
20. J. A. Pons, F. M. Walter, J. M. Lattimer, M. Prakash, R. Neuhäuser, P. An, *Astrophys. J.* **564**, 981 (2002)
21. J. J. Drake et al., astro-ph/0204159
22. J. A. McGovern, M. Birse, *Nucl. Phys. A* **506**, 367 (1990); *Nucl. Phys.* **506**, 392 (1990)
23. E. S. Fraga, R. D. Pisarski, J. Schaffner-Bielich, *Phys. Rev. D* **63**, 121702 (2001)
24. P. Haensel, J. L. Zdunik, *Nature* **340** 617 (1989)

Strangelets and Their Possible Astrophysical Origin

Luis Masperi[1]

Centro Latinoamericano de Física, Av. Venceslau Bráz 71 Fundos 22290-140 Rio de Janeiro, Brazil

Abstract. We discuss the first-order phase transition of QCD at high temperature in the universe and the possible formation of quark-matter lumps through cooling in regions of increased pressure. We show the similarity of results using confinement and breaking of chiral symmetry.

The inclusion of strange matter may give stability to small drops, or strangelets, in the colour-flavour locked phase possibly achieved in neutron stars that could reach high-altitude observatory like that of Chacaltaya as primary cosmic rays.

INTRODUCTION

The phase diagram of quantum chromo dynamics (QCD) is receiving a particular attention since it is becoming possible to achieve the quark-gluon plasma (QGP) through heavy ions collisions in accelerators at Brookhaven - $RHIC$ and CERN-SPS, and in the next future at ALICE.

The equivalent transition in the reversed direction generating hadrons in the primordial universe with the possible formation of nuggets of very dense quark-matter remnants has cosmological consequences like the baryonic matter content.

Furthermore quark matter stabilized by a strange component may exist in neutron stars which should be electrically neutral if pressure is high enough to reach the colour-flavour locked phase (CFL) with equal number of u, d and s. The collision of two neutron stars might eject small drops, strangelets, with a slight positive charge which could be absolutely stable if their equivalent atomic number is larger than a minimum.

In Section 2 we describe the first-order confinement transition through the generation of hadronic bubbles in the phase of independent quarks. When the transition is almost completed, lumps of quark matter may be pressed and survive with influence on the baryonic content necessary for primordial nucleosyn-thesis. The equivalent mechanism with vacuum which breaks the chiral symmetry is also considered.

In Section 3 the effect of stabilization due to the addition of a strange component is discussed with particular emphasys on the phases of QCD caused by the quark pairing at extremely high density.

Finally in Section 4 astrophysical conclusions are given about the possibility that strangelets reach high-altitude observatories like the Chacaltaya one and even evade the

---

[1] On leave of absence from Centro Atómico Bariloche, 8400 San Carlos de Bariloche, Argentina.

ordinary *GZK* cutoff giving another alternative for the explanation of ultra-high energy cosmic rays (*UHECR*).

FIRST-ORDER PHASE TRANSITION OF QCD

Considering that in high-temperature universe above $T_c \sim 100\ MeV$ we have non-interacting relativistic quarks, the grand-canonical partition function for fermions will give their pressure and density

$$p \simeq \frac{1}{6\pi^2} \int_0^\infty dk\, k^3\, \frac{1}{e^{\frac{1}{T}(k-\mu)}+1}\,, \tag{1}$$

$$\frac{N}{V} \simeq \frac{1}{2\pi^2} \int_0^\infty dk\, k^2\, \frac{1}{e^{\frac{1}{T}(k-\mu)}+1}\,. \tag{2}$$

For chemical potential $\mu < T$, the quark (antiquark) densities and pressures can be appreximated by

$$\frac{N_{q(\bar{q})}}{V} \simeq \frac{2.7}{\pi^2} T^3 \pm \frac{1.6}{\pi^2} T^2\, \mu_q\,, \tag{3}$$

$$p_{q(\bar{q})} \simeq \frac{3.8}{\pi^2} T^4 \pm \frac{2.7}{\pi^2} T^3\, \mu_q\,, \tag{4}$$

so that the baryon number density and total pressure are

$$\frac{N_q - N_{\bar{q}}}{V} = n_B \simeq \frac{3.2}{\pi^2} T^2\, \mu_q\,, \tag{5}$$

$$p_q + p_{\bar{q}} = p \simeq \frac{7.6}{\pi^2} T^4\,. \tag{6}$$

Therefore since the entropy density $s \sim T^3$ the normalized matter-antimatter asymmetry is

$$n_B s^{HT} \sim \frac{\mu_q}{T} \sim 10^{-10}\,, \tag{7}$$

from which $\mu_q \sim 10^{-2}\ eV$ at the transition temperature.

During the transition, when T is constant, the chemical potential is equal in both phases and in the confined one where $m_N \simeq 940\ MeV > T$ it turns out that

$$n_B s^{LT} \sim \left(\frac{m_N}{T}\right)^{\frac{3}{2}} e^{-\frac{m_N}{T}} \frac{\mu_q}{T} = \varepsilon\, n_B s^{HT}\,, \tag{8}$$

with $\varepsilon \sim 10^{-2}$, so that the matter-antimatter asymetry becomes diluted by the increase of the total entropy.

But the percolation process of hadronic bubbles ends when the universe expansion prevails on coalescence which occurs[1] for a radius of them

$$R_1 \sim \frac{1}{T_c}\left(\frac{m_{pl}}{T_c}\right)^{\frac{2}{3}} \sim 1\ cm, \qquad (9)$$

and afterwards a new stage of quark-matter bubbles starts.

If these lumps of high-temperature phase do not evaporate but keep the initial baryonic number $N_B \sim 10^{30}$ while quark-antiquark pairs annihilate, the free energy of each bubble will be

$$G \simeq R^3\,T^4 + R^3\,T^2\,\mu^2 \simeq R^3\,T^4 + \frac{N_B^2}{T^2\,R^3}. \qquad (10)$$

Since initially, for $R \simeq R_1$, $\mu \ll T$ the first term dominates and the bubble contracts, but afterwards μ inside increases and is no longer equal to the value outside. Therefore, if the process is fast enough to consider T constant, the contraction will stop when

$$\frac{\partial G}{\partial V} = 0 \simeq T^4 - \frac{N_B^2}{T^2}\frac{1}{V^2} \qquad (11)$$

so that $\bar{R} \simeq \frac{N_B^{\frac{1}{3}}}{T} \simeq 10^{-3}\ cm$ and $\mu \sim T$, i.e. we are reaching the limit of validity of Eqs.(3, 4).

For $\mu > T$ and when only quarks remain Eqs.(1, 2) can be approximated by

$$n_B \simeq \frac{\mu_q^3}{6\pi^2}, \quad p \simeq \frac{\mu_q^4}{24\pi^2}. \qquad (12)$$

It is easy to see that the decrease of volume for the annihilation of all the antiquarks in terms of the initial volume $(R \simeq R_1)$ and chemical potential (equal inside and outside the bubble) is

$$\Delta V \simeq V^{in}\left(1 - \frac{3.2}{5.4}\frac{\mu_q^{in}}{T}\right), \qquad (13)$$

so that the volume for the beginning of the case of Eq.(12) is

$$\bar{V} \sim \frac{\mu_q^{in}}{T} V^{in} \simeq 10^{-10}\,V^{in}, \qquad (14)$$

corresponding to $\bar{R} \sim 10^{-3}\ cm$. This confirms that the stabilization of the bubble occurs in the intermediate region of validity between the two cases $\mu \lesssim T$.

One thinks that confinement and breaking of chiral symmetry are related phenomena. The phase transition for the latter is understood as a change of vacuum with a corresponding decrease of energy $\Delta U_{vac}(T)$ with the limits $\Delta U_{vac}(T_c) = 0$ and $\Delta U_{vac}(0) \simeq \Lambda_{QCD}^4$. If one considers a bubble of high-temperature vacuum, it contracts like a fluid so that its free energy will be

$$G \simeq \Delta U_{vac}\,R^3 + p\left(R_1^3 - R^3\right), \qquad (15)$$

with $pR^3 \sim T$. The contraction will stop for

$$\overline{R} \simeq \left(\frac{R_1^3 T}{\Delta U_{vac}}\right)^{\frac{1}{6}} \sim 10^{-3} \ cm, \tag{16}$$

if $\Delta U_{vac} \sim T^2 \mu_q^{in\ 2}$. It is clear that with the vacuum interpretation the chemical potential has no physical meaning and is introduced for normalization.

One must remark that the chemical potential had to be zero in the epoch of baryon number non-conservation which generated the matter-antimatter asymmetry. The value $\mu_q \sim 10^{-2} \ eV$ for quarks near the confinement transition is what is needed for $\frac{n_B}{s}$ to give the primordial nucleosynthesis. But after the phase transition most of baryonic number is kept inside the bubbles which are not expected to remain stable for decreasing temperature. Since these lumps do not participate to nucleosynthesis, their lifetime may affect what is normally predicted from the evidence of this process regarding cosmology.

STABILIZED STRANGE QUARK MATTER

It is possible to see that in the approximation of negligible mass of quark s, strange quark matter may be more stable than that of only almost massless u and d. In fact for the latter case the quark number densities must be

$$\frac{N_u}{V} \sim \mu^3 \quad , \quad \frac{N_d}{V} \sim 2\mu^3, \tag{17}$$

if we wish to have neutral matter. Therefore the pressure will be

$$p \sim \mu^4 + 2^{\frac{4}{3}} \mu^4. \tag{18}$$

With the addition of the strange quark component, equal number densities of quarks u, d and s will give neutral matter and a common chemical potential $\widetilde{\mu}$ in terms of which

$$p \sim 3\widetilde{\mu}^4. \tag{19}$$

With the condition of equal pressure in both cases Eqs.(18, 19) the ratio of kinetic energy of $1\ u$, $1\ d$, $1\ s$ and $1\ u$, $2\ d$ is

$$\frac{3\widetilde{\mu}}{\mu + 2\ 2^{\frac{1}{3}} \mu} \simeq 0.89, \tag{20}$$

giving more stability to strange quark-matter.

Intuitively surface effects give a positive charge to lumps of strange quark-matter[2]. This is because the larger mass of s makes the corresponding wave function vanish on the surface decreasing its number there.

At low temperature and very high pressure QCD has a superconducting phase [3] which could lead to more stable strangelets. The appearance of this new state of matter may be understood because due to asymptotic freedom the distribution of quarks, which depends on $E - N\mu$, does not change if one is added on Fermi surface increasing the total energy E in μ and the number in 1. But including the attractive exchange of gluon between a pair produces a bound E_B which gives a corresponding decrease of $E - N\mu$ favouring the condensation and a fermionic gap.

There are two situations[4] according with the number of flavours which are taken as massless.

With two light quarks u and d, the condensate is

$$< q^{\alpha}_{iL(R)} \, C \, q^{\beta}_{jL(R)} > \; \sim \Delta \, \varepsilon_{ij} \, \varepsilon^{\alpha \beta 3}, \tag{21}$$

with colour indices α, $\beta = 1, 2, 3$ and flavour ones $i, j = 1, 2$, being the charge conjugation matrix $C = i \, \gamma^2 \, \gamma^0$. This colour superconductivity of 2 quarks ($2SC$) corresponds to a symmetry breaking

$$SU(3)_c \times SU(2)_L \times SU(2)_R \longrightarrow SU(2)_c \times SU(2)_L \times SU(2)_R$$

The chiral global symmetrry $SU(2)$ remains unbroken, the fermionic gap affects colours 1 and 2 and five gluons become massive.

The other possible condensate occurs for three light quarks u, d and s ($i, j = 1, 2, 3$)

$$< q^{\alpha}_{iL(R)} \, C \, q^{\beta}_{jL(R)} > \; \sim \Delta \, \delta^{\alpha}_i \, \delta^{\beta}_j + \Delta' \, \delta^{\alpha}_j \, \delta^{\beta}_i$$

corresponding to the breaking

$$SU(3)_c \times SU(3)_L \times SU(3)_R \times U(1)_B \times U(1)_A \longrightarrow SU(3)_{c+L+R} \times Z_2 \times Z_2$$

In this colour-flawour locking (CFL) the 8 gluons become massive and all the fermions are gapped. The global chiral $SU(3)$ symmetry is broken, as well as the baryonic $U(1)_B$ and the axial $U(1)_A$ giving therefore a total of $8 + 2 = 10$ Goldstone bosons.

This latter condensate is more suitable for stable strangelets[5].

In fact this bulk phase is electrically neutral because of the equal number of u, d and s, and more stable than ordinary quark matter due to the pairing. The surface effect decreases the s contribution and makes the CFL strangelet certainly positive avoiding catastrophic possibilities analyzed in general context[6]. This strangelet may be absolutely stable, i.e. having energy per baryon $< 938 \, MeV$, for atomic number $A > A_{\min} \sim 300$. The relation with charge is $Z \propto A^{\frac{2}{3}}$ because Z depends on surface whereas A comes from all the volume. This law is different from that of ordinary strangelets which is $Z \propto A$ for small ones and $Z \propto A^{\frac{1}{3}}$ for $A \gtrsim 150$.

ASTROPHYSICAL CONCLUSIONS

Presumably the strangelets formed at the beginning of universe were ordinary and decayed by weak interaction emitting neutrons.

On the other hand, CFL phase could be formed in neutron stars and from collision of two of them the corresponding strangelets might be emitted and reach our atmosphere.

Their initial $A \sim 10^3$ may decrease through collisions and finally they should disintegrate when becomes $< A_{min}$.

There are events in cosmic rays which can correspond to strangelets: 2 with [7] $Z \simeq 14$ and $A \simeq 350, 450$; 1 with[8] $Z \simeq 46$ and $A > 1000$; 1 with[9] $Z = 20$ and $A \simeq 460$ and the Centauro[10] of $A \sim 200$.

Strangelets, due to their large A, propagate in CMB with small loss of energy[11] and have a GZK cutoff $\sim 10^{21}$ eV, larger than that for protons, so that they might explain the $UHECR$.

If their initial atomic number is ~ 2000 they could arrive to a high-altitude observatory, like the Chacaltaya[12] one, with $A > A_{min}$ and be detected directly apart from the EAS which would have an energy spectrum of secondary hadrons softer than that caused by primary protons.

The possible detection of strangelets is also planned[13] with the AMS-02 on the International Space Station for 2005.

ACKNOWLEDGMENTS

Thanks are due to Francesco Calogero and Oscar Saavedra for extremely stimulatings comments regarding theoretical and experimental aspects of strangelets, and to Erasmo Ferreira and all the organizers of the Pan American meeting.

REFERENCES

1. E. Witten, Phys. Rev. D **30** (1984) 272.
2. J. Madsen, Phys. Rev Lett. **85** (2000) 4687.
3. K. Rajagopal and F. Wilczek, hep-ph/0011333.
4. R. Casalbuoni, hep-ph/0110107.
5. J. Madsen, hep-ph/0108036.
6. A. Dar, A. De Rujula and U. Heinz, Phys. Lett. B **470** (1999) 142; F. Calogero, to appear in Interdisciplinary Science Revieos; A. Kenz, hep-ph/0009130.
7. T. Saito, Y. Hatano and Y. Fukada, Phys. Rev. Lett. **65** (1990) 2094.
8. P.B. Price et al., Phys. Rev. D **18** (1978) 1382.
9. M. Ichimura et al., Nuovo Cim. A **106** (1993) 843.
10. C. Lattes, Phys. Rep. **65** (1980) 151.
11. M. Rybazynski and Z. Wlodarczyk, hep-ph/0109225.
12. O. Saavedra and P. Vallania, Nucl. Phys. B Suppl. Vol. **92**, No. 1-2, Feb. 2001.
13. J. Madsen, hep-ph/0112153.

Is Nuclear Multifragmentation Related with the Liquid-Gas Phase Transition?

D. P. Menezes* and C. Providência[†]

*Dep. de Física - CFM - Universidade Federal de Santa Catarina - Florianópolis - SC - CP. 476 - CEP 88.040 - 900 - Brazil
[†]Centro de Física Teórica - Dep. de Física - Universidade de Coimbra - P-3004 - 516 Coimbra - Portugal

Abstract. In this work we show a formalism based on the liquid-gas coexistence in order to calculate the caloric curve for Sm isotopes. The results are compatible with recent experimental data for heavy ion collisions where nuclear multifragmentation takes place.

The determination of the properties of the nuclear matter as functions of density, temperature and neutron-proton composition is an important problem in contemporary nuclear and astrophysics. Within the framework of relativistic models, one can investigate the above mentioned properties as well as the possibility of a liquid-gas phase transition and consequent droplet formation. The arising droplets properties as neutron skin thickness, surface energy, central density, proton and neutron radius, excitation energies, etc, can also be calculated.

The production of several intermediate mass fragments in a short time scale during heavy ion collisions is known as nuclear multifragmentation. In multifragmentation experiments, an equilibrated system is always formed and the behaviour of the energy fluctuations suggests that the system undergoes a liquid–gas phase transition. The caloric curve, which is given by the excitation energy per nucleon in terms of the thermodynamic temperature is an important quantity to be investigated in the search for a phase transition.

Droplet formation in the liquid-gas phase transition in cold [1, 2] and hot [3] asymmetric nuclear matter in the context of relativistic mean field theory, namely using the non-linear Walecka model (NLWM) [4] has been studied recently using a Thomas Fermi approximation. In this work we compare the results obtained with the above mentioned formalism with recent experimental results coming from heavy-ion collisions, where nuclear multifragmentation takes place [5, 6].

The main steps of our calculation and numerical procedure are the following: we choose a relativistic model (non-linear Walecka model) and adequate parameter sets [7, 8, 9]. The equations of motion are obtained through the Euler-Lagrange equations and we have to solve them, which are second order coupled differential equations. In order to establish appropriate boundary values, we look at which conditions phase coexistence is possible in infinite nuclear matter. Nuclear matter is studied with simple mean-field approximation (constant meson fields) and stability conditions are imposed. The binodal sections for each parametrization and each temperature is built [10]. Once the boundary

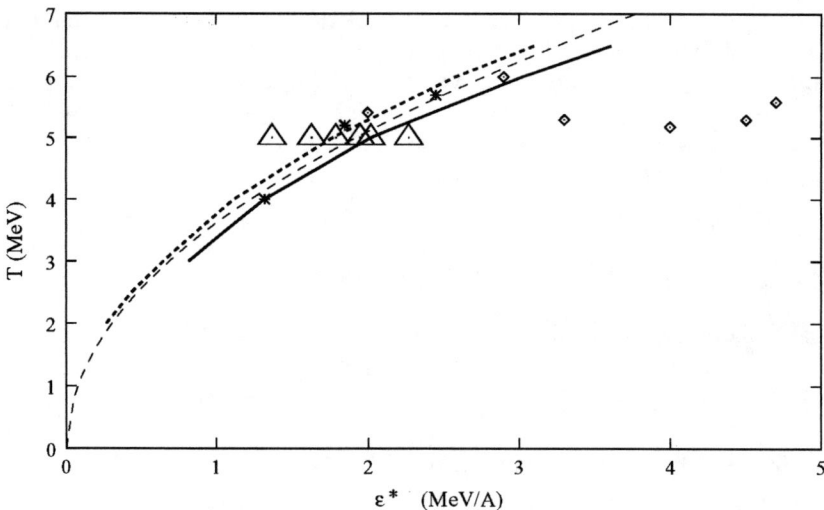

FIGURE 1. The caloric curves are shown for $^{166}_{62}Sm_{104}$ ($Y_p = 0.37$ - thick dashed line) and $^{150}_{62}Sm_{88}$ ($Y_p = 0.41$ - thick full line); the excitation energies at $T = 5$ MeV, for different proton fractions (big triangles, from the left to the right $Y_p = 0.33, 0.35, 0.37, 0.39, 0.41$ and 0.44), the experimental data from [5] (diamonds) and [6] (stars), and the Fermi-gas law [6] ($k = 13.0$ - thin dashed line) are also displayed.

conditions are found, the set of differential equations is solved. In this work we have restricted ourselves to samarium isotopes because of the available experimental data.

In table 1, we compare for the nuclei $^{150}_{62}Sm_{88}$ and $^{166}_{62}Sm_{104}$, respectively, the results obtained for the surface energy, the neutron skin thickness, the central density and the excitation energy for different parameterizations. We have considered three values of the temperature and included the Coulomb interaction. One can notice that the surface energy and the central density decrease with the increase of the temperature, while the excitation energy per particle increase.

At $T = 0$, the results given by all the parameterizations considered are quite similar. This reflects the fact that all the parameterizations have been fitted to the ground-state properties of stable (NL1) or stable and unstable (NL3 and TM1) nuclei. However, for finite temperatures the differences appear. The *neutron skin thickness*, Θ, is not sensitive to the temperature, but for a given T, it assumes smaller values for NL3 and TM1 than for NL1. This behaviour was already discussed in [11] for $T = 0$ and is related with the smaller values predicted to the symmetry energy for nuclear matter in NL3 (37.4 MeV) and in TM1 (36.9 MeV) as compared to NL1 (43.5 MeV). Also related with these values are the slightly slowlier increasing of the excitation energies with temperature for the NL3 and TM1 parameterizations.

In figure 1 we show the caloric curves obtained with the Coulomb interaction, for $^{150}_{62}Sm_{88}$ (thick full line) and $^{166}_{62}Sm_{104}$ (thick dashed line), the excitation energies at $T = 5$ MeV, for different proton fractions (big triangles, from the left to the right $Y_p = 0.33, 0.35, 0.37, 0.39, 0.41$ and 0.44), the experimental data from [5] (diamonds) and [6] (stars), and the Fermi-gas law $\varepsilon^* = 1/kT^2$, with $k = 13.0$ (thin dashed line). We conclude that the excitation energy for ^{166}Sm (thick dashed curve), proton fraction 0.37, increases slowlier with temperature than for ^{150}Sm (thick full curve), proton fraction 0.41. These two curves are consistent with data of [6] and a level density parameter A/k, $k = 13.0$ in the Fermi gas model relation. The calculation at $T = 5$ MeV, for the proton fractions represented in the figure, shows that the caloric curve is sensitive to the proton-neutron ratio in the compound nucleus.

ACKNOWLEDGMENTS

One of the authors (DPM) thanks the organizers of the PASI 2002 for the invitation to give a seminar and CNPq for partially supporting the work.

REFERENCES

1. D.P. Menezes and C. Providência, Nucl. Phys. A **650** (1999) 283.
2. G. Krein, D. P. Menezes, M. Nielsen and C. Providência, Nucl. Phys. A **674** (2000) 125.
3. D.P. Menezes and C. Providência, Phys. Rev. C **60** (1999) 024313.
4. J. Boguta and A. R. Bodmer, Nucl. Phys. A **292**, (1977) 413; A.R. Bodmer and C.E. Price, Nucl. Phys. A **505** (1989) 123; A.R. Bodmer, Nucl. Phys. A **526** (1991) 703.
5. J. Phochodzalla et al., Phys. Rev. Lett. **75** (1995) 1040.
6. K. Hagel et al., Nucl. Phys. **486** (1988) 429.
7. P.-G. Reinhard, M. Rufa, J. Maruhn, W. Greiner and J. Friedrich, Z. Phys. A **323** (1986) 13.
8. G. A. Lalazissis, J. König and P. Ring, Phys. Rev. C **55** (1997) 540.
9. K. Sumiyoshi, H. Kuwabara, H. Toki, Nucl. Phys. A **581** (1995) 725.
10. M. Barranco and J.R. Buchler, Phys. Rev. C **22** (1980) 1729.
11. Y.K. Gambir, P. Ring, A. Thimet, Ann. Phys. **198** (1990) 132; H. Berghammer, D. Vretenar and P. Ring, Nucl. Phys. A **560** (1993) 1014.

TABLE 1. **A:** Output results given by the solution of the coupled differential equations with the inclusion of the electromagnetic field, for $^{150}_{62}\text{Sm}_{88}$ (Y_p=0.41). For different temperatures, the results obtained with several parameterizations are shown.
B: Same output results for $^{166}_{62}\text{Sm}_{104}$ (Y_p=0.37).

A: $^{150}_{62}\text{Sm}_{88}$	T=0 MeV			T=5 MeV			T=6 MeV		
	NL1	NL3	TM1	NL1	NL3	TM1	NL1	NL3	TM1
σ (MeV fm^{-2})	1.000	1.007	0.987	0.810	0.857	0.874	0.650	0.725	0.762
Θ (fm)	0.16	0.11	0.09	0.16	0.12	0.10	0.15	0.11	0.10
$\rho(0)$ (fm^{-3})	0.142	0.142	0.139	0.131	0.134	0.133	0.121	0.127	0.127
$\varepsilon^*(T)$ (MeV/A)	0.00	0.00	0.00	2.02	1.93	1.87	3.01	2.79	2.67

B: $^{166}_{62}\text{Sm}_{104}$	T=0 MeV			T=5 MeV			T=6 MeV		
	NL1	NL3	TM1	NL1	NL3	TM1	NL1	NL3	TM1
σ (MeV fm^{-2})	0.920	0.933	0.917	0.660	0.723	0.742	0.530	0.615	0.640
Θ (fm)	0.32	0.25	0.22	0.32	0.26	0.23	0.31	0.25	0.22
$\rho(0)$ (fm^{-3})	0.138	0.139	0.137	0.122	0.128	0.127	0.113	0.120	0.122
$\varepsilon^*(T)$ (MeV/A)	0.00	0.00	0.00	1.79	1.74	1.68	2.73	2.54	2.45

QCD Sum Rules Applications to the J/ψ Dissociation

M. Nielsen*, F. S. Navarra*, R. S. Marques de Carvalho†, G. Krein† and M. E. Bracco**

*Instituto de Física, USP, C.P. 66318, 05315-970 São Paulo, SP, Brazil
†Instituto de Física Teórica, UNESP, Rua Pamplona 145, 01405-900 São Paulo, SP, Brazil
**Instituto de Física, UERJ, Rua São Francisco Xavier 524, Maracanã, 20559-900, Rio de Janeiro

Abstract. We use the QCD sum rules to evaluate the $D^*D\pi$ and $DD\rho$ form factors as well as the $J/\psi \pi \to \bar{D}D^*, D\bar{D}^*, \bar{D}^*D^*$ and $\bar{D}D$ cross sections. The form factors are evaluated as a function of the momentum Q^2 of the D meson and the light meson and the cross sections as a function of $\sqrt{s}$.

INTRODUCTION

Matter at very high energy density in thermal equilibrium is believed to form a Quark Gluon Plasma (QGP). This gas of almost free quarks and gluons is thought to be the proper description of matter when energy densities are larger than about 1 GeV/fm^3. This is about the energy density inside a proton or neutron and is about one order of magnitude larger than that of nuclear matter. Matter of this energy density occurs naturally in the cores of neutron stars, and is supposed to be formed in heavy-ion collisions at high energy. The central goal of the heavy-ion program at the RHIC at Brookhaven National Laboratory is to identify new forms of matter like the QGP. Various signatures of the QGP creation have been proposed, and the suppression of J/ψ mesons [1] is among the most promising ones. Experimental investigation of charmonium production in nuclear reactions has been carried out for over a decade at CERN-SPS (p+A, O+U, S+U, Pb+Pb) [2, 3, 4] and at Fermilab Tevatron (p+A) [5]. These studies have shown that the J/ψ and ψ' production on nuclear target is indeed suppressed relative to expectations from nucleon-nucleon reactions.

For p+A collisions the observed suppression of J/ψ and ψ' can be explained as absorption of a common precursor, on nucleons [6]. This systematics extends to O+U and S+U reactions for J/ψ, but for ψ' additional suppression is observed [3]. This additional suppression can be quantitatively described as absorption of the ψ' on comoving hadrons. Only recent data from Pb+Pb collision [4] have revealed the presence of an additional "anomalous" suppression mechanism for J/ψ.

However, there are calculations that reproduce the new NA50 data reasonably well up to the highest transverse energies, based on hadronic J/ψ dissociation alone. The main ingredient in the calculations based on hadronic J/ψ dissociation is the magnitude of the J/ψ absorption cross section by hadrons, which is not known experimentally. Therefore, in order to use the J/ψ suppression as a signature for the formation of the QGP in heavy ion collisions it is important to have a better knowledge on the interactions between charmonium states and hadrons. Part of these interactions happens in the early stages of the nucleus - nucleus collisions and therefore at high energies ($\sqrt{s} \simeq 10-20$ GeV), and one may try to apply perturbative QCD. However, even in this regime, nonperturbative effects may be important [7]. Interestingly, estimates using quite different methods give results clustering around the value of $3-5$ mb in this energy range. On the other hand, a significant part of the charmonium - hadron interactions occurs when other light particles have already been produced, forming a "fireball". Interactions inside this fireball happen at much lower energies ($\sqrt{s} \leq 5$ GeV) and one has to apply nonperturbative methods.

One possible nonperturbative reaction mechanism is meson exchange, which can be studied by means of effective lagrangians [8, 9]. However, the couplings and form factors needed in these effective lagrangians are not phenomenologically known.

Here we report the use of the QCD sum rule (QCDSR) method [10], based on the three-point function, to evaluate the $D^*D\pi$ and $DD\rho$ hadronic form factors and coupling constants, need in the calculation of the $J/\psi - \pi(\rho)$ cross section using meson exchange models. We also report the use of a QCD sum rule four-point function to directly evaluate the $J/\psi - \pi$ cross section.

THE QCDSR CALCULATION

Following the QCDSR formalism, the three-point function associated with a $HH'M$ vertex, where H and H' are the external mesons and M is the exchanged meson, is given by:

$$\Gamma_M(p,p') = \int d^4x d^4y \langle 0|T\{j_{H'}(x)j_M(y)j_H^\dagger(0)\}|0\rangle e^{ip'\cdot x} e^{-i(p'-p)\cdot y}, \qquad (1)$$

where j_i is the interpolating fields of the hadron i, written in terms of the quark content and with the same quantum numbers of these hadrons.

The phenomenological side of the vertex function, $\Gamma_M(p,p')$, is obtained by the consideration of H and H' state contribution to the matrix element in Eq. (1):

$$\Gamma_M^{phen}(p,p') = \lambda_H \lambda_{H'} \lambda_M g^{(M)}_{HH'M}(q^2) \Delta_{H'}(p) \Delta_H(p') \Delta_M(q), \qquad (2)$$

where Δ_i is the propagator of the meson i, $q = p - p'$, and $\lambda_H = \langle 0|j_H|H\rangle$ defines the couplings of the currents with the respective hadronic states, which is proportional to the meson decay constant. The form factor we want to determine, for an off-shell meson M, are represented by $g^{(M)}_{HH'M}(q^2)$.

The contribution of higher resonances and continuum in Eq. (2) will be taken into account as usual in the standard form of ref. [11], through the continuum thresholds s_0 and u_0, for the H and H' mesons respectively.

The QCD side, or theoretical side, of the vertex function is evaluated by performing Wilson's operator product expansion (OPE) of the operator in Eq. (1). Writing Γ_M in terms of the invariant amplitudes, we can write a double dispersion relation for each one of the invariant amplitudes, over the virtualities p^2 and p'^2 holding q^2 fixed:

$$\Gamma(p^2, p'^2, q^2) = -\frac{1}{4\pi^2} \int_{m_Q^2}^{s_0} ds \int_{m_{Q'}^2}^{u_0} du \frac{\rho(s,u,q^2)}{(s-p^2)(u-p'^2)}, \quad (3)$$

where $\rho(s,u,q^2)$ equals the double discontinuity of the amplitude $\Gamma(p^2, p'^2, q^2)$ and $m_Q(m_{Q'})$ is the mass of the heaviest quark, $Q(Q')$ in the meson $H(H')$.

As usual, our goal is to make a match between the two representations of the correlation function. To improve the matching a Borel transform [10] is performed in both sides of the sum rule in both variables $-p^2 = P^2 \to M^2$ and $-p'^2 = P'^2 \to M'^2$.

To evaluate the $J/\psi - \pi$ cross section using the QCDSR we consider the four-point function for the process $J/\psi\, \pi \to \bar{D}\, D^*$:

$$\Pi_{\mu\nu} = i \int d^4x\, d^4y\, d^4z\, e^{-ip_1 \cdot x}\, e^{-ip_2 \cdot 0}\, e^{ip_3 \cdot y}\, e^{ip_4 \cdot z}$$
$$\times \langle 0|T\{j_\pi(x) j_\nu^{D^*}(y) j_\mu^\psi(0) j_D(z)\}|0\rangle, \quad (4)$$

with the currents given by $j_\pi = \bar{d} i \gamma_5 u$, $j_\nu^{D^*} = \bar{u}\gamma_\nu c$, $j_\mu^\psi = \bar{c}\gamma_\mu c$ and $j_D = \bar{c} i \gamma_5 d$, where c, u and d are the charm, up and down quark fields respectively, and p_1, p_2, p_3 and p_4 are the four-momenta of the mesons π, J/ψ, D^* and D respectively, with $p_1 + p_2 = p_3 + p_4$.

The phenomenological side of the correlation function, $\Pi_{\mu\nu}$, is obtained by the consideration of J/ψ, π, D and D^* state contribution to the matrix element in Eq. (4):

$$\Pi_{\mu\nu}^{phen} = -\frac{m_\pi^2 f_\pi}{m_u + m_d} \frac{m_D^2 f_D}{m_c} \frac{m_{D^*} f_{D^*}\, m_\psi f_\psi\, \mathcal{M}^{\alpha\beta}}{(p_1^2 - m_\pi^2)(p_4^2 - m_D^2)}$$
$$\times \frac{g_{\mu\alpha} - p_{2\mu} p_{2\alpha}/m_\psi^2}{p_2^2 - m_\psi^2} \frac{g_{\nu\beta} - p_{3\nu} p_{3\beta}/m_{D^*}^2}{p_3^2 - m_{D^*}^2} + \text{h. r.}, \quad (5)$$

where h. r. means higher resonances and the hadronic amplitude for the process $J/\psi\, \pi \to \bar{D}\, D^*$ is given by

$$\mathcal{M} = \mathcal{M}_{\mu\nu}(p_1, p_2, p_3, p_4)\, \varepsilon_2^\mu\, \varepsilon_3^{*\nu}. \quad (6)$$

We note that one has $1/p_1^2$ pole in Eq. (5) in the limit of a vanishing pion mass. Following Reinders, Rubinstein, and Yazaki [12], and others [13, 14, 15], we can write a sum rule at $p_1^2 = 0$ and single out the leading terms in the operator product expansion (OPE) of Eq. (4) that match the $1/p_1^2$ term. The perturbative diagram does not contribute with $1/p_1^2$ and, up to dimension four, only the diagrams proportional to the quark condensate contribute. After collecting the $1/p_1^2$ terms on the theoretical side and taking the limit $p_{1\mu} \to 0$ in the residue of the pion pole, one obtains for the contribution of the quark condensate [16]:

$$\Pi_{\mu\nu}^{\langle \bar{q}q \rangle} = -\frac{2 m_c \langle \bar{q}q \rangle}{p_1^2} \frac{p_{1\nu}(p_{1\mu} + p_{2\mu} - 2 p_{3\mu}) - p_{1\mu} p_{2\nu}}{(p_3^2 - m_c^2)(p_4^2 - m_c^2)}. \quad (7)$$

Contracting the hadronic amplitude with the numerators of J/ψ and D^* propagators in Eq. (5) and comparing with Eq. (7), the structure defining $\mathscr{M}_{\mu\nu}$ in Eq. (6) is easily identified. Therefore, defining

$$\mathscr{M}_{\mu\nu} = \Lambda_{DD^*} \left(p_{1\mu} p_{1\nu} - p_{1\mu} p_{2\nu} - 2 p_{1\nu} p_{3\mu} \right), \tag{8}$$

we can write a sum rule for Λ_{DD^*} in any of the three structures appearing in Eq. (8). To improve the matching between the phenomenological and theoretical sides we make a single Borel transformation to all the external momenta (except p_1^2) taken to be equal: $-p_2^2 = -p_3^2 = -p_4^2 = P^2 \to M^2$. The problem of doing a single Borel transformation is the fact that terms associated with the pole-continuum transitions are not suppressed [17]. In ref. [17] it was explicitly shown that the pole-continuum transition has a different behavior as a function of the Borel mass as compared with the double pole contribution (triple pole contribution in our case) and continuum contribution: it grows with M^2 as compared with the contribution of the fundamental states. Therefore, the pole-continuum contribution can be taken into account through the introduction of a parameter A_{DD^*} in the phenomenological side of the sum rule [14, 15, 17].

Thus, neglecting m_π^2 in the denominator of Eq. (5) and doing a single Borel transform we get

$$\frac{\Lambda_{DD^*} + A_{DD^*} M^2}{m_{D^*}^2 - m_\psi^2} \left[\frac{e^{-m_D^2/M^2} - e^{-m_\psi^2/M^2}}{m_\psi^2 - m_D^2} - (\psi \to D^*) \right]$$
$$= -2 m_c \langle \bar{q} q \rangle \frac{e^{-m_c^2/M^2}}{M^2} \frac{m_c (m_u + m_d)}{m_\pi^2 m_D^2 m_{D^*} m_\psi f_\pi f_D f_{D^*} f_\psi}, \tag{9}$$

where we have transferred to the theoretical side the couplings of the currents with the mesons, and have introduced, in the phenomenological side, the parameter A_{DD^*} to account for possible nondiagonal transitions.

RESULTS FOR THE FORM FACTORS

The parameter values used in all calculations are $m_u + m_d = 14$ MeV, $m_c = 1.5$ GeV, $m_\pi = 140$ MeV, $m_D = 1.87$ GeV, $m_{D^*} = 2.01$ GeV, $m_\rho = 0.77$ GeV, $m_\psi = 3.1$ GeV, $f_D = 160$ MeV, $f_\pi = 131.5$ MeV, $g_\rho = 5.45$, $\langle \bar{q} q \rangle = -(0.23)^3$ GeV^3, $\langle g^2 G^2 \rangle = 0.5$ GeV^4. We parametrize the continuum thresholds as

$$s_0 = (m_H + \Delta_s)^2, \qquad u_0 = (m_{H'} + \Delta_u)^2. \tag{10}$$

We first discuss the $D^* D \pi$ form factor with an off-shell pion. In Fig. 1 we show the behavior of the perturbative and the gluon condensate contribution to the form factor $g_{D^* D \pi}(Q^2 = -q^2)$ as a function of the Borel mass M^2 (at a fixed ratio $M'^2/M^2 = (M_{H'}^2 - m_{H'}^2)/(M_H^2 - m_H^2)$, and fixed value of Q^2), using Δ_s and Δ_u given in Eq.(10) equal to 0.5 GeV. We can see that the gluon condensate, despite being small, helps the stability of the curve, as a function of M^2, providing a rather stable plateau for $M^2 \geq 3$ GeV^2.

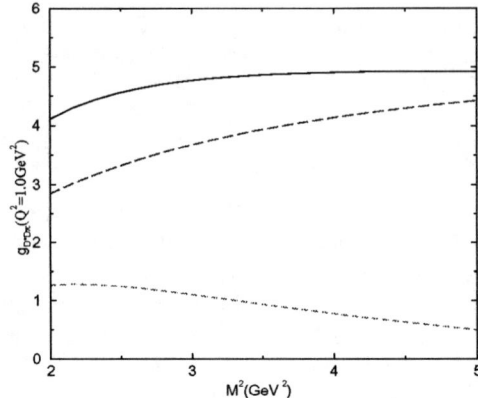

FIGURE 1. M^2 dependence of the perturbative (long-dashed line) and gluon condensate (dashed line) contributions to the $D^*D\pi$ form factor at $Q^2 = 1\,\text{GeV}^2$ (solid line) for $\Delta_s = \Delta_u = 0.5\,\text{GeV}$.

Fixing $M^2 = 3.5\,\text{GeV}^2$ we show, in Fig. 2, the momentum dependence of the form factor (squares) in the interval $2 \leq Q^2 \leq 5\,\text{GeV}$. Since the present approach cannot be used at $Q^2 = 0$, to extract the $g_{D^*D\pi}$ coupling from the form factor we need to extrapolate the curve to $Q^2 = 0$ (in the approximation $m_\pi^2 = 0$).

In ref. [15] the Q^2 dependence of the form factor, represented by the squares, was parametrized by a gaussian form (dashed line)

$$g_{D^*D\pi}^{(\pi)}(Q^2) = 5.7\, e^{-Q^4/9.17}. \tag{11}$$

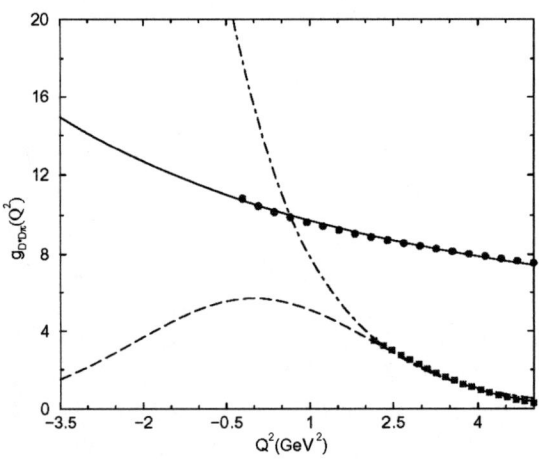

FIGURE 2. Momentum dependence of the $D^*D\pi$ form factor. The solid, dashed and dot-dashed lines give the parametrization of the QCDSR results through Eq. (14) for the circles, and Eqs. (11) and (12) for the squares.

However, as can be seen by the dot-dashed line, the Q^2 dependence of the QCDSR results for $g_{D^*D\pi}(Q^2)$ for an off-shell pion can also be well reproduced by the exponential parametrization

$$g^{(\pi)}_{D^*D\pi}(Q^2) = 15.5\, e^{-Q^2/1.48}. \tag{12}$$

Off course, the two parametrizations in Eqs. (11) and (12) lead to very different values for the $D^*D\pi$ coupling constant, defined as the value of the form factor at the pole of the off-shell meson ($Q^2 = -m_\pi^2 \sim 0$ in the case of the off-shell pion):

$$g_{D^*D\pi} = \begin{cases} 5.7 & \text{with the gaussian parametrization} \\ 15.5 & \text{with the exponential parametrization} \end{cases} \tag{13}$$

To solve this problem we analyse the $D^*D\pi$ form factor for an off-shell D meson. In Fig. 2 we also show, through the circles, the momentum dependence of the $D^*D\pi$ form factor for an off-shell D meson, in the interval $-0.5 \leq Q^2 \leq 5\,\text{GeV}$, where we expect the sum rules to be valid (since in this case the cut in the t channel starts at $t \sim m_c^2$ and thus the Euclidian region stretches up to that threshold). From this figure we can see that the Q^2 dependence of the form factor represented by the circles can be well reproduced by the monopole parametrization (solid line)

$$g^{(D)}_{D^*D\pi}(Q^2) = \frac{126.1}{Q^2 + 11.95}. \tag{14}$$

From the parametrization in Eq. (14) we can also extract the $D^*D\pi$ coupling constant, which now is defined as the value of the form factor at the D pole ($Q^2 = -m_D^2$). We get

$$g_{D^*D\pi} = 14.9, \tag{15}$$

in an excelent agreement with the exponential parametrization of $g^{(\pi)}_{D^*D\pi}(Q^2)$. Therefore, we conclude that the gaussian parametrization, despite describing very well the behaviour of the QCDSR results for $g^{(\pi)}_{D^*D\pi}(Q^2)$ in the region $2 \leq Q^2 \leq 5\,\text{GeV}$, can not be used to do the extrapolation at $Q^2 = -m_\pi^2$. By imposing that the two form factors: $g^{(\pi)}_{D^*D\pi}(Q^2)$ and $g^{(D)}_{D^*D\pi}(Q^2)$ should lead to the same coupling constant, $g_{D^*D\pi}$, we find that the exponential parametrization of $g^{(\pi)}_{D^*D\pi}(Q^2)$ is the correct parametrization of the QCDSR results. It is very interesting to notice that the obtained coupling constant, besides being still smaller than the experimental value [18]: $g_{D^*D\pi} = 17.9 \pm 0.3 \pm 1.9$, is much closer to it than other QCDSR evaluations [19].

There is another important information that we can extract from the parametrization of the QCDSR results which is the value of the cut-off. Defining the coupling constant as the value of the form factor at $Q^2 = -m_M^2$, where m_M is the mass of the off-shell meson, the monopole and the exponential parametrizations of the form factor can be written as (neglecting m_π^2):

$$g^{(D)}_{D^*D\pi}(Q^2) = g_{D^*D\pi} \frac{\Lambda_D^2 - m_D^2}{Q^2 + \Lambda_D^2}, \tag{16}$$

$$g^{(\pi)}_{D^*D\pi}(Q^2) = g_{D^*D\pi}\, e^{-\frac{Q^2}{\Lambda_\pi^2}}, \tag{17}$$

and from Eqs. (14) and (12) we get

$$\Lambda_D = 3.5\,\text{GeV}, \quad \Lambda_\pi = 1.2\,\text{GeV}. \tag{18}$$

Therefore, the form factor is harder if the off-shell meson is heavy, implying that the size of the vertex depends on the exchanged meson. This means that a heavy meson will see the vertex as pointlike, whereas a light meson will see its extension.

The same kind of behaviour is also obtained in the $DD\rho$ vertex [20]. Fixing $M'^2 = 4.7\,\text{GeV}^2$ (which corresponds to $M^2 = 0.8\,\text{GeV}^2$ for the case of off-shell D) we show, in Fig. 3, the momentum dependence of the form factor (circles for $g_{DD\rho}^{(D)}(Q^2)$ and squares for $g_{DD\rho}^{(\rho)}(Q^2)$) in the interval $0.1 \leq Q^2 \leq 5\,\text{GeV}$. In Fig. 3 we also show that the Q^2 dependence of the form factor represented by the circles can be well reproduced by the monopole parametrization (solid line)

$$g_{DD\rho}^{(D)}(Q^2) = \frac{37.5}{Q^2 + 12.12}, \tag{19}$$

and that the form factor represented by the squares can be well reproduced by the exponential parametrization (solid line)

$$g_{DD\rho}^{(\rho)}(Q^2) = 2.53 e^{-\frac{Q^2}{0.98}}. \tag{20}$$

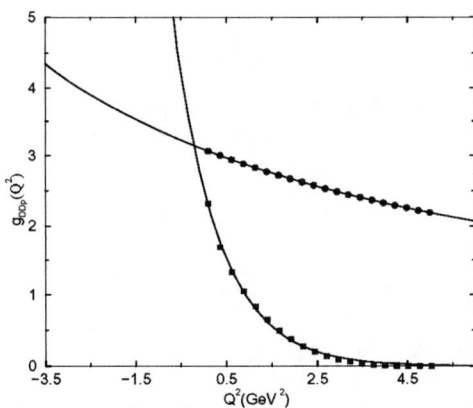

FIGURE 3. Momentum dependence of the $DD\rho$ form factor for $\Delta_s = \Delta_u = 0.5\,\text{GeV}$. The solid lines give the parametrization of the QCDSR results through Eq. (19) for the circles, and Eq. (20) for the squares.

The coupling constants and cut-offs resulting of the parametrizations in Eqs. (16) and (17), for the case of a $DD\rho$ vertex, are given in Table 1.

We see that there is a very good agreement between the two values of the coupling constant, extracted from the QCDSR results.

TABLE 1. Values of the coupling constants and cut-offs in the $DD\rho$ vertex.

	$g_{DD\rho}^{(M)}(Q^2=-m_M^2)$	Λ_M (GeV)
D off-shell	4.4	3.5
ρ off-shell	4.6	1.0

RESULTS FOR THE $J/\psi - \pi$ CROSS SECTION

The QCD sum rule results for $\Lambda_{DD^*} + A_{DD^*}M^2$, in Eq.(9), as a function of M^2 follow a straight line in the Borel region $8 \leq M^2 \leq 16 \text{ GeV}^2$. The value of the amplitude Λ is obtained by the extrapolation of the line to $M^2 = 0$ [14, 15, 17]. Fitting the QCD sum rule results to a straight line we get

$$\Lambda_{DD^*} \simeq 17.71 \text{ GeV}^{-2}. \tag{21}$$

As expected, in our approach Λ is just a number and all dependence of $\mathcal{M}_{\mu\nu}$ (Eq. (8)) on particle momenta is contained in the Dirac structure. This is a consequence of our low energy approximation.

Following the same procedure, we have also considered the processes $J/\psi \, \pi \to \bar{D} D$ and $J/\psi \, \pi \to \bar{D}^* D^*$. Similarly to the case $J/\psi \, \pi \to \bar{D} D^*$, in the OPE side the only diagrams, up to dimension four, contributing with $1/p_1^2$ are the quark condensate diagrams. Comparing the phenomenological and OPE sides of the correlators we can identify the structure defining the hadronic amplitudes [16].

The QCD sum rule results for $\Lambda_{DD} + A_{DD}M^2$ and $\Lambda_{D^*D^*} + A_{D^*D^*}M^2$ as a function of M^2 are obtained in a similar way and the amplitudes Λ_{DD} and $\Lambda_{D^*D^*}$ are extracted by the extrapolation of the line to $M^2 = 0$. We get:

$$\Lambda_{DD} \simeq 12.25 \text{ GeV}^{-1}, \quad \Lambda_{D^*D^*} \simeq 11.39 \text{ GeV}^{-3}. \tag{22}$$

Having the QCD sum rule results for the amplitude of the three processes $J/\psi \, \pi \to \bar{D} D^*$, $\bar{D} D$, $\bar{D}^* D^*$, we can evaluate the differential cross section.

We show, in Fig. 4, the cross section for the $J/\psi \, \pi$ dissociation. It is important to keep in mind that, since our sum rule was derived in the limit $p_1 \to 0$, we can not extend our results to large values of $\sqrt{s}$.

Our first conclusion is that our results show that, for values of $\sqrt{s}$ far from the $J/\psi \, \pi \to \bar{D}^* D^*$ threshold, $\sigma_{J/\psi\pi \to \bar{D}^*D^*} \geq \sigma_{J/\psi\pi \to \bar{D}D^* + D\bar{D}^*} \geq \sigma_{J/\psi\pi \to \bar{D}D}$, in agreement with the model calculations discussed, for example, in ref. [21] but in disagreement with the results obtained with the nonrelativistic quark model of ref. [22], which show that the state $\bar{D}^*D$ has a larger production cross section than $\bar{D}^*D^*$. Furthermore, our curves indicate that the cross section grows monotonically with the c.m.s. energy but not as fast, near the thresholds, as it does in the calculations in refs. [8, 9, 21]. Again, this behavior is in opposition to [22], where a peak just after the threshold followed by continuous decrease in the cross section was found.

At higher energies, due to our low energy approximation, our approach gradually looses validity. In the fiducial region, close to threshold, $4.1 \leq \sqrt{s} \leq 4.3 \text{ GeV}$, we find

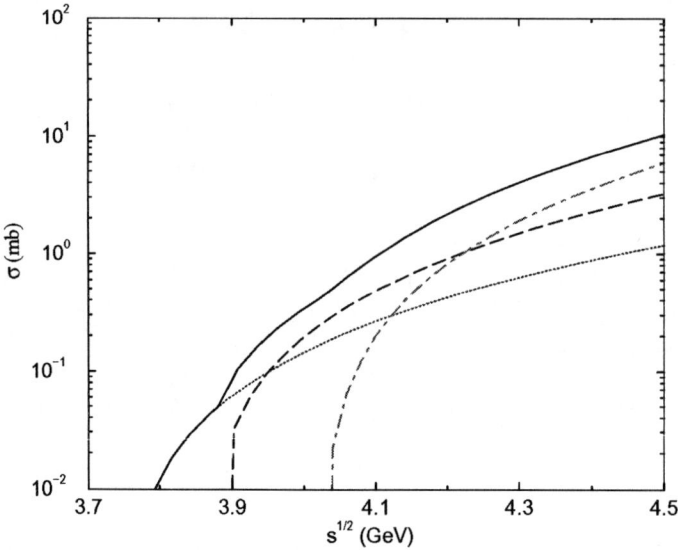

FIGURE 4. Total cross sections of the processes $J/\psi\,\pi \to \bar{D}\,D^* + D\,\bar{D}^*$ (dashed line), $\bar{D}\,D$ (dotted line) and $\bar{D}^*\,D^*$ (dot-dashed line). The solid line gives the total $J/\psi\,\pi$ dissociation cross section.

$2.5 \le \sigma \le 4.0$ mb and these values are much smaller than those obtained with the effective Lagrangians without form factors in the hadronic vertices, but agree in order of magnitude with the quark model calculations of [22].

CONCLUSIONS

We have used the method of QCD sum rules to compute form factors and coupling constants in $DD\rho$ and $D^*D\pi$ vertices. Our results for the couplings show once more that this method is robust, yelding numbers which are approximately the same regardless of which particle we choose to be off-shell and depending weakly on the choice of the continuum theshold. As for the form factors, we obtain a harder (softer) form factor when the off-shell particle is heavier (lighter). Therefore we can say that heavy mesons "see" smaller sizes in the light mesons, but the latter can not resolve small structures in the former.

We have also used the QCD sum rule approach to evaluate the hadronic amplitude of the $J/\psi\,\pi$ dissociation. From the hadronic amplitude we have evaluated the $J/\psi\,\pi \to$ charmed mesons dissociation cross section, and have obtained $2.5 \le \sigma \le 4.0$ mb at $4.1 \le \sqrt{s} \le 4.3$ GeV. In view of the uncertainties discussed above these numbers should be taken as upper limits.

It is interesting to remember that Bhanot and Peskin [23] have also used the OPE in the short distance limit to study the charmonium hadron cross section. This work was latter enlarged and updated by Kharzeev et al. [24] and also by Oh et al. [25]. In these papers the crucial assumption was made that the charmonium is very small and resolves the partonic structure of the light hadron. In our approach we do not use this assumption, and we obtain larger values for the cross section. This seems to indicate that size effects are important, and that the J/ψ cannot be considered as a nearly point like object.

ACKNOWLEDGMENTS

This work has been supported by FAPESP and CNPq - Brazil.

REFERENCES

1. T. Matsui and H. Satz, *Phys. Lett.* **B178**, 416 (1986).
2. NA38 Collaboration (C. Baglin et al.), *Phys. Lett.* B255, 459 (1991).
3. NA38 Collaboration (B. Ronceux et al.), *Phys. Lett.* B345, 617 (1995).
4. NA50 Collaboration (M.C. Abreu et al.), *Phys. Lett.* **B477**, 28 (2000).
5. E772 Collaboration (D.M. Alde et al.), *Phys. Rev. Lett.* 66, 133 (1991).
6. D. Kharzeev and H. Satz, *Phys. Lett.* **B366**, 316 (1996).
7. H.G. Dosch, F.S. Navarra, M. Nielsen and M. Rueter, *Phys. Lett.* **B466**, 363 (1999).
8. S.G. Matinyan and B. Müller, *Phys. Rev.* **C58** (1998) 2994.
9. Yongseok Oh, Taesoo Song and Su Houng Lee, *Phys. Rev.* **C63** (2001) 034901.
10. M.A. Shifman, A.I. Vainshtein and V.I. Zakharov, *Nucl. Phys.* **B120**, 316 (1977).
11. B.L. Ioffe and A.V. Smilga, *Nucl. Phys.* **B232**, 109 (1984).
12. L.J. Reinders, H. Rubinstein and S. Yazaki, *Phys. Rep.* **127**, 1 (1985).
13. S. Choe, M.K. Cheoun and S.H. Lee, *Phys. Rev.* **C53**, 1363 (1996).
14. M.E. Bracco, F.S. Navarra and M. Nielsen, *Phys. Lett.* **B454**, 346 (1999).
15. F.S. Navarra et al., *Phys. Lett.* **B489**, 319 (2000).
16. F.S. Navarra, M. Nielsen, R.S. Marques de Carvalho and G. Krein, *Phys. Lett.* **B529**, 87.
17. B.L. Ioffe and A.V. Smilga, *Nucl. Phys.* **B232** 109 (1984).
18. S. Ahmed et al., (CLEO Collaboration), hep-ex/0108013.
19. F.S. Navarra, M. Nielsen and M.E. Bracco, *Phys. Rev.* **D65**, 037502 (2002).
20. M.E. Bracco et al., *Phys. Lett.* **B521**, 1 (2001).
21. F.S. Navarra, M. Nielsen and M.R. Robilotta, *Phys. Rev.* **C64**, 021901 (R) (2001).
22. C.-Y. Wong, E. S. Swanson and T. Barnes, *Phys. Rev.* **C62**, 045201 (2000) and references therein.
23. G. Bhanot and M.E. Peskin, *Nucl. Phys.* **B156**, 391 (1979); M.E. Peskin, *Nucl. Phys.* **B156**, 365 (1979).
24. D. Kharzeev and H. Zatz, *Phys. Lett.* **B334** (1994) 155; D. Kharzeev, H. Zatz, A. Syamtomov and G. Zinovjev, *Phys. Lett.* **B389**, 595 (1996).
25. Y. Oh, S. Kim and S.H. Lee, hep-ph/0111132.

Hydrodynamical Evolution from Event-by-Event Fluctuating Initial Conditions

C. E. Aguiar*, Y. Hama[†], T. Kodama* and T. Osada[†]

*Instituto de Física/UFRJ, C.P. 68528, 21945-970 Rio de Janeiro - RJ, Brazil
[†]Instituto de Física/USP, C.P. 66318, 05389-970 São Paulo - SP, Brazil

Abstract. Effects caused by the event-by-event fluctuation of the initial conditions in hydrodynamical description of high-energy heavy-ion collisions are investigated, by applying the recently developed SPheRIO code. Non-negligible effects appear for several observable quantities, even for a fixed impact parameter $\vec{b}$. They are sensitive to the equation of state, giving different results according to the equation of state considered.

INTRODUCTION

In usual hydrodynamic treatment of high-energy heavy-ion collisions, one customarily assumes some highly symmetric and smooth initial conditions, which would correspond to some mean distributions of velocity, temperature, energy density, etc., averaged over several events. However, our systems are not large enough, so large fluctuations are expected. What are the effects of the event-by-event fluctuation of the initial conditions? Are they sizable? If so, physical information one obtains in the usual analyses may be largely distorted. Some related questions we would like to answer are whether they depend on the equation of state and which are the most sensitive variables to such an initial-state fluctuation. Here, we report preliminary results of such a study [1].

To producing fluctuating initial conditions, we use a microscopic model called NeXus [2]. Since our initial conditions do not have any symmetry nor they are smooth, and moreover we would like to consider many fluctuating initial conditions, the resolution of the hydrodynamic equations deserves some special care. We adopt the so-called smoothed-particle hydrodynamic (SPH) approach [3], first used in astrophysics and which we have previously adapted for heavy-ion collisions [4], a method flexible enough, giving a desired precision and allowing to solve with a reasonable speed. We call this version of SPS code SPheRIO[1].

In the following, we begin accounting for a brief description of the SPheRIO code in the next section. Then, in order to see the influences of the QGP phase transition, we consider in section 3 two simple equations of state [5] and predictions for some observables are shown. More realistic treatment is given in section 4, taking into account the baryon number conservation and with equations of state with many resonances. Some results are presented there. Conclusions and outlook are drawn in section 5.

---

[1] **S**moothed-**P**article **h**ydrodynamical **e**volution of **R**elativistic heavy-**IO**n collisions.

HYDRODYNAMICAL CODE SPHERIO

The main characteristic of SPH is the parametrization of the flow in terms of discrete Lagrangian coordinates attached to small volumes (called *particles*) with some conserved quantity (or quantities). In the present work, we begin by taking the entropy as our conserved quantity. Later (in Sec. 4), we shall also consider the baryon number as another conservred quantity. Then, the equations of hydrodynamics which describe the evolution of the fluid are reduced to the equations of motion of these particles. For treating problems of relativistic heavy-ion collisions, it is convenient to formulate SPH using the coordinates

$$\tau \equiv \sqrt{t^2 - z^2}, \quad x, \quad y, \quad \zeta \equiv \frac{1}{2}\ln\frac{t+z}{t-z} . \quad (1)$$

The equation[2] for the i-th particle position $\mathbf{x}_i$ is

$$\frac{d}{d\tau}\mathbf{x}_i = \left[\frac{u^x}{u^\tau}, \frac{u^y}{u^\tau}, \frac{u^\zeta}{u^\tau}\right] \equiv \mathbf{v}_i , \quad (2)$$

where u is the collective four-velocity. The equation for the i-th particle velocity $\mathbf{v}_i$ is given as

$$\frac{d}{d\tau}\left(\frac{P_i+\varepsilon_i}{s_i}\gamma_i\mathbf{v}_i\right) + \sum_j \frac{v_j}{\tau}\left[\frac{P_i}{s_i^2(u_i^0)^2} + \frac{P_j}{s_j^2(u_j^0)^2}\right]\nabla_i W[\mathbf{x}_i - \mathbf{x}_j; h] = 0 , \quad (3)$$

where s_i, P_i and ε_i are, respectively, the entropy density, pressure and energy density at $\mathbf{x}_i$. The parameter v_i is the entropy carried by the i-th particle and W is a smoothing kernel function with a scale parameter h, which defines its width. In SPH representation, the entropy density s is related to the particle entropies v_i through

$$s(\tau, \mathbf{x}) = \frac{1}{\tau u^\tau}\sum_i v_i W(\mathbf{x} - \mathbf{x}_i : h) . \quad (4)$$

Now, to generating the initial conditions $\mathbf{x}(\tau_0)$, $\mathbf{v}(\tau_0)$ and $s(\tau_0, \mathbf{x})$ in the event-by-event basis, we need some microscopic model. As stated in the Introduction, here we use the so-called NeXus event generator [2], based on the parton-based Gribov-Regge theory. Detailed explanations about the interface between NeXus and SPheRIO are found in ref. [1].

Although more elaborate description is desired [6, 7], the freeze-out process is described here by using the usual Cooper-Frye formula, adapted to SPH representation

$$E\frac{d^3N}{d\mathbf{p}^3} \approx \frac{g_\pi}{(2\pi)^3}\sum_j\left[\frac{p^\mu}{e^{u_\mu p^\mu/T_f} \mp 1}\right]_j \frac{v_j}{su^\tau}\frac{n_\mu/n^\tau}{|1+v_j^x\frac{n_x}{n_\tau}+v_j^y\frac{n_y}{n_\tau}+v_j^\zeta\frac{n_\zeta}{n_\tau}|} , \quad (5)$$

---

[2] See detailed derivation in ref. [4].

where the four-vector

$$n_\mu \equiv (n_\tau, n_x, n_y, n_\zeta) = \left(-\frac{\partial T}{\partial \tau}, -\frac{\partial T}{\partial x}, -\frac{\partial T}{\partial y}, -\frac{\partial T}{\partial \zeta}\right) \qquad (6)$$

is normal (not normalized) to the isotherms. The sum in eq. (5) should be taken for all SPH particles crossing the freeze-out surface defined by a constant temperature T_f.

STUDY WITH SIMPLE EQUATIONS OF STATE

To see the equation-of-state (EoS) dependence of the effects we are interested in, we begin considering two simple different EoS's [5]:

1. Resonance Gas (RG): $p/\varepsilon \, (= c_s^2) = 0.2$; $\qquad (7)$

2. QGP+RG: $c_s^2 = \begin{cases} 0.2, & \varepsilon < 0.28\,\text{GeV/fm}^3, \\ 0.056/\varepsilon, & \text{mixed phase}, \\ 1/3 - 4B/3\varepsilon, & \varepsilon > 1.45\,\text{GeV/fm}^3. \end{cases} \qquad (8)$

Here, we are ignoring the baryon number density of the matter. The inclusion of the latter will be done in Sec. 4.

Now, by using the combination NeXus+SPheRIO, we generated 535 and 269 minimum bias events of $Au+Au$ collisions at a center-of-mass energy of $\sqrt{s} = 130A$ GeV for RG and RG+QGP EoS, respectively. Then we computed the rapidity distributions, transverse-momentum distributions and the elliptic-flow parameter v_2, by considering only direct pion production and ρ decays.

We present in Fig. 1 the rapidity distributions for 100 events (RG+QGP EoS is used), where one can see a large event-by-event fluctuation. Below, we will show several quantities, computed by taking this fluctuation into account. A quantitative measure of this fluctuation is not included here, but some estimates have been given in ref. [1].

Centrality dependence of dN/dy

We show in Fig. 2 our preliminary results on the centrality dependence of the rapidity distributions of negative hadrons. One sees that although the shapes of all the distributions are similar, the multiplicities are larger in the QGP case, since we are starting from the same energy densities in the both cases. The STAR Collaboration reports that the height of the pseudorapidity distribution $dN/d\eta \, (\approx dN/dy)$ for the top 5% centrality events is $264\pm1\pm18$ (integrated over all p_t) [8]. Our results, 241.8 (for RG+QGP) and 227.3 (for RG), are about $10-20\%$ smaller. As mentioned above, this version of SPheRIO does not include, except ρ, the mesonic- and baryonic-resonance decays. Therefore, the discrepancy is probably due to the lack of those resonance-decay products.

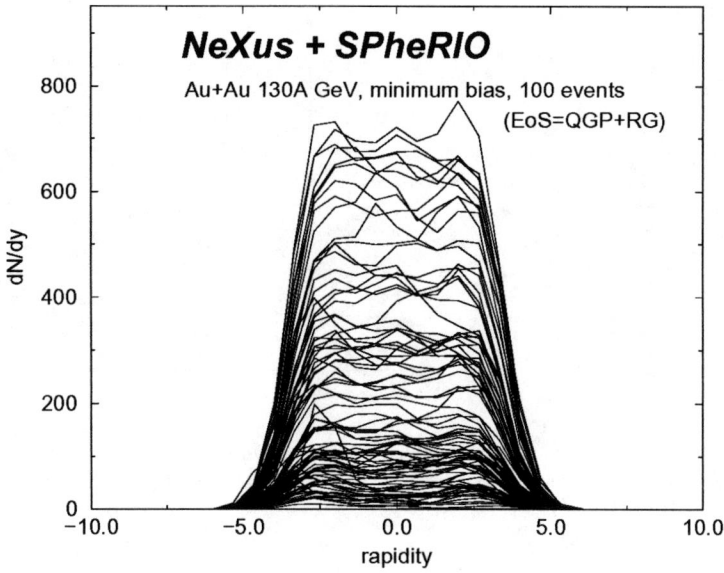

FIGURE 1. Rapidity distributions dN/dy in $Au+Au$ collisions (minimum bias) at a center-of-mass energy of $\sqrt{s} = 130A$ GeV for 100 events (RG+QGP EoS is used).

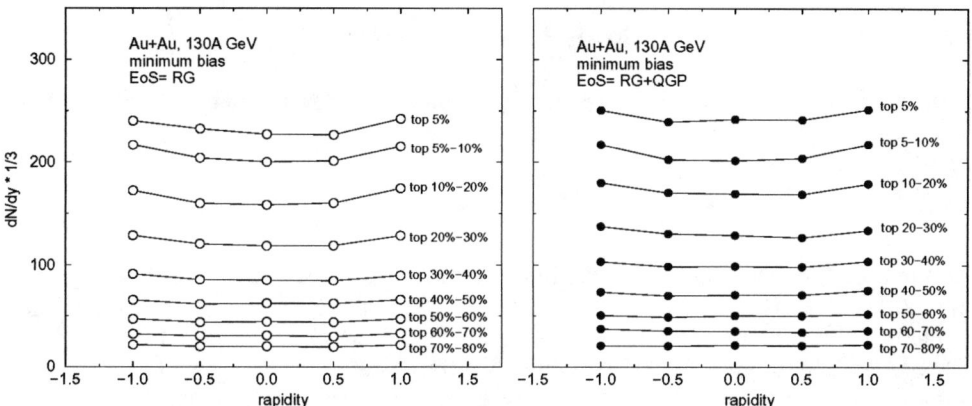

FIGURE 2. Centrality dependence of the rapidity distributions in $Au+Au$ collisions at $\sqrt{s} = 130A\ GeV$ center-of-mass energy obtained with NeXus+SPheRIO.

Centrality dependence of p_T distribution

In Fig. 3, we display the transverse-momentum distributions computed with our code. All the slopes in each case are similar, but the ones for RG+QGP EoS are steeper than those for RG. This may be understood by observing that the hydrodynamical expansion

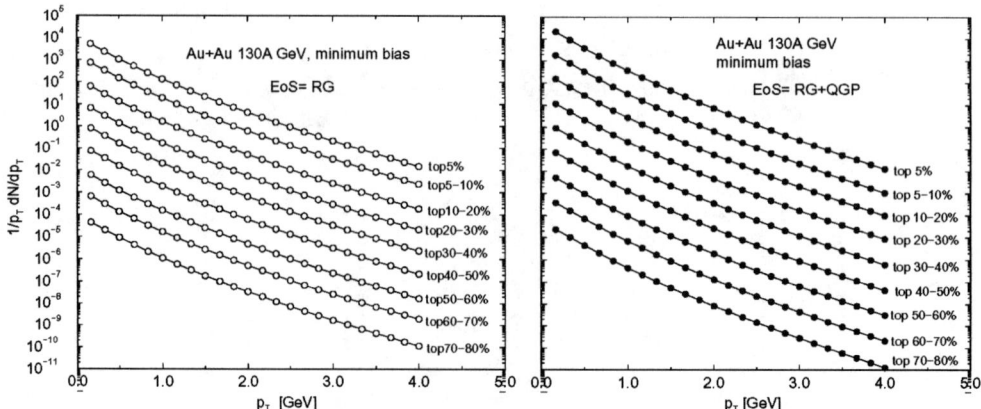

FIGURE 3. Centrality dependence of the transverse momentum distributions in $Au+Au$ collisions at $\sqrt{s}=130A$ GeV energy computed with NeXus+SPheRIO. The left panel is the result of RG EoS and the right one is those of RG+QGP EoS.

of RG+QGP is much less violent than that of RG, because during the mixed phase the expansion is acceleration free.

Elliptic flow parameter v_2

Elliptic-flow parameter v_2, defined as the second-order Fourier coefficient of the expansion

$$\frac{dN}{d\phi} \propto v_1 \cos(\phi) + v_2 \cos(2\phi) + \cdots, \tag{9}$$

is often used to study the initial-condition dependence of the transverse-flow asymmetry [9, 10]. Let us see how large is the influence of the EoS on this asymmetry. Our preliminary results for the elliptic flow parameter v_2 are shown in Fig 4 together with the STAR Collab. data [11]. As seen there, the RG results reproduce well the data for the multiplicity dependence of v_2, whereas the data on p_t dependence are reproduced better with RG+QGP EoS. Apparently there is a conflict in our results. However, we should be cautious in analyzing the results because the problem is more complex. Remark that most of the produced particles have small p_t, so that the multiplicity-dependence data correspond to a small left-side portion of the p_t-dependence data. On the other hand, when plotting p_t spectrum, most of the particles come from rather large-(n/n_{max}) region. Besides, there are effects which we have not introduced in our calculations yet, $i.e.$, baryon-number dependence of EoS and, except for ρ, baryonic- and mesonic-resonance decays in freeze out as mentioned before. Also it is not clear what effects are caused by the chemical freezeout process on the elliptic flow phenomena. To extract a right conclusion about EoS, more realistic calculations are needed.

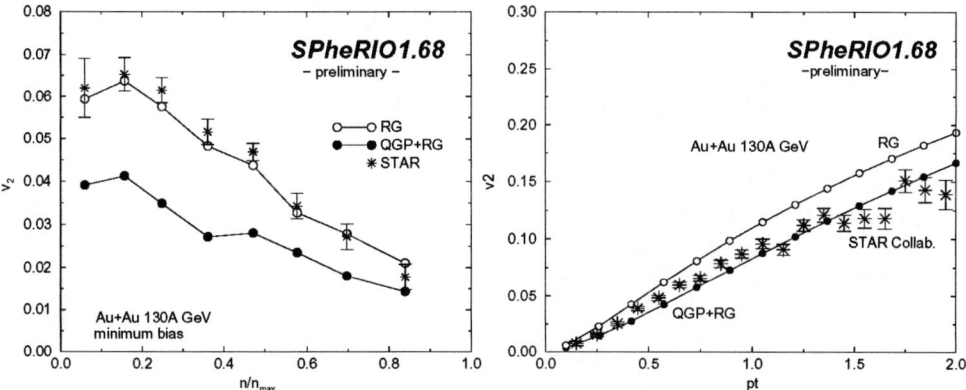

FIGURE 4. Results of multiplicity dependence (above) and p_t dependence (below) of the elliptic flow parameter v_2 compared with data [11].

EQUATION OF STATE WITH BARYON NUMBER DENSITY AND RESONANCES EXPLICITLY INCLUDED

To improve the simple EoS discussed in Sec.3, we now include the baryon number density into our equation of state and also, instead of just parameterizing EoS as in eqs. (7) and (8), explicitly computed the contributions from resonances. This new version of SPheRIO, we call SPheRIO2.

For a resonance gas which consists of N kinds of hadrons, the pressure of the gas is given by

$$P_r(T, \mu_b) = \frac{-\Omega}{V} = \mp \sum_i \frac{g_i T}{2\pi^2} \int dp \, p^2 \ln[1 \mp e^{(\mu_i - E_i)/T}] \,, \tag{10}$$

where T and μ_b are the temperature and the baryonic chemical potential. The chemical potential μ_i of the hodron i is related with μ_b and its baryon number B_i by

$$\mu_i = B_i \mu_b \,. \tag{11}$$

The thermodynamical potential Ω is defined as

$$\Omega(T, \mu_b) = \pm \sum_i^N \frac{g_i T}{2\pi^2} \int dp \, p^2 \ln[1 \mp e^{(\mu_i - E_i)/T}] \qquad \text{[Boson/Fermion]} \,. \tag{12}$$

Then, the energy and the entropy densities of the resonance gas are

$$\varepsilon_r(T, \mu_b) = \sum_i \frac{4\pi g_i}{(2\pi^3)} \int dp \, p^2 \frac{E_i(p)}{e^{(E_i(p) - \mu_i)/T} \mp 1} \,, \tag{13}$$

$$s_r(T,\mu_b) = \left(\frac{\partial P}{\partial T}\right)_{V,\mu} = -\frac{\partial}{\partial T}\left(\frac{\Omega}{V}\right)_{V,\mu} = \mp\frac{\partial}{\partial T}\sum_i\frac{g_i T}{2\pi^2}\int dp\, p^2 \ln[1\mp e^{(\mu_i-E_i)/T}]$$

$$= \mp\sum_i\frac{g_i}{2\pi^2}\int dp\, p^2\left\{\ln[1\mp e^{(\mu_i-E_i)/T}]\mp\frac{(E_i-\mu_i)/T}{e^{(E_i-\mu_i)/T}\mp 1}\right\}. \qquad (14)$$

For these thermodynamical quantities, we introduce the excluded-volume corrections [12] to all the baryons. In our calculation, almost all resonances whose mass is below 2.5 GeV, 288 baryonic and 128 mesonic resonances have been included.

Now, for the massless quark-gluon-plasma, we can write [13].

$$\varepsilon_q(T,\mu_b) = \frac{\pi^2 T^4}{15}\left(16+\frac{7}{8}6N_f\right)+\frac{3}{2}N_f\left(T^2\mu_b^2+\frac{\mu^4}{2\pi^2}\right)+B, \qquad (15)$$

$$P_q(T,\mu_b) = \frac{\pi^2 T^4}{45}\left(16+\frac{7}{8}6N_f\right)+\frac{1}{2}N_f\left(T^2\mu_b^2+\frac{\mu^4}{2\pi^2}\right)-B, \qquad (16)$$

$$s_q(T,\mu_b) = \left(\frac{\partial P}{\partial T}\right)_{V,\mu} = \frac{4\pi^2 T^3}{45}\left(16+\frac{7}{8}6N_f\right)+N_f T\mu_b^2. \qquad (17)$$

The phenomenological parameter B (bag pressure) is tuned to give $T_c = 160$ MeV at zero baryon-number density, resulting in the value of $B = 380\text{MeV/fm}^{-3}$. The phase boundary between QGP and RG phases is determined by the condition of pressure balance there,

$$P_q(T,\mu_b) = P_r(T,\mu_b). \qquad (18)$$

We show in Fig.5 the resulting phase diagram. Using this EoS, we then calculated single-particle spectra for π, p, $\bar{p}$ and so on. Preliminary results for Pb+Pb central collisions at $\sqrt{s}$=17.2A GeV energy are shown in Fig.6. As seen, our calculations roughly reproduce the experimental data obtained by NA49 [14].

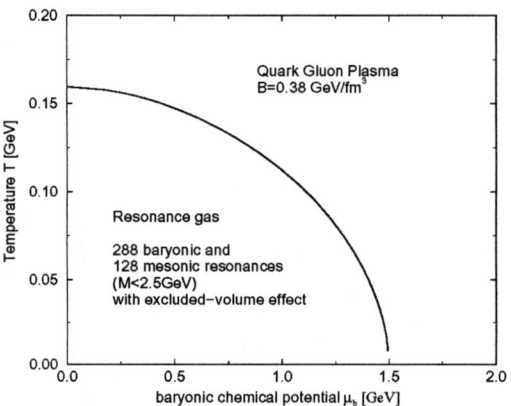

FIGURE 5. Phase boundary btween QGP and RG phases in the $T-\mu_b$ plane.

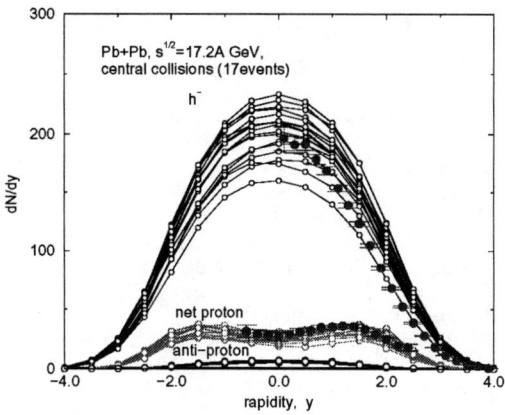

FIGURE 6. Preliminary results from NeXus+SPheRIO2. Open circles are our results of dN/dy for pion, net proton (p-$\bar{p}$) and anti proton in Pb+Pb collisions at SPS energy (central 17 events). Closed circles are data by NA49 collaboration [14].

CONCLUSIONS AND OUTLOOK

The present study shows that the effects of the event-by-event fluctuation of the initial conditions in hydrodynamics are sizable and should be considered in data analyses. We emphasize that such a fluctuation is not only due to the uncontrollable event-dependent impact-parameter, but it appears even if the impact parameter could be fixed and it is due to the finite size of our systems. We see that the propagation of such a fluctuation does depend on the equation of state, giving different final observables according to the EoS used.

The results we report here are only preliminary and could not answer some of the questions we posed at the beginning. For instance which are the most sensitive variables to such an initial-state fluctuation. This question is being examined now. Also, we have just acomplished a more realistic version of the SPheRIO code, so we are only beginning the data analyses in order to extract information from such a study.

ACKNOWLEDGMENTS

This work was supported in part by FAPESP (contract nos. 2000/04422-7 and 98/00317-2), FAPERJ (contract no. E-26/150.942/99) and PRONEX (contract no. 41.96.0886.00).

REFERENCES

1. Other results appeared in T. Osada, C.E. Aguiar, Y. Hama and T. Kodama, *Event-by-event analysis of ultra-relativistic heavy-ion collisions in smoothed particle hydrodynamics*, arXiv: nucl-th/0102011; C.E. Aguiar, Y. Hama, T. Kodama and T. Osada, *Nucl. Phys.* **A698** (2002) 639c.

2. H.J. Drescher, M. Hladik, S. Ostrapchenko, T. Pierog and K. Werner, *J. Phys.* **G25** (1999) L91; *Nucl. Phys.* **A661** (1999) 604.
3. L.B. Lucy, *Astrophys. J.* **82** (1977) 1013; R.A. Gingold and J.J. Monaghan, *Mon. Not. R. Astron. Soc.* **181** (1977) 375.
4. C.E. Aguiar, T. Kodama, T. Osada and Y. Hama, *J. Phys.* **G27** (2001) 75.
5. C.M. Hung and E.V. Shuryak, *Phys. Rev. Lett.* **75** (1995) 4003.
6. F. Grassi, Y. Hama and T. Kodama, *Phys. Lett.* **B355** (1995) 9; *Z. Phys.* **C73** (1996) 153.
7. Yu.M. Sinyukov, S.V. Akkelin and Y. Hama, *On freeze-out problem in hydro-kinetic approach to A+A collisions*, nucl-th/0203204.
8. J.W. Harris for STAR Collab., *Nucl. Phys.* **A698** (2002) 64c.
9. S.A. Voloshin and Y. Zhang, *Z. Phys. C* **70** (1996) 665; A.M. Poskanzer and S.A. Voloshin, *Phys. Rev. C* **58** (1998) 1671.
10. For a recent review, P. Danielewicz, nucl-th/0009091.
11. STAR Collaboration (K.H. Ackermann *et al.*), *Phys. Rev. Lett* **86** (2001) 402.
12. G.D. Yen, M.I. Gorenstein, W. Greiner and S.N. Yang, *Phys. Rev. C* **56**(1997), 2210.
13. C.M. Hung and E.V. Shuryak, *Phys. Rev. C* **57**(1998), 1891.
14. H. Appelshäuser et.al., NA49 Collaboration, *Phys. Rev. Lett* **82**(1999), 2471.

Finite Temperature and Density in Non-Local Chiral Quark Models

Norberto N. Scoccola* and Daniel Gomez Dumm[†]

*Physics Department, Comisión Nacional de Energía Atómica, Av.Libertador 8250, (1429) Buenos Aires, Argentina and Universidad Favaloro, Solís 453, (1078) Buenos Aires, Argentina.
[†]IFLP, Depto. de Física, Universidad Nacional de La Plata, C.C. 67, (1900) La Plata, Argentina.

Abstract. Chiral quark models with non-local covariant separable interactions at finite temperature and chemical potential are investigated. We develop a formalism in which the different quark properties are evaluated taking into account the analytic structure of the quark propagator. In this framework we study the chiral restoration phase transition for some definite non-local regulators. We find that in all cases the chiral transition is of first order for low values of T, turning into a smooth crossover at a certain "end point". Using model parameters which lead to the physical pion mass and decay constant, we find for the position of this "end point" the values $(T_E, \mu_E) \approx (60 - 70, 180 - 210)$ MeV.

INTRODUCTION

The understanding of the behaviour of strongly interacting matter at finite temperature and/or density is of fundamental interest and has important applications in cosmology, in the astrophysics of neutron stars and in the physics of relativistic heavy ion collisions. From the theory of the quark-gluon interactions, Quantum Chromo Dynamics (QCD), we believe that at zero temperature and density chiral symmetry is spontaneously broken and quarks are confined within hadrons. However, since QCD is asymptotically free, when either the temperature T or the chemical potential μ are high the effective coupling becomes small. Thus, one expects that at a certain point the system undergoes a phase transition (or crossover) to a new phase in which color is screened rather than confined and chiral symmetry is restored. Unfortunately, so far, it has not been possible to obtain detailed information about the corresponding $T - \mu$ phase diagram directly from QCD. In fact, lattice simulations, which work well for zero density and finite temperature, have serious difficulties in dealing with the complex fermion determinant that arises at finite chemical potential[1]. In this situation, different models have been used to study this sort of problems. Among them the Nambu-Jona-Lasinio model[2] is one of the most popular. In this model the quark fields interact via local four point vertices which are subject to chiral symmetry. If such interaction is strong enough the chiral symmetry is spontaneously broken and pseudoscalar Goldstone bosons appear. It has been shown by many authors that when the temperature and/or density increase, the chiral symmetry is restored[3]. Some covariant nonlocal extensions of the NJL model have been studied in the last few years[4]. Nonlocality arises naturally in the context of several of the most successful approaches to low-energy quark dynamics as, for example, the instanton liquid model[5] and the Schwinger-Dyson resummation techniques[6]. It has been also argued that nonlocal covariant extensions of the NJL model have several advantages over the local scheme. Namely, the effective interaction is finite to all orders in the loop expansion and therefore there is not need to introduce extra cut-offs, soft regulators such as Gaussian functions lead to small NLO corrections[7], etc. In addition, it has been shown[8] that a proper choice of the nonlocal regulator and the model parameters can lead to some form of quark confinement, in the sense of a quark propagator without poles at real energies. In this contribution we discuss the extension of this type of models to finite temperature and chemical potential.

FORMULATION

Our starting point is the partition function at zero T and μ,

$$Z_0 = \int D\bar{\psi}D\psi \; e^{-S_E}, \tag{1}$$

where S_E stands for the Euclidean action

$$S_E = \int d^4x \left[\bar{\psi}(x)(-i\partial + m_c)\psi(x) - \frac{G}{2} j_a(x) j_a(x) \right]. \tag{2}$$

Here m_c is the current quark mass, and the euclidean operator ∂ is defined as

$$\partial = \gamma_4 \frac{\partial}{\partial \tau} + \vec{\gamma} \cdot \vec{\nabla} \tag{3}$$

with $\gamma_4 = i\gamma_0$, $\tau = it$. The current $j_a(x)$ is given by

$$j_a(x) = \int d^4y \, d^4z \, r(y-x) \, r(x-z) \, \bar{\psi}(y) \Gamma_a \psi(z), \tag{4}$$

where $\Gamma_a = (1, i\gamma_5 \vec{\tau})$ and $r(x-y)$ is a non-local regulator function. The regulator is local in momentum space, namely

$$r(x-z) = \int \frac{d^4p}{(2\pi)^4} \, e^{-i(x-z)p} \, r(p). \tag{5}$$

In fact, Lorentz invariance implies that $r(p)$ can only be a function of p^2, hence we will use for the Fourier transform of the regulator the form $r(p^2)$ from now on.

To proceed it is convenient to deal with bosonic degrees of freedom. Let us perform a standard bosonization of the theory, introducing the sigma and pion meson fields

$$M_a(x) = (\sigma(x), \vec{\pi}(x)). \tag{6}$$

Following the usual steps we obtain a partition function equivalent to that in (1):

$$Z_0 = \int D\sigma D\pi \; \det A(M_a) \exp\left[-\frac{1}{2G} \int \frac{d^4p}{(2\pi)^4} \, M_a^2(p)\right], \tag{7}$$

where the operator A in momentum space reads

$$A(M_a) = (-\not{p}+m_c)(2\pi)^4\delta^{(4)}(p-p') + r(p^2)M_a(p-p')r(p'^2)\Gamma_a. \quad (8)$$

At this stage, we perform the mean field approximation by expanding the meson fields around their translational invariant vacuum expectation values $\bar{\sigma}$ and $\bar{\pi}_i$ and neglecting the fluctuations $\delta\sigma(x)$ and $\delta\pi_i(x)$. The mean values of the pion fields $\bar{\pi}_i$ vanish due to symmetry reasons. Within this approximation the determinant in (7) is formally given by

$$\det A = \exp\mathrm{Tr}\log A = \exp V^{(4)}\int\frac{d^4p}{(2\pi)^4}\,\mathrm{tr}\log\left[-\not{p}+m_c+\bar{\sigma}\,r^2(p^2)\right], \quad (9)$$

where tr stands for the trace over the Dirac, flavor and color indices, and $V^{(4)}$ is the four-dimensional volume of the functional integral.

Now the corresponding partition function in the grand canonical ensemble for finite temperature T and chemical potential μ can be obtained through the replacement in the integrals in (7) and (9)

$$\int\frac{d^4p}{(2\pi)^4}\,F(p_4,\vec{p}) \quad\to\quad \sumint_p F(p_4,\vec{p}) \equiv T\sum_{n=-\infty}^{n=\infty}\int\frac{d^3p}{(2\pi)^3}\,F(\omega_n-i\mu,\vec{p}), \quad (10)$$

where ω_n are the Matsubara frequencies corresponding to fermionic modes, $\omega_n = (2n+1)\pi T$. In performing this replacement we have assumed that the regulator acquires an explicit μ dependence, as it is the case e.g. in the instanton liquid model. In the same way we replace the volume $V^{(4)}$ by V/T, V being the three-dimensional volume in coordinate space. Finally, the grand canonical thermodynamic potential in the mean field approximation is given by

$$\Omega_{MF}(T,\mu) = -\frac{T}{V}\log Z_{MF}(T,\mu) = -4N_c\left(\sumint_p\log\left[p^2+\Sigma^2(p^2)\right]\right) + \frac{\bar{\sigma}^2}{2G} \quad (11)$$

with the quark selfenergy $\Sigma(p^2)$ defined as

$$\Sigma(p^2) = m_c + \bar{\sigma}\,r^2(p^2). \quad (12)$$

It can be shown that in general this quantity turns out to be divergent. However, it can be regularized by subtracting the corresponding value at zero T and μ, $\Omega_{MF}^{(r)} = \Omega_{MF} - \Omega_{MF}^{(0)}$. Now the minima of the thermodynamic potential are obtained from the solutions of

$$\frac{\partial\Omega_{MF}}{\partial\bar{\sigma}} = 0 \quad (13)$$

which leads to the following gap equation for $\bar{\sigma}$:

$$\bar{\sigma} = 8N_cG\sumint_p\frac{\Sigma(p^2)\,r^2(p^2)}{p^2+\Sigma^2(p^2)}. \quad (14)$$

Given the thermodynamic potential the expressions for all other relevant quantities can be easily derived. For each flavor the quark vacuum expectation value is given by

$$\langle \bar{q}q \rangle = \frac{\partial \Omega_{MF}}{\partial m_c} = -4N_c \fint_p \frac{\Sigma(p^2)}{p^2 + \Sigma^2(p^2)}, \qquad (15)$$

while the quark density turns out to be

$$\rho_q = -\frac{\partial \Omega_{MF}}{\partial \mu} = -4iN_c \fint_p \frac{p_4 + \Sigma(p^2)\, \partial_{p_4}\Sigma(p^2)}{p^2 + \Sigma^2(p^2)}. \qquad (16)$$

In obtaining these last two equations it should be noticed that for each quark flavor only half of the kinetic term in Eq.(11) contributes to the derivatives.

As well known[9], if the quark selfenergy is momentum independent as in the conventional NJL model, the sums over the Matsubara frequencies indicated in the previous equations can be easily carried out and one obtains rather simple expressions in terms of the usual occupation numbers. This also holds when the selfenergy only depends on the spatial components of the momentum. However, for the covariant regulators we are considering here the situation is more complicated. The main difficulty is that the analytic structure of the quark propagator $S(q) = 1/(-\slashed{q} + \Sigma(q^2))$ in the complex plane can be much richer in this case: there might be a rather complicate structure of poles and cuts.

In Minkowski space and for $\vec{q} = 0$ the positions of the poles are given by the solutions of

$$q_0^2 - \left[m_c + \bar{\sigma}\, r^2(-q_0^2)\right]^2 = 0. \qquad (17)$$

Let us denote the real and imaginary parts of these solutions with r_p and i_p respectively. For nonzero spatial momentum $\vec{q}$ it is easy to see that the poles are located at $R_p + iI_p$, where

$$R_p = \pm \varepsilon_p, \qquad I_p = \pm \frac{r_p\, i_p}{\varepsilon_p} \qquad (18)$$

with

$$\varepsilon_p = \sqrt{\frac{r_p^2 - i_p^2 + \vec{q}^{\,2} + \sqrt{(r_p^2 - i_p^2 + \vec{q}^{\,2})^2 + 4\, r_p^2\, i_p^2}}{2}}. \qquad (19)$$

On the other hand, the possible cuts of the propagator will be given in general by the cuts of the regulator as a function of q. As in the case of the poles, it is convenient to determine first their position in the q_0 plane for $\vec{q} = 0$. Then it is simple to find the position in the general case applying the above equations to all the points along the cuts.

Before going into the explicit evaluation of the Matsubara sums it is useful to analyze the pole and cut configurations in some relevant situations. One important case is that in which the regulator is such that the Minkowski quark propagator has an arbitrary set of poles but no cuts. Within this class of regulators it might exist a situation in which there are no poles along the real axis. As already mentioned, this situation might be interpreted as a realization of confinement[8]. In that case only quartets of poles located at $\alpha_p = r_p \pm i\, i_p$, $\alpha_p = -r_p \pm i\, i_p$ appear. On the other hand, if purely real poles do exist they will show up as doublets $\alpha_p = \pm r_p$. It is clear that the number and position of the poles depend on the details of the regulator. For example, if we assume it to be a step function as in the standard NJL model, only two purely real poles at $\pm M$ appear, with M being the dynamical quark mass. For a Gaussian interaction, i.e. a regulator of the form

$$r(q^2) = \exp\left(-q^2/2\Lambda^2\right), \tag{20}$$

three different situations might occur. These can be classified according to the value of $\Sigma(0)$ at zero T and μ, which we denote by $\bar{\Sigma}(0)$. For values of $\bar{\Sigma}(0)$ below a certain critical value $\Sigma(0)_{crit}$, two pairs of purely real simple poles and an infinite set of quartets of complex simple poles appear. At $\bar{\Sigma}(0) = \Sigma(0)_{crit}$, the two pairs of purely imaginary simple poles turn into a doublet of double poles with $i_p = 0$, while for $\bar{\Sigma}(0) > \Sigma(0)_{crit}$ only an infinite set of quartets of complex simple poles is obtained. There are some other physically interesting regulators as the Lorentzian regulators

$$r(q^2) = \frac{1}{1 + (q^2/\Lambda^2)^n}, \tag{21}$$

which belong to the "no cuts" group. Unfortunately some other important ones, like that arising within the instanton liquid model[5], do lead to propagators which present poles and cuts. Thus, it is necessary to consider this more general situation.

The evaluation of the sums over the Matsubara frequencies can be carried out by a convenient choice of integration paths in the complex plane and the help of Cauchy's theorem. As an example we consider the sum appearing in the gap equation Eq.(14),

$$S \equiv \int \frac{d^3q}{(2\pi)^3} \, T \sum_{n=-\infty}^{\infty} F[(\omega_n - i\mu)^2 + \vec{q}^{\,2})], \tag{22}$$

where

$$F(q^2) = \frac{\Sigma(q^2) r^2(q^2)}{q^2 + \Sigma^2(q^2)}. \tag{23}$$

Assuming that the regulator is such that the quark propagator at $\vec{q} = 0$ has an arbitrary set of simple poles located at $\pm r_p \pm i\, i_p$ and a single cut a single cut lying along the real axis in the regions $[-\infty, -r_c]$ and $[r_c, \infty]$ one gets[10]

$$S = \int \frac{d^4q}{(2\pi)^4} F(q^2) + 2 \int \frac{d^3q}{(2\pi)^3} \, \text{Re} \sum_{poles} \gamma_p \, \text{Res}\left[F(\vec{q}^{\,2} - z^2); z_p\right] \left[n_+(z_p) + n_-(z_p)\right]$$
$$- \frac{1}{\pi} \int \frac{d^3q}{(2\pi)^3} \int_{\sqrt{r_c^2 + \vec{q}^{\,2}}}^{\infty} dx \, \text{Im}\left\{F(\vec{q}^{\,2} - x^2 - i\varepsilon) \left[n_+(x) + n_-(x)\right]\right\}, \tag{24}$$

where the first term corresponds to the value of S at zero temperature an chemical potential. The coefficient γ_p is defined as $\gamma_p = 1/2$ for $\text{Im}(z_p) = 0$ and $\gamma_p = 1$ otherwise and the residues are evaluated at $z_p = \pm \varepsilon_p \pm i\, r_p\, i_p/\varepsilon_p$. The sum runs over the poles with $\text{Re}(z_p) > 0$ and $\text{Im}(z_p) \geq 0$. Finally, $n_\pm(z)$ stand for the occupation number functions

$$n_\pm(z) = \frac{1}{1 + \exp[(z \mp \mu)/T]}. \tag{25}$$

It is clear that the steps leading to Eq.(24) can be also followed to evaluate the sums appearing in e.g. the quark vacuum expectation value and the quark density, Eqs.(15) and (16) respectively, once the function $F(q^2)$ is properly redefined. In the case of the regularized thermodynamic potential $\Omega_{MF}^{(r)}$ the situation is somewhat more complicated, since the argument of the sum includes a logarithm that introduces cuts in the z plane outside the real axis. Nevertheless a similar relation can be derived to calculate the corresponding Matsubara sum.

RESULTS

Having introduced the formalism needed to deal with models with non-local separable interactions at finite temperature and chemical potential, we turn now to the results of numerical calculations carried out for some particular regulators. Here we present in detail those corresponding to the Gaussian regulator. Results for the Lorentzian and the instanton model regulators are qualitatively similar and can be found in Ref.[10]. We consider two sets of values for the parameters of the model. Set I corresponds to $G = 50$ GeV^{-2}, $m_c = 10.5$ MeV and $\Lambda = 627$ MeV, while for Set II the respective values are $G = 30$ GeV^{-2}, $m_c = 7.7$ MeV and $\Lambda = 760$ MeV. Both sets of parameters lead to the physical values of the pion mass and decay constant. For Set I the calculated value of the chiral quark condensate at zero temperature and chemical potential is $-(200 \text{ MeV})^3$, while for Set II it is $-(220 \text{ MeV})^3$. These values are similar in size to those determined from lattice gauge theory or QCD sum rules. The corresponding results for the self-energy at zero momentum are $\Sigma(0) = 350$ MeV for Set I and $\Sigma(0) = 300$ MeV for Set II. It is possible to check that Set I corresponds to a situation in which the quark propagator has no purely real poles and Set II to the case in which there are two pairs of them. Following Ref.[8], Set I might be interpreted as a confining one since quarks cannot materialize on-shell in Minkowski space.

The behaviour of the zero-momentum self-energy $\Sigma(0)$, the chiral condensate and the quark density as function of the chemical potential for some values of the temperature is shown in Fig. 1. There, we observe that at $T = 0$ there is a first order phase transition for both the confining and the non-confining sets of parameters. As the temperature increases, the value of the chemical potential at which the transition shows up decreases. Finally, above a certain value of the temperature the first order phase transition does not longer exist and, instead, there is a smooth crossover. This phenomenon is clearly shown in the left panels of Fig. 2, where we display the critical temperature at which the phase transition occurs as a function of the chemical potential (upper left panel) and as a function of the density (lower left panel). The point at which the first order phase transition ceases to exist is usually called "end point". In the chiral limit the latter turns into the so-called "tricritical point", which is the point at which the second order phase transition expected to happen in QCD with two massless quarks becomes a first order one. In fact, this is also what happens within the present model in the chiral limit, as it is shown in the right panels of Fig. 2. Some predictions about both the position of this point and its possible experimental signatures exist in the literature[11]. In our case the "tricritical point" is located at $(T_P, \mu_P) =$ (70 MeV, 130 MeV) for Set I and (70 MeV, 140 MeV)

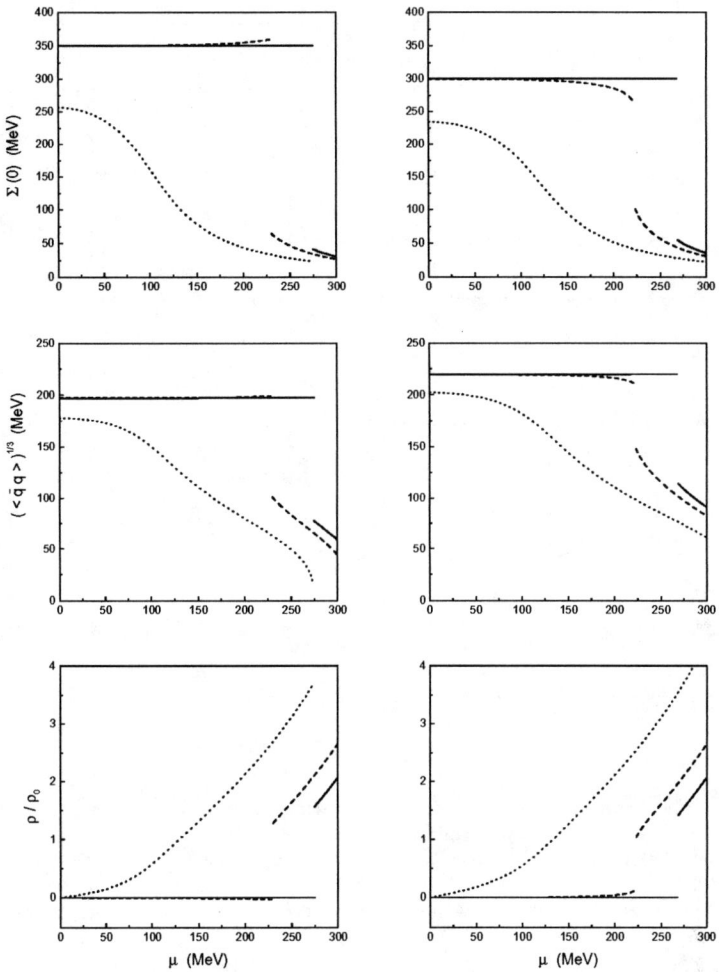

FIGURE 1. Behaviour of the self-energy, the chiral condensate and the quark density as functions of the chemical potential for three representative values of the temperature. Full line corresponds to T=0, dashed line to $T = 50$ MeV and dotted line to $T = 100$ MeV. The left panels display the results for Set I and the right panels those for Set II. The quark density ρ is given with respect to nuclear matter density $\rho_0 = 1.3 \times 10^6$ MeV^3.

for Set II, while the "end points" are placed at (T_E, μ_E) =(70 MeV, 180 MeV) and (55 MeV, 210 MeV), respectively.

It is interesting to discuss in detail the situation concerning the confining set. In this case we can find, for each temperature, the chemical potential μ_d at which confinement is lost. Following the proposal of Ref.[8], this corresponds to the point at which the self-energy at zero momentum reaches the value $\Sigma(0)_{crit}$. Using the values of m_c and Λ corresponding to Set I we get $\Sigma(0)_{crit} = 267$ MeV. For low temperatures, μ_d coincides

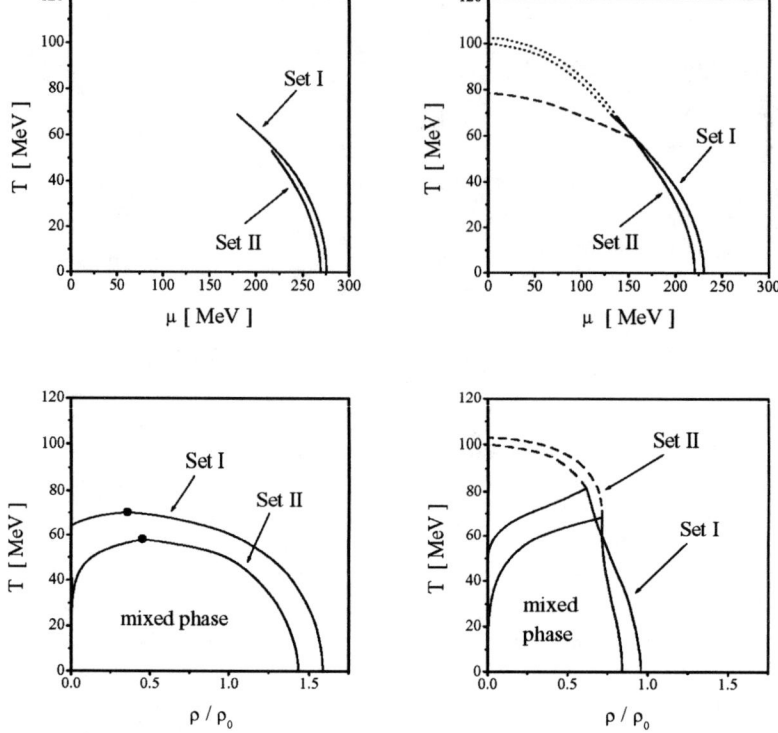

FIGURE 2. Critical temperatures as a function of the chemical potential (upper panels) and as a function of the quark density (lower panels). The right panels corresponds to the chiral limit and the left panels to the case of finite quark masses. In both panels the full line stands for a first order phase transition while in the right panel the dotted line indicates that the transition is of second order. The dashed line in the upper right panel indicates the temperature at which, for Set I, complex poles of the propagator turn into real poles.

with the the chemical potential at which the chiral phase transition takes place. However, for a temperature close enough to that of the "end point", μ_d starts to be slightly smaller than the value of μ that corresponds to the chiral restoration. Above T_E it is difficult to make an accurate comparison since, for finite quark masses, the chiral restoration proceeds through a smooth crossover. However, we can still study the situation in the chiral limit. In this case we find that, in the region where the chiral transition is of second order, deconfinement always occurs, for fixed T, at a lower value of μ than the chiral transition. The corresponding critical line is indicated by a dashed line in the upper right panel of Fig. 2. In any case, as we can see in this figure, the departure of the line of chiral restoration from that of deconfinement is in general not too large. This indicates that within the present model both transitions tend to happen at, approximately, the same point.

CONCLUSIONS

In this work we investigate the features of the chiral phase transition in some non-local chiral quark models with covariant separable interactions. We consider several types of regulators: the Gaussian regulator, the Lorentzian regulator and the instanton liquid model regulator. We find that in all the cases the phase diagram is quite similar. In particular, we obtain that for two light flavors the transition is of first order at low values of the temperature and becomes a smooth crossover at a certain "end point". Our predictions for the position of this point are very similar for both types of regulators and slightly smaller than the values in Refs.[11], $T_P \approx 100$ MeV and $\mu_P \approx 200-230$ MeV. In this sense, we remark that our model predicts a critical temperature at $\mu = 0$ of about 100 MeV, somewhat below the values obtained in modern lattice simulations which suggest $T_c \approx 140-190$ MeV. In any case, our calculation seems to indicate that μ_P might be smaller than previously expected even in the absence of strangeness degrees of freedom.

Several extensions of this work are of great interest. For example, it would be very important to investigate the impact of the introduction of strangeness degrees of freedom and flavor mixing on the main features of the chiral phase transition. Also, the competition between chiral symmetry breaking and color superconductivity at large chemical potential deserves further studies. Work along these lines is under way.

ACKNOWLEDGMENTS

The authors are fellows of CONICET, Argentina. This work was supported in part by Fundación Antorchas and Fundación Balseiro, Argentina.

REFERENCES

1. Karsch, F. *hep-lat/0106019*.
2. Nambu, Y. and Jona-Lasinio, G., Phys. Rev. **122**, 345 (1961); Phys. Rev. **124**, 246 (1961).
3. Vogl, U. and Weise, W., Prog. Part. Nucl. Phys. **27**, 195 (1991); Klevansky, S., Rev. Mod. Phys. **64**, 649 (1992); Hatsuda, T. and Kunihiro, T., Phys. Rep. **247**, 221 (1994).
4. Ripka, G., *Quarks bound by chiral fields*, Oxford University Press, Oxford, 1997.
5. Schaefer, T. and Schuryak, E., Rev. Mod. Phys. **70**, 323 (1998).
6. Roberts, C. D., and Williams, A. G., Prog. Part. Nucl. Phys. **33**, 477 (1994); Roberts, C. D., and Schmidt, S. M., Prog. Part. Nucl. Phys. **45S1**, 1 (2000).
7. Ripka, G., Nucl. Phys. **A683**, 463 (2001); Plant, R. S. and Birse, M. C., *hep-ph/0007340*.
8. Bowler, R. D. and Birse, M. C., Nucl. Phys. **A582**, 655 (1995); Plant, R. S. and Birse, M. C., Nucl. Phys. **A628**, 607 (1998).
9. Kapusta, J. I., *Finite temperature field theory*, Cambridge Univ. Press, Cambridge, 1993.
10. General, I., Gomez Dumm, D. and Scoccola, N.N., Phys. Lett. **B506**, 267 (2001); Gomez Dumm, D. and Scoccola, N.N., Phys. Rev. **D65**, 074021 (2002).
11. Berges, J. and Rajagopal, K., Nucl.Phys.**B538**, 215 (1999); Halasz, M. A., et al, Phys.Rev.**D58**, 096007 (1998); Stephanov, M., Rajagopal, K. and Shuryak, E., Phys. Rev. Lett. **81**, 4816 (1998); Phys. Rev. **D60**, 114028 (1999).

Front Form Approach to $q\bar{Q}$ Mesons with Harmonic Confinement

C. A. Z. Vasconcellos*, M. Dillig[†], E. F. Lütz*, F. G. Pilotto* and G. F. Marranghello*

*Instituto de Física, UFRGS, 91501-970 Porto Alegre, C.P. 15051, Brazil
[†]Institut für Theoretische Physik III, Universität Erlangen-Nürnberg, D 91058 Erlangen, Germany

Abstract. We investigate mesons as unequal mass $q\bar{Q}$ objects in the front form representation (equivalently, on the light cone (LC) or in the infinite momentum frame (IMF)). Our starting point is the manifestly covariant Bethe-Salpeter equation (BS) in the instant form, which we reduce to a 3-dimensional covariant LC equation by restricting the quark and sea antiquark system symmetrically off their mass shells. We discuss analytical solutions to the LC amplitude and compare our findings with current parameterizations of the distribution amplitude emphasizing the characteristic x-dependence for unequal quark masses.

INTRODUCTION

The common approach to hadronic two- and many-body systems is their formulation based either on the nonrelativistic many-body Schrödinger equation or the minimal relativistic extension towards the corresponding Dirac equation; in practical calculations, effective two-body forces and more or less severe mean-field approximations are employed to make the system tractable. The common framework for the many-body problem is the instant form: in the language of second quantization the commutators (anti-commutators) of the underlying field operators are quantized along the hyperplane with $t = const$, resulting in the standard commutation and anti-commutation relations. The advantage of this approach is the appearance of a single (unique) time-variable t for all interacting particles together with standard Feynmann contributions in perturbative expansions. As a characteristic feature the relation for a free particle $E = \pm\sqrt{\vec{p}+m^2}$ involves the intimate coupling of positive and negative energy states even without interaction i.e. the existence of a complicated "nontrivial" Dirac vacuum for fermions.

However, particularly with the advent of recent systematic experiments at high energy and/or momentum transfers, such as elastic or deep-inelastic hard scattering processes, a different frame for such processes is suggested[1]. As with increasing energy the bulk of scattering events are focussed on a small forward cone along the direction of the projectile (the beam), as characterized by $q_0^2, q_z^2 >> \vec{q}_\perp^2$, then, with the transversal momentum component being neglected, the dispersion relation is truncated to an (effectively) 2-dimensional form $q^2 = q_0^2 - q_z^2 - \vec{q}_\perp^2 \simeq (q_0-q_z)(q_0+q_z) = q_-q_+$: the

---

[1] Supported in part by the Forschungszentrum Jülich (FZJ), Contract 41445386 (COSY-62).

physics becomes 2-dimensional (we exhibit the symmetry between q_0 and q_z by introducing quantities $q_\pm$ discussed below). Though rather simplistic, already this intuitive picture suggests a different system of reference for relativistic high energy processes. Of course, observables in any field theoretical calculation should be independent of the specific frame of quantization: the quantization along the light-like hyperplane $t+z = const$ is equally acceptable. In momentum space — as intuitively envisioned — this leads to a transition to LC variables $(p_0, p_z, \vec{p}_\perp) \to (p_-, p_+, \vec{p}_\perp)$ with a rotation of $\theta = \pi/4$ around the perpendicular $\vec{p}_\perp$ axis:

$$\begin{pmatrix} p_- \\ p_+ \\ \vec{p}_\perp \end{pmatrix} = \begin{pmatrix} \sqrt{2}\cos(\pi/4) & -\sqrt{2}\sin(\pi/4) & 0 \\ \sqrt{2}\sin(\pi/4) & \sqrt{2}\cos(\pi/4) & 0 \\ 0 & 0 & 1 \end{pmatrix} \begin{pmatrix} p_0 \\ p_z \\ \vec{p}_\perp \end{pmatrix}, \tag{1}$$

ending up in a new representation of the dispersion relation of a free particle $p^2 = p_0^2 - p_z^2 - \vec{p}_\perp{}^2$. Representing p_- as the new light cone energy and p_+ as the longitudinal LC momentum, this new dispersion relation $p_- = \frac{m^2 + \vec{p}_\perp{}^2}{p_+}$ develops a unique new feature: being single-valued in the energy variable p_-, it strictly decouples particles from antiparticles and thus the free (and, as can be shown, interacting) vacuum becomes "trivial"[2]. Immediate consequences are, as shown in the pioneering work of Dirac[3], the appearance of 7 (instead of 6) kinematical generators for the Poincaré group, specifically the kinematical nature of Lorentz-boosts, as already anticipated for the standard Lorentz transformations

$$\begin{matrix} p_0' = \gamma(p_0 + \beta p_z) \\ p_z' = \gamma(\beta p_0 + p_z) \end{matrix} \to p_\pm' = \gamma(1 \pm \beta)p_\pm \tag{2}$$

(with $\beta = v$ and $\gamma = 1/\sqrt{1-\beta^2}$; as equivalently seen from $p_z \to \infty$) in the IMF) or the suppression of pair-creation out of the vacuum (which, in turn allows an appropriate decomposition of the Fock-space respecting baryon number conservation within each Fock component). Finally the corresponding Feynman rules are recovered in the old-fashioned perturbation theory basically from the substitution of the LC energies[4]

$$p_0 = \sqrt{\vec{p}^2 + m^2} \to p_- = \frac{m^2 + \vec{p}_\perp{}^2}{x} \tag{3}$$

(when we introduce the momentum fraction $x = \frac{p_+}{P_+}$ with P_+ being the total momentum of the system).

Thus going to the LC we have recovered the intuitive picture sketched at the beginning for scattering processes at high energy and momentum transfers with an effectively 2-dimensional $p_\pm$-world. As expected, a price has to be paid for this simplification: breaking the a priori spherical symmetry in the transition to the LC variables, rotational invariance is lost; similarly, the implementation of parity conservation in the 3-dimensional Hamiltonian approach, upon integrating out the time component in the LC amplitude

$$\psi(x, \vec{p}_\perp) = \int dp_- \psi(x, \vec{p}_\perp, p_-), \tag{4}$$

evidently destroys reflection symmetry $p_- \xleftrightarrow{P} p_+$, present in manifestly covariant formulations[5]. However, in spite of these deceases, getting the implementation of covariance under Lorentz boosts practically for free on the LC together with the trivial vacuum, renders the relativistic front form one of the most promising attempts to bound systems at large total 4-momenta.

In this note, we apply the covariant LC formalism to two body-systems in QCD. Quite explicitly, motivated by numerous studies of mesonic $q\bar{Q}$ systems in the recent literature, we investigate in this note the structure of mesons on the LC in the framework of Quasi-Potential (QP) equations[6]. As we are looking for analytical solutions for the corresponding LC wave-functions, we incorporate the genuine feature of QCD by a covariant scalar harmonic confinement potential (attempts in the instant form in this direction have been followed, for example, by Mitra and coworkers [7]). In view of recent activities at charm (C) and beauty (B) factories, we focus on momentum distributions both of $q\bar{q}$ mesons with equal masses and light-heavy $q\bar{Q}$ mesons with a light u, d quark and a heavy s, c, b anti-quark.

FROM THE BETHE-SALPETER TO THE QUASI-POTENTIAL EQUATION FOR THE $Q\bar{Q}$ SYSTEM ON THE LIGHT CONE

Our starting point for a quark q (with mass m_1 and momentum p_1) and an anti-quark $\bar{Q}$ (with mass m_2 and momentum p_2) is the manifestly covariant BSE in the instant form[8]

$$\Psi(p_1, p_2) = (\not{p}+m)_{(1)}(\not{p}+m)_{(2)} G(p_1, p_2)$$
$$\times \int K(p_1, p_2, k_1, k_2) \Psi(k_1, k_2) \delta(k_1 + k_2 - p_1 - p_2) dk_1 dk_2 \quad (5)$$

with the two-particle propagator

$$G(p_1, p_2) = (D_1(p_1) D_2(p_2))^{-1} = ((p_1^2 - m_1^2 - i\varepsilon)(p_2^2 - m_2^2 - i\varepsilon))^{-1}. \quad (6)$$

It is clear that the transition of the 4-dimensional BSE from the instant form to the LC is merely a formal change in the explicit representation, without modifying its physical content. Both from conceptional and practical reasons (such as to provide for example an appropriate one-body limit to the Dirac equation for phenomenological kernel, a feature which is in general not exhibited by the BS equation with truncated kernels, such as in the ladder approximation[9]) we integrate out the relative p_- component, by restricting the corresponding retardation in a covariant way. As a consequence the resulting equation is still covariant and avoids the conventional static (Salpeter) limit.

We eliminate the dependence of eq. (5) on the relative energy variable and convert the BSE into a QPE by projecting the propagation of the $q\bar{Q}$ system on the mass shell as [10]

$$G(p, P) = i\pi \frac{\delta_+(\eta D_1(p, P) - (1-\eta) D_2(p, P))}{D_1(p, P) + D_2(p, P)}, \quad (7)$$

with the η parameter restricted to the interval $[0, 1]$; above we denote the total and relative momenta for the $q\bar{Q}$ system by p and P, respectively; furthermore in eq. (7) we keep only the positive energy pole (+) in the δ function. We remark that the observables calculated should be independent of η.

Even after integrating out the dependence on the internal energy variable, the resulting equations is in its full structure involve a coupled system of 16 differential equations due to the spin of the interacting fermions and particle - anti-particle mixing, schematically

$$\psi(p_1, p_2) = \begin{pmatrix} \psi_{++} \\ \psi_{+-} \\ \psi_{-+} \\ \psi_{--} \end{pmatrix} |[j_1 j_2] J J_3 > . \tag{8}$$

Both in the instant and front form we reduce this complex system to a single radial equation by an appropriate simplification of the spin structure of the system

$$\Psi(p,P)_{\lambda\lambda' J J_3} = R(\lambda\lambda' J J_3) \Phi(p,P); \tag{9}$$

where in an helicity representation $R(\lambda\lambda' J J_3)$ is determined in the Mock representation (assuming no interaction in the spin sector [11]), or on the LC by a free Melosh rotation from the $q\bar{Q}$ rest system to the IMF [12]. Without further details we mention that we are currently exploring the detailed spin-structure of the LC by employing a one-rank separable interaction kernel, which allows to reduce the complex system of coupled differential equations to a set of 16 ordinary algebraic equations.

Upon elimination of the complexity of the spin dependence the covariant nature of the BSE allows an immediate transition to LC variables[2-5]

$$\begin{aligned} P &= (P_-, P_+, \underline{0}); P_- P_+ = M^2; P_+ = 1; \\ p &= (p, x, \underline{p}_\perp); x_i = \frac{p_i^+}{P^+}; x_1 + x_2 = 1; \end{aligned} \tag{10}$$

The resulting radial QPE for $\Phi(x, \underline{p}_\perp, M^2)$ on the LC is then obtained with the appropriate LC projection for the two-particle Green's function [13]

$$G(P_-, x, \underline{p}_\perp, M^2) \sim \delta(p_- - \frac{(1-\eta)(1-x)M^2 + \eta m_{1\perp}^2 - (1-\eta) m_{2\perp}^2}{1 - \eta + (2\eta - 1)x}), \tag{11}$$

which leads to

$$(M^2 - \frac{m_{1\perp}^2}{x} - \frac{m_{2\perp}^2}{(1-x)}) \Phi(x, \underline{p}_\perp; M^2) = \int K(x, \underline{p}_\perp, y, \underline{k}_\perp) \Phi(y, \underline{k}_\perp, M^2) dy d\underline{k}_\perp, \tag{12}$$

with M^2 being the invariant squared $q\bar{Q}$ mass and $m_{i\perp}^2 = m_i^2 + \vec{p}_\perp^2$ (we absorb all irrelevant factors from the projection of the Green's function in the kernel K).

For a given interaction kernel the solution of the equations above directly results in the light cone distribution of $q\bar{Q}$ system. As so far a detailed derivation of the LC amplitude from QCD is till lacking (only results in the large $Q^2 \to \infty$ limit or estimates from QCD sum rules exist [2, 14]), the main two goals of this short note are very modest: on the one side we would like to use a kernel which can be derived explicitly from the corresponding covariant representation in the instant form and we aim for an analytical solution for the momentum distribution. A QCD inspired kernel, which includes confinement as the unique property of QCD, is a covariant harmonic confining kernel

$$K(Z^2) \sim Z_\mu Z^\mu; \tag{13}$$

within yields, in momentum space, in the instant form

$$K(p,\kappa) = \frac{1}{g^2} \frac{\partial^2}{\partial p^2} \delta(p-\kappa) \tag{14}$$

or, equivalently, in LC coordinates

$$K(p_-, x, \vec{p}_\perp; k_-, y, \vec{k}_\perp) = \frac{1}{g^2} \left(\frac{\partial^2}{\partial p_- \partial x} - \frac{\partial^2}{\partial \vec{p}_\perp^2} \right) \delta(p_- - k_-) \delta(x-y) \delta(\vec{p}_\perp - \vec{k}_\perp) \tag{15}$$

(the strength parameter g is related to the running quark-gluon coupling constant; its value is fixed in the non-relativistic limit from meson spectroscopy). Following the same steps for the reduction to the QPE as sketched above, we obtain explicitly

$$(M^2 - \frac{m_{1\perp}^2}{x} - \frac{m_{2\perp}^2}{(1-x)}) \Phi(x, \underline{p}_\perp; M^2) = \frac{1}{g^2} (\frac{\partial}{\partial p_-} \frac{\partial}{\partial x} + \frac{\partial^2}{\partial \underline{p}_\perp^2}) \Phi(x, \underline{p}_\perp, M^2), \tag{16}$$

where the projection on the relative energy variable fixes p_- from eq. (11). Upon eliminating p_- in the last equation with

$$\frac{\partial}{\partial p_-} = \left(\frac{\partial x}{\partial p_-}\right)_{\vec{p}_\perp = const} \frac{\partial}{\partial x} + \left(\frac{\partial \vec{p}_\perp}{\partial p_-}\right)_{x = const} \frac{\partial}{\partial \vec{p}_\perp}, \tag{17}$$

from equation (16) we obtain for an arbitrary parameter η the explicit equation for $\phi_\eta(x, \vec{p}_\perp; M^2)$:

$$(M^2 - \frac{m_{1\perp}^2}{x} - \frac{m_{2\perp}^2}{1-x}) \phi_\eta(x, \vec{p}_\perp; M^2) =$$
$$\frac{1}{g^2} \left(\frac{(1-\eta+(2\eta-1)x)^2}{(\eta(\eta-1)M^2 + (1-2\eta)(\eta m_{1\perp}^2 - (1-\eta)m_{2\perp}^2)} \frac{\partial^2}{\partial x^2} \right.$$
$$\left. \frac{(1-\eta+(2\eta-1)x)}{2(2\eta-1)} \frac{1}{\vec{p}_\perp} \frac{\partial}{\partial \vec{p}_\perp} \frac{\partial}{\partial x} + \frac{\partial^2}{\partial \vec{p}_\perp^2} \right) \phi_\eta(x, \vec{p}_\perp; M^2) \tag{18}$$

As expected, for arbitrary values of η, the longitudinal and the transversal components of the light cone amplitude are intimately coupled both through the effective masses $m_{i\perp}^2 = m_i^2 + \vec{p}_\perp^2$ in the coefficient proportional to $\partial^2/\partial x^2$ and through the mixed derivative $(\vec{p}_\perp \partial/\partial \vec{p}_\perp) \frac{\partial}{\partial x}$.

In its most general form, i.e. for an arbitrary η, equation (18) admits only numerical solutions. As we in this note we are interested in analytically solvable equations, we explore the range of η to separate longitudinal and transversal components. Going back to equation (18) we immediately identify the two appropriate limits for η:

- $\eta = 1/2$ (the Blankenblecker-Sugar BBS projection [15]), yielding $p_- = (1-\eta)M^2 + \frac{1}{2}(m_1^2 - m_2^2)$ and
- $\eta = 0$ (the Gross projection [16]) in the limit $m_2 = m_Q \gg m_1 = m_q$ giving $p_- = M^2 - \frac{m_{2\perp}^2}{1-x}$ for $M^2 \sim m_2^2 \to \infty, m_{2\perp}^2 \to M^2$).

Both projections stress different physics for the off-shell propagation of the bound particles; $\eta = 1/2$ restricts the propagation of the particles symmetrically off-shell and is thus a natural choice for interacting particles with equal mass; $\eta = 0$ puts the heavy particle on its mass shell and enforces by this restriction the appropriate one-body limes (resulting in a single Dirac equation for the light particle in the limit $m_2 \to \infty$; this limit bears a strong relation to the heavy quark effective theory[17]).

In the following we focus on the BBS solution for the general case of unequal masses. We mention that with $\eta = 1/2$, but $m_q \neq m_Q$, p_- is only shifted by the difference of the squared masses and does not effect the radial solution; only the spin dependence, which we do not consider here, in detail, is modified.

The next steps are now well defined. The radial LC equation (18) in the BBS limit is explicitly given as

$$\left[\left(\frac{1}{M^2}\frac{\partial^2}{\partial x^2} + g^2 x(1-x)M^2\right) + \left(\frac{\partial^2}{\partial \vec{p}_\perp^2} - g^2 \vec{p}_\perp^2\right) - g^2 m_q^2\right]\Phi_{BBS}(x,\vec{p}_\perp,M^2) \quad (19)$$

As the longitudinal and the transversal components factorize we obtain with the ansatz (we drop the index BBS for simplicity)

$$\phi(x,\vec{p}_\perp,M^2) = \phi(x,M^2)\varphi(\vec{p}_\perp,M^2), \quad (20)$$

$$\left[\left(\frac{1}{M^2}\frac{\partial^2}{\partial x^2} + g^2 x(1-x)M^2 - E^2\right)\right]\Phi(x,M^2) = 0, \quad (21)$$

$$\left(\frac{\partial^2}{\partial \vec{p}_\perp^2} - g^2 \vec{p}_\perp^2 - \varepsilon^2\right)\varphi(\vec{p}_\perp,M^2) = 0, \quad (22)$$

and $E^2 + \varepsilon^2 = g^2 m_q^2$.

The explicit solution of the differential equation in the BBS limit is given as[18]

$$\varphi(\vec{p}_\perp,M^2) = H_{nx}(p_x)H_{ny}(p_y)e^{-\frac{g}{2}\vec{p}_\perp^2} \quad (23)$$

and

$$\Phi(x,M^2) = e^{-\frac{1}{2}\lambda x^2}\left(A\,_1F_1(a,\frac{1}{2};\lambda x^2) + B(\lambda x^2)^{\frac{1}{2}}\,_1F_1(a+\frac{1}{2},\frac{3}{2};\lambda x^2)\right), \quad (24)$$

with

$$a \equiv \frac{1}{4} - \frac{k^2}{4\lambda} = \frac{1}{4} - \frac{M^2(\frac{gM^2}{4} - E^2)}{4g^2 M^4}, \quad (25)$$

with $-1/2 \leq x \leq 1/2$. In these expressions, the functions $F(a,b,z)$ denote confluent hypergeometrical functions, $H_{ni}(p_i)$ are Hermite polynomials. The eigenvalues for the invariant mass M^2, i. e. the mesonic spectrum, are then defined by the eigenvalue conditions

$$\Phi(x = -1/2, M^2) = \Phi(x = 1/2, M^2) = 0 \qquad (26)$$

Together with the normalization condition $1 = \frac{1}{2} \int dx d\vec{p}_\perp |\Phi(x, \vec{p}_\perp, M^2)|^2$ this fixes the LC wave function completely as a function of the strength g of the harmonic kernel.

We compare our derivation with current parameterizations in the literature, where various simple analytical forms can be found[19]. In order to exhibit the influence of unequal quark masses we compare with two typical momentum distributions (for the variable x shifted to $\to x + 1/2$)

- the Lepage-Brodsky parameterization [20]

$$\Phi(x, \underline{p}_\perp, M^2) = N \exp\left\{ -\frac{Q^2}{2} \left(\frac{m_q^2 + \vec{p}_\perp^2}{x} + \frac{m_Q^2 + \vec{p}_\perp^2}{1-x} \right) \right\}; \qquad (27)$$

- the parameterization as a simple power law [21]

$$\Phi(x, M^2) = N x^{\frac{m_q}{m_Q}} (1-x)^{\frac{m_Q}{m_q}} \qquad (28)$$

(we extended the standard equal-mass parameterization to unequal quark masses). Insight into the x-dependence of the BBS solution, particularly for unequal particle masses, is obtained from a comparison of the invariant longitudinal distribution amplitude (IDA) [2, 21],

$$\Phi(x, M^2) = \int d\vec{p}_\perp |\Phi(x, \vec{p}_\perp, M^2)| \Theta(\Lambda^2 - \vec{p}_\perp^2). \qquad (29)$$

(for the parameterizations compared the transversal cut-off at a scale Λ affects only the overall normalization of the distribution amplitude; due to the Gaussian form the contribution from the perpendicular components is always finite, even without a cut-off).

Exceeding the limited scope of this brief note, a detailed discussion of our results is presented in a forthcoming publication. Here we stress only various characteristic features of the invariant amplitudes derived. As a general trend we find, that the various current parameterizations in the literature show similar gross features, but differ mainly around the end points $x \to 0$ and $x \to 1$ and in the width of the distribution around the maximum $x_{max} = m_q/m_Q$. Furthermore we confirm a systematic shift of the dominant momentum components towards $x = 0$ for $q\bar{Q}$ mesons with increasing heavy quark mass, a trend which is expected qualitatively from a minimization of the kinetic energy in the LC Hamiltonian:

$$\frac{m_1^2}{x_{max}} - \frac{m_2^2}{1 - x_{max}} \simeq 0 \qquad (30)$$

yielding $x_{max} \sim m_1^2/m_2^2$ for $m_2^2 >> m_1^2$ [22].

Let us summarize our short contribution. In this note we investigate the momentum distribution of mesons with unequal quark masses on the light cone. Without considering spin effects in detail, the radial dependence was derived in a covariant fashion starting from the BSE, by projecting it in a still covariant and physically motivated way onto the light cone. Rigorous analytical solutions were derived and compared for a covariant confining kernel for the BBS constraint of the relative energy variable of the $q\bar{Q}$ system.

We find, that our results reproduce qualitatively the trends of purely phenomenological parameterizations of LC distribution amplitudes. In detail, however, particularly in the endpoint behaviour both in the longitudinal and in the transversal components we find serious quantitative differences due to the implementation of our boundary conditions. Though very simple, our model shows some advantages. First of all it starts from a covariant interaction kernel, which can be easily extended to more realistic cases (for example, to parameterize the influence of the one-gluon exchange allowing still an analytical solution). Furthermore, the transition from the BSE to QPE allows a direct transition from the instant form to the front form for arbitrary systems, such as to two very heavy quarks in the non-relativistic limit.

Of course, the rather selective ideas presented here yield only little insight in the details of the model. Presently we are extending our approach (including additional elements of the $q\bar{Q}$ interaction and comparing different recipes for the $4 \to 3$-dim reduction); beyond that we test the model for other quantities, such as for moments of the distribution functions, decay constants and form factors [23].

REFERENCES

1. J. A. Gracey and T. Greenshaw: Proc. DIS 2000 (World Scientific, Singapoure, 2001); A. W. Thomas and W. Weise: The Structure of the Nucleon (Wiley - VCH, Berlin, 2001).
2. S. J. Brodsky and H. C. Pauli: Phys. Rep. **301** (1998) 299 (and references therein); F. Lenz, M. Thies and K. Yazaki: Phys. Rev. **D63** (2001) 045018.
3. P. A. Dirac: Rev. Mod. Phys. **21** (1949) 392; S. Weinberg: Phys. Rev. **150** (1966) 1313.
4. J. M. Namyslowski: Prog. Part. Nucl. Phys. **14** (1985) 49.
5. M. Burkardt: Adv. Nucl. Phys. **23** (1996) 1; P. G. Blunden, M. Burkardt and G. A. Miller: Phys. Rev. **C60** (1999) 2998.
6. A. Logunov and A. Tavkelidze - Nuovo Cim. **29** (1963) 380; R. Thompson - Phys. Rev. **D1** (1970) 110; H. Cohen - Phys. Rev. **D2** (1970) 1738; K. Erkelenz and K. Holinde - Nucl. Phys. **A194** (1972) 161; V. G. Kadychevsky - Nucl. Phys. **B6** (1968) 125; P. J. Hermann - Diploma Thesis, Univ. Erlangen (1991); V. Pasculatsa and J. A. Tjon: Phys. Lett. **B 435** (1998) 245. G. Ramalho, A. Arriaga and M. T. Pena: Phys. Rev. **C65** (2002) 034008.
7. A. N. Mitra and B. M. Sodermark: Nucl. Phys. **A695** (2001) 328.
8. E. E. Salpeter and H. A. Bethe: Phys. Rev. **84** (1951) 1232; N. Nakanishi: Prog. Theor. Suppl. **43** (1969) 1.
9. F. Gross, Relativistic Quantum Mechanics and Field Theory (Wiley, N.Y., 1993).
10. F. Gross - Phys. Rev. **186** (1968) 1448.
11. W. Isgur - Acta Phys. Pol. **B8** (1977) 1081. C. Hayne and N. Isgur - Phys. Rev. **D25** (1982) 1944.
12. M. V. Terent'ev - Sov. J. Nucl. Phys. **24** (1976) 106. H. J. Melosh - Phys. Rev. D 9 (1974) 1095 D. V. Ahluwalia and M. Sawicki - Phys. Rev. D 47 (1993) 5161
13. M. Schmitt: PhD-Thesis, Univ. Erlangen (1997); A. A. Belkov, M. Dillig and F. G. Pilotto: Acta Phys. Polon. **B27** (1996) 3371.
14. V. L. Chernyak and A. R. Zhitnitsky - Phys. Rep. **112** (1984) 173.
15. R. Blankenbecler and R. Sugar - Phys. Rev. **142** (1966) 1051.

16. F. Gross: CEBAF - PR - 88 - 011; Proc. Los Alamos Worksh. (1988) 455.
17. N. Isgur and M. B. Wise - Phys. Lett. **B232** (1989) 113; E. Eichten and F. L. Feinberg - Phys. Rev. Lett. **43** (1979) 1205; B. Grinstein - Annu. Rev. Nucl. Phys. **42** (1992) 101; N. Uraltsev - hep-ph/0012336.
18. E. Kamke - Differentialgleichungen I (Akad. Verlagsgesellschaft Becher & Eiler KG, Leipzig 1943); M. Abramowitz and I. A. Stegun - Handbook of Mathematical Functions (Dover Publ., New York 1965)
19. B. Q. Ma - Z. Phys. **A345** (1993) 321; O.C. Jacob and L. S. Kisslinger - Phys. Rev. Lett. **56** (1986) 225; Z. Dziembowski and L. Mankiewicz - Phys. Rev. Lett. **58** (1987) 2175; C.R. Ji, P. L. Chung and S. R. Cotanch - Phys. Rev. **D45** (1992) 4214; H. M. Choi and C. R. Ji - Phys. Rev. **D59** (1999) 074015; L. S. Kisslinger, H. M. Choi and c. R. Ji - hep-ph/0101053; M. Bauer, B. Stech and M. Wirbel - Z. Phys. **C34** (1987) 103; M. Wirbel, B. Stech and M. Bauer - Z. Phys. **C29** (1985) 637.
20. S. J. Brodsky and F. Schlumpf - Phys. Lett. **B329** (1994) 111; F. Schlumpf and S. J. Brodsky - Phys. Lett. **B360** (1995) 1.
21. G. P. Lepage and S. J. Brodsky - Phys. Rev. **D22** (1980) 2157; Phys. Rev. Lett. **43** (1979) 545; S. J. Brodsky - SLAC - PUB - 8649 (2000): At the Front. of Part. Phys. , Vol. 2 (2000) 1343 (Ed. M. Shifman).
22. M. Burkhardt: Phys. Rev. **D46** (1992) 2751.
23. M. Dillig, F. G. Pilotto, E. Lütz and C. A. Z. Vasconcellos: preprint FAU-TP3-02-10 and IF.UFRGS.

Crossing Symmetry Violation of Inverse Amplitude Method

Isabela P. Cavalcante[1]* and J. Sá Borges*

Universidade do Estado do Rio de Janeiro, Rua São Francisco Xavier, 524, Maracanã, Rio de Janeiro, Brazil.

Even though Quantum Chromodynamics (QCD) has achieved a great success in describing strong interactions, low energy hadron physics must still be modeled phenomenologically. A great theoretical improvement was made by means of chiral perturbation theory (ChPT), an effective theory derived from the basis of QCD. Here we focus on its description for pion-pion scattering. The method yields a total amplitude that respects exact crossing symmetry, however, the corresponding partial waves satisfy only approximate elastic unitarity relation. Many different methods have been proposed to improve this behaviour. Here we consider IAM [1], that allows one to access the resonance region by fixing free parameters, but violates crossing symmetry.

Our motivations is that crossing symmetry somehow interconnects the various isospin channels. In other words, if one fits IAM isospin $I=1$ amplitude to the experimental data up the the ρ resonance region, for instance, due to crossing symmetry we do not have much freedom for $I=0$ fitting. Therefore, if crossing symmetry is not fulfilled, a very constrained fit of S-wave would give meaningless result. We evaluate crossing symmetry violation (CSV) in one-loop pion-pion scattering amplitude, corrected by IAM, in the resonance region and we show that crossing symmetry is badly violated by the IAM unitarized ChPT amplitude in the resonance region.

We consider $\pi^a \pi^b \to \pi^c \pi^d$ amplitude: $\langle \pi^c \pi^d | T | \pi^a \pi^b \rangle = A(s,t,u)\delta^{ab}\delta^{cd} + B(s,t,u)\delta^{ac}\delta^{bd} + C(s,t,u)\delta^{ad}\delta^{bc}$. The total isospin defined amplitudes T_I for $I = 0, 1$ and 2 are $T_0(s,t) = 3A + B + C$, $T_1(s,t) = B - C$, and $T_2(s,t) = B + C$. Crossing symmetry implies that there is just one amplitude describing the three total isospin channels of the process, namely $A(s,t,u)$, because $B(s,t,u) = A(t,s,u)$ and $C(s,t,u) = A(u,t,s)$.

The resulting ChPT amplitudes for S-wave ($I=0$ and 2) and P-wave ($I=1$) can be written as $t_{\ell I}(s) = t_{\ell I}^{ca}(s) + t_{\ell I}^{ca\,2}(s)\bar{J}(s) + t_{\ell I}^{left}(s) + p_{\ell I}(s) = t_{\ell I}^{ca}(s) + t_{\ell I}^{1}(s)$, where $t_{\ell I}^{ca}$ are the (real) Weinberg amplitudes, $t_{\ell I}^{left}$ are the parts that bear the left-hand cuts, $\bar{J}(s)$ is an integral of phase space factor and $p_{\ell I}(s)$ are two-free-parameter polynomials.

If one wants to describe a resonant amplitude, one may wish to use Padé approximants, by writing the inverse of the partial wave. Thus, instead of the exact ChPT result $t_{\ell I}$, we use a modified amplitude $\tilde{t}_{\ell I}(s) = t_{\ell I}^{ca}(s)/(1 - t_{\ell I}^{1}(s))$. We choose the free pa-

---

[1] Present address of I.P.C.: Universidade Federal de Mato Grosso do Sul, Campo Grande, MS, Brazil.

rameters λ_1 and λ_2 in order to fit S- and P-waves to the experimental phase shifts, by using the phase shift definition. The values obtained are $\lambda_1 = -0.00345$ and $\lambda_2 = 0.01125$ [2] and the resulting phase shifts are in agreement with experimental data.

In the following we present a method to quantify CSV of the IAM in the resonance region. In general, the total amplitudes $A(s,t)$ and $B(s,t)$ can be reconstructed from the isospin defined amplitudes and the IAM modified partial waves (indicated by a tilde) can be written as

$$\tilde{A}(s,t) = \frac{1}{3}\left(\tilde{t}_{00}(s) - \tilde{t}_{02}(s)\right) + \left[A(s,t) - \frac{1}{3}\left(t_{00}(s) - t_{02}(s)\right)\right],$$

and an equivalent formula for $\tilde{B}(s,t)$. If crossing symmetry were respected, one should have $\tilde{B}(s,t) = \tilde{A}(t,s)$. Thus the difference between these two quantities is a measure of the symmetry violation. We define CSV by

$$\Delta(s, \cos\theta) = 100\% \times \frac{|\tilde{B}(s,t) - \tilde{A}(t,s)|}{|\tilde{B}(s,t)| + |\tilde{A}(t,s)|}. \tag{1}$$

We measured CSV, as a function of cms energy, for some scattering angles, first with $\lambda_1 = \lambda_2 = 0$ and as a result we obtained very small CSV at threshold, while values up to roughly 50% develop around the ρ mass region, for any scattering angle, as shown in Fig. (a). A similar trend of CSV is obtained when the parameters used to perform the fits are considered, as in Fig. (b).

In summary our results show that from one-loop ChPT it is not possible to exactly fulfill all symmetry requirements. In other words, by introducing elastic unitarity, a lot of crossing symmetry is lost, as well as keeping the latter [3] costs a big amount of the former.

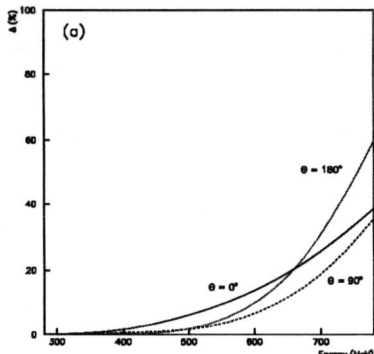
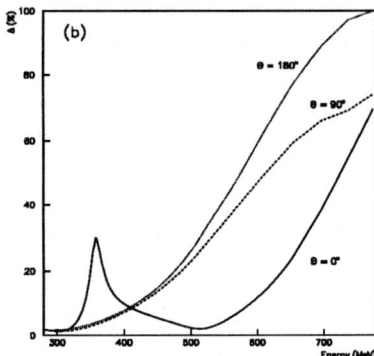

REFERENCES

1. T. N. Truong, *Phys. Lett.* **61** (1988) 2526; **67**, 2260 (1991); A. Dobado and J. R. Peláez, *Z. Phys. C* **57** (1993) 501; *Phys. Rev. D* **47** (1993) 4883; **56** (1997) 3057; T. Hannah, *Phys. Rev. D* **51** (1995) 103; **52** (1995) 4971; **54** (1996) 4648; **55** (1997) 5613.
2. I. P. Cavalcante and J. Sá Borges *Phys. Rev. D* **63** (2001) 114022.
3. J. Sá Borges, J. Soares Barbosa and V. Oguri, *Phys. Lett. B* **393** (1997) 413.

Bosons Production $W^{\pm}$ via e^-p-Collisions at CERN LEP/LHC Energies with a $W^{\pm}$ Anomalous Magnetic Moment

A. Gutiérrez-Rodríguez

Facultad de Fisica, Universidad Autónoma de Zacatecas Apartado Postal C-580, 98060 Zacatecas, Zacatecas México.

Abstract. We discuss the production of charged bosons in deep inelastic e^-p-scattering, in the context of an electroweak model, in which the vector boson self interactions may be different from those prescribed by the electroweak standard model. We show that even small deviations from the standard model value of κ ($\kappa = 1$) implies an observable deviation in the $W^{\pm}$-production rates at CERN LEP/LHC energies.

INTRODUCTION

In the present paper, we calculate charged boson production in deep inelastic e^-p-scattering in the context of an electroweak model with non-standard vector boson self interactions. Such a model was proposed by M. Kuroda *et al.* (KMSS model) [1]. In this model the trilinear vector boson coupling constants depend on only one free parameter, κ, the anomalous magnetic dipole moment of the $W^{\pm}$ [2]. The diagrams which contribute to the boson production cross section, at the quark level in the lowest order in α, contain only three vector boson self interactions, hence the boson production rates depend only on κ.

In a previous work [3], we have calculated in the context of the standard model separately the contributions to $W^{\pm}$ production from the different mechanisms: production at the leptonic vertex, at the hadronic vertex and through the boson self interaction.

We are going to work in the context of the KMSS model and perform our calculations using the parton distribution functions CTEQ5 reported by CTEQ Collaboration [4] for energies availables at LEP/LHC in the near future. In our computations for the cross section, we include besides the γ-exchange also the heavy boson exchange diagrams. We have already pointed out the importance of the contributions from heavy boson diagrams to the differential cross section of the charged boson production via e^-p collisions in a previous paper [3]. Consequently we consider the following processes:

$$e^- + p \longrightarrow e^-W^-X, \; e^-W^+X, \quad (1)$$

$$e^- + p \longrightarrow \nu_e W^-X, \; \nu_e Z^0 X. \quad (2)$$

TABLE 1. Total production of $W^\pm$ bosons according to our calculations, for an integrated luminosity of 500 pb^{-1}, at LEP/LHC energies.

Total Production of Bosons	$\kappa = 0$	$\kappa = 0.5$	$\kappa = 1$ (S.M.)	$\kappa = 1.5$	$\kappa = 2$
W^+	2950	3800	4900	6300	9300
W^-	3540	4090	5530	8685	12100

CONCLUSIONS

We have analyzed the effects of a non-standard anomalous magnetic dipole moment κ of the $W^\pm$ in the charged boson production via deep inelastic e^-p-scattering at LEP/LHC energies. We take the values of $\kappa = 0, 0.5, 1, 1.5, 2$ which do not deviate too much from the SM value of $\kappa = 1$ according to the experimental range $-1.3 < \kappa < 3.2$, reported by the CDF collaboration [5].

Taking $M_W = 80.3\ GeV$, $sin^2\theta_W = 0.223$, $E_e = 60\ GeV$, and $E_P = 7\ TeV$ (energy of the incoming electron and proton, respectively), and using the parton distributions CTEQ5, we present in the Table 1 the total production of W^+ and W^- bosons for differents values of κ. From these results we can see that it will be easier to detect effects due to $\kappa > 1$ than to $\kappa < 1$.

ACKNOWLEDGMENTS

Financial support is provided by a grant from the U.S. National Science Foundation. A.G.R. would like to thank the organizers of the Pan-American Advanced Study Institute New States of Hadronic Matter, Campos do Jordao, Sao Paulo, Brazil for their hospitality.

REFERENCES

1. J. Maalampi, D. Schildknecht, and K. H. Schwarzer, *Phys. Lett.* **166B**, 361, 1986; M. Kuroda, J. Maalampi, D. Schildknecht, and K. H. Schwarzer, *Nucl. Phys.* **B284**, 271, 1987.
2. H. Neufeld, J. D. Stroughair, and D. Schildknecht, CERN-TH.4720/87.
3. A. Gutiérrez and A. Rosado, *Phys. Rev.* **D57**, 4318, 1998.
4. CTEQ Collaboration (H. L. Lai, et al.), *Euro. Phys. J.* **C12**, 375, 2000.
5. CDF Collaboration, F. Abe et al. *Phys. Rev. Lett.* **74**, 1936, 1995.

Nucleon-Nucleon Potential: Relativistic and Heavy Baryon formalism

R. Higa

Nuclear Theory and Elementary Particle Phenomenology Group
Instituto de Física, Universidade de São Paulo
e-mail: higa@if.usp.br

Abstract. We present here preliminary results of our study of the two-pion component of the NN potential using covariant chiral symmetry restricted to just pion and nucleon degrees of freedom. In a first step we perform the full calculation using feynman parametrizations and restore the results obtained in reference [3]. Next we use several relations among feynman integrals and try to rewrite the amplitude evidencing the correct chiral scale. Finally we expand the integrals and compare it with the heavy baryon approach [5].

NN POTENTIAL - CENTRAL ISOVECTOR POTENTIAL

The main countribution to the central isovector potential is given basically by the integral $\mathscr{I}_{DD}^-$, defined as

$$\mathscr{I}_{DD}^{\pm} = -\frac{i}{2}\int \frac{d^4Q}{(2\pi)^4} \frac{[D^\pm]^{(1)}[D^\pm]^{(2)}}{(k^2-m^2)(k'^2-m^2)}, \tag{1}$$

where $+$ $(-)$ refers to the isoscalar (isovector) component and $D^\pm$ and $B^\pm$ are the usual πN amplitudes.

A set of relations among feynman integrals allow us to write it, in the local approximation, as

$$\begin{aligned}\mathscr{I}_{DD}^- &= \frac{1}{(4\pi)^2}\frac{\mu^2}{8f_\pi^4}\Bigg\{(g_A^2-1)^2\left(1-\frac{t^2}{4m^2}\right)\bar{\Pi}_{cc}^{000}\\&\quad -2g_A^2(g_A^2-1)\left(1-\frac{t}{2\mu^2}\right)\left[\Pi_{cc}^{000}-\frac{\mu}{2m}\left(1-\frac{t}{2\mu^2}\right)\Pi_{sc}^{000}\right]\\&\quad +\frac{g_A^4}{2}\left(1-\frac{t}{2\mu^2}\right)^2(\Pi_{ss}^{000}+\Pi_{reg}^{000})+g_A^4\frac{m}{\mu}\left(1-\frac{t}{2\mu^2}\right)\left(1-\frac{m}{E}\right)\Pi_{HB}\Bigg\}.\end{aligned}\tag{2}$$

In configuration space we are able to express $\bar{\Pi}_{cc}^{000}$ as

$$\bar{\Pi}_{cc}^{000} = -\frac{1}{2\pi x^4}\left[K_1(2x)+xK_0(2x)\right], \tag{3}$$

where $x = \mu r$ and K_0 and K_1 are the modified Bessel functions.

In ref. [1] Becher and Leutwyler present an expansion of the triangle integral (Π_{sc}^{000}) which preserves its analytic properties in the low energy expansion. It's represented by a heavy baryon term (Π_{HB}), a first order heavy baryon correction (Π_{HBc}) and the analytic correction to the heavy baryon expansion (Π_{th}). In configuration space one has

$$\Pi_{HB} = -\frac{e^{-2x}}{2x^2}, \quad \Pi_{HBc} = \frac{\mu}{\pi m x^2}\left[xK_0(2x) + K_1(2x)\right], \quad \Pi_{th} \approx \frac{\mu^2}{4m^2}\frac{e^{-2x}}{2x} + \mathscr{O}\left(\left(\frac{\mu}{m}\right)^3\right). \tag{4}$$

A numerical study of the integrals Π_{ss}^{000} and Π_{reg}^{000} shows that they have the same magnitude and converge to each other in the heavy baryon limit ($m \to \infty$). The expansion of Π_{ss}^{000} is written as

$$\Pi_{ss}^{000} = \frac{2}{\pi x}K_0(2x) - \frac{\mu}{2m}\frac{e^{-2x}}{x} + \frac{\mu^2}{\pi m^2}K_1(2x) + \mathscr{O}\left(\left(\frac{\mu}{m}\right)^3\right). \tag{5}$$

The above comes from a relation involving the expansion of Π_{sc}^{000}: the leading term (Π_{HB}) is canceled, the first term of the above expression comes from (Π_{HBc}) and the second one, from (Π_{th}). Ignoring for the moment terms of $\mathscr{O}((\mu/m))$ in Π_{ss}^{000} and assuming the same behaviour for Π_{reg}^{000} we have exactly the same expression of Kaiser et. al [5], up to the above order. Numerically the next term is small, but in the central isoscalar potential ($\mathscr{I}_{DD}^{+}$), due to cancelations of crossed box (Π_{ss}^{000}) and planar box (Π_{reg}^{000}) diagrams this term becomes the leading order. As this term comes essencially from Π_{th} (not presented in the heavy baryon approach) we expect a difference in this component of the potential [6].

ACKNOWLEDGMENTS

R. H. was supported by Fundação de Amparo à Pesquisa do Estado de São Paulo (FAPESP).

REFERENCES

1. T. Becher and H. Leutwyler, *Eur. Phys. J.* C **9**, 643 (1999).
2. M. R. Robilotta, pre-print nucl-th/0009001.
3. M. R. Robilotta and C. A. da Rocha, *Phys. Rev.* C **49**, 1818 (1994).
4. M. R. Robilotta and C. A. da Rocha, *Nucl. Phys.* A **615**, 391 (1997).
5. N. Kaiser, R. Brockmann and W. Weise, *Nucl. Phys.* A **625**, 758 (1997).
6. R. Higa, M. R. Robilotta and C. A. da Rocha, work in progress.

Models of Black Hole Production at LHC

R. Klippert* and C. J. Solano Salinas[†]

*Universidade Federal de Engenharia de Itajubá and ICRA-Rio, Rio de Janeiro, Brazil
[†]CBPF, Rio de Janeiro, Brazil and CINVESTAV, Mexico

Abstract. The purpose of this paper is to discuss about the different models for black hole production at the future Large Hadron Collider. In traditional scenarios the Planck scale is fundamental, and the weak scale is derived from it via some dynamical mechanism. Recently, several authors are exploring an alternative viewpoint where the weak scale is the fundamental scale of nature and the 4-dimensional Planck scale is to be derived from that [1]. These scenarios include large or warped extra dimensions, propagation of matter and gauge degrees of freedom on brane worlds, and a fundamental Planck scale of O(TeV). If the scale of quantum gravity is near TeV we will have a copious production of **mini black holes** at the Large Hadron Collider [2] and Cosmic rays interactions in the atmosphere [3]. We discussed as well other line of semi-classical models from analog gravity in nonlinear electrodynamics that can be tested as well at LHC. The possibles consequences of these models for high energy experimental physics are discussed.

Black Hole Production in Brane World. Through small-scale compactifications of D-4 dimensions an effective Planck scale $M_P \sim 10^{16} TeV$ is found. To reach a Planck scale of $O(l_{Planck})$ a large "warped volume" is required, which can be achieved by either a large warp factor [4] or a large bulk volume [5] (or possibly even both). Matter and gauge fields turn out to propagate tipically in a 4-brane, while gravitons can reach the extra dimensions.

Black Hole Production at High energy Interactions. In hadron scattering the black hole (BH) cross section is found from the partonic cross section for partons i and j to form a BH:

$$\sigma^{pp \to bh}(M_{BHmin}, s) = \sum_{i,j} \int_{\tau_m}^1 d\tau \int_\tau^1 \frac{dx}{x} f_i(x, Q^2) f_j(\frac{\tau}{x}, Q^2) \hat{\sigma}^{ij \to bh}(\tau s)$$

where $\sqrt{s}$ is the collider center of mass energy, x and τ/x are the parton momentum fractions and f_i are the parton distribution functions (PDF). M_{BHmin} is the minimum mass for a valid BH and $\tau_m = M_{BHmin}^2/s$. The Schwarzchild radius R_S of a (4+n)-dimensional BH is given by [6], assuming that the extra dimensions are large ($\gg R_S$). The total cross section is estimated by geometrical arguments[2], and is of order: $\hat{\sigma}^{ij \to bh}(\sqrt{s}) \approx \pi R_S^2$. From the theory we have the Schwarzchild radius R_S, the Hawing temperature T_H, the average inverse particle energy <1/E> and average multiplicity <N> as functions of M_P, $M_B H$ and the number of extra dimensions, n [1, 2]. From the experimental cross sections distributions, if the BH are produced, we can estimate these paramaters, including the number n of unseen extra dimensions.

Charge Asymmetries in Brane World. A possible dynamical charge asymmetry in astrophysical objects is simply sketched in [7]. They argue on brane calculations in a 5-dimensional spacetime without any compactification at all. The structure of the spectrum was found for gravitons, massless scalar fields, gauge bosons, and fermions. Only massless fields would remain localized around the brane which constitutes our conventional 3+1 world, while the massive particles are dynamically pushed apart from it. By assuming an initial asymmetry between particles and anti-particles (compensated by a symmetry between electrons and protons, such that the matter remains neutral as a whole), that mechanism would eventually drive some electrons out from the 4-D world. The same would not occur with the same efficiency with protons due to color confinement. The net result is a positively charged macroscopic visible structure. One of the devised applications of this model is to allow astrophysical objects (stars) to achieve a significant net electrical charge, the gravitational collapse of which towards a BH may provide in this way a viable engine of unbeamed gamma-ray bursts[8].

Microscopic variations of this could also be found in particle colliders. For example at the LHC (pp collisions) experiments, if these charged BHs are created, we would expect, at least, missing energy and charge asymmetry, and maybe some gamma-ray jets.

Black Hole from Analog Gravity in Nonlinear Electrodynamics. Analogous structures of black holes are also possible to occur in electrically nonlinear fluids endowed with large velocities of matter flow[9]. The authors argue on an effective modification of the speed of light due to the possible nonlinearity in the constitutive relations of Maxwell theory[10]. This admits a mathematical description in terms of an effective Riemannian geometry, and a configuration of fields and matter flow is shown to display most of the features of an event horizon when concerning to electromagnetic waves; but matter fields are not bounded by such structure. Also, this kind of black hole is not subjected to Hawking evaporation, as far as its dynamics is not determined by the mass configuration but by the distribution of electromagnetic charges. This can be studied as well at LHC in the heavy ion collider experiment (ALICE). It can be argued that when the electromagnetic energy, inside the BH, become so high it can create pairs particle-antiparticle that can cross the BH horizon, maybe in a huge explosion (jets?).

ACKNOWLEDGMENTS

Acknowledgments. This work was supported by the Brazilian agency Fundação de Amparo à Pesquisa do Estado de Minas Gerais (FAPEMIG).

REFERENCES

1. Giddings, S. B., *hep-ph/0110127* (2001).
2. Dimopoulos, S., and Landsberg, G., *Phys. Rev. Lett.*, **87**, 161602 (2001).
3. Anchordoqui, L., and Goldberg, H., *hep-ph/0109242* (2001).
4. Randall, L., and Sundrum, R., *Phys. Rev. Lett.*, **83**, 3370 (1999).
5. Arkani-Hamed, N., Dimopoulos, S., and Dvali, G., *Phys. Lett. B*, **429**, 263 (1998).
6. Myers, R., and Perry, M., *Ann. Phys. (N.Y.)*, **172**, 304 (1986).
7. Cuesta, H. M., Penna-Firme, A., and Pérez-Lorenzana, A., *hep-ph/0203010* (2002).
8. Ruffini, R., Bianco, C., Fraschetti, F., Xue, S., and Chardonnet, P., *Astrophys. J.*, **555**, L107 (2001).
9. Lorenci, V., Klippert, R., Novello, M., Bergliaffa, S. P., and Salim, J., *gr-qc/0201061* (2002).
10. Born, M., and Wolfe, E., *Principles of optics*, Cambridge, London, 1980.

Proton-Proton Total Cross Section Scenarios at Cosmic-Ray Energies

E. G. S. Luna and M. J. Menon

Instituto de Física "Gleb Wataghin", UNICAMP
13083-970, Campinas, SP, Brazil

Abstract. We parameterize pp total cross section data from accelerator to cosmic-ray energies and show that the experimental information presently available are statistically consistent with two different scenarios at the highest energies.

Proton-proton total cross sections have been determined in accelerator experiments and, presently, the highest center of mass energy reached is $\sqrt{s} = 62.5$ GeV. Some estimations are also available from cosmic-ray experiments in the interval $\sqrt{s}: 6 - 40$ TeV. In this case, as reviewed in some detail in [1], we can identify two distinct sets of estimations of the proton-proton total cross sections, one represented by the results of the Fly's Eye Collaboration and Akeno Collaboration [2], and the other by the results of Gaisser, Sukhatme, Yodh (GSY) with those by Nikolaev [3]. The discrepancies between these two sets come from both, the uncertainties in the determination of the proton-air production cross sections and the uncertainties in the connection between these cross sections and the pp total cross section, as discussed by Engel, Gaisser, Lipari and Stanev [4]: all the cosmic-ray estimations are strongly model dependent. One important point is the fact that the results by Nikolaev and GSY have never been directly critized in the literature.

In this communication we investigate in a quantitative way the behavior of the total cross section by taking into account the discrepancies that characterize the cosmic ray information. To this end, we consider *two ensembles of data and experimental information* above $\sqrt{s} = 10$ GeV with the following notation: Ensemble I: pp accelerator data + Akeno results + Fly's Eye results and Ensemble II: pp accelerator data + Nikolaev results + GSY results. We make use of the simplest parameterization for the total cross section, satisfying the Froissart-Martin bound: $\sigma^{pp}(s) = A + B\ln s + C[\ln s]^2$, where A, B, C are free parameters.

The fits have been performed through the CERN-MINUIT routine [5]. The corresponding values of the free parameters and statistical information about each fit are as follows. Ensemble I: $A = 45.78$, $B = -3.315$, $C = 0.3654$ (all in mb), $\chi^2/d.o.f. = 11.75/11$; Ensemble II: $A = 49.27$, $B = -4.435$, $C = 0.4534$ (all in mb), $\chi^2/d.o.f. = 15.04/11$. Errors from the free parameters were propagated to $\sigma_{tot}^{pp}(s)$ through the usual formulas, taking account of both variances and covariances [6]. The results with the corresponding error bands ($\pm$ one standard deviation) are shown in Fig. 1.

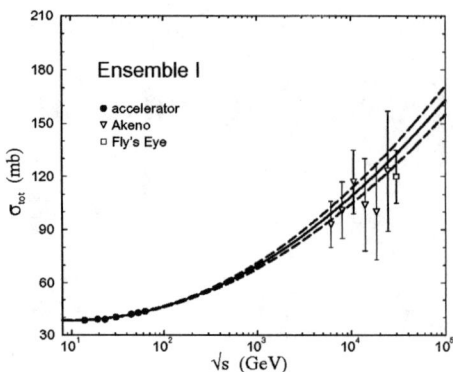 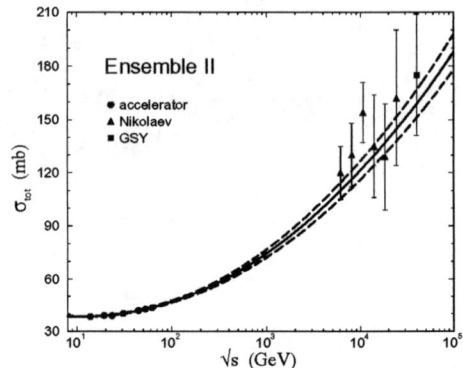

FIGURE 1. Parameterizations to Ensembles I and II. Accelerator data quoted in [1].

These results show that taking into account the error bands, the two fits become discrepant above ~ 1 TeV and they are certainly incompatible above 5 TeV. We understand this result as a *statistical evidence for the possibility of two distinct scenarios above* $\sqrt{s} \sim 5$ TeV. In particular, from each ensemble, the interpolated results at $\sqrt{s} = 14$ TeV (LHC) are $\sigma_{tot}^{pp} = 119 \pm 5$ mb (I), and $\sigma_{tot}^{pp} = 133 \pm 6$ mb (II). Ensemble I accommodates the predictions from the majority of models and in general is considered as "the cosmic-ray results". On the other hand, if the cosmic-ray estimations in Ensemble II are the correct ones, the experimental information on σ_{tot}^{pp} presently available indicates that the pp cross section rises faster with the energy than generally expected or assumed. This faster increase could be associated with the exotic phenomena observed in cosmic-ray experiments in pp collisions at $\sqrt{s} \sim 5$ TeV, explained and discussed, for example, in Ref. [7].

We are thankful to Fapesp and CNPq for financial support and to R.F. Àvila for discussions.

REFERENCES

1. R. F. Ávila, E. G. S. Luna, and M. J. Menon, Braz. J. Phys. **31**, 567 (2001), e-Print Archive hep-ph/0105065; A. F. Martini and M. J. Menon, Phys. Rev. D **56**, 4338 (1997).
2. Fly's Eye Collaboration (R. M. Baltrusaitis *et al.*), Phys Rev. Lett. **52**, 1380 (1984); Akeno Collaboration (H. Honda *et al.*), Phys. Rev. Lett. **70**, 525 (1993).
3. T.K. Gaisser, U.P. Sukhatme, and G.B. Yodh, Phys. Rev. D **36**, 1350 (1987); N.N. Nikolaev, Phys. Rev. D **48**, R1904 (1993).
4. R. Engel, T. K. Gaisser, P. Lipari, and T. Stanev, Phys. Rev. D **58**, 014019 (1998).
5. F. James, *MINUIT* - Reference Manual, Version 94.1, CERN - D506 (1994).
6. P. R. Bevington and D. K. Robinson, *Data reduction and error analysis for the physical sciences*, (McGraw-Hill, New York, 1992).
7. Chacaltaya and Pamir Collaboration, Nucl. Phys. **B370**, 365 (1992); S. L. C. Barroso et al., Nucl. Phys. B (Proc. Supp.) **75A**, 150 (1999); M. J. Menon, in *International Workshop On Hadron Physics 98*, edited by E. Ferreira, F.F.S. Cruz, S.S. Avancini(World Scientific, Singapore, 1999) p. 333, e-Print Archive hep-ph/9810508.

Charmonium–Pion Cross Section from QCD Sum Rules

F. S. Navarra*, M. Nielsen*, R. S. Marques de Carvalho[†] and G. Krein[†]

*Instituto de Física, Universidade de São Paulo, C.P. 66318, 05315-970 São Paulo, SP, Brazil
[†]Instituto de Física Teórica, Universidade Estadual Paulista, Rua Pamplona 145, 01405-900 São Paulo, SP, Brazil

Abstract. The $J/\psi \, \pi \to \bar{D} D^*$, $D \bar{D}^*$, $\bar{D}^* D^*$ and $\bar{D} D$ cross sections as a function of $\sqrt{s}$ are evaluated in a QCD sum rule calculation. We find that our results are smaller than the $J/\psi \, \pi \to$ charmed mesons cross sections obtained with models based on meson exchange, but are close to those obtained with quark exchange models.

Charmonium - hadron cross sections are of crucial importance in the context of quark-gluon plasma physics [1]. Small J/ψ - hadron dissociation cross sections may favor an interpretation of the recent Pb + Pb data in terms of the production of a new phase of matter. Part of these interactions happens in the early stages of the nucleus - nucleus collisions and therefore at high energies ($\sqrt{s} \simeq 10 - 20$ GeV) and one may try to apply perturbative QCD. However, even in this regime, nonperturbative effects may be important [2]. Interestingly, estimates using quite different methods give results clustering around the value of $3 - 5$ mb in this energy range. On the other hand, a significant part of the charmonium - hadron interactions occurs when other light particles have already been produced, forming a "fireball". Interactions inside this fireball happen at much lower energies ($\sqrt{s} \leq 5$ GeV) and one has to apply nonperturbative methods.

In this work we show some discussion an the results which we obtained in a previous paper [3] where we use the QCD sum rules (QCDSR) technique [4, 5] to study the $J/\psi - \pi$ dissociation. In view of our relatively poor understanding of J/ψ reactions in nuclear matter and considering the large discrepancies between different model estimates, we believe that our work adds to a better understanding of this important topic.

We consider all four channels $J/\psi \, \pi \to: \bar{D} D^*, D \bar{D}^*, \bar{D} D$ and $\bar{D}^* D^*$. Following [5, 6, 7, 8], we can write a sum rule at $p_1^2 = 0$ and single out the leading terms in the operator product expansion (OPE) of the correlation function of each process that match the $1/p_1^2$ term. The perturbative diagram does not contribute with $1/p_1^2$ and, up to dimension four, only the diagrams proportional to the quark condensate contribute.

Having the QCD sum rule results for the amplitude of the three processes $J/\psi \, \pi \to \bar{D} D^*$, $\bar{D} D$, $\bar{D}^* D^*$, we can evaluate the differential cross section. Using our QCD sum rule result we show, in Fig. 1, the cross section for the $J/\psi \, \pi$ dissociation. It is important to keep in mind that, since our sum rule was derived in the limit $p_1 \to 0$, we can not extend our results to large values of $\sqrt{s}$. Also, since the perturbative contribution is absent in our calculation, we were not able to properly disentangle the continuum contribution and our cross section may include contributions from higher

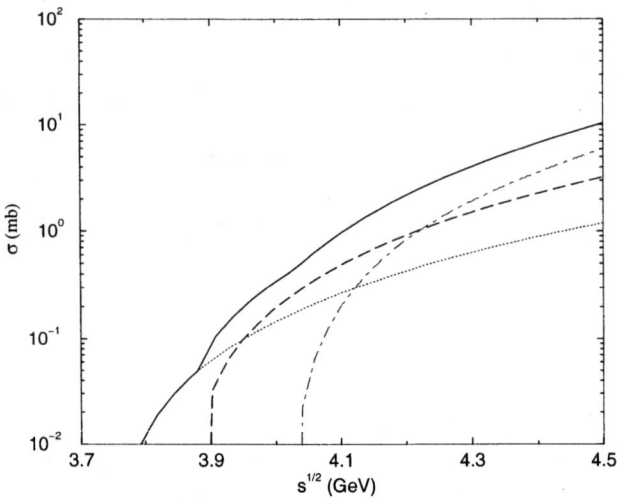

FIGURE 1. Total cross sections of the processes $J/\psi\, \pi \to \bar{D}\, D^* + D\, \bar{D}^*$ (dashed line), $\bar{D}\, D$ (dotted line) and $\bar{D}^*\, D^*$ (dot-dashed line). The solid line gives the total $J/\psi\, \pi$ dissociation cross section.

states. Whereas they are certainly not important in the case of the pion, they may give some contribution to the heavy currents. Therefore our $J/\psi\, \pi$ cross section may implicitly include (at least partially) the process $\psi'\, \pi$. For this reason, our numbers might be regarded as upper bounds.

In conclusion, we have used the QCD sum rule approach to evaluate the hadronic amplitude of the $J/\psi\, \pi$ dissociation. From the hadronic amplitude we have evaluated the $J/\psi\, \pi \to$ charmed mesons dissociation cross section, and have obtained $2.5 \leq \sigma \leq 4.0$ mb at $4.1 \leq \sqrt{s} \leq 4.3$ GeV. In view of the uncertainties discussed above these numbers should be taken as upper limits.

ACKNOWLEDGMENTS

This work was supported by CNPq and FAPESP.

REFERENCES

1. For recent reviews see B. Müller, Nucl. Phys. A 661 (1999) 272 ; T. Barnes, E.S. Swanson and C.Y. Wong, nucl-th/0006012.
2. H.G. Dosch, F.S. Navarra, M. Nielsen and M. Rueter, Phys. Lett. B 466 (1999) 363.
3. F.S. Navarra, M. Nielsen, R.S. Marques de Carvalho and G. Krein, Phys. Lett. B 529 (2002) 87.
4. M.A. Shifman, A.I. Vainshtein and V.I. Zakharov, Nucl. Phys. B120 (1977) 316.
5. L.J. Reinders, H. Rubinstein and S. Yazaki, Phys. Rep. 127 (1985) 1.
6. S. Choe, M.K. Cheoun and S.H. Lee, Phys. Rev. C53 (1996) 1363.
7. M.E. Bracco, F.S. Navarra and M. Nielsen, Phys. Lett. B454 (1999) 346.
8. F.S. Navarra, M. Nielsen, M.E. Bracco, M. Chiapparini and C.L. Schat, Phys. Lett. B489 (2000) 319.

Coulomb Gauge Quark Model, Vacuum Condensates and High Density Quark Matter

P. K. Panda and G. Krein

Instituto de Física Teórica, Universidade Estadual Paulista,
Rua Pamplona 145, 01405-900 São Paulo, SP.

Abstract. We consider here a Coulomb gauge quark model which includes an explicit construct for a nontrivial vacuum structure in QCD at finite density. Non-perturbative renormalization of ultraviolet diverges is performed by adding counterterms. The equation of state for u and d quark matter at zero temperature is calculated in the Hartree-Fock approximation.

The study of the high density and high temperature hadronic matter is of interest for the understanding of superdense stars and relativistic heavy-ion collisions. Properties of high density matter have been studied using perturbative QCD and semi-phenomenological models like bag models. Validity of a perturbative analysis is unfortunately limited to asymptotically high density and convergence of the perturbative series becomes ill behaved at densities a few time larger than the saturation density of nuclear matter. Here we sketch a calculation performed in the context of a field theoretical Coulomb-gauge quark model using a nonperturbative method with an explicit construct of a chiral symmetry breaking vacuum involving quark-antiquark condensates.

The Hamiltonian of the model is of the type of a Coulomb gauge QCD Hamiltonian [1]–[5]

$$\mathcal{H}(\vec{x}) = \psi(\vec{x})^{i\dagger}_\alpha (-i\vec{\alpha}\cdot\vec{\nabla} + \beta m_q)_{\alpha\beta} \psi(\vec{x})^i_\beta$$
$$+ \frac{1}{2}\int d\vec{y}\, \psi_\alpha(\vec{x})^{i\dagger} \psi_\beta(\vec{x})^j\, V^{ij,kl}_{\alpha\beta,\gamma\delta}(\vec{x}-\vec{y})\, \psi_\gamma(\vec{y})^{k\dagger}\psi_\delta(\vec{y})^l. \quad (1)$$

In the above i,j stand for color indices, α, β stand for the spinor indices, and V is a potential. The spinor dependence of V is such that in the zero quark mass limit the Hamiltonian is chirally symmetric. The chiral symmetry breaking vacuum state is constructed in the form

$$|\Omega\rangle = U|0\rangle \equiv \exp(B^\dagger - B)|0\rangle \quad \text{with} \quad B^\dagger = \int d\vec{k}\, \varphi(\vec{k}) c_r(\vec{k})^\dagger \chi_r^\dagger (\vec{\sigma}\cdot\hat{k}) \tilde{\chi}_s \tilde{c}_s(-\vec{k}), \quad (2)$$

where $\hat{k} = \vec{k}/|k|$, χ is the Pauli spinor and $\tilde{\chi} = -i\sigma^2\chi$. Here $\varphi(\vec{k})$ is a trial function associated with the quark-antiquark condensates and $c^\dagger$ and $\tilde{c}$ are quark and antiquark creation operators - the corresponding annihilation operators destroy the state $|0\rangle$. When $\varphi = 0$, there is no quark-antiquark condensation and the nontrivial vacuum, $|\Omega\rangle$, is equal to the trivial vacuum, $|0\rangle$. For V we use a Coulomb plus linear confining potential, which is given as $\tilde{V}(\vec{k}) = -4\pi\alpha(\vec{k}^2)/k^2 - 8\pi\sigma/(k^2)^2$ in momentum space, where $\alpha(\vec{k}^2)$ is taken from the one-loop renormalization group result and σ is the string tension. We also assume a one-loop momentum-dependent current quark mass. The energy density in condensed vacuum with respect to the trivial vacuum is given as

$$\Delta \mathscr{E} = \langle \Omega | \mathscr{H} | \Omega \rangle (\varphi) - \langle \Omega | \mathscr{H} | \Omega \rangle (\varphi = 0). \tag{3}$$

Minimization of the energy functional (3) with respect to the pairing function $\varphi(k)$ leads to an integral equation for φ, the gap equation. The Coulomb potential gives rise to ultraviolate divergencies. These are renormalized by adding appropriate counter terms to the Hamiltonian [2].

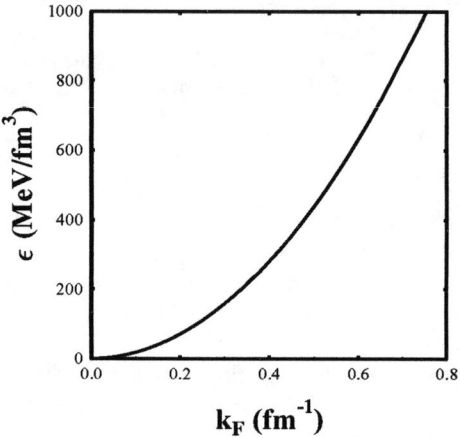

FIGURE 1. Energy density of quark matter.

The parameters of the model are fixed by fitting vacuum properties. Once these are fixed, and the counterterms at zero density are determined, the energy density of quark matter at zero temperature is calculated in the Hartree-Fock approximation. Figure 1 presents the energy density versus the quark Fermi momentum, after subtracting the nonperturbative vacuum energy - or 'bag constant', which turns out to be approximately $(178\,\text{MeV})^4$.

Work partially supported by CNPq and FAPESP (grant 99/08544-0).

REFERENCES

1. A. Amer, A. Le Yaouanc, L. Oliver, O. Pene and J.C. Raynal, Phys. Rev. Lett. 50, 87 (1983); ibid, Phys. Rev. D28, 1530 (1983).
2. S.L. Adler and A.C. Davis, Nucl. Phys. B244, 469 (1984); R. Alkofer and P. A. Amundsen, Nucl. Phys. B306, 305 (1988).
3. A. Kocić, Phys. Rev. D 33, 1785 (1986).
4. P.J.A Bicudo and J.E. Ribeiro, Phys. Rev. D 42, 1611, 1625, 1635 (1990); P.J.A Bicudo, G. Krein, J.E. Riberio and J.E. Villate, Phys. Rev D 45 1673 (1992).
5. A. Mishra, H. Mishra, S.P. Misra, Z. Phys. C 57, 241 (1993); A. Mishra, H. Mishra, V. Sheel, S.P. Misra, P.K. Panda, Int. J. Mod. Phys. E 5 93 (1996).

AUTHOR INDEX

A

Abu-Raddad, L. J., 191, 221
Aguiar, C. E., 686
Aichelin, J., 580
Altarelli, G., 70
Antinori, F., 294
Ayala, A., 620
Azevedo, E. O., 658

B

Beckmann, C., 255
Berger, J., 553
Billmeier, A., 294
Bracco, M. E., 676
Brenner Mariotto, C., 631

C

Cardamone, D. M., 191
Castillo, J., 371
Cavalcante, I. P., 713
Csernai, L. P., 428
Cuautle, E., 620

D

Di Giacomo, A., 168
Dillig, M., 704

E

Eichmann, U., 553
Elze, H.-T., 229

F

Fachini, P., 325, 358
Ferreira, E., 615
Fiolhais, M., 658

G

Gans, J., 255
Gastineau, F., 580
Gay Ducati, M. B., 631, 637
Gomez Dumm, D., 695
Grassi, F., 646
Gulbrandsen, K. H., 255
Gutiérrez-Rodríguez, A., 715

H

Hama, Y., 686
Harris, J. W., 255
Herrera, G., 620
Higa, R., 717

I

Ingelman, G., 631

K

Karsch, F., 112
Kharzeev, D. E., 27
Klippert, R., 719
Kodama, T., 3, 686
Kraus, I., 377
Krein, G., 676, 723, 725

L

Letessier, J., 142, 460
Liu, F. M., 422
Luna, E. G. S., 721
Lütz, E. F., 704

M

Machado, M. V. T., 637, 655
Malheiro, M., 658
Markert, C., 533

Marques de Carvalho, R. S., 676, 723
Marranghello, G. F., 704
Masperi, L., 666
Menezes, D. P., 672
Menon, M. J., 721
Montaño, L. M., 620

N

Navarra, F. S., 676, 723
Newby, J., 377
Nielsen, M., 676, 723
Nuss, L. G., 658

O

Osada, T., 686

P

Panda, P. K., 725
Pilotto, F. G., 704
Providência, C., 672

R

Rafelski, J., 142, 460, 533
Raufeisen, J., 27, 525
Roland, C., 349

S

Sá Borges, J., 713
Šafařík, K., 377
Salur, S., 553
Scherer, S., 553
Scoccola, N. N., 695
Solano Salinas, C. J., 719
Sorensen, P., 366, 377
Stöcker, H., 553

T

Taurines, A., 658
Thews, R. L., 490
Torrieri, G., 533, 580
Trainor, T. A., 599

V

van Kolck, U., 191
Vasconcellos, C. A. Z., 704

W

Werner, K., 397, 422

Z

Zajc, W. A., 325
Zaranek, J., 294
Zschiesche, D., 553